AF763162

Handbook of Geometry and Topology of Singularities VIII

José Luis Cisneros-Molina • Lê Dũng Tráng •
José Seade
Editors

Handbook of Geometry and Topology of Singularities VIII

Springer

Editors
José Luis Cisneros-Molina
Instituto de Matemáticas, Unidad Cuernavaca
Universidad Nacional Autónoma de México
Cuernavaca, Mexico

Lê Dũng Tráng
Marseille, France

José Seade
Instituto de Matemáticas, Unidad Cuernavaca
Universidad Nacional Autónoma de México
Cuernavaca, Mexico

ISBN 978-3-031-99570-5 ISBN 978-3-031-99571-2 (eBook)
https://doi.org/10.1007/978-3-031-99571-2

Mathematics Subject Classification: 32SXX, 14BXX, 14PXX, 53CXX, 57R20, 58KXX

© The Editor(s) (if applicable) and The Author(s), under exclusive license to Springer Nature Switzerland AG 2026

This work is subject to copyright. All rights are solely and exclusively licensed by the Publisher, whether the whole or part of the material is concerned, specifically the rights of translation, reprinting, reuse of illustrations, recitation, broadcasting, reproduction on microfilms or in any other physical way, and transmission or information storage and retrieval, electronic adaptation, computer software, or by similar or dissimilar methodology now known or hereafter developed.
The use of general descriptive names, registered names, trademarks, service marks, etc. in this publication does not imply, even in the absence of a specific statement, that such names are exempt from the relevant protective laws and regulations and therefore free for general use.
The publisher, the authors and the editors are safe to assume that the advice and information in this book are believed to be true and accurate at the date of publication. Neither the publisher nor the authors or the editors give a warranty, expressed or implied, with respect to the material contained herein or for any errors or omissions that may have been made. The publisher remains neutral with regard to jurisdictional claims in published maps and institutional affiliations.

This Springer imprint is published by the registered company Springer Nature Switzerland AG
The registered company address is: Gewerbestrasse 11, 6330 Cham, Switzerland

If disposing of this product, please recycle the paper.

Dedicated to Lê Dũng Tráng, who passed away in November 2025, leaving behind a remarkable legacy of friendship, generosity, and outstanding mathematics.

Preface

This is the eighth and tentatively last volume of the Handbook of Geometry and Topology of Singularities, a series with contributions by more of one hundred world leading mathematicians.

Singularity theory dates back to Newton, Bézout, Puiseux, Klein, Severi and many others. Yet, it emerged as a field of mathematics in itself in the early 1960s, thanks to pioneering work by R. Thom, O. Zariski, H. Whitney, H. Hironaka, J. Milnor, E. Brieskorn, C. T. C. Wall, V. I. Arnold, J. Mather and many others. Singularities are ubiquitous in mathematics, appearing naturally in a wide range of different areas of knowledge. Their scope is vast, their purpose is multifold and its potential for applications in other areas of mathematics and of knowledge in general is unlimited, and so are its possible sources of inspiration. This is a meeting point where many areas of mathematics, and science in general, come together.

This wideness and amplitude of its scope can make singularity theory hard to grasp for graduate students and researchers in general: what are and what have been the major lines of development in the last decades, what is known and where to find it, what is the current state of the art in its many branches, the various directions into which this theory is flourishing, its interaction with other areas of current research in mathematics. Those are questions that gave birth to this project, the "Handbook of Geometry and Topology of Singularities". This handbook covers a wide scope of singularity theory, presenting articles on various aspects of the theory and its interactions with other areas of mathematics. The authors are world experts; the various articles deal with both classical material and modern developments. They are addressed to graduate students and newcomers into the theory, as well as to specialists that can use these as guidebooks.

Professor Lê Dũng Tráng passed away in November 2025, shortly after sending this Volume VIII to press. He was among the greatest singularists ever, and his mathematical legacy will endure, inspiring generations of young mathematicians to come.

The first four volumes of this collection gathered together both, foundational as well as "cutting-edge" aspects of the theory. The topics studied so far include:

- The topology of plane curves and surface singularities
- The analytic classification of plane curve singularities
- Introductions to four classical methods for studying singularities: resolutions, deformations, stratifications and slicing the spaces *à la* Lefschetz
- The geometry and topology of analytic maps near a critical point; Milnor's fibration theorem, the monodromy, vanishing cycles and Lê numbers
- Morse theory for stratified spaces and constructible sheaf complexes
- Simple Lie algebras and simple singularities
- Limits of tangents to a complex analytic variety and the Nash blow up
- Zariski's equisingularity and intersection homology
- The interplay between analytic and topological invariants of complex surface singularities and their relation with modern 3-manifold invariants
- Quasi-projective varieties, complements of plane curves and hypersurfaces in projective space
- Chern classes, Segre classes and indices of vector fields and 1-forms on singular varieties. The motivic Hirzebruch class for singular varieties
- The space of arcs in algebraic varieties
- Determinantal singularities
- Lipschitz geometry in singularity theory
- The Hilbert-Samuel multiplicity
- Grothendieck's and other comparison theorems. Free divisors and D-modules

and many other important topics.

Volumes V and VI of this collection focus on singular holomorphic foliations, a whole area of mathematics in itself with close connections with singularity theory, as well as with dynamical systems, geometry and topology. For instance:

1. The integral lines of a vector field on a manifold, or the orbits of a smooth flow, determine a foliation with singularities at the fixed points of the flow. Similarly, the orbits of a holomorphic action of a complex Lie group determine a holomorphic foliation with singular set the union of all orbits with special isotropy.
2. Differential 1-forms on a manifold determine a codimension-one distribution (a sub-bundle of the tangent bundle), except at the points where the 1-form vanishes. Under a certain "integrability condition", this gives rise to a codimension-one foliation with singularities at the points where the 1-form vanishes. If the 1-form is holomorphic, we get a holomorphic foliation.
3. If we consider a holomorphic function f from an $(n+k)$-manifold M into an n-manifold N which is a submersion (like a projection locally) at most points, and we consider the fibers $f^{-1}(y)$ of all points in $y \in N$, we get a codimension-n holomorphic foliation on M, with singularities at the critical points of f.
4. Open-books are a special class of foliations, with codimension-one leaves (the pages) and a codimension-two singular set (the binding). The Milnor fibrations

associated with isolated critical points of holomorphic maps are examples of open-books.

The theory of holomorphic foliations has its origins in the study of differential equations on the complex plane by C. F. Gauss, A. L. Cauchy, B. Riemann and others. We are happy to have in volumes V and VI important contributions from several of the best research schools worldwide. Volume VI ends with an epilogue by Jean-Pierre Ramis, one of the main world leaders in the theory of complex foliations, which abounds in deep thoughts and reflections concerning the Stokes phenomena, with plenty of historical roots and a vast bibliography.

Volume VII consists of fourteen chapters, and it begins with an exposition on Jim Damon's amazing legacy, full of brilliant ideas and results on both, the foundations of singularity theory and its applications, particularly to medical image computing. Other topics included in volume VII are as follows:

- Singularities of real functions and their bifurcation sets
- Perturbation theory of polynomials and linear operators
- C^∞ and complex analytic frontal singularities
- Global singularity theory of real and complex functions
- Singularity theory from the perspective of symplectic geometry and Floer-type homology theories
- Derivation Lie algebras of isolated singularities
- Three-dimensional rational isolated complete intersection singularities
- Stability of KLT singularities, log terminal singularities and Stable Degeneration Conjecture
- V-filtrations, a theory which springs from the work of Bernstein and Kashiwara on $\mathcal{D}$-modules
- Hodge, toric and motivic methods in the study of Milnor fibers in singularity theory, both from the local and the global points of view
- The Monodromy Conjecture, a longstanding open problem in singularity theory that lies in the crossroads of number theory, algebra, analysis, geometry and topology
- Recent developments in the study of the algebraic complexity of concrete optimization problems in applied algebraic geometry and algebraic statistics

This Volume VIII consists of twelve chapters and begins with a chapter by Norbert A'Campo and Pablo Portilla on "Plane curve singularities via divides", a concept introduced in the 1970s independently by A'Campo and Sabir Gusein-Zade. Divides are generic relative immersions of compact one-dimensional manifolds in the closed unit disk, where relative means that if one or more components of the 1-manifold are segments, then the boundary is mapped to the boundary of the disk.

Chapter 2, by Ilia Itenberg, is about Viro's method in the topology of real algebraic varieties. This is a powerful construction proposed by O. Viro in the late 1970s, originally designed as an approach to obtain topologically controlled smoothings of real singularities and to produce real algebraic varieties with prescribed topology. The Viro method also has deep connections with tropical geometry. The main purpose of this chapter is to give a brief introduction to the Viro method and its connections with tropical geometry.

Chapter 3 is an introduction to local tropicalization, by Patrick Popescu-Pampu and Dmitry Stepanov. This chapter explains four viewpoints on the local tropicalization of formal subgerms of toric germs, which is a local analog of the global tropicalization of subvarieties of algebraic tori. The chapter first illustrates some of those viewpoints in the case of plane curve singularities, then it moves to arbitrary dimensions.

Chapter 4 by Enrique Artal concerns superisolated singularities in $\mathbb{C}^3$, and their "friends". Superisolated surface singularities in $\mathbb{C}^3$ were introduced by I. Luengo to prove that the μ-constant stratum may be singular. The main feature of this family is that it can bring information from the projective plane curves (global setting but smaller dimension) into surface singularities and complicated enough to offer a variety of properties. The Lê-Yomdin singularities are a generalization which offers a wider catalog of examples. The chapter studies properties of these singularities and introduces another family with similar properties in the quasi-homogeneous setting.

Chapter 5 by José Luis Cisneros and Agustín Romano studies the McKay correspondence and generalizations of it. This establishes a bijection between the non-trivial irreducible representations of finite subgroups Γ of $\mathrm{SL}(2, \mathbb{C})$ and the irreducible components of the exceptional divisor in the minimal resolution of the associated Kleinian singularity, $X = \mathbb{C}^2/\Gamma$. The chapter begins by reviewing the classical McKay correspondence and then describes the reinterpretation by Artin and Verdier using reflexive $\mathcal{O}_X$-modules. Then the authors discuss the generalization for rational singularities and the classification of reflexive modules of rank one over rational and minimally elliptic singularities. Finally, they look at special reflexive modules on Gorenstein singularities and the deformation theory of full sheaves.

Chapter 6 by David Massey is on characteristic cycles of complex of sheaves. This is a concept that has its origins in the interplay between algebraic geometry, representation theory and microlocal analysis. Its modern formulation is much indebted to the work of Kashiwara and Shapira, on the one hand, and to Beilinson and Bernstein, on the other hand. The Riemann-Hilbert correspondence provides a deep connection between characteristic cycles and Lagrangian intersection theory. In this chapter, the author provides results on how to begin with the characteristic cycle (possibly in a graded, enriched form) of a given complex of sheaves and then calculate the characteristic cycles of related complexes, such as the nearby cycles and vanishing cycles of the original complex.

Chapter 7 by Sabir Gusein-Zade is on monodromy zeta functions in singularity theory. Consider, for instance, the germ of a holomorphic function $f : (\mathbb{C}^n, 0) \to (\mathbb{C}, 0)$. One has the classical Milnor fibration over a small circle C_δ in $\mathbb{C}$ with fibre $V_f = f^{-1}(t) \cap B_\varepsilon$. Let $h = h_f : V_f \to V_f$ be a monodromy transformation of this fibration. We can assume that h is a diffeomorphism, and it is well defined up to isotopy. The monodromy operators are the actions h^* of h on the (co)homology groups $H^q(V_f;\mathbb{Z})$. Then the monodromy zeta function of f is the rational function in t defined by

$$\zeta_f(t) = \prod_{q\geq 0}\left(\det(\mathrm{id} - t\cdot h^*_{|H^q(V_f;\mathbb{R})})\right)^{(-1)^q}.$$

This chapter discusses notions of monodromy zeta functions, their properties, methods of computation and some modifications of them.

In Chap. 8, Christopher Hacon surveys results related to the singularities of the minimal model program of complex quasi-projective varieties. One of the central problems in algebraic geometry is to classify algebraic varieties up to birational isomorphism. The main approach to this classification is the minimal model program which has yielded great progress for varieties defined over the complex numbers $\mathbb{C}$. The goal is to construct a birational model which is as simple as possible, of any complex projective variety. This chapter focuses on quasi-projective varieties defined over the complex numbers. A key features of this program is that even if one is primarily interested in smooth varieties, singularities naturally appear. Minimal models have terminal singularities, and in fact, a key step in the proof of the minimal model program for threefolds achieved in the 1980s, by Sh. Mori et al., was a comprehensive classification of terminal threefold singularities.

Chapter 9 by Toru Ohmoto is an introduction to a general theory of universal polynomials associated with classification of map-germs, called Thom polynomials. The theory was originated by René Thom in the 1950s and has since been evolved in various aspects by many authors. This is about intersection theory on certain moduli spaces, say classifying spaces of mono/multi-singularities of maps, which provides consistent and deep insights into both classical and modern enumerative geometry with many potential applications.

Chapter 10 by Maurício Corrêa and José Seade intertwines the theories of complex analytic singular varieties and singular holomorphic foliations on complex manifolds, two different settings with many points in common. The chapter studies invariants of vector fields and holomorphic foliations which are defined in the two settings, with the Baum-Bott residues permeating the whole chapter. The study of indices and residues is of utmost importance in the local and global study of both, singular holomorphic foliations in complex manifolds and singular complex analytic varieties. The chapter uses as spearhead the GSV-index and surveys the remarkable work of Marco Brunella on the subject. It is also a gentle introduction to the Baum-Bott residues.

In Chap. 11, entitled Monodromy in Integral Geometry and PDE, Victor Vassiliev discusses two classical problems concerning functions with integral representations and shows how the monodromy considerations can assist in the solution of these problems. The problems considered are the algebraicity of functions given by volumes cut off by hyperplanes from spatial bodies and the problem of sharp fronts in the theory of hyperbolic partial differential equations.

The final Chap. 12 is by Masahiko Yoshinaga, and it focuses on the topological theory of Hyperplane Arrangements, particularly in the case where the hyperplanes are defined over the reals. The theory of hyperplane arrangements has undergone significant development in the last decades, connecting various branches of mathematics, including Singularity Theory, Combinatorics, Algebraic Geometry, Geometric Group Theory and more. The chapter is an introduction to the subject, and it focuses, mainly, on the relationship between topology and the real structure, such as adjacent relations of chambers and stratifications related to real structures.

This collection is addressed to graduate students and newcomers to the theory, as well as to specialists who can use it as a guidebook, and it provides an accessible account of the state of the art in several aspects of the subject, its frontiers and its interactions with other areas of research. These eight volumes cover together a wide spectrum of singularity theory and its interactions with other related areas of mathematics.

For completeness, we include the tables of content of Volumes I–VII.

Extrapolating Bernard Teissier's words in the preface to Volume I of this handbook, the various topics presented in this collection are lineaments of the single landscape that goes under the name "singularity theory". Our heartfelt gratitude to all the many authors and referees that have contributed to this collection. And very special thanks to Bernard Teissier, for his great support for making this project, the Handbook of Geometry and Topology of Singularities, for being always a friend and an example. Our thanks also to Springer Verlag for publishing this collection.

Cuernavaca, Mexico — José Luis Cisneros-Molina
Marseille, France — Lê Dũng Tráng
Cuernavaca, Mexico — José Seade
March 2025

Contents of Volumes I to VII

Contents of Volume I

Contents of Volume II

Contents of Volume III

Contents of Volume IV

Contents of Volume V

Contents of Volume VI

Contents of Volume VII

Contents

Contributors

Norbert A'Campo University of Basel, Basel, Switzerland

E. Artal Departamento de Matemáticas, IUMA, Universidad de Zaragoza, Zaragoza, Spain

José Luis Cisneros-Molina Instituto de Matemáticas, Unidad Cuernavaca, Universidad Nacional Autónoma de México and International Research Laboratory CNRS, Laboratorio Solomon Lefschetz, Cuernavaca, Mexico

Maurício Corrêa Università degli Studi di Bari, Bari, Italy

Sabir M. Gusein-Zade Moscow State University, Faculty of Mechanics and Mathematics, Moscow Center for Fundamental and Applied Mathematics, GSP-1, Moscow, Russia
National Research University "Higher School of Economics", Moscow, Russia

Christopher Hacon Department of Mathematics, University of Utah, Salt Lake City, UT, USA

Ilia Itenberg Sorbonne Université and Université Paris Cité, CNRS, IMJ-PRG, Paris, France

David B. Massey Department of Mathematics, Northeastern University, Boston, MA, USA

Toru Ohmoto Department of Pure and Applied Mathematics, Waseda University, Okubo, Shinjuku, Tokyo, Japan

Patrick Popescu-Pampu Univ. Lille, CNRS, UMR 8524 - Laboratoire Paul Painlevé, Lille, France

Pablo Portilla Cuadrado Universidad Politécnica de Madrid, Madrid, Spain

Agustín Romano-Velázquez Instituto de Matemáticas, Unidad Cuernavaca, Universidad Nacional Autónoma de México, Cuernavaca, Morelos, Mexico

José Seade Instituto de Matemáticas, Unidad Cuernavaca, Universidad Nacional Autónoma de México and International Research Laboratory CNRS, Laboratorio Solomon Lefschetz, Cuernavaca, Mexico

Dmitry Stepanov Department of Higher Mathematics and Center of Fundamental Mathematics, Moscow Institute of Physics and Technology, Dolgoprudny, Moscow, Russia

Lê Dũng Tráng Marseille, France

V. A. Vassiliev Weizmann Institute of Science, Rehovot, Israel

Masahiko Yoshinaga Department of Mathematics, Osaka University, Toyonaka, Osaka, Japan

Chapter 1
Plane Curve Singularities via Divides

Norbert A'Campo and Pablo Portilla Cuadrado

Contents

N. A'Campo
University of Basel, Basel, Switzerland
e-mail: norbert.acampo@unibas.ch

P. Portilla Cuadrado (✉)
Universidad Politécnica de Madrid, Madrid, Spain
e-mail: p.portilla@upm.es

© The Author(s), under exclusive license to Springer Nature Switzerland AG 2026

J. L. Cisneros-Molina et al. (eds.), *Handbook of Geometry and Topology of Singularities VIII*, https://doi.org/10.1007/978-3-031-99571-2_1

Abstract Generic relative immersions of compact one-manifolds in the closed unit disk, i.e. divides, provide a powerful combinatorial framework, and allow a topological construction of fibered classical links, for which the monodromy diffeomorphism is explicitly given as a product of Dehn twists. Complex isolated plane curve singularities provide a classical fibered link, the Milnor fibration, with its Milnor monodromy, monodromy group, and vanishing cycles.

This surveys puts together much of the work done on divides and their role in the topology of isolated plane curve singularities. We review two complementary approaches for constructing divides: one via embedded resolution techniques and controlled real deformations, and another via Chebyshev polynomials, which yield explicit real morsifications. A combinatorial description of the Milnor fiber is developed, leading to an explicit factorization of the geometric monodromy as a product of right-handed Dehn twists. We further explore the structure of reduction curves that arise from the Nielsen description of quasi-finite mapping classes and from iterated cabling operations on divides.

The interplay between the geometric and integral homological monodromies is analyzed, with special attention to symmetries induced by complex conjugation and strong invertibility phenomena. In particular, the integral homological monodromy for isolated plane curve singularities can be computed effectively. In contrast, for complex hypersurface singularities in higher dimensions no method of computation of the integral homology monodromy is known.

Connections with mapping class groups, contact and symplectic geometry, and Lefschetz fibrations are also discussed. We conclude by outlining several open problems and conjectures related to the characterization of divides among fibered links, the presentation of geometric monodromy groups, and the existence of symplectic fillings compatible with the natural fibration structures.

1.1 Introduction

The concept of divides (or partage in French) has become a significant tool in the study of isolated plane curve singularities and their associated monodromy groups. Introduced in the 1970s, independently by the first author [3] and Sabir Gusein-Zade [22, 23], divides provided a novel combinatorial approach to understanding the topology of singularities and their deformations. Both researchers were deeply influenced by the broader developments in singularity theory initiated by John Milnor's foundational work on the Milnor fibration [37], which showed that isolated hypersurface singularities could be understood through their associated fibration structure. Divides emerged as a geometric tool that could encode the topological behavior of these singularities by associating them with fibered links in three-dimensional spaces.

The first author's pioneering work on divides tied tightly these objects to the monodromy of singularities. His work throughout the years drew a big picture that placed these objects as an useful and novel tool to understand plane curves. In particular, he showed that a divide associated with a plane curve singularity encoded every topological piece of information associated with the plane curve singularity: the link, a model of the Milnor fiber, the geometric monodromy as a product of Dehn twists around vanishing cycles and even the reduction curves of the Nielsen-Thurston decomposition of the monodromy. This work also demonstrated that divides could serve as a bridge between knot theory and the study of singularities, particularly by offering a combinatorial framework for understanding the monodromy representation of some fibered link that share a lot of similarities with those coming from singularity theory.

In parallel, Gusein-Zade developed an alternative approach to divides, focusing on the real morsifications of singularities and using Chebyshev polynomials to generate divides for specific classes of real plane curves. This approach offered a more algebraic route, allowing the explicit construction of divides from polynomial models. Gusein-Zade's contributions enriched the theory of divides, showing that they could be systematically constructed for a wide variety of singularities, further connecting the geometric properties of divides with algebraic invariants like Puiseux pairs and Newton polygons.

Divides also exhibit strong symplectic properties, connecting them to the study of symplectic geometry and symplectic fillings. Research into the symplectic properties of divides has led to advances in the understanding of how these objects interact with higher-dimensional spaces, further expanding their applications.

This survey explores the modern theory of divides, focusing on their combinatorial and geometric properties, as well as their applications to the study of singularities. We discuss the role of divides in generating fibered links, describe their connections to the Milnor fiber and geometric monodromy, and highlight recent advances in the symplectic and topological study of these objects. Additionally, we state and discuss some open questions, such as the broader symplectic properties of divides.

Through this work, we aim to provide a comprehensive overview of the current state of the theory of divides, demonstrating its central role in singularity theory, knot theory, and beyond. Much of the content comes directly from the classical work of the first author. Some proofs have been revisited or expanded with some details added and new detailed examples are considered. The order in which the material is told, does not necessarily respect the chronological order in which it appeared but rather an expository one.

1.1.1 Organization of the Paper

In Sect. 1.2 We fix notation and introduce some basic concepts about the theory of mapping class groups and the theory of plane curves. Among others we introduce the Nielsen-Thurston decomposition and the important concepts of geometric vanishing cycle and geometric monodromy group associated with an isolated plane curve singularity.

In Sect. 1.3 we give the abstract definition of divide and we associate to it a link in the three-sphere. Furthermore, in Lemma 1.3.9 we prove the existence of adapted functions (Morse functions that define the divide) for each divide.

In Sect. 1.4 we prove a general fibration theorem (Theorem 1.4.1): the link that we associated to each divide is a fibered link. The fibration is given by a very explicit map which can be constructed from an adapted function.

In Sect. 1.5 we start the study of divides coming from plane curve singularities. In particular we introduce the concept of totally real plane curve singularity and we explain two methods to produce divides for totally real plane curves. The first method, presented in Sect. 1.5.2, is due to the first author and relies on choosing a resolution and then perform a series of controlled perturbations of the strict transform and contractions of exceptional divisors with self intersection -1 alternately. The second method, explained in Sect. 1.5.3 is due to Gusein-Zade and it relies on the fact that Chebyshev polynomials can be used to parametrize the 0 set of Brieskorn-Pham polynomials. This fact together with an iterative technique, allows one to produce divides for any topological type of plane curve realized by a particular real model.

In Sect. 1.6 we describe combinatorially the Milnor fiber of the function constructed in Sect. 1.4. This combinatorial description is crucial in the description of the monodromy given in Sect. 1.7. In the first subsection we describe the monodromy as a product of right handed Dehn twists around the geometric vanishing cycles of a distinguished basis. In the second subsection we deal with special properties of the geometric monodromy associated with a plane curve singularity, more concretely we explore the property of being strongly invertible and we discuss the state of the art in higher dimensions.

In Sect. 1.8, we explain how one can, from a divide, to easily produce an explicit model for the Milnor fiber together with a set of geometric vanishing cycles associated with a distinguished basis. Furthermore, if the divide is produced by the previously introduced cabling technique, we explain how to visualize in this model, the reduction curves of the geometric monodromy associated with the divide.

Finally, in Sect. 1.9 we state a few open questions related to divides associated with plane curves together with some other properties that are not fully explored in this work.

1.2 Preliminary Theory on Mapping Class Groups and Plane Curves

We turn now our attention to the theory of mapping class groups. The purpose of this section it to collect the results on mapping class group that we use throughout the rest of the text as well as to fix notations and conventions. For more insight on this topic we refer to the book by B. Farb and D. Margalit [19].

1.2.1 Mapping Class Groups

Let F be a compact oriented connected surface with boundary.

Definition 1.2.1 The *mapping class group* of F is defined as

$$\mathrm{Mod}(F) := \pi_0(\mathrm{Diff}^+(F, \partial F)).$$

That is, elements of $\mathrm{Mod}(F)$ are isotopy classes of oriented diffeomorphisms of F that fix the boundary point-wise. Each of these classes is called a *mapping class*.

Relative diffeomorphisms α, β of a surface $(F, \partial F)$ are in fact in the same mapping class if relatively homotopic.

Next we define the basic elements that form the mapping class group: the Dehn twists.

Notation Given a simple closed curve $\gamma \subset \Sigma$ disjoint from the boundary on an oriented surface we denote by

$$T_\gamma : \Sigma \to \Sigma$$

a **right-handed Dehn twist** around γ or its mapping class in the mapping class group of Σ relative to its boundary. The support of T_γ is concentrated, by definition, in a tubular neighborhood of γ. Also, the mapping class of T_γ only depends on the isotopy class of the simple closed curve γ (Fig. 1.1).

Now we turn to our attention to another construction, very much related to the one above, that will turn out to be very useful later on this text. We introduce the notions: minimal positive pair of Dehn twists, half twist, right half Dehn twist and left half Dehn twist.

Definition 1.2.2 Let F be an oriented surface and let $z : S^1 \to F$ be a simple closed parametrized curve on F. Let F' be the surface obtained from the surface F by cutting F along the image Z of z and by gluing back with the diffeomorphism that identifies images of opposite points of S^1; denote by $Z' \subset F'$ the image of Z; we call this diffeomorphism a *half twist*. The surfaces F and F' are of course diffeomorphic and $F \setminus Z$ and $F' \setminus Z'$ are equal as sets. A *minimal positive pair of*

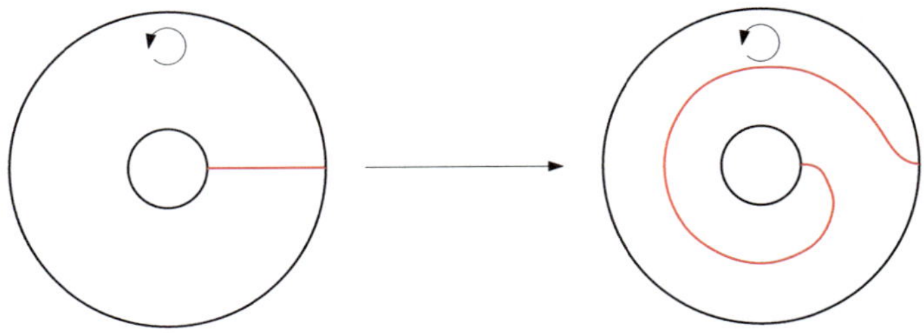

Fig. 1.1 A right-handed Dehn twist acting on a properly embedded segment. The orientation of the annulus is indicated by the curved arrows

Dehn twists from F to F' is a pair of diffeomorphisms (p, q) from F to F' such that the following holds:

(a) The composition $q^{-1} \circ p : F \to F$ is a right Dehn twist with respect to the orientation of F having the curve Z as core. In addition $p(Z) = q(Z) = Z'$ holds
(b) There exists a regular collar neighborhood N of Z in F such that both p and q coincide with the identity of $F \setminus Z = F' \setminus Z'$ outside N.
(c) For some volume form ω on N, which we think of as a symplectic structure, we have $p^*\omega = q^*(\omega) = \omega$, and the sum of the Hofer distances [28, Chapter 5] to the identity of the restrictions of p and q to $N \setminus z$ is minimal.

Minimal positive pairs of Dehn twists exist and are well defined up to isotopy. For a minimal positive pair (p, q) of Dehn twists, the member p is called *positive or right half Dehn twist* and the member q is called *negative or left half Dehn twist*.

1.2.1.1 Nielsen-Thurston Decomposition

The following decomposition result is a landmark in mapping class group theory.

Theorem 1.2.3 (See [46] and Corollary 13.3 from [19]) *Let $\phi : \Sigma \to \Sigma$ be an orientation preserving homeomorphism that restricts to the identity on $\partial\Sigma$. Then there exists ϕ' isotopic to ϕ and a collection $\mathcal{C}$ of non-null-homotopic disjoint simple closed curves (called reduction curves) including all boundary components such that:*

1. *The collection of curves is invariant by ϕ', i.e. $\phi'(\mathcal{C}) = \mathcal{C}$.*
2. *The homeomorphism ϕ' restricted to the union of components of each ϕ'-orbit of connected components of $\Sigma \setminus \mathcal{C}$ is isotopic either to a periodic or to a pseudo-Anosov homeomorphism.*

Definition 1.2.4 Any collection of curves $\mathcal{C}$ satisfying 1 above is called a *reduction system*. That is, a collection of simple closed curves that is invariant (up to isotopy) by ϕ is a reduction system.

The decomposition given by Theorem 1.2.3 is called *Nielsen-Thurston decomposition* and is unique up to isotopy if the reduction system of curves $\mathcal{C}$ is minimal by inclusion. We assume whenever necessary that the representatives of the mapping classes used in this work satisfy 1 and 2 from the previous theorem. This decomposition leads to the following definition.

Definition 1.2.5 Let $\phi : \Sigma \to \Sigma$ be a representative of a mapping class, we say that ϕ is **pseudo-periodic** if only periodic pieces appear in its Nielsen-Thurston decomposition.

Pseudo-periodic homeomorphisms are of special importance in complex singularity theory, as all geometric monodromies of holomorphic germs of functions on isolated complex surface singularities are of this kind. In particular, monodromies of isolated plane curve singularities are of this type. In general, local geometric monodromies of complex hypersurface singularities are dynamically restricted: the geometric monodromy can be realized by a distal map, which topological entropy vanishes. This shows that in the case of curve singularities the reduction of the geometric monodromy does not have pseudo-Anosov components. See also [2].

1.2.2 Isolated Plane Curve Singularities

In this section we review some basic notions on plane curve singularities that are mentioned and used throughout this text.

The algebra of convergent power series $\mathbb{C}\{x, y\}$ is a unique factorization domain and so, up to multiplication by a unit, a series f defining an isolated plane curve singularity, can be uniquely expressed as $f = f_1 \cdot \cdots \cdot f_r$ where each f_i is an irreducible convergent power series. Each f_i is called a *branch* of f and if $r = 1$ we say that f is *irreducible*.

1.2.2.1 The Milnor Fibration

In [37], Milnor proved that for a holomorphic map $f : \mathbb{C}^{n+1} \to \mathbb{C}$ with an isolated singularity at the origin,

$$\frac{f}{|f|} : S_\epsilon^{2n+1} \setminus K \to S^1$$

is a locally trivial fibration. Here $S_\epsilon^{2n+1} \subset \mathbb{C}^{n+1}$ denotes a suitably small sphere and the *link* K is defined by $K := f^{-1}(0) \cap S_\epsilon^{2n+1}$. If we denote by $D_\delta \subset \mathbb{C}$ a small disk of radius δ centered at 0 and by $B_\epsilon \subset \mathbb{C}^n$ a ball of radius ϵ, it follows from Ehresmann's fibration lemma that

$$f|_{f^{-1}(\partial D_\delta) \cap B_\epsilon} : f^{-1}(\partial D_\delta) \cap B_\epsilon \to \partial D_\delta \tag{1.1}$$

is also a locally trivial fibration for ϵ small enough and δ small with respect to ϵ. In this case, we say that B_ϵ is a *Milnor ball* and that ϵ is a Milnor radius. Milnor proved in [37] that these two fibrations are essentially equivalent.

The fibers of second fibration are connected compact $2n$-manifolds with non-empty boundary $(F, \partial F)$, and the fibers of first fibration are diffeomorphic to the interior $F \setminus \partial F$ of the fibers of second fibration which carry a complex structure.

Any fiber F of these fibrations is called *the Milnor fiber* of f. When $n = 1$, the second is a connected oriented compact surface with non-empty boundary. The Milnor fiber has r boundary components, where r is the number of branches of f. Its first Betti number $b_1(F)$ coincides with

$$\dim_{\mathbb{C}} \frac{\mathbb{C}\{x, y\}}{(\partial f/\partial x, \partial f/\partial y)}$$

and any of these quantities is called *the Milnor number of* f and denoted by μ_f or simply by μ if there is no ambiguity. The topological information of a plane curve singularity is carried by the r-component oriented classical link $\partial F \subset \partial B_\epsilon$ in its oriented ambient 3-sphere.

1.2.2.2 Puiseux Pairs and Intersection Multiplicities

There are a lot of different ways of codifying by numerical invariants the topological information of a plane curve singularity. We cite [14, 49] or [18] as standard references on this topic.

Next, we briefly recall what Puiseux pairs are, since they appear several times on this work. In particular, Puiseux pairs codify topologically uni-branch plane curve singularities. We give an axiomatic approach to this theory.

Definition 1.2.6 A finite sequence Pp of pairs of integers $(a_1, b_1), \ldots, (a_k, b_k)$ is a sequence of *essential Puiseux pairs* if and only if

$$2 \leq a_i < b_i,$$

$$b_i/(a_1 a_2 \cdots a_i) < b_{i+1}/(a_1 a_2 \cdots a_{i+1}) \text{ and}$$

$$\gcd(b_i, a_1 \cdots a_i) = 1$$

for all $i = 1, \ldots, k$.

A sequence of essential Puiseux pairs Pp defines a family of topologically equivalent singularities. A specific member $f_{Pp}(x, y)$ of this family is obtained from a Puiseux expansion with fractional and strictly increasing exponents

$$y = x^{b_1/a_1} + x^{b_2/a_1 a_2} + \cdots + x^{b_k/a_1 a_2 \cdots a_k}$$

by the rule, which takes into account the ramification of $x^{1/a_1 a_2 \cdots a_k}$,

$$f_{Pp}(x, y) = \prod_{\theta} (y - \theta^{a_2 \cdots a_k b_1} - \theta^{a_3 \cdots a_k b_2} - \dots - \theta^{b_k}),$$

where θ runs over the $a_1 a_2 \cdots a_k$ roots of $z^{a_1 a_2 \dots a_k} - x = 0$ in the algebraic closure of the field $\mathbb{C}((x))$. The coefficients of the polynomial $f_{Pp}(x, y)$ are integers.

For example, the Puiseux expansion $y = x^{3/2} + x^{7/4}$ with strictly increasing exponents for $Pp = ((2, 3), (2, 7))$ leads to the polynomial $f_{(2,3),(2,7)} = (y^2 - x^3)^2 - 4x^5 y - x^7$ and the Puiseux expansion $y = x^{3/2} + x^{11/6}$ to the polynomial $f_{(2,3),(3,11)} = (y^2 - x^3)^3 - 6x^7 y^2 - 2x^{10} - x^{11}$.

There is a second equivalent set of numerical invariants that are more suitable for certain matters: *Newton pairs*. These are a finite sequence Np of coprime integers (a_i, λ_i) that can be computed from the Puiseux pairs by the recursive formula

$$\begin{aligned} \lambda_1 &:= b_1, \\ \lambda_{i+1} &:= b_{i+1} - b_i a_{i+1} + \lambda_i a_{i+1} a_i . \end{aligned} \tag{1.2}$$

Instead of using the above Puiseux expansion one uses for given Newton pairs Np Newton's Ansatz

$$y = x^{\lambda_1/a_1}(1 + x^{\lambda_2/a_1 a_2}(1 + x^{\lambda_3/a_1 a_2 a_3}(\dots)))$$

which expands to the corresponding Puiseux expansion.

Now suppose that $f = f_1 \cdots f_r$ is the factorization of a reducible isolated singularity f in branches f_i. In this case, Puiseux or Newton pairs of each of the branches are not enough to determine the topology of the singularity. One need one extra piece of information, for example *intersection multiplicities*. For each pair $i, j \in \{1, \dots, r\}, i \neq j$, we define the *intersection multiplicity* between the branches f_i and f_j as

$$\dim_{\mathbb{C}} \mathbb{C}\{x, y\}/(f_i, f_j)$$

which is in fact the linking number of the oriented knots corresponding to these branches. For an interpretation of the numbers λ_i of Eq. 1.2 in terms of intersection numbers, see Sect. 1.5.3.

For instance, Puiseux pairs and Newton pairs can be truncated: denote by Pp_j or $Np_j, 0 \leq j \leq r$, the initial j entries of given equivalent Pp, Np and let $f_{Pp_j}(x, y)$ define a sequence of plain curve singularities. Here, Pp_0, Np_0 will be the empty set of Puiseux pairs, Newton pairs corresponding to $f_{\emptyset}(x, y) = y$. The equations $f_{Pp_j}(x, y) = 0, j = 0, \dots, r$, define oriented knots $K_j \subset S^3_{\epsilon}$. The Newton pairs get a topological interpretation: The knot K_j is the (a_j, λ_j) cable knot on the knot $K_{j-1}, j = 1, 2, \dots, r$. See the book by David Eisenbud and Walter Neumann [18] for references and historical remarks.

1.2.3 The Versal Deformation Space and the Geometric Monodromy Group

1.2.3.1 The Versal Unfolding

We briefly recall here the notion of the *versal unfolding* of an isolated singularity; see [10, Chapter 3] for more details. Recall the algebra

$$A_f = \frac{\mathbb{C}\{x, y\}}{(\partial f/\partial x, \partial f/\partial y)};$$

Let

$$g_1, \ldots, g_\mu \in \mathbb{C}[x, y]$$

be polynomials that project to a basis of A_f, assume (we can always do so) that $g_1 = 1$. For $\lambda = (\lambda_1, \ldots, \lambda_\mu) \in \mathbb{C}^\mu$, define the function f_λ by

$$f_\lambda = f + \sum_{i=1}^{\mu} \lambda_i g_i.$$

The *base space of the versal unfolding* of f is the parameter space of all λ which is naturally isomorphic to $\mathbb{C}^\mu$. The *discriminant locus* is the subset

$$\text{Disc} = \{\lambda \in \mathbb{C}^\mu \mid f_\lambda^{-1}(0) \text{ is not smooth}\}.$$

It can be shown that Disc is an algebraic hypersurface. The discriminant locus admits different stratifications; for example, it is stratified according to the sum of the Milnor numbers of the singular points of the singular fibers lying above. In this sense, the top-dimensional stratum (that is, the smooth part of Disc) parameterizes curves with a single node. Denote by V_f a small closed ball in $\mathbb{C}^\mu$ centered at the origin. Define

$$X_f = \{(\lambda, (x, y)) \mid (x, y) \in f_\lambda^{-1}(0),\ \lambda \notin \text{Disc}\}. \tag{1.3}$$

Then, for V_f small enough and after intersecting X_f with a sufficiently small closed polydisk, this family has the structure of a smooth surface bundle with base $V_f \setminus \text{Disc}$ and fibers diffeomorphic to the Milnor fiber F of Eq. 1.1. We fix a point in $V_f \setminus \text{Disc}$ and we denote, also by F, the fiber with boundary lying over it.

1.2.3.2 The Geometric Monodromy Group

Recall that a mapping class (Definition 1.2.1) is an isotopy class of orientation preserving diffeomorphisms of F that restricts to the identity on the boundary, where the isotopies are required to fix the boundary *point-wise*. Let $f : \mathbb{C}^2 \to \mathbb{C}$ define an isolated singularity at the origin, with Milnor fiber F.

Definition 1.2.7 The *geometric monodromy group* is the image in $\mathrm{Mod}(F)$ of the monodromy representation

$$\rho : V_f \setminus \mathrm{Disc} \to \mathrm{Mod}(F)$$

of the universal family X_f of Eq. 1.3.

Definition 1.2.8 A *quadratic vanishing cycle* or *geometric vanishing cycle* or, for the purposes of this work, simply a *vanishing cycle* is a simple closed curve $c \subset F$ that gets contracted to a point when transported to the nodal curve lying over a smooth point of the discriminant Disc.

Let B_ϵ be a Milnor ball for an isolated plane curve singularity f. Then, for any linear form $\ell : \mathbb{C}^2 \to \mathbb{C}$ generic with respect to f and $\eta > 0$ small enough, the map

$$\tilde{f} := f + \eta\ell : B_\epsilon \to \mathbb{C}$$

only has Morse-type singularities and the corresponding critical values $c_1, \ldots, c_\mu$ are all distinct and close to $0 \in \mathbb{C}$. The holomorphic map $\tilde{f}$ is usually called a *morsification* of f.

Remark 1.2.9 Two isolated plane curve singularities with the same topological type can be connected by a 1-parameter μ-constant family and thus, they have the same geometric monodromy group. This last statement follows from the following: in [22, Theorem 3] it is proven that two plane curve singularities with the same topological type can be joined by a μ-constant family or, equivalently, by family of singularities with the same topological type. As a consequence, the intersection matrices (of distinguished basis) of these singularities coincide. Also, by a result of Hamm and Lê [25], the geometric monodromy group o a plane curve singularity can be computed from any morsification. As a consequence of this discussion, the geometric monodromy group (Definition 1.2.7) of an isolated plane curve singularity is a topological invariant.

1.3 Definition of Divide and Some Combinatorial Properties

In this section we give the first definitions and properties of divides. In particular, we prove the existence of *adapted functions* to divides (Lemma 1.3.9) which are Morse functions with very particular properties.

1.3.1 Definition of Divide

We start with the definition of the central object of this work.

Definition 1.3.1 A *divide* is a generic relative immersion $\alpha : J \to D$ of a compact 1-dimensional manifold J into a disk D. Usually D is a disk in the Euclidean plane $\mathbb{R}^2$ or the Gaussian plane $\mathbb{C}$.

Notation A divide is usually identified with the image of the immersion and denoted by P, that is, with the notation of the above definition $P = \alpha(J)$.

Remark 1.3.2

- *relative* means that $\alpha(\partial J) \subset \partial D$ and $\alpha^{-1}(\partial D) = \partial J$.
- *generic* means that the restriction of α to ∂J is injective, that $\alpha(J)$ is transverse to ∂D, and that the curve $\alpha(J)$ has only ordinary double points in the interior of D (this leads to the following Definition 1.3.3)
- When not important, the radius of the disk will not be specified. If we need to specify the radius of the disk, we will write D_ρ for a disk of radius ρ centered at the origin.
- In certain parts of this survey, we only consider divides which are immersions of disjoint union of segments and, in other parts, we also consider divides which are immersions of disjoint union of segments and circles. It will be pointed out when necessary. If no clarification is made, we assume the second and more general version.

It will be very useful to deal with a more *relaxed* version of the notion of divide which is one where we allow ordinary singularities other than double points.

Definition 1.3.3 A *pre-divide* is a relative immersion $\alpha : J \to D$ that is generic near ∂D as above and such that for all distinct $p, q \in J$ with $\alpha(p) = \alpha(q)$ the images of the tangent spaces

$$(D\alpha)_p(T_p J) \neq (D\alpha)_q(T_q J)$$

in $T_{\alpha(p)} D$.

Note that the only difference between a divide and a pre-divide is that in a pre-divide we admit singularities which are multiple crossing points (see Fig. 1.2).

We need to prepare the construction of a classical link $L(P)$ in the 3-sphere from a divide. Let $D = D_\rho$ be a disk (for convenience centered at the origin) of radius ρ in the Euclidean plane $\mathbb{R}^2$ with norm $||x||^2 = |x_1|^2 + |x_2|^2$. We equip the tangent space $T\mathbb{R}^2 = \mathbb{R}^2 \times \mathbb{R}^2$ with the Euclidean norm $||(x, u)||^2 = |x_1|^2 + |x_2|^2 + |u_1|^2 + |u_2|^2$, where we denote the tangent vector $u \in T_x\mathbb{R}^2$ by (x, u). The 4-ball B^4_D of radius ρ over D is the subset in TD defined by $B^4_D = \{(x, u) \in TD \mid ||(x, u)|| \leq \rho\}$. The 3-sphere S^3_D that will contain the link $L(P)$ is the boundary ∂B^4_D. Observe that $\partial D_\rho \times \{0\}$ is identified with the set of tangent vectors $\{(x, 0) \in TD_\rho \mid ||x|| = \rho\}$.

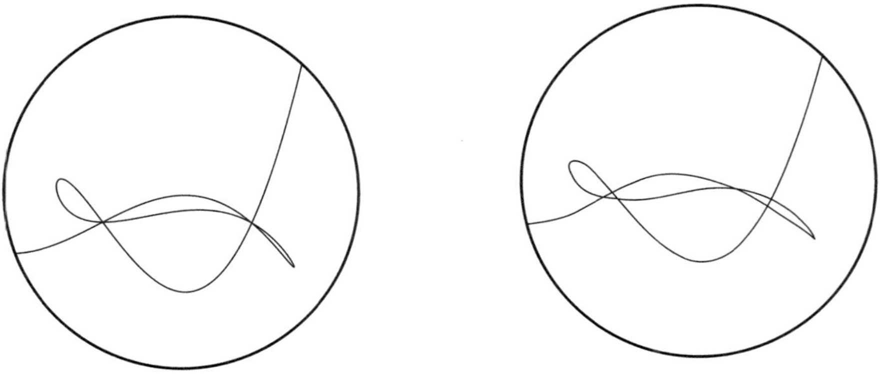

Fig. 1.2 On the left we see a divide where there are two ordinary singularities of order 3. On the right we see a divide that is obtain from the first one after a small perturbation

Let P be a divide or pre-divide in D. We denote by $T_D P$ the set of all vectors $(x, u) \in TD$ with $x \in P, u \neq 0$, or $x \in P \cap \partial D$ that are tangent to one of the branches of P.

In view of the study of complex plane curve singularities, we think $T\mathbb{R}^2$ as the complex plane $\mathbb{C}^2$ via the identification

$$(x, u) \in T\mathbb{R}^2 = \mathbb{R}^2 \times \mathbb{R}^2 \mapsto (x_1 + u_1 i, x_2 + u_2 i) \in \mathbb{C}^2$$

and choose to orient $T\mathbb{R}^2$ by the "complex" orientation given by the frame $(\frac{\partial}{\partial x_1}, \frac{\partial}{\partial u_1}, \frac{\partial}{\partial x_2}, \frac{\partial}{\partial u_2})$. As a consequence, the 4-ball B^4_D and its boundary S^3_D inherit orientations. Also the smooth part of $T_D P$ becomes an oriented surface.

With this setting we are ready to define the link associated to a given divide.

Definition 1.3.4 Given a divide $P \subset D$ we define the *associated link* $L(P)$ as

$$L(P) := \{(x, u) \in T_D P \mid \|(x, u)\| = \rho\} = T_D P \cap S^3_D \subset S^3_D. \tag{1.4}$$

We will later see that not all links come from divides. We will also see that links that come from divides are fibered and that, even not all fibered links come from divides. For example, in Remark 1.7.14, it is explained why the figure eight knot (Fig. 1.3) cannot appear as the knot associated to a divide.

The projection $TD \to D$ restricts to a map $S^3_D \to D$ having a circle as preimage of non boundary points of D. Those preimages shrink to circles of radius 0 as the point moves to the boundary. The restriction to $L(P)$ is generically $2:1$ over the smooth points of P not in ∂D and $1:1$ above ∂D. It is $4:1$ over the double points of P.

As example, let $\gamma : [-1, 1] \to D$ define a divide P in the disk D with end points $A = \gamma(-1), B = \gamma(+1)$ on $\partial D \subset S^3_D$. The divide P is also defined by $\gamma^*(t) = \gamma(-t)$. The knot $L(P)$ is the union of two oriented arcs on S^3_D, one running

Fig. 1.3 The figure eight knot

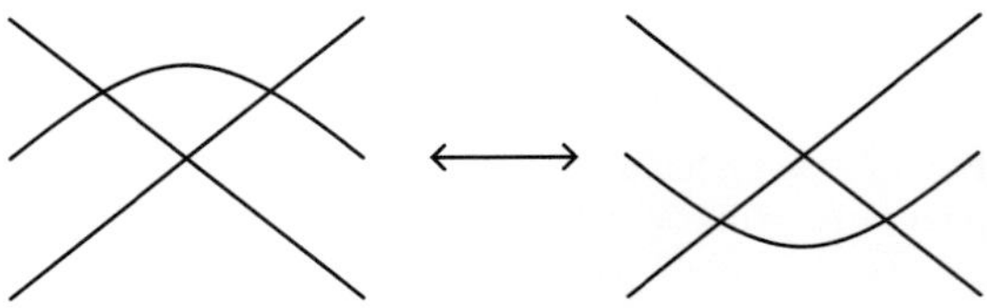

Fig. 1.4 The movement caused by an admissible isotopy near an intersection point

from A to B, the other running from B to A. The first arc, using γ is parametrized by

$$t \in [-1, +1] \mapsto (\gamma(t), \frac{\dot{\gamma}(t)}{||\dot{\gamma}(t)||}(1 - ||\gamma(t)||^2)^{1/2}) \in S^3_D$$

The second arc, similarly, uses γ^*. Both oriented arcs A, B fit together as an oriented knot.

A variant is following construction: First extend γ to an immersion Γ : $[-\tau, +\tau] \times [-1, +1] \to D$ of a rectangular thickening of the interval $[-1, +1]$. Next, for small τ', restrict Γ to a rectangle $S = [-\tau', +\tau'] \times^* [-1 + \tau', +1 - \tau']$ where $\times^*$ means that we smooth the corners. Let $\partial^+ S$ be the boundary of S with any chosen orientation. The restriction $\partial^+ \Gamma$ is a generic immersion of an oriented copy of S^1 into D. The lifting of the oriented speed vector of $\partial^+ \Gamma$ yields an oriented knot which is isotopic to the oriented knot $L(P)$. So knots coming from generic immersed intervals, come also from generically immersed oriented circles. William Gibson and Masaharu Ishikawa [21] have proven that every oriented knot can be obtained by lifting a immersion of an oriented circle in D.

Naturally, a small perturbation of a divide, produces an isotopic links in B^4_D. We can if necessary make a divide by a small isotopy more "Euclidean friendly" without changing its link: for instance we may assume without restricting generality that the divide P is near its double points an orthogonal intersection of segments. Furthermore, we can consider more involved (and no longer necessarily small) perturbations that still produce isotopic links. This leads to the following definition and lemma.

Definition 1.3.5 We say that an isotopy of a divide P is admissible if it is generated by Reidemeister moves of type III only. See Fig. 1.4.

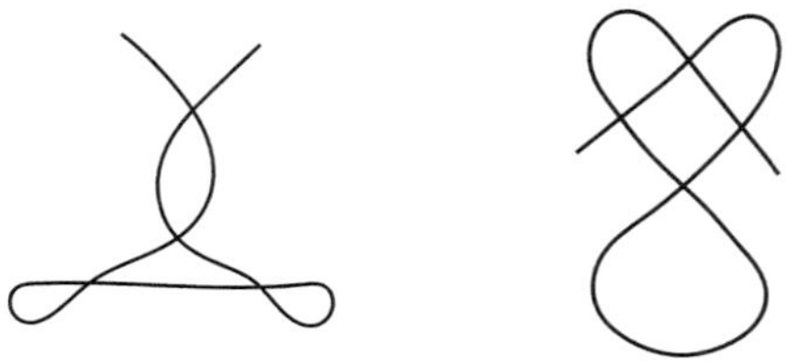

Fig. 1.5 Two divides that yield the same knot but that are not equivalent through admissible isotopies

Lemma 1.3.6 *If P and P' are related by an admissible isotopy, then $L(P)$ and $L(P')$ are isotopic in ∂B_ρ.*

Proof Following the definition Eq. 1.4, it is possible to explicitly lift an admissible isotopy as the one described in Fig. 1.4 to an isotopy of links of divides as defined by Definition 1.3.4. □

Example 1.3.7 Note that admissible isotopies do not yield a complete set of moves to go from one divide P to another one P' with $L(P) \sim L(P')$. Indeed it can be verified that the links from the figure yield isotopic knots but one cannot go from one to another only admissible isotopies, see Gibson-Ishikawa, two divides for knot 10_{139} not related by III-moves. See also Fig. 1.5.

Knots or links $L(P)$ for divides P are very special. For instance let $L(P)$ be the knot of a divide $P \subset D$. The symmetry given by the involution $(x, u) \mapsto (x, -u)$ fixes point-wise $\partial D \subset S^3_D$ and preserves $L(P)$ globally, with two fixed point on $L(P)$. So the knot $L(P)$ is strongly invertible, see Makoto Sakuma [42]. Divides links $L(P)$ are closures of quasi positive braids by Tomomi Kawamura's result [32].

1.3.2 Adapted Morse Function f_P on D and Its Extension θ_P on B^4_D

Definition 1.3.8 Let $P \subset D$ be a divide. We say that a smooth function

$$f_P : D \to \mathbb{R}$$

is *adapted* to P if

1. f_P is locally generic, i.e. it only has non-degenerate critical points
2. each bounded region has exactly one non-degenerate maximum or minimum. in its interior with critical values ± 1
3. the saddle points of f_P are exactly at the double points of P and $P = f_P^{-1}(0)$.
4. each contractible unbounded region has exactly one non-degenerate maximum or minimum on its intersection with ∂D.
5. if $P \cap \partial D = \emptyset$ then the restriction of f_P to ∂D is constantly ± 1.

Note that f_P is a Morse function in a disk slightly bigger than D where we allow different critical points to have the same critical values.

Lemma 1.3.9 *Let $P \subset D$ be a connected divide. Then there exist adapted functions $f_P : D \to \mathbb{R}$.*

Proof We may change the divide $P \subset D \subset \mathbb{R}^2$ by a small isotopy in order to make it "Euclidean friendly", i.e. at its double points P will consist of a pair of Euclidean orthogonal segments. Every connected component of $D \setminus P$, that we also call region, does not meet ∂D or meets ∂D in a connected set since by hypothesis P is connected. We mark each such connected set by one point if it is a contractible arc in the boundary of D. We color the regions by signs $+$ and $-$ in a chequerboard way, i.e. two regions separated by a subinterval in P will have opposite signs. This is possible since D is simply connected and P as planar graph has only vertices of valency 4. Next, we mark each region that not meets the boundary of D with a point. Let $r > 0$ be a real such that the Euclidean balls of radius r in $\mathbb{R}^2$ with center a marked point are disjoint from P, from ∂D if its center is not on ∂D, and from each other. Assume also that the intersection of P with a ball of radius r and center a double point of P is the union of two Euclidean segments. We denote by B_p the ball of radius r with center p a marked point or double point of P.

Now we can construct the function f_P. Define f_P at a marked point p or double point of P to be standard Euclidean expression defining a local maximum, local minimum, or saddle point according sign or to be a double point of P. So on B_p define $f_P(q) = 1 - ||q - p||^2$ if p is in a $+$ region, $f_P(q) = -1 + ||q - p||^2$ or $f_P(q) = L(q)K(q)$ where L, K are local signed linear equations of Euclidean norm 1 for the segments of P at p. For a marker point $p \in \partial D$ we define f_P on B_p as a norm 1 linear boundary maximum or minimum with value 1 or -1. If P does not meet ∂D we define $f_P(q) = \pm 1, q \in \partial D$ according the sign of the adjacent region. Finally put $f_P(q) = 0$ for $q \in P$. The so partially defined function extends in a up to isotopy unique way to a smooth function $f_P : D \to \mathbb{R}$ that has no singularities except at the marked points and double points of P. □

Let $\beta : \mathbb{R} \to \mathbb{R}$ be a non negative smooth bump function that evaluates to 1 for $t \in [0, r/2]$ and to 0 for $t \geq r$. Let $\chi : D \to \mathbb{R}$ be the smooth function that evaluates to 0 on the complement of the union $\cup B_p$ and to $\beta(||p - q||), q \in B_p$.

We define a complex function $\theta_{P,\eta} : B^4_D \to \mathbb{C}$ that depends on a real small parameter $\eta > 0$ by

$$\theta_{P,\eta}(x, u) := f_P(x) + i\eta\, df_P(x)(u) - \frac{1}{2}\eta^2\chi(x)H_{f_P}(x)(u, u) \tag{1.5}$$

The function $\theta_P = \theta_{P,\eta}$ satisfies the Cauchy-Riemann equation at points in D and points in 4-balls of radius $r/2$ in $T\mathbb{R}^2 = \mathbb{C}^2$ and center a Morse critical point of f_P. These properties help to prove in next section a fibration theorem and compute the geometric monodromy, mainly following pioneering work of Milnor and Brieskorn.

More precisely, the argument function $\pi_{P,\eta} = \frac{\theta_{P,\eta}}{|\theta_{P,\eta}|}$ defines on S^3_P an oriented open book structure with smooth binding like the argument function $\frac{f}{|f|}$ does on the Milnor sphere of an isolated hypersurface singularity $f = 0$. More over $\theta_{P,\eta}$ behaves like a morsification of the singularity f, which allows a computation of the monodromy of that open book following:

1.4 General Fibration Theorem

1.4.1 A Connected Divide Yields a Fibered Link

A main results from [5] about divides is:

Theorem 1.4.1 *Let P be a connected divide. For $\eta > 0$ and sufficiently small, the map $\pi_P := \pi_{P,\eta}$ induces an oriented open book decomposition of S^3_P with binding $L(P)$, that is:*

1. *π_P is a fibration of the complement of $L(P)$ over S^1,*
2. *For all $\alpha \in S^1$, the boundary of each fiber (page of the open book) is the binding, i.e., $\partial\pi_P^{-1}(\alpha) = L(P)$*

Proof There exists a regular product tubular neighborhood N of $L(P)$, such that the map $\pi_{P,\eta}$ for any $1 \geq \eta > 0$ is on $N \setminus L(P)$ a fibration over S^1, for which near $L(P)$ the fibers look like the pages of a book near its back. It is crucial to observe that in the intersection of the link $L(P)$ with the support of the function

$$(x, u) \in S^3 \mapsto \chi(x) \in \mathbb{R}$$

the kernel of the Hessian of $\theta_{P,\eta}$ and the kernel of the differential of the map

$$(x, u) \in S^3 \mapsto f_P(x) \in \mathbb{R}$$

coincide. For an alternative proof of this step, look at 1.7.1 and the discussion before.

For any $\eta > 0$, the map $\pi_{P,\eta}$ is regular at each point of $U' := \{(x, u) \in S^3 | x \in U\}$ were U is a small neighborhood of the critical points of f_P. Take now local real coordinates (x, y, u, v) around a maximum $p \in D$ so that $f_P(x, y) = 1 - x^2 - y^2$ and $p = (0, 0)$. In these coordinates, seeing $\mathbb{C} \simeq \mathbb{R}^2$, we have that θ_P looks like

$$\theta_P(x, y, u, v) = \left(\left(u^2 + v^2\right)\eta^2 - x^2 - y^2 + 1,\ -2\eta(ux + vy)\right)$$

and

$$\pi_P(x,y,u,v) = \left(\frac{(u^2+v^2)\eta^2 - x^2 - y^2 + 1}{\sqrt{\left((u^2+v^2)\eta^2 - x^2 - y^2 + 1\right)^2 + 4\eta^2(ux+vy)^2}}, \right.$$

$$\left. -\frac{2\,(\eta ux + \eta vy)}{\sqrt{\left((u^2+v^2)\eta^2 - x^2 - y^2 + 1\right)^2 + 4\eta^2(ux+vy)^2}} \right)$$

Let $(0, 0, u, v)$ be a point in U' that projects to p. Then, the differential of π_P evaluated at that point equals:

$$\begin{pmatrix} 0 & 0\,0\,0 \\ -\frac{2\,\eta u}{(u^2+v^2)\eta^2+1} \; -\frac{2\,\eta v}{(u^2+v^2)\eta^2+1} & 0\,0 \end{pmatrix}$$

and the two non-zero entries of the matrix can't be 0 at the same time for a point of the form $(0, 0, u, v) \in S^3$ since $u^2 + v^2 = 1$. This shows that the jacobian of π_P has rank 1 on those points and hence, π_P is a submersion on a neighborhood U'. The computations for a minimum or a saddle point are very similar.

Finally, since the norm of the derivative of χ can be bounded, a similar computation yields that there exists $\eta_0 > 0$ such that for any η with $0 < \eta < \eta_0$, the map $\pi_{P,\eta}$ is regular on $S^3 \setminus (N \cup U')$. Hence, for η sufficiently small the map $\pi_{P,\eta}$ is a submersion, so since already a fibration near $L(P)$, it is a fibration by Ehresmann fibration theorem. □

1.5 Divides for Plane Curve Singularities

In this section we construct a divide P associated to a given isolated plane curve singularity in the sense that $L(P)$ is a model for the link of the plane curve singularity. What we actually produce is a *real morsification* of a totally real plane curve singularity; and this real morsification yields a divide. We do so, following two different programs: the first one Sect. 1.5.2 gives an algorithm to construct such a divide from an embedded resolution of the plane curve, the second one, in Sect. 1.5.3 gives a closed formula that produces a divide from a complete set of topological invariants of a plane curve.

1.5.1 Totally Real Models of Plane Curve Singularities

For both methods, it is necessary to start with a *totally real* model of our isolated plane curve singularity.

Definition 1.5.1 Let $f \in \mathbb{C}\{x, y\}$ be a converging power series and let $f = f_1 \cdots f_r$ be a factorization into irreducible factors. We say that f is *real* or that it defines a *real plane curve singularity* if it is given by a real equation, that is, if $f \in \mathbb{R}\{x, y\}$. We say that f is *totally real* if $f_i \in \mathbb{R}\{x, y\}$ for all $i = 1, \ldots, r$.

Equivalently, f is real if the curve $(f^{-1}(0), 0)$ is invariant under complex conjugation and f is totally real if each of its branches is real.

Once one has a real plane curve singularity, one can try to perturb it in a controlled way with the hope that all the critical points (as a complex function) are actually real and as simple as possible: Morse. It turns out that if one is able to do this, the real picture of the deformed curve tells us a lot about the complex singularity. Actually, this contains all the topological information of the plane curve singularity. In particular one is able to recover a distinguished basis of geometric vanishing cycles. These special types of deformations are called real morsifications:

Definition 1.5.2 Let $f : \mathbb{C}^2 \to \mathbb{C}$ define an isolated real plane curve singularity, let $B_\epsilon \subset \mathbb{R}^2$ be a Milnor ball for f and let D_δ be a Milnor disk for f. We say that a family of convergent power series $\{f_t\}_{t\in[0,\tau)}$ is a *real morsification* of f if

1. it is a deformation of f, that is, $f_0 = f$,
2. f_t is real for all t,
3. for all $t \in (0, \tau)$ the map f_t only has Morse critical points and they all lie in B_ϵ. Furthermore, the number of Morse critical points does not depend on t.
4. all its critical points are real, that is, they lie in $\mathbb{R}^2 \subset \mathbb{C}^2$.
5. all saddle points lie over 0.

In the next two sections we prove, in two different ways, that there exist real morsifications for totally real isolated plane curve singularities. It is still an open question if there exist real morsifications for all isolated real plane curve singularities. See [35] for a partial solution to this question and see [20] for the implications of a solution of this problem in other conjectures.

Remark 1.5.3 It follows from the topological classification of isolated plane curve singularities that, given a topological type, it is always possible to find a totally real isolated plane curve singularity realising that topological type (see [50–52] for a classical reference or [49, Corollary 5.3.3 and Lemma 2.3.1] for a more recent and concrete reference). See also the previous related Remark 1.2.9.

1.5.2 Divides from an Embedded Resolution

In this section we reproduce the proof of one of the main results (Théorème 1.) from [3] which proves the existence of divides for plane curve singularities. More concretely, the proof gives an algorithm to construct a polynomial real deformation of a plane curve singularity under the hypothesis that the singularity is given by a totally real polynomial.

The topology of special types of divides obtained by the first author's method were described in the works [43] and [16] of Schulze-Röbbecke and Castellini, for branches and for arbitrary complex topological types.

Theorem 1.5.4 *Let $f(x, y)$ be a totally real polynomial defining an isolated plane curve singularity. Then there exists a real polynomial deformation $f(x, y; t)$, $t \in \mathbb{R}$, such that $f(x, y; 0) = f(x, y)$ and that for all $t \neq 0, t \in \mathbb{R}$ and small enough, we have*

1. *the real curve $C_t = \{(x, y) \in D_\varepsilon | f(x, y; t) = 0\}$ is a divide with r branches in a small disk D_ε near the origin.*
2. *the number k of double points of C_t verifies $2k - r + 1 = \mu$, where μ is the Milnor number of f at O.*

Next we explain the main construction on which its based the proof of Theorem 1.5.4 which consists in iteratively blowing up until, exactly, the strict transform is resolved and consists of smooth branches meeting transversely (but not in normal crossings) the exceptional divisor. At this point, we translate slightly the strict transforms along some exceptional divisor and blow-down back all the exceptional divisors, proving that the translation yields a deformation of the original plane curve in $\mathbb{C}^2$. Performing this operation as many times as necessary, yields a pre-divide (recall Definition 1.3.3) which can be further polynomially deformed into an actual divide.

Next we explain this process in more detail. Let $f(x, y)$ be a totally real polynomial defining an isolated plane curve singularity C. Let

$$\pi_n : X_n \to X_{n-1}, \quad 1 \leq n \leq N$$

be a sequence of transformations that yield an embedded resolution of C_0, where

1. $X_0 = \mathbb{C}^2$
2. $\pi_1 : X_1 \to X_0$ is the blow-up at the origin.
3. $\pi_k : X_k \to X_{k-1}$ for $i = 2, \ldots, N - 1$ is a blow-up at a point in the exceptional divisor E_{k-1} of $\pi_1 \circ \cdots \circ \pi_{k-1}$
4. $H_{N-1} = \{z \in X_N | f \circ \pi_1 \circ \cdots \circ \pi_{N-1}(z) = 0\}$ is a union of pairwise transversal smooth curves.
5. $\pi_N : X_N \to X_{N-1}$ is a composition of blow ups that resolves all the ordinary singularities of H_{N-1} where two or more components of the strict transform of C meet
6. $H_N = \{z \in X_N | f \circ \pi_1 \circ \cdots \circ \pi_N(z) = 0\}$ is a hypersurface in X_N having as singularities only normal crossings.

In other words, $\pi_1, \ldots, \pi_{N-1}$ is an usual sequence of blow ups of the minimal resolution of the plane curve defined by f just until the strict transform consists of smooth curves not meeting in normal crossings, and π_N is the composition of all the remaining blow ups that give an embedded resolution of the plane curve.

For $1 \leq n \leq N$, we denote by

$$H_n = \{z \in X_n | f \circ \pi_1 \circ \cdots \circ \pi_n(z) = 0\}$$

the total transform in X_n, by

$$E_n = \{z \in X_n | \pi_1 \circ \cdots \circ \pi_n(z) = 0\} \subset H_n \subset X_n$$

the exceptional divisor in X_n, and by B_n the component of E_n, which is the exceptional divisor corresponding to the blow-up π_n.

Let $[x_n, y_n]$ be a homogeneous coordinate system on B_n such that the affine chart $A_n = \{y_n = 1\}$ contains the intersection points of B_n with the other branches of H_n. Let $V_n = \{(\lambda_n, x_n)\}, \lambda_n \in \mathbb{C}, x_n \in A_n$ be a chart of X_n such that we have $B_n \cap V_n = A_n$ and $B_n \cap V_n = \left\{\lambda_n^{m_n} = 0\right\}$ with m_n the multiplicity of B_n in X_n.

Since $f(x, y)$ is totally real by hypothesis, we can choose the blow-ups $\pi_n : X_n \to X_{n-1}$ for $1 \leq n < N$ having as center a real point of X_{n-1}. So we can also choose the coordinates $[x_n, y_n]$ and (x_n, λ_n) to be real at each step of the resolution process. The function

$$f_{N-1} = f \circ \pi_1 \circ \cdots \circ \pi_{N-1}$$

is written in the chart V_{N-1} like

$$f_{N-1}(x_{N-1}, \lambda_{N-1}) = \lambda_{N-1}^{m_{N-1}} \tilde{f}_{N-1}(x_{N-1}, \lambda_{N-1}).$$

We decompose $\tilde{f}_{N-1} = h + g$ with h the initial homogeneous part. Assume that B_M, with $1 \leq M \leq N-2$, is the component of E_{N-1} that intersects B_{N-1}. Then, h is of degree $m_{N-1} - m_M$. So the set

$$\{(x_{N-1}, \lambda_{N-1}) \in V_{N-1} | h(x_{N-1}, \lambda_{N-1}) = 0\}$$

is the union of the tangents to the non-exceptional branches of H_{N-1} at the points of $B_{N-1} \cap H_{N-1}$. For $t_{N-1} \in \mathbb{R}$ we put

$$\begin{aligned} \bar{f}_{N-1}(x_{N-1}, \lambda_{N-1}, t_{N-1}) = \; & h(x_{N-1}, \lambda_{N-1} - t_{N-1}) \\ & \times \left[1 + t_{N-1} \cdot \prod_{i=1}^{l} (x_{N-1} - x_i) \cdot \exp(x_{N-1})\right] \\ & + g(x_{N-1}, \lambda_{N-1} - t_{N-1}) \end{aligned}$$

where $x_1, \ldots, x_l$ are the roots of $h(x_{N-1}, 0) = 0$. For $t_{N-1} \in \mathbb{R}$ and $t_{N-1} \neq 0$ small enough, the equation

$$\bar{f}_{N-1}(x_{N-1}, 0, t_{N-1}) = 0$$

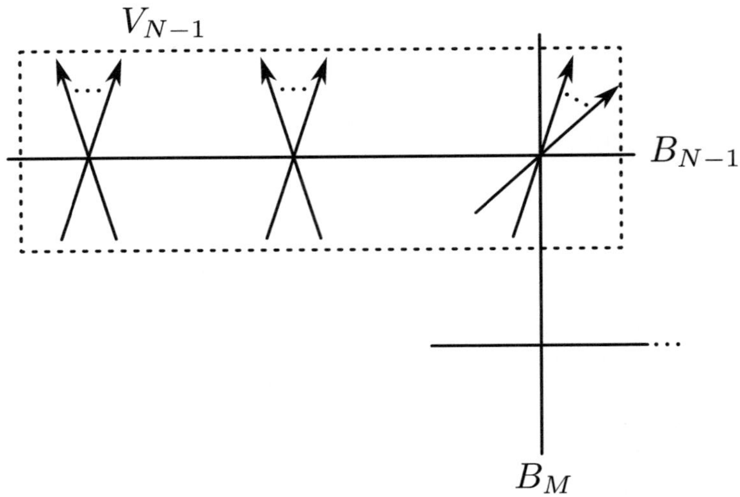

Fig. 1.6 H_{N-1} near B_{N-1}

has exactly $m_{N-1} - m_M$ distinct roots on $B_{N-1} \cap V_{N-1}$ (and not in B_M) and

$$\bar{f}_{N-1}(\infty, 0, t_{N-1}) = \infty \quad \text{at the point} \quad \infty \in B_{N-1} \setminus V_{N-1}.$$

We can see in Fig. 1.6 the branches of H_{N-1} that cross B_{N-1} and B_M. For t_{N-1} real, non-zero and small, Fig. 1.7 is the actual real drawing of

$$H_{N-1,t_{N-1}} \cap V_{N-1} = \left\{(x_{N-1}, \lambda_{N-1}) \in V_{N-1} | \lambda_{N-1}^{m_{N-1}} \bar{f}_{N-1}(x_{N-1}, \lambda_{N-1}, t_{N-1}) = 0\right\}$$

Neighbouring points on the B_{N-1} branch are translated out and next to B_{N-1}. So $H_{N-1,t_{N-1}}$ meets B_{N-1} in normal crossings. The presence of the factor $\lambda_{N-1}^{m_{N-1}}$ in $f_{N-1,t_{N-1}}$ allows us to push the function $f_{N-1,i_{N-1}}$ to a function $f(x, y, t_{N-1})$ on $\mathbb{C}^2$. We observe that by construction,

$$f(x, y; 0) = f(x, y)$$

and for $t_{N-1} \neq 0$, t_{N-1} real and small enough, the hypersurface

$$H_{N-1}(t_{N-1}) = \left\{(x, y) \in \mathbb{C}^2 | f(x, y, t_{N-1}) = 0\right\} = \pi_1 \circ \cdots \circ \pi_{N-1}\left(H_{N-1,t_{N-1}}\right)$$

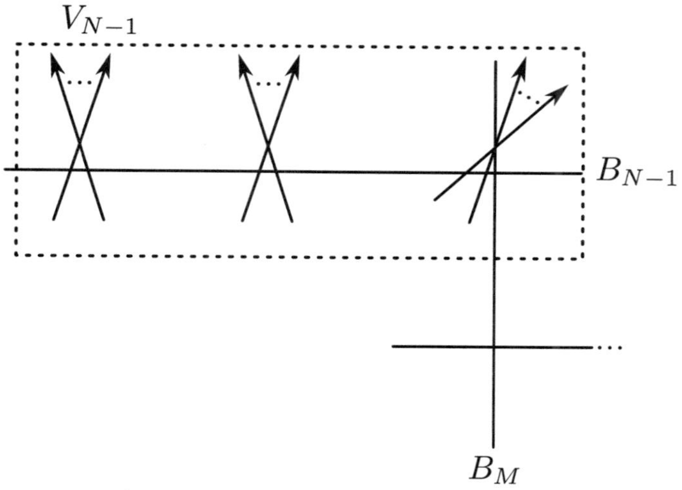

Fig. 1.7 We can see $H_{N-1,t_{N-1}}$ in the chart V_{N-1}

has in B_ε the singularities:

(a) an isolated singularity in $0 \in B_\varepsilon$,
(b) if $N_i^{N-1}(t_{N-1}), 1 \le i \le l_{N-1}$ are the translates of neighbouring points on B_{N-1}; then $H_{N-1}(t_{N-1})$ has at points

$$w_i^{N-1}(t_{N-1}) = \pi_1 \circ \cdots \circ \pi_{N-1}\left(N_i^{N-1}(t_{N-1})\right)$$

isolated ordinary singularities. That is, transversal intersections of smooth branches.

The blow-ups $\pi_1, \ldots, \pi_{N-2}$ resolve the singularity at $0 \in B_\varepsilon$ of $H_{N-1}(t_{N-1})$ so that the branches of $H_{N-1,t_{N-1}} = (\pi_1 \circ \cdots \circ \pi_{N-2})^{-1} - (H_{N-1}(t_{N-1}))$ meet each other transversely. We can therefore start the construction again and obtain a deformation of $f(x, y, t_{N-1})$.

$$f(x, y, t_{N-1}, t_{N-2}), t_{N-2} \in \mathbb{R}$$

such that the hypersurface $\{f(x, y, t_{N-1}, t_{N-2}) = 0\} \subset \mathbb{C}^2$ has an isolated singularity in $0 \in B_\varepsilon$. In addition, one must take care so that

$$f(x, y, t_{N-1}) \equiv f(x, y, t_{N-1}, t_{N-2})$$

at a high enough order at the points $w_i^{N-1}(t_{N-1}), 1 \le i \le l_{N-1}$. This operation can be iterated $N-1$ times to obtain a family of functions on $\mathbb{C}^2$.

$$f(x, y, t_{N-1}, t_{N-2}, \ldots, t_1), (t_{N-1}, \ldots, t_1) \in \mathbb{R}^{N-1}$$

We finally define

$$\bar{f}(x, y, t) = f(x, y, t, t, \ldots, t), \quad t \in \mathbb{R}$$
$$\bar{H}_t = \left\{x, y \in \mathbb{C}^2 | \bar{f}(x, y, t) = 0\right\}$$

From this construction, the following lemma follows:

Lemma 1.5.5 *For $t \in \mathbb{R}, t \neq 0$ and small enough, the curve $\bar{C}_t = \bar{H}_t \cap D_\varepsilon$ is an immersion $\bar{\alpha} : J \to D_\varepsilon$ such that*

(a) The number of components of J equals the number r of branches of f,
(b) $\bar{\alpha}(\partial J) \subset \partial D_\varepsilon$, and the set $\bar{\alpha}(J) \subset D_\varepsilon, \alpha(J)$ is connected.
(c) $\bar{\alpha}(J) = \bar{C}_t$ has as its only singularities, ordinary singularities.

In particular, (b) and (c) say that $\bar{\alpha}$ is a connected pre-divide (recall Definition 1.3.3).

The number of double points appearing near a multiple point with r_i branches of $\bar{C}_t$, is

$$\frac{1}{2} r_i (r_i - 1).$$

So the number of double points of any divide $\alpha : J \to D_\varepsilon$ next to $\bar{\alpha} : J \to D_\varepsilon$ is

$$\frac{1}{2} \sum_{1 \leq i \leq l} r_i (r_i - 1). \tag{1.6}$$

In the next theorem we complete the proof of Theorem 1.5.4, by proving that $\bar{\alpha}$ can be polynomially perturbed into a divide and we also show that the δ invariant of the singularity coincides with the number of double points of any divide coming from a totally real singularity is we only need to prove.

Lemma 1.5.6 *Let $\alpha : J \to D_\varepsilon$ be a divide resulting from a small perturbation of $\bar{\alpha} : J \to D_\varepsilon$. Then there exists a real polynomial $M(x, y, t)$ so that the function*

$$f(x, y; t) = \bar{f}(x, y, t) + M(x, y, t), \quad t \in \mathbb{R}$$

is a deformation of $f(x, y)$, such that its divide $C_t \subset D_\varepsilon$ has the configuration of the divide defined by α and $C_t \subset D_\varepsilon$ is close to the image of $\bar{\alpha}$.

Moreover, if $w_i(t)$, $1 \leq i \leq l$ are the crossing points (ordinary singularities) of $\bar{C}_t$, and r_i is the number of branches of $\bar{C}_t$ at points $w_i(t)$. Then we have

$$\sum_{1 \leq i \leq l} r_i (r_i - 1) - r + 1 = \mu.$$

Equivalently, the number of double points of the divide D coincides with the δ invariant of the singularity and

$$D = \delta = \frac{1}{2} \sum_{1 \le i \le l} r_i \left(r_i - 1\right)$$

Proof Let $(x_1(t), y_1(t)), \ldots, (x_e(t), y_e(t)) \in D_\varepsilon$ the crossing points (double or multiple) of of $\bar{C}_t$. Assume that $w_1(t) = (x_1(t), y_1(t))$ is a multiple crossing point since if there are only double points, the lemma is proven because our deformation $\bar{C}_t$ is given by a polynomial.

Let r_1 be the number of branches at the point $w_1(t)$. Let, for $1 \le i \le r_1$,

$$l_i(x, y, t) = 0$$

the equations of the tangents at $w_1(t)$ to the branches of $\bar{C}_t$. Any configuration of a divide which is close to $\bar{C}_t$ can be described near $w_1(t)$ by an equation

$$L_1(x, y, t) = \prod_{1 \le i \le r_1} t \left(l_i(x, y, t) - t a_i\right) = 0$$

where $a_i \in \mathbb{R}$. Thus, the equation

$$\bar{f}(x, y, t) + L_1(x, y, t) = 0 \tag{1.7}$$

gives for $t \neq 0$ and small enough the same configuration. Let let $w_{1,1}(t), \ldots, w_{1,s}(t)$ be the $s = 1/2 r_1 \left(r_1 - 1\right)$ double points that appear near $w_1(t)$ for Eq. 1.7. Let $P_1(x, y; t)$ be a polynomial such that for a high enough order we have

$$\begin{aligned} P_1\left(w_{1,j}(t); t\right) &\equiv 1, \quad 1 \le j \le s, \\ P_1\left(x_i(t), y_i(t); t\right) &\equiv 0, \quad 2 \le i \le e. \end{aligned}$$

Then

$$\bar{f}_{(1)}(x, y, t) = \bar{f}(x, y, t) + P_1(x, y; t) L_1(x, y, t)$$

is a deformation of $f(x, y)$ such that the curve

$$C_{(1)}(t) = \left\{(x, y) \in D_\varepsilon \,|\, \bar{f}_{(1)}(x, y, t) = 0\right\}$$

has near $w_1(t)$ the desired configuration and $(x_i(t), y_i(t))$, $2 \le i \le e$, are still multiple points of $\bar{C}_{(1)}(t)$. Repeat the same construction for another multiple crossing point until there are no more. This proves the first part of the statement.

The second part follows from the following observation: if one has an ordinary real singularity consisting of r_i branches, after a small perturbation of these

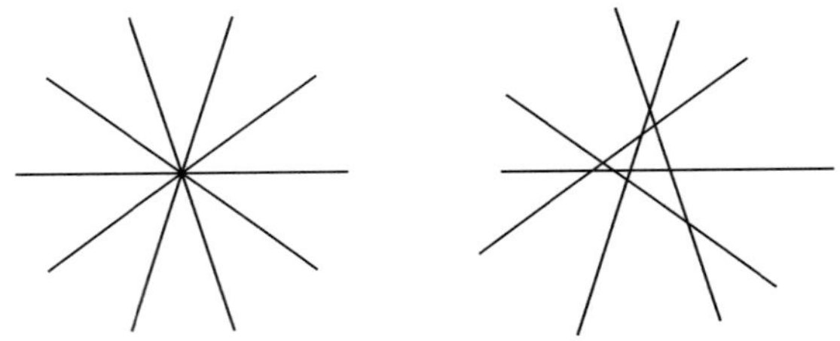

Fig. 1.8 On the left 5 local branches and on the right the perturbed branches that lead to 10 double points

branches, there appear $\frac{1}{2}r_i(r_i - 1)$ double points (Fig. 1.8). Let's denote

$$D = \frac{1}{2} \sum_{1 \le i \le l} r_i \, (r_i - 1)$$

the number of double points of the divide. We first recall Milnor's formula [37] relating the Milnor number μ, the delta invariant δ and the number of branches $\mu = 2\delta - r + 1$. Let b be the number of bounded regions of the divide. Since on each bounded region there lies a critical point of $f(x, y; t)$ and all the other critical points are the nodes of the divide, we have

$$b + D = \mu. \tag{1.8}$$

The connectedness of the divide implies that the union of the closure of the bounded regions is a topological disk, hence a direct calculation of the Euler characteristic yields

$$b - (2D - r) + D = 1 \quad \Rightarrow \quad b - D = 1 - r \tag{1.9}$$

Taking Eq. 1.8 minus Eq. 1.9 yields $2D = \mu - 1 + r$ which, after Milnor's formula implies $\delta = D$. □

Example 1.5.7 This is an example of a divide using a similar technique as the one explained in Theorem 1.5.4. Note that the perturbations chosen here are not exactly the same.

Start with the isolated plane curve singularity defined by the polynomial $f(x, y) = -x^8 - x^7 - 3x^5y + y^3$.

After four blow ups we arrive at the situation of Fig. 1.9. In a local chart, near the intersection point of the divisors E_4 and E_2, the total transform is defined by the expression

$$f_4(x, y) = -\Big(x^2y^2 + 3xy + x - y\Big)x^6y^{14}$$

We can perturb the strict transform by considering the polynomial:

$$f_{4,t}(x, y) = -\Big(x^2y^2 + 3xy + t + x - y\Big)x^6y^{14}$$

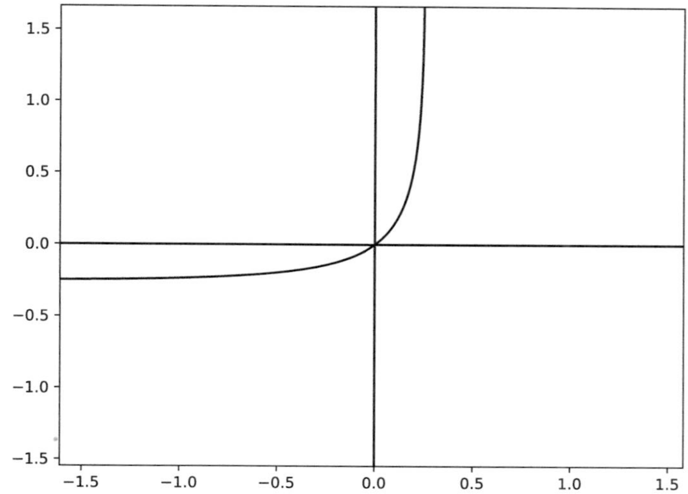

Fig. 1.9 Situation after the first 4 blow ups of the minimal embedded resolution

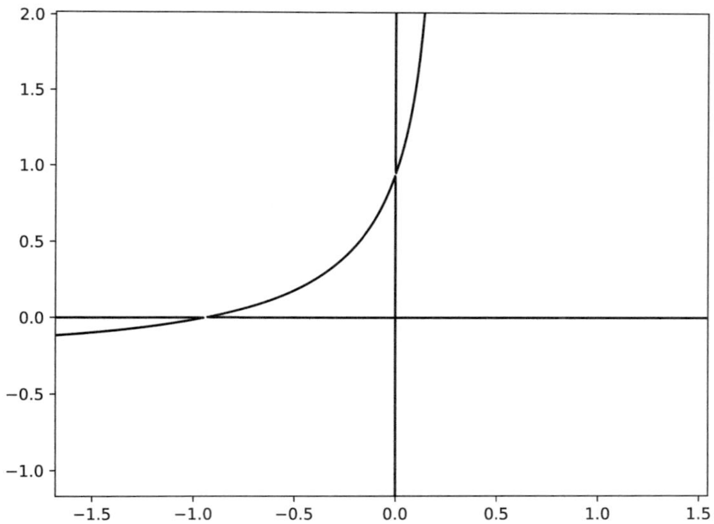

Fig. 1.10 Situation after perturbing the strict transform near E_4 and E_2

which, for $t > 0$ defines the picture of Fig. 1.10. Now, we observe that the contraction of E_4, yields the equation

$$\hat{f}_{4,t}(x, y) = -\left(x^2y + ty + 3xy - y^2 + x\right)x^6y^7$$

which, for $t > 0$ describes the situation depicted in Fig. 1.11 Again, we perturb the strict transform in order to achieve normal crossings by considering the following

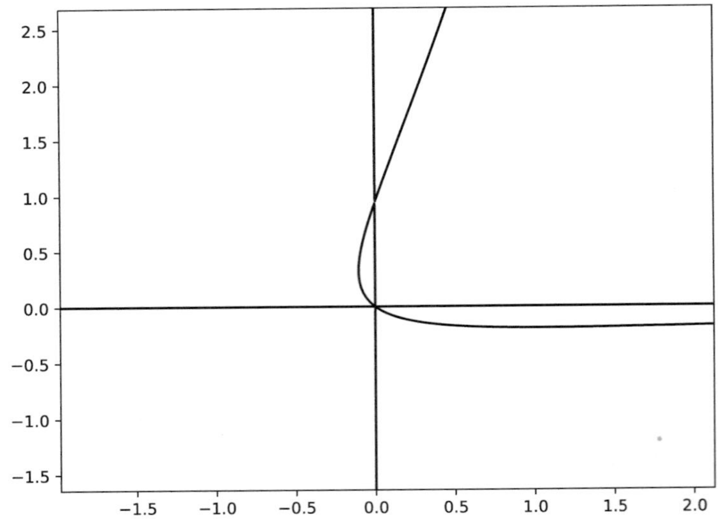

Fig. 1.11 The divisor E_4 is contracted and the strict transform is only perturbed by the previously defined perturbation. We see that the strict transform and the divisor do not meet in normal crossings

polynomial

$$f_{3,t}(x, y) = -\left(x^2y + ty + 3xy - y^2 + t + x\right)x^6y^7.$$

We arrive to the situation of To avoid cumbersome notation, we have used the same parameter t for the perturbation. One can think of this as two steps: first one considers a perturbation over a two-parameter family; then one considers the perturbation over a disk contain in this family. By the choice of our perturbations, we are considering the perturbation over the diagonal of this two-parameter family. Note that Theorem 1.5.4 proposes, a priori different perturbations to ensure that they behave well with respect to the blow-down process. In our case, we verify this at each step even though we make different choices that the ones proposed by the theorem. In the above equation, $y = 0$ defines locally E_3 and $x = 0$ defines E_2 (Fig. 1.12). After the contraction of E_3 we arrive to the polynomial

$$\hat{f}_{3,t}(x, y) = -\left(ty^2 - y^3 + x^2 + ty + 3xy + x\right)x^6$$

which, for $t > 0$ defines the situation depicted in Fig. 1.13. In this case, the strict transform already intersects the divisor E_2 transversely. So we don't perform a perturbation at this stage and we directly contract E_2 to arrive to the equation

$$\hat{f}_{2,t}(x, y) = -\left(x^5 + x^4 + tx^2y + 3x^3y + txy^2 - y^3\right)x^3.$$

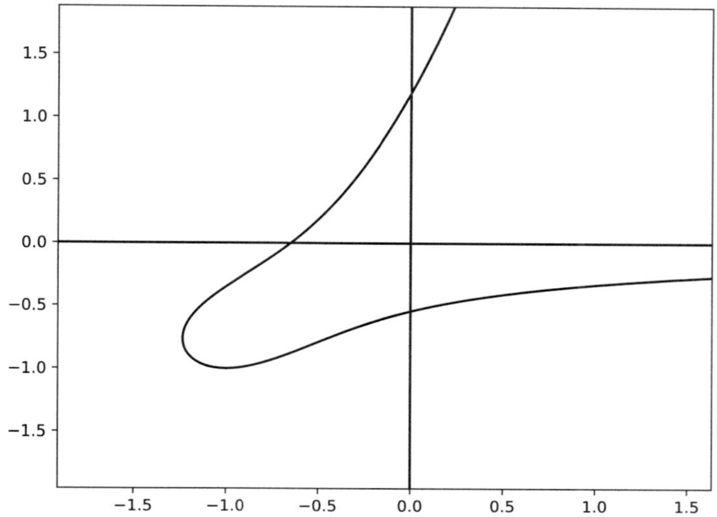

Fig. 1.12 After a small perturbation of the strict transform we arrive at a normal crossings situation

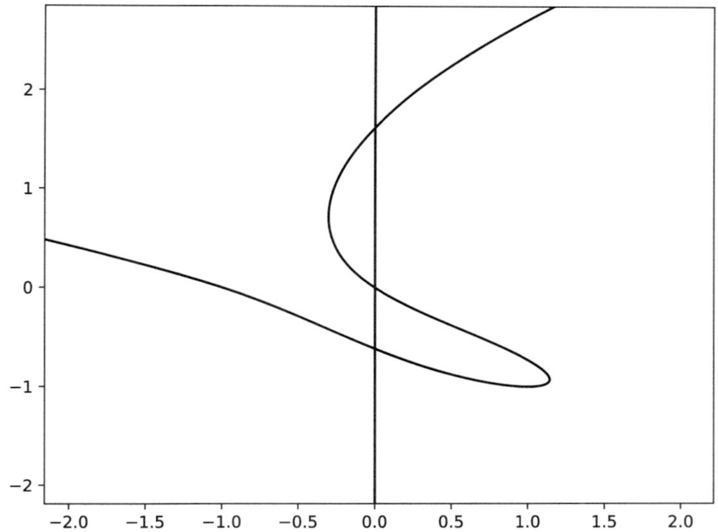

Fig. 1.13 The divisor E_3 is contracted and the strict transform is perturbed by the previously defined perturbation. In this case we automatically achieve normal crossings

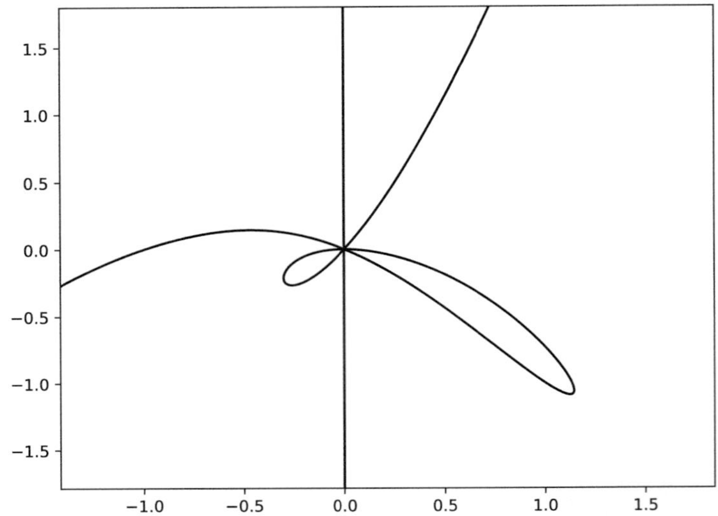

Fig. 1.14 The divisor E_2 is contracted

which describes the local situation depicted at Fig. 1.14. As we see, the (perturbed) strict transform does not meet in normal crossings the divisor E_1. So we need to perturb it a little bit. We do so by considering the polynomial

$$\begin{aligned} f_{1,t}(x, y) = -\Big(& t^{10} + 5\,t^8 x + t^8 + 10\,t^6 x^2 + 4\,t^6 x + 10\,t^4 x^3 + 3\,t^6 y + 6\,t^4 \\ & x^2 + 5\,t^2 x^4 + t^5 y + 9\,t^4 xy + 4\,t^2 x^3 + x^5 + 2\,t^3 xy + 9\,t^2 x^2 y \\ & + t^3 y^2 + x^4 + tx^2 y + 3\,x^3 y + txy^2 - y^3\Big)\, x^3. \end{aligned}$$

which results from substituting x by $x + t^2$ in the strict transform of the equation defined by $\hat{f}_{2,t}(x, y)$. For $t > 0$ the corresponding situation is depicted in Fig. 1.15.

We are ready to contract the divisor E_1 to arrive to the polynomial

$$\begin{aligned} \hat{f}_{1,t}(x, y) = & -t^{10} x^3 - 5\,t^8 x^4 - t^8 x^3 - 10\,t^6 x^5 - 4\,t^6 x^4 - 10\,t^4 x^6 - 6\,t^4 x^5 \\ & - 5\,t^2 x^7 - 3\,t^6 x^2 y - 4\,t^2 x^6 - x^8 - t^5 x^2 y - 9\,t^4 x^3 y - x^7 - 2\,t^3 x^3 y \\ & - 9\,t^2 x^4 y - tx^4 y - 3\,x^5 y - t^3 xy^2 - tx^2 y^2 + y^3. \end{aligned}$$

This is already a polynomial defined in the (original) $\mathbb{C}^2$ that is a deformation of $f(x, y) = f_{1,0}(x, y)$ and that defines a pre-divide (recall Definition 1.3.3) which is depicted in Fig. 1.16. After performing a generic perturbation as explained in Lemma 1.5.6 we arrive at a divide like the one in Fig. 1.17 which is a divide for our original singularity.

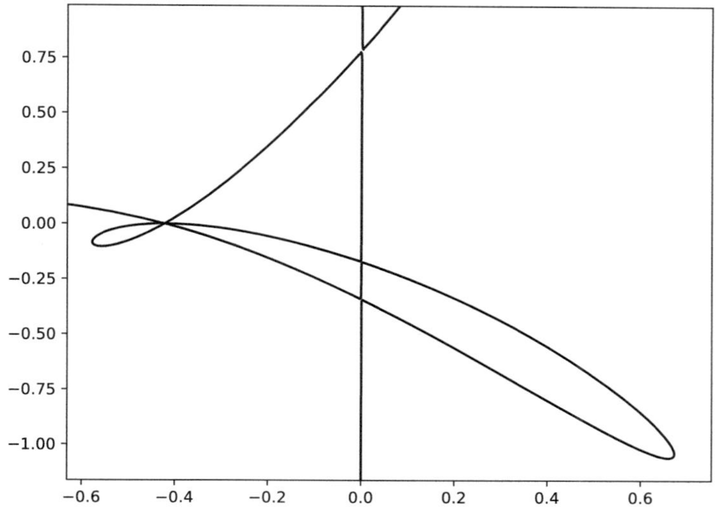

Fig. 1.15 A small translation is performed to achieve normal crossings

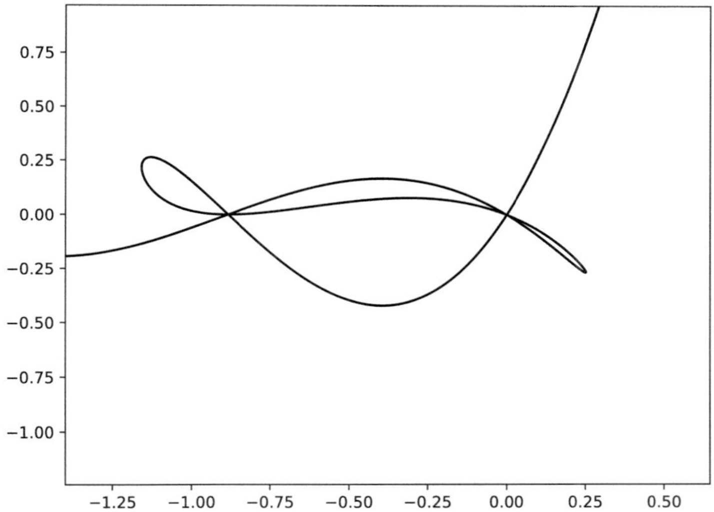

Fig. 1.16 A small translation is performed to achieve normal crossings

1.5.3 Divides from Chebyshev Polynomials

In this section we explain a different way to obtain divides for plane curves. This was envisaged by Sabir Gusein-Zade in [22, 23] and was also further explored by the first author.

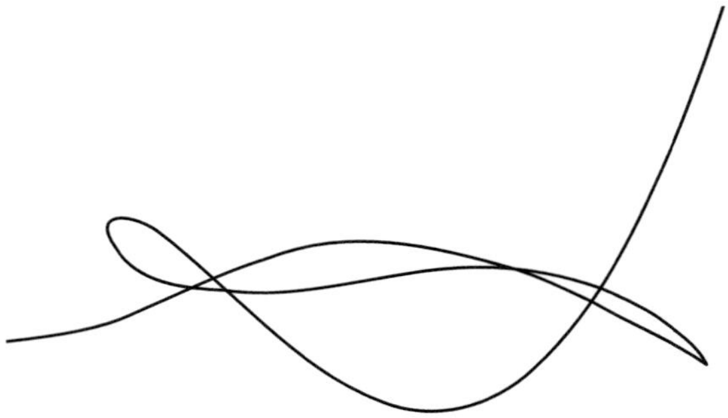

Fig. 1.17 A small perturbation can be performed to get rid of the ordinary singularities and get only double points

It was observed by René Thom that the Chebyshev polynomials $T : \mathbb{C} \to \mathbb{C}$ up to affine equivalence of functions are precisely the polynomial mappings from $\mathbb{C}$ to $\mathbb{C}$ with two or less critical values and with only quadratic singularities [45]. The standard Chebyshev polynomial $T(p, z)$ is a polynomial in the variable z of degree p and has the following properties:

1. its critical are values $+1, -1$,
2. it has the symmetry $T(p, z) = (-1)^p T(p, -z)$, and
3. the coefficient of z^p is 2^{p-1}.

As a consequence of these properties we find that, for example, the map $T(1, z)$ has no critical values and the map $T(2, z)$ has only the critical value -1. Furthermore, the Chebyshev polynomial $T(p, z)$ satisfies the identity

$$T(p, \cos(x)) = \cos(px),$$

and its restriction to $[-1, 1]$ is defined by

$$T(p, t) = \cos(p \arccos(t)).$$

Sabir Gusein-Zade [22, 23] constructed real morsifications for real plane curve singularities with Chebyshev polynomials. The building block for his construction is the real morsification for the map $f(x, y) = 2^{p-1}x^p - 2^{q-1}y^q$ given by

$$f_s(x, y) = s^{pq}(T(p, x/s^q) - T(q, y/s^p)), \quad s \in [0, 1].$$

In particular, it is a direct computation that for each $s \in]0, 1]$, the function $f_s : \mathbb{C}^2 \to \mathbb{C}$ has $\mu_f = (p-1)(q-1)$ quadratic singularities all at points with real coordinates, its critical values are contained in $\{-2s^{pq}, 0, 2s^{pq}\}$ and

$$\lim_{\substack{s \to 0 \\ s > 0}} f_s = f.$$

If the exponents p and q are relatively prime to each other, the level set $\{f(x, y) = 0\}$ can be parametrized by the monomial map

$$t \in \mathbb{C} \mapsto (t^q/2^{p-1}, t^p/2^{q-1}) \in \mathbb{C}^2.$$

The Chebyshev composition identity (see [41])

$$T(p, T(q, z)) = T(pq, z),$$

implies that in this case, the Chebyshev polynomials can also be used to parametrize the level sets $\{f_s(x, y) = 0\}$ as well. Indeed, the map

$$t \in \mathbb{C} \mapsto (s^q T(q, t/s), s^p T(p, t/s)) \in \mathbb{C}^2$$

parametrizes the level set $\{f_s(x, y) = 0\}$. Equivalently, one can verify that

$$\lim_{s \to 0} \frac{s^p T(p, t/s)}{2^{p-1}} = t^q \tag{1.10}$$

1.5.3.1 Gusein-Zade Method

As a consequence of this, Gussein-Zade [22] showed that if

$$t \mapsto (t^m, \sum \lambda_k t^k))$$

is a parametrization of a branch with $\lambda_k \in \mathbb{R}$, then

$$(t, s) \mapsto \left(\frac{s^m T(m, t/s)}{2^{m-1}}, \sum \frac{\lambda_k s^p T(k, t/s)}{2^{k-1}} \right) \tag{1.11}$$

defines, for $s \neq 0$, a divide for that branch. Observe that any real branch admits a real parametrization.

Example 1.5.8 We compare now this method with the method described in the previous Sect. 1.5.2. In particular, we consider the plane curve singularity defined by the polynomial

$$f(x, y) = -x^8 - x^7 - 3x^5 y + y^3$$

which is the one used in Example 1.5.7. A direct computation yields that the map

$$(t^3, t^7 + t^8)$$

is a parametrization of the branch $\{f(x, y) = 0\}$ near the origin. And so the formula Eq. 1.11, yields that the following describes a divide for each $s \neq 0$:

$$
\begin{aligned}
(t, s) \mapsto & \left(\frac{s^3 T(3, t/s)}{2^2}, \frac{s^7 T(7, t/s)}{2^6} + \frac{s^8 T(8, t/s)}{2^7} \right) \\
= & \left(-\frac{3}{4} s^2 t + t^3, \right. \\
& \left. -\frac{7}{64} s^6 t + \frac{7}{8} s^4 t^3 - \frac{7}{4} s^2 t^5 + t^7 + \frac{1}{128} s^8 \right).
\end{aligned}
$$

Where the equality is just the evaluation of the Chebyshev polynomials and simplification of the expressions. For $s \neq 0$ we obtain the divide depicted in Fig. 1.18. Compare with that of Fig. 1.17.

It follows from the defining property $T(x, cos\theta) = \cos(n\theta)$ that the curve defined by the above parametrization has

$$
\delta(P_{p,q}) = (p-1)(q-1)/2 \tag{1.12}
$$

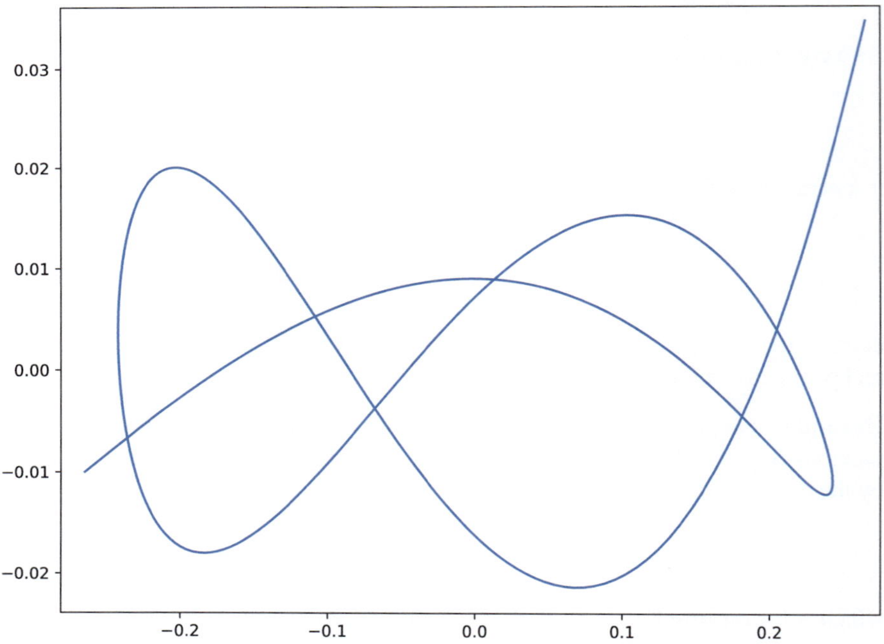

Fig. 1.18 Divide for the plane curve defined by $-x^8 - x^7 - 3x^5 y + y^3$ using Gusein-Zade method

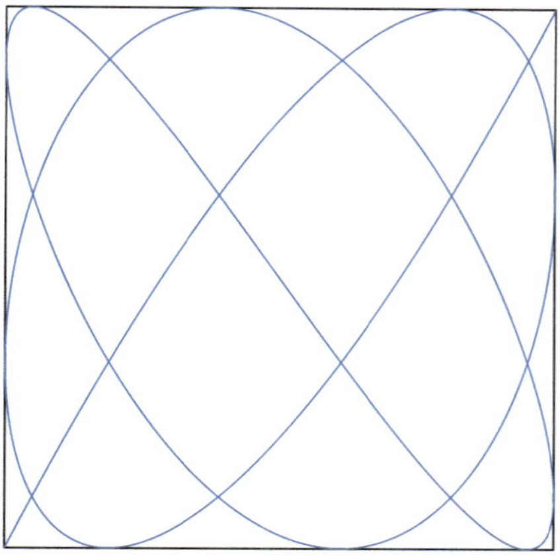

Fig. 1.19 Divide in box $[-1, 1] \times [-1, 1]$ for $2^6x^7 - 2^4y^5 = 0$

double points near the origin. Since f_s has $(p-1)(q-1)$ critical points (all of them Morse), we have proven that the previous parametrization yields a divide. More concretely, for each $s \in]0, 1]$ the intersection

$$P_{p,q;s} := \{f_s(x, y) = 0\} \cap D$$

is a divide for the Brieskorn-Pham singularity defined by the polynomial $2^{p-1}x^p - 2^{q-1}y^q$ at $0 \in \mathbb{C}^2$. The curve $P_{p,q} := \{f_1(x, y) = 0\}$ can be drawn in a rectangular box as in Fig. 1.19. As a first type of building block we will need the box $B := [-1, 1] \times [-1, 1]$ with the curve $P_{p,q}$. If $(p, q) = 1$ holds, the curve $P_{p,q}$ is the image of

$$\begin{aligned} T_{p,q} : [-1, 1] &\longrightarrow B \\ t &\mapsto (T(p, t), T(q, t)) \end{aligned}$$

which leaves the box through the corners. In general, the immersed curve has several components, which are immersions of the interval or of the circle. At most two components are immersions of the interval, which leave the box through the corners.

1.5.3.2 Cabling Divides

Let P be any divide having one branch given by an immersion $\gamma : [-1, 1] \to D$. We assume, that the speed vector $\dot{\gamma}(t)$ and the position vector $\gamma(t)$ are proportional at $t = \pm 1$, i.e. the divide P meets ∂D at right angles. Let

$$N\gamma : [-1, 1] \times [-1, 1] \to D$$

be an immersion of a rectangular box around P, that is, the restriction

$$N\gamma|_{[-1,1]\times\{0\}}$$

is the immersion γ and the image of $N\gamma$ is in a small tubular neighborhood of P. For instance, for a small value of the parameter $\eta \in \mathbb{R}_{>0}$ the following expression defines such an immersion $N\gamma : B \to \mathbb{R}^2$ of the rectangular box $B := [-1, 1] \times [-1, 1]$:

$$N\gamma(s,t) := \gamma(t) + s\eta \frac{J(\dot{\gamma}(t))}{\|\dot{\gamma}(t)\|},$$

where J is the rotation of $\mathbb{R}^2$ over $\pi/2$, equivalently, J is multiplication by i if we see the divide as a subset of $\mathbb{C}$. The four corners $N\gamma(\pm 1, \pm 1)$ are on the circle of radius $\sigma := \sqrt{1+\eta^2}$. We finally define

$$N\gamma(s,t) := N_\eta\gamma(s,t) := \frac{1}{\sigma}\left(\gamma(t) + s\eta \frac{J(\dot{\gamma}(t))}{\|\dot{\gamma}(t)\|}\right),$$

that is an immersion $N\gamma : B \to D$ mapping the corners of the box B into ∂D.

Definition 1.5.9 Let P be a divide and let p, q be two natural numbers with $\gcd(p, q)$, we denote by

$$P_{p,q} * P$$

the divide in D, which is the image by $N\gamma : B \to D$ of $P_{p,q} \subset B$. We call $P_{p,q} * P$ the *(p, q)-iterated divide around P*.

Note that by definition, an iterated $*$-composition of divides has to be evaluated from the right to the left.

The number of double points $\delta(P_{p,q} * P)$ of $P_{p,q} * P$ is computed inductively from the number of double points $\delta(P)$ of P by:

$$\delta(P_{p,q} * P) = (p-1)(q-1)/2 + \delta(P)p^2. \tag{1.13}$$

Indeed, observe that the divide has:

1. By construction, at least as many crossings as $P_{p,q}$ had (recall Eq. 1.12), and hence the first summand.
2. Near each intersection point of P, there appear after the cabling operation, p lines crossing with other p lines, and hence the second summand.

Let $R_\eta\gamma$ be the union of the image of $N_\eta\gamma$ with the two chordal caps at the endpoints of γ. One can think of $R_\eta\gamma$ as a thick version of γ. The connected components of $D\setminus R_\eta\gamma$ correspond via inclusion to the connected components of $D\setminus P$. We declare a connected component of $D\setminus P_{p,q} * P$ to be signed by $+$, if the component contains a component of $D \setminus R_\eta\gamma$, that corresponds to a $+$ component of $D \setminus P$. In this case we call the connected component of $D \setminus R_\eta\gamma$ a *P_+-component*. Observe that there exists a chess board sign distribution for the components of $D \setminus R_\eta\gamma$ that makes P_+-components indeed to $+$ components.

The field $\Phi_{p,q}$ of cones on the box $B \subset \mathbb{R}^2$ is the subset in the tangent space of TB given by:

$$\Phi_{p,q} := \{(x, u) \in TB \mid | < u, e_1 >_{\mathbb{R}^2} | \geq \cos(\alpha(x))\|u\|\}$$

where $e_1 = (1, 0) \in \mathbb{R}^2$ and where $\alpha : B \to \mathbb{R}$ is a function, such that for every $(x, u) \in TB$ with $x \in P_{p,q}$ and $u \in T_x P_{p,q}$ we have the equality

$$| < u, e_1 >_{\mathbb{R}^2} | = \cos(\alpha(x))\|u\|.$$

Moreover, α has the boundary values $\alpha(\pm 1, t) = 0$ and $\alpha(s, \pm 1) = \pi/2$. We interpolate the function α on B by upper and lower convexity, i.e such that $\frac{\partial^2}{\partial t}\alpha < 0$ and $\frac{\partial^2}{\partial s}\alpha > 0$. The definition of $\alpha(x)$ seems to be conflicting at the double points of the curve $P_{p,q}$; at a double point $x = (x_1, x_2)$ of the curve $P_{p,q}$ the two tangents lines to $P_{p,q}$ have opposite slopes $\tan(\alpha(x))$ and $-\tan(\alpha(x))$, since the curve $P_{p,q}$ is defined by the equation

$$T(q, x_1) - T(p, x_2) = 0$$

that separates the variables. For example, a nice such function α is given by:

$$\alpha(x_1, x_2) := \arctan(\frac{q\sqrt{1 - x_2^2}}{p\sqrt{1 - x_1^2}}).$$

The interest of the field $\Phi_{p,q}$ comes from the following lemma, that follows from the definitions.

Lemma 1.5.10 *Let the image of $\gamma : [-1, 1] \to D$ be a divide P, that meets ∂D at right angles. For $\eta > 0$ small enough, the intersection of $S^3 \subset T\mathbb{R}^2$ with the image in $T\mathbb{R}^2$ of the field of sectors $\Phi_{p,q}$ under the differential of $N_\eta\gamma$ is a tubular neighborhood of the knot $L(P)$. The composition of $P_{p,q} : [-1, 1] \to B := [-1, 1] \times [-1, 1]$ and of $N\gamma : B \to D$ is again a divide, whose knot is a torus cable knot of type (p, q) of the knot $L(P)$.*

The image of the field of sectors $\Phi_{p,q}$ of B under the differential of $N_\eta\gamma$ will be denoted by $\Phi_{\eta,p,q}\gamma$ and for small η contains those vectors, that have feet near P and form a small angle with the tangent vectors of the divide P.

1.5.3.3 Divides for Plane Branches

Let $\{f_{a,b}(x, y) = 0\}$ be a singularity having one branch and with essential Puiseux pairs (recall Sect. 1.2.2.2) $(a_i, b_i)_{1\le i\le n}$. The theorem of S. Gusein-Zade [23] very efficiently describes a divide for the singularity $\{f_{a,b}(x, y) = 0\}$ in a closed form, namely the iteratively composed divide

$$P_{a_n,b'_n} * \cdots * P_{a_2,b'_2} * P_{a_1,b_1},$$

where the numbers $b'_2, \ldots, b'_n$ can be computed recursively, as we will show here below. We denote by S_k, $1 \le k \le n$ the divide

$$P_{a_k,b'_k} * P_{a_{k-1},b'_{k-1}} \cdots * P_{a_2,b'_2} * P_{a_1,b_1}$$

and let $f_k(x, y)$ be a specific equation for a singularity with essential Puiseux pairs $(a_i, b_i)_{1\le i\le k}$.

Remember, that the product $a_1a_2 \ldots a_k$ is the multiplicity at 0 of the curve $\{f_k(x, y) = 0\}$ and that the linking number λ_k of $L(S_k)$ and $L(S_{k-1})$ in S^3 can be computed recursively by:

$$\lambda_1 = b_1,\ \lambda_{k+1} = b_{k+1} - b_k a_{k+1} + \lambda_k a_k a_{k+1}.$$

The linking number λ_k is equal to the intersection multiplicity

$$\dim \mathbb{C}[[x, y]]/(f_k(x, y), f_{k-1}(x, y))$$

at 0 of the curves $\{f_k(x, y) = 0\}$ and $\{f_{k-1}(x, y) = 0\}$ (recall the definition of Newton pairs Eq. 1.2). We have also for the linking number λ_k an interpretation in terms of divides (see the next section for the first equality) which leads to the definition of the numbers b'_k:

$$\lambda_k = \#(S_k \cap S_{k-1}) = 2a_k\delta(S_{k-1}) + b'_k.$$

Remembering, that we already have computed recursively the numbers $\delta(S_k)$ and $\#(S_k \cap S_{k-1})$, we conclude that b'_k too can be computed recursively.

For example, for the Puiseux expansion $y = x^{3/2} + x^{7/4}$ we have: $\lambda_1 = 3$, $\lambda_2 = 7-3\cdot 2+3\cdot 2\cdot 2 = 13$, $b'_2 = 13-2\cdot 2\cdot 1 = 9$. Hence, the divide for the irreducible singularity with Puiseux expansion $y = x^{3/2} + x^{7/4}$ is the divide $P_{2,9} * P_{2,3}$, see Fig. 1.20. For the Puiseux expansion $y = x^{3/2} + x^{11/6}$ we found: $\lambda_1 = 3$, $\lambda_2 = 11 - 3\cdot 3 + 3\cdot 2\cdot 3 = 20$, $b'_2 = 20 - 2\cdot 3\cdot 1 = 14$. Hence, the divide for its singularity $\{(y^2 - x^3)^3 - 6x^7y^2 - 2x^{10} - x^{11} = 0\}$ is $P_{3,14} * P_{2,3}$, see Fig. 1.21.

Using the contents of the (independent) Sect. 1.8 (cf. [5, 7]), we can read off from this divide the Milnor fibration of the singularity $\{f_{a,b}(x, y) = 0\}$. In particular we can describe the Milnor fiber with a distinguished basis of quadratic vanishing

Fig. 1.20 On the upper left part we see the divide $P_{2,3}$, on the upper right part the divide $P_{2,9}$ and on the lower part, the divide $P_{2,9} * P_{2,3}$

Fig. 1.21 The divide $P_{3,14} * P_{2,3}$ for $(y^2 - x^3)^3 - 6x^7y^2 - 2x^{10} - x^{11}$

cycles. Using the above iterated cabling construction, we will also be able to read off from the divide, the reduction system (Definition 1.2.4) of the geometric monodromy of an irreducible plane curve singularity (see Sect. 1.8.3), as described in [1]. For instance, intersection numbers in the sense of Nielsen of quadratic vanishing cycles and reduction cycles can be computed.

1.5.3.4 Divides for All Real Plane Curve Singularities

In general, for an isolated singularity of a real polynomial $f(x, y)$ having several local branches, the divide $\{f_1(x, y) = 0\} \cap D$ of a real morsification $f_t(x, y)$ may have immersed circles as components. The above cabling construction $P_{p,q} * P$

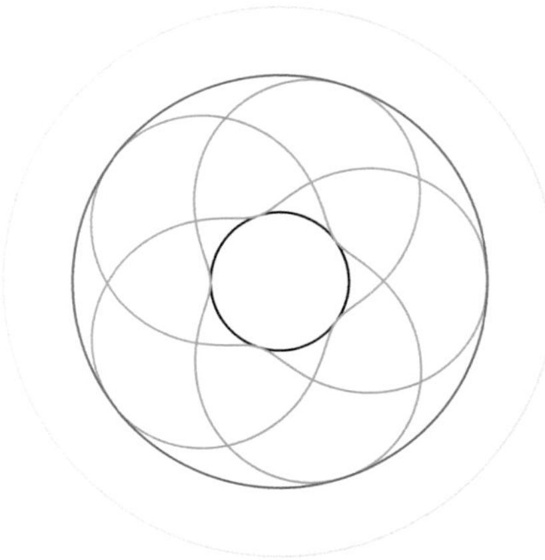

Fig. 1.22 The building block $L_{3,5}$ of Lissajous type

does not work if the divide P consists of an immersed circle. Of course, if one is willing to change the equation of the singularity to an equation, which defines a topologically equivalent singularity and which has only real local branches, one will only have to deal with divides consisting of immersed intervals. If we do not want to change the real equation, we conjecture how a different type of block would produce real morsifications for these real singularities. Note that the question whether any real singularity admits a morsification or not is still open. See [35] for some partial results to answer the question in the positive and see [20] for other recent conjectures that relies on this open question.

We conjecture that, if the real plane curve contains pairs of complex conjugate branches, then we could still produce divides using a second type of building blocks for a cabling construction, see Fig. 1.22.

These building blocks are the divides $L_{p,q}$ in the annular region $A := \{(x, y) \in D | 1/4 \leq \sqrt{x^2 + y^2} \leq 3/4\}$. If for the integers $(p, q) = 1$ holds, the divide $L_{p,q}$ is the Lissajous curve

$$s \in [0, 1] \mapsto (1/2 + 1/4 \sin(2\pi q s))(\sin(2\pi p s), \cos(2\pi p s))$$

in $A \subset D$. The curve $L_{p,q}$ has p-fold rotational symmetry. If $(p, q) = r > 1$ the divide $L_{p,q}$ is defined defined as the union of r rotated copies of $L_{p/r,q/r}$ with rotations of angles $2\pi k/p, k = 0 \ldots r-1$, of D. Again, the system of curves $L_{p,q}$ has a q-fold rotational symmetry.

The star-product $L_{p,q} * P$ can be defined as above if the divide P consists of one immersed circle. The two types of building blocks $P_{p,q}$ and $L_{p,q}$ together with the star-products $P_{p,q} * P$ and $L_{p,q} * P$ will allow one to describe the iterated cablings of real plane curve singularities in general.

1.5.3.5 Natural Orientations

The link of a divide P is naturally oriented by the following recipe. Let $\gamma :]0,1[\to D$ be a local regular parametrization of P. The orientation of $L(P)$ is such that the map $t \in]0,1[\mapsto (\gamma(t), \lambda(t)\dot{\gamma}(t)) \in L(P)$ is oriented. Here $\lambda(t)$ is a positive scalar function, which ensures that the map takes its values in $L(P)$. For a connected divide P we orient its fiber surface F_P such that the oriented boundary of $(F_P \cup L(P), L(P))$ coincides with the orientation of $L(P)$.

Vanishing cycles δ_c where c is a critical point of f_P do not carry a natural orientation, since, for example, the third power of the geometric monodromy of the singularity $\{x^3 - y^2 = 0\}$ reverses the orientations on the vanishing cycles.

We orient the tangent space $TD = D \times \mathbb{R}^2$ so that the orientation of its unit sphere S^3 as the boundary of its unit ball gives that the linking numbers of $L(P_1)$ and $L(P_2)$ are positive for generic pairs of divides P_1 and P_2. In fact, the orientation TD is opposite to its orientation as tangent space. With this convention, we have $Lk_{S^3}(L(P_1), L(P_2)) = \#(P_1 \cap P_2)$, a fact which was already used in the previous section.

1.6 Description of the Milnor Fiber

In this section, we give a description of the Milnor fibers associated to a divide that lie over the points ± 1 and $\pm i$. The description is done in terms of more simple pieces. This decomposition is useful in the forthcoming sections where we describe the geometric monodromy associated with the divide.

1.6.1 Description of the Milnor Fibers over $+1$ *and* -1

Let P be a connected divide and let $\pi_P : S^3 \setminus L(P) \to S^1$ be its fibration of Theorem 1.4.1. In this subsection, we show how to read off geometrically the fibers $F_1 := \pi_P^{-1}(1)$ and $F_{-1} := \pi_P^{-1}(-1)$. For our construction we assume the disk $D \subset \mathbb{C}$ oriented. We think of its orientation as an orthogonal complex structure $J : TD \to TD$. Define

$$P_+ := \{x \in D \setminus \partial D | f_P(x) > 0, df_P(x) \neq 0\}$$

The level curves of f_P define a oriented foliation F_+ on P_+, where a tangent vector u to a level of f_P at $x \in P_+$ is oriented so that $df_P(x)(Ju) > 0$. Put

$$P_{+,+} := \{(x,u) \in S^3 | x \in P_+,\ u \in T(F_+)\}$$

and

$$P_{+,-} := \{(x, u) \in S^3 | x \in P_+,\ u \in T(F_-)\},$$

where F_- is the foliation with the opposite orientation. Put

$$F_M := \{(x, u) \in S^3 | x = M\}$$

for a maximum M, and

$$F_m := \{(x, u) \in S^3 | x = m\}$$

for a minimum m of f_P. For saddle points s of f_P (equivalently, double point of P), define

$$F_{s,+} := \{(x, u) \in S^3 | x = s,\ H_{f_P}(x)(u, u) < 0\}$$

and

$$F_{s,-} := \{(x, u) \in S^3 | x = s,\ H_{f_P}(x)(u, u) > 0\}$$

Observe that the angle in between $u, v \in F_m$ or $u, v \in F_M$ is a natural distance function on F_m or F_M, which allows us to identify F_m and F_M with a circle. Finally, put

$$\partial D_+ := \{x \in \partial D | f_P(x) > 0\}$$

Let $p_{\mathbb{R}} : S^3 \to D$ be the projection $(x, u) \mapsto x$. The projection $p_{\mathbb{R}}$ maps each of the sets $P_{+,+}$ and $P_{+,-}$ homeomorphically to P_+. The sets F_m or F_M are homeomorphic to S^1, if M or m is a maximum or minimum of f_P respectively. And the sets $F_{s,\pm}$ are homeomorphic to a disjoint union of two open intervals if s is a crossing point of P. The set ∂D_+ is homeomorphic to a disjoint union of open intervals. We have the following decomposition of F_1:

$$F_1 = P_{+,+} \cup P_{+,-} \cup \partial D_+ \cup \bigcup_{s \in P_d} F_{s,+} \cup \bigcup_{M \in P_+} F_M.$$

Where P_d denotes the set of double points of the divide. Observe that for $(x, u) \in P_{+,+} \cup P_{+,-}$ we have $\theta_P(x, u) \in \mathbb{R}_{>0}$ (recall Eq. 1.5) since $f_P(x) > 0$, $df_P(x)(u) = 0$ and $\chi(x)H_{f_P}(u, u) \leq 0$. So $P_{+,+} \cup P_{+,-}$ is an open and dense subset in F_1. Accordingly, with the obvious changes of signs, we get a similar description for F_{-1}:

$$F_{-1} = P_{-,+} \cup P_{-,-} \cup \partial D_- \cup \bigcup_{s \in P_d} F_{s,-} \cup \bigcup_{m \in P_-} F_m.$$

1.6.1.1 Combinatorial Description

Forming the closure of $P_{+,+} \cup P_{+,-}$ in F_1 leads to the following combinatorial description of the above decomposition. First, we add to the open surface F_1 its boundary and get

$$\bar{F}_1 := F_1 \cup L(P).$$

Now let R be a connected component of P_+. The inverse image $p_{\mathbb{R}}^{-1}(R) \cap \bar{F}_1$ in $\bar{F}_1$ are two disjoint open cells or cylinders $R_+ \subset P_{+,+}$ and $R_- \subset P_{+,-}$ which are in fact subsets of F_1. The closure of R_+ in $\bar{F}_1$ is a surface $\bar{R}_+$ with boundary and corners. The set F_M is a common boundary component without corners of $\bar{R}_+$ and $\bar{R}_-$ if M is a maximum in R. If there is no maximum in R the closures $\bar{R}_+$ and $\bar{R}_-$ meet along the component of ∂D_+ which lies in the closure of R. Let S, R be connected components of P_+ such that the closures of R and S have a crossing point s in common. The closures of R_+ and S_- in F_1 meet along one of the components of $F_{s,+}$ and the closures of R_- and S_+ in F_1 meet along the other component of $F_{s,+}$. The closure of $F_{s,+} \cap \bar{R}_+$ in $\bar{R}_+$ intersects $L(P)$ in 2 corners, that are also corners of the closure of $F_{s,+} \cap \bar{S}_-$ in $\bar{S}_-$ (see Fig. 1.23). Notice that the foliation $\mathcal{F}_+$ on P_+ does not lift to a foliation, which extends to an oriented foliation on F_1.

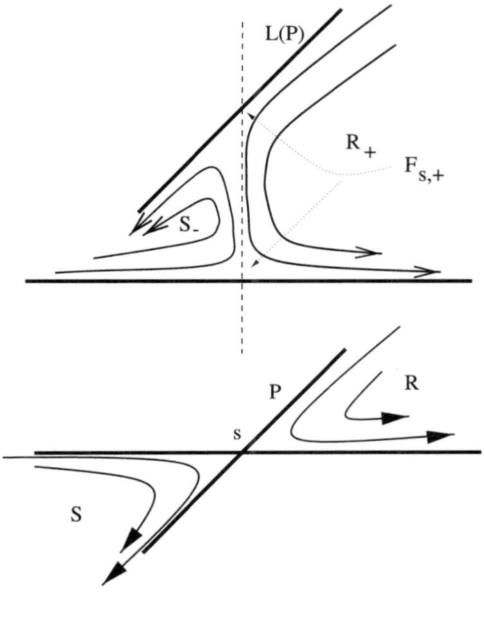

Fig. 1.23 Gluing of the lifts of R with $+$ and S with $-$ foliation to F_1

1.6.2 Description of the Milnor Fibers over $+i$ and $-i$

Now we quickly work out the fibers $F_i := \pi_P^{-1}(i)$ and $F_{-i} := \pi_P^{-1}(-i)$. Observe that F_i and F_{-i} are projected to a subset of $P \cup \mathrm{supp}(\chi)$ by $p_{\mathbb{R}}$. Put

$$F_{i,P} := \{(x,u) \in S^3 | x \in P,\ \chi(x) = 0,\ df_P(x)(u) > 0\}.$$

For a crossing point c of P we put

$$F_{i,c} := \{(x,u) \in S^3 | \chi(x) > 0,\ df_P(x)(u) > 0,\ f_P(x) - \frac{1}{2}\eta^2\chi(x)H_{f_P}(c)(u,u) = 0\}.$$

Observe that

$$f_P(x) - \frac{1}{2}\eta^2\chi(x)H_{f_P}(c)(u,u) = \mathrm{Re}(\theta_{P,\eta}(x,u))$$

and recall (Eq. 1.5). In order to get nice sets it is necessary to choose a nice bump function χ as before. The set

$$F_{i,P} \cup F_{i,c}$$

is an open and dense subset in F_i and forming its closure results in

$$\bar{F}_i := F_i \cup L(P)$$

leads to a combinatorial description of F_i similar to the one given above.

We would like to direct the interested reader to some beautiful pictures in the works of Sebastian Baader, Pierre Dehornoy and Livio Liechti [11, 17] which complement the ones contained in this work and which exemplify how the Milnor fiber(s) are recovered from the divide. The referee called our attention to these works and we are thankful for that.

1.7 Descriptions of the Monodromy

In this section we give a complete description of the geometric monodromy of the fibered link $L(P)$.

In Sect. 1.7.1 we give the monodromy as a composition of right-handed Dehn twists. This factorization, certainly completely determines the monodromy but it is general not trivial to give the Nielsen-Thurston decomposition from a factorization. In Sect. 1.7.2 we take care of this and give the minimal reduction system of curves of the Nielsen-Thurston decomposition of the monodromy.

1.7.1 Monodromy as a Product of Dehn Twists

We will use the integral curves of the distribution $J(\ker(df_P))$, which pass through the crossing points of the divide P. In a connected component R of $D \setminus P$, those integral curves of $J(\ker(df_P))$ meet at the critical point of f_P in the component R with distinct tangents, or they go to distinct points of ∂D.

We denote by P' the union of the integral curves of $J(\ker(df_P))$, which pass through the crossing points of P. The complement in D of the union $P' \cup P \cup \partial D$ is a disjoint union of tiles, which are homeomorphic to open squares or triangles

Notation We call a pair (A, B) of tiles *opposite*, if $A \neq B$ and the closures of A and B in D have a segment of P in common. For an opposite pair of tiles (A, B) let $A|B$ be the interior in D of the union of the closures of A and B in D.

The set is foliated by the levels of f_P and also by the integral lines of the distribution $J(\ker(df_P))$. Both foliations are non-singular and meet in a J-orthogonal way (see Fig. 1.24).

Notation For a pair (A, B) of opposite we put,

$$F_{1,A|B} := \{(x, u) \in F_1 | x \in A|B\}$$

and

$$F_{-1,A|B} := \{(x, u) \in F_{-1} | x \in A|B\}$$

The sets $F_{1,A|B}$ and $F_{-1,A|B}$ each have two connected components:

$$F_{1,A|B} = F_{1,+,A|B} \cup F_{1,-,A|B}$$

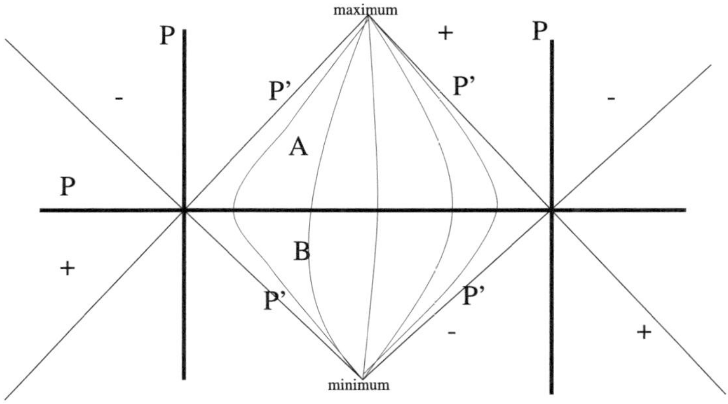

Fig. 1.24 Two tiles with the $J(\ker(df_P))$ foliation

where

$$F_{1,+,A|B} := \{(x,u) \in F_1 | x \in A|B,\ df_P(Ju) > 0\}$$

$$F_{-1,+,A|B} := \{(x,u) \in F_{-1} | x \in A|B,\ df_P(Ju) > 0\}$$

and

$$F_{1,-,A|B} := \{(x,u) \in F_1 | x \in A|B,\ df_P(Ju) < 0\}$$

$$F_{-1,-,A|B} := \{(x,u) \in F_{-1} | x \in A|B,\ df_P(Ju) < 0\}$$

The closures of $F_{1,\pm,A|B}$ in $\bar{F}_1$ and of $F_{-1,\pm,A|B}$ in $\bar{F}_{-1}$ are polygons with 6 edges: let M, c, c' be the vertices of the triangle A; the six edges of the closure H of $F_{1,+,A|B}$ in $\bar{F}_1$ are

$$\begin{aligned}&\{(x,u) \in H \mid x = M\},\ \{(x,u) \in H \mid x \in [c, M]\},\\ &\{(x,u) \in H \mid x = c\},\ \ \{(x,u) \in H \mid x \in [c, c']\},\\ &\{(x,u) \in H \mid x = c'\},\ \{(x,u) \in H \mid x \in [c', M]\},\end{aligned}$$

where $[M, c]$ and $[M, c']$ are segments included in P' and $[c, c']$ is a segment in P. Next, we define two diffeomorphisms

$$S_{i,A|B} : F_{1,A|B} \to F_{-1,A|B}$$

and

$$S_{-i,A|B} : F_{1,A|B} \to F_{-1,A|B}$$

for each pair of opposite tiles (A, B). To do so we choose the adapted function $f_P : D \to \mathbb{R}$ (recall Definition 1.3.8 and Lemma 1.3.9) such that the maxima are of value 1 and the minima of value -1. Moreover, we modify the function f_P at the boundary ∂D such that along each of the integral lines of the foliation given by the distribution $J(\ker(df_P))$ the function f_P takes all values in an interval $[-m, m]$ with $1 \geq m > 0$. The latter modification of f_P is useful if the tile A or B meets ∂D. We also need the rotations $J_\theta : T(D) \to T(D)$ about the angle $\theta \in [-\pi, \pi]$. Recall that the complex structure J is precisely $J_{\pi/2}$.

We are now ready to define the map $S_{i,A|B}$. Let $(x, u) \in F_1$ with $x \in A|B$ and do as follows:

1. let $y \in A|B$ be the point in the opposite tile on the integral line of the distribution $J(\ker(df_P))$ with $f_P(x) = -f_P(y)$;
2. now move x to y along the integral curve $\gamma_{x,y}(t), t \in [f_P(x), f_P(y)]$ which connects x and y with the parameterization $f_P(\gamma_{x,y}(t)) = t$;

3. consider the rotation angle function $\theta(x,t) := \frac{(|f_P(x)|-t)\pi}{2|f_P(x)|}$ and move the vector u along the path

$$\left(\gamma_{x,y}(t), U_{x,y}(t)\right) := \left(\gamma_{x,y}(t), s(x,t)(J_{\theta(x,t)}(|f_P(x)|u_{x,y}(t)/2) + u_{x,y}(t))\right),$$

where $(\gamma_{x,y}(t), u_{x,y}(t)) \in S^3$ is the continuous vector field along $\gamma_{x,y}(t)$ such that $df_P(u_{x,y}(t)) = 0$, we have the equality $u_{x,y}(f_P(x)) = U_{x,y}(f_P(x)) = u$, and the stretching factor $s(x,t) \geq 1$ is chosen such that $(\gamma_{x,y}(t), U_{x,y}(t)) \in S^3$ holds.
4. We finally define

$$S_{i,A|B}((x,u)) := \left(y, u_{x,y}(f_P(y))\right) = \left(y, U_{x,y}(f_P(y))\right)$$

The definition of $S_{-i,A|B}$ is analogous, but uses rotations in the sense of $-J$.

The names $S_{i,A|B}$ or $S_{-i,A|B}$ indicate that the flow lines $(\gamma_{x,y}(t), u_{x,y}(t))$ pass through the fiber F_i or F_{-i} respectively. The flow lines defining $S_{i,A|B}$ or $S_{-i,A|B}$ are different. However, the maps $S_{i,A|B}$ and $S_{-i,A|B}$ are equal. The system of paths $(\gamma_{x,y}(t), U_{x,y}(t)) \in S^3$ is local near the link $L(P)$, i.e. for every neighborhood N in S^3 of a point $(x', u') \in L(P)$ there exists a neighborhood M of (x', u') in S^3 such that each path $(\gamma_{x,y}(t), U_{x,y}(t))$ with $(x,u) \in F_1 \cap M$ stays in N. It will follow that the flow lines of the monodromy vector field are meridians of the link $L(P)$ in its neighborhood.

Remark 1.7.1 After verifying that the constructed flow lines form a monodromy flow for the π_P, this last fact proves that the map $\pi_{P,\eta}$ is a fibration near the link $L(P)$. Proving this was a step in the proof of Theorem 1.4.1 and this is an alternative proof.

The partially defined diffeomorphisms $S_{i,A|B}$ and $S_{-i,A|B}$ glue to diffeomorphisms

$$S_i, S_{-i} : \bigcup_{A|B} F_{1,A|B} \to \bigcup_{A|B} F_{-1,A|B}$$

where the sum ranges over all opposite pairs of tiles (A, B) with $A \subset P_+$. The gluing poses no problem since those unions are disjoint, but the diffeomorphisms S_i and S_{-i} do not extend continuously to F_1. We will see that the discontinuities, which are the obstruction for extending S_i and S_{-i}, can be compensated by a composition of right half Dehn twists (recall Definition 1.2.2).

At a maximum $M \in D$ of f_P each vector $(M,u) \in S^3$ belongs to F_1. Let a and b be the integral curves of $J(\ker(df_P))$ with one endpoint at M and orthogonal to u. We assume that neither a nor b passes through a crossing point of P (see Fig. 1.25) and that a and b belong to different pairs of opposite tiles. A continuous extension of the maps S_i or S_{-i} has to map the vector (M,u) to two vectors based at the other

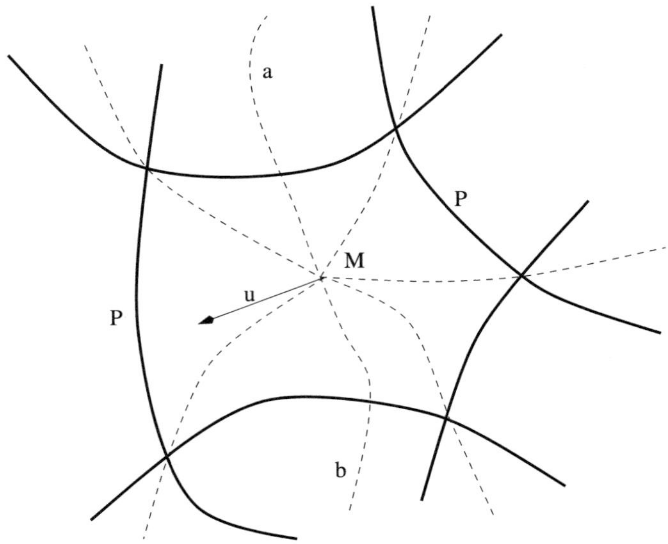

Fig. 1.25 The discontinuity at F_M

endpoint of a and b. Since these endpoints differ in general, a continuous extension is impossible.

In order to allow a continuous extension at the common endpoint of a and b we make a new surface F_1' by cutting F_1 along the cycles F_M, where M runs through all the maxima of f_P and by gluing back after a rotation of angle π of each of the cycles F_M. In the analogous manner, we make the surface F_{-1}' in doing the half twist along F_m, where m runs through the minima of f_P. The subsets $F_{1,A|B}$ do not meet the support of the half twists, so they are canonically again subsets of F_1', which we denote by $F_{1,A|B}'$. Analogously, we have subsets $F_{-1,A|B}'$ in F_{-1}'. A crucial observation is that the partially defined diffeomorphisms

$$S_i', S_{-i}' : \bigcup_{A|B} F_{1,A|B}' \to \bigcup_{A|B} F_{-1,A|B}'$$

have less discontinuities, which are the obstruction for a continuous extension. We denote by a' and b' the arcs on F_1', which correspond to the arcs a and b on F_1. Indeed, the continuous extension at the end points of a' and b' is now possible.

Let s be a crossing point of P and let $I_{s,+}$ be the segment of P', which passes through s and lies in P_+. The inverse image of $Z_s^\circ := p_{\mathbb{R}}^{-1} I_{c,+} \cap F_1$ is not a cycle, except if both endpoints of $I_{s,+}$ lie on ∂D. If a maximum M of f_P is an endpoint of $I_{s,+}$, the inverse image $p_{\mathbb{R}}^{-1}(M) \cap F_1$ consists of 2 points on F_M, which are antipodal. On the new surface F_1' the inverse image $p_{\mathbb{R}}^{-1}(I_{s,+}) \cap F_{-1}'$ is a cycle. An extension of S_i' and S_{-i}' will be discontinuous along this cycle (see Fig. 1.26). We now observe that the partially defined diffeomorphisms S_i' and S_{-i}'

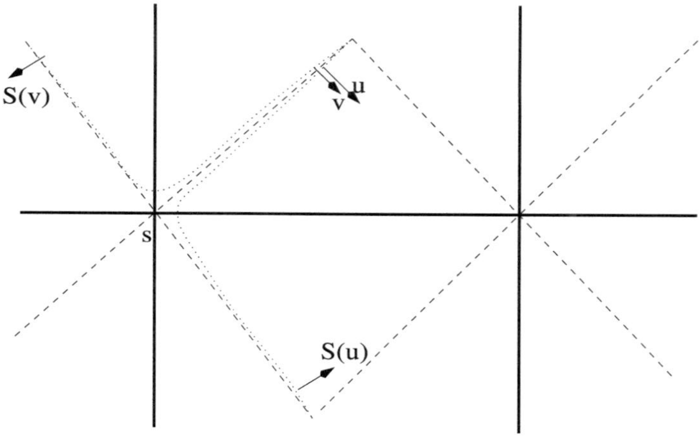

Fig. 1.26 The discontinuity along Z_s

have discontinuities along the cycle $p_{\mathbb{R}}^{-1}(I_{s,+}) \cap F'_{-1}$, which can be compensated by half twists along the inverse images $p_{\mathbb{R}}^{-1}(I_{s,-}) \cap F'_{-1}$, where s runs through the crossing points of P. Note that for a crossing point s of P the curve $Z'_{s,-1} := p_{\mathbb{R}}-1(I_{s,-}) \cap F'_{-1}$ is in fact a simply closed curve on F'_{-1}.

For a crossing point s of the divide P we now define a simply closed curve on F_1, by putting:

$$Z_s := Z_s^\circ \cup \bigcup_{M \in \partial I_{s,+}} F_{s,M}$$

where for an endpoint M of $I_{s,+}$, which is a maximum of f_P, the set $F_{s,M}$ is the simple arc of F_M, which connects the two points of $Z_s^\circ \cap F_M$ and contains an inward tangent vector of $I_{s,+}$ at M. As we already have noticed the set $Z_s^\circ \cap F_M$ has only one element if $M \in \partial D$, so we define $F_{s,M} := \emptyset$ in that case.

We have the inclusion $F_m \subset F_{-1}$. We now define the cycle $Z_m \subset F_1$. Define for a minimum m of f_P the region

$$B_m := \bigcup_{A|B, m \in \bar{B}} F_{-1,A|B}$$

Let $B_{m,\epsilon}$ be the level curve

$$B_{m,\epsilon} := \{(x,u) \in B_m | f_P(x) = -\epsilon\}$$

For a small ϵ the set

$$(S_i)^{-1}(B_{m,\epsilon} \cap \bigcup_{A|B} F_{-1,A|B})$$

is a union of copies of an open interval and is not a cycle but nearly a cycle. The union closes up to a cycle by adding small segments which project to the integral lines through the crossing points of P. We denote this cycle by $Z_m \subset F_1$.

We are now able to state the main theorem.

Theorem 1.7.2 *Let P in D be a connected divide. Let $\pi_P : S^3 \setminus L(P) \to S^1$ be the fibration of Theorem 1.4.1 The counter clockwise monodromy of the fibration π_P is the composition of right handed Dehn twists*

$$T := T_- \circ T_. \circ T_+ : F_1 \to F_1,$$

where T_- is the product of the right handed twists along Z_m, m running through the minima of f_P, $T_.$ is the product of the right handed Dehn twists along the cycles Z_s, s running through the crossing points of P, and T_+ is the product of the right handed twists along F_M, M running through the maxima of f_P.

Proof We need to introduce one more surface. Let F''_{-1} be the surface obtained from the surface F'_{-1} by cutting F'_{-1} along the cycles $Z'_{s,-1}$ and by gluing back after a half twist along each $Z'_{s,-1}$, with s running through the crossing points of P. We still have partially defined diffeomorphisms

$$S''_i, S''_{-i} : \bigcup_{A|B} F'_{1,A|B} \to \bigcup_{A|B} F''_{-1,A|B}$$

since the cutting was done in the complement of $\bigcup_{A|B} F'$. By a direct inspection we see that the diffeomorphisms extend continuously to

$$S''_i, S''_{-i} : F'_1 \to F''_{-1}$$

Let

$$(p_+, q_+) : F_1 \to F'_1$$

$$(p_., q_.) : F''_{-1} \to F'_{-1}$$

$$(p_-, q_-) : F'_{-1} \to F_{-1}$$

be minimal positive pairs of Dehn twists (recall Definition 1.2.2). A direct inspection shows that the composition

$$p_+ \circ S_i' \circ p_. \circ p_- \circ (q_+ \circ S_{-i}' \circ q_. \circ q_-)^{-1} : F_1 \to F_1$$

is the monodromy of the fibration π_P. This composition evaluates to

$$T_- \circ T_. \circ T_+ : F_1 \to F_1$$

□

1.7.1.1 Some Consequences

Next we state some immediate consequences of the previous constructions.

Remark 1.7.3 We list some special properties of the monodromy of links and knots of divides. The number of Dehn twists of the above decomposition of the monodromy equals the first betti number $\mu = 2\delta - r + 1$ of the fiber, and the total number of intersection points among the core curves of the involved Dehn twists is less then 5δ. This means that the *complexity* of the monodromy is bounded by a function of μ. For instance, the coefficients of the Alexander polynomial of the link of a divide are bounded by a quantity, which depends only on the degree of the Alexander polynomial. This observation suggests the following definition for the complexity C of an element of the mapping class group ϕ of a surface: the minimum of the quantity $L + I$ over all decompositions as product of Dehn twists of ϕ, where L is the number of factors and I is the number of mutual intersections of the core curves. We do not know properties of this exhaustion of the mapping class group. Notice, that the function $(\phi, \psi) \mapsto C(\psi^{-1} \circ \phi) \in \mathbb{N}$ defines a left invariant distance on the mapping class group.

Remark 1.7.4 It can be seen (see next Sect. 1.7.2) that for any link of a divide the monodromy diffeomorphism and its inverse are conjugate by an orientation reversing element in the mapping class group. In our previous notations this conjugation is given by the map

$$(x, u) \in F_1 \mapsto (x, -u) \in F_1,$$

which moreover realizes geometrically the symmetry of G. Torres [47]

$$t^\mu \chi(1/t) = (-1)^\mu \chi(t)$$

for the Alexander polynomial $\chi(t)$ of knots.

Remark 1.7.5 In fact the proof of Theorem 1.7.2 shows that the fibration of the link of a connected divide P can be filled with a singular fibration in the 4-ball, which has 3 singular fibers with only quadratic singularities, as in the case of a divide of the singularity of a complex plane curve. The filling has only two singular fibers if the function f_P has no maxima or no minima. By this construction from a connected divide we obtain a contractible 4-dimensional piece with a Lefschetz pencil. This is part of what is usually called *Hurwitz equivalence*. For more on this topic, see the classical references by Kas [31] or Matsumoto [36], or the more recent by Baykur and Hayano [13].

We make the following important observation that we have not lost generality by considering only divides which are generic immersions of intervals.

Remark 1.7.6 It is important to note that Theorems 1.4.1 and 1.7.2 remain true for generic immersions of disjoint unions of intervals and circles in the 2-disk. It is also possible to start with a generic immersion of a 1-manifold I in an oriented compact connected surface with boundary S. The pair (S, I) defines a link $L(S, I)$ in the 3-manifold

$$M_S := T^+(S)/\mathrm{zip},$$

where $T^+(S)$ is the space of oriented tangent directions of the surface S and where zip is the identification relation, which identifies $(x, u), (y, v) \in T^+(S)$ if and only if $x = y \in \partial S$ or if $(x, u) = (y, v)$. In order to get a fibered link, the topological pair $(R, R \cap \partial S)$ has to be contractible for each connected component R of $S \setminus I$ and moreover, the complement $S \setminus I$ has to allow a chess board coloring in positive and negative regions. For more details on this construction, see [30].

1.7.2 Other Decompositions of Monodromy

In this subsection we study yet a decomposition of the *algebraic* monodromy in terms of involutions. The contents of this section come mainly from [9].

1.7.2.1 Digression in Higher Dimensions

We start with a discussion about geometric monodromies of isolated hypersurface singularities in general dimension (recall the discussion of Sect. 1.2.2.1 on page 1.2.2.1).

Let $f : \mathbb{C}^{n+1} \to \mathbb{C}$ be a map defined by a polynomial. We assume that $f(0) = 0$ and that $0 \in \mathbb{C}^{n+1}$ is an isolated critical point of f. For $p \in \mathbb{C}^{n+1}$ let

$$\|p\| = \sqrt{|z_0(p)|^2 + |z_1(p)|^2 + \cdots + |z_n(p)|^2}.$$

Let B_ϵ be a Milnor ball for the singularity of f and let

$$\text{Tube}_{\epsilon,\delta} := \{p \in B_\epsilon \mid |f(p)| \leq \delta\}, 0 < \delta << \epsilon,$$

be a regular tubular neighborhood of $\{p \in B_\epsilon \mid f(p) = 0\}$ in B_ϵ. A *monodromy vector field* X for the singularity is a smooth vector field

$$p \in \text{Tube}_{\epsilon,\delta} \mapsto X_p \in \mathbb{C}^{n+1}$$

such that we have the following properties for $p \in \text{Tube}_{\epsilon,\delta}$

- $(df)_p(X_p) = 2\pi i f(p)$,
- X_p is tangent to ∂B_ϵ if $p \in \partial B_\epsilon$,
- trajectories of X starting at $p \in \partial B_\epsilon$ are periodic with period 1 and are the boundary of a smooth disk in ∂B_ϵ, that is transversal to the function f. For this property to be satisfied, the hypothesis that f has a critical point is crucial.

Using partition of unity, one can construct monodromy vector fields. The flow at time 1 of a monodromy vector field X defines a monodromy diffeomorphism $T = T_X \colon F \to F$, where the manifold with boundary $(F, \partial(F)) := \{p \in B_\epsilon \mid f(p) = \delta\}$ is the Milnor fiber of the singularity. The relative isotopy class of the diffeomorphism T is independent from the chosen monodromy vector field and is called the *geometric monodromy* of the singularity. The geometric monodromy is a topological invariant of the singularity (see [34, Theorem 3] for $n \neq 2$ and [38] for $n = 2$).

From now on we will assume in addition, that the polynomial f is real meaning that its coefficients are real numbers. Let $c : \mathbb{C}^{n+1} \to \mathbb{C}^{n+1}$ denote the involution on complex space given by the complex conjugation of coordinate values. Hence with the above notations, we have $c(\text{Tube}_{\epsilon,\delta}) = \text{Tube}_{\epsilon,\delta}$ and $c(F) = F$. We denote by $c_F : F \to F$ the restriction of the involution c to F.

Let $X \colon \text{Tube}_{\epsilon,\delta} \to \mathbb{C}^{n+1}$ be a monodromy vector field for the isolated singularity of f. We may assume that we have constructed the vector field X with more care near the boundary of the Milnor ball in order to achieve that for some $\epsilon' < \epsilon$ we have the symmetry

$$c(X_p) = -c(X_{c(p)}), \text{ for } p \in \text{Tube}_{\epsilon,\delta}, \text{ with } \|p\| > \epsilon'.$$

Since f is real, we have

$$c((df)_p(X_p)) = (df)_{c(p)}(c(X_p)) = -2\pi i f(c(p))$$

hence, we see (by substituting q for $c(p)$ and accordingly $c(q)$ for p) that the vector field X^c defined by:

$$q \in \text{Tube}_{\epsilon,\delta} \mapsto X^c_q := -c(X_{c(q)}) \in \mathbb{C}^{n+1},$$

is a monodromy vector field too. Let $Y : \mathrm{Tube}_{\epsilon,\delta} \to \mathbb{C}^{n+1}$ be the vector field

$$Y := \frac{X + X^c}{2},$$

which due to the extra care is also a monodromy vector field. We have $Y^c = Y$. The following is an important symmetry of the geometric monodromy:

Lemma 1.7.7 *Let T_Y be a monodromy diffeomorphism, which has been computed with a monodromy vector field Y satisfying $Y^c = Y$. We have the symmetry*

$$c_F \circ T_Y \circ c_F = T_Y^{-1}.$$

The geometric monodromy T satisfies (up to relative isotopy) the symmetry

$$c_F \circ T \circ c_F = T^{-1}.$$

Proof The restriction of complex conjugation $c_{\mathrm{Tube}_{\epsilon,\delta}} : \mathrm{Tube}_{\epsilon,\delta} \to \mathrm{Tube}_{\epsilon,\delta}$ maps the monodromy vector field Y to $-Y$ and F to F. Hence, since $c_{\mathrm{Tube}_{\epsilon,\delta}}$ reverses the orientations of the trajectories, we have

$$T_Y^{-1} = (c_F)^{-1} \circ T_Y \circ c_F = c_F \circ T_Y \circ c_F.$$

Since the geometric monodromy T is in the relative mapping class represented by T_Y, we have the symmetry $c_F \circ T \circ c_F = T^{-1}$ at the mapping class group level. □

Symmetries of monodromies as in the lemma can occur in the more general context of so-called *strongly invertible knots*, see for instance [26, 33, 44].

See also the work of Sabir Gusein-Zade [24], in which he shows among other results, that the integral homological monodromy of an isolated complex hypersurface singularity with real defining equation f is the composition of two involutions, both conjugated to the action of complex conjugation on the homology of a real regular fiber.

Remark 1.7.8 The symmetry property $c_F \circ T \circ c_F = T^{-1}$ expresses that the geometric monodromy T of a complex hypersurface with real defining equation is conjugate in the mapping class group by an element of order 2 to its inverse T^{-1}. This is a statement in the mapping class group of the Milnor fiber and does not refer to any complex conjugation, so it can be stated for any complex hypersurface singularity with complex defining equation. We say that the singularity is *strongly invertible* if its geometric monodromy diffeomorphism T is conjugate by an element of order 2 in the relative mapping class group of the Milnor fiber to its inverse T^{-1}. Thus, the property of strong invertibility is a topological property for hypersurface singularities.

We can rewrite the symmetry property as follows: $T_Y \circ c_F \circ T_Y \circ c_F = \mathrm{Id}_F$. We see that $T_Y \circ c_F : F \to F$ is an involution of F. It follows the

Corollary 1.7.9 *The geometric monodromy T of an isolated complex hypersurface singularity, which is defined by a real equation, is the composition of two involutions of the fiber $F_\delta, \delta \in \mathbb{R}$, namely: $T = (T \circ c_F) \circ c_F$, where c_F is the restriction of the complex conjugation.*

For $k \in \mathbb{Z}$ we also have the relation $T_Y^k \circ c_F \circ T_Y^k \circ c_F = \mathrm{Id}_F$, which shows that $T_Y^k \circ c_F : F \to F, k \in \mathbb{Z}$, is a sequence of involutions of F.

The above observations can be applied to plane curve singularities in general, since it follows from the Theory of Puiseux Pairs that every plane curve singularity is topologically equivalent to a singularity given by a real equation and hence plane curve singularities are strongly invertible.

For complex hypersurface singularities of higher dimension the situation seems to be opposite. In $\mathbb{C}^{n+1}$, $n > 1$, there exist isolated hypersurface singularities which are not topologically equivalent to a singularity with a real defining equation [44]. We expect that in general the geometric monodromy T of a complex hypersurface singularity fails to be strongly invertible.

Mathias Schulze communicated to the first author a proof of strong invertibility of the homological monodromy with real coefficients (this was never published and stayed as a personal communication). The following is a strengthening of his result to the case of homology with rational coefficients.

Theorem 1.7.10 *The rational homological monodromy of a complex hypersurface singularity is strongly invertible.*

Proof Let $E = \bigoplus_i E_i$ be a finest possible direct sum decomposition in T_*-invariant $\mathbb{Q}$-subspaces of E. The characteristic polynomial of the restriction $A_i : E_i \to E_i$ of T_* to a summand E_i is a power $\psi(t) = \phi(t)^L$ of a cyclotomic polynomial $\phi(t)$. We have $\psi(A_i) = 0$ by the Hamilton-Cayley theorem and we have $\psi(A_i^{-1}) = 0$ since a power of cyclotomic polynomial satisfies $\psi(1/t) = \pm t^{\mathrm{degree}(\psi)}\psi(t)$.

Since the decomposition has no refinement, we may choose a vector $e_1 \in E_i$, that is cyclic for A_i and for A_i^{-1}. The systems

$$e_1, e_2 := A_i(e_1), \cdots, e_{\dim E_i} := A^{\dim E_i - 1}(e_1),$$

$$f_1 := e_1, f_2 := A_i^{-1}(e_1), \cdots, f_{\dim E_i} := A^{-\dim E_i + 1}(e_1)$$

are basis for the space E_i. Let $b_i : E_i \to E_i$ be the linear map defined by $b_i(e_j) = f_j, 1 \le j \le \dim E_i$.

We have $A_i^{-1}b_i(e_j) = b_i A_i(e_j), 1 \leq j < \dim E_i$. The polynomial $\psi_1(t) := -\psi(t) + t^{\dim E_i}$ satisfies $\psi_1(A_i^{\pm 1}) = A_i^{\pm \dim E_i}$. Since the degree of the polynomial $\psi_1(t)$ is less than $\dim E_i$, we have

$$A_i^{-1}b_i(e_{\dim E_i}) = A^{-\dim E_i}b_i(e_1) = \psi_1(A_i^{-1})b_i(e_1) =$$

$$b_i\psi_1(A_i)(e_1) = b_i A_i^{\dim E_i}(e_1) = b_i A_i(e_{\dim E_i}).$$

We conclude that $A_i^{-1}b_i = b_i A_i$ and $A_i^{-1} = b_i A_i b_i^{-1}$ hold.

The polynomial $\psi_0(t) := \frac{\psi(t)-\psi(0)}{-t\psi(0)}$ satisfies $A_i^{-1} = \psi_0(A_i)$ and $A_i = \psi_0(A_i^{-1})$. We deduce $b_i A_i^{-1} = b_i\psi_0(A_i) = \psi_0(A_i^{-1})b_i = A_i b_i$ and conclude $A_i^{-1} = b_i^{-1}A_i b_i$. We observe at this point that both the conjugates of A_i by b_i and by b_i^{-1} are equal to the inverse A_i^{-1}.

For $0 \leq j < \dim E_i$ we have (remember $e_1 = f_1 = b_i(e_1) = b_i(f_1)$)

$$b_i(f_{j+1}) = b_i A_i^{-j}(f_1) = A_i{}^j b_i(f_1) = A_i{}^j(e_1) = e_{j+1}.$$

Hence b_i is of order two, which shows that A_i is strongly invertible over $\mathbb{Q}$. The direct sum $b := \bigoplus b_i$ is a rational strong inversion for T_*. □

Question Is the integral homological monodromy of a complex hypersurface singularity strongly invertible? For example, take the isolated surface singularity

$$f(x, y, z) = L_1 L_2^2 L_3^3 L_4^4 L_5^5 L_6^6 L_7^7 L_8^8 L_9^9 + x^{46} + y^{46} + z^{46}$$

where $L_j, 1 \leq j \leq 9$ are linear forms on $\mathbb{C}^3$, such that the 9 lines $\{L_j = 0\}$ in the complex projective plane span the 9_3 configuration of flex tangents to a nonsingular cubic. No real equation for this singularity can exist, since the configuration 9_3 cannot be realized in the real projective plane, see [27, 44]. Is its monodromy strongly invertible?

1.7.2.2 Going Back to Plane Curves and Divides

Now we focus on the study of complex conjugation on the topology of plane curve singularities via divides. First we show that the fiber of the link of a connected divide carries naturally a cellular decomposition with tri-valent 1-skeleton.

Let $f : \mathbb{C}^2 \to \mathbb{C}$ define an isolated plane curve singularity at $0 \in \mathbb{C}^2$ given by a real convergent power series $f \in \mathbb{R}\{x, y\}$. Let $f = f_1 f_2 \cdots f_r$ be the factorization in local branches. The topology of a plane curve singularity f is completely encoded in a divide for f as we have seen in Sect. 1.4 or in the previous section where we have given a factorization of the monodromy from the divide (cf. [7, 12]), see also [3, 4, 23].

We state the results more generally for links of connected divides, since the involution given by complex conjugation on real isolated plane curve singularities corresponds to a natural involution of links of divides as explained below, (cf. [5, 7]).

Denote by δ_P the number of double points of the divide P. The local topology of a plane curve singularity is obtained from a divide for the singularity. More precisely, the link $L_f \subset \partial B_\epsilon$ of an isolated plane curve singularity $\{f = 0\}$, $f \in \mathbb{R}\{x, y\}$, is equivalent to the link L_P of a divide P, see Sect. 1.5 (cf. [7]).

We recall, that for a totally real (Definition 1.5.1) plane curve singularity f one can obtain a divide P by performing a small real deformation $f_s, 0 \leq s \leq 1$, of the singularity, where for $0 < s \leq 1$ the restriction of f_s to the euclidean disk $D_\epsilon := B_\epsilon \cap \mathbb{R}^2$ of radius ϵ in $\mathbb{R}^2$ is a Morse function with $\mu(f)$ critical points and such that the 0-level is connected and contains all the saddle points (this is the content of Theorem 1.5.4). The divide P for f is the curve $\{p \in D_\epsilon \mid f_1(p) = 0\}$, which we rescale by the factor $\frac{1}{\epsilon}$ from D_ϵ into the unit disk D (Fig. 1.27).

The homology $H_1(F_P, \mathbb{Z})$ can be described combinatorially in terms of the divide P as a direct sum $H_1(F_P, \mathbb{Z}) = E_- \bigoplus E_0 \bigoplus E_+$, where E_-, E_0 and E_+ are the subspaces in $H_1(F, \mathbb{Z})$, which are freely generated as follows:

$$
\begin{aligned}
E_+ &:= [\delta_1, \cdots, \delta_{\mu_+}], \\
E_0 &:= [\delta_{\mu_+ + 1}, \cdots, \delta_{\mu_+ + \mu_0}], \quad \text{and} \\
E_- &:= [\delta_{\mu_+ + \mu_0 + 1}, \cdots, \delta_{\mu_- + \mu_0 + \mu_+}],
\end{aligned}
$$

where $(\delta_i)_{1 \leq i \leq \mu}$ is the oriented system of vanishing cycles of the divide P with positive upper triangular Seifert form

$$S : H_1(F, \mathbb{Z}) \to H^1(F, \mathbb{Z}) = \mathrm{Hom}(H_1(F, \mathbb{Z}), \mathbb{Z}),$$

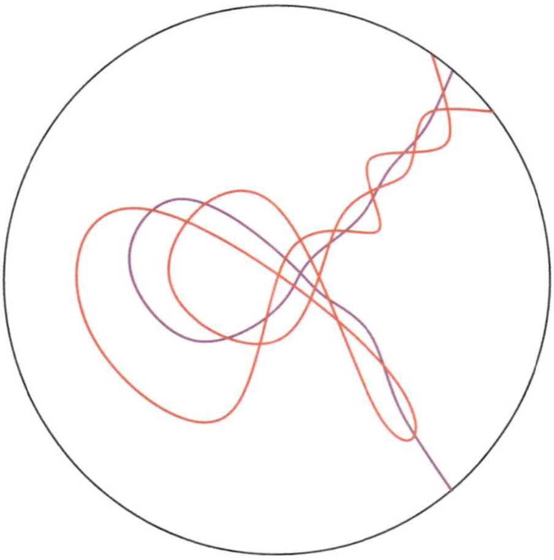

Fig. 1.27 The divide corresponding to the singularity $(x^3 - y^2)((x^3 - y^2)^2 - 4x^8 y)$

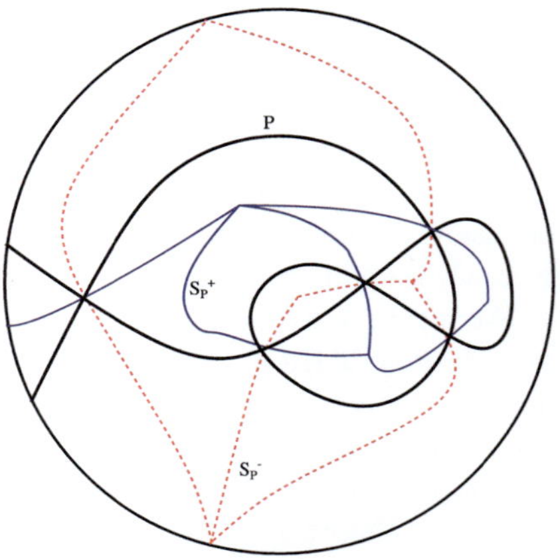

Fig. 1.28 Divide and graph of divide

see. We define $N := S - \text{Id}$, which is upper-triangular nilpotent matrix. The monodromy $T_* : H_1(F, \mathbb{Z}) \to H_1(F, \mathbb{Z})$ is given by:

$$T_* = (S^t)^{-1} \circ S.$$

For the rest of this section, recall the construction of the fiber surface F_P of a divide P in the unit euclidian disk D. In particular, recall the notation from Sects. 1.3 and 1.5.

Let S_P be a system of gradient lines of f, which connect saddle points of f with local or relative maxima or minima of f. See Fig. 1.28. In fact, near the boundary of D the lines of the system S_P are only gradient-like, but end in a relative critical point of f. We call the set S_P the graph of the divide P. By a slalom construction, see [6] (not included in this work), one can reconstruct from each of the sets $S_P^{\pm} := \{x \in S_P \mid \pm f(x) \geq 0\}$ the divide P. We assume that at the maxima and minima of f, that different arcs of S_P have different unoriented tangent directions. Let Σ_P be the following subset of the fiber F_P over 1 of the natural fibration over S^1 of the complement of the link of P: the subset Σ_P is the closure in the tangent space TD of D of

$$\Sigma'_P := \{(x, u) \in TD \mid x \in S_P,\ f(x) > 0,\ (df)_x(u) = 0,\ \|x\|^2 + \|u\|^2 = 1\}.$$

We recall that the union of fiber F_P with the link L_P is the closure in TD of

$$F'_P := \{(x, u) \in TD \mid x \in D,\ f(x) > 0,\ (df)_x(u) = 0,\ \|x\|^2 + \|u\|^2 = 1\},$$

which is a surface of genus $g(P)$ equal to the number of double points in P.

The system of cycles δ_i, $1 \leq i \leq \mu$, can be drawn on Σ_P as follows. The vanishing cycle δ_i, $\mu_- + \mu_0 < i \leq \mu$, which corresponds to a local maximum M of f is the circle $\{(M, u) \mid u \in T_M D,\ \|M\|^2 + \|u\|^2 = 1\}$ oriented counter clock-wise. The vanishing cycle δ_i, $\mu_- < i \leq \mu_- + \mu_0$, which corresponds to a saddle point z

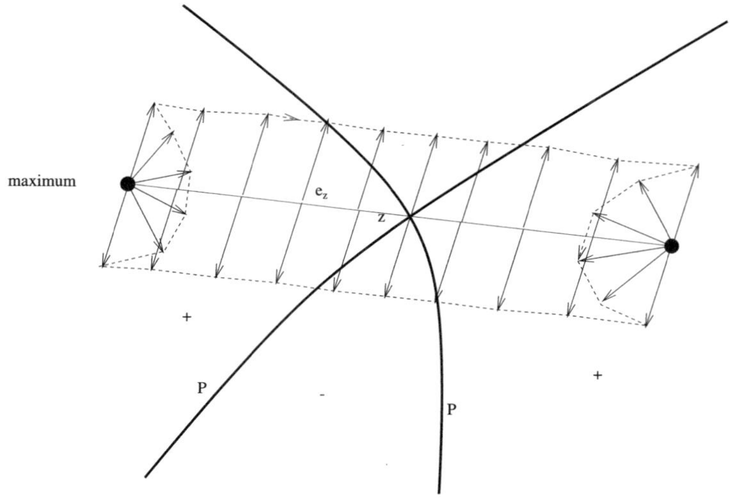

Fig. 1.29 Vanishing cycle δ_z for a saddle point

of f is a curve in the set

$$E_z := \{(x,u) \in TD \mid x \in e_z, u \in \operatorname{Ker}(df)_{e_z}, \|x\|^2 + \|u\|^2 = 1\},$$

as drawn in Fig. 1.29. The curve δ_i is a piecewise smooth embedded copy of S^1 in F_P with image in E_z and with non-constant projection to the disk D. The orientation is chosen such that at both ends the orientation agrees with the oriented vanishing cycle of the maximum. Moreover, the inward tangent vectors (x, u) to e_z at end points $x \in e_z$ do belong to δ_i. The vanishing cycle $\delta_i, 1 \leq i \leq \mu_-$, which corresponds to a local minimum of f projects to an oriented circuit $e_1, e_2, \cdots, e_k$ of edges of the graph S_P. The circuit surrounds the region in counter clock-sense to which the minimum corresponds. The circuit is a polygon and bounds a cell in D. The curve δ_i is a subset of $\{(x,u) \in TD \mid x \in \cup_{1\leq j\leq k} e_j, u \in \operatorname{Ker}(df)_x, \|x\|^2 + \|u\|^2 = 1\}$ and is the image of a piecewise smooth embedding of S^1 as drawn in Fig. 1.30. The vectors (x, u), which belong to δ_i, point out of the cell of the circuit. Smooth representatives for the system of vanishing cycles can be obtained using *tears* as in [8] (see also Sect. 1.8.1).

The involution $c_{T\mathbb{R}^2} : (x,u) \in T\mathbb{R}^2 \mapsto (x, -u) \in T\mathbb{R}^2$ induces an involution c_{S^3} on $S^3 \subset T\mathbb{R}^2$ that preserves the link L_P of any divide P and that induces involutions on the fiber surfaces above ± 1. If the divide P is a divide for a real plane curve singularity f, the involutions induced by $c_{T\mathbb{R}^2}$ or by the complex conjugation c on the triples (S^3, L_P, F_P) and on $(\partial B_\epsilon, L_f, F)$ correspond to each other by the homeomorphism of pairs of the main theorem of (cf. [7]).

Theorem 1.7.11 *Let $P \subset D$ be a connected divide consisting of the image of a relative generic immersion of a compact one dimensional manifold in the unit disk*

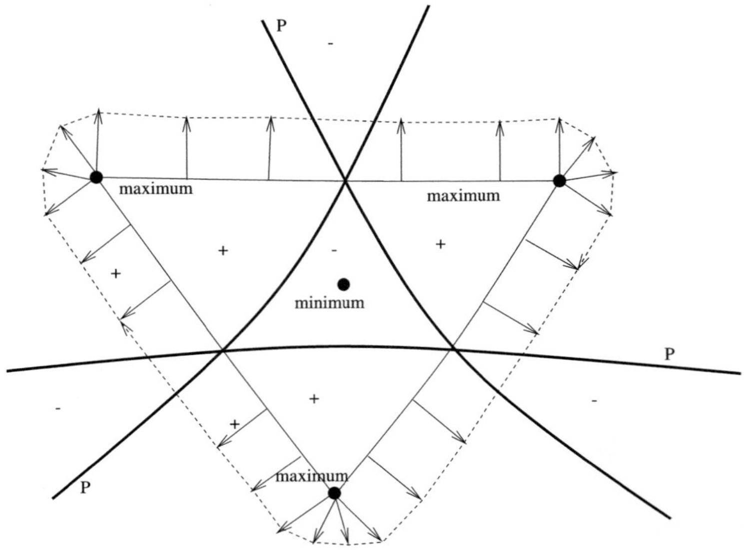

Fig. 1.30 Vanishing cycle δ_i for a minimum

$D \subset \mathbb{R}^2$. Let E_-, E_0 and E_+ be the summands in $H_1(F, \mathbb{Z})$ as above. The involution $c_{F*} : H_1(F, \mathbb{Z}) \to H_1(F, \mathbb{Z})$ fixes point-wise the summand E_+, in particular for $\delta_i \in E_+$ we have

$$c_{F*}(\delta_i) = \delta_i.$$

For $\delta_i \in E_0$ we have

$$c_{F*}(\delta_i) = -\delta_i + \sum_{1 \le j \le \mu_+} \langle N(\delta_j), \delta_i \rangle.$$

For $\delta_i \in E_-$ we have

$$c_{F*}(\delta_i) = \delta_i + \sum_{1 \le j \le \mu_+} \langle N(\delta_j), \delta_i \rangle - \sum_{\mu_+ < j \le \mu_+ + \mu_0} \langle N(\delta_j), \delta_i \rangle \delta_j.$$

The trace of the involution $c_{F*} : H_1(F, \mathbb{Z}) \to H_1(F, \mathbb{Z})$ is given by:

$$\mathrm{Trace}(c_{F*}) = \mu_- - \mu_0 + \mu_+.$$

Proof A vanishing cycle $\delta_i \in E_+$ corresponds to a maximum $M \in D$ of f, hence as set we have $\delta_i = \{(M, u) \in TD \mid \|M\|^2 + \|u\|^2 = 1\}$. The involution $(x, u) \mapsto (x, -u)$ induces on δ_i the antipodal map, which is orientation preserving. It follows $c_{F*}(\delta_i) = \delta_i$ in homology. A vanishing cycle $\delta_i \in E_0$ corresponds to a saddle point $z \in D$ of f. Working with the tear model of (cf. [8]), see also (the independent) Sect. 1.8.1, we see that the involution reverses the orientation and that at the endpoints, which are maxima of f, we have outward instead of inward vectors. Hence,

$$c_{F*}(\delta_i) = -\delta_i + \sum m_{i,j}\delta_j,$$

where in the sum j runs through the maxima of f in the interior of D which are connected by gradient lines of S_P to the saddle point i. The coefficient $m_{i,j}$ equals 1 or 2 depending on whether the connection by gradient lines is simple or double. Finally one gets $c_{F*}(\delta_i) = -\delta_i + \sum_{1\leq j\leq\mu_+}\langle N(\delta_j), \delta_i\rangle$. For a vanishing cycle $\delta_i \in E_-$ one can work with the model of (cf. [8]) and get $c_{F*}(\delta_i) = \delta_i + \sum_{1\leq j\leq\mu_+}\langle N(\delta_j), \delta_i\rangle - \sum_{\mu_+<j\leq\mu_++\mu_0}\langle N(\delta_j), \delta_i\rangle$. Since N is strictly upper-triangular, the matrix of c_{F*} is upper-triangular with ± 1 on the diagonal. We get $\mathrm{Trace}(c_{F*}) = \mu_- - \mu_0 + \mu_+$. □

We can present the Seifert form and homological monodromy with block-matrices S, T as in [24]. On $H_1(F, \mathbb{Z})$, E_-, E_0, E_+ and $H^1(F, \mathbb{Z})$ we work with the basis or dual basis given by the system δ_i, $1 \leq i \leq \mu$. The matrix in block form of the Seifert form is

$$S = \begin{pmatrix} \mathrm{Id}_{\mu_+} & A & G \\ O & \mathrm{Id}_{\mu_0} & B \\ O & O & \mathrm{Id}_{\mu_-} \end{pmatrix}$$

where the block G equals the block matrix product $1/2(A \circ B)$. The matrix coefficients of $A \circ B$ and G have interpretations in terms of the divide P or in terms of the Morse function f_P on the disk D. The matrix coefficient $(A \circ B)_{(i,j)}$, $1 \leq i \leq \mu_+, \mu_+ + \mu_0 < j \leq \mu$, counts the number of sector adjacencies, that has the $+$-region i of P with the $-$-region j, while the coefficient $G_{(i,j)}$ counts the number of common boundary segments of the regions i and j. The coefficient $(A \circ B)_{(i,j)}$ also counts the number of saddle connections via gradient lines of f_P in between the minimum j and maximum i, while the coefficient $G_{(i,j)}$ counts the number of components of regular gradient line connections from the minimum j to the maximum i. This explains the above factor $1/2$, since a segment of P is twice incident with a saddle point of f_P.

The matrix of the action of complex conjugation on $H_1(F, \mathbb{Z})$ is

$$C = \begin{pmatrix} \mathrm{Id}_{\mu_+} & A & G \\ O & -\mathrm{Id}_{\mu_0} & -B \\ O & O & \mathrm{Id}_{\mu_-} \end{pmatrix}$$

and is obtained from the matrix S by multiplying the middle row of blocks by -1. It is interesting to compute the matrix of the involution $T \circ c_F$ on $H_1(F, \mathbb{Z})$

$$TC = \begin{pmatrix} \mathrm{Id}_{\mu_+} & O & O \\ -{}^t A & -\mathrm{Id}_{\mu_0} & O \\ {}^t G & {}^t B & \mathrm{Id}_{\mu_-} \end{pmatrix}$$

The matrix $T \circ C$ is the transgradient of the matrix of C.

It turns out that the combinatorial property $G = 1/2(A \circ B)$ for divides is equivalent to $C \circ C = \mathrm{Id}_\mu$ or $T \circ C \circ T \circ C = \mathrm{Id}_\mu$.

For an isolated plane curve singularity at $0 \in \mathbb{C}^2$, which is given by a real equation $\{f = 0\}$, $f \in \mathbb{R}\{x, y\}$, $f(0) = 0$, we will denote by $\delta_{\mathbb{R}}(f)$ the number of double points of a divide P for the singularity. Hence $\delta_{\mathbb{R}}(f)$ is the maximal number of local real saddle points in some level, that can occur for a small real deformation of f. Observe, that one has $\delta_{\mathbb{R}}(f) \le \delta(f)$, where $\delta(f)$ is the maximal number of local critical points in some level, that can occur for a small deformation of f. We recall the formula $\mu(f) = 2\delta(f) - r + 1$ of Milnor [37]. As example for $f = x^4 + Kx^2y^2 + y^4$, $-2 \neq K \neq 2$, one has $\delta_{\mathbb{R}}(f) = 4, \delta(f) = 6, r = 4$ and very surprisingly, Callahan shows that for $-2 < K < 2$ there exists a small real deformation with 5 local minima in the same level [15].

Theorem 1.7.12 *For an isolated plane curve singularity at* $0 \in \mathbb{C}^2$, *which is given by a real equation* $\{f = 0\}$, $f \in \mathbb{R}\{x, y\}$, $f(0) = 0$, *we have*

$$\mu(f) = 2\delta_{\mathbb{R}}(f) + \mathrm{Trace}(c_{F*})$$

Proof We have $\delta_{\mathbb{R}} = \dim E_0 = \mu_0$. Hence $\mu(f) = \mu_- + \mu_0 + \mu_+ = \mathrm{Trace}(c_{F*}) + 2\mu_0 = 2\delta_{\mathbb{R}}(f) + \mathrm{Trace}(c_{F*})$. □

Curves δ_i that correspond to maxima of f_P are invariant by the involution c. A curve δ_i that corresponds to a saddle point or minimum of f_P is in general not invariant by c. The union Σ_P of the curves δ_i is invariant by c. We get

Theorem 1.7.13 *Let P be a connected divide. The pair (F_P, Σ_P) defines a tri-valent cellular decomposition with $r(P)$ punctured cells of the fiber surface with boundary F_P. The involution $c : (x, u) \to (x, -u)$ acts on the map (F_P, Σ_P).*

Proof The union Σ_P of the vanishing cycles δ_i is a tri-valent graph Σ_P that is invariant by c. The inclusion $\Sigma_P \subset F_P$ induces an isomorphism $H_1(\Sigma_P, \mathbb{Z}) \to H_1(F_P, \mathbb{Z})$, hence, Σ_P is a spine for the surface F_P. □

Question In particular if $r(P) = 1$ the triple (F_P, Σ_P, c) is a so-called *maximal unicell map* of genus $g(P)$ with orientation reversing involution c with $r(P) = 1$ fixed points on the graph of the map. The involution c_F has a unique fixed point on Σ_P, which corresponds to the intersection of the folding curve $F_P \cap \partial D$ of c_F with the graph Σ_P. Maximal here means that the number of edges of the graph of the map is maximal, i.e. $3g$.

It would be interesting to compute the generating series

$$\mathrm{divide}(t) := \sum_{g \in \mathbb{N}} d(g) t^g = 1 + t + 2t^2 + 8t^3 + 36t^4 + \cdots ,$$

where $d(g)$ denotes the number of simple, relative and generic immersions of the interval [0, 1] with g double points in the disk D, counted up to homeomorphism in the source and image. We have taken the coefficients up to the term t^4 of divide(t) from the listings of simple, relative, free or oriented divides by Masaharu Ishikawa [29]. We ask to compare the numbers $d(g)$ with the numbers $m^+(g)$ of maximal maps of genus g with orientation reversing involution having 1 fixed point on its graph.

1.7.2.3 Involutions Induced by π-Rotations

Next we study the involutions that appear in the decomposition of the geometric monodromy of plane curve singularities. We show that they lift to π-rotations about an axis in the universal cyclic covering of the complement of the link.

Our next goal is to visualize the two involutions c_F and $T \circ c_F$. We assume that the monodromy diffeomorphism T is chosen in its isotopy class such that $T \circ c_F$ is an involution of the fiber surface F_P. In fact, we assume that the monodromy T was given by a monodromy vector field X, which satisfies $X = X^c$.

We assume for simplicity, that the divide P meets the boundary of the disk D. The involution c_F is an orientation reversing diffeomorphism of $F_1 = F_P$, which fixes point-wise the system of arcs $a := F_P \cap \partial D$. The Lefschetz number of the orientation reversing involution $T \circ c_F$ is equal to the Lefschetz number of c_F by Theorem 1.7.10. Hence the involution $T \circ c_F$ also fixes point-wise exactly a system of arcs b on the fiber surface F_P with $\chi(a) = \chi(b)$, i.e. the systems a and b consist of the same number of arcs.

Let $z : Z \to S^3 \setminus K_P$ be the infinite cyclic covering of the knot complement. Let X^Z be the lift of the vector field X. Let $T_{\frac{1}{2}} : Z \to Z$ be the flow diffeomorphism of the vector field X^Z with stopping time $\frac{1}{2}$. Let $F_1' \subset Z$ be a lift of the fiber surface F_1 and let F_{-1}' be the lift $T_{\frac{1}{2}}(F_1')$ of the fiber surface F_{-1}. The involution c_F can first be lifted unambiguously to an involution of F_1' and then extended unambiguously to an involution A of Z such that A maps the vector field X^Z to its opposite. The involution A is a π-rotation about a lift of the system of arcs a into F_1'. The involution $(x, u) \mapsto (x, -u)$ induces an involution $c_{F_{-1}}$ of the fiber surface F_{-1} above -1, and, as we have done for the involution c_F, the involution $c_{F_{-1}}$ lifts unambiguously to a π-rotation $B : Z \to Z$ about a system of arcs $b'' \subset F_{-1}'$ that satisfies $z(b'') = b' := \partial D \cap F_{-1}$. We have $B \circ A = T_{\frac{1}{2}} \circ T_{\frac{1}{2}} = T_1$. Since T_1 is a lift of the monodromy, it follows from $(T \circ c_F) \circ c_F = T$, that the involution $T \circ c_F$ fixes the system of arcs $b := z(T_{\frac{1}{2}}^{-1}(b''))$.

In Sect. 1.7.1 (cf. [5]) we have computed the monodromy diffeomorphism $T_+ : F_1 \to F_{-1}$ for which $T_{\frac{1}{2}}$ is a lift to Z as a product of half Dehn twists for a divide. From the above we deduce, that the involution $T \circ c_F : F_1 \to F_1$ is the composition $T_+^{-1} \circ c_{F_{-1}} \circ T_+$. We see that $T \circ c_F$ fixes the arc $b = T_+^{-1}(\partial D \cap F_{-1})$. We also see that the involution $c_F \circ T$ of F_1 fixes the system of arcs $c_F(b) = T_+(\partial D \cap F_{-1})$. Both arc systems b and $c_F(b)$ have equal projections β into D. See Figs. 1.31 and 1.32 for examples.

The composition of the orientation reversing involutions, that fix point-wise the arc systems a, b of Fig. 1.31 on the cylinder surface, is indeed a Dehn twist. To see

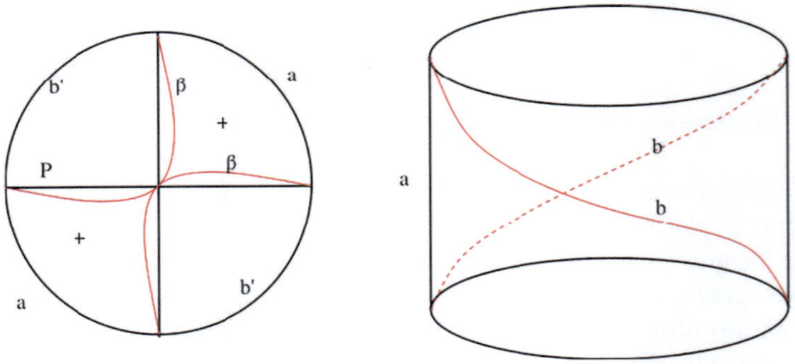

Fig. 1.31 Dehn twist as composition of two involutions

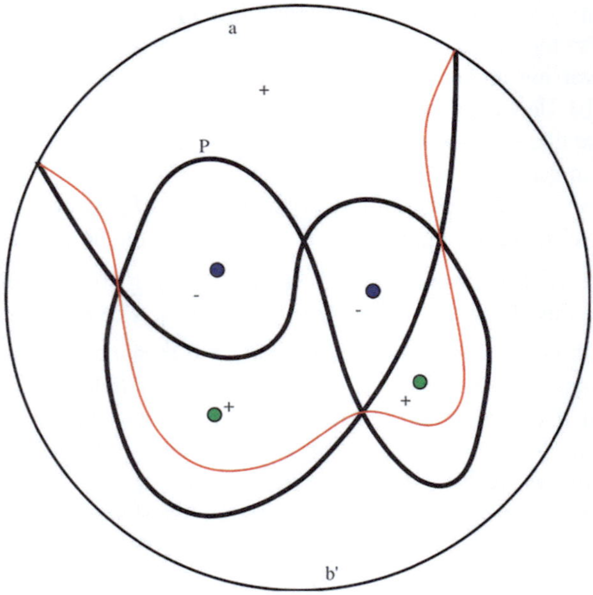

Fig. 1.32 Divide for the E_8 singularity and the curve β in red

this, think of the cylinder as $[0, \pi] \times S^1$; the two involutions induce on the circle $\{\theta\} \times S^1$ reflections about diameters that make an angle θ, so the composition is a rotation of angle 2θ of the circle $\{\theta\} \times S^1$.

In Fig. 1.32 the curve β is drawn for a more complicated divide for the plane curve singularity E_8 with equation $y^3 - x^5 = 0$.

We conclude with a remark we promised that shows that uses the results of this section to prove that not all fibered links come from divides.

Remark 1.7.14 The figure eight knot (see Fig. 1.3) is not the knot of a divide. The figure eight knot's complement fibers over the circle having as fiber the punctured torus and as monodromy the isotopy class of the linear diffeomorphism given by a matrix in $SL(2, \mathbb{Z})$ of trace 3. Such a matrix M is not the product of two unipotent matrices, which are conjugate in $SL(2, \mathbb{Z})$ and the matrices M and M^{-1} are not conjugated by an integral matrix of determinant -1. So according to the remarks this section, the figure eight can not be the knot of a divide. Another argument to rule out the figure eight as the knot of a divide goes as follows. The first betti number of the fiber of the figure eight knot is 2. But only two connected divides have fibers with betti number 2 and these two have monodromies with trace equal to 1 or 2.

1.8 Abstract Visualization of the Milnor Fiber and the Monodromy

In this section we describe an easy and fast graphical algorithm of visualizing the Milnor fiber together with a distinguished basis of vanishing cycles directly from the divide [7].

1.8.1 Visualization of the Vanishing Cycles for a Divide

We start by visualizing the vanishing cycles.

Let P be a connected divide and let $\pi_P : S^3 \setminus L(P) \to S^1$ be the fibration of the complement of the link $L(P)$ over S^1 as in Sect. 1.4 (cf. [5, 7]). Recall that the fibration map is given with the help of an auxiliary Morse function $f_P : D \to \mathbb{R}$. The fiber F_P above $1 \in S^1$ projects to the positive components of the complement of P in D. One has that the closure of

$$\{(x, u) \in TD | f_P(x) > 0, (df_P)_x(u) = 0, \|x\|^2 + \|u\|^2 = 1\}$$

in $S^3 \setminus L(P)$ is the fiber surface F_P.

To each critical point of $f_P : D \to \mathbb{R}$ corresponds a vanishing cycle on the surface F_P. In the case, where the divide P is a divide of a singularity, the surface

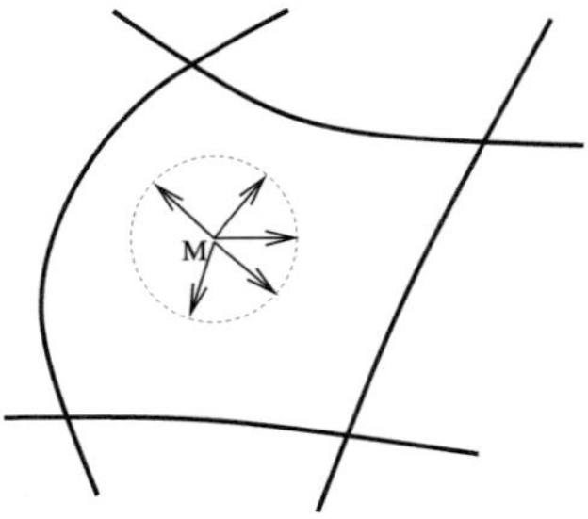

Fig. 1.33 The vanishing cycle δ_M for a maximum

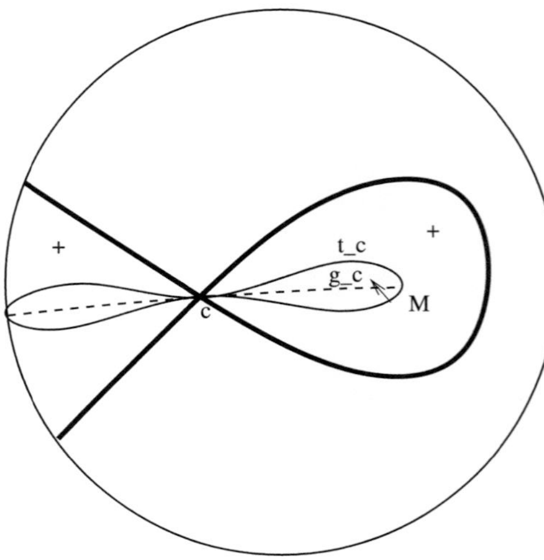

Fig. 1.34 Tear for vanishing cycle δ_c for a saddle point

F_P is a model for the Milnor fiber and the system of vanishing cycles on F_P is a model for a distinguished system of quadratic vanishing cycles of the singularity.

Let M be a maximum of f_P. The vanishing cycle δ_M is the non-oriented simply closed curve on F_P, see Fig. 1.33,

$$\delta_M := \{u \in T_M D \mid \|M\|^2 + \|u\|^2 = 1\}.$$

Let c be a crossing point of P. The point c is a saddle type singularity of F_P. The vanishing cycle is the non-oriented simply closed curve δ_c on F_P that results from the following construction. Let $g_c \subset P_+ \cup \{c\}$ be the singular gradient line through c, for which the endpoints are a maximum of f_P or a point in ∂D. We splice g_c and get a double *tear* $t_c \subset P_+ \cup \{c\}$ as in Fig. 1.34. The tear t_c is a closed curve, that has at c a non-degenerate tangency with g_c from both sides. Moreover, t_c is perpendicular to g_c at the endpoints of g_c, if the endpoint is a maximum of f_P and else t_c has a tangency with ∂P. The vanishing cycle δ_c is the closure in F_P of the set

$$\{(x, u) \in F_P | x \in t_c,\ (df_P)_x \neq 0,\ u\, \textit{points to the inside of the tear } t_c\}.$$

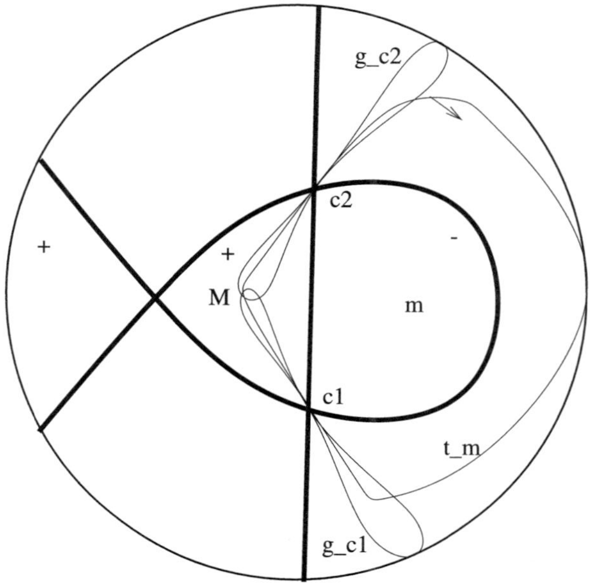

Fig. 1.35 The vanishing cycle δ_M for a minimum

Let m be a minimum of f_P. The following is a description of the vanishing cycle δ_m on F_P. The projection of δ_m in D is a simply closed curve t_m in $P_+ \cup \{c_1, c_2, \ldots, c_k\}$, where $\{c_1, c_2, \ldots, c_k\}$ is the list of the double points of P that lie in the closure of the region of m, see Fig. 1.35. The curve t_m and the singular gradient line g_{c_i}, $1 \leq i \leq k$ coincide in a neighborhood of c_i. Moreover, if the endpoint of g_{c_i} is a maximum M, the curve t_m leaves transversally the tear t_{c_i} at M and enters transversally the next tear $t_{c_{i\pm1}}$. If the endpoint of g_{c_i} is on the boundary of D, the curve leaves the tear t_{c_i}, becomes tangent to the boundary of D and enters the next tear, see Fig. 1.35. The vanishing cycle δ_m is the non-oriented simply closed curve on F_P, that is the closure of

$$\{(x, u) \in F_P | x \in t_m,\ u\, \textit{points inwards to the disk bounded by}\ \ t_m\}.$$

1.8.2 *Visualization of the Milnor Fiber Associated with a Divide*

Now we explain how to visualize the Milnor fiber together with the divide. Think of the divide as a road network which has δ junctions, and replace every junction by a roundabout, which leads you to a new road network with 4δ T-junctions. Realize now every road section in between two T-junctions by a strip with a half twist. Do the same for every road section in between a T-junction and the boundary of

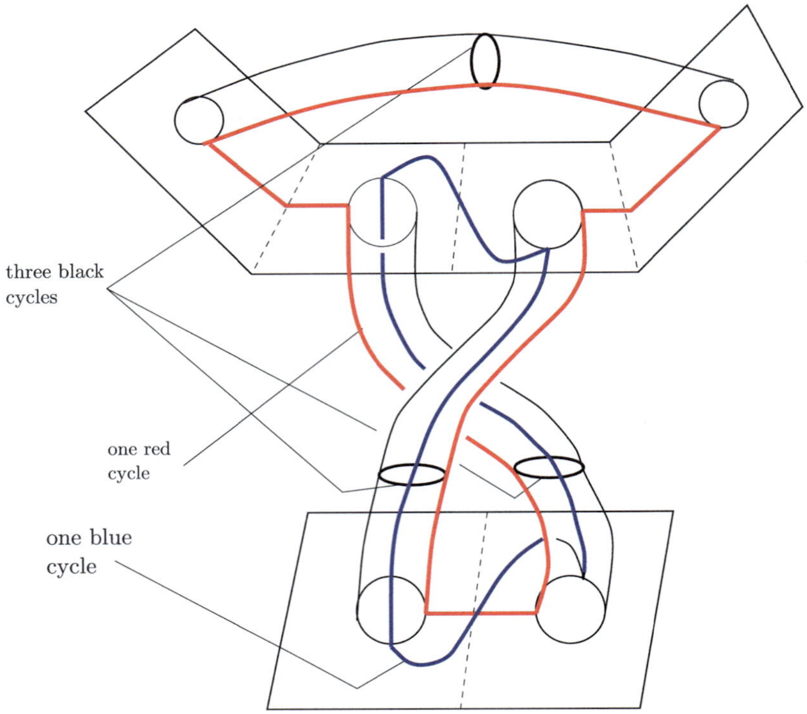

Fig. 1.36 Milnor fiber with vanishing cycles for D_5

the divide. Altogether you will need $6\delta + r$ strips. Now, applying what we have seen in the previous Sect. 1.8.1, the core line of the four strips of a roundabout is a black vanishing cycle, the strips corresponding to boundary edges and corners of a positive or negative region have as core line a red or blue vanishing cycle (see Fig. 1.36 or 1.37 for color scheme).

In Fig. 1.37 is worked out the singularity with two Puiseux pairs and $\mu = 16$, where we used a divide equivalent (up to an admissible isotopy) to that of Fig. 1.20. We have drawn for convenience in Fig. 1.37 only one red, black, or blue cycle. We have also indicated the position of the arc α, which will play a role in the next section.

1.8.3 Reduction Cycles and Reduction Tori

We consider a divide of the form $Q := P_{p,q} * P$, where the divide P is given by an immersion $\gamma : [-1, 1] \to D$. So, the divide Q is the image of $N\gamma \circ T_{p,q}$. If we change the immersion by a reparametrization $\gamma_1 := \gamma \circ \phi$, where ϕ is an oriented diffeomorphism of $[-1, 1]$, the divide $Q_1 := N\gamma_1 \circ \phi([-1, 1])$ is isotopic

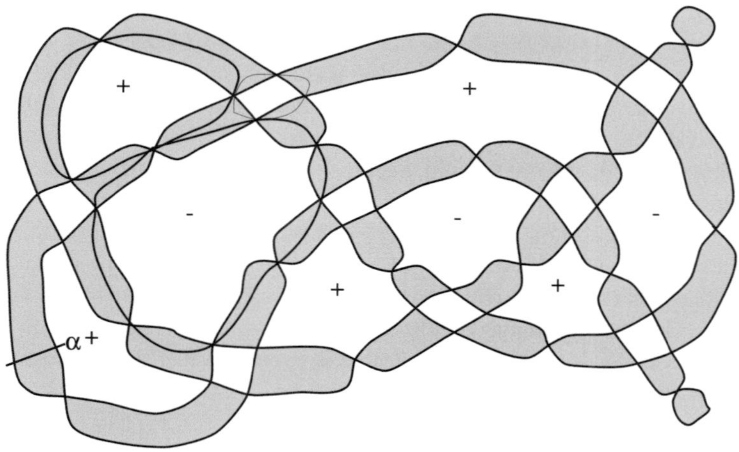

Fig. 1.37 Milnor fiber with vanishing cycles for $y^4 - 2y^2x^3 + x^6 - x^7 - 4yx^5$

Fig. 1.38 Manhattan: crossing of the box with 4 by 4 strands

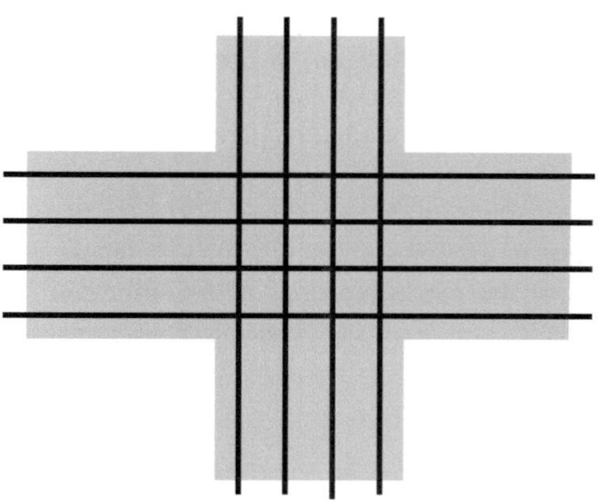

to Q by an transversal isotopy, which does not change the type of its knot. By choosing ϕ appropriately and η small, one can achieve that to each double point of P correspond p^2 double points of the divide Q_1, which look like the intersection points of a system of p almost parallel lines with an other system of p almost parallel lines, see Fig. 1.38.

We may assume, that the divide Q for each double point of P already has a grid of p^2 intersections.

We will construct reduction curves for the monodromy of the knot $L(Q)$ following the ideas of [1]. The reduction curves of the cabling $P_{p,q} *_\eta P$ are the intersection of the fiber F_Q over $1 \in S^1$ of the fibration on the complement of the knot $L(Q)$ with the boundary of a regular tubular neighborhood U of the closed tubular neighborhood V of the knot $L(P)$ for which $L(Q) \subset \partial V$ holds. The intersection $(F_Q \cap \partial U) \subset F_Q$ is indeed a system of reduction curves provided that the torus ∂U is transversal to the fibration of the knot $L(Q)$.

Assume that the tubular neighborhood V was constructed with the field $\Phi_{p,q}$ and a particular value of the parameter η. The same field of sectors, but a slightly bigger parameter value η' yields a tubular neighborhood U of V in S^3. The construction of the fibration will be done as in Sect. 1.4 (cf. [5]). The main choice for the construction of the fibration for the knot $L(Q)$ is a carefully chosen adapted Morse function (recall Definition 1.3.8) $f_Q : D \to \mathbb{R}$ with $f_Q^{-1}(0) = Q$. For our purpose here, where we must achieve the above transversality so we will choose f_Q as follows. First, after applying a regular transversal small isotopy, we may assume that the divide P has perpendicular rectilinear crossings. Next, we consider an adapted function $f_P : D \to \mathbb{R}$ for the divide P, that is euclidian near its crossings. Let the fibration on the complement of the knot $L(P)$ be $\pi_{P,\eta} : S^3 \setminus L(P) \to S^1$ where we recall that $\pi_P(x,u) := \theta_P(x,u)/|\theta_P(x,u)|$ and

$$\theta_{P,\lambda}(x,u) := f_P(x) + i\lambda^{-1}\, df_P(x)(u) - \frac{1}{2}\lambda^{-2}\chi(x)H_{f_P}(x)(u,u).$$

Here, the function $\chi : D \to \mathbb{R}$ is a bump function at the crossing points of P and λ is a big real parameter (compare with Eq. 1.5). We now choose a small positive real number v, such that $\{x \in D \mid |f_P(x)| \leq v\}$ is a regular tubular neighborhood of P, that meets each component of $\{x \in D \mid \chi(x) = 1\}$. Next we choose $\eta' > 0$ such the corners of $N_{\eta'}\gamma(B)$ are in $\{|f_P(x)| = v\}$ i.e. $\eta'^2 = v$. We construct the torus knot $L(Q)$ with $Q := P_{p,q} *_\eta P$ where $0 < \eta < \eta'$. Since $Q \subset \{|f_P(x)| < v\}$ holds, we can construct a Morse function $f_Q : D \to \mathbb{R}$ for the divide Q, such that on $\{|f_P(x)| \geq v\}$ the function f_Q is constant along the level sets of f_P.

The following theorem follows directly from [1, Lemme 2, pg. 153] and the above construction.

Theorem 1.8.1 *The torus $\partial\Phi_{\eta',p,q}\gamma$ is transversal to the fibration π_Q induced by f_Q on the complement of the knot $L(Q)$. The intersection*

$$\partial\Phi_{\eta',p,q}\gamma \cap F_Q$$

is a system of p closed curves on the fiber F_Q, which is a reduction of the monodromy of $L(Q)$.

With a few examples, we now explain how to inductively depict in the Milnor fiber a distinguished system of vanishing cycles and the reduction curves for the monodromy of a singularity, for which a divide of the form $Q = P_{p,q} * P$ is given.

For a (p,q) cabling $(p < q)$ the reduction system consists of p simply closed curves on the fiber F_Q. Each of them cuts out from F_Q a surface diffeomorphic to the fiber F_P of the divide P. The p copies of F_P in F_Q are cyclicly permuted by the monodromy T_Q. This description follows from the construction of the monodromy of an irreducible plane curve singularity given in [1].

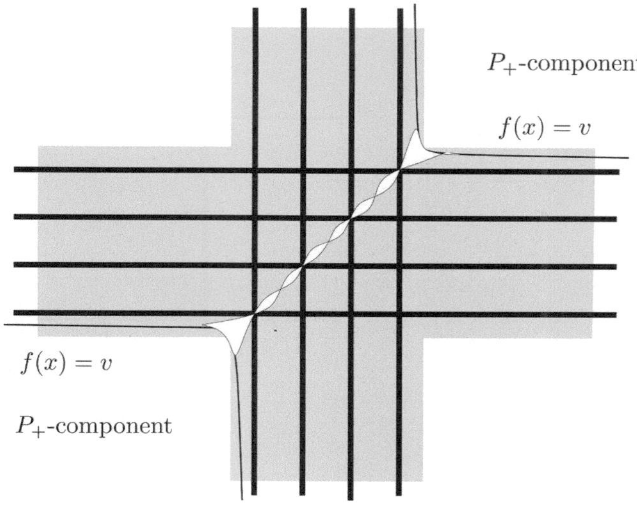

Fig. 1.39 Bridge through Manhattan

One of those copies can be visualized more easily as follows. Let $\{x \in D | f_P(x) \geq v\}$. For each double point c of P we connect the two components of $\{x \in D | f_P(x) \geq v\}$, that are incident with the double point c, by a special bridge which projects diagonally through the Manhattan part of the divide Q, that corresponds to c. The projection of the bridge is a twisted strip S_c in D, that realizes a boundary connected sum of the P_+-components (see Fig. 1.39). The twist points of the strip S_c are precisely the critical points of f_Q, that lie on the diagonal. The boundary of S_c consists of two smooth curves, that intersect each other transversally and that also intersect the divide Q transversally.

Let C be the union of the projections S_c of the bridges with $\{x \in D | f_P(x) \geq v\}$, see Fig. 1.39. In Fig. 1.40 we have zoomed out one block to show more details. The copy $F_{P,Q}$ of the fiber of the knot $L(P)$ is the closure in the fiber F_Q of the knot $L(Q)$ of the set

$$\{(x, u) \in TD | x \in C, (df_Q)_x \neq 0, (df_Q)_x(u) = 0\} \cap S^3.$$

The first reduction curve is the boundary

$$R := \partial F_{P,Q}$$

of the surface $F_{P,Q} \subset F_Q$. The reduction system is the orbit

$$\{R, T_Q(R), T_Q^2(R), \dots\}$$

under the monodromy T_Q of the singularity with divide Q.

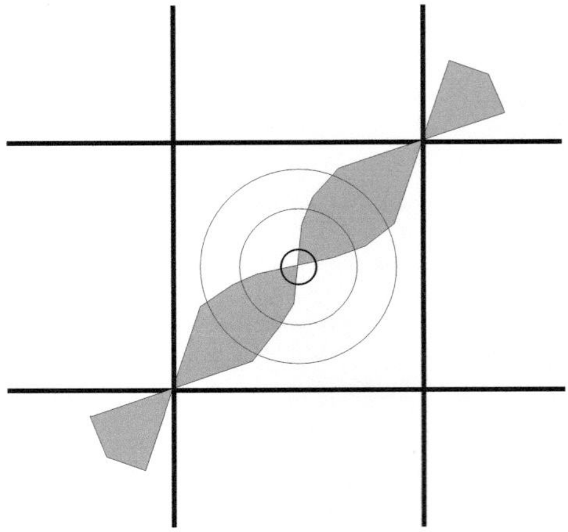

Fig. 1.40 Bridge through a block of Manhattan

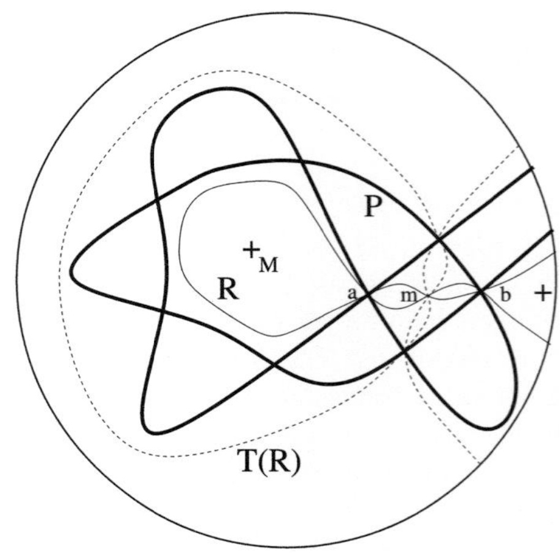

Fig. 1.41 Divide $Q = P_{2,9} * P_{2,3}$ with reduction curves R and $T(R)$ (dotted)

Example 1.8.2 Our first example is the singularity with two essential Puiseux pairs $(x^3 - y^2)^2 - 4x^5y - x^7$, whose link is a two stage iterated torus knot. For a detailed description of the link, the monodromy and the Milnor fiber of this singularity, follow the construction of [1]. Its divide $Q = P_{2,9} * P_{2,3}$, (see Fig. 1.20), has two P_+-components, where $P = P_{2,3}$ (recall Fig. 1.41 and contents of that section), where the projection of the reduction curve R is drawn). In this case Manhattan consists of one block. The reduction curve R is the pre-image in the fiber F_Q of its projection $\text{proj}(R) \subset D$ under the map $(x, u) \mapsto x$ (see Fig. 1.41). That means R is the closure in F_Q of the set

$$\{(x, u) \in TD | x \in \text{proj}(R), (df_Q)_x \neq 0, (df_Q)_x(u) = 0, \|x\|^2 + \|u\|^2 = 1\}.$$

The reduction system is $\{R, T_Q(R)\}$.

The curve R is homologically trivial in F_Q. It turns out that the power T_Q^{156} of the monodromy is the composition of the right Dehn twists, whose core curves are $\{R, T_Q(R)\}$. The power T_Q^{156} is a product of $2496 = 16 \times 156$ Dehn twists, since T_Q is the product of those Dehn twists whose core curves are the system of distinguished quadratic vanishing cycles of the real morsification with divide Q. It turns out that the expression as product of Dehn twists is far from being as short as possible. In fact, the right Dehn twist Δ_R with core curve R can be written as a product of 36 right Dehn twists that have core curves coming from the morsification with divide Q. More precisely, the Dehn twist Δ_R factors as

$$\Delta_R = (\Delta_M \circ \Delta_b \circ \Delta_m \circ \Delta_a \circ \Delta_m^{-1} \circ \Delta_b^{-1})^6,$$

The factors are right Dehn twists whose core curves are among the quadratic vanishing cycles $\delta_m, \delta_a, \delta_M, \delta_b$ of the divide Q as indicated in Fig. 1.40, δ_M is the vanishing cycle of a P_+-region, δ_m of the maximum of Manhattan, and δ_a, δ_b of street corners of Manhattan. It follows that T_Q^{156} can also be written as a composition of 72 Dehn twists with core curves among the vanishing cycles of the divide Q. The composition $\Delta_b \circ \Delta_m \circ \Delta_a \circ \Delta_m^{-1} \circ \Delta_b^{-1}$ is the Dehn twist with core curve $\bar{a} := \Delta_b(\Delta_m(a))$.

The reduction curve R cuts off from F_Q a piece $F_{P,Q}$ of genus one, which corresponds to a copy of the Milnor fiber of the Brieskorn-Pham singularity y^2+x^3 (corresponding to the first Puiseux pair). The Dehn twists Δ_M and $\Delta_{\bar{a}}$ act only on this piece, since the curves δ_M and $\bar{a}$ lie entirely in this piece; in this piece, that is a copy of the fiber F_P, they generate the geometric monodromy group of the accompanying singularity $x^3 - y^2 = 0$ with divide $P_{2,3}$.

Example 1.8.3 Our second example is the *double cusp*, that is, the singularity with two branches $(x^3 - y^2)(y^3 - x^2)$. Its homological monodromy is of infinite order by [1]. Each branch is a torus knot. Again Manhattan consists of one block. In Fig. 1.42 we have drawn the projections of the curves R, R' and S, S', that together are the boundary components of the two diagonals through Manhattan. In this case the curves R and R' are isotopic to each other, as are the curves S and S'. A complete reduction system for the geometric monodromy is the system $\{R, S\}$. Each component of this system carries a non-trivial homology class. The isotopy classes of the curves R and S are permuted by the monodromy T_P, hence the system $\{R, S\}$ is invariant under the monodromy.

Let h be the action of T_P on the homology $H_1(F_P, \mathbb{Z})$ of the fiber F_P. Let $\delta_a, \delta_b, \delta_c, \delta_d$ be the vanishing cycles of the double points, that are the corners of Manhattan of P, let δ_m be the vanishing cycle of the maximum in the center of Manhattan.

If one chooses the orientations appropriately, on has

$$[R] = [\delta_a]+[\delta_m]+[\delta_c],\ [S] = [\delta_b]+[\delta_m]+[\delta_d],\ h([R]) = -[S],\ h([S]) = -[R],$$

Fig. 1.42 Divide P for $(x^3 - y^2)(y^3 - x^2)$ with reduction system $R \cup S$

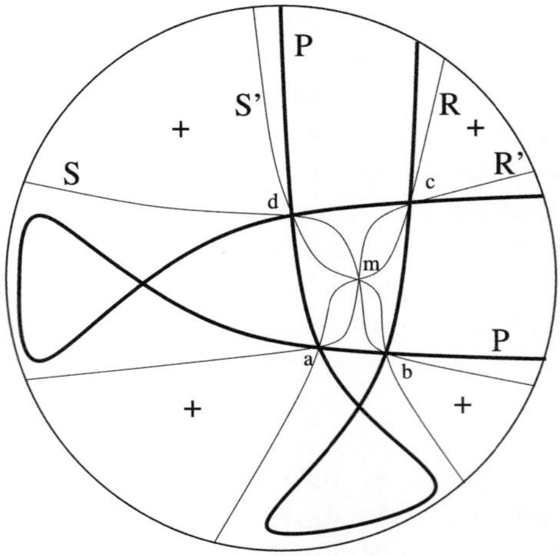

Fig. 1.43 Divide P for $(x^3 - y^2)(y^3 - x^2)$ with reduction system $A \cup B$

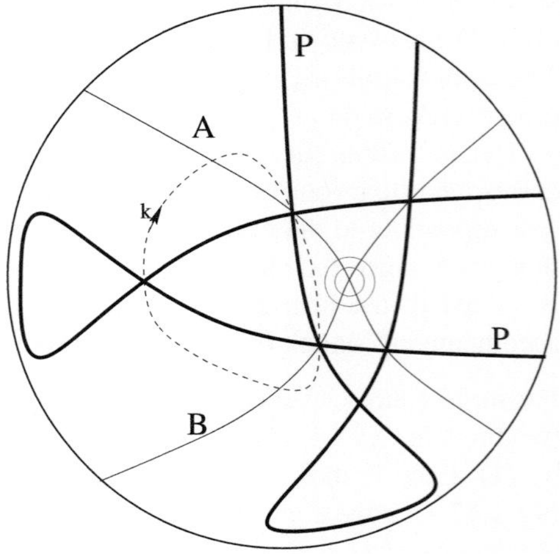

hence also $h([R]-[S]) = [R]-[S]$. Let $[k]$ be any cycle on F_P, that is carried by a simple oriented curve k and intersects the curves R and S each transversally in one point. One has $h^{10}([k]) = [k] \pm ([R]+[S])$, which shows that the homological monodromy h is not of finite order. We have drawn in Fig. 1.42 the oriented projection of such a cycle k, that intersects the curves A and B. The curve A is halfway in between the curves R and R' on the cylinder they cut out. Let B be the curve halfway in between S and S'. The curves A and B are the reduction curves of Figure 3 on page 167 of [1]. The reduction system A, B is much easier to draw, see Fig. 1.43, where are drawn the projections in D. The projections meet transversally at the maximum in Manhattan of f_P. The curve δ_m intersects transversally in two points each curve R and S. One has $h^{10}([\delta_m]) = [\delta_m] \pm 2([R]+[S])$.

The power T_P^{10} of the geometric monodromy, that is a word of length 110 in the Dehn twists of the divide P, is equal to the composition of those right Dehn twists, whose core curves are R and S. So, the power T_P^{10} also can be written as the much shorter word

$$\Delta_c \circ \Delta_m \circ \Delta_a \circ \Delta_m^{-1} \circ \Delta_c^{-1} \circ \Delta_d \circ \Delta_m \circ \Delta_b \circ \Delta_m^{-1} \circ \Delta_d^{-1}.$$

Remark 1.8.4 The curves A and $\Delta_m(\Delta_c(\delta_a))$ are isotopic, where Δ_m and Δ_c are the right Dehn twists with core curves δ_m and δ_c. It follows that the reduction system A, B consists of quadratic vanishing cycles of the singularity of $\{(x^3 - y^2)(y^3 - x^2) = 0\}$ with two branches. In contrast, a reduction curve of a singularity with only one branch can not be a quadratic vanishing cycle, since all reduction curves are zero in the homology.

1.8.4 Geometric Monodromy Group and Reduction System

Let the polynomial $f_{(a,b)}$ be an equation for an irreducible plane curve singularity with n essential Puiseux pairs $(a_i, b_i)_{1 \le i \le n}$ with $a_i < b_i$. The number of simply closed curves contained in a complete reduction system R for the monodromy of f is

$$a_n a_{n-1} \cdot \ldots \cdot a_2 + a_n a_{n-1} \cdot \ldots \cdot a_3 + \cdots + a_n a_{n-1} + a_n.$$

Indeed, this follows from the construction of the Milnor fiber in [1]. Let

$$\Gamma_{f,red} < \Gamma_f$$

be the subgroup of the geometric monodromy group (recall Definition 1.2.7) of Γ_f of f of those elements $\gamma \in \Gamma_f$ that up to isotopy fix each component of R. Let $\Gamma^0_{f,red}$ be the subgroup of Γ_f which is generated by the Dehn twist whose core curves are quadratic vanishing cycles and do not intersect any component of R. Obviously, one has $\Gamma^0_{f,red} \subset \Gamma_{f,red}$, but we do not know if this inclusion is strict. A component of $F \setminus R$ is called a top-component if its closure in F meets only one component of R. Let $\Gamma_{f,top}$ be the subgroup of Γ_f of those monodromy transformations, which induce the identity in each component of $F \setminus R$ that is not a top-component. Let $\Gamma^0_{f,top}$ be the intersection $\Gamma_{f,top} \cap \Gamma^0_{f,red}$. We have

Theorem 1.8.5 *Let $f = f_{(a,b)}$ be an irreducible singularity with $n \ge 2$ essential Puiseux pairs $(a_i, b_i)_{1 \le i \le n}$. Let $g = f_{(a',b')}$ be a singularity with the $n-1$ essential Puiseux pairs $(a', b') = (a_i, b_i)_{1 \le i \le n-1}$. The group Γ_f contains the product of a_n copies of the group Γ_g.*

Theorem 1.8.6 *Let $f_{(a,b)}$ be an irreducible singularity with $n \ge 2$ essential Puiseux pairs. The group $\Gamma^0_{f,top}$ is isomorphic to the product of $a_n a_{n-1} \cdot \ldots \cdot a_2$ copies of the geometric monodromy group of the singularity $y^{a_1} - x^{b_1} = 0$.*

Proof of Theorem 1.8.5 Let P be the divide $P_{a_{n-1},b'_{n-1}} * \cdots * P_{a_2,b'_2} * P_{a_1,b_1}$ for the singularity of g and let $Q = P_{a_n,b'_n} * P$ be the divide for the singularity of f. A copy $F_{P,Q}$ of the fiber F_P is constructed as a subset of the fiber F_Q. Remember, that F_P is obtained by connecting with strips the sets $\{(x,u) \in TD | f_P(x) > 0, (df_P)_x(u) = 0\}$, where $f_P : D \to \mathbb{R}$ is a Morse function for the divide P. For each double point of P there are two connecting strips. To each $+$-component of P corresponds a P_+-component of Q with the same topology and to each double point of P corresponds a Manhattan grid of Q, in which we have drawn diagonally the projection of the strips that connect $\{(x,u) \in TD | x \in Q_{P,+}, (df_Q)_x(u) = 0\}$. Here, $Q_{P,+}$ denotes the union of the P_+-components of the complement of the divide Q. From the divide P is deduced a distinguished basis of quadratic vanishing cycles for the singularity of f. Let B_P be the union of the curves of this basis. This basis can be drawn on the fiber F_P, see Sect. 1.8.1 for an algorithmic method to visualize distinguished basis of vanishing cycles.

In order to prove the theorem, we will construct inside $F_{P,Q}$ a system of simply closed curves with union $B_{P,Q}$, each of them being a quadratic vanishing cycle for the singularity g, such that the pairs (F_P, B_P) and $(F_{P,Q}, B_{P,Q})$ are diffeomorphic. This finishes the proof, since the Dehn twist, whose cores are the quadratic vanishing cycles of $B_{P,Q}$, generate a copy of Γ_g in Γ_f. By acting with the geometric monodromy T of the singularity f, one obtains a_n commuting copies of Γ_g in Γ_f.

To each $+$-region of P corresponds one P_+-region of Q. The maximum of f_P, say at M in the region, is also a maximum of f_Q. The quadratic vanishing cycle $\delta_M := \{(M,u) \in TD | \|M\|^2 + \|u\|^2 = 1\}$ of F_Q lies in F_P and also in $F_{P,Q}$. For each double point c of P the quadratic vanishing cycle $\delta_c \subset F_P$ projects in D to a tear splicing the gradient line g_c of f_P through c. The endpoints of g_c are maxima of f_P or points on ∂D. The function f_Q has exactly one gradient line $g_{Q,c}$ that has the same endpoints as g_c and coincides with g_c in a neighborhood of the common endpoints. The gradient line $g_{Q,c}$ runs along a diagonal through the Manhattan grid corresponding to c. Let $g_{Q,c}$ be the simply closed curve on F_Q, that projects to a tear $t_{Q,c}$ equal to $g_{Q,c}$, except above a neighborhood of its endpoints where $t_{Q,c}$ equals t_c. We remark that $\delta_{Q,c}$ is a cycle in $F_{P,Q}$. Let $c_1, \ldots, c_p$ be the $p := a_n$ double points of Q that occur along the $g_{Q,c}$ and let $M_2, \ldots, M_p$ along $g_{Q,c}$ be the maxima. Let δ_{Q,c_1} be the quadratic vanishing cycle of the singularity f that corresponds to c_1. One verifies that the cycles $\delta_{Q,c}$ and

$$\Delta_{c_p} \circ \Delta_{M_p} \circ \cdots \circ \Delta_{c_2} \circ \Delta_{M_2}(\delta_{Q,c_1})$$

are isotopic. Here Δ_{c_i} or Δ_{M_i} stands for the right Dehn twist of F_Q whose core curve is the quadratic vanishing cycle δ_{c_i} or δ_{M_i} of the singularity f. Hence $\delta_{Q,c} \subset F_{P,Q}$ is a quadratic vanishing cycle for the singularity f. So far, we have constructed for each maximum and for each saddle point of f_P a simply closed curve on $F_{P,Q}$ that is a quadratic vanishing cycle of the singularity f. These cycles intersect on $F_{P,Q}$ as do the corresponding quadratic vanishing cycles of the singularity g on F_P.

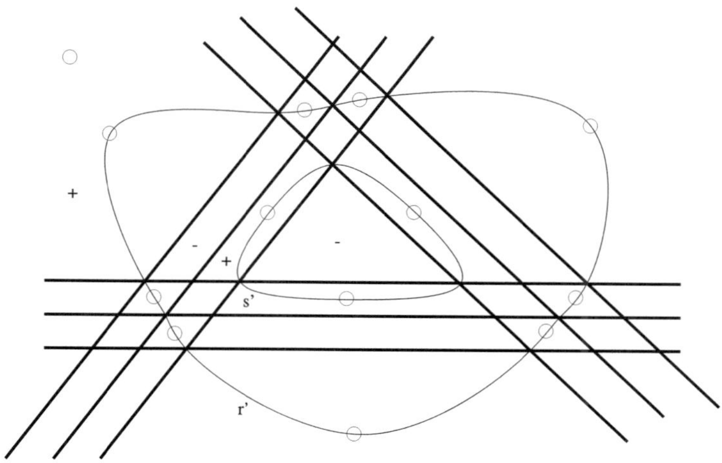

Fig. 1.44 Vanishing cycle s on F_Q from a minimum of f_P and cycle r on $F_{P,Q}$

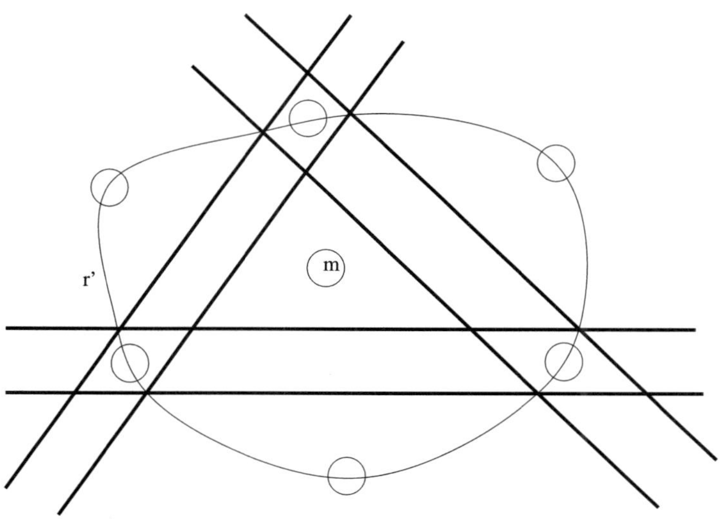

Fig. 1.45 Vanishing cycle $\delta_{Q,m}$ on F_Q from a minimum of f_P and cycle r on $F_{P,Q}$

We now wish to construct for each minimum of f_P a vanishing cycle on $F_{P,Q}$. We have to handle two cases: p odd, see Fig. 1.44, and p even, see Fig. 1.45.

If p is odd, a minimum m of f_P will also be a minimum of f_Q. Let $\delta_{Q,m}$ be the vanishing cycle on F_Q corresponding to m, see Fig. 1.44. The projection of $\delta_{Q,m}$ into D is a smooth simply closed curve s' transversal to Q, that surrounds the $-$ region of m through the its neighboring $+$ regions of Q. One needs to take care that in each neighboring $+$ component the projection runs through the maximum of f_Q in that region. The points of s correspond to pairs (x, u) with $x \in s'$ and u pointing inwards to m. Let r be a simply closed cycle on $F_{P,Q}$ that projects into D upon the curve r', which now surrounds the $-$ region of m through the P_+-components of Q, see Fig. 1.44. In the Manhattan grids r' is just a diagonal, again r' runs through the maxima of the regions or touches ∂D. On r we only allow pairs (x, u) where u points inwards to m. It is clear that the cycle r on $F_{P,Q}$ intersects the cycles of the previous construction as the vanishing cycle to the minimum of f_P intersects the vanishing cycles of the critical points of f_P. It remains however to check that the cycle r is a quadratic vanishing cycle of the singularity of g. By applying to $r \subset F_Q$ the Dehn twist corresponding to the critical points of f_Q that are in between the curves r' and s', one can transform the isotopy class of the curve $\delta_{Q,m}$ to the class of the curve r. This proves that r is indeed a quadratic vanishing cycle of the singularity of g.

If p is even, a minimum m of f_P will be a maximum of f_Q. Let $\delta_{Q,M}$ be the vanishing cycle on F_Q corresponding to maximum $M := m$, see Fig. 1.45. Its projection into D is the point $M := m$. Let r be a simply closed cycle on $F_{P,Q}$ that projects into D upon the curve r' which now surrounds the $-$ region of $M := m$ through the P_+-regions of Q, see Fig. 1.45. In the Manhattan grids r' is just a diagonal, again r' runs through the maxima of the regions. On r we only allow pairs (x, u) where u points inwards to m. It is clear that the cycle r on $F_{P,Q}$ intersects the cycles of the previous construction as the vanishing cycle to the minimum of f_P intersects the vanishing cycles of the critical points of f_P. By applying to $\delta_{Q,M} \subset F_Q$ the Dehn twist corresponding to the critical points of f_Q that are in between the curve r' and the point $M := m$, one can transform the isotopy class of the curve $\delta_{Q,M}$ to the class of the curve r, and proves that r is indeed a quadratic vanishing cycle of the singularity of g. As explained, this terminates the proof. □

Proof of Theorem 1.8.6 The proof of Theorem 1.8.5 constructs a copy $\Gamma_{P,Q}$ of the monodromy group Γ_f of the singularity f as subgroup in the monodromy group Γ_g of the singularity g. This copy acts with support in a copy $F_{P,Q}$ of the fiber F_P. The the first $a_n - 1$ iterates of the monodromy T_Q of the singularity g constructs a_n copies of F_P in F_Q. By conjugation with T_Q one gets a_n copies from $\Gamma_{P,Q}$. We end the proof by repeating this argument. One gets $a_n a_{n-1} \cdots a_2$ commuting copies of the geometric monodromy group of the singularity $x^{b_1} - y^{a_1}$ in Γ_g. □

1.9 Some Other Questions and Properties

We conclude by stating some other open questions and properties related to divides associated with plane curve singularities.

1.9.1 Connected Sum and Characterization of Plane Curve Divides

The connected sum of two divides (D_1, P_1) and (D_2, P_2) is done by making a boundary connected sum of D_1 and D_2 such that a boundary point of P_1 matches with a boundary point of P_2. For divides with one branch we have the formula:

$$L(P_1 \# P_2) = L(P_1) \# L(P_2)$$

A relative immersion $i : I \to D$ of a copy of $[0, 1]$ in D, such that at selftangencies the velocities are with opposite orientation, defines an embedded and oriented arc I' in S^3 by putting:

$$I' := \{(x, u) \in S^3 | x \in i(I), (di^{-1})_x(u) \geq 0\}$$

Let $j : I \to D$ be a relative immersion with only transversal crossings and opposite selftangencies, such that the endpoints of i and j are tangent with opposite orientations and that all tangencies of i and j are generic and have opposite orientations. The union $I' \cup J'$ is the oriented knot of the pair (i, j). A divide P defines pairs (i_P, j_P) of relative immersions with opposite orientations by taking both orientations. Those pairs (i_P, j_P) have a special 2-fold symmetry. For instance the complex conjugation realizes this 2-fold symmetry for a divide, which arises as a real deformation of a real plane curve singularity. It is interesting to observe that this symmetry acts on F_1 with as fixed point set the intersection $D \cap F_1$, which is a collection of r disjointly embedded arcs in F_1. The quotient of $\bar{F_1}$ by the symmetry is an orbifold surface with exactly $2r$ boundary $\frac{\pi}{2}$-singularities. Any link of singularity of a plane curve can be obtained as the link of a divide by Sect. 1.5 (cf. [3, 7, 23]). It is an interesting problem to characterize links of singularities among links of divides.

1.9.2 Symplectic Properties

The link of a divide is transversal to the standard contact structure in the 3-sphere. This can be seen explicitly by the following computation, where we use the multiplication of quaternions. Let P be a divide in the unit disk. We assume that the part of P, which lies in the collar of ∂D with inner radius $\frac{1}{\sqrt{2}}$, consists of radial line segments. We think of the branches of P as parametrized curves $\gamma(t) = (a(t), b(t)), -A \leq t \leq A$, where the parameter speed is adjusted such that $a^2 + b^2 + \dot{a}^2 + \dot{b}^2 = 1$. To the branch γ correspond two arcs Γ^+ and Γ^- on the sphere of quaternions of unit length:

$$\Gamma^+(t) := a(t) - \dot{a}(t)i + b(t)j + \dot{b}(t)k$$

$$\Gamma^-(t) := a(-t) + \dot{a}(-t)i + b(t)j - \dot{b}(-t)k$$

The left invariant speed of Γ^+ at time t is

$$v(t) := \Gamma^+(t)^{-1}\frac{d}{dt}\Gamma^+(t)$$

We have

$$v = a\dot{a} + \dot{a}\ddot{a} + b\dot{b} + \dot{b}\ddot{b} + [-a\ddot{a} + \dot{a}^2 - b\ddot{b} + \dot{b}^2]i + v_j j + v_k k$$

The coefficient $v_0 := a\dot{a} + \dot{a}\ddot{a} + b\dot{b} + \dot{b}\ddot{b}$ vanishes, since $\frac{d}{dt}\Gamma(t)$ is perpendicular to $\Gamma(t)$, and hence we can rewrite the coefficient v_i of i in v as

$$v_i = -a\ddot{a} + \dot{a}^2 - b\ddot{b} + \dot{b}^2 =< (a + \dot{a}, b + \dot{b}) \mid (\dot{a}, \dot{b}) >$$

Outside of the collar neighborhood of ∂D we have $v_i > 0$, since $a^2+b^2 < 1/2 < \dot{a}^2 + \dot{b}^2$. In the collar we also have $v_i > 0$ by a direct computation. Since the left invariant contact structure on the unit sphere in the skew field of the quaternions is given by the span of the tangent vectors j and k at the point 1, we conclude that Γ^+ with its orientation is in the positive sense transversal to the left invariant contact structure S^3.

For the link of a divide we now will construct a polynomial, hence symplectic, spanning surface in the 4-ball. For $\lambda \in \mathbb{R}$, $\lambda > 0$ put

$$B_\lambda := \{p + ui \in \mathbb{C}^2 | p, u \in \mathbb{R}^2, \|p\|^2 + \lambda^{-2}\|u\| \leq 1\}$$

We have $B_\lambda \cap \mathbb{R}^2 = D$ and B_λ is a strictly holomorphically convex domain with smooth boundary in $\mathbb{C}^2$. The map $(p, u) \mapsto ((p, u/\lambda)$ identifies ∂B_λ with the unit 3-sphere of $\mathbb{C}^2$.

Theorem 1.9.1 *Let P be a connected divide in the disc D with δ double points and r branches. There exist $\lambda > 0, \eta > 0$ and there exists a polynomial function $F : B_\lambda \to \mathbb{C}$ with the following properties:*

(a) the function F is real, i.e. $F(\overline{p + ui}) = \overline{F(p + ui)}$,
(b) the set $P_0 := \{p \in D | F(p) = 0\}$ is a divide, which is C^1 close to the divide P, and hence the divides P and P_0 are combinatorially equivalent,
(c) the function F has only non degenerate singularities, which are all real,
(d) for all $t \in \mathbb{C}$, $|t| \leq \eta$ the intersection $K_t := \{(p + iu) \in B_\lambda | F(p + iu) = t\} \cap \partial B_\lambda$ is transversal and by a small isotopy equivalent to the link $L(P)$,

(e) *for the link K_η the surface $\{(p + iu) \in B_\lambda | F(p + iu) = \eta\}$ is a connected smooth symplectic spanning surface of genus $\delta - r + 1$ in the 4-ball.*

Proof Let the divide P be given by smooth parametrized curves $\gamma_l : [-1, 1] \to \mathbb{R}^2, 1 \leq l \leq r$. Using the Weierstrass Approximation Theorem, we can construct polynomial approximations $\gamma_{l,0} : [-1, 1] \to \mathbb{R}^2$ being C^2 close to γ_l and henceforth give a divide P_0 with the combinatorics of the divide P. We may choose $\gamma_{l,0}$ such that $\gamma_{l,0}(s) \notin D, |s| > 1$. Let $F : \mathbb{C}^2 \to \mathbb{C}$ be a real polynomial map such $F = 0$ is a regular equation for the union of the images of $\gamma_{l,0}$. Let S^3_λ be the sphere $S^3_\lambda := \{p+iu \in \mathbb{C}^2 | \|p\|^2 + \lambda^{-2}\|u\|^2 = 1\}$. For a sufficiently small $\lambda > 0$ we have that the 0-level of F on S^3_λ is a model for the link $L(P)$. For $t \in \mathbb{C}, t \neq 0$, and t sufficiently small, say $|t| \leq \eta$, the surface $X_t := \{p+iu \in B^4_\lambda | F(p+iu) = t\}$ is connected and smooth of genus $\delta - r + 1$, and has a polynomial equation, hence is a symplectic surface in the 4-ball B_λ equipped with the standard symplectic structure of $\mathbb{C}^2$. The intersection $K_\eta := X_\eta \cap \partial B_\lambda$ is also a model for the link $L(P)$ and has hence a symplectic filling with the required properties. □

Remark 1.9.2 Unfortunately, it is not the case that the restriction of F to B_λ is a fibration with only quadratic singularities, such that for some $\eta > 1$ the fibers $f_0^{-1}(t), t \in \mathbb{C}, |t| < \eta$, are transversal to the boundary of B_λ. So, we do not know, as it is the case for divides coming from plane curve singularities, if it is possible to fill in with a Picard-Lefschetz fibration, which is compatible with the contact and symplectic structure.

1.9.3 Presentation of the Geometric Monodromy Group

We like to state the problem of presenting the geometric monodromy group of plane curve singularities with generators and relations. It would be particulary nice to express the presentation in terms of a divide of the singularity. The same problem can also be stated for the homological monodromy group of plane curve singularities, but we think that the problem for the geometric monodromy group is more tractable, since all reduction curves can be taken into account Theorems 1.8.5 and 1.8.6. However, an important missing piece in this program is a presentation with generators and relations of the geometric monodromy group of the singularities $y^p - x^q = 0$ for $3 \leq p \leq q, 7 \leq p+q$. The fundamental group of the complement of the discriminant in the unfolding of the singularity $y^2 - x^q = 0$ is the braid group B_{q-1}. Bernard Perron and Jean-Pierre Vannier [39] have proved for the singularities $y^2 - x^q = 0$ that the geometric monodromy group is a faithful image of the braid group B_{q-1} and that a similar result holds for the D singularities $x(y^2 - x^q) = 0$. The fundamental group of the complement of the discriminant in the unfolding of the singularity $y^3 - x^6 = 0$ is the Artin AE_6 group of the Dynkin diagram E_6. Bronislaw Wajnryb [48] has proved that the geometric monodromy representation of E_6 into the mapping class group of the Milnor fiber of the singularity $y^3 - x^6 = 0$ is not faithful. Recently, Nick Salter and the second author [40, Corollary C] have proven that the geometric monodromy representation of a plane curve singularity that is not of type A_n or D_n is never faithful if the genus of the Milnor fiber is at least 7.

Acknowledgments This publication has been partially funded by the grant (holded by the second author) RYC2022-035158-I, funded by MCIN/AEI/10.13039/501100011033 and by el FSE+

References

1. A'Campo, N.: Sur la monodromie des singularites isolees d'hypersurfaces complexes. Invent. Math. **20**, 147–169 (1973). https://doi.org/10.1007/BF01404063
2. A'Campo, N.: La fonction zeta d'une monodromie. Comment. Math. Helv. **50**, 233–248 (1975). https://doi.org/10.1007/BF02565748
3. A'Campo, N.: Le groupe de monodromie du déploiement des singularités isolées de courbes planes. I. Math. Ann. **213**, 1–32 (1975). https://doi.org/10.1007/BF01883883
4. A'Campo, N.: Le groupe de monodromie du déploiement des singularités isolées de courbes planes. II. Proc. int. Congr. Math., Vancouver 1974, Vol. 1, 395–404 (1975).
5. A'Campo, N.: Generic immersions of curves, knots, monodromy and Gordian number. Publ. Math., Inst. Hautes Étud. Sci. **88**, 151–169 (1998). https://doi.org/10.1007/BF02701769
6. A'Campo, N.: Planar trees, slalom curves and hyperbolic knots. Publ. Math., Inst. Hautes Étud. Sci. **88**, 171–180 (1998). https://doi.org/10.1007/BF02701770
7. A'Campo, N.: Real deformations and complex topology of plane curve singularities. Ann. Fac. Sci. Toulouse, Math. (6) **8**(1), 5–23; erratum: 8, no.2, 343 (1999). https://doi.org/10.5802/afst.918
8. A'Campo, N.: Quadratic vanishing cycles, reduction curves and reduction of the monodromy group of plane curve singularities. Tôhoku Math. J. (2) **53**(4), 533–552 (2001). https://doi.org/10.2748/tmj/1113247799
9. A'Campo, N.: Monodromy of real isolated singularities. Topology **42**(6), 1229–1240 (2003). https://doi.org/10.1016/S0040-9383(02)00099-X
10. Arnold, V.I., Guseĭn-Zade, S.M., Varchenko, A.N.: Singularities of differentiable maps. Vol. II, *Monographs in Mathematics*, vol. 83. Birkhäuser Boston, Inc., Boston, MA (1988). https://doi.org/10.1007/978-1-4612-3940-6. Monodromy and asymptotics of integrals, Translated from the Russian by Hugh Porteous, Translation revised by the authors and James Montaldi
11. Baader, S., Dehornoy, P.: Minor theory for surfaces and divides of maximal signature. Preprint, arXiv:1211.7348 [math.GT] (2012). URL https://arxiv.org/abs/1211.7348
12. Balke, L., Kaenders, R.: On a certain type of Coxeter-Dynkin diagrams of plane curve singularities. Topology **35**(1), 39–54 (1996). https://doi.org/10.1016/0040-9383(95)00010-0
13. Baykur, R., Hayano, K.: Hurwitz equivalence for Lefschetz fibrations and their multisections. In: Real and complex singularities. XIII international workshop on real and complex singularities, São Carlos, Brazil, July 27–August 8, 2014 in honor of María del Carmen Romero Fuster's 60th birthday. Proceedings, pp. 1–24. Providence, RI: American Mathematical Society (AMS); Madrid: Real Sociedad Matemática Española (RSME) (2016). https://doi.org/10.1090/conm/675/13581
14. Brieskorn, E., Knörrer, H.: Plane algebraic curves. Transl. from the German by John Stillwell, reprint of the hardback ed. 1986 edn. Mod. Birkhäuser Class. Basel: Birkhäuser (2012). https://doi.org/10.1007/978-3-0348-0493-6
15. Callahan, J.: The double cusp has five minima. Math. Proc. Camb. Philos. Soc. **84**, 537–538 (1978). https://doi.org/10.1017/S0305004100055341

16. Castellini, R.: The topology of A'Campo deformations of singularities: an approach through the lotus. Theses, Université de Lille 1 (2015). URL https://theses.hal.science/tel-01207005
17. Dehornoy, P., Liechti, L.: Divide monodromies and antitwists on surfaces. Preprint, arXiv:1910.00851 [math.DS] (2019) (2019). URL https://arxiv.org/abs/1910.00851
18. Eisenbud, D., Neumann, W.: Three-dimensional link theory and invariants of plane curve singularities, *Ann. Math. Stud.*, vol. 110. Princeton University Press, Princeton, NJ (1985). https://doi.org/10.1515/9781400881925
19. Farb, B., Margalit, D.: A primer on mapping class groups, *Princeton Math. Ser.*, vol. 49. Princeton, NJ: Princeton University Press (2011)
20. Fomin, S., Pylyavskyy, P., Shustin, E., Thurston, D.: Morsifications and mutations. J. Lond. Math. Soc. (2) **105**(4), 2478–2554 (2022). https://doi.org/10.1112/jlms.12566
21. Gibson, W., Ishikawa, M.: Links and gordian numbers associated with generic immersions of intervals. Topology Appl. **123**(3), 609–636 (2002). https://doi.org/10.1016/S0166-8641(01)00224-3
22. Gusein-Zade, S.M.: Dynkin diagrams for singularities of functions of two variables. Funct. Anal. Appl. **8**, 295–300 (1974). https://doi.org/10.1007/BF01075485
23. Gusein-Zade, S.M.: Intersection matrices for certain singularities of functions of two variables. Funct. Anal. Appl. **8**, 10–13 (1974). https://doi.org/10.1007/BF02028301
24. Gusein-Zade, S.M.: Index of a singular point of a gradient vector field. Funct. Anal. Appl. **18**, 6–10 (1984). https://doi.org/10.1007/BF01076355
25. Hamm, H.A., Tráng, L.D.: Un théorème de Zariski du type de Lefschetz. Ann. Sci. Éc. Norm. Supér. (4) **6**, 317–355 (1973). https://doi.org/10.24033/asens.1250
26. Heil, W., Tollefson, J.L.: Deforming homotopy involutions of 3-manifolds to involutions. Topology **17**, 349–365 (1978). https://doi.org/10.1016/0040-9383(78)90003-4
27. Hilbert, D., Cohn-Vossen, S.: Anschauliche Geometrie., *Grundlehren Math. Wiss.*, vol. 37. Berlin, J. Springer (1932)
28. Hofer, H., Zehnder, E.: Symplectic invariants and Hamiltonian dynamics. Basel: Birkhäuser (1994)
29. Ishikawa, M.: Links of plane curves, fiberedness, quasipositivity and Jones polynomials. Theses, Universität Basel (2001)
30. Ishikawa, M.: Tangent circle bundles admit positive open book decompositions along arbitrary links. Topology **43**(1), 215–232 (2004). https://doi.org/10.1016/S0040-9383(03)00040-5
31. Kas, A.: On the handlebody decomposition associated to a Lefschetz fibration. Pac. J. Math. **89**, 89–104 (1980). https://doi.org/10.2140/pjm.1980.89.89
32. Kawamura, T.: Quasipositivity of links of divides and free divides. Topology Appl. **125**(1), 111–123 (2002). https://doi.org/10.1016/S0166-8641(01)00264-4
33. Kawauchi, A.: A survey of knot theory. Transl. from the Japan. by the author. Basel: Birkhäuser (1996)
34. King, H.C.: Topological type of isolated critical points. Ann. of Math. (2) **107**(2), 385–397 (1978). https://doi.org/10.2307/1971121
35. Leviant, P., Shustin, E.: Morsifications of real plane curve singularities. J. Singul. **18**, 307–328 (2018). https://doi.org/10.5427/jsing.2018.18p
36. Matsumoto, Y.: Lefschetz fibrations of genus two – a topological approach. In: Topology and Teichmüller spaces. Proceedings of the 37th Taniguchi symposium, Katinkulta, Finland, July, 24–28, 1995, pp. 123–148. Singapore: World Scientific (1996)
37. Milnor, J.W.: Singular points of complex hypersurfaces, *Ann. Math. Stud.*, vol. 61. Princeton University Press, Princeton, NJ (1968). https://doi.org/10.1515/9781400881819
38. Perron, B.: Conjugaison topologique des germes de fonctions holomorphes à singularité isolée en dimension trois. Invent. Math. **82**, 27–35 (1985). https://doi.org/10.1007/BF01394777
39. Perron, B., Vannier, J.P.: Geometric monodromy groups of simple singularities. Math. Ann. **306**(2), 231–245 (1996). https://doi.org/10.1007/BF01445249
40. Portilla Cuadrado, P., Salter, N.: Vanishing cycles, plane curve singularities and framed mapping class groups. Geom. Topol. **25**(6), 3179–3228 (2021). https://doi.org/10.2140/gt.2021.25.3179

41. Rivlin, T.J.: The Chebyshev polynomials. Pure and Applied Mathematics. New York etc.: John Wiley & Sons, a Wiley- Interscience Publication. VI, 186 p. £ 8.60 (1974). (1974)
42. Sakuma, M.: On strongly invertible knots. In: Algebraic and topological theories. Papers from the symposium dedicated to the memory of Dr. Takehiko Miyata held in Kinosaki, October 30-November 9, 1984, pp. 176–196. Tokyo: Kinokuniya Company Ltd. (1986)
43. Schulze-Röbbecke, T.: Algorithmen zur Auflösung und Deformation von Singularitäten ebener Kurven. Dipl.-Arbeit, *Bonn. Math. Schr.*, vol. 96. Mathematisches Institut der Universität Bonn, Bonn (1977)
44. Teissier, B.: Un exemple de classe d'équisingularité irrationnelle. (An example of irrational equisingularity class). C. R. Acad. Sci., Paris, Sér. I **311**(2), 111–113 (1990)
45. Thom, R.: L'équivalence d'une fonction différentiable et d'un polynôme. Topology **3**, 297–307 (1965). https://doi.org/10.1016/0040-9383(65)90079-0
46. Thurston, W.P.: On the geometry and dynamics of diffeomorphisms of surfaces. Bull. Am. Math. Soc., New Ser. **19**(2), 417–431 (1988). https://doi.org/10.1090/S0273-0979-1988-15685-6
47. Torres, G.: On the Alexander polynomial. Ann. Math. (2) **57**, 57–89 (1953). https://doi.org/10.2307/1969726
48. Wajnryb, B.: Artin groups and geometric monodromy. Invent. Math. **138**(3), 563–571 (1999). https://doi.org/10.1007/s002220050353
49. Wall, C.T.C.: Singular points of plane curves, *Lond. Math. Soc. Stud. Texts*, vol. 63. Cambridge: Cambridge University Press (2004)
50. Zariski, O.: Studies in equisingularity. I. Equivalent singularities of plane algebroid curves. Amer. J. Math. **87**, 507–536 (1965). https://doi.org/10.2307/2373019
51. Zariski, O.: Studies in equisingularity. II. Equisingularity in codimension 1 (and characteristic zero). Amer. J. Math. **87**, 972–1006 (1965). https://doi.org/10.2307/2373257
52. Zariski, O.: Studies in equisingularity. III. Saturation of local rings and equisingularity. Amer. J. Math. **90**, 961–1023 (1968). https://doi.org/10.2307/2373492

Chapter 2
Viro Method in Topology of Real Algebraic Varieties

Ilia Itenberg

Contents

Abstract The main purpose of this article is to give a brief introduction to the Viro method of construction of real algebraic varieties with prescribed topology.

2.1 Introduction

The Viro method is a powerful construction, originally designed as an approach to obtain topologically controlled smoothings of real singularities and to produce real algebraic varieties with prescribed topology (see [36, 39], and [28]). It was proposed by O. Viro in the late 1970s, and it consists in a reduction of problems on polynomials with an arbitrary Newton polytope to similar problems on polynomials

I. Itenberg (✉)
Sorbonne Université and Université Paris Cité, CNRS, IMJ-PRG, Paris, France
e-mail: ilia.itenberg@imj-prg.fr

© The Author(s), under exclusive license to Springer Nature Switzerland AG 2026

J. L. Cisneros-Molina et al. (eds.), *Handbook of Geometry and Topology of Singularities VIII*, https://doi.org/10.1007/978-3-031-99571-2_2

with smaller Newton polytopes. The *combinatorial patchworking* (see Sect. 2.3), which is a particular case of the Viro method, allows one to construct real algebraic hypersurfaces in toric varieties in a combinatorial way, gluing hypersurfaces from pieces that, essentially, are hyperplanes (a close link of this construction to the combinatorics of Newton polytopes of discriminants in the spaces of hypersurfaces in toric varieties is presented in Chapter 11 of [12]). The general Viro construction is discussed in Sect. 2.5. It should also be mentioned that the Viro method can be adapted for a construction of complete intersections in toric varieties; such an extension was initiated by B. Sturmfels in [33].

One of the first successful applications of the Viro method is an isotopy classification, obtained by Viro in 1979, of non-singular curves of degree 7 in the real projective plane. Another example of an application of the Viro method (more precisely, of the combinatorial patchworking) is the construction of counter-examples to the Ragsdale conjecture (see [15, 16, 22]).

It should be underlined that almost all recent constructions in topology of real algebraic varieties were realized using the Viro method. At the same time, it has become clear that the potential field of application of the Viro method is much broader than topology of real algebraic varieties. One can mention, for example, a series of results obtained by E. Shustin [29–31] (in particular, he adapted the Viro method to the study of possible collections of types of critical points of multivariate polynomials of fixed degree) and a construction of polynomial vector fields in $\mathbb{R}^2$ with many limit cycles (see [19]).

The Viro method has deep connections to tropical geometry (for an introduction to tropical geometry, see, for example, [9]). In fact, the combinatorial patchworking can be seen as one of the roots of tropical geometry (see [38]). Relations between the combinatorial patchworking and tropical hypersurfaces in $\mathbb{R}^n$ are presented in Sect. 2.4.4.

2.2 Topology of Algebraic Hypersurfaces in $\mathbb{R}P^n$

2.2.1 Hilbert 16*-th Problem*

The questions concerning the topology of real algebraic hypersurfaces in the n-dimensional real projective space $\mathbb{R}P^n$ (where n is a positive integer) were included by D. Hilbert in the 16-th problem of his famous list [14]. By an *algebraic hypersurface* in $\mathbb{R}P^n$ we mean a real homogeneous polynomial in $n + 1$ variables which is considered up to multiplication by a non-zero real constant. Such a polynomial has a zero locus $\mathbb{R}X \subset \mathbb{R}P^n$ and a zero locus $\mathbb{C}X \subset \mathbb{C}P^n$. The former (respectively, latter) zero locus is called the *real part* (respectively, the *complex part*) of the hypersurface. The *degree* of an algebraic hypersurface in $\mathbb{R}P^n$ is the degree of a defining polynomial. All hypersurfaces considered in this text are *non-singular*, that is, their defining polynomials do not have critical points in $\mathbb{C}^{n+1} \setminus \{0\}$.

Fix positive integers n and d; what kind of topology can have the real part $\mathbb{R}X$ of a non-singular hypersurface X of degree d in $\mathbb{R}P^n$? In the case of curves ($n = 2$), this question was solved by A. Harnack [13]. Each connected component of a non-singular curve of degree d in $\mathbb{R}P^2$ is a circle smoothly embedded in $\mathbb{R}P^2$. According to the Harnack theorem, the number of these connected components can take any integer value between 0 (respectively, 1) and $(d-1)(d-2)/2+1$ if d is even (respectively, odd). The upper bound

$$b_0(\mathbb{R}X) \leq \frac{(d-1)(d-2)}{2} + 1,$$

where $b_0(\mathbb{R}X)$ is the number of connected components of the real part $\mathbb{R}X$ of a non-singular curve X of degree d in $\mathbb{R}P^2$ is called the *Harnack inequality*. The question about possible topology of the pair $(\mathbb{R}P^2, \mathbb{R}X)$, where X is a non-singular curve of degree d in $\mathbb{R}P^2$, is much more difficult; the answer is known only for $d \leq 7$ (see, for example, [37]).

In the case $n = 3$, the answer to the question on possible topology of the real part of a non-singular surface of degree d is known only for $d \leq 4$ (in fact, for non-singular surfaces of degree $d \leq 4$, several finer classifications are available, namely, the isotopy classification and the deformation classification). For non-singular surfaces of degree 5 in $\mathbb{R}P^3$, even the maximal possible number of connected components of the real part is not known (this maximal number is at least 23 and at most 25; see [18, 25]).

More generally, one can fix a deformation family of complex algebraic varieties and ask for topological or equivariant deformation classification of real algebraic varieties in the family (a *real algebraic variety* is a complex algebraic variety Y equipped with a *real structure*, that is, an antiholomorphic involution $c : Y \to Y$; the fixed point set of the involution c is the *real part* of the real algebraic variety (Y, c)). In this paper, we confine ourselves to the case of algebraic hypersurfaces in toric varieties. Each toric variety has its standard real structure induced by the complex conjugation in the dense complex torus.

2.2.2 *Betti Numbers*

A natural question, a priori simpler than the one formulated above, concerns the possible values of the Betti numbers of the real part of a non-singular hypersurface of degree d in $\mathbb{R}P^n$. In this paper, we consider Betti numbers with $\mathbb{Z}/2\mathbb{Z}$-coefficients: by the p-th Betti number of a topological space $\mathcal{M}$ we mean $b_p(\mathcal{M}) = \dim_{\mathbb{Z}/2\mathbb{Z}} H_p(\mathcal{M}; \mathbb{Z}/2\mathbb{Z})$.

A generalization of the Harnack inequality is the *Smith-Thom inequality* for total Betti numbers (see [10, 32], and [34]): for any algebraic hypersurface X in $\mathbb{R}P^n$, one has

$$\sum_{p=0}^{n-1} b_p(\mathbb{R}X) \leq \sum_{p=0}^{2n-2} b_p(\mathbb{C}X).$$

In fact, the inequality is valid for any (quasi-projective) real algebraic variety of dimension $n-1$ (and even in a more general situation); see, for example, [7] and [40]. A real algebraic variety realizing equality in the Smith-Thom inequality is said to be *maximal*.

Sharp upper bounds for individual Betti numbers of the real part of a non-singular hypersurface of degree d in $\mathbb{R}P^n$ are, in general, difficult to obtain. O. Viro [37] proposed the following conjecture (related to the so-called Ragsdale conjecture for real algebraic curves; see [26]): for any smooth projective real algebraic surface X with simply connected complex point set $\mathbb{C}X$, one has

$$b_1(\mathbb{R}X) \leq h^{1,1}(\mathbb{C}X),$$

where, as before, $\mathbb{R}X$ denotes the real part of X. Counter-examples to the Ragsdale conjecture [15] give rise to counter-examples to the Viro conjecture, in particular, among algebraic surfaces in $\mathbb{R}P^3$ (see [5, 6, 17]). Nevertheless, the Viro conjecture is true for the *primitive T-surfaces* in $\mathbb{R}P^3$ (see [17]), i.e., for algebraic surfaces in $\mathbb{R}P^3$ constructed using the primitive combinatorial patchworking (see Sect. 2.3); these hypersurfaces are close to non-singular tropical limit (see, for example, [9]).

2.3 Combinatorial Patchworking

2.3.1 Construction

The special class of real algebraic hypersurfaces we are interested in is described by the *combinatorial patchworking*. This construction is a particular case of the *Viro method*; we present the general construction in Sect. 2.5.

Let n and d be positive integers, and let $T^n(d)$ be the simplex in $\mathbb{R}^n$ with vertices $(0, 0, \dots, 0)$, $(0, 0, \dots, 0, d)$, $(0, \dots, 0, d, 0)$, $\dots$, $(d, 0, \dots, 0)$. We shorten the notation of $T^n(d)$ to T, when n and d are unambiguous and call $T^n(d)$ the *standard n-simplex of size d*. Take a triangulation τ of T with straight edges and integer vertices (i.e., vertices having integer coordinates). Suppose that a distribution of signs at the vertices of τ is given. The sign (plus or minus) at the vertex with coordinates $(i_1, \dots, i_n)$ is denoted by $\sigma_{i_1,\dots,i_n}$.

Denote by T_* the union of all symmetric copies of T under reflections and compositions of reflections with respect to coordinate hyperplanes. Extend the triangulation τ to a symmetric triangulation τ_* of T_*, and the distribution of signs $\sigma_{i_1,\dots,i_n}$ to a distribution at the vertices of the extended triangulation by the following rule: passing from a vertex to its mirror image with respect to a coordinate hyperplane we preserve the sign if the distance from the vertex to the hyperplane is even, and change the sign if the distance is odd.

If an n-simplex of the triangulation of T_* has vertices of different signs, select a piece of hyperplane being the convex hull of the middle points of the edges having endpoints of opposite signs. Denote by Γ the union of the selected pieces. It is a piecewise-linear hypersurface contained in T_*. It is not a simplicial subcomplex of T_*, but can be deformed by an isotopy preserving τ_* to a subcomplex Υ of the first barycentric subdivision τ'_* of τ_*. Each n-simplex of τ'_* has a unique vertex belonging to τ_*. Denote by τ_*^+ the union of all the n-simplices of τ'_* containing positive vertices of τ_* and by τ_*^- the union of all the remaining n-simplices. The subcomplex Υ is the intersection of τ_*^+ and τ_*^-.

Identify by the symmetry with respect to the origin the faces of T_*. The quotient space $\widetilde{T}$ is homeomorphic to the real projective space $\mathbb{R}P^n$. Denote by $\widetilde{\Gamma}$ the image of Γ in $\widetilde{T}$. (An example for $n = 2$ and $d = 10$ is shown in Fig. 2.1.)

A triangulation (or, more generally, a polyhedral subdivision) τ of T is said to be *convex* if there exists a convex piecewise-linear function $\nu : T \longrightarrow \mathbb{R}$ whose domains of linearity coincide with the n-simplices (n-polytopes) of τ. One can show that, if a convex piecewise-linear function certifying the convexity of a subdivision

Fig. 2.1 Construction of a counter-example to the Ragsdale conjecture

does exist, the function can be chosen in such a way that all its values at integer points of T are non-negative integers.

Theorem 2.3.1 (O. Viro (See [36, 39])) *If τ is convex, there exists a non-singular hypersurface X of degree d in $\mathbb{R}P^n$ and a homeomorphism $\mathbb{R}P^n \to \widetilde{T}$ mapping the real part $\mathbb{R}X$ of X onto $\widetilde{\Gamma}$.*

Consider a family (depending on positive real parameter t) of polynomials

$$P^{(t)}(x_0, \dots, x_n) = \sum_{(i_1,\dots,i_n)\in V} \sigma_{i_1,\dots i_n} t^{-\nu(i_1,\dots i_n)} x_0^{d-i_1-\dots-i_n} x_1^{i_1} \dots x_n^{i_n},$$

where V is the set of vertices of τ and ν is a convex piecewise-linear function certifying that the triangulation τ is convex. If t is big enough, the polynomial $P^{(t)}$ is called *Viro polynomial* and defines a hypersurface that satisfies the properties described in Theorem 2.3.1. The hypersurface X defined by a Viro polynomial is called a *T-hypersurface*. If the triangulation τ is *primitive* (that is, each n-simplex of τ is of volume $\frac{1}{n!}$), then X is called a *primitive T-hypersurface*.

We discuss a proof of Theorem 2.3.1 in Sect. 2.4.4.

2.3.2 Topology of T-Hypersurfaces

The construction of the combinatorial patchworking is very powerful in the case of curves: T-curves (even primitive ones) can have various interesting and rich topologies. In higher dimensions there are strong restrictions on the topology of T-hypersurfaces (and, in particular, on the topology of primitive T-hypersurfaces).

A smooth projective real algebraic variety X is said to be *of type* $\chi = \sigma$ if the Euler characteristic of the real part $\mathbb{R}X$ of X is equal to the signature of the complex point set $\mathbb{C}X$ of X. If X is of even (complex) dimension, then by the signature of $\mathbb{C}X$ we mean the signature of the intersection form in the middle homology; if the dimension of X is odd, we say that the signature of $\mathbb{C}X$ is 0. An example of real algebraic varieties of type $\chi = \sigma$ is provided by primitive T-hypersurfaces in $\mathbb{R}P^n$ (of course, this statement is immediate if n is even).

Theorem 2.3.2 (B. Bertrand) *Let n and d be positive integers. Then, any primitive T-hypersurface of degree d in $\mathbb{R}P^n$ is of type $\chi = \sigma$.*

In the case $n = 3$ this statement was proved in [17]. The general statement was proved by B. Bertrand [4] (see also [1]). A much more simple proof can be found in [8].

The Betti numbers (with $\mathbb{Z}/2\mathbb{Z}$-coefficients) of real parts of primitive T-hypersurfaces in $\mathbb{R}P^n$ satisfy important inequalities in terms of Hodge numbers of the complex point sets of the hypersurfaces (notice that the Hodge numbers of a non-singular algebraic hypersurface in $\mathbb{C}P^n$ are completely determined by the degree of the hypersurface). The following upper bounds (which adapt the Viro conjecture to the case of primitive T-hypersurfaces in $\mathbb{R}P^n$) was proved recently by A. Renaudineau and K. Shaw [27].

Theorem 2.3.3 (A. Renaudineau and K. Shaw) *For any integer* $0 \leq p \leq n-1$ *and any primitive* T*-hypersurface* X *of degree* d *in* $\mathbb{R}P^n$*, we have*

$$b_p(\mathbb{R}X) \leq \begin{cases} h^{p,n-1-p}(\mathbb{C}X), & \text{if } p = (n-1)/2, \\ h^{p,n-1-p}(\mathbb{C}X) + 1, & \text{otherwise.} \end{cases}$$

The case $n = 3$ is treated in [17]. A proof of an asymptotic version of the above inequalities for primitive T-hypersurfaces can be found in [23].

Remark 2.3.4 The inequalities proved by Renaudineau and Shaw imply that for any **maximal** primitive T-hypersurface X of degree d in $\mathbb{R}P^n$, one has

$$b_p(\mathbb{R}X) = \begin{cases} h^{p,n-1-p}(\mathbb{C}X), & \text{if } p = (n-1)/2, \\ h^{p,n-1-p}(\mathbb{C}X) + 1, & \text{otherwise.} \end{cases}$$

Such hypersurfaces exist for any positive integers n and d as shown in [21].

The inequalities of Theorem 2.3.3 are far from being true for all non-singular hypersurfaces of degree d in $\mathbb{R}P^n$ (and even for all not necessarily primitive T-hypersurfaces in $\mathbb{R}P^n$). For example, one can easily construct a non-singular cubic surface in $\mathbb{R}P^3$ such that the real part of the surface has two connected components (slightly perturbing the union of a plane and a non-singular quadric surface whose real part is homeomorphic to a sphere and does not intersect the plane), though the upper bound in Theorem 2.3.3 for $b_0(\mathbb{R}X)$ in the case of a primitive T-surface of degree 3 in $\mathbb{R}P^3$ is 1. For quartic surfaces in $\mathbb{R}P^3$ the maximal number of connected components of the real part is equal to 10, see [24] (the upper bound in Theorem 2.3.3 is 2). Examples for the first Betti number are provided by counter-examples to the Viro conjecture (see [5, 6, 17]).

The equalities in Remark 2.3.4 are also far from being true for all maximal hypersurfaces in $\mathbb{R}P^n$. For example, there exist maximal surfaces in $\mathbb{R}P^3$ whose real parts have much more connected components (respectively, much smaller first Betti number) than the value given in Remark 2.3.4; see [35]. However, among maximal surfaces in $\mathbb{R}P^3$, there are no known examples with less connected components (respectively, bigger first Betti number) than the value given in Remark 2.3.4. More generally, there are no known examples of maximal hypersurfaces in $\mathbb{R}P^n$, where n is odd (respectively, even), such that the $\frac{n-1}{2}$-th (respectively, $\frac{n}{2}$-th) Betti number of the real part is bigger than the value given in Remark 2.3.4.

Another intriguing question concerns possible modifications of the inequalities of Theorem 2.3.3 in the case of not necessarily primitive T-hypersurfaces in $\mathbb{R}P^n$. Some results in this direction for T-surfaces in $\mathbb{R}P^3$ can be found in [20].

2.4 Tropical Hypersurfaces in $\mathbb{R}^n$

The construction of combinatorial patchworking is one of the sources of tropical geometry (for tropical reformulations of the combinatorial patchworking, see [38], [9] and [27]). We recall here several basic tropical notions.

2.4.1 *Tropical Operations*

Tropical geometry can be seen as an algebraic geometry based on so-called *tropical numbers*. The set $\mathbb{T}$ of tropical numbers is the set $\mathbb{R}$ of real numbers completed with negative infinity: $\mathbb{T} = \mathbb{R} \cup \{-\infty\}$. Thus, $\mathbb{T}$ has a natural topology and is homeomorphic to a (closed) half-line. The tropical numbers are equipped with two arithmetic operations: *tropical addition* and *tropical multiplication*. The tropical addition is the operation of taking maximum, and the tropical multiplication is the ordinary addition:

$$\text{“}a + b\text{”} = \max\{a, b\}, \quad \text{“}a \cdot b\text{”} = a + b.$$

(We use the quotation marks when operations are tropical.) The set $\mathbb{T}$ equipped with these two operations is a *semi-field*, which means that

- $(\mathbb{T}, \text{“}+\text{”})$ is a commutative monoid (with the neutral element $-\infty$),
- $(\mathbb{T}^\times, \text{“}\cdot\text{”})$, where $\mathbb{T}^\times = \mathbb{T} \setminus \{-\infty\}$, is an abelian group (with neutral element 0),
- and the tropical multiplication is distributive with respect to the tropical addition:

$$\text{“}a \cdot (b + c)\text{”} = \text{“}(a \cdot b) + (a \cdot c)\text{”}$$

for any $a, b, c \in \mathbb{T}$.

This semi-field is called *tropical*.

2.4.2 *Maslov's Dequantization*

The tropical arithmetic operations can be seen as results of deformation of standard addition and multiplication. For any real number $t > 1$, consider the base t logarithmic map

$$\log_t : \mathbb{R}_{\geq 0} \to \mathbb{T}.$$

This map establishes a bijection between the set $\mathbb{R}_{\geq 0}$ of non-negative real numbers and the set $\mathbb{T}$. Thus, $\log_t$ allows one to introduce new arithmetic operations on $\mathbb{T}$:

$$a \oplus_t b = \log_t(t^a + t^b), \quad a \odot_t b = a + b.$$

These operations degenerate to the tropical operations when t tends to $+\infty$.

2.4.3 Tropical Hypersurfaces in $\mathbb{R}^n$

Let n be a positive integer. In the tropical framework, the space $\mathbb{R}^n = (\mathbb{T}^\times)^n$ can be seen as tropical analog of the complex torus $(\mathbb{C}^\times)^n$, where $\mathbb{C}^\times = \mathbb{C} \setminus \{0\}$.

Let $V \subset (\mathbb{Z}_{\geq 0})^n$ be a finite set, where $\mathbb{Z}_{\geq 0}$ is the set of non-negative integers. Consider a *tropical polynomial*

$$p(u_1, \dots, u_n) = \text{``} \sum_{(i_1, \dots, i_n) \in V} a_{i_1, \dots, i_n} u_1^{i_1} \dots u_n^{i_n} \text{''},$$

where the coefficients $a_{i_1, \dots, i_n}$ are tropical numbers (assume that all these coefficients are different from $-\infty$). Such a tropical polynomial gives rise to a *tropical polynomial function*

$$f_p : (u_1, \dots, u_n) \mapsto \max_{(i_1, \dots, i_n) \in V} \{a_{i_1, \dots, i_n} + i_1 u_1 + \dots + i_n u_n\}.$$

This is a convex piecewise-linear function. It is defined on $\mathbb{R}^n = (\mathbb{T}^\times)^n$ and it takes values in $\mathbb{R} = \mathbb{T}^\times$.

Denote by $\mathcal{X}_p$ the *corner locus* in $\mathbb{R}^n$ of the function f_p, that is, the subset of $\mathbb{R}^n$ formed by the points where the function f_p is not locally affine-linear.

The graph of f_p has a natural structure of polyhedral complex, so the corner locus $\mathcal{X}_p$ is also a polyhedral complex. It defines a subdivision Φ_p of $\mathbb{R}^n$. This subdivision is dual to a certain subdivision Ψ_p of the convex hull Δ_p of $V \subset \mathbb{R}^n$; the convex hull Δ_p is called the *Newton polytope* of p; this definition makes sense for any field or semi-field of coefficients. The latter subdivision can be described as follows. For any strata δ of Φ_p, take a point $(u_1^0, \dots, u_n^0)$ in the relative interior of δ and evaluate at $(u_1^0, \dots, u_n^0)$ the affine-linear functions $a_{i_1, \dots, i_n} + i_1 u_1 + \dots + i_n u_n$, $(i_1, \dots, i_n) \in V$. The convex hull of the points of V that correspond to the affine-linear functions $a_{i_1, \dots, i_n} + i_1 u_1 + \dots + i_n u_n$ realizing the maximum is the polytope dual to δ. If δ is of dimension ℓ, then the dual polytope is of dimension $n - \ell$. The dual polytopes form the subdivision Ψ_p of Δ_p (see, for example, [9] for details).

The *tropical hypersurface* defined by p in $\mathbb{R}^n$ is the polyhedral complex $\mathcal{X}_p$ whose $(n-1)$-dimensional facets are equipped with positive integer weights equal to integer lengths of the dual segments (the *integer length* of a segment with integer endpoints is the number of its integer points diminished by 1). We use the

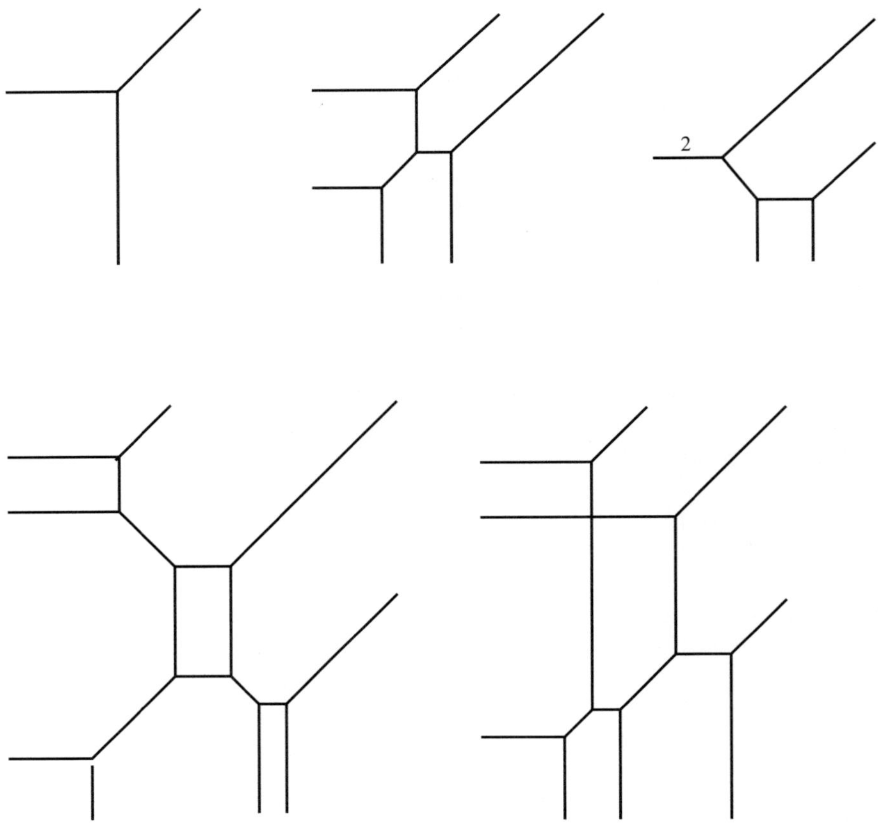

Fig. 2.2 Examples of tropical curves of degrees 1, 2 et 3 in $\mathbb{R}^2$

same notation $\mathcal{X}_p$ for this tropical hypersurface. The *degree* and *Newton polytope* of $\mathcal{X}_p$ are, respectively, the degree and Newton polytope of the defining tropical polynomial p. Figure 2.2 contains examples of tropical curves of small degrees in $\mathbb{R}^2$ (only the weights different from 1 are indicated); Fig. 2.3 presents examples of dual subdivisions.

The subdivision Ψ_p can be described in a different way. Consider the convex hull in $\mathbb{R}^{n+1}$ of the graph of the function $(i_1, \ldots, i_n) \mapsto -a_{i_1,\ldots,i_n}$ defined on V. The projection to Δ_p of the lower part of the boundary of this convex hull gives the subdivision Ψ_p. In particular, Ψ_p is convex in the sense of Sect. 2.3.1: there exists a convex piecewise-linear function $\nu : \Delta_p \longrightarrow \mathbb{R}$ whose domains of linearity coincide with the n-dimensional polytopes of Ψ_p. The tropical polynomial function

$$f_p : (u_1, \ldots, u_n) \mapsto \max_{(i_1,\ldots,i_n)\in V} \{a_{i_1,\ldots,i_n} + i_1u_1 + \ldots + i_nu_n\}$$

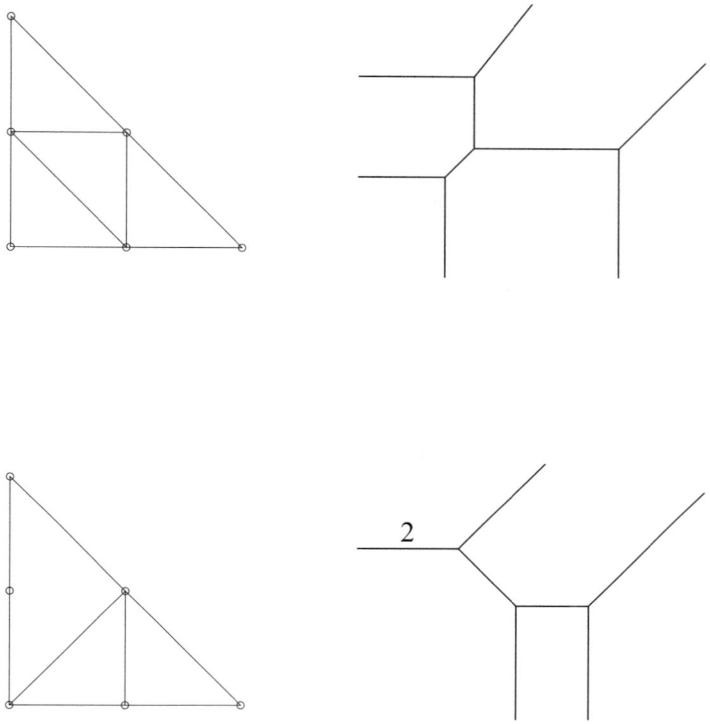

Fig. 2.3 Examples of tropical conics and their dual subdivisions

is the *Legendre transform* of the function $(i_1, \ldots, i_n) \mapsto -a_{i_1,\ldots,i_n}$ defined on the set V.

The tropical hypersurface $\mathcal{X}_p$ is said to be *non-singular*, if the dual subdivision Ψ_p is a primitive triangulation. In this case, the weights of all $(n-1)$-dimensional facets of $\mathcal{X}_p$ are equal to 1, but the latter property does not guarantee that the tropical hypersurface is non-singular.

2.4.4 Combinatorial Patchworking Revisited

A proof of the combinatorial patchworking theorem (Theorem 2.3.1) can be obtained along the following lines (see [38] for details). Consider a Viro polynomial

$$P^{(t)}(x_0, \ldots, x_n) = \sum_{(i_1,\ldots,i_n)\in V} \sigma_{i_1,\ldots i_n} t^{-\nu(i_1,\ldots i_n)} x_0^{d-i_1-\ldots-i_n} x_1^{i_1} \ldots x_n^{i_n},$$

where t is sufficiently big (we use the notation of Sect. 2.3.1). The statement is proved orthant by orthant, so let us study the zero set of $P^{(t)}$ in the open positive orthant of the affine chart $x_0 \neq 0$ with the affine coordinates

$$y_1 = x_1/x_0, \quad \ldots, \quad y_n = x_n/x_0.$$

(The other orthants can be treated using the obvious coordinate changes.) So, from now on, we assume that $y_1 > 0$, ..., $y_n > 0$. The equality $P^{(t)}(y_1, \ldots, y_n) = 0$ can be rewritten as

$$P_+^{(t)}(y_1, \ldots, y_n) = P_-^{(t)}(y_1, \ldots, y_n),$$

where $P_+^{(t)}$ (respectively, $P_-^{(t)}$) is formed by the monomials of $P^{(t)}$ that have positive (respectively, negative) coefficients.

When t tends to $+\infty$, the graphs of the polynomials $P_+^{(t)}$ and $P_-^{(t)}$ in the logarithmic coordinates

$$u_1 = \log_t y_1, \quad \ldots, \quad u_n = \log_t y_n, \quad u_{n+1} = \log_t y_{n+1}$$

tend, respectively, to the graphs

$$u_{n+1} = p_+(u_1, \ldots, u_n) \quad \text{and} \quad u_{n+1} = p_-(u_1, \ldots, u_n)$$

of the tropical polynomials

$$p_\pm(u_1, \ldots, u_n) = \text{“} \sum_{(i_1, \ldots, i_n) \in V_\pm} -\nu(i_1, \ldots i_n) u_1^{i_1} \ldots u_n^{i_n} \text{''},$$

where V_+ (respectively, V_-) is the set of vertices of τ that have positive (respectively, negative) sign.

The projection that forgets the coordinate u_{n+1} sends the intersection of the graphs of p_+ and p_- to a part $\mathcal{X}_{\text{positive}}$ of the tropical hypersurface $\mathcal{X}$ defined in $\mathbb{R}^n$ (with the coordinates u_1, ..., u_n) by the tropical polynomial

$$p(u_1, \ldots, u_n) = \text{“} \sum_{(i_1, \ldots, i_n) \in V} -\nu(i_1, \ldots i_n) u_1^{i_1} \ldots u_n^{i_n} \text{''}.$$

To describe $\mathcal{X}_{\text{positive}}$, recall that each connected component of the complement of $\mathcal{X} \subset \mathbb{R}^n$ corresponds to a vertex of τ. Since the vertices of τ have signs, the connected components of the complement of $\mathcal{X} \subset \mathbb{R}^n$ are divided into two classes: positive and negative. The part $\mathcal{X}_{\text{positive}}$ is the common boundary of the union of positive components and the union of negative components.

Since $P_+^{(t)}$ and $P_-^{(t)}$ tend, respectively, to the graphs of the tropical polynomials p_+ and p_-, it can be shown that the projection (that forgets the coordinate u_{n+1}) of the intersection of the graphs of $P_+^{(t)}$ and $P_-^{(t)}$ is isotopic (in the stratified sense) to $\mathcal{X}_{\text{positive}}$. Furthermore, under a natural identification of the n-dimensional tropical projective space $\mathbb{T}P^n$ and the simplex T, the closure of $\mathcal{X}_{\text{positive}}$ in $\mathbb{T}P^n$ is isotopic (again in the stratified sense) to the intersection of the piecewise-linear hypersurface Γ with the simplex T (see Sect. 2.3.1 for notation).

2.5 General Viro Method

In this section, we discuss the general version of the Viro method.

2.5.1 *Notations*

Put $\mathbb{R}^\times = \mathbb{R} \setminus \{0\}$ and $\mathbb{R}_+ = \{x \in \mathbb{R} \mid x > 0\}$. Fix a positive integer n. Further on, the term *polytope* in $\mathbb{R}^n$ means a convex (possibly degenerate) polytope in $\mathbb{R}^n$. If all vertices of a polytope are integer, then it is said to be *lattice*.

Given a complex polynomial $P(z_1, \dots, z_n) = \sum_{(i_1,\dots,i_n)\in V} A_{i_1,\dots,i_n} z_1^{i_1} \dots z_n^{i_n}$ in n variables, where $V \subset (\mathbb{Z}_{\geq 0})^n \subset (\mathbb{R}_{\geq 0})^n$ is a finite set, we denote by $\Delta(P)$ the Newton polytope of P. The *truncation* of P on a face δ of $\Delta(P)$ is the polynomial $P^\delta(z_1, \dots, z_n) = \sum_{(i_1,\dots,i_n)\in V\cap\delta} A_{i_1,\dots,i_n} z_1^{i_1} \dots z_n^{i_n}$. A polynomial $P \in \mathbb{C}[z_1, \dots, z_n]$ is called *non-degenerate*, if P and any truncation P^δ of P on a proper face δ of $\Delta(P)$ has a non-singular zero set in $(\mathbb{C}^\times)^n$, cf. [36].

2.5.2 *Main Statement*

The initial data of the general Viro construction is a convex subdivision

$$\Delta = \Delta_1 \cup \dots \cup \Delta_N$$

of a non-degenerate lattice polytope $\Delta \subset \mathbb{R}^n$ into non-degenerate lattice polytopes Δ_1, ..., Δ_N, and a collection $A_{i_1,\dots,i_n}$ of real or complex numbers indexed by the integer points $(i_1, \dots, i_n)$ of Δ such that $A_{i_1,\dots,i_n}$ is non-zero for any vertex $(i_1, \dots, i_n)$ of the subdivision. The polynomials

$$P_k(z_1, \dots, z_n) = \sum_{(i_1,\dots,i_n)\in\Delta_k} A_{i_1,\dots,i_n} z_1^{i_1} \dots z_n^{i_n}, \quad k = 1, \dots, N,$$

are often supposed to be non-degenerate.

The result of the construction is an algebraic hypersurface X with Newton polytope Δ in the toric variety $\mathrm{Tor}_{\mathbb{C}}(\Delta)$ associated with Δ. The topology of the complex part $\mathbb{C}X$ of X (and of the real part $\mathbb{R}X$ of X if the polynomials $P_1, \dots, P_N$ are real) is described in terms of the topology of zero sets of $P_1, \dots, P_N$. Namely, the complex (respectively, real) part of X is, in a sense, "glued" out of the complex (respectively, real) zero sets of $P_1, \dots, P_N$.

We restrict ourselves in this text to a detailed description of the Viro construction only in the real case, assuming in addition that the polynomials $P_1, \dots, P_N$ are non-degenerate. For such a description, we need the notion of a *chart* of a real polynomial.

Let $\nabla \subset (\mathbb{R}_{\geq 0})^n$ be a non-degenerate lattice polytope. Denote by V_∇ the set of vertices of ∇, and consider the *moment map* $\mu_\nabla : (\mathbb{C}^\times)^n \to \overset{\circ}{\nabla}$ (see [2, 3, 11]), where $\overset{\circ}{\nabla}$ is the relative interior of ∇ :

$$\mu_\nabla(z_1, \dots, z_n) = \frac{\sum_{(i_1,\dots,i_n)\in V_\nabla} |z_1|^{i_1} \dots |z_n|^{i_n} \cdot (i_1, \dots, i_n)}{\sum_{(i_1,\dots,i_n)\in V_\nabla} |z_1|^{i_1} \dots |z_n|^{i_n}} \, . \tag{2.1}$$

Consider a real polynomial P in n variables (the most interesting situation for the definitions below is the case where ∇ is the Newton polytope of P). Denote by $\overset{\circ}{Z}$ the hypersurface defined by P in $(\mathbb{C}^\times)^n$, and denote by Z the corresponding hypersurface in the toric variety $\mathrm{Tor}_\mathbb{C}(\nabla)$ associated with ∇. The variety $\mathrm{Tor}_\mathbb{C}(\nabla)$ has the tautological real structure

$$\mathrm{conj} : \mathrm{Tor}_\mathbb{C}(\nabla) \to \mathrm{Tor}_\mathbb{C}(\nabla),$$

and the fixed point set of this involution is the real part $\mathrm{Tor}_\mathbb{R}(\nabla)$ of $\mathrm{Tor}_\mathbb{C}(\nabla)$. Denote by $\mathbb{R}Z \subset \mathrm{Tor}_\mathbb{R}(\nabla)$ the real part of Z, and denote by $\mathbb{R}\overset{\circ}{Z}$ the real part of $\overset{\circ}{Z}$ (in other words, $\mathbb{R}\overset{\circ}{Z}$ is the intersection of $\mathbb{R}Z$ with $(\mathbb{R}^\times)^n \subset \mathrm{Tor}_\mathbb{R}(\nabla)$). For each open orthant $Q \subset (\mathbb{R}^\times)^n$, put $\mathbb{R}Z_Q = \mathbb{R}Z \cap Q$.

Consider the group S generated by the orthogonal symmetries with respect to coordinate hyperplanes in $\mathbb{R}^n$. Any element of this group, as well as any orthant of $(\mathbb{R}^\times)^n$ is naturally identified with an n-tuple of elements of $\{1, -1\}$. We label the elements of S either by such n-tuples, or by orthants (for example, if Q is an orthant of $(\mathbb{R}^\times)^n$, then $s_Q \in S$ is the symmetry such that $s_Q((\mathbb{R}_+)^n) = Q$). For any orthant $Q \subset (\mathbb{R}^\times)^n$, the ∇-*chart* $Ch_{\nabla,Q}(P)$ in Q of the polynomial P is the closure in ∇_Q of the subset $(s_Q \circ \mu_\nabla)(\mathbb{R}Z_Q) \subset \nabla_Q$, where $\nabla_Q = s_Q(\nabla)$. (By abuse of notation, we use the same symbol Q to denote an orthant in $(\mathbb{R}^\times)^n \supset \nabla$ and the corresponding orthant in $(\mathbb{R}^\times)^n \supset \mathbb{R}\overset{\circ}{Z}$.) This definition is a key ingredient of the Viro construction (see [28, 36, 39]). The ∇-*chart* $Ch_\nabla(P)$ of P is the union

$$\cup_Q \, (Ch_{\nabla,Q}(P))$$

over all orthants Q of $(\mathbb{R}^\times)^n$.

Recall that the real part $\mathrm{Tor}_\mathbb{R}(\nabla)$ of $\mathrm{Tor}_\mathbb{C}(\nabla)$ can be obtained as a quotient of the disjoint union $\sqcup_Q \nabla_Q$ (as before, the union is taken over all orthants of $(\mathbb{R}^\times)^n$): for every orthant Q and every codimension one face δ_Q of ∇_Q, we identify δ_Q with its image under the symmetry $s_{((-1)^{j_1},\dots,(-1)^{j_n})}$, where $(j_1, \dots, j_n)$

is a primitive integer normal vector of δ_Q (here, *primitive integer vector* means a non-zero integer vector which is not a multiple of an integer vector with the proportionality coefficient bigger than 1).

In the case where ∇ is the standard n-simplex $T^n(d)$ of size d for a certain positive integer d (cf. Sect. 2.3), the above procedure produces a copy of $\mathbb{R}P^n$.

The image of the chart $Ch_\nabla(P)$ of P in $\mathrm{Tor}_\mathbb{R}(\nabla)$ is called the *toric ∇-chart* of P and is denoted by $Ch_{\mathrm{Tor}(\nabla)}(P)$. If ∇ coincides with the Newton polytope of P, we speak about the *toric chart* of P.

Theorem 2.5.1 (O. Viro (See [36, 39])) *Let $\nu : \Delta \to \mathbb{R}$ be a convex piecewise-linear function certifying that the subdivision $\Delta = \Delta_1 \cup \ldots \cup \Delta_N$ is convex and such that all values of ν at integer points of Δ are non-negative integers. Put*

$$P^{(t)}(z_1, \ldots, z_n) = \sum_{(i_1,\ldots,i_n)\in\Delta} A_{i_1,\ldots,i_n} z_1^{i_1} \ldots z_n^{i_n} t^{-\nu(i_1,\ldots,i_n)},$$

where t is a positive real parameter. Then, for sufficiently big values of t and for any orthant Q of $(\mathbb{R}^\times)^n$, the Δ-chart $Ch_{\Delta,Q}(P^{(t)})$ in Q of the polynomial $P^{(t)}$ is isotopic (in the stratified sense) to the union

$$\bigcup_{k=1}^{N} (Ch_{\Delta_k,Q}(P_k))$$

of Δ_k-charts in Q of the polynomials P_k, $k = 1, \ldots, N$. Furthermore, for sufficiently big values of t, the toric chart $Ch_{\mathrm{Tor}(\Delta)}(P^{(t)})$ of polynomial $P^{(t)}$ is isotopic (in the stratified sense) to the image in $\mathrm{Tor}_\mathbb{R}(\Delta)$ of the union

$$\bigcup_{k=1}^{N} (Ch_{\Delta_k}(P_k))$$

of Δ_k-charts of the polynomials P_k, $k = 1, \ldots, N$.

Theorem 2.3.1 is a particular case of Theorem 2.5.1. Indeed, in the case where, for each index $k = 1, \ldots, N$, the polytope Δ_k is a simplex and all non-zero monomials of P_k correspond to the vertices of Δ_k, the (stratified) isotopy types of the Δ_k-charts of P_k in all orthants Q of $(\mathbb{R}^\times)^n$ depend only on the signs of monomials of P_k and are given by the procedure described in Sect. 2.3. Thus, in this case, the statement of Theorem 2.3.1 provides a description, up to (stratified) isotopy, of the toric chart $Ch_{\mathrm{Tor}(\Delta)}(P^{(t)})$ of the polynomial $P^{(t)}$ for sufficiently big values of t.

We present here only a sketch of a proof of Theorem 2.5.1. Assume that Q is the positive orthant of $(\mathbb{R}^\times)^n$, and consider a cut epigraph

$$\Lambda = \{(x_1, \ldots, x_n, x_{n+1}) \in \mathbb{R}^{n+1} \mid (x_1, \ldots, x_n) \in \Delta \text{ and } \nu(x_1, \ldots, x_n) \leq x_{n+1} \leq M\}$$

of ν and the toric variety $\mathrm{Tor}_{\mathbb{C}}(\Lambda)$ associated with the polytope Λ (here M is an integer which is bigger than all values of ν). The family

$$P^{(1/t')}(z_1, \ldots, z_n) = \sum_{(i_1,\ldots,i_n)\in\Delta} A_{i_1,\ldots,i_n} z_1^{i_1} \ldots z_n^{i_n} (t')^{\nu(i_1,\ldots,i_n)}$$

can be seen as a real polynomial F in $n+1$ variables z_1, ..., z_n, and t'. One can show that, when $t'_0 > 0$ tends to 0, the Λ-chart in Q of the polynomial $t' - t'_0$ tends the graph of ν. For each index $k = 1, \ldots, N$, the projection $\mathbb{R}^{n+1} \to \mathbb{R}^n$ that forgets the last coordinate identifies $Ch_{\Lambda,Q}(F) \cap \widetilde{\Delta}_k$ with the Δ_k-chart in Q of P_k, where $\widetilde{\Delta}_k = \nu(\Delta_k)$. Details can be found in [36] and [39].

References

1. Ch. Arnal, *Patchwork combinatoire et topologie d'hypersurfaces algébriques réelles*, Université Pierre et Marie Curie, Paris, 2017.
2. M. F. Atiyah, *Convexity and commuting hamiltonians*, Bull. Lond. Math. Soc. **14** (1982), 1–15.
3. ———, *Angular momentum, convex polyhedra and algebraic geometry*, Proc. Edinburgh Math. Soc. **26** (1983), 121–138.
4. B. Bertrand, *Euler characteristic of primitive T-hypersurfaces and maximal surfaces*, J. Inst. Math. de Jussieu **9** (2010), no. 1, 1–27.
5. F. Bihan, *Une sextique de l'espace projectif réel avec un grand nombre d'anses*, Revista Matematica Complutense **14** (2001), no. 2, 439–461.
6. ———, *Asymptotic behaviour of betti numbers of real algebraic surfaces*, Comm. Math. Helv. **78** (2003), 227–244.
7. G. Bredon, *Introduction to compact transformation groups*, Academic Press, 1972.
8. E. Brugallé, *Euler characteristic and signature of real semi-stable degenerations*, J. Inst. Math. de Jussieu (2022), no. 5, 1–8.
9. E. Brugallé, I. Itenberg, G. Mikhalkin, and K. Shaw, *Brief introduction to tropical geometry*, GGT, Proceedings of the Gökova Geometry-Topology Conference 2014 (2015), 1–75.
10. E. E. Floyd, *On periodic maps and the Euler characteristics of associated spaces*, Trans. AMS **72** (1952), 138–147.
11. W. Fulton, *Introduction to toric varieties*, Ann. Math. Studies, vol. 131, Princeton Univ. Press, Princeton NJ, 1993.
12. I. Gelfand, M. Kapranov, and A. Zelevinsky, *Discriminants, resultants, and multidimensional determinants*, Birkhäuser, 1994.
13. Axel Harnack, *Ueber die Vieltheiligkeit der ebenen algebraischen Curven*, Math. Ann. **10** (1876), no. 2, 189–198. MR 1509883
14. D. Hilbert, *Mathematische probleme*, Nachrichten von der Gesellschaft der Wissenschaften zu Göttingen, 1900.
15. I. Itenberg, *Contre-exemples à la conjecture de Ragsdale*, C. R. Acad. Sci. Paris **317** (1993), no. 3, 277–282.
16. ———, *Counter-examples to Ragsdale conjecture and T-curves*, Cont. Math. **182** (1995), 55–72, Proceedings, Michigan 1993.
17. ———, *Topology of real algebraic T-surfaces*, Rev. Mat. Univ. Complut. Madrid **10** (1997), 131–152, Real algebraic and analytic geometry, Segovia, 1995.
18. I. Itenberg and V. Kharlamov, *Towards the maximal number of components of a nonsingular surface of degree* 5 *in* RP^3, Topology of real algebraic varieties and related topics, Ser. 2, vol. 173, Amer. Math. Soc., Providence, RI, 1996, pp. 111–118.

19. I. Itenberg and E. Shustin, *Singular points and limit cycles of planar polynomial vector fields*, Duke Math. Journal **102** (2000), no. 1, 1–37.
20. ———, *Critical points of real polynomials and topology of real algebraic T-surfaces*, Geom. Dedicata **101** (2003), no. 1, 61–91.
21. I. Itenberg and O. Viro, *Maximal real algebraic hypersurfaces of projective space*, in preparation.
22. ———, *Patchworking algebraic curves disproves the Ragsdale conjecture*, Math. Intelligencer **18** (1996), no. 4, 19–28.
23. ———, *Asymptotically maximal real algebraic hypersurfaces of projective space*, Proceedings of Gökova Geometry-Topology Conference 2006 (GGT) (2007), no. 4, 91–105.
24. V. Kharlamov, *The maximum number of components of a surface of degree* 4 *in* RP^3, Funkc. Anal. i Prilozen. **6** (1972), no. 4, 101.
25. S. Orevkov, *Real quintic surface with* 23 *components*, C. R. Acad. Sci. Paris **333** (2001), no. 2, 115–118.
26. V. Ragsdale, *On the arrangement of the real branches of plane algebraic curves*, Amer. J. Math. **28** (1906), 377–404.
27. A. Renaudineau and K. Shaw, *Bounding the Betti numbers of real hypersurfaces near the tropical limit*, Ann. Sci. ENS **56** (2023), no. 3, 945–980.
28. J.-J. Risler, *Construction d'hypersurfaces réelles* [*d'après Viro*], 1992.
29. E. Shustin, *Real plane algebraic curves with prescribed singularities*, Topology **32** (1993), no. 4, 845–856.
30. ———, *Surjectivity of the resultant map: a solution to the inverse Bézout problem*, Comm. in Algebra **23** (1995), no. 3, 1145–1164.
31. ———, *Critical points of real polynomials, subdivisions of Newton polyhedra and topology of real algebraic hypersurfaces*, Amer. Math. Soc. Transl. (2) **173** (1996), 203–223.
32. P. A. Smith, *Transformations of finite period*, Ann. of Math. (2) **39** (1938), no. 1, 127–164. MR 1503393
33. B. Sturmfels, *Viro's theorem for complete intersections*, Ann. Scuola Norm. Sup. Pisa **21** (1994), no. 3, 377–386.
34. René Thom, *Sur l'homologie des variétés algébriques réelles*, Differential and Combinatorial Topology (A Symposium in Honor of Marston Morse), Princeton Univ. Press, Princeton, NJ, 1965, pp. 255–265. MR 0200942
35. O. Viro, *Construction of multicomponent real algebraic surfaces*, Dokl. Akad. Nauk SSSR **248** (1979), no. 2, 279–282.
36. ———, *Gluing of algebraic hypersurfaces, smoothing of singularities and construction of curves*, Proc. Leningrad Int. Topological Conf. Leningrad, Nauka, 1983, pp. 149–197 (in Russian).
37. ———, *Progress in the topology of real algebraic varieties over the last six years*, Rus. Math. Surv. **41** (1986), no. 3, 55–82.
38. ———, *Dequantization of real algebraic geometry on logarithmic paper*, European Congress of Mathematics, Vol. I, Barcelona, vol. 201, Birkhäuser, Basel, 2001, pp. 135–146.
39. ———, *Patchworking real algebraic varieties*, 2018, Appendix in G.-M. Greuel, Ch. Lossen, E. Shustin, "Singular algebraic curves".
40. G. Wilson, *Hilbert's sixteenth problem*, Topology **17** (1978), no. 1, 53–73.

Chapter 3
An Introduction to Local Tropicalization

Patrick Popescu-Pampu and Dmitry Stepanov

Contents

Abstract In this paper we explain four viewpoints on the *local tropicalization* of formal subgerms of toric germs, which is a local analog of the global tropicalization of subvarieties of algebraic tori. We start by illustrating some of those viewpoints for plane curve singularities, then we pass to arbitrary dimensions. We conclude by describing several variants and extensions of the notion of local tropicalization presented in this paper.

P. Popescu-Pampu (✉)
University of Lille, CNRS, UMR 8524 - Laboratoire Paul Painlevé, Lille, France
e-mail: patrick.popescu-pampu@univ-lille.fr

D. Stepanov
Department of Higher Mathematics and Center of Fundamental Mathematics, Moscow Institute of Physics and Technology, Moscow, Russia
e-mail: stepanov.da@phystech.edu

© The Author(s), under exclusive license to Springer Nature Switzerland AG 2026

J. L. Cisneros-Molina et al. (eds.), *Handbook of Geometry and Topology of Singularities VIII*, https://doi.org/10.1007/978-3-031-99571-2_3

3.1 Introduction

The aim of this paper is to present various viewpoints on the notion of *local tropicalization*, introduced by the authors in their 2013 paper [39] as a local analog of the global notion of tropicalization of a subvariety of an algebraic torus. We assume that the reader is familiar with the basics of singularity theory and toric geometry. For details about global tropicalization, one may consult either the introductory papers [20, 22, 42] of Kajiwara, Teissier and Katz or Maclagan and Sturmfels' textbook [28].

Let us explain briefly what does it mean to tropicalize locally a reduced complex formal germ $Y \hookrightarrow (\mathbb{C}^n, 0)$ whose irreducible components are not contained in the union $\partial\, \mathbb{C}^n$ of coordinate hyperplanes of $\mathbb{C}^n$. The initial definitions of [39, Definitions 6.6, 6.7] involved spaces of valuations, which allowed us to apply them both to local and global situations, for arbitrary commutative rings, not necessarily containing a field.

But the shortest definition of the local tropicalization of the embedding $Y \hookrightarrow (\mathbb{C}^n, 0)$ is most likely as *the topological closure inside $\mathbb{R}^n$ of the union of rays $\mathbb{R}_{\geq 0} w$ generated by the weight vectors $w = (w_1, \ldots, w_n)$ which appear as initial exponent vectors of the formal arcs*

$$t \mapsto (c_1 t^{w_1} + \cdots , \ \ldots \ , c_n t^{w_n} + \cdots)$$

contained in Y but not contained in $\partial\, \mathbb{C}^n$ (that is, such that $c_j \in \mathbb{C}^*$ and $w_j \in \mathbb{Z}_{>0}$ for every $j \in \{1, \ldots, n\}$). For instance, if Y is a plane curve singularity defined by a series $f(x, y) \in \mathbb{C}[[x, y]]$, then its local tropicalization is the union of the rays orthogonal to the compact edges of the Newton polygon of f (see Corollary 3.3.10).

For the geometrically-inclined reader, perhaps the most suggestive viewpoint on local tropicalization is the following *toric-geometrical* one. Seen as an affine toric variety, $\mathbb{C}^n$ has the set $\mathcal{F}_0$ of faces of the non-negative orthant $\sigma_0 := (\mathbb{R}_{\geq 0})^n$ of its vector space $\mathbb{R}^n$ of weight vectors as associated fan. Note that σ_0 also belongs to $\mathcal{F}_0$, as its nonproper face. Let $\mathcal{F}$ be a fan whose cones are contained in σ_0. There exists a canonical birational morphism

$$\pi_{\mathcal{F}} : \mathcal{X}_{\mathcal{F}} \to \mathbb{C}^n$$

from the toric variety $\mathcal{X}_{\mathcal{F}}$ associated to the fan $\mathcal{F}$ to the affine toric variety $\mathbb{C}^n = \mathcal{X}_{\mathcal{F}_0}$. This birational morphism is a *modification*, that is, it is moreover *proper*, if and only if the *support* $|\mathcal{F}|$ of $\mathcal{F}$ (that is, if the union of its cones) is equal to σ_0. Denote by $Y_{\mathcal{F}}$ the strict transform of Y by $\pi_{\mathcal{F}}$ and let

$$\pi : Y_{\mathcal{F}} \to Y$$

be the restriction of $\pi_{\mathcal{F}}$ to $Y_{\mathcal{F}}$. *It may happen that π is a modification without $\pi_{\mathcal{F}}$ being so* (see Example 3.5.2).

Question Under which condition is the morphism $\pi : Y_{\mathcal{F}} \to Y$ proper?

Answer Precisely when the support $|\mathcal{F}|$ of $\mathcal{F}$ contains the local tropicalization of $Y \hookrightarrow (\mathbb{C}^n, 0)$! That is, this local tropicalization may be defined as the intersection of the supports of the fans $\mathcal{F}$ such that $\pi : Y_{\mathcal{F}} \to Y$ is proper.

This is the announced toric-geometrical interpretation of the notion of local tropicalization. In this paper we will present two more definitions of local tropicalization for embeddings $Y \hookrightarrow (\mathbb{C}^n, 0)$ as above and we will explain the equivalences of the four definitions (see Theorem 3.6.2). We will indicate also related works (see Sect. 3.9).

Let us describe the structure of the paper. Each time a notation is introduced, it is included in a $\boxed{\text{box}}$ and each time a terminology is presented, it is written in **boldface characters**. We start by explaining basic ideas about local tropicalization in dimension two, for plane curve singularities. Namely, in Sect. 3.2 we illustrate the above toric interpretation of local tropicalization using a cuspidal cubic curve in the affine plane $\mathbb{C}^2$. In Sect. 3.3 we recall the notion of *Newton polygon* of a plane curve singularity $Y \hookrightarrow (\mathbb{C}^2, 0)$ in a way suitable for our purposes and we explain in this context the equivalence of the above two definitions of local tropicalization. Then we pass to arbitrary dimensions. In Sect. 3.4 we explain our notations and basic facts about *toric geometry* and *valuation theory*. In Sect. 3.5 we turn the toric-geometric viewpoint on local tropicalization into a definition of local tropicalization for certain formal subgerms Y of a germ of affine toric variety $(\mathcal{X}_\sigma, 0)$ and we explain how to associate weight vectors to certain arcs included in Y or to semivaluations of the local ring of Y. In Sect. 3.6 we formulate Theorem 3.6.2, stating that various definitions of the local tropicalization of an embedding of Y into a toric germ $(\mathcal{X}_\sigma, 0)$ are equivalent and we explain intuitively some ingredients of its proof. In Sect. 3.7 we describe the local tropicalization of a principal effective divisor on a toric germ in terms of its Newton polyhedron. In Sect. 3.8 we describe our work done in collaboration with Cueto on the local tropicalizations of a large and important class of surface singularities, Neumann and Wahl's *splice type singularities*. Finally, in Sect. 3.9 we present variants and extensions of the notion of local tropicalization explained in this paper, mentioning papers of Kajiwara, Payne, Ulirsch, Esterov, J. Giansiracusa, N. Giansiracusa and Lorscheid.

3.2 Motivating Considerations About the Cusp Singularity

In this section we illustrate the toric-geometrical interpretation of local tropicalization explained in Sect. 3.1 on the example of a plane cubic curve with a cusp singularity at the origin. For this purpose, we recall basic facts about toric surfaces, their fans and their orbits (see points (1)–(6) below).

Throughout the paper, we denote by $\boxed{Z(f)}$ the principal Cartier divisor defined by a rational function f on a normal complex analytic variety, or the germ of such a divisor defined by an element of a local ring of such a variety. We denote by $\boxed{\mathbb{C}^2_{x,y}}$ the complex affine plane with coordinates x and y.

Definition 3.2.1 A **modification** of an irreducible complex analytic variety X is a proper bimeromorphic morphism $\pi : X_\pi \to X$.

An **embedded resolution** of a curve contained in a smooth surface X is a modification π of the surface such that X_π is smooth, the total transform of the curve is a normal crossings divisor and π induces an isomorphism between the complement of the exceptional set of π in X_π and the complement of the singular locus of the curve in X. There exists a **minimal embedded resolution**, through which all other resolutions factor. It is obtained by blowing up successively the points of the curve and of its total transforms where one does not have locally normal crossings (see [50, Section 3.4]).

In order to illustrate this fact, let $\boxed{Y} \hookrightarrow \mathbb{C}^2_{x,y} =: \boxed{S_0}$ be the plane cubic curve $Z(f)$, where:

$$f(x, y) := y^2 - 2x^3 + x^2 y. \tag{3.1}$$

It has only one singular point, at the origin $\boxed{p_0} := (0, 0)$ of S_0, as may be seen by solving the system of equations $f = \partial_x f = \partial_y f = 0$. The germ of Y at p_0 is a **cusp singularity**. That is, it is formally isomorphic to the germ at the origin of the semicubical parabola $Z(u^2 - v^3) \hookrightarrow \mathbb{C}^2_{u,v}$. This fact may be shown explicitly by first completing the square in the polynomial (3.1):

$$f(x, y) = \left(y + \frac{x^2}{2}\right)^2 - 2x^3 \left(1 + \frac{x}{8}\right)$$

and by performing then the formal (even analytic) change of variables:

$$u := y + \frac{x^2}{2}, \quad v := 2^{1/3} x \left(1 + \frac{x}{8}\right)^{1/3},$$

in which the last cubic root is expanded using Newton's binomial series expansion formula for $(1 + t)^{1/3}$.

One may reach the minimal embedded resolution of Y by blowing up successively three points p_0, p_1, p_2 of Y and of its strict transforms, as illustrated on the upper part of Fig. 3.1. For each $i \in \{0, 1, 2\}$, denote by

$$\boxed{\pi_i} : S_{i+1} \to S_i$$

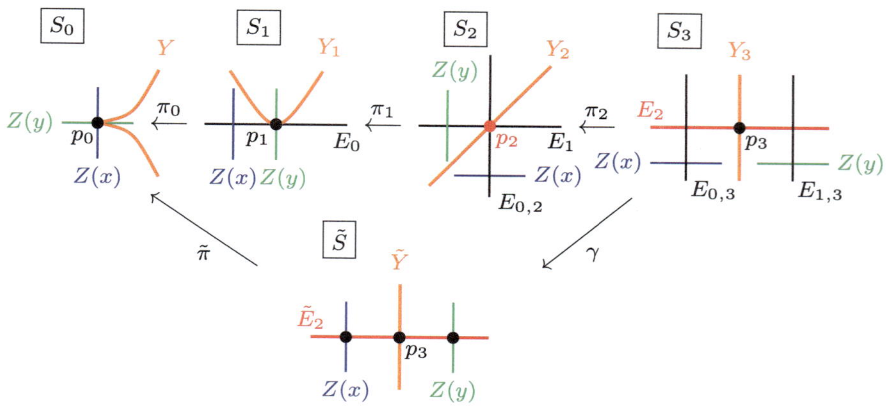

Fig. 3.1 The minimal embedded resolution $\pi_0 \circ \pi_1 \circ \pi_2$ and a toric partial resolution $\tilde{\pi}$ of the cuspidal cubic Y contained in the plane $S_0 = \mathbb{C}^2$. The exceptional curve created by blowing up the point p_i is denoted by E_i and its strict transform on the surface S_j is denoted by $E_{i,j}$. The strict transforms of $Z(x)$ and $Z(y)$ are still denoted by $Z(x)$ and $Z(y)$. The strict transform of Y on the surface S_j is denoted by Y_j and that on $\tilde{S}$ is denoted by $\tilde{Y}$

the blowup morphism of the surface $\boxed{S_i}$ at the point p_i and by $\boxed{E_i} \hookrightarrow S_{i+1}$ the exceptional divisor of π_i. It is a smooth rational curve of self-intersection -1. For each pair (i, j) such that $j \geq i+2$, we denote by $\boxed{E_{i,j}}$ the strict transform of E_i on the surface S_j and by $\boxed{Y_j}$ the strict transform of the curve Y on the same surface. For instance, $Y_0 = Y$.

Each time one blows up a smooth point of a compact curve contained in a smooth surface, the self-intersection number of its strict transform is one less than the initial self-intersection number (see [25, Lemma 4.3] or [50, Lemma 8.1.6]). Therefore, one has the following self-intersection numbers on the final surface S_3:

$$E_{0,3}^2 = -3, \ E_{1,3}^2 = -2, \ E_2^2 = -1.$$

As $E_{0,3}$ and $E_{1,3}$ have negative self-intersections, they may be contracted to normal singular points (see [25, Theorem 4.9]). Denote by $\boxed{\tilde{S}}$ the resulting surface and by

$$\boxed{\gamma} : S_3 \to \tilde{S}$$

the contraction morphism (see the right diagonal arrow of Fig. 3.1).

The composed morphism $\pi_0 \circ \pi_1 \circ \pi_2 : S_3 \to S_0$ may be factored as

$$\pi_0 \circ \pi_1 \circ \pi_2 = \tilde{\pi} \circ \gamma$$

where

$$\boxed{\tilde{\pi}} : \tilde{S} \to S_0$$

contracts the image $\boxed{\tilde{E}_2} := \gamma(E_2) \hookrightarrow \tilde{S}$ to the initial point p_0 (see the left diagonal arrow of Fig. 3.1). Note that the restriction of γ to the curve E_2 induces an isomorphism from E_2 to $\tilde{E}_2$. For this reason, we still denote by p_3 the image $\gamma(p_3)$.

In Fig. 3.1 we have also represented the axes of coordinates $Z(x)$, $Z(y) \hookrightarrow \mathbb{C}^2 = S_0$ and their strict transforms on the various surfaces appearing in the drawing. For simplicity, we denote those strict transforms in the same way. Note that the morphisms π_i blow up singular points of the total transforms of the divisor $Z(xy)$. This means that those morphisms are *toric*, if one endows the affine plane $\mathbb{C}^2$ with its standard toric structure with dense torus $(\mathbb{C}^*)^2$. As the morphisms π_i are also birational, it means that they may be described by successive subdivisions of **fans**, that is, finite sets of strictly convex rational cones inside the **weight plane** $\mathbb{R}^2$ of the toric surface $\mathbb{C}^2$, closed under the operation of taking faces and such that the intersection of any two of its cones is a common face. The cones of such a fan may be of dimension *two*, of dimension *one*—in which case they are **rays** spanned by well-defined **primitive vectors** of $\mathbb{Z}^2$, that is, vectors of coprime coordinates—or of dimension *zero*—that is, equal to the origin of the weight plane. A two-dimensional cone has three proper faces—two rays and the origin—while a one-dimensional cone has only one proper face, namely, the origin.

Let us recall in which way a fan $\mathcal{F}$ contained in the weight plane $\mathbb{R}^2$ encodes a **toric surface** $\boxed{\mathcal{X}_{\mathcal{F}}}$ (see [7, 13, 33] or [14, Section 1.3]):

1. There is an inclusion-preserving bijection between the cones τ of the fan and the affine open toric subsurfaces $\boxed{\mathcal{X}_\tau}$ of $\mathcal{X}_{\mathcal{F}}$. The set of $\mathbb{C}$-valued points of X_τ is the set of all *monoid* homomorphisms from $\tau^\vee \cap \mathbb{Z}^2$ to $\mathbb{C}$, where $\tau^\vee$ is the dual cone of τ (see Sect. 3.4) and $\mathbb{C}$ is viewed here as a *multiplicative* monoid. Each such surface $\mathcal{X}_\tau$ contains a unique closed orbit $\boxed{O_\tau}$, which is canonically a complex algebraic torus of dimension $2 - \dim \tau$. It is also the unique smallest dimensional orbit of $\mathcal{X}_\tau$. As a set, O_τ consists of those monoid homomorphisms which send $(\tau^\vee \setminus \tau^\perp) \cap \mathbb{Z}^2$ to $0 \in \mathbb{C}$, where $\tau^\perp$ is the orthogonal complement (or the annihilator) of τ in the dual vector space.
2. To the trivial cone 0 reduced to the origin of the weight plane $\mathbb{R}^2$ corresponds an algebraic torus $\mathcal{X}_0$ isomorphic to $(\mathbb{C}^*)^2$. It is Zariski-dense in $\mathcal{X}_{\mathcal{F}}$.
3. To a one-dimensional cone, that is, a ray $\mathbb{R}_{\geq 0}w$ spanned by a non-zero weight vector w with integral coordinates, corresponds a toric affine open surface $\mathcal{X}_{\mathbb{R}_{\geq 0}w}$ isomorphic to $\mathbb{C} \times \mathbb{C}^*$.
4. To a **regular** two-dimensional cone, that is, to a cone spanned by a basis of the weight lattice $\mathbb{Z}^2$, corresponds an affine open surface isomorphic to $\mathbb{C}^2$.
5. To a two-dimensional cone which is not regular corresponds an affine open surface which is normal and has one singular point, namely, its 0-dimensional orbit.

6. If $(p, q) \in \mathbb{Z}^2$ is an integer weight vector, then its corresponding **one parameter subgroup**

$$\begin{array}{rcl} \mathbb{C}^* & \to & (\mathbb{C}^*)^2 \\ t & \mapsto & (t^p, t^q) \end{array}$$

has a limit inside $\mathcal{X}_{\mathcal{F}}$ when $t \to 0$ if and only if (p, q) belongs to the **support** $\boxed{|\mathcal{F}|}$ of $\mathcal{F}$, that is, to the union of its cones. This limit is the identity element of the orbit O_τ seen as a group, where τ is the unique cone of $\mathcal{F}$ whose relative interior contains (p, q).

Note that if τ is a strictly convex rational cone of the weight plane $\mathbb{R}^2$ and $\mathcal{F}$ is the fan consisting of its faces, then one has $\mathcal{X}_{\mathcal{F}} = \mathcal{X}_\tau$.

The process of successive subdivisions of fans corresponding to the toric morphisms of Fig. 3.1 is represented in Fig. 3.2. The starting fan $\boxed{\mathcal{F}_0}$, corresponding to the surface $S_0 = \mathbb{C}^2$, consists of the non-negative quadrant $\boxed{\sigma_0}$ of the weight plane $\mathbb{R}^2$ of $\mathbb{C}^2$ and of its faces. Near each ray τ we write the name of the irreducible component of the total transform of $Z(xy)$ which is the closure of the corresponding one-dimensional orbit O_τ. That irreducible component is isomorphic either to $\mathbb{C}$ (in the case of the strict transforms of $Z(x)$ and $Z(y)$) or to $\mathbb{CP}^1$ (in the case of an exceptional divisor $E_{i,j}$). On the upper horizontal line of Fig. 3.2, each blowup is represented combinatorially by the subdivision of a regular cone τ along a ray. Inside each such cone τ we indicate the name of the associated point, which is the

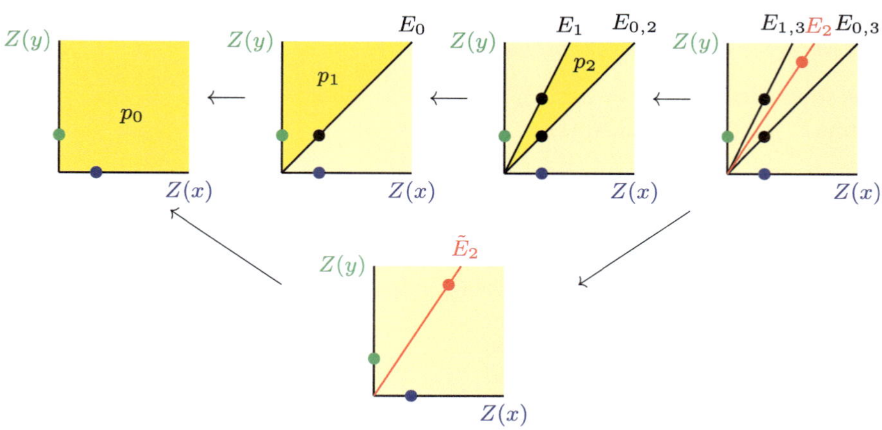

Fig. 3.2 The fans corresponding to the toric surfaces of Fig. 3.1 and their subdivisions corresponding to the toric morphisms of the same figure. We write the name of the corresponding irreducible component of the total transform of $Z(xy) \hookrightarrow S_0$. Inside each regular two-dimensional cone which gets subdivided, we indicate the name of the corresponding 0-dimensional orbit. The right diagonal arrow is the toric modification which blows down the orbit closures $E_{0,3}$ and $E_{1,3}$. The left diagonal arrow blows then down the orbit closure $\tilde{E}_2$

0-dimensional orbit of $\mathcal{X}_\tau$. The subdividing ray is spanned by the sum $w_i + w_j$ of the primitive weight vectors w_i and w_j spanning the edges of τ. All those primitive vectors are drawn as small dots.

Let us consider inside each surface S_1, S_2, S_3 appearing in the upper part of Fig. 3.1 the affine toric open set $\mathcal{X}_\tau$ isomorphic to $\mathbb{C} \times \mathbb{C}^*$ which corresponds to the ray τ which was added to get the surface. The strict transforms of the curve Y inside those affine toric surfaces $\mathcal{X}_\tau$ are the intersections of $\mathcal{X}_\tau$ with the curves Y_1, Y_2, Y_3, that is, $Y_1 \setminus p_1, Y_2 \setminus p_2$ and Y_3 respectively. Indeed, only Y_3 does not pass through a singular point of the total transform of the divisor $Z(xy)$, that is, through a 0-dimensional orbit of the ambient toric surface, and $\mathbb{C} \times \mathbb{C}^*$ does not contain such an orbit. Restricting the toric birational morphisms $S_i \to S_0$ to those three strict transforms, we get the restrictions:

$$\pi_1| : Y_1 \setminus p_1 \to Y, \;\; \pi_1 \circ \pi_2| : Y_2 \setminus p_2 \to Y, \;\; \pi_1 \circ \pi_2 \circ \pi_3| : Y_3 \to Y.$$

Only the last morphism is *proper*, similarly to what happens for the morphism $\tilde{\pi}$: $\tilde{S} \to S_0$ (see the bottom of Fig. 3.1). The curve Y_3 is the strict transform of Y in the affine toric surface $\mathcal{X}_{\mathbb{R}_{\geq 0}(2,3)}$ determined by the ray $\mathbb{R}_{\geq 0}(2, 3)$ spanned by the weight vector $(2, 3)$. If we want to also turn the first two morphisms into proper ones, we need to add the orbits p_1 and p_2 respectively, that is, to look inside the affine toric surfaces $\mathcal{X}_\tau$, where τ is one of the two-dimensional regular cones decorated by p_1 and p_2 in Fig. 3.2. Note that both cones contain the ray $\mathcal{X}_{\mathbb{R}_{\geq 0}(2,3)}$.

In fact, *for every fan $\mathcal{F}$ whose support is contained in σ_0, the restriction $\pi : Y_\pi \to Y$ of the toric morphism $\pi_\mathcal{F} : \mathcal{X}_\mathcal{F} \to \mathcal{X}_{\sigma_0} = \mathbb{C}^2$ to the strict transform Y_π in $\mathcal{X}_\mathcal{F}$ is proper if and only if $|\mathcal{F}|$ contains the ray* $\mathbb{R}_{\geq 0}(2, 3)$. Therefore, according to the toric-geometrical interpretation of local tropicalization explained in Sect. 3.1, *this ray is the local tropicalization of the germ* (Y, p_0).

Question How is the ray $\mathbb{R}_{\geq 0}(2, 3)$ related to the defining series $f(x, y)$ of the germ (Y, p_0) (see formula (3.1))?

Answer *It is the ray orthogonal to the unique compact edge of the Newton polygon of f.*

Proposition 3.3.12 below states the analogous result for all plane curve singularities without branches contained in the coordinate axes.

3.3 The Newton Polygon of a Plane Curve Singularity

In this section we relate the two viewpoints on local tropicalization explained in the introduction—via weight vectors of parametrizations and toric geometry—in the case of plane curve singularities $Y \hookrightarrow (\mathbb{C}^2, 0)$ defined by series $f(x, y) \in \mathbb{C}[[x, y]]$ which are divisible neither by x nor by y. The object allowing to make this connection precise is the *Newton polygon* $\mathcal{N}(f)$ of f (see Definition 3.3.4,

Corollary 3.3.10 and Proposition 3.3.12). We end the section by introducing a third viewpoint on the local tropicalization, using the *initial forms* $\text{in}_{(p,q)} f$ of f (see Corollary 3.3.14).

Consider a **plane curve singularity** $Y \hookrightarrow (\mathbb{C}^2, 0)$, which for us means a reduced formal germ of dimension one. Since the local ring $\mathbb{C}[[x, y]]$ is factorial, we know that Y is a principal divisor, that is, there exists $f(x, y) \in \mathbb{C}[[x, y]] \setminus \{0\}$ satisfying $f(0, 0) = 0$, such that $Y = Z(f)$. The series $f(x, y)$ is well-defined up to multiplication by a unit of $\mathbb{C}[[x, y]]$, that is, by a series with non-zero constant term. The germ Y consists of a finite number of **branches**, that is, irreducible germs of curves.

We assume in the sequel that *no branch of Y is contained in the union of coordinate axes $Z(x)$ and $Z(y)$ of $\mathbb{C}^2$*, which amounts to say that f is not divisible by x or y. Note that in terms of Definition 3.4.1 below, this means that Y is an *internal subtoric germ* of $(\mathbb{C}^2, 0)$.

A basic property of plane *branches* B is that their normalizations are *smooth*, which means that they are described by parametrizations $t \mapsto (x(t), y(t))$, with $x(t), y(t) \in \mathbb{C}[[t]]$. Special kinds of such normalizing parametrizations may be obtained by the classical iterative method of Newton (see [16, pages 59-60] or [14, Section 1.2.6]). They satisfy the property that $x(t) = t^n$, where $n \in \mathbb{Z}_{>0}$ is the intersection number at 0 of B with the branch $(Z(x), 0)$. Each step of Newton's method uses a suitable *Newton polygon*.

Let us explain how Newton polygons pop up naturally in this context. Instead of looking for *Newton-Puiseux parametrizations* of the form $t \mapsto (t^n, y(t))$, we will consider an arbitrary parametrization $t \mapsto (x(t), y(t))$ of a branch of $Y = Z(f)$ and we will look for constraints on its *initial terms*. As no branch of Y is contained in the coordinate axes of $\mathbb{C}^2$, both series $x(t)$ and $y(t)$ are non-zero. We have the relation

$$f(x(t), y(t)) = 0 \tag{3.2}$$

in the ring $\mathbb{C}[[t]]$. Let us write

$$x(t) = \sum_{i=p}^{\infty} \boxed{x_i}\, t^i, \quad y(t) = \sum_{j=q}^{\infty} \boxed{y_j}\, t^j, \quad x_i, y_j \in \mathbb{C}, \tag{3.3}$$

where $p > 0, q > 0, x_p \neq 0, y_q \neq 0$. Therefore $x_p t^p$ and $y_q t^q$ are the **initial terms** of $x(t)$ and $y(t)$.

Definition 3.3.1 The **support** $\boxed{\text{supp}\, f}$ of $f \in \mathbb{C}[[x, y]] \setminus \{0\}$ is the set of exponents $(\alpha, \beta) \in \mathbb{Z}^2_{\geq 0}$ such that the coefficient $f_{\alpha,\beta}$ of the monomial $x^\alpha y^\beta$ in the expansion of f does not vanish. With this notation we can write

$$f(x, y) = \sum_{(\alpha,\beta) \in \text{ supp } f} \boxed{f_{\alpha,\beta}}\, x^\alpha y^\beta, \quad f_{\alpha,\beta} \in \mathbb{C}^* \text{ for all } (\alpha, \beta) \in \text{supp } f. \tag{3.4}$$

Substituting (3.3) into (3.4) and rearranging the terms, the identity (3.2) becomes

$$\left(\sum_{(\alpha,\beta)\in \delta(p,q)} f_{\alpha,\beta}\, x_p^{\alpha} y_q^{\beta} \right) \cdot t^{m(p,q)} + \sum_{m>m(p,q)} d_m t^m = 0, \tag{3.5}$$

for some coefficients $d_m \in \mathbb{C}$. The set $\delta(p,q) \subseteq \operatorname{supp} f$ and the positive integer $m(p,q)$ appearing in (3.5) are defined as follows, recalling that σ_0 denotes the non-negative quadrant of $\mathbb{R}^2$:

Definition 3.3.2 Consider $(p,q) \in \sigma_0^\circ \cap \mathbb{Z}^2 = \mathbb{Z}^2_{>0}$ and $f \in \mathbb{C}[[x,y]] \setminus \{0\}$. The **basis** $\boxed{\delta_{(p,q)}(f)}$ **relative to** (p,q) of the support $\operatorname{supp} f$ of f is the subset of $\operatorname{supp} f$ on which the linear function

$$\begin{array}{rccc} \boxed{(p,q)}: & \mathbb{R}^2 & \to & \mathbb{R}, \\ & (\alpha,\beta) & \mapsto & p\alpha + q\beta \end{array}$$

achieves its minimum $\boxed{m(p,q)} \in \mathbb{Z}_{\geq 0}$ when restricted to $\operatorname{supp} f$.

As (α,β) is the *vector of exponents* of the monomial $x^\alpha y^\beta$, the pair (p,q) may be interpreted as a *weight vector* relative to the variables (x,y). We chose the name of *basis of* $\operatorname{supp} f$ for $\delta_{(p,q)}(f)$ as we think about it as the basis on which $\operatorname{supp} f$ lies when it is studied using the *height* function $(p,q) : \mathbb{R}^2 \to \mathbb{R}$.

The following classical result explains how convex geometry introduces itself at this precise moment:

Proposition 3.3.3 *Let F be a finite subset of the real vector space V and let $\ell : V \to \mathbb{R}$ be a linear form. Then, the minimum of the restriction of ℓ to F is achieved precisely on the intersection of F with a face of its own convex hull. Moreover, all the faces of the convex hull of F appear when ℓ varies in the dual vector space of V.*

For a summary of facts from convex geometry used in the theory of toric varieties see Oda's [33, Appendix].

In our context, we have to apply Proposition 3.3.3 to $F := \operatorname{supp} f$. As f is a series, its support $\operatorname{supp} f$ may be *infinite*. Nevertheless, it is easy to see that if $p > 0, q > 0$, that is, if (p,q) varies in the interior σ_0° of the non-negative quadrant σ_0 of the plane $\mathbb{R}^2$ of weight vectors, then the restriction of the linear form (p,q) of Definition 3.3.2 to $\operatorname{supp} f$ achieves its minimum on a *finite* set. Indeed, the possible such sets are the intersections of $\operatorname{supp} f$ with the *compact* faces of a variant of the convex hull of $\operatorname{supp} f$, which is called the *Newton polygon* of f. This polygon is defined as follows:

Definition 3.3.4 The **Newton polygon** $\boxed{\mathcal{N}(f)}$ **of the series** $f \in \mathbb{C}[[x, y]] \setminus \{0\}$ is the convex hull in $\mathbb{R}^2$ of the set

$$\text{supp } f + \mathbb{R}^2_{\geq 0} = \bigcup_{(\alpha,\beta)\in \text{ supp } f} \left((\alpha, \beta) + \mathbb{R}^2_{\geq 0}\right).$$

The sum of the left-hand side of the previous equality is the so-called **Minkowski sum** of subsets of a real vector space, defined by:

$$A + B := \{a + b,\ a \in A, b \in B\}.$$

Our assumption that f is not divisible by x or y means that its Newton polygon $\mathcal{N}(f)$ intersects both axes of the plane of exponents of monomials. Its compact faces are either *vertices* or *edges*.

Example 3.3.5 Let us consider again the series $f(x, y) := y^2 - 2x^3 + x^2y$ of formula (3.1). Figure 3.3 shows the Minkowski sum supp $f + \mathbb{R}^2_{\geq 0}$ of Definition 3.3.4 and its convex hull, which is by definition the Newton polygon $\mathcal{N}(f)$.

Remark 3.3.6 Assume that $f \in \mathbb{C}[x, y]$, as in Example 3.3.5. If instead of weight vectors with positive coordinates we had considered arbitrary non-zero vectors (p, q) of the weight plane then, seen as linear forms on the plane of exponents of monomials, they would have achieved their minima on the intersections of supp f with the faces of the *convex hull of this support*. This convex hull—a triangle in Example 3.3.5—is by definition the *global Newton polygon* of the polynomial $f(x, y)$, which gives information about the parametric curves contained in $Z(f)$ which are described by *Laurent power series* (with possibly negative exponents) instead of power series with non-negative exponents. In this paper we are concerned only with parametric curves *passing through the origin of* $\mathbb{C}^2$ *or, more generally, of* $\mathbb{C}^n$. That is why we define the Newton polygon of f as in Definition 3.3.4. In order to differentiate it from the convex hull of the support, one may call it the *local Newton polygon*. Note that the term *Newton diagram* is also met in the literature,

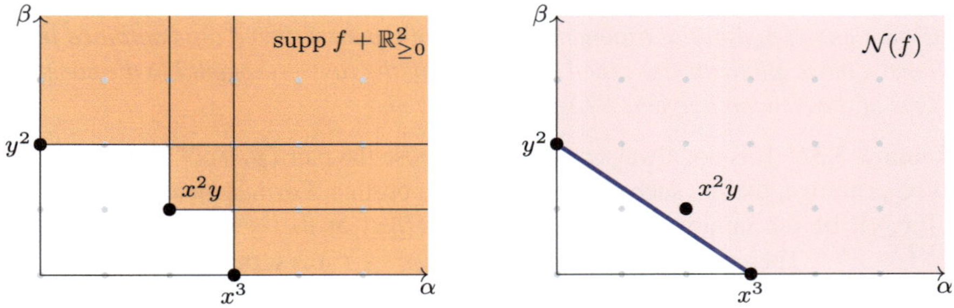

Fig. 3.3 The Minkowski sum supp $f + \mathbb{R}^2_{\geq 0}$ of Definition 3.3.4 (on the left) and the Newton polygon $\mathcal{N}(f)$ of the series $f(x, y) = y^2 - 2x^3 + x^2y$ (on the right)

sometimes meaning the strict subset of the local Newton polygon which is the union of its compact faces (see for instance [46]).

The map associating to each non-zero series its Newton polygon behaves like the logarithm function, in the sense that it transforms products into Minkowski sums (see [14, Proposition 1.6.13]):

Proposition 3.3.7 *Assume that* $f, g \in \mathbb{C}[[x, y]] \setminus \{0\}$. *Then:*

$$\mathcal{N}(fg) = \mathcal{N}(f) + \mathcal{N}(g).$$

Proposition 3.3.7 implies that $\mathcal{N}(fg) = \mathcal{N}(f)$ whenever g is a unit of $\mathbb{C}[[x, y]]$. Therefore the Newton polygon does not depend of the defining series of a given plane curve singularity $Y \hookrightarrow \mathbb{C}^2$. This allows to simply speak of the **Newton polygon** $\boxed{\mathcal{N}(Y)}$ **of** Y. Even if we drop it for simplicity from the notation, it is not an invariant of the abstract germ Y, since it depends strongly on the chosen embedding of Y in $(\mathbb{C}^2, 0)$ (see [35] or [15]).

Let us come back to the equality (3.5). For this equality to be possible, it is necessary that *the coefficient*

$$\boxed{d_{m(p,q)}} := \sum_{(\alpha,\beta)\in\, \delta(p,q)} f_{\alpha,\beta}\, x_p^{\alpha} y_q^{\beta}$$

of $t^{m(p,q)}$ *contains at least two summands*. That is, *the face* $\delta(p, q)$ *must be a compact edge of* $\mathcal{N}(f)$ *and not a vertex*. If this is the case, then, since $\mathbb{C}$ is algebraically closed, it is indeed possible to choose the initial coefficients in the series $x(t)$ and $y(t)$ so that $d_{m(p,q)} = 0$. Moreover, as shown for instance in [49, Section IV.3.2] or [10, Section 2.1], one can organize in this way an iterative process which computes parametrizations of all the branches of $Z(f)$.

To say that $(p, q) \in \sigma_0^\circ$ *achieves its minimum* on the edge $\delta(p, q)$ of $\mathcal{N}(f)$ means that the ray $\mathbb{R}_{\geq 0}(p, q)$ is *orthogonal* to $\delta(p, q)$. We may thus summarize the previous explanations as follows:

Theorem 3.3.8 *For a given plane curve singularity* $Y \hookrightarrow (\mathbb{C}^2, 0)$ *without branches contained in the coordinate axes, there exists only a finite number of possibilities for the quotient* q/p *of initial exponents* p *and* q *of a parametrized arc contained in* Y. *Namely, those quotients are exactly the slopes of the rays orthogonal to the compact edges of the Newton polygon* $\mathcal{N}(Y)$ *of* Y.

Remark 3.3.9 It is not always possible to choose the pair (p, q) of Theorem 3.3.8 to be primitive, that is, such that p and q are coprime. For instance, let $f(x, y) \in \mathbb{C}[[x, y]]$ be the minimal polynomial in $\mathbb{C}[[x]][y]$ of the Newton-Puiseux series $x^{3/2} + x^{7/4}$. Therefore, $(x(t) := t^4,\ y(t) := t^6 + t^7)$ is a parametrization of the branch $Y \hookrightarrow \mathbb{C}^2$ defined by f. The weight vector $(4, 6)$ of initial exponents of the series $(x(t), y(t))$ is not primitive and $(2, 3)$ is the unique primitive weight vector with positive coordinates which is proportional to it. Assume that a pair

$(x_1(t), y_1(t)) \in \mathbb{C}[[t]]$ has $(2, 3)$ as weight vector of initial exponents. Then, one may perform a change of variable, replacing t by $u \in \mathbb{C}[[t]]$ such that $x_1(t) = u^2$. The series $y_1(t)$ becomes a series $y_2(u) \in \mathbb{C}[[u]]$. The pair $(u^2, y_2(u))$ parametrizes the same branch $B \hookrightarrow \mathbb{C}^2$ as $(x_1(t), y_1(t))$ therefore, as 2 is a prime number, this branch has at most one *characteristic exponent* (recall that the *characteristic exponents of a Newton-Puiseux series* are those exponents of its support which cannot be written as a fraction whose denominator is the lowest common denominator of the exponents strictly lower than them, see [14, Definition 1.6.2]). This shows that B cannot be equal to the branch Y, which has two characteristic exponents, namely $3/2$ and $7/4$.

Using the first definition of the *local tropicalization* of $Y \hookrightarrow (\mathbb{C}^2, 0)$ given in the introduction, in terms of exponent vectors of arcs contained in Y, Theorem 3.3.8 implies:

Corollary 3.3.10 *Let $Y \hookrightarrow (\mathbb{C}^2, 0)$ be a plane curve singularity without branches contained in the coordinate axes. Then its local tropicalization is the union of the rays orthogonal to the compact edges of the Newton polygon $\mathcal{N}(Y)$ of Y.*

The local tropicalization of the series of Example 3.3.5 is represented in Fig. 3.4. One recognizes the diagram of the bottom of Fig. 3.2. This is not a coincidence. Indeed, one has the following property, which is illustrated in Figs. 3.1 and 3.2:

Proposition 3.3.11 *Let $Y \hookrightarrow (\mathbb{C}^2, 0)$ be a plane curve singularity without branches contained in the coordinate axes. Let $\mathcal{F}$ be a fan with support contained in the non-negative quadrant σ_0 of the weight plane $\mathbb{R}^2$. Consider the strict transform $Y_{\mathcal{F}}$ of Y by the toric morphism $\pi_{\mathcal{F}} : \mathcal{X}_{\mathcal{F}} \to \mathbb{C}^2 = \mathcal{X}_{\sigma_0}$. Then, $Y_{\mathcal{F}}$ avoids the 0-dimensional orbits of the toric surface $\mathcal{X}_{\mathcal{F}}$ if and only if the rays orthogonal to the compact edges of the Newton polygon $\mathcal{N}(Y)$ of Y are cones of the fan $\mathcal{F}$.*

Proof We add cones to the fan $\mathcal{F}$ until we obtain a fan $\mathcal{F}'$ whose support is σ_0. Then, we subdivide the two-dimensional cones of $\mathcal{F}'$ using the rays orthogonal to the compact edges of the Newton polygon $\mathcal{N}(Y)$ of Y, obtaining a fan $\hat{\mathcal{F}}$. By Barroso et al. [14, Proposition 1.4.18], the strict transform $Y_{\hat{\mathcal{F}}}$ of Y by the toric morphism $\pi_{\hat{\mathcal{F}}} : \mathcal{X}_{\hat{\mathcal{F}}} \to \mathbb{C}^2 = \mathcal{X}_{\sigma_0}$ intersects the toric boundary $\partial\mathcal{X}_{\hat{\mathcal{F}}}$ only at points of the orbits of dimension one which correspond to the rays orthogonal to the compact edges of

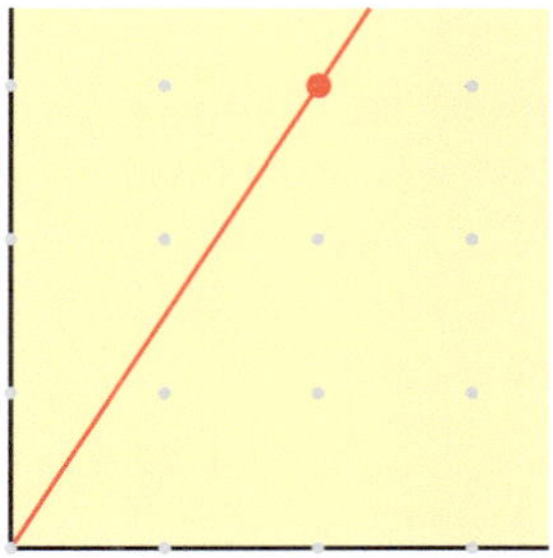

Fig. 3.4 The local tropicalization of the series $f(x, y) = y^2 - 2x^3 + x^2y$ is the ray spanned by the primitive vector (2, 3), which is indicated with a filled dot

$\mathcal{N}(Y)$. As a consequence, the strict transform $Y_{\mathcal{F}}$ intersects the toric boundary $\partial \mathcal{X}_{\mathcal{F}}$ only at points of the images of those orbits by the toric morphism $\hat{\pi} : \mathcal{X}_{\hat{\mathcal{F}}} \to \mathcal{X}_{\mathcal{F}}$. Those images are precisely the orbits of $\mathcal{X}_{\mathcal{F}}$ which correspond to the cones σ of $\mathcal{F}$ whose relative interiors contain rays orthogonal to the compact edges of $\mathcal{N}(Y)$. To say that no such orbit is of dimension 0 means that no such cone σ is of dimension two. This is equivalent to the fact that the rays orthogonal to the compact edges of the Newton polygon of Y are cones of the fan $\mathcal{F}$. □

The previous proof yields also the following result:

Proposition 3.3.12 *Let $Y \hookrightarrow (\mathbb{C}^2, 0)$ be a plane curve singularity without branches contained in the coordinate axes. Let $\mathcal{F}$ be a fan with support contained in the non-negative quadrant σ_0 of the weight plane $\mathbb{R}^2$. Consider the strict transform $Y_{\mathcal{F}}$ of Y by the toric morphism $\pi_{\mathcal{F}} : \mathcal{X}_{\mathcal{F}} \to \mathbb{C}^2 = \mathcal{X}_{\sigma_0}$. Then the restriction $\pi : Y_{\mathcal{F}} \to Y$ of $\pi_{\mathcal{F}}$ to $Y_{\mathcal{F}}$ is proper if and only if the support of $\mathcal{F}$ contains the rays orthogonal to the compact edges of the Newton polygon $\mathcal{N}(Y)$ of Y.*

This proves the equivalence of the two characterizations of local tropicalization given in Sect. 3.1 in the case of plane curve singularitues $Y \hookrightarrow (\mathbb{C}^2, 0)$ without branches contained in the coordinate axes.

We have worked above with initial terms of univariate series (see the equalities (3.3)). There is a corresponding notion for bivariate series, this time *relative to a weight vector*:

Definition 3.3.13 Let $(p, q) \in \sigma_0^\circ \cap \mathbb{Z}^2 = \mathbb{Z}^2_{>0}$ and $f \in \mathbb{C}[[x, y]] \setminus \{0\}$. The (p, q)-**initial form** of the series f is the (p, q)-weighted homogeneous polynomial

$$\boxed{\text{in}_{(p,q)} f} := \sum_{(\alpha,\beta)\in\delta_{(p,q)}(f)} f_{\alpha,\beta}\, x^\alpha y^\beta \quad \in \mathbb{C}[x, y].$$

Here $\delta_{(p,q)}(f) \subseteq \text{supp}\, f$ denotes the basis of $\text{supp}\, f$ relative to (p, q), introduced in Definition 3.3.2.

This notion allows to rephrase Theorem 3.3.8 using a defining series of the given plane curve singularity, instead of parametrized arcs contained in it:

Corollary 3.3.14 *Let $(p, q) \in \mathbb{Z}^2_{>0}$ and $f \in \mathbb{C}[[x, y]] \setminus \{0\}$. Then, (p, q) belongs to the local tropicalization of $Z(f) \hookrightarrow (\mathbb{C}^2, 0)$ if and only if the initial form* $\text{in}_{(p,q)} f$ *is not a monomial.*

Proof By Corollary 3.3.10, (p, q) belongs to the local tropicalization of $Z(f) \hookrightarrow (\mathbb{C}^2, 0)$ if and only if (p, q) is orthogonal to a compact edge of the Newton polygon $\mathcal{N}(f)$, that is, if and only if the basis $\delta_{(p,q)}(f)$ has at least two elements. In turn, by Definition 3.3.13, this is equivalent to the fact that the initial form $\text{in}_{(p,q)} f$ is not a monomial. □

Corollary 3.3.14 may be turned into a third definition of the local tropicalization of $Z(f) \hookrightarrow (\mathbb{C}^2, 0)$, as the union of rays generated by the vectors $(p, q) \in \mathbb{Z}^2_{>0}$

such that the initial form $\mathrm{in}_{(p,q)} f$ *is not a monomial.* We will see below that a slight modification of this definition may be given for germs of higher dimension embedded in germs of toric varieties (see Definition 3.5.7) and that it is equivalent to the two definitions presented in the introduction (see Theorem 3.6.2). Before that, we need to explain our notations from toric geometry and valuation theory in arbitrary dimensions.

3.4 Basic Facts About Toric Geometry and Valuation Theory

In this section we explain elementary facts as well as our notations and conventions about *toric geometry* and *valuation theory*. A basic relation between the two is that any *subtoric arc* in the sense of Definition 3.4.3 defines a *semivaluation* (see Example 3.4.7).

Let us start with toric geometry. Here we work in arbitrary dimension, by contrast with Sects. 3.2 and 3.3, in which we considered only toric surfaces. The readers needing more details may consult the textbooks [7, 13, 33]. Those who want quick introductions to the subject may consult [2, 6] or [14, Section 1.3].

From now on, we replace the field $\mathbb{C}$ of the previous sections with an arbitrary algebraically closed field $\boxed{\mathbb{K}}$. Let $\boxed{N}$ and $\boxed{M}$ be two **lattices**, that is, two free abelian groups of finite rank. We assume that they are **dual** of each other, which means that we fix a unimodular bilinear pairing

$$\begin{aligned} N \times M &\to \quad \mathbb{Z} \\ (w, m) &\mapsto \boxed{w \cdot m} \end{aligned}.$$

This pairing produces identifications

$$M = \mathrm{Hom}(N, \mathbb{Z}), \;\; N = \mathrm{Hom}(M, \mathbb{Z}).$$

We interpret the two lattices differently: M is the **lattice of exponents of monomials** and N is the corresponding **weight lattice**. That is, each vector $m \in M$ gives rise to a **monomial** $\boxed{\chi^m} \in \mathbb{K}[M]$, where $\boxed{\mathbb{K}[M]}$ denotes the $\mathbb{K}$-algebra of the lattice M, and each vector $w \in N$ attributes the **weight** $w \cdot m \in \mathbb{Z}$ to the monomial χ^m.

Once a basis $(e_1, \ldots, e_n)$ of M is fixed, one gets a dual basis $(e_1^\vee, \ldots, e_n^\vee)$ of N. One may write then $m = m(1)e_1 + \cdots + m(n)e_n$ and $w = w(1)e_1^\vee + \cdots + w(n)e_n^\vee$, where all the coefficients $m(i)$ and $w(i)$ are integers. This allows to write $\chi^m = x_1^{m(1)} \cdots x_n^{m(n)}$, where $\boxed{x_j} := \chi^{e_j}$ for every $j \in \{1, \ldots, n\}$ and w endows the variable x_j with the weight $w(j)$. That is, $\mathbb{K}[M]$ is an intrinsic way of writing the $\mathbb{K}$-algebra of **Laurent polynomials** in n variables and N is an intrinsic way of looking at all the ways to endow the variables of such polynomials with integer weights.

Let τ be a **strictly convex rational polyhedral cone** inside the real vector space $\boxed{N_{\mathbb{R}}} := N \otimes_{\mathbb{Z}} \mathbb{R}$ generated by N. That is, τ is a cone $\boxed{\mathbb{R}_{\geq 0}\langle w_1, \ldots, w_k\rangle}$ generated by a finite set of vectors $w_j \in N$ and it is assumed not to contain vector subspaces of positive dimension. We denote by $\boxed{\tau^\circ}$ the interior of τ in the topological space $N_{\mathbb{R}}$ and by

$$\boxed{\tau^\vee} := \{m \in M_{\mathbb{R}},\ w \cdot m \geq 0\ \forall\, w \in \tau\} \subseteq M_{\mathbb{R}}$$

its **dual cone**. The intersection $\tau^\vee \cap M$ is a submonoid of $(M, +)$, generating a subalgebra $\mathbb{K}[\tau^\vee \cap M]$ of $\mathbb{K}[M]$. Its spectrum is by definition the **affine toric variety** $\mathcal{X}_\tau$ **defined by the cone** τ **over the field** $\mathbb{K}$:

$$\boxed{\mathcal{X}_\tau} := \mathrm{Spec}(\mathbb{K}[\tau^\vee \cap M]).$$

If τ is the origin 0 of $N_{\mathbb{R}}$, one gets therefore $\mathcal{X}_0 = \mathrm{Spec}(\mathbb{K}[M])$. This space is also denoted by $\boxed{T_N}$ and is called the **algebraic torus** defined over $\mathbb{K}$ and corresponding to the weight lattice N.

The set $\boxed{\mathcal{X}_\tau(\mathbb{K})}$ of $\mathbb{K}$**-valued points of** $\mathcal{X}_\tau$ is identified naturally with

$$\mathrm{Hom}((\tau^\vee \cap M, +), (\mathbb{K}, \cdot)),$$

where the morphisms are taken in the category of morphisms of monoids. This identification is obtained by associating to each $\mathbb{K}$-valued point $p \in \mathcal{X}_\tau(\mathbb{K})$ the evaluation map which sends every $m \in \tau^\vee \cap M$ to $\chi^m(p) \in \mathbb{K}$. A basis $(e_1, \ldots, e_n)$ of M allows to identify the set $T_N(\mathbb{K})$ of $\mathbb{K}$-valued points of T_N with $(\mathbb{K}^*)^n$.

The variety $\mathcal{X}_\tau$ is *smooth* if and only if the cone τ is **regular**, that is, $\tau = \mathbb{R}_{\geq 0}\langle w_1, \ldots, w_k\rangle$, where $(w_1, \ldots, w_k)$ may be extended to a basis of the lattice N. A choice of such an extension identifies $\mathcal{X}_\tau(\mathbb{K})$ with $\mathbb{K}^k \times (\mathbb{K}^*)^{n-k}$. Particular cases of this fact were listed as points (2), (3) and (4) of Sect. 3.2.

A **fan** is a finite set $\mathcal{F}$ of strictly convex rational polyhedral subcones of the weight vector space $N_{\mathbb{R}}$, satisfying the following properties:

- if $\tau \in \mathcal{F}$, then all the faces of τ also belong to $\mathcal{F}$;
- if $\tau_1, \tau_2 \in \mathcal{F}$, then $\tau_1 \cap \tau_2$ is a face of both τ_1 and τ_2.

The **support** $\boxed{|\mathcal{F}|} \subseteq N_{\mathbb{R}}$ of the fan $\mathcal{F}$ is the union of its cones. All the fans of Fig. 3.2 have the same support, equal to the non-negative quadrant σ_0 of $\mathbb{R}^2$.

Every fan $\mathcal{F}$ determines a toric variety $\boxed{\mathcal{X}_{\mathcal{F}}}$, obtained by gluing the affine toric varieties $\mathcal{X}_\tau$ defined by the cones τ of $\mathcal{F}$. Each variety $\mathcal{X}_\tau$ becomes a Zariski-open subvariety of $\mathcal{X}_{\mathcal{F}}$. The torus $T_N = \mathcal{X}_0$ acts on $\mathcal{X}_{\mathcal{F}}$ by an extension of the canonical translational action on itself. This action partitions $\mathcal{X}_{\mathcal{F}}$ into orbits. Each open subset $\mathcal{X}_\tau$, for $\tau \in \mathcal{F}$, contains a unique orbit $\boxed{O_\tau}$ which is closed in $\mathcal{X}_\tau$. It is also its unique smallest-dimensional orbit, and one has the equality:

$$\dim(O_\tau) + \dim(\tau) = \mathrm{rk}(N).$$

This generalizes point (1) of Sect. 3.2. The **toric boundary** $\boxed{\partial \mathcal{X}_\tau}$ of $\mathcal{X}_\tau$ is the complement of the dense torus of $\mathcal{X}_0$. That is, it is the union of orbits of codimension at least one in $\mathcal{X}_\tau$.

If $|\mathcal{F}| \subseteq \sigma$, then there is a canonical toric birational morphism

$$\boxed{\pi_{\mathcal{F}}} : \mathcal{X}_{\mathcal{F}} \to \mathcal{X}_\sigma$$

which identifies the dense tori of $\mathcal{X}_{\mathcal{F}}$ and $\mathcal{X}_\sigma$. This morphism is a modification in the sense of Definition 3.2.1 (that is, it is moreover *proper*) if and only if $|\mathcal{F}| = \sigma$.

In the rest of the paper, we assume that $\boxed{\sigma} \subset N_{\mathbb{R}}$ **denotes a strictly convex rational polyhedral cone with non-empty interior** σ°. This means that the closed orbit O_σ of $\mathcal{X}_\sigma$ has dimension zero. For simplicity, we denote it by $\boxed{0}$ and the monoid $(\sigma^\vee \cap M, +)$ by $\boxed{\Gamma}$. Let $\boxed{(\mathcal{X}_\sigma, 0)}$ be the formal germ of $\mathcal{X}_\sigma$ at its closed orbit 0. Its local ring $\boxed{\mathcal{O}_{\mathcal{X}_\sigma,0}}$ is equal to the $\mathbb{K}$-algebra $\boxed{\mathbb{K}[[\Gamma]]}$ of formal power series with coefficients in $\mathbb{K}$ and exponents in the monoid Γ. If $f \in \mathbb{K}[[\Gamma]]$, its **support** $\boxed{\mathrm{supp}\, f} \subseteq \Gamma$ may be defined as in the two-dimensional regular case (see Definition 3.3.1).

In Definition 3.6.3 we will explain what does it mean to tropicalize locally *interior subtoric germs*, in the following sense:

Definition 3.4.1 We say that a formal subgerm $Y \hookrightarrow (\mathcal{X}_\sigma, 0)$ is a **subtoric germ**. If it is reduced and none of its irreducible components is contained in the toric boundary $\partial\mathcal{X}_\sigma$, we call it an **interior subtoric germ**.

Remark 3.4.2 Denote by $\boxed{I_Y} \subseteq \mathbb{K}[[\Gamma]]$ the defining ideal of the subtoric germ $Y \hookrightarrow (\mathcal{X}_\sigma, 0)$. Note that Y is contained in the toric boundary $\partial\mathcal{X}_\sigma$ if and only if I_Y contains monomials χ^m, with $m \in \Gamma$. Therefore, an irreducible subtoric germ $Y \hookrightarrow (\mathcal{X}_\sigma, 0)$ is interior if and only if its defining ideal I_Y does not contain monomials.

Note that already in Sect. 3.1 we considered only *interior* subtoric germs of $(\mathbb{C}^n, 0)$, without using this terminology. The plane curve singularities of Sect. 3.3 were also assumed to be interior subtoric. Let us introduce an analogous terminology for *arcs*:

Definition 3.4.3 An **arc of a formal germ** Y is a formal morphism of the form

$$\mathrm{Spec}(\mathbb{K}[[t]]) \to Y.$$

A **subtoric arc** is an arc of $(\mathcal{X}_\sigma, 0)$. An **interior subtoric arc** is a subtoric arc of $(\mathcal{X}_\sigma, 0)$ whose image is not included in the toric boundary $\partial\mathcal{X}_\sigma$.

We want to explain now how each arc of Y defines a *semivaluation* of its local ring $\mathcal{O}_{Y,0}$ (see Example 3.4.7). We first explain basic facts about valuations and semivaluations. In this paper, we always assume that the *semivaluations* are real-valued:

Definition 3.4.4 Let R be a commutative ring with $1 \neq 0$. A **valuation centered in** R is a map $\nu \colon R \to [0, +\infty]$ such that:

- $\nu(x) = +\infty$ if and only if $x = 0$;
- $\nu(1) = 0$;
- $\nu(xy) = \nu(x) + \nu(y)$ for all $x, y \in R$;
- $\nu(x + y) \geq \min\{\nu(x), \nu(y)\}$ for all $x, y \in R$.

A **semivaluation centered in** R is defined similarly, excepted that the first condition is relaxed to $\nu(0) = +\infty$. That is, ν is allowed to take the value $+\infty$ on elements different from 0.

For instance, the map

$$\boxed{\nu_t} : \mathbb{K}[[t]] \to [0, +\infty]$$

associating to each non-zero formal power series its initial exponent and to 0 the symbol $+\infty$ is a valuation of the $\mathbb{K}$-algebra $\mathbb{K}[[t]]$, called the **t-adic valuation**.

Remark 3.4.5 The classical notion of *Krull valuation* has a field instead of a ring as source and an arbitrary totally ordered abelian group enriched with the symbol $+\infty$ as target (see [51, Chapter VI.8]). This notion was later extended to rings, replacing fields by rings as sources (see for instance [45, Section 2]) and the notion of *semivaluation* was obtained by allowing non-zero elements to take the value $+\infty$. In particular, real-valued semivaluations of a ring R take values in $\mathbb{R} \cup \{+\infty\}$. Instead, the *centered* semivaluations of Definition 3.4.4 take only non-negative values. **In this paper we will only consider centered semivaluations, that is why we suppress the qualificative *centered*.** A centered semivaluation has a *center* in $\mathrm{Spec}(R)$, defined by the prime ideal of elements $x \in R$ such that $\nu(x) > 0$. We will not use this notion.

It follows immediately from Definition 3.4.4 that if ν is a semivaluation on R, then the set

$$\boxed{\mathfrak{p}_\nu} := \{x \in R \mid \nu(x) = +\infty\}$$

is a prime ideal of R.

Definition 3.4.6 Let ν be a semivaluation of the ring R. We say that the closed subscheme of $\mathrm{Spec}(R)$ defined by the prime ideal $\mathfrak{p}_\nu$ is the **support** of ν.

The name of *support* alludes to the fact that the semivaluation ν centered in R is determined by a valuation centered in $R/\mathfrak{p}_\nu$. Indeed, the semivaluation ν descends to a valuation ν' of $R/\mathfrak{p}_\nu$, therefore it may be expressed as the composition:

$$R \to R/\mathfrak{p}_\nu \xrightarrow{\nu'} [0, +\infty].$$

More generally, whenever $\varphi : R \to R'$ is a morphism of rings and $\nu' : R' \to [0, +\infty]$ is a valuation, the following composition is a semivaluation:

$$R \xrightarrow{\varphi} R' \xrightarrow{\nu'} [0, +\infty].$$

Example 3.4.7 For instance, from the algebraic viewpoint an arc of Y is defined by a morphism of rings $\varphi : \mathcal{O}_{Y,0} \to \mathbb{K}[[t]]$, therefore the t-adic valuation $\nu_t : \mathbb{K}[[t]] \to [0, +\infty]$ induces a semivaluation:

$$\nu_t \circ \varphi : \mathcal{O}_{Y,0} \to [0, +\infty].$$

This is the announced explanation of the way an arc determines a semivaluation. Note that this semivaluation is never a valuation when Y is of dimension at least two, because then there are non-zero elements h of $\mathcal{O}_{Y,0}$ satisfying $\varphi(h) = 0$.

In Definitions 3.4.1 and 3.4.3 we introduced the notions of *interior subtoric germs* and *interior arcs* respectively. Let us define an analogous notion for *semivaluations*, in the sense of Definition 3.4.4:

Definition 3.4.8 Let $\nu : \mathbb{K}[[\Gamma]] \to [0, +\infty]$ be a semivaluation. We call it an **interior semivaluation** if its support in the sense of Definition 3.4.6 is not contained in the toric boundary of $\mathcal{X}_\sigma$.

A semivaluation ν of the ring $\mathcal{O}_{Y,0}$ of a subtoric germ $Y \hookrightarrow (\mathcal{X}_\sigma, 0)$ is called **interior**, if the semivaluation of $\mathbb{K}[[\Gamma]]$ obtained by composing ν with the canonical projection $\mathbb{K}[[\Gamma]] \to \mathcal{O}_{Y,0}$ is so.

The reader can check that a semivaluation $\nu : \mathbb{K}[[\Gamma]] \to [0, +\infty]$ is interior if and only if $\nu(\chi^m) \neq +\infty$ for all $m \in \Gamma$. Moreover, the semivaluation associated to an arc of a subtoric germ in the sense of Example 3.4.7 is interior if and only if the arc is itself interior.

3.5 Four Sets of Weight Vectors Associated to an Interior Subtoric Germ

In this section we generalize the ideas of Sect. 3.3 from plane curve singularities without branches contained in the coordinate axes to arbitrary interior subtoric germs in the sense of Definition 3.4.1. Namely, to every interior subtoric germ

$Y \hookrightarrow (\mathcal{X}_\sigma, 0)$ we associate four kinds of vectors contained in the weight cone σ (see Definitions 3.5.1, 3.5.5, 3.5.7, 3.5.8), which will be used in Theorem 3.6.2 to give four different characterizations of a conic subset of σ. It is this subset which we call the local tropicalization of $Y \hookrightarrow (\mathcal{X}_\sigma, 0)$ (see Definition 3.6.3).

First, let us turn the toric-geometric explanations of Sect. 3.1 into the following definition:

Definition 3.5.1 Let $Y \hookrightarrow (\mathcal{X}_\sigma, 0)$ be an interior subtoric germ, in the sense of Definition 3.4.1. A **compactifying fan** of Y is a fan $\mathcal{F}$ such that $|\mathcal{F}| \subseteq \sigma$ and such that the restriction $\pi : Y_\mathcal{F} \to Y$ of the toric birational morphism $\pi_\mathcal{F} : \mathcal{X}_\mathcal{F} \to \mathcal{X}_\sigma$ to the strict transform $\boxed{Y_\mathcal{F}}$ of Y by $\pi_\mathcal{F}$ is *proper*. The **compactifying cone** of Y is the intersection of the supports of all its compactifying fans.

Example 3.5.2 Let us give a simple example of a fan $\mathcal{F}$ such that $\pi_\mathcal{F}$ is not proper but π is proper. Choose $n = 2$ and let Y be the line $Z(y - x)$ of the affine plane $\mathbb{K}^2_{x,y}$ with coordinates x, y. Let $\pi_\mathcal{F}$ be the toric birational morphism $\mathbb{K}^2_{u,v} \to \mathbb{K}^2_{x,y}$ defined by the equations $x = u$, $y = uv$, that is, one of the two charts of the blowup of the origin of $\mathbb{K}^2_{x,y}$. The corresponding fan $\mathcal{F}$ consists of the faces of the cone $\mathbb{R}_{\geq 0}\langle(1, 1), (0, 1)\rangle$ spanned by the vectors $(1, 1)$ and $(0, 1)$. The strict transform of Y by $\pi_\mathcal{F}$ is $Y_\mathcal{F} = Z(v - 1)$. The toric morphism $\pi_\mathcal{F}$ is obviously not proper, as the preimage $\pi_\mathcal{F}^{-1}(0) = Z(u) \hookrightarrow \mathbb{K}^2_{u,v}$ of the origin of $\mathbb{K}^2_{x,y}$ is not complete. But the restriction $\pi : Z(v - 1) \to Z(y - x)$ of $\pi_\mathcal{F}$ to $Y_\mathcal{F}$ is an isomorphism, hence it is proper.

Let us look at the compactifying cone of an interior subtoric branch $B \hookrightarrow (\mathcal{X}_\sigma, 0)$, in the sense of Definition 3.5.1. A normalisation morphism $\mathrm{Spec}(\mathbb{K}[[t]]) \to B$ is an interior subtoric arc $\gamma : \mathrm{Spec}(\mathbb{K}[[t]]) \to (\mathcal{X}_\sigma, 0)$, in the sense of Definition 3.4.3. Each such arc has an associated weight vector $w(\gamma)$, defined as follows. Denote by

$$\boxed{\gamma^*} : (\mathbb{K}[[\Gamma]], \cdot) \to (\mathbb{K}[[t]], \cdot).$$

the corresponding local morphism of $\mathbb{K}$-algebras seen as a morphism of multiplicative monoids. By composing γ^* at the source with the natural injection of monoids $\boxed{i_\Gamma} : (\Gamma, +) \hookrightarrow (\mathbb{K}[[\Gamma]], \cdot)$ and at the target with the t-adic valuation seen as a morphism $\nu_t : (\mathbb{K}[[t]], \cdot) \to (\mathbb{Z}_{\geq 0} \cup \{+\infty\}, +)$ of monoids, we get a morphism of monoids $\nu_t \circ \gamma^* \circ i_\Gamma : (\Gamma, +) \to (\mathbb{Z}_{\geq 0}, +)$, that is, an element of the monoid $(\sigma \cap N, +)$. As the arc γ is assumed to be subtoric, one has $(\nu_t \circ \gamma^* \circ i_\Gamma)(\Gamma \setminus \{0\}) \subseteq \mathbb{Z}_{>0}$, which shows that $\nu_t \circ \gamma^* \circ i_\Gamma \in \sigma^\circ \cap N$. We may formulate:

Definition 3.5.3 Let $\gamma : \mathrm{Spec}(\mathbb{K}[[t]]) \to (\mathcal{X}_\sigma, 0)$ be an interior subtoric arc. Its **weight vector** $\boxed{w(\gamma)}$ is the morphism of monoids

$$\nu_t \circ \gamma^* \circ i_\Gamma : (\Gamma, +) \to (\mathbb{Z}_{\geq 0}, +),$$

seen as an element of $\sigma^\circ \cap N$.

For instance, if $\mathcal{X}_\sigma(\mathbb{K}) = \mathbb{K}^2$ and γ is defined by $t \mapsto (x(t), y(t))$ as in formula (3.3), one gets $w(\gamma) = (p, q) = (\nu_t(x(t)), \nu_t(y(t)))$.

One has the following property, which shows that *in the case of interior subtoric branches*, the two definitions of local tropicalization given in Sect. 3.1 are equivalent:

Proposition 3.5.4 *Let* $B \hookrightarrow (\mathcal{X}_\sigma, 0)$ *be an interior subtoric branch and let* γ : $\mathrm{Spec}(\mathbb{K}[[t]]) \to (\mathcal{X}_\sigma, 0)$ *be an arc whose image is B. Then, the ray* $\mathbb{R}_{\geq 0} w(\gamma) \subset \sigma$ *generated by its weight vector in the sense of Definition 3.5.3 is independent of the choice of the arc* γ *and is equal to the compactifying cone of B, in the sense of Definition 3.5.1.*

Proof Consider the affine toric variety $\mathcal{X}_{\mathbb{R}_{\geq 0} w(\gamma)}$. One may check that the arc γ may be lifted to an arc $\gamma' : \mathrm{Spec}(\mathbb{K}[[t]]) \to \mathcal{X}_{\mathbb{R}_{\geq 0} w(\gamma)}$ which sends the origin of $\mathrm{Spec}(\mathbb{K}[[t]])$ to a point of the closed orbit $O_{\mathbb{R}_{\geq 0} w(\gamma)}$ of $\mathcal{X}_{\mathbb{R}_{\geq 0} w(\gamma)}$. This proves that the restriction $\pi' : B' \to B$ of the toric morphism $\pi_{\mathbb{R}_{\geq 0} w(\gamma)} : \mathcal{X}_{\mathbb{R}_{\geq 0} w(\gamma)} \to \mathcal{X}_\sigma$ to the strict transform B' of B is proper. As a consequence, the restriction $\pi : B_{\mathcal{F}} \to B$ of $\pi_{\mathcal{F}}$ to the strict transform $B_{\mathcal{F}}$ of B by $\pi_{\mathcal{F}}$ is proper whenever $\mathbb{R}_{\geq 0} w(\gamma) \subseteq |\mathcal{F}|$.

Indeed, consider a subdivision $\mathcal{F}'$ of $\mathcal{F}$ containing $\mathbb{R}_{\geq 0} w(\gamma)$ among its rays. Then $\mathcal{X}_{\mathcal{F}'}$ contains the affine toric variety $\mathcal{X}_{\mathbb{R}_{\geq 0} w(\gamma)}$ as a Zariski-open subset. The restriction of the toric morphism $\pi_{\mathcal{F}'}$ to this subset is equal to $\pi_{\mathbb{R}_{\geq 0} w(\gamma)}$: $\mathcal{X}_{\mathbb{R}_{\geq 0} w(\gamma)} \to \mathcal{X}_\sigma$. This shows that the restriction of $\pi_{\mathcal{F}'}$ to the strict transform $B_{\mathcal{F}'}$ of B on $\mathcal{X}_{\mathcal{F}'}$ is proper. As $\pi_{\mathcal{F}'}$ factors through $\pi_{\mathcal{F}}$, this implies that $\pi : B_{\mathcal{F}} \to B$ is also proper. □

Let us turn now to arbitrary interior subtoric germs. As explained in Sect. 3.1, we may look at all the interior subtoric arcs contained in them and at the corresponding weight vectors:

Definition 3.5.5 Let $Y \hookrightarrow (\mathcal{X}_\sigma, 0)$ be an interior subtoric germ, in the sense of Definition 3.4.1. An **arcwise weight vector** of Y is the weight vector of an interior subtoric arc contained in Y, in the sense of Definition 3.5.3.

Instead of looking at the arcs $(\mathrm{Spec}(\mathbb{K}[[t]]), 0) \to (\mathcal{X}_\sigma, 0)$ whose image is contained in Y, we may look dually at the formal germs of functions on $(\mathcal{X}_\sigma, 0)$ vanishing on Y, that is, at the elements of the ideal $I_Y \subseteq \mathbb{K}[[\Gamma]]$ defining Y in $(\mathcal{X}_\sigma, 0)$ (see Remark 3.4.2). Then, one may again define preferred weight vectors (see Definition 3.5.7). Note first that Definition 3.3.13 readily generalizes to an arbitrary weight vector $w \in \sigma^\circ$ and a formal series $f \in \mathbb{K}[[\Gamma]]$, respectively an ideal I of $\mathbb{K}[[\Gamma]]$:

Definition 3.5.6 Consider $w \in \sigma^\circ$ and $f = \sum_{m \in \text{ supp } f} f_m \chi^m \in \mathbb{K}[[\Gamma]]$. The **basis** $\boxed{\delta_w(f)} \subseteq \text{supp } f$ **of** supp f **relative to** w is the locus where the restriction of the linear form $w : M_{\mathbb{R}} \to \mathbb{R}$ to the support supp f of f achieves its minimum. The

w-weighted homogeneous polynomial

$$\boxed{\mathrm{in}_w f} := \sum_{m \in \delta_w(f)} f_m \chi^m$$

is called the w-**initial form** of the series f.

If $I \hookrightarrow \mathbb{K}[[\Gamma]]$ is an ideal, then its w-**initial ideal** $\boxed{\mathrm{in}_w I}$ is the ideal of $\mathbb{K}[[\Gamma]]$ generated by the w-initial forms $\mathrm{in}_w f$, for all $f \in I$.

Let us distinguish the weight vectors which define initial ideals without monomials (recall Remark 3.4.2), generalizing the two-dimensional situation considered in Corollary 3.3.14:

Definition 3.5.7 Let $Y \hookrightarrow (\mathcal{X}_\sigma, 0)$ be an interior subtoric germ and $I_Y \hookrightarrow \mathbb{K}[[\Gamma]]$ be its defining ideal. A weight vector $w \in \sigma^\circ$ is called an **initial weight vector of** Y if the initial ideal $\mathrm{in}_w I_Y \hookrightarrow \mathbb{K}[[\Gamma]]$ does not contain monomials.

Finally, let us also define weight vectors associated to semivaluations:

Definition 3.5.8 Let $Y \hookrightarrow (\mathcal{X}_\sigma, 0)$ be an interior subtoric germ. Let ν be an interior semivaluation of the ring $\mathcal{O}_{Y,0}$, in the sense of Definition 3.4.8. We look at it as a morphism of monoids $\nu : (\mathcal{O}_Y, \cdot) \to ([0, +\infty], +)$. Denote by $\boxed{r_Y}$: $(\Gamma, +) \to (\mathcal{O}_Y, \cdot)$ the composition of the morphism $i_\Gamma : (\Gamma, +) \hookrightarrow (\mathbb{K}[[\Gamma]], \cdot)$ of Definition 3.5.3 and of the restriction morphism $(\mathbb{K}[[\Gamma]], \cdot) \to (\mathcal{O}_Y, \cdot)$. By composing ν and r_Y we get a morphism of monoids

$$\nu \circ r_Y : (\Gamma, +) \to ([0, +\infty], +),$$

that is, an element $\boxed{w(\nu)}$ of σ. We call it the **weight vector of the interior semivaluation** ν. The weight vector of an interior semivaluation of ν is called a **valuative weight vector of** Y.

One may check that:

- as ν was assumed to be interior, one has $w(\nu) \in \sigma^\circ$;
- the weight vector of the semivaluation associated to an interior arc γ of Y as explained in Example 3.4.7 is equal to the weight vector $w(\gamma)$ of γ in the sense of Definition 3.5.3.

3.6 Four Viewpoints on Local Tropicalization

This section presents the main result of this paper, Theorem 3.6.2, stating that given an interior subtoric germ, four subsets of the real weight space constructed using the notions introduced in Sect. 3.5 coincide. This provides four different interpretations of the *local tropicalization* of an interior subtoric germ (see Definition 3.6.3).

We conclude the section with a structure result about local tropicalizations (see Theorem 3.6.5).

In the sequel we will use the following notion:

Definition 3.6.1 Let Σ be a subset of a finite-dimensional real vector space V. Its **cone-closure** is the closure $\overline{\mathbb{R}_{\geq 0}\Sigma}$ inside V of the non-negative cone $\mathbb{R}_{\geq 0}\Sigma$ over Σ, consisting of all the rays $\mathbb{R}_{\geq 0}w$ generated by the elements w of Σ.

The following result confirms that for each interior subtoric germ, the cone-closures of four different subsets of the weight cone of the ambient affine toric variety agree:

Theorem 3.6.2 *Let $Y \hookrightarrow (\mathcal{X}_\sigma, 0)$ be an interior subtoric germ, in the sense of Definition 3.4.1. Then, the following subsets of the weight cone $\sigma \subset N_{\mathbb{R}}$ coincide:*

1. *the compactifying cone of Y, in the sense of Definition 3.5.1;*
2. *the cone-closure of the set of arcwise weight vectors of Y, in the sense of Definition 3.5.5;*
3. *the cone-closure of the set of initial weight vectors of Y, in the sense of Definition 3.5.7;*
4. *the cone-closure of the set of valuative weight vectors of Y, in the sense of Definition 3.5.8.*

Proof

• When the field $\mathbb{K}$ has characteristic 0, the equality of the sets defined in *2.* and *3.* is a consequence of [41, Corollary 4.3]. Here we present a sketch of a different proof valid in arbitrary characteristic. In fact, more is true: both starting cones agree, not only their cone-closures. Let us explain this fact. Fix a ray $\lambda \subset \sigma$ which is spanned by an element of $\sigma^\circ \cap N$. One proves that each of the following statements is equivalent to the next one:

(a) there exists an arcwise weight vector w of Y spanning λ;
(b) there exists an interior subtoric arc of Y whose strict transform on the affine toric variety $\mathcal{X}_\lambda$ intersects its unique closed orbit O_λ;
(c) the strict transform Y_λ of Y on $\mathcal{X}_\lambda$ is such that $Y_\lambda \cap O_\lambda$ is non-empty;
(d) $\lambda \setminus \{0\}$ is included in the set of initial weight vectors of Y.

Therefore, the non-negative cone over the set of arcwise weight vectors of Y coincides with the non-negative cone over the set of initial weight vectors of Y.

The equivalence (b) $\iff$ (c) uses the fact that in any characteristic, branches admit parametrizations, that is, are images of arcs (see [10, Section 4.1] for an argument using their resolution by blowups and [5, Section 1.3.2] for an argument using the regularity of their normalizations).

The equivalence (c) $\iff$ (d) results from the fact that if $w \in N$ spans the ray λ, then the initial ideal $\mathrm{in}_w I_Y$ of the defining ideal $I_Y \subset \mathbb{K}[[\Gamma]]$ of Y, seen as an ideal of $\mathbb{K}[M]$ instead of $\mathbb{K}[[\Gamma]]$ as in Definition 3.5.6, defines a subvariety of the dense torus T_N whose closure in $\mathcal{X}_\lambda$ intersects O_λ at a point of the intersection $Y_\lambda \cap O_\lambda$.

• The equality of the sets defined in *1.* and *3.* is a consequence of [9, Proposition 3.19]. Let us explain the main ideas of the proof, generalizing again the situation from the case $\mathbb{K} = \mathbb{C}$ of [9] to the case of arbitrary field $\mathbb{K}$. Denote by $\mathcal{F}$ a fan whose support is contained in σ and by $\boxed{\mathfrak{W}(Y)}$ the set of initial weight vectors of Y in the sense of Definition 3.5.7. Choose a fan Σ subdividing the cone σ and such that $\mathcal{F} \subseteq \Sigma$. Then, we get the following commutative diagram

$$\begin{array}{ccccccc} Y_\Sigma & \hookrightarrow & \mathcal{X}_\Sigma & \hookleftarrow & \mathcal{X}_\mathcal{F} & \hookleftarrow & Y_\mathcal{F} \\ & & \pi_\Sigma \searrow & & \swarrow \pi_\mathcal{F} & & \swarrow \pi \\ & & & \mathcal{X}_\sigma & \hookleftarrow & Y & \end{array}$$

where $Y_\mathcal{F}$ and Y_Σ are the strict transforms of Y in $\mathcal{X}_\mathcal{F}$ and $\mathcal{X}_\Sigma$ respectively, $\pi_\mathcal{F}$ and π_Σ are the natural toric birational morphisms, and π is the restriction of $\pi_\mathcal{F}$ to $Y_\mathcal{F}$. Note that π is *proper* (that is, $\mathcal{F}$ is a compactifying fan of Y) if and only if Y_Σ is contained in $\mathcal{X}_\mathcal{F}$.

The equality of the sets defined in *1.* and *3.* is therefore a consequence of the following equivalence, for every cone τ of Σ:

$$O_\tau \cap Y_\Sigma \neq \emptyset \Longleftrightarrow \tau^\circ \cap \mathfrak{W}(Y) \neq \emptyset, \tag{3.6}$$

where O_τ is the corresponding toric orbit of $\mathcal{X}_\Sigma$.

Let Us First Prove the Implication $\Longrightarrow$ **of** (3.6). Choose $y_0 \in O_\tau \cap Y_\Sigma$. Let $\gamma : \mathrm{Spec}(\mathbb{K}[[t]]) \to (Y_\Sigma, y_0)$ be an arc passing through y_0. Denote by $w \in N$ the weight vector (in the sense of Definition 3.5.3) of the interior subtoric arc $\pi_\Sigma \circ \gamma : \mathrm{Spec}(\mathbb{K}[[t]]) \to Y$. The fact that $y_0 \in O_\tau \cap Y_\Sigma$ means that the weight vector w lies in $\tau^\circ \cap N \subseteq \tau^\circ$. To show that $w \in \mathfrak{W}(Y)$, we let B be the branch contained in Y which is the image of the arc $\pi_\Sigma \circ \gamma$. Since $B \subseteq Y$, the ideal $I_Y \subset \mathbb{K}[[\Gamma]$ of Y is contained in the ideal $I_B \subset \mathbb{K}[[\Gamma]$ of B. Thus, $\mathfrak{W}(B) \subseteq \mathfrak{W}(Y)$. But, by arguments similar to those used in the proof of Proposition 3.5.4, we see that $w \in \mathfrak{W}(B)$.

Let Us Now Prove the Implication $\Longleftarrow$ **of** (3.6). Consider a primitive lattice vector $w \in \tau^\circ \cap \mathfrak{W}(Y)$ and a refinement Σ_w of Σ such that the ray $\lambda_w := \mathbb{R}_{\geq 0} \cdot w$ is a cone of the fan Σ_w. By construction, the orbit O_{λ_w} is mapped via the toric morphism $\pi_w : \mathcal{X}_{\Sigma_w} \to \mathcal{X}_\Sigma$ to the orbit O_τ. The intersection of the strict transform Y_{Σ_w} of Y in $\mathcal{X}_{\Sigma_w}$ with the orbit O_{λ_w} is determined by the w-initial ideal $\mathrm{in}_w I(Y)$ of the ideal $I(Y)$ defining Y, viewed in the Laurent polynomial ring $\mathbb{K}[M]$. As $w \in \mathfrak{W}(Y)$, this initial ideal is monomial free. Therefore, one has $Y_{\Sigma_w} \cap O_{\lambda_w} \neq \emptyset$. Since $O_{\tau_w} \subset \mathcal{X}_{\Sigma_w}$, the map π_w ensures that $Y_\Sigma \cap O_\tau \neq \emptyset$ as well.

• The equality of the sets defined in *3.* and *4.* is a consequence of [39, Theorem 11.2]. In fact, the inclusion of the set of *4.* inside the set of *3.* follows from essentially the same argument as described before Theorem 3.3.8. Namely, if $f \in I_Y \subset \mathbb{K}[[\Gamma]]$ is a series of the defining ideal of Y and $\nu : \mathbb{K}[[\Gamma]]/I_Y \to \mathbb{R} \cup \{+\infty\}$ is an interior semivaluation, then ν lifts to an interior semivaluation ν' of $\mathbb{K}[[\Gamma]]$ which necessarily takes value $+\infty$ on f. Since ν' does not take infinite value on any monomial χ^m with $m \in \Gamma$, it follows that f must involve at least two monomials where ν' attains its minimal value. Thus $\mathrm{in}_w(f)$, where w is the weight vector corresponding to ν', is not a monomial. The inverse inclusion is based on subtler results from the theory of valuations, see [39, Theorem 11.2] and references therein. □

Theorem 3.6.2 provides four equivalent definitions of the *local tropicalization* of an interior subtoric germ:

Definition 3.6.3 Let $Y \hookrightarrow (\mathcal{X}_\sigma, 0)$ be an interior subtoric germ, in the sense of Definition 3.4.1. The set defined by any of the four equivalent conditions of Theorem 3.6.2 is called the **local tropicalization** of the subtoric germ Y and is denoted by $\boxed{\mathrm{Trop}_{\mathrm{loc}} Y} \subset \sigma$.

The local tropicalization $\mathrm{Trop}_{\mathrm{loc}} Y$ depends strongly on the embedding $Y \hookrightarrow (\mathcal{X}_\sigma, 0)$ (think, for instance, about the various embeddings $t \mapsto (t, t^n)$ of a smooth branch inside $(\mathbb{K}^2, 0)$, whose local tropicalizations are, by Proposition 3.5.4, the rays spanned by $(1, n)$). Even if we drop it from the notation for simplicity, one has to keep this dependence in mind.

Each of the four viewpoints on local tropicalization presented in Theorem 3.6.2 is adapted to particular contexts. For instance:

- viewpoint *1.* is useful when one wants to construct modifications of Y starting from a particular embedding in a germ of toric variety;
- viewpoint *2.* is useful when one wants to think in terms of the arc space of the germ Y;
- viewpoint *3.* is useful when one deals with concrete defining equations of Y, as in our work [9] discussed in Sect. 3.8;
- viewpoint *4.* is useful when one wants to think in terms of the space of semivaluations associated to Y; for instance, our most general definitions of tropicalization given in [39, Section 6] used this space.

In fact, in our paper [39] we introduced two notions of local tropicalization, a *positive* and a *nonnegative* one (see [39, Definitions 6.6 and 6.7]), both being *extended* local tropicalizations in the sense explained in Sect. 3.9. For simplicity, we preferred to take here as definition of local tropicalization only the intersection of the nonnegative local tropicalization of [39] with the cone σ.

Remark 3.6.4 Combining Proposition 3.5.4 with Theorem 3.6.2, we see that the local tropicalization of an interior subtoric branch of $(\mathcal{X}_\sigma, 0)$ is the ray spanned by the weight vector of any arc parametrizing the branch. Maurer had introduced such weight vectors in his 1980 paper [29] for curve singularities $Y \hookrightarrow (\mathbb{C}^n, 0)$ with several branches, under the name of *critical tropisms*. For this reason, we may see Maurer's paper as a precursor of viewpoint *2.* on local tropicalization. The semantic proximity of the noun *tropism* with the adjective *tropical* is interesting. Lejeune-Jalabert and Teissier had already used the expression *tropisme critique* in [26] (see [14, end of Section 1.4.5]).

Viewpoint *2.* implies that the local tropicalization of the union of two interior subtoric germs of $(\mathcal{X}_\sigma, 0)$ is the union of their two local tropicalizations. Therefore, every local tropicalization is a finite union of local tropicalizations of irreducible interior subtoric germs. Our main structure theorem about the local tropicalizations of such germs becomes (see [39, Theorem 11.9], [9, Proposition 3.11]):

Theorem 3.6.5 *Let $Y \hookrightarrow (\mathcal{X}_\sigma, 0)$ be an irreducible interior subtoric germ with defining ideal $I_Y \subset \mathbb{K}[[\Gamma]]$. Then, the local tropicalization* $\mathrm{Trop}_{\mathrm{loc}} Y$ *is the support of a fan $\mathcal{F}$ which satisfies the following conditions:*

- *the dimension of each maximal cone of $\mathcal{F}$ is equal to the dimension of Y;*
- *the maximal cones of $\mathcal{F}$ all have non-empty intersections with σ°;*
- *given any cone τ of $\mathcal{F}$, the w-initial ideal of I_Y is independent of the choice of $w \in \tau^\circ$.*

Viewpoint *3.* implies that:

Proposition 3.6.6 *The local tropicalization of an interior subtoric germ is equal to the intersection of the local tropicalizations of the interior principal effective divisors containing it.*

This fact, which is a local version of one of the definitions of global tropicalization of a subvariety of an algebraic torus (see [28, Definition 3.2.1]), explains the importance of understanding the local tropicalizations of effective principal divisors on $(\mathcal{X}_\sigma, 0)$. Next section is dedicated to this topic.

3.7 The Local Tropicalization of an Effective Principal Subtoric Divisor

In this section we describe the local tropicalizations of the interior subtoric germs which are *effective principal divisors*. They are determined by the Newton polyhedra of their defining series in a way which generalizes the case of plane curve singularities described in Corollary 3.3.10 (see Theorem 3.7.4).

We continue using the notations of Sect. 3.4. We assume that the interior subtoric germ $Y \hookrightarrow (\mathcal{X}_\sigma, 0)$ is an effective principal divisor on $\mathcal{X}_\sigma$. Therefore $Y = Z(f)$, where $f \in \mathbb{K}[[\Gamma]]$. The fact that no irreducible component of Y is contained in the toric boundary $\partial \mathcal{X}_\sigma$ is equivalent to the condition that the series f is not contained in any of the ideals of $\mathbb{K}[[\Gamma]]$ generated by the sets of monomials $(\sigma^\vee \setminus \tau^\perp) \cap M$, where τ is an edge of σ. For example, if σ is the positive orthant $\mathbb{R}^n_{\geq 0}$, which allows to write $\mathbb{K}[[\Gamma]] = \mathbb{K}[[x_1, \ldots, x_n]]$, this simply says that f is not divisible by any of the variables $x_1, \ldots, x_n$.

The following definition is a straightforward generalization of Definition 3.3.4:

Definition 3.7.1 Let $f = \sum_{m \in \text{supp } f} f_m \chi^m \in \mathbb{K}[[\Gamma]]$ (therefore $f_m \in \mathbb{K}^*$ for every $m \in \text{supp } f$). Its **Newton polyhedron** $\boxed{\mathcal{N}(f)} \subseteq \sigma^\vee$ is the convex hull of the set

$$\bigcup_{m \in \text{supp } f} (m + \sigma^\vee) \subset \sigma^\vee.$$

Note that, as in the two-dimensional case examined in Sect. 3.3, for every $w \in \sigma^\circ$ the basis $\delta_w(f)$ appearing in Definition 3.5.6 is the intersection of supp f with a compact face of $\mathcal{N}(f)$. This face is the locus where the linear form $w : M_{\mathbb{R}} \to \mathbb{R}$ achieves its minimum in restriction to the polyhedron $\mathcal{N}(f)$.

There is a duality between the faces of the Newton polyhedron $\mathcal{N}(f)$ and a certain fan with support σ. Namely, if κ is a face of $\mathcal{N}(f)$, its **dual cone** $\boxed{\sigma(\kappa)} \subset \sigma$ is the closure of the set of weight vectors $w \in \sigma$ whose restriction $w : \mathcal{N}(f) \to \mathbb{R}_{\geq 0}$ achieves its minimum exactly on the face κ. Then, as results from [36, Chapter II.3.(A)], one has the following statement:

Lemma 3.7.2 *The set of cones $\sigma(\kappa)$, where κ varies over the faces of the Newton polyhedron $\mathcal{N}(f)$ of a series $f \in \mathbb{K}[[\Gamma]]$, forms a fan with support σ.*

Definition 3.7.3 The fan $\boxed{\mathcal{F}(f)}$ with support σ described in Lemma 3.7.2 is called the **Newton fan** of the series f.

For instance, the subdivision of the non-negative quadrant shown in Fig. 3.4 is the Newton fan of the series $f(x, y) = y^2 - 2x^3 + x^2 y \in \mathbb{C}[[x, y]]$, whose Newton polygon is shown in Fig. 3.3.

By an argument similar to those explained for plane curve singularities after Proposition 3.3.7, the Newton polyhedron of a defining series f of the interior subtoric principal divisor Y is independent of the choice of this series. Therefore, one may speak simply of the **Newton fan** $\boxed{\mathcal{F}(Y)}$ of $Y \hookrightarrow (\mathcal{X}_\sigma, 0)$.

The connection of the Newton fan $\mathcal{F}(Y)$ with the local tropicalization of Y (see Theorem 3.6.2), is given by the following result of [39, Proposition 11.8]:

Theorem 3.7.4 *Let $Y \hookrightarrow (\mathcal{X}_\sigma, 0)$ be an interior effective principal divisor. Then, the local tropicalization* $\text{Trop}_{\text{loc}} Y$ *of Y coincides with the union of the cones of the Newton fan $\mathcal{F}(Y)$ of Y which are dual to the compact edges of $\mathcal{N}(Y)$.*

One may prove Theorem 3.7.4 similarly to the proofs of Theorem 3.3.8 and Corollary 3.3.10, starting from characterization (2) in Theorem 3.6.2 of the local tropicalization $\text{Trop}_{\text{loc}} Y$ of Y.

Example 3.7.5 Consider a **Pham-Brieskorn surface singularity** (a name which alludes mainly to the papers [40] and [3], as explained in [4, Pages 42–49]). It is by definition a hypersurface singularity in $\mathbb{C}^3$ defined by an equation of the form

$$x_1^\alpha + x_2^\beta + x_3^\gamma = 0, \tag{3.7}$$

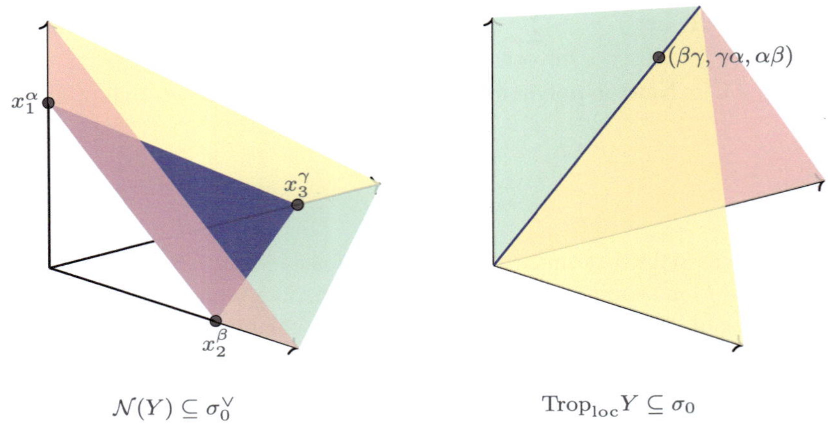

Fig. 3.5 The Newton polyhedron $\mathcal{N}(Y)$ of the Pham-Brieskorn singularity $Y \hookrightarrow (\mathbb{C}^3, 0)$ defined by equation $x_1^\alpha + x_2^\beta + x_3^\gamma = 0$ (on the left) and its local tropicalization (on the right)

where α, β, γ are pairwise coprime positive integers. Its Newton polyhedron $\mathcal{N}(Y)$ and its local tropicalization $\mathrm{Trop}_{\mathrm{loc}} Y$ are represented in Fig. 3.5. The weight cone σ_0 of $\mathbb{C}^3$ seen as an affine toric variety is the non-negative octant of the weight space $\mathbb{R}^3$. The compact faces of $\mathcal{N}(Y)$ are the triangle with vertices $(\alpha, 0, 0)$, $(0, \beta, 0)$, $(0, 0, \gamma)$, its edges and its vertices. By Theorem 3.7.4, $\mathrm{Trop}_{\mathrm{loc}} Y$ is the union of the three two-dimensional cones dual to the edges of the previous triangle. Those cones are spanned by the vector $(\beta\gamma, \gamma\alpha, \alpha\beta) \in \sigma_0$ orthogonal to the triangle and by each of the three vectors of the canonical basis of the weight space $\mathbb{R}^3$.

Note that András Némethi and Baldur Sigurðsson described in [30] the local tropicalizations of effective but not necessarily principal divisors on toric germs.

3.8 The Local Tropicalization of Splice Type Surface Singularities

Theorem 3.7.4 shows that the local tropicalization of an interior effective principal subtoric divisor is determined by the associated Newton polyhedron. By contrast, if an interior subtoric germ $Y \hookrightarrow (\mathcal{X}_\sigma, 0)$ has a non-principal defining ideal I_Y, then it is much more difficult to determine its local tropicalization. The reason is that $\mathrm{Trop}_{\mathrm{loc}} Y$ is in general not the intersection of the local tropicalizations of the members of a generating family of I_Y, even if this generating family describes Y as a complete intersection. In this section we explain how we described in [9], in collaboration with María Angélica Cueto, the local tropicalizations of *splice type singularities*, which form an important class of isolated complete intersection surface singularities.

Splice type surface singularities were introduced in [31] and [32] by Walter Neumann and Jonathan Wahl as examples of complex isolated complete intersection surface singularities whose links are integral homology spheres. One may refer to Wahl's papers [47] and [48] for a description of their motivations for defining them.

This class of singularities contains the class of Pham-Brieskorn surface singularities (see Example 3.7.5). The following system defines also a splice type surface singularity in $\mathbb{C}^4$ (see [8, Examples 2.27, 2.34, 2.40]):

$$\begin{cases} x_1^2 - x_2^3 + x_3 x_4 = 0, \\ x_3^5 - x_4^2 + x_1 x_2^4 = 0. \end{cases} \tag{3.8}$$

In general, a splice type surface singularity is determined up to a special kind of equisingular deformation by a **splice diagram**. By this, one means a pair (Γ, w), where $\boxed{\Gamma}$ is a tree whose vertices have valencies either 1 (**leaves**) or ≥ 3 (**nodes**) and w is a function that assigns to each pair (u, e) of a node u and adjacent edge e a positive integer (**weight**) $\boxed{w_{u,e}}$. Thus, each edge connecting nodes u and v carries two weights – one near the node u and one near v. In the context of integral homology sphere links, these data must satisfy three more properties, called *coprime, edge determinant* and *semigroup conditions*:

- The *coprime condition* states that *for each node, the weights of adjacent edges are* ≥ 2 *and pairwise coprime.*
- For each node u and vertex p (which may be either a node or a leaf), define $\boxed{\ell_{u,p}}$ to be the product of all the weights adjacent to, but not lying on, the shortest path from u to p. In particular, when $p = u$ one gets the product $\boxed{d_u}$ of weights adjacent to u. The *edge determinant condition* states that *for every pair of adjacent nodes u and v, one has the inequality* $d_u d_v > \ell_{u,v}^2$.
- Denote by $1, \ldots, n$ the leaves of Γ. For each node u, consider the **weight vector**

$$\boxed{w_u} = (\ell_{u,1}, \ldots, \ell_{u,n}) \in \mathbb{Z}^n \subset \mathbb{R}^n.$$

  If e is an edge adjacent to u, then it makes sense to speak about the leaves of Γ seen from u in the direction e. If $i_1, \ldots, i_s$ are these leaves, then the *semigroup condition* says that *the number* d_u *must be contained in the sub-semigroup of* $(\mathbb{Z}_{\geq 0}, +)$ *generated by* $\ell_{u,i_1}, \ldots, \ell_{u,i_s}$, *for all pairs* (u, e) *of a node u and an adjacent edge e.*

Examples of splice diagrams satisfying the three conditions are shown in Fig. 3.6.

To get from such data a singularity with integral homology sphere link, one should apply the following procedure developed by Neumann and Wahl. First of all, associate a variable x_i to every leaf i. Then, let u be a node, e an adjacent edge, and $i_1, \ldots, i_s$ the leaves seen from u in the direction of e. Choose a representation

$$d_u = \sum_{k=1}^{s} n_k \ell_{e,i_k},$$

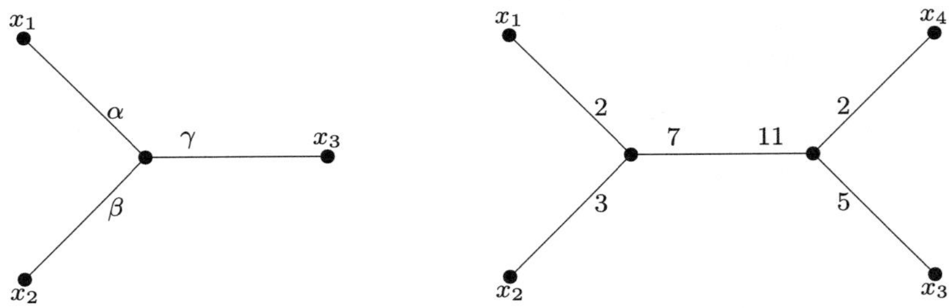

Fig. 3.6 The splice diagrams corresponding to the singularities defined by Eq. (3.7) (on the left) and by the system (3.8) (on the right)

such that $n_k \in \mathbb{Z}_{\geq 0}$ for every $k \in \{1, \ldots, s\}$. Such a representation exists, by the semigroup condition. Define the **admissible monomial** $\chi_{u,e}$ as

$$\boxed{\chi_{u,e}} := x_{i_1}^{n_1} \cdots x_{i_s}^{n_s}.$$

If the node u has valency $r_u \geq 3$ and $e_1, \ldots e_{r_u}$ are the adjacent edges, it is more convenient to denote the corresponding admissible monomials by $\chi_{u,1}, \ldots, \chi_{u,r_u}$. For each node u, choose a complex matrix $\boxed{(a_{u,i,j})_{i,j}}$ of size $(r_u - 2) \times r_u$, such that all its maximal minors are non-zero. The **splice type system** determined by the splice diagram (Γ, w) and a choice of admissible monomials $\chi_{u,j}$ and coefficients $a_{u,i,j}$ is the system of equations

$$\sum_{j=1}^{r_u} a_{u,i,j} \chi_{u,j} = 0, \quad \text{for all } i \in \{1, \ldots, r_u - 2\}, \tag{3.9}$$

where u varies among all the nodes of Γ. It is an easy combinatorial exercise to check that the system (3.9) has $n - 2$ equations. Finally, by definition, a **splice type singularity** is the germ at the origin of the variety in $\mathbb{C}^n$ defined by the system (3.9) or by any system obtained by adding to (3.9) terms of heigher w_u-weight to the equations corresponding to the node u. The interested reader can check that splice diagrams shown in Fig. 3.6 indeed correspond to the singularities defined by Eq. (3.7) and by the system (3.8).

A system of equations corresponding to a big splice diagram can be rather complicated, and *it is far from obvious that the singularity defined by such a system is an irreducible isolated complete intersection of dimension* 2. Indeed, the proof of this fact spans several pages in [31] and [32], where the authors proceed by direct calculations with partial resolutions of the germ. In turn, the theory of local tropicalization allows for a more transparent and conceptual proof of these properties. We developed this strategy in full detail in [9]. Here we present a brief summary of our method.

Our main point in [9] was to give a previously unknown tropical interpretation of the splice diagram. We associated above a weight vector w_u to each node u of a splice diagram Γ. If we additionally associate the i-th vector e_i of the canonical basis of $\mathbb{R}^n$ to the ith leaf of Γ, we proved in [9, Theorem 5.11] that once conveniently normalized, *those weight vectors determine a piecewise linear embedding of the splice diagram Γ into the simplex Δ_{n-1} whose vertices are the points e_i.*

Further, by analyzing by purely combinatorial means (no calculations with resolutions being needed) the algorithm by which Neumann and Wahl constructed a system of equations from a given splice diagram Γ, we showed in [9, Theorem 1.2] that:

Theorem 3.8.1 *The local tropicalization of a splice type surface singularity $Y \hookrightarrow (\mathbb{C}^n, 0)$ corresponding to the splice diagram Γ is the non-negative cone in $\mathbb{R}^n$ over the tree Γ embedded in Δ_{n-1}.*

For instance, note that by intersecting the local tropicalization of the Pham-Brieskorn singularity represented on the right side of Fig. 3.5 with the triangle Δ_2, one gets a tree isomorphic to the corresponding splice diagram, which is represented on the left side of Fig. 3.6.

Since $\Gamma \subset \Delta_{n-1}$ is a 1-dimensional simplicial complex, the local tropicalization $\text{Trop}_{\text{loc}} Y$ (endowed with appropriate fan structure) has maximal cones of dimension 2. Note that we started with integral weight vectors corresponding to the vertices of Γ, thus $\text{Trop}_{\text{loc}} Y$ is indeed the support of a *rational* polyhedral fan, as stated in Theorem 3.6.5. From the same theorem, we got $\dim Y = 2$. The splice system corresponding to Γ has the correct number of equations (namely $n - 2$), so it follows at once that Y is indeed a complete intersection and that all its irreducible components have dimension 2.

Recall that throughout this paper we assumed for simplicity that all the subtoric germs were *interior* in the sense of Definition 3.4.1. However, the theory of local tropicalization may be extended to subtoric germs which are not interior by adding *strata at infinity* to the local tropicalization (see Sect. 3.9.1 for an introduction and [39, Sections 4 and 6] for details). In [9] we performed the corresponding analysis and we showed that if Y is a splice type surface singularity, then its local tropicalization $\text{Trop}_{\text{loc}} Y$ has no strata at infinity except the closures of its finite strata, and thus Y has no irreducible components in the toric boundary $\partial\mathbb{C}^n$.

As a byproduct of our techniques, we got a relatively easy proof of the following new result (see [9, Theorem 1.1]):

Theorem 3.8.2 *Any splice type surface singularity is a Newton nondegenerate complete intersection.*

Combining this fact with Theorem 3.6.2 and the notion of *embedded toric resolution* (see, e.g., [24]), we proved that a resolution of singularities of Y could be obtained by a toric morphism corresponding to a convenient fan supported by $\text{Trop}_{\text{loc}} Y$. But since the entire local tropicalization (with the exception of the coordinate rays $\mathbb{R}_{\geq 0}\, e_i$) is contained in the positive orthant $\mathbb{R}^n_{>0}$ of $\mathbb{R}^n$, this toric morphism does not blow up anything except the origin in $\mathbb{C}^n$. This implies that *the splice type singularity Y is isolated.*

Finally, *Y must be irreducible* because if a complete intersection singularity of dimension ≥ 2 has two or more irreducible components, then it cannot be isolated (this is a special case of a theorem of Hartshorne, as explained in the proof of [39, Corollary 7.13]).

As a generalization of Goldin and Teissier's [19, Corollary of Theorem 6.1], which concerned only *branches*, de Felipe, González Pérez and Mourtada showed in [12] that any plane curve singularity may be reembedded explicitly in a higher dimensional space $\mathbb{C}^n$ as a Newton non-degenerate germ and they expressed the local tropicalization of the reembedded germ in terms of invariants of the starting plane curve singularity. In [9, Corollary 7.14] we obtained a second proof of part of their reembedding theorem as a particular case of our general results about splice type singularities.

3.9 Variants and Extensions of Local Tropicalization

In this final section we describe briefly a natural extension of the notion of local tropicalization to the case of a general subtoric germ, which can possess components in the toric boundary (see Sect. 3.9.1), as well as other generalizations by several authors (see Sects. 3.9.2 and 3.9.3).

3.9.1 Extended Local Tropicalization

Till now we have always assumed that the subtoric germ $Y \hookrightarrow (\mathcal{X}_\sigma, 0)$ was *interior* in the sense of Definition 3.4.1, that is, without irreducible components contained in the toric boundary $\partial\mathcal{X}_\sigma$. However, it is sometimes necessary to relax this condition. It is the fourth definition of the local tropicalization, based on valuative weight vectors (see point *4.* of Theorem 3.6.2), that most naturally can be generalized.

To illustrate the difficulty that arises when the subtoric germ is not interior, let us start with an example. Let again $\mathbb{K}$ be an algebraically closed field and $\mathcal{X}_\sigma = \operatorname{Spec}(\mathbb{K}[x_1, \ldots, x_n])$ be the affine space considered as a toric variety. Assume that the subtoric germ $Y \hookrightarrow (\mathcal{X}_\sigma, 0)$ is contained in the coordinate hyperplane defined by $x_1 = 0$, but in no other coordinate hyperplane. This means that $x_1 \in I_Y$, but $x_i \notin I_Y$ for any $i \neq 1$, where $I_Y \subset \mathbb{K}[[x_1, \ldots, x_n]]$ is the defining ideal of Y in $\mathcal{O}_{\mathcal{X}_\sigma,0}$. Now, if we try to compute the local tropicalization of Y according, say, to point *4.* of Theorem 3.6.2, we see that since $r_Y(x_1) = 0 \in \mathcal{O}_Y$, we must have $\nu \circ r_Y(x_1) = +\infty$ for *each* semivaluation ν of Y, where the morphism of monoids

$$r_Y : (\Gamma, +) \to (\mathcal{O}_Y, \cdot)$$

was introduced in Definition 3.5.8. Thus there are no interior semivaluations and the local tropicalization is empty. To get something nontrivial, we have to allow the semivaluation ν to take infinite values on Γ, but then none of the corresponding valuative weight vectors of Y can be interpreted as an element of the real vector space $N_{\mathbb{R}}$.

To see how to deal with this issue, let us denote by Γ' the submonoid of $\Gamma = \mathbb{Z}^n_{\geq 0}$ generated by the exponent vectors of the monomials $x_2, \ldots, x_n$. Observe that despite the fact that any semivaluation ν of Y takes value $+\infty$ on x_1, such a semivaluation can be finite on $r_Y(\Gamma')$. If this is the case, then ν defines a linear form on the space $M'_{\mathbb{R}}$, where M' is the sublattice of M generated by Γ'.

Recall the simple fact from linear algebra that if U is a linear subspace of a finite dimensional vector space V, then a linear form on U is the same thing as an element of the quotient space $V^\vee/U^\perp$, i.e., there exists a canonical isomorphism

$$U^\vee \simeq V^\vee/U^\perp,$$

where $V^\vee$ is the dual space of V and $U^\perp \subseteq V^\vee$ is the orthogonal of U. It follows that we can interpret the general valuative weight vectors of Y as elements of appropriate quotients of the space $N_{\mathbb{R}}$. Following an approach initiated by Ash, Mumford, Rapoport and Tai in [1, Section I.1] and developed by Kajiwara [20] and Payne [37], we regard those quotients as *strata at infinity* that can be adjoined to the space $N_{\mathbb{R}}$.

The formal definition proceeds as follows. Let as before σ be a strictly convex rational polyhedral cone in $N_{\mathbb{R}}$. If τ is a face of σ, denote $\boxed{N_\tau} := N \cap \mathbb{R}\tau$, where $\mathbb{R}\tau$ is the real vector subspace of $N_{\mathbb{R}}$ spanned by τ. Since the abelian group N_τ is a *saturated* sublattice of N, then the quotient N/N_τ is also a lattice. Consider the disjoint union:

$$\boxed{L(\sigma, N)} := \bigsqcup_{\tau \preceq \sigma} (N/N_\tau)_{\mathbb{R}}, \tag{3.10}$$

where τ varies in the set of faces of the polyhedral cone σ.

It is a formal exercise to see that $L(\sigma, N)$ can be identified with the set of all monoid homomorphisms from $(\Gamma := \sigma^\vee \cap M, +)$ to $(\boxed{\overline{\mathbb{R}}} := \mathbb{R} \cup \{+\infty\}, +)$. Moreover, using the standard topology on $\overline{\mathbb{R}}$, for which a basis of neighborhoods of $+\infty$ is the set of all intervals of the form $(a, +\infty]$, we can endow $L(\sigma, N)$ with the topology of pointwise convergence. Then, (3.10) represents $L(\sigma, N)$ as a disjoint union of locally closed subsets, which we call its **strata**.

Each stratum $(N/N_\tau)_{\mathbb{R}}$ carries naturally an integral lattice N/N_τ. The cone $\sigma \subset N_{\mathbb{R}}$ has images in all the strata $(N/N_\tau)_{\mathbb{R}}$, by the canonical projections $N_{\mathbb{R}} \to (N/N_\tau)_{\mathbb{R}}$. We say that the disjoint union of σ with all those images is the **extended cone** $\boxed{\overline{\sigma}}$. It is a closed topological subspace of $L(\sigma, N)$ which can be alternatively defined as the closure of the cone σ in $L(\sigma, N)$. The topological spaces $L(\sigma, N)$ had appeared several times in the literature, but since there seemed to be no standard name for them, we called them **linear varieties** in [39, Section 4]. We chose that name by analogy with that of *toric variety*. Indeed, similarly to the fact that a toric variety is endowed with an action of an algebraic torus embedded in it as a dense open set, a linear variety is endowed with an action of a linear space which is also embedded in it as a dense open set (see [39, Remark 4.4]).

Definition 3.9.1 Let $Y \hookrightarrow (\mathcal{X}, 0)$ be a subtoric germ, not assumed to be interior. Then, its **extended local tropicalization** is defined as the closure in the extended cone $\overline{\sigma}$ of all valuative weight vectors of Y of the form $\nu \circ r_Y$ (see Definition 3.5.8), where ν ranges over all (not necessarily interior) semivaluations of Y.

The notion of rational polyhedral fan has its extended version, and an analog of the structure Theorem 3.6.5 holds for extended local tropicalization. In fact, the main finiteness theorem is formulated and proven in [39, Theorem 11.9] for extended local tropicalization (which we called there *non-negative local tropicalization*).

In the case when Y is not contained in the toric boundary of $\mathcal{X}$, Definition 3.9.1 generalizes Definition 3.6.3 of local tropicalization, in the sense that $\text{Trop}_{\text{loc}} Y$ is obviously contained in the extended local tropicalization. Then, the question arises whether the extended and non-extended tropicalizations are essentially different from each other. The answer is *no*, and this again follows formally from [39, Theorem 11.9]:

Proposition 3.9.2 *If $Y \hookrightarrow (\mathcal{X}_\sigma, 0)$ is an interior subtoric germ, then the extended local tropicalization of Y is the closure in $L(\sigma, N)$ of the local tropicalization* $\text{Trop}_{\text{loc}} Y$ *in the sense of Definition 3.6.3.*

Remark 3.9.3 We have to warn the reader wishing to continue studying the local tropicalization from [39] about a terminological difference with the present paper. In [39, Definitions 6.6 and 6.7], we distinguished between the *positive* and *nonnegative local tropicalization*, whereas here we speak simply about *local tropicalization*. In fact, the main Definitions 3.6.3 and 3.9.1 of this paper correspond to the nonnegative version of [39]. The difference with the positive local tropicalization can be explained as follows. Let Y be a subgerm of an affine toric germ $(\mathcal{X}_\sigma, 0)$. Then we get the positive local tropicalization if in Definition 3.9.1 we consider only the semivaluations whose *center* is exactly the unique closed orbit 0 of $\mathcal{X}_\sigma$, and the closure operation is performed in the *interior* of the extended cone $\overline{\sigma}$. If we allow the semivaluations whose center only *contains* the unique closed orbit (as is implicitly done in Definition 3.9.1, see also the very Definition 3.4.4 of semivaluation used in this paper), then we get the nonnegative local tropicalization. The two local tropicalizations determine each other, see [39, Theorems 11.2, 11.9]. In this introductory text we felt no need to overload the reader's attention with such technical details.

To give the reader a better feeling of what is the extended tropicalization, let us consider one more example. In all our definitions, it creates no problem to take the germ Y to be *equal* with the germ $(\mathcal{X}_\sigma, 0)$ of affine toric variety itself. *Let us see what are the local tropicalization of* $(\mathcal{X}_\sigma, 0)$ *and its extended version.* Assume that the base field $\mathbb{K}$ is the field $\mathbb{C}$ of complex numbers. First recall from Sect. 3.4 that the set of $\mathbb{C}$-valued points $\mathcal{X}_\sigma(\mathbb{C})$ of the toric variety $\mathcal{X}_\sigma$ corresponding to a cone $\sigma \subset N_\mathbb{R}$ can be identified with the set of *monoid homomorphisms* $\sigma^\vee \cap M \to \mathbb{C}$, where $\mathbb{C}$ is considered as a monoid with respect to multiplication. If we replace $(\mathbb{C}, \cdot)$ by $(\mathbb{R}, \cdot)$, we get in a similar way the set of real points $\mathcal{X}_\sigma(\mathbb{R})$ of $\mathcal{X}_\sigma$, and, if we replace $(\mathbb{C}, \cdot)$ by the monoid $(\mathbb{R}_{\geq 0}, \cdot)$ of nonnegative real numbers, we get its set $\mathcal{X}_\sigma(\mathbb{R}_{\geq 0})$ of nonnegative real points. The group $(\mathbb{R}_{>0}, \cdot)$ is isomorphic via the continuous isomorphism $\boxed{\log}$ to the group $(\mathbb{R}, +)$, and if we extend $\log$ by setting formally $\log 0 := -\infty$, we get an isomorphism

$$- \log \colon (\mathbb{R}_{\geq 0}, \cdot) \to (\overline{\mathbb{R}}, +)$$

of monoids. Note that the subset $[0, +\infty]$ of $\overline{\mathbb{R}}$ corresponds under this isomorphism to the segment $[0, 1]$ of $\mathbb{R}_{\geq 0}$.

By composing any monoid morphism $\rho \colon (\sigma^\vee \cap M, +) \to (\mathbb{R}_{\geq 0}, \cdot)$ with $-\log$, we get a monoid morphism

$$(-\log) \circ \rho \colon (\sigma^\vee \cap M, +) \to (\overline{\mathbb{R}}, +).$$

If such a morphism does not take the value $+\infty$, then it is simply an element of $\sigma \subset N_\mathbb{R}$. Otherwise, it is an element of $\overline{\sigma} \subset L(\sigma, N)$ belonging to a stratum at infinity, that is, different from $N_\mathbb{R}$. It is easy to see that any such element can be realized by a semivaluation on the formal local ring $\mathbb{C}[[\Gamma]]$ of $(\mathcal{X}_\sigma, 0)$. For example, if $w \in \sigma^\circ$, we can define the associated semivaluation (which is in fact a valuation) by:

$$\nu_w(f) := \min_{m \in \operatorname{supp} f} w \cdot m, \quad \text{for every } f \in \mathbb{C}[[\Gamma]].$$

We conclude that:

Proposition 3.9.4 *One has* $\operatorname{Trop}_{\text{loc}}(\mathcal{X}_\sigma, 0) = \sigma$*, whereas the extended local tropicalization of* $(\mathcal{X}_\sigma, 0) \hookrightarrow (\mathcal{X}_\sigma, 0)$ *is the extended cone* $\overline{\sigma}$*.*

Example 3.9.5 In Fig. 3.7 are shown the structures of the extended local tropicalizations of formal germs $(\mathcal{X}_\sigma, 0)$, for a 2-dimensional cone σ on the left and a three-dimensional one with four rays on the right (see also [39, Figures 3 and 4]). If ζ is a face of the cone σ, we denote by σ_ζ the image of σ by the quotient linear map $N_\mathbb{R} \to (N/N_\zeta)_\mathbb{R}$. It is the intersection of the extended cone $\overline{\sigma}$ with the stratum $(N/N_\zeta)_\mathbb{R}$. We denote by λ_1, λ_2 the edges of the left-hand example and by λ, τ an edge and a 2-dimensional face respectively of the right-hand 3-dimensional example. In both drawings we have indicated by a filled disk the 0-dimensional stratum $(N/N_\sigma)_\mathbb{R}$ at infinity.

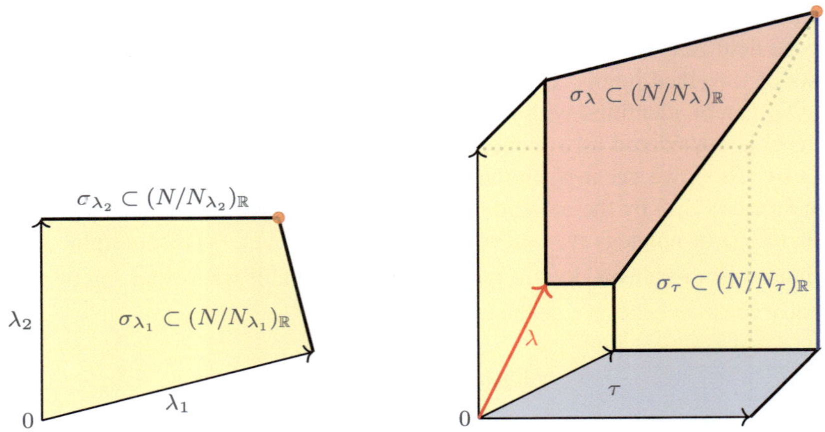

Fig. 3.7 The extended local tropicalizations $\overline{\sigma}$ of a germ of affine toric variety $(\mathcal{X}_\sigma, 0)$ of dimension 2 on the left and dimension 3 on the right (see Example 3.9.5)

In [39, Section 12] we have compared local tropicalizations with the global tropicalization of subvarieties of algebraic tori. We showed in particular (see [39, Theorem 12.11]) that if $Y \hookrightarrow \mathcal{X}_\sigma$ is an algebraic subvariety of the affine toric variety $\mathcal{X}_\sigma$ with dense torus T_N, whose germ Y is an interior toric subgerm, then the local tropicalization of Y is equal to the intersection of the cone σ with the global tropicalization of $Y \cap T_N \hookrightarrow T_N$.

Local tropicalization has also *functorial properties*, described in [39, Proposition 6.12]. This allowed us to generalize in [39, Section 13] the definition of local tropicalization to germs of subvarieties of toric (not necessarily affine) varieties, as well as to subvarieties of toroidal embeddings in the sense of [23]. However, a much stronger generalization is possible, as we explain in Sect. 3.9.2.

3.9.2 Ulirsch's Tropicalization of Logarithmic Schemes

The most basic piece of data which gives rise to a local tropicalization is a morphism of monoids

$$\gamma : (\Gamma, +) \to (\mathcal{O}, \cdot) \tag{3.11}$$

where $\Gamma := \sigma \cap M^\vee$ as before and $\mathcal{O}$ is a local ring, such that γ sends all the non-units of Γ to the maximal ideal of $\mathcal{O}$. If, from a general point of view, the choice of monoids Γ of that particular form seems not motivated enough, note that they can be characterized by abstract properties as the monoids which are simultaneously *commutative, finitely generated, cancellative, torsion free*, and *saturated*. The monoids satisfying the first three properties are also called *fine*.

Morphisms of the form (3.11) were the basic building blocks used in [39] for the development of various generalizations of the local tropicalizations, trying to stick to elementary methods. The theory of *logarithmic structures*, initiated by Fontaine and Illusie and first described by Kato in his 1989 paper [21], incorporates monoid morphisms of this sort in a very general framework. For details on this framework, one may consult Ogus' textbook [34].

A **logarithmic structure** on a ringed space $(X, \mathcal{O}_X)$ is a sheaf $\mathcal{M}_X$ of monoids together with a morphism $\alpha\colon \mathcal{M}_X \to (\mathcal{O}_X, \cdot)$ of sheaves of monoids, such that α induces an isomorphism $\alpha^{-1}(\mathcal{O}_X^*) \simeq \mathcal{O}_X^*$.

An important class of logarithmic structures comes from pairs (X, D), where X is a normal scheme and D is a hypersurface of it. The associated **divisorial logarithmic structure** on X is defined by

$$\mathcal{M}_X(U) := \{f \in \mathcal{O}_X(U) \mid f|_{U-D} \in \mathcal{O}_X^*(U)\},$$

for every open subset U of X. That is, $\mathcal{M}_X$ is the sheaf of regular functions on X which do not vanish outside D. Toric varieties and toroidal embeddings possess therefore canonical logarithmic structures, which are the divisorial logarithmic structures induced by their boundaries. Moreover, to each sufficiently good logarithmic scheme (namely a fine and saturated one), it is possible to associate a *conical complex*, both in an extended and non-extended variants. Informally, a conical complex is like a fan, but its maximal cones do not have to be contained in a fixed vector space. In [44], Martin Ulirsch developed a theory of tropicalization of logarithmic schemes and their subvarieties. In the special case where X is an affine toric variety, it recovers the local tropicalization described at the end of Sect. 3.9.1. In [43], Ulirsch proved an analog of (1) of Theorem 3.6.2 for the case of so called *log-regular varieties*.

We would like to mention one more aspect of the theory of logarithmic structures which is of particular interest to singularity theory. Namely, this theory provides also a general framework for the operation of *rounding*, or *real oriented blowing up*, of algebraic varieties and analytic spaces (see [38]). This gives a new way of approaching some problems connected to Milnor fibrations, as explained in [8, Section 3.1].

3.9.3 Other Generalizations of Local Tropicalization

In [11, Section 2], Alexander Esterov defined a notion of *local tropical fan* as an independent object, i.e., not necessarily coming from a tropicalization. Then, in [11, Definition 3.4] he defined the local tropicalization of an analytic germ $Y \hookrightarrow \mathcal{X}_\sigma$ of pure dimension k essentially as in point *1.* of Theorem 3.6.2, but endowing it with an additional structure, namely a *weight function* that assigns a positive integral weight to each maximal cone of a particular kind of fan $\mathcal{F}$ whose support is the local tropicalization. If Y is of pure dimension k, then the weight of a k-dimensional cone τ equals the intersection number of Y with the orbit O_τ corresponding to τ (see also [9, Definition 3.20]). Moreover, in order to develop the intersection theory of local tropical fans and of local tropical characteristic classes he showed that one can associate to a k-dimensional subtoric germ not only the tropicalization above, which is a k-dimensional local tropical fan, but a sequence of tropical fans of dimension $0, 1, \ldots, k$, where the support of the k-dimensional fan is the local tropicalization, and which reflects more information about the germ. Esterov applied this extension of the theory of local tropicalization to the study of the monodromy of complex analytic functions.

We would like also to mention two recent developments in the theory of tropicalization. They are however not so much concerned with the local tropicalization as with the general question of determining which are the most general objects which may be tropicalized.

In [17] and [18], Jeffrey and Noah Giansiracusa investigated the connections between *tropicalization*, $\mathbb{F}_1$*-schemes* (i.e., schemes over the so-called *field with one element*), and *Berkovich analytification* of algebraic varieties. A $\mathbb{K}$-scheme **with a model over** $\mathbb{F}_1$ is a scheme over a field $\mathbb{K}$ which admits an affine covering where the coordinate ring of each affine piece is a monoid ring and their gluing is induced by localizations of monoids. In the affine case $X = \text{Spec}(A)$, where A is a $\mathbb{K}$-algebra and $\mathbb{K}$ is a valued field, the underlying set of the **Berkovich analytification** X^{an} of X is the set of all semivaluations of A extending the given valuation on $\mathbb{K}$. In particular, $\mathbb{K}$ can be trivially valued, that is, $0 \in \mathbb{K}$ has value $+\infty$ and all the other elements of $\mathbb{K}$ have value 0. The pioneering work on this subject was the 2009 paper [37] in which Sam Payne considered all possible embeddings of an affine algebraic variety X into different affine spaces $\mathbb{K}^n$, viewed as toric varieties. He noted that a monomial map between such embeddings induces a natural map between the corresponding extended global tropicalizations, and he showed that the associated category-theoretic limit of all such tropicalizations seen as sets can be identified with the analytification X^{an}.

In [17], J. and N. Giansiracusa generalized the construction of tropicalization from embeddings of some scheme X over $\mathbb{K}$ to toric varieties to the embeddings of X to a $\mathbb{K}$-scheme equipped with a model over $\mathbb{F}_1$. Also, they endowed this tropicalization with a kind of scheme structure. In [18], they constructed an embedding of a $\mathbb{K}$-scheme X to an $\mathbb{F}_1$-scheme $\widehat{X}$ which is *universal* in the sense that its tropicalization admits a morphism to the tropicalization of X with respect to any other embedding. Moreover, they proved that the Berkovich analytification X^{an} can be recovered as the tropicalization of this universal embedding $X \hookrightarrow \widehat{X}$. Finally, they reproved and refined Payne's limit result.

The paper [27] of Oliver Lorscheid is an attempt to determine even a deeper structure than that of an $\mathbb{F}_1$-scheme that underlies all the instances of tropicalization. The author proposed for that role his theory of *ordered blueprints*. Also this theory enhances the tropicalization with a scheme structure. One may find in [27] the technical definitions of the ordered blueprint and the *ordered blue scheme* as well as the references to other works of Lorscheid on the theory of blueprints. Here we only draw the reader's attention to the fact that Lorscheid's work succeeds in interpreting Berkovich analytification, Kajiwara-Payne tropicalization [20], J. and N. Giansiracusa tropicalization, Ulirsch tropicalization and some other approaches to tropical geometry via his theory of ordered blueprints.

Acknowledgments This research was funded, in whole or in part, by l'Agence Nationale de la Recherche (ANR), project SINTROP (ANR-22-CE40-0014) and Labex CEMPI (ANR-11-LABX-0007-01). We thank Jose Luis Cisneros Molina, Dũng Tráng Lê and Jose Seade for the invitation to contribute to this volume of the Handbook of Geometry and Topology of Singularities. We are grateful to María Angélica Cueto for everything we understood about local tropicalization during our collaboration with her and for her careful reading of the first version of this paper. We thank also Evelia García Barroso, Arthur Renaudineau, Bernard Teissier and the anonymous referee for their remarks on previous versions of this text.

References

1. A. Ash, D. Mumford, M. Rapoport and Y.-S. Tai, *Smooth compactification of locally symmetric varieties.* Math Sci Press, 1975. Second edition in 2010, Cambridge Univ. Press.
2. J.-P. Brasselet, Introduction to toric varieties. Publicações Matemáticas do IMPA. 23o Colóquio Brasileiro de Matemática. IMPA, Rio de Janeiro, 2001.
3. E. Brieskorn, Rationale Singularitäten komplexer Flächen. Invent. Math. **4** (1968), 336–358.
4. E. Brieskorn, *Singularities in the work of Friedrich Hirzebruch.* Surv. Differ. Geom. **7**, International Press, 2000, 17–60.
5. A. Campillo, J. Castellanos, *Curve singularities.* Hermann, 2005.
6. D. Cox, What is a toric variety? In *Topics in algebraic geometry and geometric modeling*, 203–223, Contemp. Math. **334**, Amer. Math. Soc., 2003.
7. D. Cox, J. Little and H. Schenck, *Toric varieties.* Springer, 2011.
8. M. A. Cueto, P. Popescu-Pampu and D. Stepanov, The Milnor fiber conjecture of Neumann and Wahl, and an overview of its proof. Chapter of the book *Essays in Geometry, Dedicated to Norbert A'Campo*, A. Papadopoulos ed., EMS Publishing House, Berlin, 2023, 629–709.
9. M. A. Cueto, P. Popescu-Pampu and D. Stepanov, Local tropicalizations of splice type surface singularities. With an appendix by J. Wahl. Math. Ann. **390** (2024), 811–887.
10. S. D. Cutkosky, *Resolution of Singularities.* Graduate Studies in Maths. **63**, AMS, 2004.
11. A. Esterov, The ring of local tropical fans and tropical nearby monodromy eigenvalues. arXiv:1807.00609v3. 21 September 2021.
12. A. B. de Felipe, P. González Pérez and H. Mourtada, Resolving singularities of curves with one toric morphism. Math. Ann. **387** (2023), 1853–1902.
13. W. Fulton, *Introduction to toric varieties.* Princeton Univ. Press, 1993.
14. E. García Barroso, P. González Pérez and P. Popescu-Pampu, The combinatorics of plane curve singularities. How Newton polygons blossom into lotuses. In *Handbook of Geometry and Topology of Singularities I.* J.-L. Cisneros Molina, D.T. Lê, J. Seade eds., 1–150. Springer, 2020.
15. E. García Barroso, A. Lenarcik and A. Płoski, Newton diagrams and equivalence of plane curve germs. J. Math. Soc. Japan **59** (2007), no. 1, 81–96.
16. É. Ghys, A singular mathematical promenade. ENS Éditions, Lyon, 2017.
17. J. Giansiracusa and N. Giansiracusa, Equations of tropical varieties. Duke Math. J. **165** (2016), no. 18, 3379–3433.
18. J. Giansiracusa and N. Giansiracusa, The universal tropicalization and the Berkovich analytification. Kybernetika (Prague) **58** (2022), no. 5, 790–815.
19. R. Goldin and B. Teissier, Resolving singularities of plane analytic branches with one toric morphism. In *Resolution of singularities* (Obergurgl), Progr. Math. **181**, 315–340. Birkhäuser, 2000.

20. T. Kajiwara, *Tropical toric geometry.* In *Toric topology*, 197–207, Contemp. Math. **460**, Amer. Math. Soc., 2008.
21. K. Kato, Logarithmic structures of Fontaine-Illusie. In *Algebraic analysis, geometry, and number theory (Baltimore, MD, 1988)*, 191–224, Johns Hopkins Univ. Press, 1989.
22. E. Katz, A tropical toolkit. Expo. Math. **27** (2009), no. 1, 1–36.
23. G. Kempf, F. F. Knudsen, D. Mumford and B. Saint-Donat, *Toroidal embeddings I.* Lect. Notes in Maths. **339**, Springer-Verlag, Berlin, 1973.
24. A. G. Khovanskii, Newton polyhedra (resolution of singularities). Journ. Soviet Math., **27** (1984), 2811–2830.
25. H. B. Laufer, *Normal two-dimensional singularities.* Princeton Univ. Press, 1971.
26. M. Lejeune-Jalabert and B. Teissier, Transversalité, polygone de Newton, et installations. In *Singularités à Cargèse*, 75–119. Astérisque **7–8**, Soc. Math. France, Paris, 1973.
27. O. Lorscheid, A unifying approach to tropicalization. Trans. Amer. Math. Soc. **376** (2023), no. 5, 3111–3189.
28. D. Maclagan and B. Sturmfels, *Introduction to tropical geometry.* Grad. Stud. in Maths. **161**. Amer. Math. Soc., Providence, RI, 2015.
29. J. Maurer, Puiseux expansion for space curves. Manuscripta Math. **32** (1980), 91–100.
30. A. Némethi and B. Sigurðsson, Local Newton nondegenerate Weil divisors in toric varieties. arXiv:2102.02948.
31. W. Neumann and J. Wahl, Complete intersection singularities of splice type as universal abelian covers. Geom. Topol. **9** (2005), 699–755.
32. W. Neumann and J. Wahl, Complex surface singularities with integral homology sphere links. Geom. Topol. **9** (2005), 757–811.
33. T. Oda, *Convex bodies and algebraic geometry.* Springer, 1988.
34. A. Ogus, *Lectures on logarithmic algebraic geometry.* Cambridge Studies in Advanced Mathematics **178**. Cambridge University Press, 2018.
35. M. Oka, *On the stability of the Newton boundary.* Proc. Symp. Pure Math. **40**, American Math. Soc., 1983, 259–268.
36. M. Oka, *Non-degenerate complete intersection singularity.* Hermann, 1997.
37. S. Payne, Analytification is the limit of all tropicalizations. Math. Res. Lett. **16** (2009), no. 3, 543–556.
38. P. Popescu-Pampu, *Introduction to real oriented blowups in toric, toroidal and logarithmic geometries.* To appear.
39. P. Popescu-Pampu and D. Stepanov, Local tropicalization. In *Algebraic and Combinatorial aspects of Tropical Geometry.* Proceedings Castro Urdiales 2011, E. Brugallé, M.A. Cueto, A. Dickenstein, E.M. Feichtner and I. Itenberg editors, Contemp. Math. **589**, AMS, 2013, 253–316.
40. F. Pham, Formules de Picard-Lefschetz généralisées et ramification des intégrales. Bull. Soc. Math. France **93** (1965), 333–367.
41. D. Stepanov, Universal valued fields and lifting points in local tropical varieties. Comm. Algebra **45** (2017), no. 2, 469–480.
42. B. Teissier, Amibes non archimédiennes. In *Géométrie tropicale*, 85–114, Ed. Éc. Polytechnique, 2008.
43. M. Ulirsch, Tropical compactification in log-regular varieties. Math. Zeitschrift **280** (2015), 195–210.
44. M. Ulirsch, Functorial tropicalization of logarithmic schemes: the case of constant coefficients. Proc. Lond. Math. Soc. (3) **114** (2017), no. 6, 1081–1113.
45. M. Vaquié, *Valuations.* In *Resolution of singularities (Obergurgl, 1997)*, 539–590, Progr. Math. **181**, Birkhäuser, 2000.
46. A. Varchenko, *Zeta-Function of monodromy and Newton's diagram.* Invent. Math. **37** (1976), 253–262.
47. J. Wahl, Topology, geometry, and equations of normal surface singularities. In *Singularities and computer algebra*, 351–371, London Math. Soc. Lecture Note Ser. **324**, Cambridge Univ. Press, Cambridge, 2006.

48. J. Wahl, Splice diagrams and splice-quotient surface singularities. Celebratio Mathematica, article no. **1030**, Math. Sci. Publishers, 2022.
49. R. J. Walker, *Algebraic curves*. Princeton Univ. Press, 1950.
50. C. T. C. Wall, *Singular points of plane curves.* London Math. Soc. Student Texts **63**. Cambridge Univ. Press, 2004.
51. O. Zariski and P. Samuel, *Commutative algebra.* Vol. II. Reprint of the 1960 edition. Graduate Texts in Mathematics **29**. Springer-Verlag, 1975.

Chapter 4
Superisolated Singularities and Friends

Enrique Artal Bartolo

Contents

Abstract Superisolated surface singularities in $(\mathbb{C}^3, 0)$ were introduced by I. Luengo to prove that the μ-constant stratum may be singular. The main feature of this family is that it can bring information from the projective plane curves (global setting but smaller dimension) into surface singularities. They are simple enough to allow to retrieve information and complicated enough to offer a variety of properties. The so-called Lê-Yomdin singularities are a generalization which offers a wider catalog of examples. We study some properties of these singularities, mainly topological and related with the monodromy, and we introduce another family which exploits the same properties but in the quasi-homogeneous setting.

E. Artal Bartolo (✉)
Departamento de Matemáticas, IUMA, Universidad de Zaragoza, Zaragoza, Spain
e-mail: artal@unizar.es

© The Author(s), under exclusive license to Springer Nature Switzerland AG 2026

J. L. Cisneros-Molina et al. (eds.), *Handbook of Geometry and Topology of Singularities VIII*, https://doi.org/10.1007/978-3-031-99571-2_4

4.1 Introduction

This work is devoted to the relationship between two subjects: projective plane curves and germs of surface singularities in $\mathbb{C}^3$. A projective plane curve is defined by a homogeneous polynomial in three variables, and hence it defines a germ of (homogeneous) surface singularity in $(\mathbb{C}^3, 0)$. This germ will be in general non-isolated (actually it will be isolated if and only if the curve is smooth). On the other side, a germ of surface singularity in $(\mathbb{C}^3, 0)$ has associated a projective plane curve, its tangent cone.

Probably the first approach to this idea was done by Y. Yomdin [60] (see also [34]). Starting from a reduced homogeneous function $F_d(x, y, z) \in \mathbb{C}[x, y, z]$ of degree d, Yomdin proposes an *asymptotical* approach using *generic* functions $F_{d+k}(x, y, z)$, $k \in \mathbb{N}$, such that the zero locus V of $F_d + F_{d+k}$ has an isolated singularity at the origin (there is a more concrete condition which will be made explicit in the text).

Some years later, I. Luengo used similar ideas in [37] to study the smoothness of the μ-constant stratum of surface singularities in $\mathbb{C}^3$. He defined the family of *superisolated singularities* as the singularities which are (abstractly) resolved with one blow-up. Actually, they coincide with the Yomdin (Lê-Yomdin from now on) singularities for $k = 1$. While being a particular case, they can be defined in a more intrinsic way and he made a thorough study of their properties.

The main idea of Luengo is to translate properties of the projective plane curve C_d into properties of a superisolated singularity whose tangent cone is C_d, allowing to derive properties of singularities (local, dimension 2 embedded in dimension 3) from properties of plane curves (global, dimension 1 embedded in dimension 2).

As a consequence, Luengo proved that the μ-constant stratum of a germ of isolated singularities in $\mathbb{C}^3$ may be non smooth; note that the μ-constant stratum of a germ of an isolated singularity in $\mathbb{C}^2$ is smooth (see e.g. [57]). Examples of singular strata were found among superisolated singularities, and the key point of the proof came from the properties of the tangent cones as projective plane curves.

In the eighties, S.S.-T. Yau [59] stated the following conjecture. The abstract topology of the link of a germ of isolated singularity in $\mathbb{C}^3$ and the characteristic polynomial of the monodromy determines the embedded topology of the germ, see Conjecture 4.5.3 for a precise statement. The conjecture was verified in important infinite families of singularities. In the nineties, an embedded resolution of superisolated singularity was found in [5]. With this tool the link of the singularity V can be computed; actually it was already done [37] and it can be read from C_d. The geometric monodromy of the Milnor fibration of V can also be described as some of its properties. Namely the characteristic polynomial of the action of the monodromy on the cohomology of the Milnor fibration (previously computed by J. Stenvens [54]) and also its Jordan form. How the properties of C_d are related to the above properties of V can be used to discard the conjecture. The purpose of this work is to show this parallelism in order to present how they benefit from each other, and also to present problems in this field which are open since the nineties.

In the nineties, W.D. Neumann and J. Wahl studied the classic Casson's conjecture and refined it in the case of links of germs of isolated singularities in $\mathbb{C}^3$, supporting the conjectures by large families of examples. These conjectures were generalized by A. Némethi and L.I. Nicolaescu to conjectures involving finer invariants such as Seiberg-Witten's. These conjectures are related to oriented closed 3-manifolds whose link is a $\mathbb{Q}$-homology sphere. In [27] characterized the superisolated singularities of surface whose link is a $\mathbb{Q}$-homology sphere in terms of the tangent cone, leading mainly to rational cuspidal curves. As for the previous problems, the properties of those curves furnished counterexamples to these conjectures and gave key ideas to reformulate problems.

These are three concrete examples of how superisolated singularities have helped to the study of more general singularities of surface. There is survey in 2006 [15], which develop also these ideas, mainly from an algebraic point of view. In this work we recall some of these properties but we focus specially on topological properties of the singularities coming from topological properties of the tangent cones, and we extend the results to Lê-Yomdin singularities, since having a larger family it is likely that more problems can be attacked in the future.

We may study this problem in higher dimensions, but since there are quite interesting problems as it is, we will focus on surface singularities. In this work we will present some other generalizations. As in the homogeneous case, a *weighted* projective plane curves is defined by a quasi-homogeneous function in three variables, and hence it defines also a germ of singularity in $(\mathbb{C}^3, 0)$. In most cases, this germ has non isolated singularities. Adding *generic* terms of weighted degree greater than the original one, the singularity becomes isolated and under this genericity conditions, its properties depend on the weighted projective curve. These ideas lead in [14] to the definition of the *weighted Lê-Yomdin singularities*.

In Sect. 4.2, we introduce the main concepts, notations and first properties of superisolated and Lê-Yomdin singularities. In Sect. 4.3 is devoted to the other object of interests, the projective plane curves; we define the notions of combinatorics and Zariski pairs. The combinatorics of the tangent cone of a superisolated singularity is the main ingredient to determine its link; this is recalled in Sect. 4.4, where the links of superisolated and Lê-Yomdin singularities are described. In Sect. 4.5 we introduce the main topological invariant for germs of hypersurface singularities, the Milnor fibration and its monodromy; we recall the classical theory and introduce the necessary notations. Monodromy have been defined, but in general its computation is a challenge; embedded resolutions, described in Sect. 4.6, provide a model for the monodromy. Steenbrink used an embedded resolution to endow the cohomology of the Milnor fiber with a Mixed Hodge Structure. One ingredient of this Mixed Hodge Structure is the weight filtration which is topological nature; Sect. 4.6 ends with the description of this weight filtration and its relationship with the monodromy, some algebraic aspects have been pushed to Sect. 4.13.

There are two main goals in Sect. 4.7. The first one is to describe an embedded resolution for superisolated and Lê-Yomdin singularities; this is very long for superisolated singularities and actually undone for weighted Lê-Yomdin singularities and this is the reason of the second goal, the definition of $\mathbb{Q}$-resolutions which allow to simplify its presentation for superisolated singularities and to effectively construct one such embedded $\mathbb{Q}$-resolutions for Lê-Yomdin singularities. Moreover, we see that Steenbrink theory is also valid for $\mathbb{Q}$-resolutions. This section needs the material in Sects. 4.11 and 4.12.

The next sections are more devoted to problems than two results. The previous study of the monodromy only holds for fields of characteristic 0; Sect. 4.8 deals with a method of M. Escario to compute the monodromy over $\mathbb{Z}$ and relates it to a question of N. A'Campo. In Sect. 4.9, we state some open topological problems for superisolated singularities. We end with Sect. 4.10 introducing the weighted Lê-Yomdin singularities where projective plane curves are replaced by weighted projective plane curves. There is a still some work to do about these singularities, so they are only introduced; they have been helpful to solve problems where superisolated and Lê-Yomdin singularities fail, so they enlarge the options of future work.

Summarizing, all these type of singularities are simple enough to allow a thorough study and complicated enough to enlighten deep properties of singularities.

4.2 Definitions and First Properties

Definition 4.2.1 (Luengo [37]) Let $(V, 0) \subset (\mathbb{C}^3, 0)$ be a germ of isolated surface singularity. We say that V is a *superisolated surface singularity* (SIS) if it is resolved after one blow-up.

More precisely, let $\sigma : (\hat{\mathbb{C}}^3, E) \to (\mathbb{C}^3, 0)$ be the blow-up of the origin, i.e.,

$$\hat{\mathbb{C}}^3 := \{(p, \ell) \in \mathbb{C}^3 \times \mathbb{P}^2 \mid p \in \ell\}, \quad E := \{0\} \times \mathbb{P}^1,$$

and σ is the restriction of the first projection; we identify $\mathbb{P}^2$ with the set of lines through 0. The *strict transform* of V is $\hat{V} := \overline{\sigma^{-1}(V \setminus \{0\})}$, and the restriction $\hat{\sigma} : \hat{V} \to V$ is the blow-up of V. Hence V is superisolated if $\hat{V}$ is smooth. There is a simple characterization of SIS.

Proposition 4.2.2 ([37]) *Let us suppose that V is the zero locus of $F(x, y, z) \in \mathbb{C}\{x, y, z\}$. Let*

$$F(x, y, z) = F_d(x, y, z) + \sum_{m>d} F_m(x, y, z), \qquad F_d \neq 0, \tag{4.1}$$

be the decomposition in homogeneous forms (the subscript stands for the degree). Let $C_m := V_{\mathbb{P}}(F_m) \subset \mathbb{P}^2$ be the projective zero locus of F_m (a maybe non-reduced curve if $F_m \neq 0$).

Then, V is SIS if and only if $\operatorname{Sing}(C_d) \cap C_{d+1} = \emptyset$.

This proposition relates SIS and Lê-Yomdin singularities.

Definition 4.2.3 Let $(V, 0) \subset (\mathbb{C}^3, 0)$ be a germ of isolated surface singularity. We say that V (zero-locus of $F(x, y, z) \in \mathbb{C}\{x, y, z\}$) is a *Lê-Yomdin singularity* of order k ($k-$LYS) if

$$F(x, y, z) = F_d(x, y, z) + \sum_{m \geq d+k} F_m(x, y, z), \qquad F_d \neq 0,$$

and $\mathrm{Sing}(C_d) \cap C_{d+k} = \emptyset$.

Remark 4.2.4 Hence SIS is equivalent to 1-LYS. It is interesting to study the above proposition and definition in local coordinates. Let us assume that $P = [0 : 0 : 1]$ is a point of the tangent cone C_d. In particular $[0 : 0 : 1] = ((0, 0, 0), [0 : 0 : 1]) \in \hat{V} \cap E$ and it is the image of the origin of $\mathbb{C}^3$ by the chart

$$\begin{aligned} \mathbb{C}^3 &\xrightarrow{\Phi} \hat{\mathbb{C}}^3 \\ (x, y, z) &\longmapsto ((xz, yz, z), [x : y : 1]). \end{aligned} \tag{4.2}$$

Under these conditions E has $z = 0$ as equation while the total transform $\sigma^* V$ has as equation

$$z^d(f_d(x, y, 1) + z^k(f_{d+k}(x, y, 1) + z f_{d+k+1}(x, y, 1) + \ldots)) = 0.$$

If we replace z by the product z_1 of z and a suitable unit at the origin, the local equation of E is still $z_1 = 0$ while the local equation of $\hat{V}$ is $f_d(x, y, 1) + z_1^k = 0$. For later use, let us denote $f_d(x, y, 1)$ as $f_{d,P}$. Then,

(Loc1) If P is a smooth point of C_d, then P is a smooth point of $\hat{V}$ and $E \pitchfork \hat{V}$.
(Loc2) If P is a singular point of C_d and $k = 1$, then P is a smooth point of $\hat{V}$ and $E \not\pitchfork \hat{V}$.
(Loc3) If P is a singular point of C_d and $k > 1$, then P is a singular point of $\hat{V}$.
(Loc4) In all these cases, the local topological type at P of $E \cup \hat{V}$ is determined by the topological type of (C_d, P) and k.

Remark 4.2.5 As a first consequence the tangent cone of a LYS is a reduced curve. The idea behind k-LYS is to think about them as series with $k \gg 0$, while a SIS can be interpreted as a *generic* isolated singularity with a fixed tangent cone. In the case of non-reduced curves, A. Melle [42] defined the t-singularities with the same ideas, both in the definition of *generic* singularities with fixed tangent cone and the definition of an infinite series of singularities. The main feature is that one can stratify the tangent cone such that (Loc4) is essentially true and the topological type is constant at each stratum.

Let us come back to the discussion of Remark 4.2.4. The local singularity of $(\hat{V}, P)$ is an interesting type of singularity which has been thoroughly studied.

Definition 4.2.6 An isolated germ of hypersurface in $\mathbb{C}^3$ is a *k-suspension* if it admits an equation of the type $z^k - f(x, y) = 0$, where f defines an isolated germ of curve singularity.

Note that any isolated singularity of multiplicity 2 is a 2-suspension, which have been thoroughly studied [32, 33]. It is not the case for higher multiplicities.

4.3 Combinatorics and Zariski Pairs

A key component of SIS and k-LYS is the tangent cone which is a reduced projective plane curve. Let us introduce some concepts which will be useful later.

Let $C = C^1 \cup \cdots \cup C^r$ be the irreducible decomposition of a reduced projective plane curve; we will denote d its degree, and d_i the degree of C^i, i.e., $d = d_1 + \cdots + d_r$. A consequence of Zariski-van Kampen theory is that

$$H_1(\mathbb{P}^2 \setminus C; \mathbb{Z}) \cong \mathbb{Z}^{r-1} \oplus (\mathbb{Z}/\gcd(d_1, \ldots, d_r)).$$

Consider the minimal embedded resolution $\rho : Y \to \mathbb{P}^2$ of the points in $\operatorname{Sing}'(C)$, the singular points which are not nodes belonging to distinct irreducible components of C.

Definition 4.3.1 The *combinatorics* of C is the dual graph Γ_C of $\rho^{-1}(C)$ decorated by the self-intersections and where the vertices corresponding to the strict transforms are marked.

For an irreducible curve the combinatorics determines and is determined by the degree and the topological types of the singularities. A more sophisticated example is in Fig. 4.1.

Lemma 4.3.2 *The genera of the irreducible components corresponding to the vertices can be recovered from the combinatorics.*

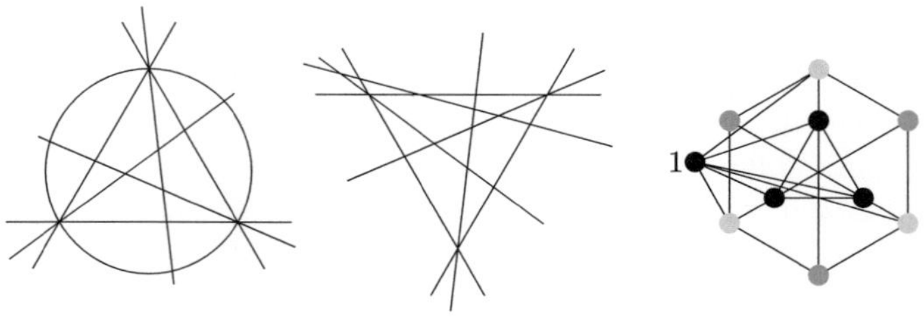

Fig. 4.1 Curves C^1, C^2 with common (unmarked) graph. The self-intersection of the undecorated black vertices is 0; the self-intersection of the gray vertices is -1

Proof The non-marked vertices correspond to rational curves. Let v be a marked vertex corresponding to the strict transform $\hat{C}_i$ of C_i. Let us consider the connected components of non-marked vertices. They correspond to the points in $\operatorname{Sing}'(C)$. Knowing $\hat{C}_i^2$ and the contractions we can recover C_i^2 which is d_i^2. Note that $\operatorname{Sing}'(C) \cap C_i \supset \operatorname{Sing}(C_i)$; we can compute for $p \in \operatorname{Sing}(C_i)$

$$\delta_p(C_i) = \frac{\mu_p(C_i) - r_p(C_i) + 1}{2}$$

where $\mu_p(C_i)$ is the Milnor number and $r_p(C_i)$ is the number of local branches. Then, the genus of C_i equals

$$\frac{(d_i - 1)(d_i - 2)}{2} - \sum_{p \in \operatorname{Sing}(C_i)} \delta_p(C_i)$$

and the result follows. □

Proposition 4.3.3 *Let us C', C'' be two plane curves. They have the same combinatorics if and only if there are regular neighbourhoods $N(C'), N(C'')$ and a homeomorphism of pairs $(N(C'), C') \to (N(C''), C'')$.*

Proof Let us consider the graph Γ of the combinatorics of a curve C. Let us construct a model for $(N(C), C)$ from Γ. Consider the minimal resolution Γ. We can pick up a tubular neighbourhood $T(C_v)$ for each irreducible component of $\rho^{-1}(C)$ where v is the corresponding vertex. This is an oriented disk-bundle onto an oriented closed Riemann surface of genus g_v and Euler number e_v, i.e., determined by Γ. We can choose these tubular neighbourhoods such that for each double point of $\rho^{-1}(C)$ the intersection of the two tubular neighbourhoods is a polydisk in $\mathbb{C}^2$ and the curves are the intersection with the axes. With this construction, the boundary of the the union of the tubular neighbourhoods is a plumbing manifold [45] with plumbing graph Γ. The contraction of the singular points provide $N(C)$.

If two curves C', C'' have the same graph the above construction provides a homeomorphism $(N(C'), C') \to (N(C''), C'')$. If such a homeomorphism exist, Waldhausen's theory of graph manifolds [58] combined with Neumann's plumbing calculus show that the marked graphs must coincide. □

Example 4.3.4 The combinatorics of an irreducible curve is equivalent to the degree and the topological type of the curves. In general, the marking is not necessary but we have to take it into account because of Cremona transformations. Consider the curves

$$C^1 : xyz(xy + yz + xz)(x + 2y)(y + 2z)(z + 2x),$$

$$C^2 : xyz(x + y + z)(2x + y)(2y + z)(2z + x).$$

We pass from one to the other via the Cremona transformation $[x : y : z] \mapsto [yz : xz : xy]$. The black vertices are always marked; for C^1 (resp. C^2) the other marked vertices are the darker (resp. lighter) gray ones.

We can interpret Proposition 4.3.3 as follows: two curves with the same combinatorics have the same *local topology*.

Example 4.3.5 In [61] O. Zariski studied sextic curves with six ordinary cusps. These are irreducible curves. He studied a subfamily with a special property, the six cusps were on a conic and showed that for such a curve C' one has that $\pi_1(\mathbb{P}^2\setminus C') \cong \mathbb{Z}/2 * \mathbb{Z}/3$. In this paper he showed also that if for such a sextic C'' the cusps were not in a conic, then there was no epimorphism $\pi_1(\mathbb{P}^2 \setminus C'') \to \mathfrak{S}_3$, in particular, there is no homeomorphism $(\mathbb{P}^2, C') \to (\mathbb{P}^2, C'')$.

Definition 4.3.6 ([6]) Two reduced projective plane curves form a *Zariski pair* if they have the same combinatorics but there is no homeomorphism of $\mathbb{P}^2 \to \mathbb{P}^2$ sending one to the another.

Remark 4.3.7 Actually, in [61] Zariski did not show the existence of a Zariski pair, since he did not prove the existence of a curve C''. He found such sextics in [63] using deformation arguments (no explicit example) and he proved that his examples have abelian fundamental group. Several decades later, explicit examples appeared in [6, 47].

Construction 4.3.8 Given a curve $C = \{f(x, y, z) = 0\}$ of degree d, let

$$X_d := \{[x : y : z : t] \in \mathbb{P}^3 \mid t^d - f(x, y, z) = 0\}, \qquad \begin{array}{rcl} X_d & \xrightarrow{\tau} & \mathbb{P}^2 \\ [x : y : z : t] & \longmapsto & [x : y : z] \end{array}$$

be the d-cyclic cover of $\mathbb{P}^2$ ramified along C. The monodromy of the cover is generated by

$$\begin{array}{rcl} X_d & \xrightarrow{\eta} & X_d \\ [x : y : z : t] & \longmapsto & [x : y : z : \exp \frac{2i\pi}{d} t] \end{array}$$

Let us recall that if X'_d is any birational smooth model of X_d, then $H^1(X_d; \mathbb{C})$ and $H^1(X'_d; \mathbb{C})$ are isomorphic; moreover, this isomorphism commutes with the morphisms induced by η. This construction is closed to the one of Alexander polynomial in knot theory.

Definition 4.3.9 (Libgober [36]) The *Alexander polynomial* of C is the characteristic polynomial of the map induced by η on H^1.

In modern terms, Zariski proved in [62] that the Alexander polynomials of C', C'' were different, $t^2 - t + 1$ and 1, respectively. The Alexander polynomial can be obtained from the fundamental groups of their complements (which are not isomorphic). The interesting point is that Zariski computes the Alexander polynomial from some geometric properties of the curves and no fundamental group computation is needed. The difference in the Alexander polynomials comes from the geometric property of the cusps being or not in a conic.

Definition 4.3.10 An *Alexander-Zariski* pair is a Zariski pair where the Alexander polynomials of the curves are distinct.

Zariski's example is an Alexander-Zariski pair. We know now a lot of Zariski pairs, not all of them are Alexander-Zariski, sometimes there are geometric properties which characterize them but do not have an effect on the Alexander polynomial. There are also the so-called *arithmetic Zariski pairs* where the equations of the members of the pairs live in a number field and they are related by a Galois automorphism of the field, compare with the discussion in Sect. 4.9. A more detailed description can be found in [13], though many new features have appeared since then.

Let us summarize some of the dependence between combinatorics and topological properties.

Properties 4.3.11 Let C be a reduced projective planes curve.

(a) $H_*(\mathbb{P}^2 \setminus C; \mathbb{Z})$ depends only on the combinatorics.
(b) $\pi_1(\mathbb{P}^2 \setminus C)$ does not depend only on the combinatorics.
(c) The Alexander polynomial is not a combinatorial invariant.

4.4 Abstract Topology

Let us come back to surface singularities. The main difference between SIS and k-LYS for $k > 1$ is that the first blowing up does not solve the singularities of the surface. Nevertheless, since the tangent cone is reduced, whenever it is smooth the strict transform and the exceptional component are transversal. Let us elaborate these ideas.

Let V as in Definition 4.2.3. The total transform $\sigma^* V$ of V by σ is a divisor which decomposes as $\hat{V} + dE$. Recall that $E = \{0\} \times \mathbb{P}^2 \equiv \mathbb{P}^2$ is a projective plane and the tangent cone C_d is $E \cap \hat{V}$. Any irreducible compact divisor D of a variety W has an *Euler class* (or *self-intersection*) $e_W(D)$; this is true also when W is not singular, and in this case $e_W(D)$ can be a rational number. It is a linear class of divisors in D corresponding to the line bundle $\mathcal{O}(D)$ restricted to D. Sometimes it is also denoted as $E \cdot E = E^2 = (E^2)_W$ since it is actually an intersection (the notation for the ambient variety will drop if no ambiguity is expected).

For the exceptional divisor $E \subset \hat{\mathbb{C}}^3$ of σ the Euler class can be detected by the following property. Let H be any divisor in $\mathbb{C}^3$; then $E \cdot \sigma^* H = 0$.

If H is a hyperplane through the origin we have that $\sigma^* H = E + \hat{H}$, where $\hat{H}$ is the strict transform. It is clear that $\hat{H} \cap E = L$ which is a line (namely $\mathbb{P}(H)$). Hence:

$$0 = E \cdot \sigma^* H = E^2 + E \cdot \hat{H} = E^2 + L \Longrightarrow e(E) = E^2 = -L.$$

The intersection properties of $C_d = \hat{V} \cap E$ in $E \equiv \mathbb{P}^2$ come from Bézout Theorem. We use the Euler class to compute the intersection properties of C_d in V.

Proposition 4.4.1 *Let V be a k-LYS. Let $C_d := C^1 + \cdots + C^r$ the decomposition of C_d in irreducible divisors of degrees $d_1, \ldots, d_r$ with $d = d_1 + \cdots + d_r$. Then,*

$$(C_i \cdot C_j)_{\hat{V}} = \frac{d_i \cdot d_j}{k} \text{ if } i \neq j, \quad (C^i \cdot C^i)_{\hat{V}} = -\frac{d_i \cdot (d - d_i + k)}{k}.$$

Proof (Guidelines) The ideas for the proof are inspirated from [37] (resp. [10]) for the SIS (resp. k-LYS) case. Since $\hat{V}$ is generically transversal to E, we recall that then $(C^j \cdot C_d)_{\hat{V}} = (C^j \cdot e(E))_E = (C^j \cdot L)_E = -d_j$.

Let us start with the case $i \neq j$. In the SIS case, Luengo [37] showed that the intersections of C^i, C^j, $i \neq j$ are isomorphic in both E and $\hat{V}$, and then $(C^i \cdot C^j)_{\hat{V}} = d_i \cdot d_j$. In [10], using properties of intersections and coverings, we see that in general $(C^i \cdot C^j)_{\hat{V}} = \frac{d_i \cdot d_j}{k}$. Combining with the first formula, we obtain the statement. □

By the very definition of SIS, $\hat{\sigma} : \hat{V} \to V$ is a resolution of V. Note that from Proposition 4.4.1, there is no (-1)-curve, hence, $\hat{\sigma}$ is the minimal resolution of V. A minimal log resolution $\tilde{\sigma} : \tilde{V} \to V$ is obtained combining $\hat{\sigma}$ with the minimal embedded resolution of $\operatorname{Sing}'(C_d) \subset \hat{V}$. The divisor $\tilde{\sigma}^*(0)$ is a simple normal crossing divisor. A regular neighbourhood of $\tilde{\sigma}^*(0)$ in $\tilde{V}$ can be constructed as in Proposition 4.3.3 and its neighbourhood is also a graph manifold associated to the dual graph $\tilde{\Gamma}$; if Γ is the graph obtained in that proposition, note that they are equal if we forget the decorations. More precisely, the decorations of the non-marked vertices coincide for both graphs. If e_i (resp. $\tilde{e}_i$) is the Euler number of the vertex in Γ (resp. $\tilde{\Gamma}$) corresponding to C_i, then

$$e_i - \tilde{e}_i = d_i^2 - (-d_i(d - d_i + 1)) = d_i(d + 1),$$

since we have to perform the same blow-ups in order to get simple normal crossing divisors in both cases.

Proposition 4.4.2 ([10]) *If V is a k-LYS ($k > 1$), a log resolution $\tilde{\sigma} : \tilde{V} \to V$ is obtained from $\hat{\sigma}$ by performing the embedded resolution of the function $f_{d,P} : (\hat{V}, P) \to \mathbb{C}$, $\forall P \in \operatorname{Sing}(C_d)$.*

In particular, for $k \geq 1$, the topology of the link of V depends only on k and on the combinatorics of C_d.

Proof The discussion previous to the statement serves for $k = 1$. Let us consider now the case $k > 1$. Recall that $\operatorname{Sing}(C_d)$ and $\operatorname{Sing}(\hat{V})$ are the same sets.

Let $P \in \operatorname{Sing}(C_d)$, and let us assume that we are in the situation of Remark 4.2.4 where the local equation of $\hat{V}$ at $P = (0, 0, 0)$ is $z_1^k - f_{d,p}(x, y, 1) = 0$. Let us consider a (minimal) embedded resolution of the function $f_{d,P}$ in this singularity; its dual graph is completely determined by k and the topological type of C_d at P.

The union of the exceptional components for all $P \in \operatorname{Sing}(C_d)$ with the strict transform of f is a simple normal crossing divisor and we obtain a log resolution of $\hat{V}$. Moreover, the decorated dual graph is completely determined by the combinatorics of C_d. □

Remark 4.4.3 There is a simple way to produce a log-resolution of the k-suspension of $f(x, y) = 0$. Note that the link of the singularity is the cyclic cover of $\mathbb{S}^3$ ramified along the link of $f(x, y) = 0$. There is a general procedure by C. Safont [50] to obtain the plumbing graph of a cover of a graph manifold, which can be extended to the case where it is ramified along fibers of the blocks. When $k = 2, 3$, this procedure can be simplified. When $k > 3$, one needs to use the theory of quotient singularities, see Sect. 4.11. The normalization Z of the pull-back of the suspension and an embedded resolution of $f = 0$ is a singular surface. The singular points are the preimages of the double points of the divisor of the embedded resolution, and they are the normalizations of singularities with equations $z^k - u^a v^b = 0$ which are disjoint unions of quotient singularities, see Example 4.11.9. How to deal with the intersections numbers in the singular space Z is done in [17], and for its resolution in Proposition 4.11.8.

Remark 4.4.4 There is a global version of the previous remark. Since the topology of a k-LYS does not depend on the particular $f_m, m \leq d+k$, as far as f_{d+k} satisfies the condition in Definition 4.2.3, we may assume that $F = f_d(x, y, z)+z^{d+k}$ where $f_d(x, y, 0)$ does not have multiple factors. Under this hypothesis $(x, y, z) \mapsto \zeta \cdot (x, y, z)$ preserves V if $\zeta^k = 1$. This action lifts to $\hat{V}$, the quotient $\hat{V}_k$ is smooth, the preimage of C_d is isomorphic to C_d and if we keep the notation for the irreducible components of C_d we have

$$(C_i \cdot C_j)_{\hat{V}_k} = d_i \cdot d_j \text{ if } i \neq j, \quad (C^i \cdot C^i)_{\hat{V}_k} = -d_i \cdot (d - d_i + k).$$

Since $\hat{V}$ is a k-cyclic cover of $\hat{V}_k$ ramified along C_d, a log-resolution of V can be effectively obtained from an embedded resolution of $C_d \subset \hat{V}_k$.

As we stated in the introduction, singularities whose link is a $\mathbb{Q}$-homology sphere is an important family. SIS and k-LYS with this property can be characterized.

Proposition 4.4.5 *Let V be a k-LYS with tangent cone C_d. If $k = 1$, the link of V is a $\mathbb{Q}$-homology sphere if and only if all these irreducible components are cuspidal rational curves and if C_d is reducible, then all the components intersect at only one point.*

If $k > 1$, besides the above conditions, $\forall P \in \operatorname{Sing}(C_d)$ the link of its k-suspension is also a $\mathbb{Q}$-homology sphere.

Proof Recall that the condition on the link is equivalent to ask that all the irreducible components are rational and the dual graph is a tree. □

4.5 On the Monodromy

One of the main invariants in singularity theory is the monodromy. Let $F : (\mathbb{C}^{n+1}, 0) \to (\mathbb{C}, 0)$ be the defining equation of a germ of isolated singularity of a hypersurface. Following Milnor [43] we know that there exists $\varepsilon > 0$ such that F is defined in a neighbourhood of $\overline{\mathbb{B}}_\varepsilon^{2n+2}$, the origin is its unique singular point, and $V := F^{-1}(0)$ intersects transversally $\mathbb{S}_{\varepsilon_0}^{2n+1}$, $\forall \varepsilon_0 \in (0, \varepsilon]$. By arguments of compactness, there exists $\eta > 0$ such that $F^{-1}(t) \pitchfork \mathbb{S}_\varepsilon^{2n+1}$ for all $t \in \overline{\mathbb{D}}_\eta^2$.

The Milnor fibration has as representative the map $F_{\varepsilon,\eta}^*$

$$\begin{array}{ccc} \overline{\mathbb{B}}_{\varepsilon,\eta}^* := \overline{\mathbb{B}}_\varepsilon^{2n+2} \cap F^{-1}(\overline{\mathbb{D}}_\eta^2 \setminus \{0\}) & \xrightarrow{F_{\varepsilon,\eta}^* := F_|} & \overline{\mathbb{D}}_\eta^2 \setminus \{0\} \\ \downarrow & & \downarrow \\ \overline{\mathbb{B}}_{\varepsilon,\eta} := \overline{\mathbb{B}}_\varepsilon^{2n+2} \cap F^{-1}(\overline{\mathbb{D}}_\eta^2) & \xrightarrow{F_{\varepsilon,\eta} := F_|} & \overline{\mathbb{D}}_\eta^2 \end{array} \tag{4.3}$$

and its topological type does not depend on the particular choices. Its homotopy type is the same if we replace $\overline{\mathbb{D}}_\eta^2 \setminus \{0\}$ by $\mathbb{S}_\eta^1$ and we know that fibrations on the circle are completely determined by the so-called *geometric monodromy*. If

$$\Sigma := F^{-1}(\eta) \cap \overline{\mathbb{B}}_\varepsilon^{2n+2} \quad \text{(the Milnor fiber)}$$

then the monodromy $\Phi : \Sigma \to \Sigma$ is a homeomorphism, only well-defined up to isotopy (actually it can be chosen a diffeomorphism), such that this fibration

$$[0, 1] \times \Sigma/(1, z) \sim (0, \Phi(z)) \longrightarrow [0, 1]/0 \sim 1 \equiv \mathbb{S}^1.$$

is isomorphic to $F_{\varepsilon,\eta}^*$.

Moreover Milnor showed that it has the homotopy type of a wedge of μ n-spheres, where μ (the Milnor number) is actually the codimension of the Jacobian ideal of F. Then, $H_n(\Sigma; \mathbb{Z})$ is the only non trivial reduced homology group of Σ and $\Phi_* : H_n(\Sigma; \mathbb{Z}) \to H_n(\Sigma; \mathbb{Z})$ is a well-defined automorphism called the *homological monodromy*. In the same way we can define its dual $\Phi^* : H^n(\Sigma; \mathbb{Z}) \to H^n(\Sigma; \mathbb{Z})$, the *cohomological monodromy*. Most of the time we will write only *monodromy* and the context will determine the one we refer to.

Though Φ_* loses a lot of information from Φ, it provides more effective invariants. The first one is its characteristic polynomial Δ. Resolutions, see Sects. 4.6 and 4.7, provide tools to compute a representative of Φ. In particular, Δ can be retrieved from the famous A'Campo's formula [2]. This polynomial can be computed for k-LYS.

Proposition 4.5.1 *Let V be a k-LYS singularity with tangent cone C_d. For $k = 1$*

$$\Delta = \frac{(t^d - 1)^{d^2-3d+3-\mu(C_d)}}{t-1} \prod_{p \in \mathrm{Sing}(C_d)} \Delta_p(t^{d+1}), \tag{4.4}$$

where Δ_p is the characteristic polynomial of the monodromy Φ_p of (C_d, p).

If $k > 1$, then

$$\Delta = \frac{(t^d - 1)^{d^2-3d+3-\mu(C_d)}}{t-1} \prod_{P \in \mathrm{Sing}(C_d)} \Delta_P^{(k)}(t^{d+k}), \tag{4.5}$$

where $\Delta_P^{(k)}$ is the characteristic polynomial of Φ_P^k. In particular, $\mu = (d-1)^3 + k\mu(C_d)$.

Formula (4.4) was stated without proof in [54]; it was restated and proved in [5], using an embedded resolution and A'Campo's formula. Formula (4.5) was proved in [30], using an A'Campo-like formula for partial resolutions, and reproved in [40].

Corollary 4.5.2 *The characteristic polynomial of the monodromy of a k-LYS depends only on k and on the combinatorics of its tangent cone.*

This corollary brings to our attention a conjecture of S.S.-T. Yau cited in the Introduction.

Conjecture 4.5.3 ([59]) Let $(V_i, 0) \subset (\mathbb{C}^3, 0)$, $i = 1, 2$, two germs of isolated singularity of surface. Let K_1, K_2 be their links and let Δ_1, Δ_2 be the characteristic polynomials of their monodromies. If K_1 is homeomorphic to K_2 (they have the *same abstract topology*) and $\Delta_1 = \Delta_2$, then they have the *same embedded topology*, i.e., there is a germ $\Psi : (\mathbb{C}^3, 0) \to (\mathbb{C}^3, 0)$ of homeomorphism such that $\Psi(V_1) = V_2$.

The author supported this conjecture since it holds for large families of examples. The above discussion shows that if we have two k-LYS whose tangent cones have the same combinatorics, then they satisfy the hypotheses of the conjecture. The characteristic polynomial of the monodromy is an invariant of the conjugacy class of the action of the monodromy on cohomology. In order to study the conjecture, we want study finer invariants of the monodromy, namely the Jordan form.

Note also that another consequence of Proposition 4.5.1 is that all the eigenvalues are roots of unity. Actually a cornerstone in singularity theory gives a sharper statement, the Monodromy Theorem. Let us introduce some notation in order to give a shorter statement. We need some concepts from Sect. 4.13, namely the weight filtrations of nilpotent endomorphisms.

Definition 4.5.4 Let V a $\mathbb{Q}$-vector space of finite dimension and let $h : V \to V$ be an automorphism such that its characteristic polynomial is a product of cyclotomic polynomials. Let $m \in \mathbb{Z}_{>0}$ such that the characteristic polynomial of h^m is a power of $t - 1$. Let W be the weight filtration of the nilpotent endomorphism $\mathbf{1} - h^m$, centered at 0. For $k \geq 0$, the *k-characteristic polynomial of h* is the characteristic polynomial $\Delta^{[k]}$ of the isomorphism induced by h on gr_k^W (which coincides with the one on gr_{-k}^W).

The reader may skip in a first approach Sect. 4.13. What it is relevant, from Propositions 4.13.5*(c)*, is that the degree of $\Delta^{[k]}$ is the number of Jordan blocks of size $(k + 1)$, and the polynomial itself gives their eigenvalues.

Theorem 4.5.5 ([19], [29, pp. 10–13]) *The eigenvalues of Φ are roots of unity; moreover, $\Delta^{[m]} = 1$, if $m > n$, and 1 is not a root of $\Delta^{[n]}$, i.e., the maximal size for Jordan blocks is $n + 1$ (and n for the eigenvalue 1).*

4.6 Embedded Resolutions

In order to prove (4.4) an embedded resolution of a SIS is needed, and this is the main result of [5]. Since an embedded resolution provides a model for the geometric monodromy, much more information can be retrieved. We recall the general definition.

Let $(V, 0) \subset (\mathbb{C}^{n+1}, 0)$ be a germ of isolated hypersurface singularity defined by $F \in \mathbb{C}\{x_0, x_1, \dots, x_n\}$.

Definition 4.6.1 An *embedded resolution* of V is a proper birational map $\pi : (X, E) \to (\mathbb{C}^{n+1}, 0)$, X smooth, which is an isomorphism outside the origin and such that $\pi^* V$ is a simple normal crossing divisor.

Under these conditions, $E = \pi^{-1}(0)$ is also a simple normal crossing divisor with irreducible components $E_1, \dots, E_r$. The strict transform of V is $\hat{V} := \overline{\pi^{-1}(V \setminus \{0\})}$ and the total transform is the simple normal crossing divisor

$$\pi^* V = \hat{V} + \sum_{j=1}^{r} N_j E_j. \tag{4.6}$$

Let us recall A'Campo's formula. For the sake of simplicity let us denote $E_0 := \hat{V}$. We set for $J \subset \{0, 1, \dots, r\}$:

$$E_J = \bigcap_{j \in J} E_j, \qquad \overset{\circ}{E}_J := E_J \setminus \bigcup_{j \notin J} E_j.$$

Let h be a homeomorphism of an n-dimensional manifold. We define the zeta function of h as

$$\zeta_h(t) := \prod_{j=0}^{n} \det\left(\mathbf{1} - th^{*,H^j(X;\mathbb{C})}\right)^{(-1)^j},$$

i.e., the alternating product of the characteristic polynomials of h acting on cohomology. When V is an isolated singularity of hypersurface, ζ_V denotes the zeta function of the monodromy.

Theorem 4.6.2 (A'Campo [3]) *Let V is an isolated singularity of hypersurface. With the above notation,*

$$\zeta_V(t) = \prod_{j=1}^{r} (1 - t^{N_j})^{\chi(\mathring{E}_j)}, \tag{4.7}$$

where $E_j := E_{\{j\}}$.

In this formula, only the *smooth part* of $\pi^{-1}(V)$ matters, i.e. only subsets J with one element, so apparently we have introduced useless notation. It is not the case, since, starting from an embedded resolution J. Steenbrink constructs in [52, 53] (using all the above spaces) a spectral sequence $\mathcal{E}$ (see Construction 4.6.4 and the discussion after Theorem 4.6.5 for details) with the following properties:

1. The level $\mathcal{E}_1$ is composed by the cohomology of some cyclic covers of the spaces E_J, $\#J \leq 1$.
2. It converges to the cohomology of the Milnor fiber.
3. It degenerates at the level $\mathcal{E}_2$.
4. It provides a Mixed Hodge Structure to the cohomology of the Milnor fiber.

Mixed Hodge Structures are related to two filtrations, Hodge (analytic) and weight (topological), and we are mainly interested in this second one, since it contains the information about the Jordan form of the cohomology of the fiber. We are going to explain how to construct the spectral sequence $\mathcal{E}$, how to obtain the weight filtration and how to relate it to the monodromy, and in particular to its Jordan form.

In order to construct the spectral sequence we need to start with the semistable normalization, see [53]. Let N be any common multiple of all N_i's and let us consider the following diagram

$$\begin{array}{ccccc} Z & \xrightarrow{\nu} & Y & \xrightarrow{g} & X \\ & & \downarrow & & \downarrow {\scriptstyle F\circ\pi} \\ & & \mathbb{C} & \xrightarrow[t\mapsto t^N]{} & \mathbb{C} \end{array}$$

where Y is the pull-back and ν the normalization. While X is smooth, it is not the case for neither Y nor Z, but Z is normal and its singularities are of abelian quotient type. It is not hard to see that g is an isomorphism from $g^{-1}(E)$ to E, but ν is a (piecewise) cyclic cover $(g \circ \nu)^{-1}(E) \to E$.

Let $D := (g \circ \nu)^{-1}(E)$, and let us denote $D_j := (g \circ \nu)^{-1}(E_j)$; note that it may be reducible and not connected. We define D_J and $\mathring{D}_J$ in the same way.

Proposition 4.6.3 *The map $\nu_J := \nu_{|D_J} : D_J \to E_J$ is a cyclic ramified cover, unramified along $\mathring{E}_J$, with $N_J := \gcd(N_j,\ j \in J)$ sheets, and the ramifications indices are defined by the other N_j's.*

Proof Let us pick-up a point $P \in \mathring{D}_J$. We can choose local coordinates $x_1, \ldots, x_{n+1}$ centered at P such that if $J = \{j_1, \ldots, j_s\}$, then the local equation of D_{j_i} is $x_i = 0$. Let us denote $P_Y \in Y$ the only point in $g^{-1}(P)$. The local equation of Y on P_Y is $t^N - \prod_{i=1}^s x_i^{N_i} = 0$. The normalization process has two parts: separate the local irreducible components and *add* holomorphic forms. Only the first one affects to the topology of D_J. Since N is a multiple of all the N_j's, Y has N_J irreducible components and there is an obvious cyclic monodromy. The ramifications are produced along $D_{J'}$, with $J \subsetneq J'$, and the number of points depend on N_j for $j \in J'$. □

Construction 4.6.4 Steenbrink spectral sequence is perfectly explained in the original paper [53]. Let us recall what we need about spectral sequences and weight filtrations. Let us consider a complex of vector spaces $\{(C_n, d_n)\}_{n\in\mathbb{Z}}$, $d_n : C_n \to C_{n+1}$, and let us consider a decreasing filtration $L_k(C_n)$ such that $d_n(L_k(C_n)) \subset L_k(C_{n+1})$.

This filtration induces a graduation and we can define

$$C^0_{p,q} := \frac{L_q(C_{p+q})}{L_{q+1}(C_{p+q})} = \operatorname{gr}^L_q C_{p+q} \text{ (the 0-level of the spectral sequence),}$$

which is also a complex with differentials $d^0_{p,q} : E^0_{p,q} \to E^0_{p+q,q}$.

Let us define for $r \geq 1$:

$$Z^r_{p,q} := \frac{d^{-1}_{p+q}(L_{q+r}(C_{p+q+1})) + L_{q+1}(C_{p+q})}{L_{q+1}(C_{p+q})},$$

$$B^r_{p,q} := \frac{d_{p+q-1}(L_{q-r+1}(C_{p+q-1})) + L_{q+1}(C_{p+q})}{L_{q+1}(C_{p+q})},$$

$$\mathcal{E}^r_{p,q} := \frac{Z^r_{p,q}}{B^r_{p,q}}, \text{ the } r\text{-level of the spectral sequence.}$$

The actual cocycles are the elements in $\ker d$; the elements of $Z^r_{p,q}$ are represented by the elements in $L_q(C_{p+q})$ such that their images by d_{p+q} live in $L_{q+r}(C_{p+q+1})$, i.e., the image by the differential *approaches* to 0. Something similar can be said for the coboundaries and the elements in $B^r_{p,q}$: they are represented by the images by d_{p+q-1} of elements in $L_{q-r+1}(C_{p+q-1})$ and their images *approach* to the total set of boundaries.

There is a natural map $d^r_{p,q} : \mathcal{E}^r_{p,q} \to \mathcal{E}^r_{p-r+1,q+r}$ induced by the original differential. By its very definition, $\mathcal{E}^{r+1}_{p,q}$ is exactly the cohomology defined by the differentials d^r, i.e., the $r+1$-level is the cohomology of the r-level. We can also define,

$$Z^\infty_{p,q} := \frac{\ker d_{p+q} + L_{q+1}(C_{p+q})}{L_{q+1}(C_{p+q})},$$

$$B^\infty_{p,q} := \frac{d_{p+q-1}(C_{p+q-1}) + L_{q+1}(C_{p+q})}{L_{q+1}(C_{p+q})},$$

$$\mathcal{E}^\infty_{p,q} := \frac{Z^\infty_{p,q}}{B^\infty_{p,q}}, \text{ the } \infty\text{-level of the spectral sequence.}$$

It is important to notice that $\mathcal{E}^\infty_{p,q}$ is the associated graded of $H^{p+q}(C^\bullet)$ by the filtration induced by L. Hence, we can interpret a spectral sequence as a cohomological path starting from the cohomology of the induced graded complex and ending in the associated graded of the cohomology of the complex; we say that the spectral sequence converges to the cohomology of the complex C. Note that if after some r-level all the differentials vanish then $\mathcal{E}^r = \mathcal{E}^{r+1} = \cdots = \mathcal{E}^\infty$ and we say that the spectral sequence *degenerates* at $\mathcal{E}^r$.

Theorem 4.6.5 ([53]) *There is a spectral sequence $\mathcal{E}$ converging to $H^*(\Sigma; \mathbb{Q})$, degenerating at $\mathcal{E}^2$.*

Since its description is cumbersome we will present it only for the cases $n = 1, 2$, which are the cases we are interested in. The generalization to the general case can be deduced with a little effort. Let us fix the following notation:

$$D^{(\ell)} = \coprod_{\#J=\ell+1} D_J, \quad D^{(\ell)}_0 = \coprod_{\#J=\ell+1}^{0\notin J} D_J,$$

where $\coprod$ stands for disjoint union; $D^{(\ell)}$ is also the normalization of the intersection of $\ell+1$ divisors. For $n = 1$ we present the non-trivial $\mathcal{E}^1_{p,q}$ spaces and the non trivial differentials.

$$
\begin{array}{ccc}
H^0(D_0^{(1)};\mathbb{Q}) & & \\
\uparrow & & \\
H^0(D_0^{(0)};\mathbb{Q}) & H^1(D_0^{(0)};\mathbb{Q}) & H^2(D_0^{(0)};\mathbb{Q}) \\
 & & \uparrow \\
 & & H^0(D_0^{(1)};\mathbb{Q})
\end{array}
\tag{4.8}
$$

We do the same thing for $n = 2$ (we omit the coefficient ring):

$$
\begin{array}{ccccc}
H^0(D_0^{(2)}) & & & & \\
\delta_{0,1}\uparrow & & & & \\
H^0(D_0^{(1)}) & H^1(D_0^{(1)}) & H^2(D_0^{(1)}) & & \\
\delta_{0,0}\uparrow & \delta_{1,0}\uparrow & \delta_{2,0}\uparrow & & \\
 & & H^0(D^{(2)}) & & \\
 & & \oplus & & \\
H^0(D_0^{(0)}) & H^1(D_0^{(0)}) & H^2(D_0^{(0)}) & H^3(D_0^{(0)}) & H^4(D_0^{(0)}) \\
 & & \delta_{2,-1}\uparrow & \delta_{3,-1}\uparrow & \delta_{4,-1}\uparrow \\
 & & H^0(D^{(1)}) & H^1(D^{(1)}) & H^2(D^{(1)}) \\
 & & & & \delta_{4,-2}\uparrow \\
 & & & & H^0(D^{(2)})
\end{array}
\tag{4.9}
$$

The *vertical* differentials at $\mathcal{E}^1$ are *simplicial* combinations of restriction and Gysin maps between projective varieties with quotient singularities (V-varieties in the language of Steenbrink) which are Hodge morphisms. The differentials at $\mathcal{E}^2$ vanish and it is the reason of the convergence of the spectral sequence and the Mixed Hodge Structure can be somewhat visualized.

In general, the only interesting elements of $\mathcal{E}^2 = \mathcal{E}^\infty$ correspond to the case $p+q = n$. We are going to define the *vertical filtration* $W^{\mathcal{E}}$ of $H^n(\Sigma;\mathbb{Q})$ associated to the Steenbrink spectral sequence as the one such that

$$
\operatorname{gr}_p^{W^{\mathcal{E}}} H^n(\Sigma;\mathbb{Q}) \cong \mathcal{E}^2_{p,n-p}.
$$

This is actually the filtration described earlier with a change of indices: $L_{n-p} = W_p^{\mathcal{E}}$ (in particular, it is increasing). In Sect. 4.13 we describe the weight filtration of a nilpotent morphism. We show how Steenbrink relates these two filtrations. From Theorem 4.5.5, we have that $N_\Phi := \Phi^N - \mathbf{1}$ is a nilpotent morphism.

Theorem 4.6.6 ([53]) *Let $H^n_{\neq 1}(\Sigma;\mathbb{Q})\oplus H^n_{=1}(\Sigma;\mathbb{Q})$ be the Φ-invariant decomposition of $H^n(\Sigma;\mathbb{Q})$ such that:*

(i) *$H^n_{\neq 1}(\Sigma;\mathbb{Q})$ does not have 1 as eigenvalue;*
(ii) *Φ is unipotent on $H^n_{=1}(\Sigma;\mathbb{Q})$.*

Let us denote:

(a) *$W^{\neq 1}$: the weight filtration of N_Φ restricted to $H^n_{\neq 1}(\Sigma;\mathbb{Q})$ and centered at n;*
(b) *$W^{=1}$: the weight filtration of N_Φ restricted to $H^n_{=1}(\Sigma;\mathbb{Q})$ and centered at $n+1$.*

Then, the vertical *filtration of $\mathcal{E}$ coincides with $W^{\neq 1}\oplus W^{=1}$.*

We obtain some consequences of this result.

Corollary 4.6.7 *The Jordan form of Φ is determined by the spectral sequence, more precisely, by the action of Φ on $\mathcal{E}^2_{p,n-p}$.*

We are going to be more concrete for $n = 1, 2$, though the same ideas can be generalized for arbitrary dimension.

Corollary 4.6.8 *Let $(V,0)\subset(\mathbb{C}^2,0)$ be a germ of isolated singularity and let Φ be its monodromy.*

(i) *The first column of* (4.8) *is the cochain complex of the dual graph of D.*
(ii) *The first column on $\mathcal{E}^2$ is the cohomology of this graph; in particular, the graph is connected.*
(iii) *There is a natural action of Φ on this dual graph, which determines the number of* 2*-Jordan blocks for each eigenvalue ($\neq 1$). More precisely,*

$$\Delta^{[1]}(t)=\frac{(t-1)\Delta_{\Phi,H^0(D^{(1)})}(t)}{\Delta_{\Phi,H^0(D^{(0)})}(t)}.$$

Corollary 4.6.9 *Let $(V,0)\subset(\mathbb{C}^3,0)$ be a germ of isolated singularity and let Φ be its monodromy.*

(i) *The first column of* (4.9) *is the cochain complex of the dual complex of D.*
(ii) *The first column on $\mathcal{E}^2$19 the cohomology of this complex; in particular, the* 2*-complex must be connected and with vanishing H^1.*
(iii) *There is a natural action of Φ on this dual complex, which determines the number of* 3*-Jordan blocks for each eigenvalue ($\neq 1$). More precisely,*

$$\Delta^{[2]}(t)=\frac{\Delta_{\Phi,H^0(D^{(2)})}(t)\Delta_{\Phi,H^0(D^{(0)})}(t)}{(t-1)\Delta_{\Phi,H^0(D^{(1)})}(t)}.$$

(iv) *The maps $\delta_{2,0}, \delta_{3,-1}, \delta_{4,-1}$ are surjective; the maps $\delta_{1,0}, \delta_{2,-1}$ are injective. These data determine the Jordan blocks of size 2:*

$$\Delta^{[1]}(t) = \frac{(t-1)^{\dim E^2_{4,-2}} \Delta_{\Phi, H^1(D^{(1)})}(t)}{(t-1)^{\dim E^2_{0,2}} \Delta_{\Phi, H^1(D^{(0)})}(t)}.$$

The condition on the Jordan blocks of maximal size can be expressed for any dimension. We consider the dual n-complex Γ_D of the divisor D. We have that $\tilde{H}^j(\Gamma_D) = 0$, if $j \neq n$ and the characteristic polynomial of Φ on $\tilde{H}^n(\Gamma_D)$ equals $\Delta^{[n]}$. The fact that this maximal bound is reached appeared in the same number of *Invent. Math.* in two papers of N. A'Campo [1] and B. Malgrange [38].

4.7 Embedded $\mathbb{Q}$-resolutions

Thanks to Steenbrink's results, in order to obtain the Jordan form of the monodromy of an isolated hypersurface singularity we need an embedded resolution of the hypersurface. The main goal of [5] was to provide such an embedded resolution of a SIS. A. Melle and me tried to use the same techniques to obtain an embedded resolution of a k-LYS but we encountered technical difficulties.

Note that the main ingredient of Steenbrink spectral sequence is the (non-smooth) space of the semistable normalization. Actually the main goal in [53] is to extend the results of [52] to the case of V-varieties. Having these facts in mind, J. Martín-Morales [40] provided what is called an embedded $\mathbb{Q}$-resolution for a k-LYS, where the Steenbrink spectral sequence can be constructed with the same properties. For SIS, his construction follows the same basic ideas of [5] but the output is combinatorially much more simple.

Some background about cyclic abelian singularities can be found in Sect. 4.11.

Definition 4.7.1 Let X be a V-variety of dimension $n+1$, i.e. a complex variety with cyclic abelian singularities. Let D be a divisor in X. We say that D is a *$\mathbb{Q}$-simple normal crossing divisor* if $\forall P \in X$ there is a *chart* $(\mathbb{C}^{n+1}, 0) \to (X, P)$ realizing the quotient singularity type of X such that the preimage of D is contained in the coordinate hyperplanes of $\mathbb{C}^{n+1}$.

Definition 4.7.2 Let $(V, 0) \subset (\mathbb{C}^{n+1}, 0)$ be a germ of isolated hypersurface singularity. An *embedded $\mathbb{Q}$-resolution* of V is a proper birational map $\pi : (X, E) \to (\mathbb{C}^{n+1}, 0)$ which is an isomorphism outside the origin and such that X is a V-variety and $\pi^* V$ is a $\mathbb{Q}$-simple normal crossing divisor.

The semistable normalization is defined in the same way and it is still a V-variety and D is a $\mathbb{Q}$-simple normal crossing divisor [41]. The main feature is that its combinatorics are much simpler that the ones of an embedded resolution.

It is a well-known result that the embedded resolution of plane curve singularities can be done by blowing-up points. We have stated in Theorem 4.12.7 that an embedded $\mathbb{Q}$-resolution can be done using weighted blowing-up on points, see Sect. 4.12 for details about weighted blowing-ups.

For hypersurface singularities in dimension 2, in general, one needs to blow up along curves. The hypersurface singularities that can be solved by blowing-up points are called *absolutely isolated singularities* (AIS). They have been studied specially in problems related with complex singular foliations. They have been also useful to check some conjectures, e.g., A. Melle proved Durfee's conjecture [24] for AIS in [42]. This conjecture states that 6 times the geometric genus is not greater than the Milnor number.

It is very rare for k-LYS to be AIS; actually only when $\operatorname{Sing} C_d = \emptyset$. If it is not the case, the embedded resolution for SIS of [5] needs blowing-ups along points and smooth rational curves. Nevertheless, the method of [40] provides both for SIS and k-LYS, a $\mathbb{Q}$-resolution process where only weighted blowing-ups of points are needed. It seems reasonable to say that k-LYS are *$\mathbb{Q}$-absolutely isolated singularities* and to study further features of these singularities.

Let us sketch the $\mathbb{Q}$-resolution process for k-LYS [40]. The first step is to blow-up the origin in $\mathbb{C}^3$. This way, we obtain an exceptional divisor $E_1 \cong \mathbb{P}^2$, and the singularities of $\hat{V}$ are in $\operatorname{Sing} C_d$. We have seen this in Remark 4.2.4. The surfaces $\hat{V}$ and E_1 are not transversal exactly at $\operatorname{Sing} C_d$.

Let us fix $P \in \operatorname{Sing} C_d$, which we assume to be $[0 : 0 : 1]$, in order to describe the part of the $\mathbb{Q}$-resolution process over P. Let us recall that there is a chart where E_1 has equation $z = 0$ and $\hat{V}$ has equation $z^k + f_{d,P}(x, y) = 0$.

To simplify the notations the strict transform of a divisor under a weighted blow-up will be denoted as the divisor, the ambient space will make the difference from the divisor and its strict transform.

As stated in Theorem 4.12.7, there is a sequence of weighted blowing-ups which provide the minimal embedded $\mathbb{Q}$-resolution of (C_d, P). Let us assume that we start with a (p, q)-weighted blowing-up $\sigma_1^P : (Z_1^P, E_1^P) \to (E_1, P)$. Let $P_1, \dots, P_r$ the points in E_1^P for which $(\sigma_1^P)^{-1}(C_d)$ is not a $\mathbb{Q}$-normal crossing divisor. Let m_1^P the (p, q)-weighted multiplicity of C_d at P.

Let us consider now the situation in dimension 3 around $P \in E \subset \hat{\mathbb{C}}^3$. We perform a (kp, kq, m_1^P)-weighted blowing-up $\tilde{\sigma}_1^P : (\tilde{Z}, \tilde{E}_1^P) \to (\hat{\mathbb{C}}^3, P)$ with the above coordinates (x, y, z). Actually:

- $E_1 \cap \tilde{E}_1^P = E_1^P$,
- $E_1 \cap \hat{V} = C_d$, and
- $E_1 \cap \tilde{E}_1^P \cap \hat{V} = \{P_1, \dots, P_r\}$.

These are the points in where E_1, $\tilde{E}_1^P$, $\hat{V}$ do not form a $\mathbb{Q}$-normal crossing divisor (as in dimension 2).

We iterate the process, though we may find some technical difficulties, since $\tilde{Z}$ (and the subsequent spaces) may not be smooth, but weighted blowing-up are still valid. This is one of the consequences.

Proposition 4.7.3 *The topology of the part of the resolution process of V after the first blowing-up is completely determined by the combinatorics of C_d, including the multiplicities of the divisor.*

The next step is to compare the divisors of the $\mathbb{Q}$-resolution and its semistable normalization for both $(V, 0)$ and $(C_d, \operatorname{Sing} C_d)$. There is a special irreducible component in the divisor of the semistable normalization, namely D_1 which is a d-cyclic cover of E_1 ramified along the intersection with the other irreducible components of the resolution. As stated in the previous proposition, the rest of the components are completely determined by the combinatorics. We can summarize this discussion.

Theorem 4.7.4 *All the terms in* (4.9) *(including its action by the monodromy) are determined by the combinatorics except the terms $H^j(D_1)$, factor of $H^j(D^{(0)})$, for* $j = 1, 2, 3$.

Let us look closer at the V-variety D_1 and its cohomology, together with the monodromy action. The first important remark is that V-varieties have also Poincaré duality over $\mathbb{Q}$. Recall that E_1 has been obtained by several blow-ups of the original exceptional component of the first blowing-up σ which was also denoted by E_1. Since it was naturally isomorphic to the complex projective plane, and in order to avoid confusion, it is identified with $\mathbb{P}^2$. The following commutative diagram holds:

$$\begin{array}{ccc} D_1 & \xrightarrow{\tilde{\sigma}_D} & X_d \\ \downarrow{\scriptstyle \tau_D} & & \downarrow{\scriptstyle \tau} \\ E_1 & \xrightarrow{\tilde{\sigma}} & \mathbb{P}^2 \end{array} \tag{4.10}$$

where

(CD1) τ is the d-cyclic cover of $\mathbb{P}^2$ ramified along C_d, see Construction 4.3.8;
(CD2) $\tilde{\sigma}$ is the restriction to E_1 of the composition of all the weighted blowing-ups of the $\mathbb{Q}$-resolution of V, except the blowing-up of the origin. It coincides with the minimal embedded $\mathbb{Q}$-resolution of C_d in $\mathbb{P}^2$;
(CD3) the diagram coincides with the pull-back of τ and $\tilde{\sigma}$.

We can derive some conclusions from this diagram. The monodromy η of X_d defined in Construction 4.3.8 lifts to the monodromy of τ_D induced by the monodromy Φ of V. We are going to break the computation of the monodromy of η_D on cohomology in several steps. Since η_D^d is the identity, it is enough to compute the characteristic polynomials $\Delta^j(t) := \det\left(\mathbf{1} - th^{*,H^j(D_1;\mathbb{Q})}\right)$ and combine them with the zeta function

$$\zeta_{\eta_D}(t) = \prod_{j=0}^{4} (-1)^j \Delta^j(t).$$

This is the first consequence of the diagram.

Step 4.7.5 *The V-variety D_1 is birational to X_d, actually it is a model with only cyclic quotient singularities.*

Since X_d is connected we obtain the following step.

Step 4.7.6 *The V-variety D_1 is connected, i.e.* $\dim H^0(D_1) = \dim H^4(D_1) = 1$ *and*

$$\Delta^0(t) = \Delta^4(t) = 1 - t.$$

Step 4.7.7 *The ζ function ζ_{η_D} is combinatorial, so it is also the case for $\chi(D_1)$.*

Proof Since $\tilde{\sigma}^{-1}(C_d)$ is a $\mathbb{Q}$-normal crossing divisor, E_1 admits a stratification where over each stratum D the map τ_D is an unramified cyclic cover (over each stratum the number of sheets is a divisor of d). Hence, there is a formula for $\chi(D_1)$ which is purely combinatorial. Actually, a combinatorial formula also exists for ζ_{η_D}. □

Step 4.7.8 *The polynomial Δ^1 is the Alexander polynomial Δ_{C_d} of C_d.*

Proof Hodge theory is also valid on projective V-varieties; in particular $H^1(X_d; \mathbb{C}) \cong H^1(X_d; \mathcal{O}_{X_d}) \oplus H^0(X_d; \Omega^1_{X_d})$ and $\overline{H^1(X_d; \mathcal{O}_{X_d})} \cong H^0(X_d; \Omega^1_{X_d})$. Hence, $H^1(X_d; \mathbb{C})$ is a birational invariant and Δ^1 is also the characteristic polynomial of η on $H^1(X_d; \mathbb{Q})$, which is, by Definition 4.3.9, the Alexander polynomial of C_d. □

Step 4.7.9 *The combinatorics and the Alexander polynomial of C_d completely determines the characteristic polynomials of η_D on cohomology.*

Proof Using Poincaré duality we have that $\Delta^1(t) = \Delta^3(t)$. Hence,

$$\frac{\Delta^2(t)}{\Delta^1(t)^2} = \frac{\zeta_{\eta_D}(t)}{(1-t)^2}.$$

The right hand-side is combinatorial, so the it is enough to compute $\Delta^1(t) = \Delta_{C_d}$. □

We end this section applying the previous discussion to the determination of the Jordan form of the monodromy of Φ for a k-LYS. Let us start with the blocks of maximal size, i.e. 3 (only eigenvalues different from 1). The details can be found in [5, Théorème 4.3.1] and they follow from Corollaries 4.6.8 and 4.6.9.

Proposition 4.7.10 *For a SIS the structure of Jordan blocks is a combinatorial invariant, given $m \gg 1$*

$$\Delta^{[2]}(t) = \gcd\left((t-1)^m, \prod_{P \in \operatorname{Sing} C_d} \Delta_P^{[1]}(t)\right).$$

The same result holds for k-LYS as it can be deduced from [40, 41]. Note also that a necessary condition for the existence of Jordan blocks of size 3 is that there are singularities in C_d with Jordan blocks of size 2. Let us consider now the Jordan blocks of size 2 of a k-LYS.

Theorem 4.7.11 *Let V a k-LYS and let $\Delta_{C_d}(t)$ be the Alexander polynomial of the tangent cone.*

$$\Delta^{[1]}(t) = \frac{\Delta^{\mathrm{Comb},k}(t)}{\Delta_{C_d}(t)},$$

where $\Delta^{\mathrm{Comb},k}(t)$ is a polynomial which depends only on the combinatorics and k.

Remark 4.7.12 A precise formula appears in [5, Théorème 5.4.2 and Corollaire 5.5.4]. Let us recall that up to a power of $t-1$ (which is combinatorially determined), we know that $\Delta^{[1]}(t)$ equals

$$\frac{\Delta_{\Phi,H^1(D^{(1)})}(t)}{\Delta_{\Phi,H^1(D^{(0)})}(t)}.$$

It is not complicated to compute the numerator once one has the data from the embedded $\mathbb{Q}$-resolution. The connected components of $D^{(1)}$ are compact Riemann surfaces and the corresponding characteristic polynomials using ζ functions and Riemann-Hurwitz theory.

For the denominator, there are two kind of components. On one side we have D_1, and we have seen in Step 4.7.8 that its characteristic polynomial is $\Delta_{C_d}(t)$. To have the specific formula we need to compute the characteristic polynomials for the other components.

We will not enter in the details but the key point for SIS is the following. Using [5, Proposition 5.3.2] we find that the each the maps $D_j \to E_j$, $j > 1$ are birationally equivalent to maps $T \to \mathbb{P}^1 \times \mathbb{P}^1$ which are ramified only along vertical and horizontal factors. Since on H^1 all these invariants do not depend on the birational model we can compute the characteristic polynomials in a relatively easy way.

There is a parallel work for k-LYS in [40] but in this case $\mathbb{P}^1 \times \mathbb{P}^1$ is replaced by a special ruled surface with quotient singularities; the ramification locus is composed by fibers and pairwise disjoint multi-sections. The computation of these characteristic polynomials is more involved and it is the goal of [12]. A concrete formula is expected in a forthcoming paper of J. Martín-Morales.

Theorem 4.7.13 *There are singularities of hypersurface having the same link, the same characteristic polynomial of the monodromy but which do not have the same embedded topology, i.e., Conjecture* 4.5.3 *does not hold.*

Proof It is enough to consider two SIS, or k-LYS with the same k, whose tangent cones form an Alexander-Zariski pair, see [6] for the existence. Since the link and the characteristic polynomial are combinatorial invariants they are the same. Since the Alexander polynomials of the tangent cones differ, the structure of the Jordan form of the monodromy is different, and then the embedded topology is different. □

4.8 Monodromy over $\mathbb{Z}$

In the previous section, we have described the action of the monodromy of the Milnor fibration on the cohomology of the Milnor fiber with coefficients in a field of characteristic 0, say $\mathbb{Q}$ or $\mathbb{C}$. We have used an embedded resolution (or more precisely an embedded $\mathbb{Q}$-resolution) to obtain a model of the Milnor fiber and the geometric monodromy. With these ingredients and Steenbrink's and Milnor's theories, these computations have been achieved.

This method does not hold if we want to consider $\mathbb{Z}$ as ring of coefficients, much more finer ingredients are needed. There is a theoretic classic method which involves morsifications, Picard-Lefschetz theory, and the intersection form of the Milnor fiber. There is an excellent exposition in [4, Part I]. Let us sketch this method. Let us start with a defining equation $F : (\mathbb{C}^{n+1}, 0) \to (\mathbb{C}, 0)$ of V and let us consider a *good* representative $F_{\varepsilon,\eta} : \overline{\mathbb{B}}_{\varepsilon,\eta} \to \overline{\mathbb{D}}^2_\eta$ as in (4.3), which is a fibration outside the origin.

Definition 4.8.1 A *morsification* of F is a holomorphic map $\mathfrak{F}$ defined on an open neighbourhood $\mathfrak{U}$ of $\overline{\mathbb{B}}_{\varepsilon,\eta} \times \overline{\mathbb{D}}^2_r$, $r > 0$, such that, if we define for $\lambda \in \overline{\mathbb{D}}^2_r$

$$\overline{\mathbb{B}}_{\varepsilon,\eta,\lambda} := \{\mathbf{p} \in \mathbb{C}^{n+1} \mid (\mathbf{p}, \lambda) \in \mathfrak{U}, |\mathfrak{F}(\mathbf{p}, \lambda)| \leq \eta\}, \qquad \begin{array}{rcl} \overline{\mathbb{B}}_{\varepsilon,\eta,\lambda} & \xrightarrow{\mathfrak{F}} & \overline{\mathbb{D}}^2_\eta \\ \mathbf{p} & \longmapsto & \mathfrak{F}(\mathbf{p}, \lambda) \end{array}, \text{ then}$$

(a) $F^0 = F_{\varepsilon,\eta}$

(b) If $\lambda \neq 0$ then F^λ is a Morse function, i.e., all the singular points are non-degenerate and the singular values are pairwise distinct (and in the open disk $\mathbb{D}^2_\lambda$).

From classical Morse theory and standard compactness arguments we have the following statement.

Proposition 4.8.2 *Morsifications of F do exist. Moreover, after eventually restricting $r > 0$:*

(1) *If $\lambda \neq 0$, then F^λ has exactly μ singular points, where μ is the Milnor number of F.*

(2) *If $t \in \overline{\mathbb{D}}^2_\eta$ is a regular value of F^λ, then $(F^\lambda)^{-1}(t)$ is diffeomorphic to the Milnor fiber Σ of F.*

(3) *Each F^λ defines a locally trivial fibration over $\overline{\mathbb{D}}^2_\eta \setminus \operatorname{disc} F^\lambda$ where* disc *stands for the set the critical values.*
(4) *The restrictions of these fibrations to $\mathbb{S}^1_\eta$ are all equivalent.*

This proposition shows that we can compute the monodromy of F by computing the monodromy of F^λ over $\overline{\mathbb{D}}^2_\eta \setminus \operatorname{disc} F^\lambda$ and then restrict its result to $\mathbb{S}^1_\eta$. This is where Picard-Lefschetz theory enters, see [4, Lemmas 1.3 and 1.4].

Proposition 4.8.3 *Let V be a non-degenerate singularity (i.e., after a change of coordinates, $F = x_0^2 + \cdots + x_n^2$). Then, Σ is diffeomorphic to a closed regular neighbourhood of the zero-section of the tangent vector bundle over $\mathbb{S}^n$. The image by σ of the zero section in Σ is called a* vanishing cycle *and $(\sigma \cdot \sigma)^2_\Sigma = (-1)^{\frac{n(n-1)}{2}}(1 + (-1)^n)$.*

Fix a value λ, $0 < |\lambda| \leq r$. Let $\operatorname{disc} F^\lambda = \{t_1, \dots, t_\mu\}$ and let $t_0 \in \mathbb{S}^1_\lambda$. We can choose a small radius $\delta > 0$ such that the closed disks of radius δ centered at t_i are pairwise disjoint and contained in the open disk $\mathbb{D}^2_\eta$ as in Fig. 4.2. Let us choose points $s_i \in \mathbb{S}^1_\delta(t_i)$. In the fibers $(F^\lambda)^{-1}(s_i)$ we find a vanishing cycle $\hat{\sigma}_i$ corresponding to the critical point t_i. Fix a system of simple paths γ_i in $\overline{\mathbb{D}}^2_\eta \setminus \bigcup_{j=1}^{\mu} \mathbb{D}^2_\delta(t_i)$, whose only intersection is their common origin t_0, which end at s_i and such that from top to bottom, near t_0 they are ordered as $\gamma_1, \dots, \gamma_\mu$, see Fig. 4.2.

The fiber bundles over γ_i are trivial, and then it is possible to transport $\hat{\sigma}_i \subset (F^\lambda)^{-1}(s_i)$ to $\sigma_i \subset (F^\lambda)^{-1}(t_0) =: \Sigma_\lambda$; recall that Σ_λ is homeomorphic to Σ. Following [4] we have the following definition.

Definition 4.8.4 The tuple $(\sigma_1, \dots, \sigma_\mu)$ is a *distinguished basis of vanishing cycles* of the homology space $H_n(\Sigma_\lambda; \mathbb{Z})$.

The name of basis is justified by the following result.

Theorem 4.8.5 ([4, Theorem 2.1]) *A distinguished basis of vanishing cycles is a basis of the abelian group $H_n(\Sigma_\lambda; \mathbb{Z})$.*

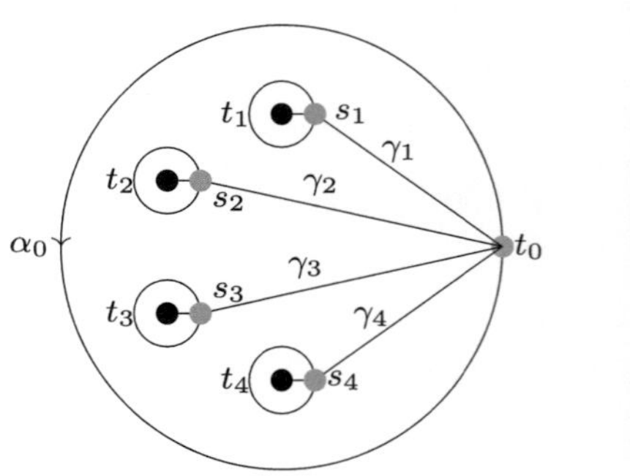

Fig. 4.2 Paths from the base value to the critical values, $\mu = 4$

In Proposition 4.8.3, we pay attention to the intersection form in the Milnor fiber. Though this is a priori a weak invariant, we are going to see that this is not really the case. Let us come back to the target of F^λ. Note that

$$\pi_1\left(\overline{\mathbb{D}}^2_\eta \setminus \bigcup_{j=1}^{\mu} \mathbb{D}^2_\delta(t_i); t_0\right) = \langle \alpha_1, \ldots, \alpha_\mu \mid \_ \rangle$$

where $\alpha_i := \gamma_i \cdot \hat{\alpha}_i \cdot \gamma_i^{-1}$, and $\hat{\alpha}_i$ is the cycle based at s_i that runs along $\mathbb{S}^1_\delta(t_i)$ counterclockwise. The previous choices guarantee that, up to homotopy, $\alpha_0 := \alpha_1 \cdot \ldots \cdot \alpha_\mu$ is the cycle based at t_0 that runs along $\mathbb{S}^1_\eta$ counterclockwise. The fact that F^λ is a locally trivial fibration associates to each path α_i a monodromy diffeomorphism $\Phi_i : \Sigma_\lambda \to \Sigma_\lambda$ which induces an automorphism $\Phi_{i,*}$ of $H_n(\Sigma_\lambda; \mathbb{Z})$.

Proposition 4.8.6 (Picard-Lefschetz Formula [4, Corollaries in page 26]) *The monodromy Φ_0 (on Σ_λ) is conjugate to the monodromy Φ on Σ. Moreover, $\Phi_0 = \Phi_{\mu,*} \circ \ldots \cdot \Phi_{1,*}$,*

$$\Phi_{i,*}(\sigma_j) = \sigma_j + (-1)^{\frac{(n+1)(n+2)}{2}} (\sigma_j \cdot \sigma_i)_{\Sigma_\Lambda} \sigma_i,$$

and, hence, Φ_* can be computed in terms of the intersection form on a distinguished basis of $H_n(\Sigma_\lambda, \mathbb{Z})$.

In [4], one can learn much more, including the properties of the variation operator and the Seifert form. The main problem about this method is that actual computations are really hard. The fact that there is a strong geometric knowledge of the Milnor fibration in the case $n = 1$ makes very interesting the approach of A.M. Gabrièlov in [28], since he proposed an inductive method on the dimension.

After a choice of good coordinates, the input of Gabrièlov's method are the invariants of a *generic* polar curve of V and the intersection form of $(V_\beta, 0) \subset (\mathbb{C}^n, 0)$, where V_β is the zero locus of $F(\beta, x_1, \ldots, x_n)$, for $0 < \beta \ll 1$. We are not going to enter in the details, the polar curve is a classical invariant of hypersurface singularities, see e.g. [55]. Starting from a distinguished basis of vanishing cycles for V_β and its intersection form, using the polar invariants Gabrièlov constructs a distinguished basis of vanishing cycles for V and computes its intersection form.

In [26], M. Escario uses this method to provide an effective way to compute the intersection form of a k-LYS. Let us assume that the equation $f_d(x, y, z) = 0$ of C_d is given in some generic coordinates. In particular, the homogeneous polynomial $f_d(x, y, 0)$ has no multiple roots. As we pointed out, it is enough to compute the intersection form for $F(x, y, z) = f_d(x, y, z) + z^{d+k}$. Escario's procedure is quite technical but we want to emphasize that the input is the monodromy of the polynomial $f(x, y, 1)$. This polynomial is a locally trivial fibration outside the critical values, and this is the monodromy needed for Gabrièlov's method.

In [25] using a so-called *discriminant method*, he computes the braid monodromy of the discriminant of $f(x, y, 1) - t$ (with respect to the variable y). This is principal part of the input in Gabrièlov's method. The relevant point is that this computation can be made effective using the work of J. Carmona [20] and the joint work of M. Marco and M. Rodríguez [39].

We are going to see how this method applies to k-LYS. The first step is to present a problem stated (and solved) by O. Saeki.

Problem 4.8.7 ([49]) Are there germs of isolated hypersurface singularities such that in its μ-constant stratum there is no germ admitting real equations?

It is well-known that the answer is no if $n = 1$. For $n = 2$, the first step would be to study the problem for the tangent cone (and looking for singularities where the combinatorics of the tangent cone is constant in the μ-constant stratum, which is not always the case, see [14]).

Saeki started from the curve $(x^3-y^3)(y^3-z^3)(z^3-x^3) = 0$. Of course, this is not a right example, since the previous equation is actually *real*. Nevertheless, it is not hard to see that in the space of curves with the same combinatorics, it is not possible to get always real equations for the irreducible components $\ell_1 = 0, \dots, \ell_9 = 0$, all linear.

The idea of O. Saeki is to consider a non-reduced tangent cone $g_0 := \ell_1^{n_1} \cdot \ldots \cdot \ell_9^{n_9}$, where the n_i are positive and pairwise distinct. Of course, there is no SIS with this tangent cone, but Melle's t-singularities [42] can do the job. Let us consider a *generic* homogeneous polynomial g_1 of degree $n_1 + \cdots + n_9 + 1$, namely, the projective curve $\{g_1 = 0\}$ intersects transversally the reduced curve $\{\ell_1 \cdot \ldots \cdot \ell_9 = 0\}$. Under these conditions, the singularity defined by $F := g_0 + g_1$ is isolated and for any $\tilde{F}$ defining a singularity in its μ-constant stratum, $\tilde{F}$ does not admit real equations, since it can be proved that the tangent cone is fixed.

Let us relate this problem with another one stated by N. A'Campo. Let us recall that from the Monodromy Theorem 4.5.5, the characteristic polynomial of Φ_* over $H^n(\Sigma; \mathbb{Q})$ is a product of cyclotomic polynomials. In particular, Φ^* and $(\Phi^*)^{-1}$ are conjugate since they have the same Jordan blocks.

Problem 4.8.8 ([3]) Let V be a germ of an isolated singularity of hypersurface. Let Φ be its monodromy and let $\Phi_*^{\mathbb{Z}}$ be the automorphism defined on $H_n(\Sigma; \mathbb{Z})$. Are $\Phi_*^{\mathbb{Z}}$ and $(\Phi_*^{\mathbb{Z}})^{-1}$ conjugate?

The relationship between the two problems can from the following reasoning. Let V be a germ of an isolated singularity of hypersurface with real equations. One can take the Milnor fiber over a real value. Note that the monodromy of V is taken running counterclockwise a circle about the origin. The above condition implies that V is preserved by the complex conjugation, which sends Φ to Φ^{-1}. Hence, the answer to Problem 4.8.8 is yes if V is defined by a real equation, or more precisely, in its μ-constant stratum there is a singularity with real equations, e.g., if $n = 1$.

Remark 4.8.9 Actually in the previous discussion, one can give a sharper argument. If V and $\overline{V}$ are in the same μ-constant stratum (even if no element of this stratum has real equations), then the answer to Problem 4.8.8 is also yes.

Note that in Saeki's example V, the singularities V and $\overline{V}$ are not in the same μ-constant stratum, so, it would be interesting to check its answer to Problem 4.8.8. The main problem is the computation of the monodromy over the integers. There is an aside problem; even if this monodromy can be computed, the Milnor number of V is huge, so it will be difficult to check the conjugacy. To be honest, in a private conversation, N. A'Campo ensured that he had effective methods for this checking, even for large ranks.

Thanks to M. Escario's work an example of a k-LYS (or SIS) not satisfying the condition in Remark 4.8.9 could lead to an answer to Problem 4.8.8. In [15] it was proved that two SIS were in the same μ-constant stratum if and only if their tangent cones had the same combinatorics and they were in the same connected component of the space of curves with that combinatorics.

There is a combinatorics in [8, Example 4.2], such that no curve with this combinatorial type has real equations and the space of curves with that combinatorics has two connected components, which are *conjugate*. The degree of the combinatorics is 6. Independently, M. Namba [44] and A.I. Degtyarëv [22] proved that the space of curves of degree ≤ 5 with a given combinatorics is always connected. Hence, the degree of this example is minimal.

Proposition 4.8.10 *The space of irreducible sextic curves having three singular points of types $\mathbb{E}_8$ ($v^3 + u^5 = 0$), $\mathbb{E}_7$ ($v^3 + uv^3 + u^5 = 0$), and $\mathbb{A}_4$ ($v^2 + u^5 = 0$) has two conjugate connected components. Moreover, any such curve is projectively equivalent to either C or $\overline{C}$ where C is defined by the polynomial $f_6(x, y, z)$:*

$$(8i + 4)\, x^3y^3 + (-24i - 12)\, x^3y^2z + (-24i - 12)\, x^2y^3z + (24i + 12)\, x^3yz^2+$$
$$(48i + 4)\, x^2y^2z^2 + (24i + 12)\, xy^3z^2 + (-8i - 4)\, x^3z^3 + (-24i + 3)\, x^2yz^3+$$
$$(-24i + 28)\, xy^2z^3 + (-8i - 4)\, y^3z^3 + 5x^2z^4 + 20ixyz^4 - 20y^2z^4.$$

Proof In [8] we start with such a curve C. It is easy to prove that there is a unique projective automorphism such that the singular points are $[0 : 1 : 0]$, $[1 : 0 : 0]$, $[0 : 0 : 1]$ and that the tangent lines at the first two singular points are $x - z = 0$ and $y - z = 0$. We obtain then the equation must be either $f_6 = 0$ or $\overline{f}_6 = 0$. Moreover, the tangent line at the last singular point has no real equation. If the curve would admit real equations, the coordinates of the three singular points are real (they are of different type, hence they cannot be complex conjugate), as the equations of the tangent lines. Then, the projective automorphism must be real and it is not possible. □

Remark 4.8.11 It is obvious that these curves $C, \overline{C}$ do not form a Zariski pair, since the complex conjugation in $\mathbb{C}$ defines a homeomorphism $(\mathbb{P}^2, C) \to (\mathbb{P}^2, \overline{C})$. Note that this homeomorphism respects the natural orientation of $\mathbb{P}^2$ but does not respect the natural orientation of $C, \overline{C}$; see e.g. [4, p. 8] for the notion of complex orientation of a complex variety. These curves are natural candidates for the notion of *oriented*

Zariski pairs, i.e., curves with the same combinatorics with no homeomorphism of pairs preserving orientations.

Unfortunately we do not know if $C, \overline{C}$ form an oriented Zariski pair. In [8] we prove that if C' is the curve of equation $xz(125x - 4(2+11i)z) = 0$, then $C \cup C', \overline{C} \cup \overline{C'}$ form an oriented Zariski pair.

The Milnor number of a SIS having C as tangent cone is 144. It is not known if it is minimal along the singularities of surface not satisfying the condition in Remark 4.8.9 but most likely the minimal Milnor number will be in the same order of magnitude. There is a lot of work for these computations, but all the tools are available.

These ideas fall into the main features of SIS: they provide sophisticated examples which have computable features.

4.9 SIS, *k*-LYS, and Zariski Pairs Not of Alexander Type

After the work of I. Luengo, where he used SIS to prove that μ-constant strata may not be smooth, using properties of projective plane curves, the second main contribution of SIS was to provide counterexamples to Yau's Conjecture 4.5.3. The idea is the same one: use the properties of the tangent cones. The counterexamples shown in Theorem 4.7.13 come from the existence of Alexander-Zariski pairs, exhibited in [6] and coming from the works of O. Zariski.

Let us try to isolate the geometric properties that led to these results. A model of the Milnor fibration is obtained from divisor of the semistable normalization of an embedded $\mathbb{Q}$-resolution. Let us consider to k-LYS V', V'' with tangent cones C', C'' having the same combinatorics. We obtain two embedded $\mathbb{Q}$-resolutions with the same number r of irreducible components and a natural bijection

$$\begin{aligned}\{E'_J \mid \emptyset \neq J \subset \{0,1,\dots,r\}\} &\longleftrightarrow \{E''_J \mid \emptyset \neq J \subset \{0,1,\dots,r\}\}\\ E'_J &\longleftrightarrow E''_J.\end{aligned}$$

This bijection induces homeomorphisms $h_J : E'_J \to E''_J$ if $J \neq \{1\}$. Moreover, these homeomorphisms are naturally compatible for distinct sets of indices.

As a consequence, if we consider the divisors D', D'' of the semistable normalizations, the bijection also induces compatible homeomorphisms $\tilde{h}_J$, $J \neq \{1\}$, such that the following diagram is commutative:

$$\begin{array}{ccc} D'_J & \xrightarrow{\tilde{h}_J} & D''_J \\ \downarrow & & \downarrow \\ E'_J & \xrightarrow{h_J} & E''_J. \end{array}$$

From the definition of Zariski pairs, we obtain the sufficient and necessary condition to obtain this diagram also for $J = \{1\}$.

Lemma 4.9.1 *There are homeomorphisms $h_1 : E_1' \to E_1''$ and $\tilde{h}_1 : D_1' \to D_1''$ extending the compatibility if and only the tangent cones do* not *form a Zariski pair.*

We can interpret these results as the existence of a decomposition of the Milnor fibration. If the tangent cones are not in a Zariski pair, it is easy to see that both Milnor fibrations are equivalent, and the singularities have the same embedded topology.

On the other side, if the tangent cones form a Zariski pair, then the decompositions are not equal and we cannot deduce that the Milnor fibrations are equivalent. The decomposition of the Milnor fibration in terms of the semistable normalization have been thorough studied for $n = 1$. Actually, the induced the decomposition of the Milnor fiber coincides with Nielsen-Thurston [46, 56] decomposition of the monodromy, and this decomposition is unique up to isotopy!

If the unicity of this decomposition would hold in higher dimensions, we would deduce that if C', C'' form a Zariski pair, then the k-LYS V', V'' would have different embedded topology. Actually, with some extra-work, if V is a SIS with tangent cone the curve $C \cup C'$ in Remark 4.8.11, it would provide a negative answer for Problem 4.8.8.

Even without this unicity, Theorem 4.7.13 works if the tangent cones form an *Alexander-Zariski pair*, since the Jordan form of the monodromies are distinct, and the same happens for the embedded topology.

Problem 4.9.2 Let V', V'' be k-LYS whose tangent cones C', C'' form a Zariski pair which is not of Alexander type. What can be said about the embedded topology?

The original examples of Zariski pairs were of Alexander type. The first Zariski pairs not of Alexander type appeared in [7]. They correspond to combinatorics of reducible curves. Actually, putting some weights on the connected components, they are distinguished by the Alexander polynomial of some non-reduced curves. The problem is that up to now, we have not found how to translate this invariant into an invariant of a SIS or a k-LYS.

Much more types of Zariski pairs have been found, with different invariants that distinguish them, see a survey [13] of 2008, to check much more of them. Actually, after this survey, qualitative different kinds of Zariski pairs have been found.

Let us point out several particular cases:

(ZP1) Zariski pairs C_1, C_2 such that $\pi_1(\mathbb{P}^2 \setminus C_1) \cong \pi_1(\mathbb{P}^2 \setminus C_2)$.

(ZP2) Zariski pairs C_1, C_2 where $\mathbb{P}^2 \setminus C_1 \cong \mathbb{P}^2 \setminus C_2$ but no homeomorphism can be extended to $\mathbb{P}^2$.

(ZP3) Zariski pairs C_1, C_2 defined by equations $F_i = 0$, such that $F_i \in \mathbb{K}[x, y, z]$, where $\mathbb{K}$ is a number field, and there exists $\varphi \in \operatorname{Gal}(\mathbb{K}, \mathbb{Q})$ such that $F_2 = F_1^{\varphi}$. They are called *arithmetic Zariski pairs* and they share all the algebraic properties. In the same way, we can construct conjugated k-LYS V_1, V_2 which share all the algebraic properties.

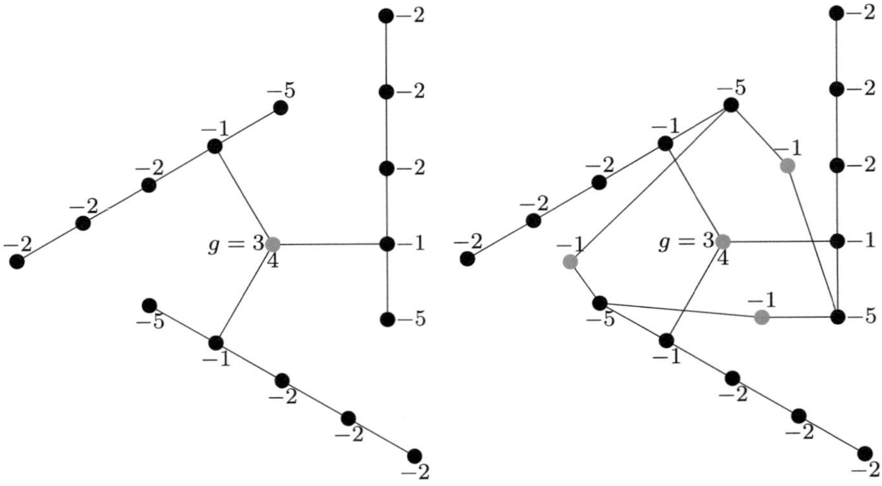

Fig. 4.3 Comb_4 (right) and $\text{Comb}_4^{\text{Irr}}$ (left)

All the above Zariski pairs are not of Alexander type. There are many examples in the literature, we sketch a family which serves for (ZP1) and (ZP3).

Example 4.9.3 ([11]) Let us consider the combinatorics Comb_n, $n \geq 3$, defined by the following properties (see the dual graph for $n = 4$ in Fig. 4.3, right). There are four irreducible components, three lines and a curve of degree $2n$. This curve has three singular points all of them with the topological type of $v^n - u^{n+1} = 0$, and the lines pairwise connect the singular points. We prove in [11] that after a projective change of coordinates the equation of such a curve can be written as $xyzg_{2n}(x, y, z) = 0$ with the following properties:

(a) The monomials $x^a y^b z^c$ in g_{2n} with non-vanishing coefficient satisfy the condition $\min(a + b, b + c, c + a) \geq n$.
(b) The sum of the terms corresponding to the monomials $x^n y^b z^c$ equals $(y + z)^n$.
(c) The sum of the terms corresponding to the monomials $x^a y^n z^c$ equals $(x + z)^n$.
(d) The sum of the terms corresponding to the monomials $x^a y^b z^n$ equals $(x + \zeta y)^n$, where $\zeta^n = 1$.
(e) The polynomial is *generic* satisfying the above conditions.

Two curves in Comb_n are in the same connected component of Comb_n if and only if the roots in (d) are either equal or conjugate.

For $n = 3$, we have two connected components, associated to $1, \exp \frac{2i\pi}{3}$. They form a Zariski pair since the fundamental groups of the complement are not isomorphic. For $n > 3$, the fundamental groups of the complements are always abelian. There is a sort of linking invariant that shows they form Zariski pairs (actually tuples). For $n > 3$, we are in (ZP1), and for $n > 4$, we have also examples of (ZP3).

Remark 4.9.4 If we consider the combinatorics $\mathrm{Comb}_n^{\mathrm{Irr}}$ of the irreducible curves in Comb_n, we have also a disconnected space (see the dual graph for $n = 4$ in Fig. 4.3, left). For $n = 3$, we have actually Alexander-Zariski pairs. For $n > 3$, it is unknown if they form a Zariski pair. It would be interesting if the corresponding SIS may have different embedded topology and this property could be translated into a topological property of the curves in order to show that they form a Zariski pair.

4.10 Weighted Lê-Yomdin Singularities

Classical k-LYS allow to transform knowledge from complex projective plane curves to a family of germs of isolated singularities in $(\mathbb{C}^3, 0)$ and the other way around. Following this idea in [14] weighted Lê-Yomdin singularities play the same role where complex projective plane curves are replaced by projective curves in a weighted projective plane.

The idea is similar. Fix a weight $\omega := (p, q, r)$, $\gcd \omega = 1$. Let $(V, 0)$ be a germ of isolated singularity in $(\mathbb{C}^3, 0)$ defined by $F(x, y, z) \in \mathbb{C}\{x, y, z\}$. As in (4.1), for some $k > 0$, we can consider

$$F(x, y, z) = F_d(x, y, z) + \sum_{m \geq d+k} F_m(x, y, z), \qquad F_d \neq 0, \tag{4.11}$$

but now instead of the decomposition of F in homogeneous forms, we consider the one in ω-weighted homogeneous forms. The subindex is the ω-degree. We can consider in this case d as the *ω-multiplicity* of F or V.

We consider $C_m := V_{\mathbb{P}^\omega}(F_m)$ which is a closed curve in the *orbifold* $\mathbb{P}^2_\omega$, see Sect. 4.12. This is a *weighted projective plane curve.*

We want to consider a family of germs with the following property. Let $\sigma_\omega : (\hat{\mathbb{C}}^3_\omega, E) \to (\mathbb{C}^3, 0)$ be the ω-weighted blow-up of the origin, see Sect. 4.12. Recall that $E \cap \hat{V} = C_d$, where $E \equiv \mathbb{P}^2_\omega$ is the exceptional divisor and $\hat{V}$ is the strict transform. We want that outside $\mathrm{Sing}(C_d)$ this intersection has $\mathbb{Q}$-normal crossings, and at $\mathrm{Sing}\, C_d$ we have a situation similar to (Loc4).

Following Definition 4.2.3, we may consider the condition $\mathrm{Sing}(C_d) \cap C_{d+k} = \emptyset$. This is basically what we want but since the total space of the ω-weighted blowup is singular, we have to be more precise, as in [10].

Definition 4.10.1 Let $\omega := (p, q, r)$, $\gcd \omega = 1$, and $k > 0$. Let $(V, 0)$ be a germ of singularity in $(\mathbb{C}^3, 0)$ defined by $F(x, y, z) \in \mathbb{C}\{x, y, z\}$ which can be decomposed as in (4.11). Then we say that $(V, 0)$ is an *(ω, k)-weighted Lê-Yomdin singularity* ((ω, k)-WLYS) if the following points do not belong to C_{d+k}.

(i) $P \in \mathrm{Sing}(C_d) \setminus \mathrm{Sing}(\hat{\mathbb{C}}^3_\omega)$.
(ii) $P = [0 : 0 : 1]$, $r > 1$, and the multiplicity of $F(x, y, 1)$ is > 1. Something similar holds for the other vertices.

(iii) $P \in C_d \cap \{z = 0\}$, $P \neq [1 : 0 : 0], [0 : 1 : 0]$, $\gcd(p, q) > 1$, and the intersection is not transversal at P. Something similar holds for the other axes.

As for k-LYS, it is not hard to prove that an (ω, k)-WLYS is isolated.

Remark 4.10.2 There is a geometric interpretation for (ii). Let us consider the minimal resolution of $(\mathbb{P}^2_\omega, P)$ of Proposition 4.11.8, with its exceptional divisor $E := E_1 \cup \cdots \cup E_r$. Let $\hat{C}_d$ be the strict transform of C_d. The multiplicity of $F(x, y, 1)$ is 1 if and only if

(w) $\hat{C}_d \cap E$ has only one point $\hat{Q}$ which is smooth for both $\hat{C}_d$ and E, with distinct tangent directions, and $Q \in E_1 \cup E_r$.
(x) If Q is in the strict transform of $x = 0$, then $\gcd(q, r) = 1$.
(y) If Q is in the strict transform of $y = 0$, then $\gcd(p, r) = 1$.

For (iii), if $\gcd(p, q) > 1$, then for each $P = [x_0 : y_0 : 0]_\omega$, for which either

(a) $x_0 y_0 \neq 0$, or
(b) $P = [0 : 1 : 0]_\omega$ and $\gcd(p, r) = 1$, or
(c) $P = [1 : 0 : 0]_\omega$ and $\gcd(q, r) = 1$, then

The singularity $(\hat{\mathbb{C}}^3_\omega, P)$ is cyclic quotient by a group of order $\gcd(p, q)$ and this is why the transversality condition is needed.

Remark 4.10.3 In the non-weighted case, given a curve C_d we can construct k-LYS with C_d as tangent cone for any $k > 0$. In the weighted case when there are singular points of $\mathbb{P}^2_\omega$ in C_d, it may happen that some k cannot happen, since for some degrees a singular point of E can be forced to be in any curve of that degree.

Example 4.10.4 Let $\omega := (2, 2, 3)$ and $C_{16} := \{y^2 z^4 - x^5 z^2 + y^8 = 0\}$. Following (4.12), there is an *isomorphism*

$$\begin{array}{ccc} \mathbb{P}^2_\omega & \xrightarrow{\psi} & \mathbb{P}^2_\eta \\ [x : y : z]_\omega & \longmapsto & [x : y : z^2]_\eta \end{array}$$

where $\eta = (1, 1, 3)$. Note that $\Phi(C_{16}) = C_8 = \{y^2 z^2 - x^5 z + y^8 = 0\}$. The only singular point of $\mathbb{P}^2_\eta$ is $[0 : 0 : 1]_\eta$ and $C_8 \cap \{z = 0\} = [1 : 0 : 0]_\eta$. If we consider the affine chart $x = 1$, the equation is $y^2 z^2 - z + y^8 = 0$, i.e, C_8 is smooth at $[1 : 0 : 0]_\eta$ but it is tangent to $z = 0$.

Let us come back to the weight ω. The space $(\hat{\mathbb{C}}^3_\omega$ is singular at $P := [0 : 0 : 1]_\omega \in C_{16}$ and at $\{z = 0\} \subset E$. We have that $C_{16} \cap \{z = 0\} = [1 : 0 : 0]_\omega =: Q$. If we want to construct a (ω, k)-WLYS, we need to take some ω-weighted projective curve $C_{16+k} = \{f_{16+k}(x, y, z) = 0\}$ of ω-degree $16 + k$ such that $P, Q \notin C_{16+k}$.

This means that there are powers of z and x with non-zero coefficients in $f_{16+k}(x, y, z)$, i.e., $16 + k$ is a multiple of 6. Hence the smallest k equals 2. We can take $f_{18} = x^9 + z^6$. Then, $\{y^2 z^4 - x^5 z^2 + y^8 + x^9 + z^6 = 0\}$ is an $(\omega, 2)$-WLYS.

Example 4.10.5 Note that a singularity can be a (ω, k)-WLYS for different values of (ω, k). This example was provided by I. Luengo. Let $F_t(x, y, z) = x^5 + y^3 z + z^{11} + t x^2 y^2$, $t \in \mathbb{C}$.

We can consider it as a semiquasihomogeneous deformation of F_0. Note that F_0 defines a smooth curve in $\mathbb{P}^2_{33,50,15}$ of degree 165. The weight of $x^2 y^2$ is 166, hence the corresponding singularity is a $((33, 50, 15), 1)$-WLYS.

Consider the weight $(2, 3, 1)$. The weighted homogeneous polynomial $x^5 + y^3 z + t x^2 y^2$ has weighted degree 10 and then the corresponding singularity is a $((2, 3, 1), 1)$-WLYS. It is also an example of μ-constant family which is not μ^*-constant family as the well-known Briançon-Speder example [18].

We can study (ω, k)-WLYS using the same techniques as for k-LYS. In particular, an embedded $\mathbb{Q}$-resolution is obtained as for k-LYS; it is enough to replace the first standard blowing-up by an ω-weighted blowing-up of the origin in $\mathbb{C}^3$. Of course, the total space has quotient singularities, but we already encountered this situation in the process to build an embedded $\mathbb{Q}$-resolution for k-LYS. The numerical invariants of these resolution are trickier than for k-LYS, but the idea is the same one.

As one can define combinatorics of weighted projective curves, (ω, k)-WLYS with ω-*weighted* tangent cones with the same combinatorics have embedded $\mathbb{Q}$-resolutions which may differ only in the first divisor and the results of Sect. 4.7 remain true with slights modifications. Actually, the same happens for the results of other sections. The notion of Zariski pairs (and Alexander-Zariski pairs) can be also defined, see [10], providing also counterexamples to Yau's conjecture 4.5.3.

Why (ω, k)-WLYS may be interesting? They provide a bridge between singularity theory and weighted projective plane curves and this bridge is larger than the one provided by k-LYS. Let us state some cases.

(W1) Examples of μ-constant deformations which are not μ^*-constant do exist among (ω, k)-WLYS and not among k-LYS. Moreover, it provides simple proofs for the topological triviality of these families when the family of ω-*weighted* tangent cone is also topologically trivial.

(W2) A singularity of plane curve has a unique minimal embedded $\mathbb{Q}$-resolution, and most likely the same happens for most k-LYS. The fact that a singularity can be (ω_i, k_i)-WLYS, $i = 1, 2$, for $(\omega_1, k_1) = (\omega_2, k_2)$, may imply the existence of several minimal embedded $\mathbb{Q}$-resolutions.

(W3) There are k-LYS whose link is a $\mathbb{Q}$-homology sphere, see Proposition 4.4.5. In [10] we prove that there is no k-LYS whose link is a $\mathbb{Z}$-homology sphere, but there are among (ω, k)-WLYS.

(W4) Monodromy conjecture for the topological and Igusa zeta functions holds for SIS [9], and following the same ideas it should hold for k-LYS. There are no yet formulas for the characteristic polynomial Δ and for the topological zeta function Z of an (ω, k)-WLYS but using [30] (for Δ) and [35] (for Z) the computation of such formulas is an affordable task, and monodromy conjecture could be confronted.

4.11 Quotient Singularities

In this section we may a short presentation of abelian quotient singularities. There are several key references, e.g. [21, 48]. Roughly speaking, an (abelian) quotient singularity is the quotient of $(\mathbb{C}^n, 0)$ by the action of a finite abelian subgroup G of analytic isomorphisms. To be more precise, such a G acts also on $\mathcal{O}_{\mathbb{C}^n,0} = \mathbb{C}\{x_1, \ldots, x_n\}$ and the fixed ring $\mathcal{O}^G_{\mathbb{C}^n,0}$ is normal and defines a germ of complex variety, namely the corresponding quotient singularity.

Let us add a word of warning. This is closely related to *orbifolds*, where one focuses not only on $\mathcal{O}^G_{\mathbb{C}^n,0}$, but also on the action. We are mostly interested in the varieties, but the quotient and orbifold language simplifies the exposition. We can simplify the exposition assuming that G is a linear subgroup [21, Théorème 4].

Example 4.11.1 One could start with quotient singularities in dimension 1. It is not hard to see that a faithful action must be cyclic and it will be by multiplication of roots of unity. If it is the action of $\mu_d := \{\zeta \in \mathbb{C} \mid \zeta^d = 1\}$, then it is clear that $\mathbb{C}\{x\}^{\mu_d} = \mathbb{C}\{x^d\}$ and the corresponding *quotient singularity* is actually smooth.

D. Prill [48] took care of this situation and defined the notion of *small linear group*, the only reflection about a hyperplane is the identity. He showed that it is enough to consider small groups to get all the quotient singularities. We borrow some notation from [17] in order to describe these singularities.

Step 4.11.2 The quotient singularity of $(\mathbb{C}^n, 0)$ defined by the action of the group $\mu_{d_1} \times \cdots \times \mu_{d_r}$ given by

$$(\zeta_1, \ldots, \zeta_r) \cdot (x_1, \ldots, x_n) := \left(x_1 \prod_{j=1}^{r} \zeta_j^{a_{j1}}, \ldots, x_n \prod_{j=1}^{r} \zeta_j^{a_{jn}} \right)$$

will be denoted as

$$X \left(\begin{array}{c|ccc} d_1 & a_{11} & \cdots & a_{1n} \\ \vdots & \vdots & \ddots & \vdots \\ d_r & a_{r1} & \cdots & a_{rn} \end{array} \right).$$

If $d_1 = \cdots = d_r = d$, we may denote it also by $\frac{1}{d}A$ where A is the above $(a_{i,j})$ matrix of size $r \times n$.

This notation also serves for the orbifolds. Different symbols can represent the same singularity. Let us recall some operations which keep invariant the type of the quotient singularity.

Lemma 4.11.3 ([16, Lemma 1.2]) *The following operations do not change the isomorphism type of* $X(\mathbf{d}|A)$.

(1) *Permutation of the columns of* A *(the variables are correspondingly permuted).*
(2) *Permutation of the rows of* $(\mathbf{d}|A)$.
(3) *Multiplication of the last row of* $(\mathbf{d}|A)$ *by a positive integer.*
(4) *Multiplication of the last row of* A *by an integer coprime with* d_r.
(5) *Replace* $a_{i,j}$ *by* $a_{i,j} + kd_i$, $k \in \mathbb{Z}$.
(6) *If* $d_r = 1$, *eliminate the last row.*

All these operations do not change the orbifold.

Example 4.11.4 With this lemma we can revisit Example 4.11.1. It is not hard to check that we can assume $r = 1$, and that $\frac{1}{d}(a) = \frac{\gcd(a,d)}{d}(1)$. As orbifold, it is normalized, but as singularity, we end with the smooth germ.

There is a last operation which do not apply to orbifolds.

Lemma 4.11.5 ([16, Lemma 1.2]) *The following operation does not change the isomorphism type of* $X(\mathbf{d}|A)$.

(7) *If* e *divides* $\gcd(d_1, a_{1,1}, \ldots, a_{1,n-1})$ *and* $\gcd(e, a_{1,n}) = 1$ *then replace* $a_{i,n}$ *by* $ea_{i,n}$.

In this case the isomorphism onto the new type is given by

$$[(x_1, \ldots, x_{n-1}, x_n)] \mapsto [(x_1, \ldots, x_{n-1}, x_n^e)].$$

With these operations, we can normalize 2-dimensional quotient singularities, and we see that they coincide with Jung-Hirzebruch singularities, see [31] for a nice description.

Lemma 4.11.6 *Any* 2*-dimensional quotient singularity is of the type* $\frac{1}{d}(a, b)$, $\gcd(d, a) = \gcd(d, b) = 1$. *Moreover, there* $\alpha, \beta \in \mathbb{Z}$, $0 < \alpha, \beta < d$, *and coprime with* d *such that* $\frac{1}{d}(a, b) = \frac{1}{d}(\alpha, 1) = \frac{1}{d}(1, \beta)$.

Sometimes it is useful to present the singularities not in normal form to avoid cumbersome notation. A minimal resolution of such a quotient singularity can be expressed using continuous fractions.

Definition 4.11.7 Let $q \in \mathbb{Q}_{>1}$. The *continuous fraction decomposition* of q is a tuple defined recursively as follows. If $q \in \mathbb{Z}$, the tuple is $[q]$. If not, let $r := \lceil q \rceil - q$. If $[b_1, \ldots, b_r]$ is the continuous fraction decomposition of $r_1^{-1} > 1$, then $[\lceil q \rceil, b_1, \ldots, b_r]$ is the continuous fraction decomposition of q.

8

For example, the continuous fraction decomposition of $\frac{n+1}{n}$ is $[2, \ldots, 2]$ with length n.

Fig. 4.4 $\rho^{-1}(Y)$

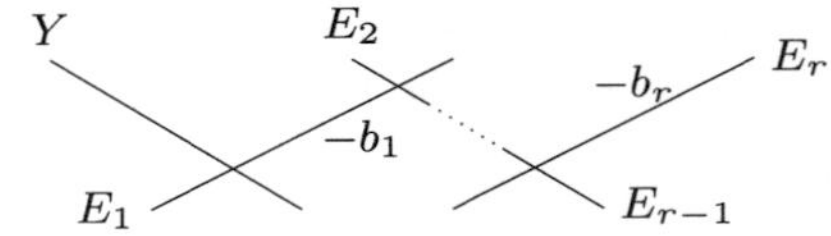

Proposition 4.11.8 ([31]) *Let (Z, P) be a quotient singularity of type $\frac{1}{d}(1, \beta)$, where $\gcd(d, \beta) = 1$. Let $[b_1, \ldots, b_r]$ be the continuous fraction decomposition of $\frac{d}{\beta}$.*

Let $\rho : (\hat{Z}, E) \to (Z, P)$ be the minimal resolution of (Z, P). Then $E = \rho^{-1}(P)$ is a linear divisor with r irreducible components having as self-intersections $-b_1, \ldots, -b_r$ as in Fig. 4.4.

Let us assume that $Y \subset Z$ is a curve such that the local equation of (Y, P) is $x_2 = 0$. Let us denote also by Y the strict transform of Y by ρ. Then, Y intersects transversally E at a point in E_1 and $(Y \cdot Y)_{\hat{Z}} = (Y \cdot Y)_Z - \frac{\beta}{d}$.

Example 4.11.9 In Remark 4.4.3, we stated that the normalization of the singularity defined by $z^k - u^a v^b = 0$ is a disjoint union of quotient singularities, as many as $\gcd(k, a, b)$. Let us assume for simplicity that $\gcd(k, a, b) = 1$. In that case it has the same normalization as $z^{k_1} - u^{a_1} v^b = 0$, where $k_1 = \frac{k}{\gcd(a,k)}$ and $a_1 = \frac{a}{\gcd(a,k)}$.

Hence, again for the sake of simplicity we can assume that $\gcd(k, a) = \gcd(k, b) = 1$. If $0 < c < k$ and $ac \equiv b \bmod k$, then $z^k - u^a u^b = 0$ and $z^k - uv^c = 0$ have the same normalization and the quotient singularity is $\frac{1}{k}(1, k - c)$.

4.12 Weighted Blow-ups

The weighted blow-ups are a generalizations of standard blow-ups, where the standard projective space is replaced by a weighted projective space, see [23], and are the bricks of embedded $\mathbb{Q}$-resolutions. We restrict ourselves to the dimensions we are interested in.

Let $\omega = (p_0, p_1, \ldots, p_n) \in \mathbb{Z}^{n+1}_{>0}$ such that $\gcd \omega = 1$. The ω-weighted projective space $\mathbb{P}^n_\omega$ is the quotient of $\mathbb{C}^{n+1} \setminus \{0\}$ by the following action of $\mathbb{C}^*$:

$$t \cdot (x_0, x_1, \ldots, x_n) := (t^{p_0} x_0, t^{p_1} x_1, \ldots, t^{p_n} x_n)$$

with a natural structure of normal variety (sometimes an orbifold structure is preferred); the equivalence classes are denoted as $[x_0 : x_1 : \cdots : x_n]_\omega$ which can be identified with the closure of the orbit in $\mathbb{C}^{n+1}$.

Example 4.12.1 If $\gcd(p, q) = 1$, then $\mathbb{P}^1_{(p,q)}$ is naturally isomorphic to $\mathbb{P}^1$, via $[x : y] \mapsto [x^p : y^q]$. This isomorphism no longer holds if we treat $\mathbb{P}^1_{(p,q)}$ as orbifold.

Remark 4.12.2 As a generalization let $\omega = (p_0, p_1, \ldots, p_n) \in \mathbb{Z}^{n+1}_{>0}$, $\gcd\omega = 1$, and such that $\gcd(p_1, \ldots, p_n) = d > 1$. Let $\eta := \left(p_0, \frac{p_1}{d}, \ldots, \frac{p_n}{d}\right)$. Then there is a natural isomorphism

$$\begin{aligned} \mathbb{P}^n_\omega &\longrightarrow \mathbb{P}^n_\eta \\ [x_0 : x_1 : \cdots : x_n]_\omega &\longmapsto [x_0^d : x_1 : \cdots : x_n]_\eta. \end{aligned} \tag{4.12}$$

The weighted blow-ups can be defined using incidence varieties for the previous action as it is done for usual blow-ups.

Definition 4.12.3 Let $\omega = (p_1, \ldots, p_n) \in \mathbb{Z}^n_{>0}$ such that $\gcd\omega = 1$. The ω-*weighted blow-up* of $(\mathbb{C}^n, 0)$ is the map

$$\begin{aligned} \{(\mathbf{p}, \mathbf{q}) \in \mathbb{C}^n \times \mathbb{P}^{n-1}_\omega \mid \mathbf{p} \in \mathbf{q}\} =: \hat{\mathbb{C}}^n_\omega &\xrightarrow{\sigma_\omega} \mathbb{C}^n \\ (\mathbf{p}, \mathbf{q}) &\longmapsto \mathbf{p}. \end{aligned}$$

Remark 4.12.4 If $\gcd\omega := d > 1$, the $\frac{\omega}{d}$-weighted blowing-up is also called an ω-weighted blow-up.

As it happens with the usual blow-up, it is an isomorphism outside the origin, and the *exceptional divisor* $E := \sigma_\omega^{-1}(0)$ is isomorphic to $\mathbb{P}^{n-1}_\omega$. The main difference is that the total space is no longer smooth, cyclic quotient singularities may arise.

Remark 4.12.5 In a similar way, weighted blow-ups of quotient singularities can be defined, see [16, 17]. If $Z = X(\mathbf{d}|A)$ is a quotient singularity, $\mathbf{d} \in \mathbb{Z}^r_{>0}$, $< A \in$ $\mathrm{Mat}(n \times r; \mathbb{Z})$, and we perform an ω-weighted blow-up of $(\mathbb{C}^n, 0)$, the group acts naturally on it source, and its quotient defines an ω-weighted blow-up of Z. In this case the singularities of the total space may be non-cyclic.

We describe singularities and intersections for some cases.

Proposition 4.12.6 ([17, Theorem 4.3 and p. 14]) *Let* $Z := \frac{1}{d}(p, q)$, *with* d, p, q *pairwise coprime. Let* $\sigma_{(p,q)} : \hat{Z} \to Z$ *be the* (p, q)*-weighted blow-up of* Z *and* E *its exceptional divisor. Then* $E^2 = -\frac{d}{pq}$ *and* $\hat{Z}$ *has (at most) two singular points both in* E *of type* $\frac{1}{p}(-d, q)$ *and* $\frac{1}{p}(q, -d)$*; moreover the local equation of* E *at these singular points corresponds to the coordinate for the weight* $-d$.

We end the case of dimension 2, with the weighted version of resolution of singularities by blow-ups.

Theorem 4.12.7 *Let* $(C, 0) \subset (\mathbb{C}^2, 0)$ *be a germ of isolated singularity. Let* $\sigma :$ $(X, E) \to (\mathbb{C}^2, 0)$ *be the minimal resolution of* $(C, 0)$. *Let* r *be the number of branching components in* E *(i.e. those who intersect at least three other irreducible components of* $\sigma^{-1}(C)$*). Then, there exists a composition* $\sigma_w : (X_w, E_w) \to (\mathbb{C}^2, 0)$ *of a sequence of* r *weighted blow-ups which is a* $\mathbb{Q}$*-embedded resolution of* C.

In dimension 3, though the idea is the same the statements are cumbersome. To give some light, we describe the weighted blow-ups of smooth points.

Proposition 4.12.8 *Let* $\omega := (p, q, r)$*, where* $\gcd \omega = 1$. *Let* $\sigma_\omega : (\hat{Z}, E) \to (\mathbb{C}^3, 0)$ *be the* ω*-weighted blow and* E *its exceptional divisor. Then* $\hat{Z}$ *may have the following singularities in* $E \equiv \mathbb{P}^2_\omega$

(a) $[1:0:0]_\omega : \frac{1}{p}(-1, q, r)$;
(b) $[0:1:0]_\omega : \frac{1}{q}(p, -1, r)$;
(c) $[0:0:1]_\omega : \frac{1}{r}(p, q, -1)$;
(d) *at* $x = 0$ *if* $\gcd(q, r) > 1$;
(e) *at* $y = 0$ *if* $\gcd(p, r) > 1$;
(f) *at* $z = 0$ *if* $\gcd(q, r) > 1$.

4.13 Weight Filtration of a Nilpotent Morphism

The conjugacy class of a nilpotent morphism $N : V \to V$, V a vector space of finite dimension, is completely determined by its Jordan form. There is an equivalent and useful invariant called the *filtration!weight*, see e.g. [51].

Definition 4.13.1 Let $v \in V$. The *α-weight, β-weight, and γ-weight* of v are defined as $-\infty$ for $v = 0$ and for $v \neq 0$:

$$\alpha(v) := \min\{\alpha \in \mathbb{Z}_{\geq 0} \mid N^\alpha(v) \neq 0\}, \quad \beta(v) := -\max\{\beta \in \mathbb{Z}_{\geq 0} \mid v \in N^\beta(V)\},$$

and $\gamma(v) := \alpha(v) + \beta(v)$.

The following lemma is straightforward.

Lemma 4.13.2 *For* $v_1, v_2 \in V$ *and* $\omega = \alpha, \beta, \gamma$ *we have*

$$\omega(v_1 + v_2) \leq \max(\omega(v_1), \omega(v_2)).$$

Moreover, if $N(v_1) \neq 0$*, then*

$$\alpha(N(v_1)) = \alpha(v_1) - 1, \quad \beta(N(v_1)) = \beta(v_1) - 1, \quad \gamma(N(v_1)) = \gamma(v_1) - 2.$$

Definition 4.13.3 The *weight filtration* W *centered at* 0 of N is defined as

$$W_k = W_k(V) := \{v \in V \mid \gamma(v) \leq k\}.$$

It is clear that W_k is a subspace and it defines an increasing filtration. In order to understand better this filtration let us check it over a Jordan block.

Example 4.13.4 Let us assume that $\dim V = r$ and there is a basis $v_1, \dots, v_r$ such that $N(v_1) = 0$ and $N(v_i) = v_{i-1}$, if $i > 1$, i.e., N has a Jordan block of size r. From the definitions:

$$\alpha(v_i) = i - 1, \quad \beta(v_i) = i - r, \quad \gamma(v_i) = 2i - (r + 1).$$

We have $(\gamma(v_1), \gamma(v_1), \dots, \gamma(v_r)) = (-r - 1, -r + 1, \dots, r + 1)$, i.e., it is a symmetric arithmetic sequence of ratio 2.

This example is the key for the following properties.

Proposition 4.13.5 *The weight filtration W and the endomorphism N satisfy:*

(a) $N(W_k) \subset W_{k-2}$.
(b) *For* $k \geq 0$, *the map* $\operatorname{gr}_k^W \to \operatorname{gr}_{-k}^W$ *induced by* N^k *is an isomorphism.*
(c) *For* $k \geq 0$, $\dim \operatorname{gr}_k^W = \dim \operatorname{gr}_{-k}^W$ *is the number of Jordan blocks of size k in Jordan form of N.*

Sometimes it is convenient to shift this filtration.

Definition 4.13.6 Let $m \in \mathbb{Z}$; the *weight filtration* $W^{(m)}$ *centered at m* of N is defined as $W_k^{(m)} := W_{k+m}$.

Acknowledgments Partially supported by MCIN/AEI/10.13039/501100011033 (grant code: PID2020-114750GB-C31) and by Departamento de Ciencia, Universidad y Sociedad del Conocimiento del Gobierno de Aragón (grant code: E22_20R: "Álgebra y Geometría").

References

1. A'Campo, N.: Sur la monodromie des singularités isolées d'hypersurfaces complexes. Invent. Math. **20**, 147–169 (1973). https://doi.org/10.1007/BF01404063
2. A'Campo, N.: La fonction zêta d'une monodromie. Comment. Math. Helv. **50**, 233–248 (1975). https://doi.org/10.1007/BF02565748
3. A'Campo, N.: Monodromy of real isolated singularities. Topology **42**(6), 1229–1240 (2003). https://doi.org/10.1016/S0040-9383(02)00099-X
4. Arnol'd, V., Guseĭn-Zade, S., Varchenko, A.: Singularities of differentiable maps. Vol. II, *Monographs in Mathematics*, vol. 83. Birkhäuser Boston Inc., Boston, MA (1988). Translated from the Russian by Hugh Porteous, Translation revised by the authors and James Montaldi
5. Artal, E.: Forme de Jordan de la monodromie des singularités superisolées de surfaces. Mem. Amer. Math. Soc. **109**(525), x+84 (1994). https://doi.org/10.1090/memo/0525
6. Artal, E.: Sur les couples de Zariski. J. Algebraic Geom. **3**(2), 223–247 (1994)
7. Artal, E., Carmona, J.: Zariski pairs, fundamental groups and Alexander polynomials. J. Math. Soc. Japan **50**(3), 521–543 (1998). 10.2969/jmsj/05030521
8. Artal, E., Carmona, J., Cogolludo, J.: Effective invariants of braid monodromy. Trans. Amer. Math. Soc. **359**(1), 165–183 (2007)
9. Artal, E., Cassou-Noguès, P., Luengo, I., Melle, A.: Monodromy conjecture for some surface singularities. Ann. Sci. École Norm. Sup. (4) **35**(4), 605–640 (2002). https://doi.org/10.1016/S0012-9593(02)01100-X

10. Artal, E., Cogolludo, J., Martín-Morales, J.: Cremona transformations of weighted projective planes, Zariski pairs, and rational cuspidal curves. In: J. Fernández de Bobadilla, T. Laszlo, A. Stipsicz (eds.) Singularities and Their Interaction with Geometry and Low Dimensional Topology, Trends in Mathematics. Birkhäuser, Basel (2020). https://doi.org/10.1007/978-3-030-61958-9_7
11. Artal, E., Cogolludo, J., Martín-Morales, J.: Triangular curves and cyclotomic Zariski tuples. Collect. Math. **71**(3), 427–441 (2020). https://doi.org/10.1007/s13348-019-00269-y
12. Artal, E., Cogolludo, J., Martín-Morales, J.: Cyclic coverings of rational normal surfaces which are quotients of a product of curves. Publ. Mat. **68**(2), 359–406 (2024). https://doi.org/10.5565/publmat6822402
13. Artal, E., Cogolludo, J., Tokunaga, H.: A survey on Zariski pairs. In: Algebraic geometry in East Asia—Hanoi 2005, *Adv. Stud. Pure Math.*, vol. 50, pp. 1–100. Math. Soc. Japan, Tokyo (2008). 10.2969/aspm/05010001
14. Artal, E., Fernández de Bobadilla, J., Luengo, I., Melle, A.: Milnor number of weighted-Lê-Yomdin singularities. Int. Math. Res. Not. IMRN (22), 4301–4318 (2010). https://doi.org/10.1093/imrn/rnq041
15. Artal, E., Luengo, I., Melle, A.: Superisolated surface singularities. In: Singularities and computer algebra, *London Math. Soc. Lecture Note Ser.*, vol. 324, pp. 13–39. Cambridge Univ. Press, Cambridge (2006). https://doi.org/10.1017/CBO9780511526374.005
16. Artal, E., Martín-Morales, J., Ortigas, J.: Cartier and Weil divisors on varieties with quotient singularities. Internat. J. Math. **25**(11), 1450100, 20 (2014). https://doi.org/10.1142/S0129167X14501006
17. Artal, E., Martín-Morales, J., Ortigas, J.: Intersection theory on abelian-quotient V-surfaces and $\mathbf{Q}$-resolutions. J. Singul. **8**, 11–30 (2014). 10.5427/jsing.2014.8b
18. Briançon, J., Speder, J.P.: La trivialité topologique n'implique pas les conditions de Whitney. C. R. Acad. Sci. Paris Sér. A-B **280**(6), Aiii, A365–A367 (1975)
19. Brieskorn, E.: Über die Auflösung gewisser Singularitäten von holomorphen Abbildungen. Math. Ann. **166**, 76–102 (1966)
20. Carmona, J.: Monodromía de trenzas de curvas algebraicas planas. Ph.D. thesis, Universidad de Zaragoza (2003)
21. Cartan, H.: Quotient d'un espace analytique par un groupe d'automorphismes. In: Algebraic geometry and topology. A symposium in honor of S. Lefschetz, pp. 90–102. Princeton Univ. Press, Princeton, NJ (1957)
22. Degtyarëv, A.: Isotopic classification of complex plane projective curves of degree 5. Algebra i Analiz **1**(4), 78–101 (1989)
23. Dolgachev, I.: Weighted projective varieties. In: Group actions and vector fields (Vancouver, B.C., 1981), *Lecture Notes in Math.*, vol. 956, pp. 34–71. Springer, Berlin (1982)
24. Durfee, A.: The signature of smoothings of complex surface singularities. Math. Ann. **232**(1), 85–98 (1978). https://doi.org/10.1007/BF01420624
25. Escario, M.: Discriminant method for the homological monodromy of tame polynomials. arXiv e-prints p. arXiv0602.297 (2006)
26. Escario, M.: Método del discriminante para la monodromía homológica de singularidades aisladas. Ph.D. thesis, Universidad de Zaragoza (2006)
27. Fernández de Bobadilla, J., Luengo, I., Melle, A., Némethi, A.: On rational cuspidal projective plane curves. Proc. London Math. Soc. (3) **92**(1), 99–138 (2006). https://doi.org/10.1017/S0024611505015467
28. Gabrièlov, A.: Polar curves and intersection matrices of singularities. Invent. Math. **54**(1), 15–22 (1979). https://doi.org/10.1007/BF01391174
29. Grothendieck, A.: Résumé des premiers exposés de A. Grothendieck rédigé par P. Deligne. In: Groupes de monodromie en géométrie algébrique. I, Lecture Notes in Mathematics, Vol. 288, pp. viii+523. Springer-Verlag, Berlin (1972). Séminaire de Géométrie Algébrique du Bois-Marie 1967–1969 (SGA 7 I), Dirigé par A. Grothendieck. Avec la collaboration de M. Raynaud et D. S. Rim

30. Gusein-Zade, S., Luengo, I., Melle, A.: Partial resolutions and the zeta-function of a singularity. Comment. Math. Helv. **72**(2), 244–256 (1997). 10.1007/S000140050014
31. Hirzebruch, F., Neumann, W., Koh, S.: Differentiable manifolds and quadratic forms. Marcel Dekker Inc., New York (1971). Appendix II by W. Scharlau, Lecture Notes in Pure and Applied Mathematics, Vol. 4
32. Laufer, H.: Normal two-dimensional singularities. Princeton University Press, Princeton, N.J.; University of Tokyo Press, Tokyo (1971). Annals of Mathematics Studies, No. 71
33. Laufer, H.: On normal two-dimensional double point singularities. Israel J. Math. **31**(3-4), 315–334 (1978). https://doi.org/10.1007/BF02761497
34. Lê D.T.: Ensembles analytiques complexes avec lieu singulier de dimension un (d'après I. N. Iomdine). In: Seminar on Singularities (Paris, 1976/1977), *Publ. Math. Univ. Paris VII*, vol. 7, pp. 87–95. Univ. Paris VII, Paris (1980)
35. León-Cardenal, E., Martín-Morales, J., Veys, W., Viu-Sos, J.: Motivic zeta functions on $\mathbb{Q}$-Gorenstein varieties. Adv. Math. **370**, 107192, 34 (2020). https://doi.org/10.1016/j.aim.2020.107192
36. Libgober, A.: Alexander polynomial of plane algebraic curves and cyclic multiple planes. Duke Math. J. **49**(4), 833–851 (1982). 10.1215/S0012-7094-82-04941-9
37. Luengo, I.: The μ-constant stratum is not smooth. Invent. Math. **90**(1), 139–152 (1987). https://doi.org/10.1007/BF01389034
38. Malgrange, B.: Letter to the editors. Invent. Math. **20**, 171–172 (1973). https://doi.org/10.1007/BF01404064
39. Marco, M., Rodríguez, M.: `SIROCCO`: A library for certified polynomial root continuation. In: Mathematical Software - ICMS 2016, *Lecture Notes in Comput. Sci.*, vol. 9725, pp. 191–197. Springer-Verlag, Berlin (2016). https://doi.org/10.1007/978-3-319-42432-3_24
40. Martín-Morales, J.: Embedded $\mathbb{Q}$-resolutions for Yomdin-Lê surface singularities. Israel J. Math. **204**(1), 97–143 (2014). https://doi.org/10.1007/s11856-014-1078-z
41. Martín-Morales, J.: Semistable reduction of a normal crossing $\mathbb{Q}$-divisor. Ann. Mat. Pura Appl. (4) **195**(5), 1749–1769 (2016). https://doi.org/10.1007/s10231-015-0546-3
42. Melle, A.: Milnor numbers for surface singularities. Israel J. Math. **115**, 29–50 (2000). https://doi.org/10.1007/BF02810579
43. Milnor, J.: Singular points of complex hypersurfaces, *Annals of Mathematics Studies*, vol. 61. Princeton University Press, Princeton, N.J. (1968)
44. Namba, M.: Geometry of projective algebraic curves. Marcel Dekker Inc., New York (1984)
45. Neumann, W.: A calculus for plumbing applied to the topology of complex surface singularities and degenerating complex curves. Trans. Amer. Math. Soc. **268**(2), 299–344 (1981)
46. Nielsen, J.: Surface transformation classes of algebraically finite type. Danske Vid. Selsk. Mat.-Fys. Medd. **21**(2), 89 (1944)
47. Oka, M.: Symmetric plane curves with nodes and cusps. J. Math. Soc. Japan **44**(3), 375–414 (1992). 10.2969/jmsj/04430375
48. Prill, D.: Local classification of quotients of complex manifolds by discontinuous groups. Duke Math. J. **34**, 375–386 (1967). 10.1215/S0012-7094-67-03441-2
49. Saeki, O.: Theory of super-isolated singularities and its applications. Sūrikaisekikenkyūsho Kōkyūroku (Kyoto) **764**, 23–36 (1991)
50. Safont, C.: Coverings of S^3 branched over iterated torus links. Rev. Mat. Univ. Complut. Madrid **3**(2-3), 181–210 (1990)
51. Schmid, W.: Variation of Hodge structure: the singularities of the period mapping. Invent. Math. **22**, 211–319 (1973). https://doi.org/10.1007/BF01389674
52. Steenbrink, J.: Limits of Hodge structures. Invent. Math. **31**(3), 229–257 (1975/76). https://doi.org/10.1007/BF01403146
53. Steenbrink, J.: Mixed Hodge structure on the vanishing cohomology. In: Real and complex singularities (Proc. Ninth Nordic Summer School/NAVF Sympos. Math., Oslo, 1976), pp. 525–563. Sijthoff and Noordhoff, Alphen aan den Rijn (1977)
54. Stevens, J.: On the μ-constant stratum and the V-filtration: an example. Math. Z. **201**(1), 139–144 (1989). https://doi.org/10.1007/BF01162001

55. Teissier, B.: Variétés polaires. I. Invariants polaires des singularités d'hypersurfaces. Invent. Math. **40**(3), 267–292 (1977). https://doi.org/10.1007/BF01425742
56. Thurston, W.: On the geometry and dynamics of diffeomorphisms of surfaces. Bull. Amer. Math. Soc. (N.S.) **19**(2), 417–431 (1988). https://doi.org/10.1090/S0273-0979-1988-15685-6
57. Wahl, J.: Equisingular deformations of normal surface singularities. I. Ann. of Math. (2) **104**(2), 325–356 (1976). https://doi.org/10.2307/1971049
58. Waldhausen, F.: Eine Klasse von 3-dimensionalen Mannigfaltigkeiten. I, II. Invent. Math. **3**, 308–333; ibid. **4** (1967), 87–117 (1967). https://doi.org/10.1007/BF01402956
59. Yau, S.T.: Topological type of isolated hypersurface singularities. In: Recent developments in geometry (Los Angeles, CA, 1987), *Contemp. Math.*, vol. 101, pp. 303–321. Amer. Math. Soc., Providence, RI (1989)
60. Yomdin, Y.: Complex surfaces with a one-dimensional set of singularities. Sibirsk. Mat. Ž. **15**, 1061–1082, 1181 (1974)
61. Zariski, O.: On the Problem of Existence of Algebraic Functions of Two Variables Possessing a Given Branch Curve. Amer. J. Math. **51**(2), 305–328 (1929). https://doi.org/10.2307/2370712
62. Zariski, O.: On the irregularity of cyclic multiple planes. Ann. of Math. (2) **32**(3), 485–511 (1931). https://doi.org/10.2307/1968247
63. Zariski, O.: The topological discriminant group of a Riemann surface of genus p. Amer. J. Math. **59**, 335–358 (1937). https://doi.org/10.2307/2371416

Chapter 5
A Review on the McKay Correspondence

José Luis Cisneros-Molina and Agustín Romano-Velázquez

Contents

Abstract The McKay correspondence is a deep and beautiful connection that links several areas of mathematics, including group theory, algebraic geometry, and representation theory. Discovered by John McKay in 1979, it has since been generalized and extensively studied. The original form of the correspondence relates finite subgroups of $\mathrm{SL}(2, \mathbb{C})$ to Kleinian singularities. Specifically, it establishes a bijection between the non-trivial irreducible representations of a finite subgroup

J. L. Cisneros-Molina (✉) · A. Romano-Velázquez
Instituto de Matemáticas, Unidad Cuernavaca, Universidad Nacional Autónoma de México, Cuernavaca, Morelos, Mexico
e-mail: jlcisneros@im.unam.mx; agustin.romano@im.unam.mx

© The Author(s), under exclusive license to Springer Nature Switzerland AG 2026

J. L. Cisneros-Molina et al. (eds.), *Handbook of Geometry and Topology of Singularities VIII*, https://doi.org/10.1007/978-3-031-99571-2_5

Γ of $\mathrm{SL}(2,\mathbb{C})$ and the irreducible components of the exceptional divisor in the minimal resolution of the associated Kleinian singularity, $X=\mathbb{C}^2/\Gamma$. This chapter begins by reviewing the classical McKay correspondence for Kleinian singularities, then introduces the necessary definitions and results to describe the reinterpretation given by Artin and Verdier through the Riemann-Hilbert correspondence. In this reinterpretation, the McKay correspondence becomes a bijection between non-trivial indecomposable reflexive $\mathcal{O}_X$-modules and the irreducible components of the exceptional divisor in the minimal resolution of X. Next, we explore the generalization of the McKay correspondence to rational singularities. We continue with the classification of reflexive modules of rank one over rational and minimally elliptic singularities. Finally, we examine the classification of special reflexive modules on Gorenstein singularities and discuss the deformation theory of full sheaves.

5.1 Introduction

The McKay correspondence is a deep and beautiful connection that links several different areas of mathematics, such as group theory, algebraic geometry, and representation theory. It was discovered by John McKay in 1979 and has since been generalized and studied extensively. The original form of the correspondence relates the finite subgroups of $\mathrm{SL}(2,\mathbb{C})$ to Kleinian singularities. In his now classical book [33] Klein gives the classification of finite subgroups of $\mathrm{SL}(2,\mathbb{C})$: they are isomorphic to cyclic groups, binary dihedral groups, or binary polyhedral groups. Any finite subgroup Γ of $\mathrm{SL}(2,\mathbb{C})$ acts naturally on $\mathbb{C}^2$ with a unique fixed point at the origin. Klein also proved that the quotient $\mathbb{C}^2/\Gamma$ can be embedded in $\mathbb{C}^3$ as a complex hypersurface with an isolated singularity corresponding to the fixed point. Such singularities are called *Klenian singularities*. Later, Du Val [11] noticed that there is a relationship between Klenian singularities and Dynkin diagrams of simple Lie algebras of type A, D and E. He showed that the dual graph Δ_Γ of the minimal resolution of a Klenian singularity $\mathbb{C}^2/\Gamma$ coincides with a Dynkin diagram of the aforementioned type. As time went by, it became clear that there was a relationship between the properties of Klenian singularities and ADE classification. Quoting Arnol'd (1976) in [6]: "The problem is to find a common origin of all the ADE classification theorems, and to substitute a priori proofs to a posterior verifications of the parallelism of the classifications."

In 1979, McKay [43] provided a bijection between finite subgroups of $\mathrm{SL}(2,\mathbb{C})$ and (extended) Dynkin diagrams of type ADE. Given a finite subgroup Γ of $\mathrm{SL}(2,\mathbb{C})$, McKay constructed a graph $\mathfrak{G}_\Gamma$ using the representation theory of Γ, where the vertices correspond to the irreducible representations of Γ. The graph $\mathfrak{G}_\Gamma$ turned out to be an extended Dynkin diagram of type A, D or E and McKay made the remarkable observation that removing from $\mathfrak{G}_\Gamma$ the vertex corresponding to the trivial representation, one obtains the Dynkin diagram corresponding to the dual graph Δ_Γ of the minimal resolution of the corresponding Kleinian singularity

$\mathbb{C}^2/\Gamma$. This establishes a bijection, now known as the *McKay correspondence*, between the non-trivial irreducible representations of a finite subgroup of SL(2, $\mathbb{C}$) and the irreducible components of the exceptional divisor of the minimal resolution of the associated Kleinian singularity. McKay noticed the correspondence by a case-by-case study using the classification of finite subgroups of SL(2, $\mathbb{C}$). A geometrical construction of this correspondence was done in a series of papers by Gonzalez-Sprinberg and Verdier [18], by Artin and Verdier [2] for the correspondence at the level of vertices and by Esnault and Knörrer [17] at the level of edges.

Let us summarize the correspondence at the level of vertices: let Γ be a finite subgroup of SL(2, $\mathbb{C}$) and denote by $X = \mathbb{C}^2/\Gamma$ the corresponding Kleinian singularity. Let $\pi : \tilde{X} \to X$ be the minimal resolution of X. To a non-trivial irreducible representation ρ of Γ one can associate a non-trivial indecomposable reflexive $\mathcal{O}_X$-module M via the Riemann-Hilbert correspondence. Note that the module π^*M/tor is locally free, and its first Chern class $c_1(\pi^*M/\text{tor})$ is the Poincaré dual of a curvette hitting transversely an unique irreducible component of the exceptional divisor. Such a component is the image of the representation ρ under McKay correspondence. Artin and Verdier proved that the first Chern class determines the module M along with the representation ρ. Conversely, for any irreducible component of the exceptional divisor there is a representation and a module realizing it. We do not spell the correspondence at the level of edges since the main concern of this chapter is a wide generalization of the results of Artin and Verdier at the level of vertices. The classification of reflexive modules has been studied by several people in different cases: Esnault [16] introduced full sheaves and improved the results of Artin and Verdier [2] for rational surface singularities, and classified rank one reflexive modules for quotient singularities. Wunram [58] classified the family of *special* reflexive modules for rational surface singularities, nevertheless the complete classification of reflexive modules on rational singularities remained open. For a non-rational singularity, Kahn [28] classified all the reflexive modules in the simply elliptic case using Atiyah's classification of vector bundles on elliptic curves. For a general normal Gorenstein singularity, Fernández de Bobadilla and Romano [4] classified the family of *cohomological special* reflexive modules. For a general surface singularity, the classification problem for reflexive $\mathcal{O}_X$-modules remains open. In the particular case of reflexive modules of rank one, Nemethi and Romano [49] classified all the reflexive modules over rational and minimally elliptic singularties.

The theory, study and applications of reflexive modules has different approaches, for example, Iyama and Wemyss [26] generalized the notion of speciality as the vanishing of $\text{Ext}^1_{\mathcal{O}_X}(M, \mathcal{O}_X)$. Using this algebraic notion of speciality, they classified all the non-trivial indecomposable special modules on quotient surface singularities in a combinatorial way. For this, they use the vanishing of the first Ext group to reduce the classification problem to a combinatorial problem on the Auslander-Reiten quiver. Another use of the McKay correspondence is provided by Hara et al. [22] and the results of Gustavsen and Ile [21]. Hara, Sawada and Yasuda prove that the F-blowup of a rational surface singularity of positive characteristic is dominated by the minimal resolution and has only rational singularities. Gustavsen and Ile prove the analogous statement in characteristic zero, i.e., they prove that the blow-up at any maximal Cohen-Macaulay module in a rational surface singularity of characteristic zero is a partial resolution dominated by the minimal resolution.

It is important to recall that this theory has applications to other areas, for example, due to a result of Eisenbud [14], in the hypersurface case reflexive modules correspond to matrix factorizations. Following an idea of Kontsevich, the work of Kapustin and Li (see [29]), and Orlov (see [50, 51]) showed that matrix factorizations have applications to the study of Landau-Ginzburg models appearing in string theory, and to the study of Kontsevich's homological mirror symmetry. By Khovanov and Rozansky [30, 31], reflexive modules have interesting applications to link invariants.

In this survey, we decided to focus on the different generalizations of the McKay correspondence following the ideas of Artin and Verdier. For that reason, we will only consider the classification problem of reflexive modules over normal surface complex singularities. Since the theory, study and applications of reflexive modules is a really big area having a lot of interesting results and important open problems, we decided to focus ourselves in this survey just in a small part of this theory. The reader may check the previous references to see different flavors of this problem.

This survey is organized as follows: In Sect. 5.2, we review the classical McKay correspondence, in Sect. 5.3, we introduce all the definitions and notions about normal surface singularities, later in Sect. 5.4 we present the Riemann-Hilbert correspondence and results on reflexive modules and full sheaves. In Sect. 5.5 we review the McKay correspondence for rational singularities following the idea of Artin and Verdier, later in Sect. 5.6 we show the classification of rank one reflexive modules on rational and minimally elliptic singularities. In Sect. 5.7, we show the classification of special modules on Gorenstein surface singularities.

5.2 The McKay Correspondence

In this section, we review the classical McKay correspondence for Kleinian singularities in an expository way, delaying the formal definitions for the following sections.

5.2.1 Kleinian Singularities

Kleinian singularities are an important class of surface singularities embedded in $\mathbb{C}^3$. In [12] Durfee presents fifteen characterizations of this class of singularities.

Let $\langle 2, q, r\rangle$ denote the group given by the presentation,

$$\langle 2, q, r\rangle = \langle b, c \mid (bc)^2 = b^q = c^r\rangle \tag{5.1}$$

and let $\zeta_n = e^{\frac{2\pi i}{n}}$ be a n-th primitive root of unity. The classification of the finite subgroups Γ of $\mathrm{SL}(2, \mathbb{C})$ is well known (see [36]), they are conjugate to the finite subgroups of $\mathrm{SU}(2)$. Up to conjugacy, they are the following :

1. Cyclic groups C_n of order n with $n \in \mathbb{N}$ and $n \geq 2$. In this case,
$$\Gamma = \left\langle \begin{pmatrix} \zeta_n & 0 \\ 0 & \zeta_n^{-1} \end{pmatrix} \right\rangle.$$
2. Binary dihedral groups $\mathsf{BD}_{2n} = \langle 2, 2, n \rangle$, $n \geq 2$ of order $4n$. In this case,
$$\Gamma = \left\langle \begin{pmatrix} 0 & 1 \\ (-1) & 0 \end{pmatrix}, \begin{pmatrix} \zeta_{2n} & 0 \\ 0 & \zeta_{2n}^{-1} \end{pmatrix} \right\rangle.$$
3. Binary tetrahedral group $\mathsf{BT} = \langle 2, 3, 3 \rangle$, of order 24. In this case,
$$\Gamma = \left\langle \begin{pmatrix} 0 & 1 \\ -1 & 1 \end{pmatrix}, \begin{pmatrix} 0 & -\zeta_3 \\ \zeta_3^2 & 1 \end{pmatrix} \right\rangle.$$
4. Binary octahedral group $\mathsf{BO} = \langle 2, 3, 4 \rangle$ of order 48. In this case,
$$\Gamma = \left\langle \frac{1}{2} \begin{pmatrix} 1+\sqrt{2}i & -i \\ -i & 1-\sqrt{2}i \end{pmatrix}, \frac{1}{2} \begin{pmatrix} -\sqrt{2}-i & -1 \\ 1 & -\sqrt{2}+i \end{pmatrix} \right\rangle.$$
5. Binary icosahedral group $\mathsf{BI} = \langle 2, 3, 5 \rangle$ of order 120. In this case,
$$\Gamma = \left\langle \frac{1}{2} \begin{pmatrix} 1+i & 1+i \\ -1+i & 1-i \end{pmatrix}, \frac{1}{2} \begin{pmatrix} \frac{\sqrt{5}+1}{2} + \frac{\sqrt{5}-1}{2} i & 1 \\ -1 & \frac{\sqrt{5}+1}{2} - \frac{\sqrt{5}-1}{2} i \end{pmatrix} \right\rangle.$$

If Γ is a finite subgroup of $\mathrm{SL}(2, \mathbb{C})$ it acts naturally on $\mathbb{C}^2$ by matrix multiplication. The action on $\mathbb{C}^2 \setminus \{0\}$ is free and the origin is a fixed point, thus, the orbit space $(\mathbb{C}^2 \setminus \{0\})/\Gamma$ is a complex manifold of dimension 2 and the orbit space $\mathbb{C}^2/\Gamma$ has a singularity at the orbit of the origin.

Definition 5.2.1 Let $\Gamma \leq \mathrm{SL}(2, \mathbb{C})$ be a finite subgroup. The *Klenian singularity* obtained by Γ is the quotient $X_\Gamma := \mathbb{C}^2/\Gamma$.

If Γ and Γ' are conjugate finite subgroups of $\mathrm{SL}(2, \mathbb{C})$, then $X_\Gamma = X_{\Gamma'}$, therefore, Kleinian singularities are classified by the conjugacy classes of finite subgroups of $\mathrm{SL}(2, \mathbb{C})$, the reader may see [36, Chapter 2] for more details.

Klein in [32] proved that X_Γ is a complex algebraic surface with an isolated singularity at the origin. This is done by embedding X_Γ in $\mathbb{C}^3$ and showing that the image is the zero set of a polynomial $f_\Gamma \colon \mathbb{C}^3 \to \mathbb{C}$ as follows.

Let Γ be a finite subgroup of SL(2, $\mathbb{C}$). The action of Γ on $\mathbb{C}^2$ induces an action on the polynomial ring $\mathbb{C}[u, v]$ as follows:

$$\begin{aligned} \Gamma \times \mathbb{C}[u, v] &\to \mathbb{C}[u, v] \\ (\gamma, f(u, v)) &\mapsto f(\gamma \cdot (u, v)^T) \end{aligned} \tag{5.2}$$

where $-^T$ means transpose. As usual, denote by $\mathbb{C}[u, v]^\Gamma$ the polynomials that are invariant under the action of Γ. The invariant polynomials $\mathbb{C}[u, v]^\Gamma$ form a subalgebra and Klein proved that this subalgebra is generated by three invariant polynomials X, Y and Z which satisfy only one non-trivial polynomial equation $f_\Gamma(X, Y, Z) = 0$. Using the generators of the invariant algebra $\mathbb{C}[u, v]^\Gamma$ one defines the map

$$F : \mathbb{C}^2 \to \mathbb{C}^3$$
$$F(z_1, z_2) = (X(z_1, z_2), Y(z_1, z_2), Z(z_1, z_2)) .$$

The map F is invariant under the action of Γ, so it factors over the quotient

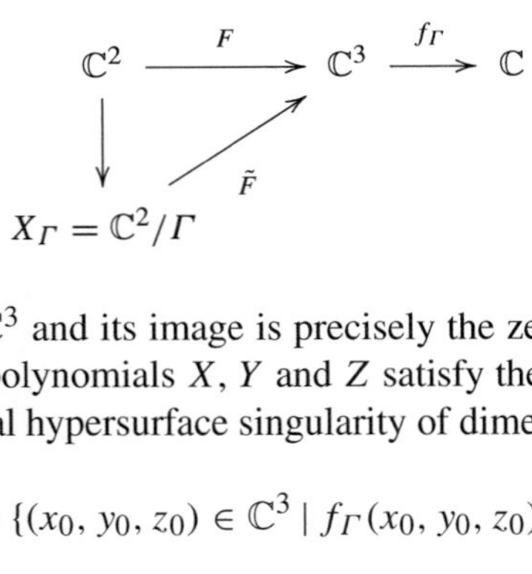

thus $\tilde{F}$ embeds X_Γ in $\mathbb{C}^3$ and its image is precisely the zero-set of the polynomial f_Γ, since the invariant polynomials X, Y and Z satisfy the relation $f_\Gamma(X, Y, Z) = 0$. Hence, X_Γ is a normal hypersurface singularity of dimension two, i.e.,

$$X_\Gamma = \{(x_0, y_0, z_0) \in \mathbb{C}^3 \mid f_\Gamma(x_0, y_0, z_0) = 0\},$$

with an isolated singularity at the origin, i.e., the origin is the unique point where all the partial derivates of f_Γ vanish. The polynomials f_Γ for the Klenian singularities are shown in Table 5.1.

In the following example, we show how to obtain the polynomial f_Γ in the case $\Gamma = \mathsf{C}_n$. The reader may see [36, Chapter 2] for the remaining cases.

Table 5.1 Polynomials defining Klenian singularities

Γ	C_n	BD_{2n}	BT	BO	BI
f_Γ	$xy - z^n$	$x^2 + zy^2 + z^{n+1}$	$x^2 + y^3 + z^4$	$x^2 + zy^3 + z^3$	$x^2 + y^3 + z^5$

Example 5.2.2 Let $\Gamma \leq \mathrm{SL}(2, \mathbb{C})$ be a finite cyclic subgroup. We have

$$\Gamma = \left\langle \begin{pmatrix} \zeta_n & 0 \\ 0 & \zeta_n^{-1} \end{pmatrix} \right\rangle,$$

where ζ_n is a primitive root of the unity of order n. The action (5.2) in this case is given by

$$\Gamma \times \mathbb{C}[u, v] \to \mathbb{C}[u, v]$$
$$\left(\begin{pmatrix} \zeta_n^j & 0 \\ 0 & \zeta_n^{-j} \end{pmatrix}, f(u, v) \right) \mapsto f\left(\begin{pmatrix} \zeta_n^j & 0 \\ 0 & \zeta_n^{-j} \end{pmatrix} \cdot (u, v)^T \right) = f(\zeta_n^j u, \zeta_n^{-j} v).$$

Therefore,

$$\mathbb{C}[u, v]^\Gamma = \{ f \in \mathbb{C}[u, v] \mid f(\zeta_n^j u, \zeta_n^{-j} v) = f(u, v) \text{ for any } j \}.$$

We have that $X = u^n$, $Y = v^n$ and $Z = uv$, thus, $\mathbb{C}[u, v]^\Gamma = \mathbb{C}[u^n, v^n, uv]$. Now consider the morphism

$$\phi \colon \mathbb{C}[x, y, z] \to \mathbb{C}[u^n, v^n, uv]$$

given by $\phi(x) = u^n$, $\phi(y) = v^n$ and $\phi(z) = uv$. This morphism is onto and its kernel is given by the ideal generated by $xy - z^n$, since $\phi(xy - z^n) = u^n v^n - (uv)^n = 0$. Thus,

$$\mathbb{C}[x, y, z]/\langle xy - z^n \rangle \cong \mathbb{C}[u^n, v^n, uv].$$

Therefore, $f_\Gamma = xy - z^n$. This finishes the example.

Kleinian singularities are characterized by their minimal resolution. Let X_Γ be a Klenian singularity. Denote by $\pi : \tilde{X}_{\Gamma_{\min}} \to X_\Gamma$ the minimal resolution of X_Γ (see Sect. 5.3.3 below). Du Val proved in [11] that the exceptional fiber $E = \pi^{-1}(0)$ is the connected union of smooth rational curves with self-intersection -2, i.e.,

$$E = \pi^{-1}(0) = E_1 \cup E_2 \cup \ldots \cup E_r,$$

where E_j is isomorphic to the projective line $E_j \cong \mathbb{P}^1_{\mathbb{C}}$ with $E_j \cdot E_j = -2$ for any j. The components E_j meet transversely, not three of them meet at a point, its pairwise intersection is a single point or empty. This information is encoded associating to the exceptional set E a graph Δ_Γ, called the dual graph of the minimal resolution: there is a vertex for each E_j and two vertices are connected by an edge if the corresponding components intersect. The dual graphs of the minimal resolution of Kleinian singularities are showed in Table 5.2 and they exactly correspond to the Dynkin diagrams of the simple Lie algebras of type A, D and E. This property also characterizes Kleinian singularities, for this reason they are also

Table 5.2 Dual graph of the minimal resolution of Kleinian singularities

Kleinian singularity	Dual graph	Dynkin diagram
X_{C_n}		A_{n-1}
$X_{\mathsf{BD}_{2n}}$	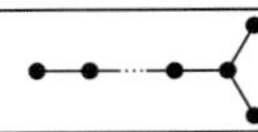	D_{n+2}
X_{BT}	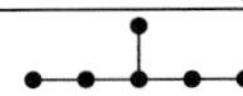	E_6
X_{BO}	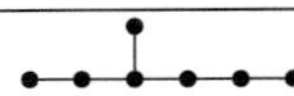	E_7
X_{BI}	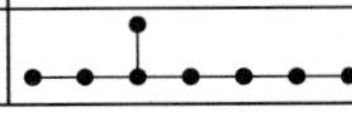	E_8

called *ADE-singularities* and one refers to a particular Kleinian singularity by its Dynkin diagram: "the A_n-singularity", "the E_8-singularity", etc. To see the relation between Kleinian singularities and Lie algebras see [8].

As we can see from the above, with the aid of the Kleinian singularities and their minimal resolutions one can establish a one-to-one correspondence between:

1. Conjugacy classes of finite subgroups of SL(2, $\mathbb{C}$).
2. Dynkin diagrams of type ADE.

This is summarized in the following diagram where all the arrows are bijections:

$$\begin{array}{ccc} \{\text{Finite subgroups } \Gamma \leq \mathrm{SL}(2,\mathbb{C}) \text{ up to conjugacy}\} & & \{\text{Dynkin diagram } \Delta_\Gamma \text{ of type ADE}\} \\ \updownarrow & & \updownarrow \\ \{\text{Klenian singularities } X_\Gamma = \mathbb{C}^2/\Gamma\}/\sim & \longleftrightarrow & \{\text{Minimal resolution } \tilde{X}_\Gamma\}/\sim \end{array}$$

In this diagram, the symbol $\sim$ means up to biholomorphism.

5.2.2 McKay Correspondence

In 1979, McKay in [43] gave the bijection

$$\{\text{Finite subgroups } \Gamma \leq \mathrm{SL}(2,\mathbb{C}) \text{ up to conjugacy}\}$$
$$\leftrightarrow \{\text{ Dynkin diagram } \Delta_\Gamma \text{ of type ADE}\}$$

constructing the Dynkin diagram directly from the group Γ using representation theory as follows. Let $\Gamma \leq \mathrm{SL}(2,\mathbb{C})$ be a finite subgroup and denote by

$$\mathrm{Irr}\,\Gamma := \{\rho_0, \rho_1, \ldots, \rho_r\},$$

a complete set of irreducible complex representations of Γ, where ρ_0 is the trivial representation. Denote by Q the natural representation given by the inclusion $\Gamma \subset$

SL(2, $\mathbb{C}$). By Maschke's theorem (see [27, Chapter 8]) we can decompose the tensor products $\rho_i \otimes Q$ as direct sum of elements of Irr Γ

$$\rho_i \otimes Q \cong \bigoplus_j a_{ij} \rho_j. \tag{5.3}$$

Using the decompositions (5.3) McKay constructed a directed graph $\mathfrak{G}_\Gamma$ as follows:

1. One vertex for each irreducible representation.
2. Given the decomposition (5.3), we draw a_{ij} arrows from the vertex v_i (that corresponds to the representation ρ_i) to the vertex v_j (that corresponds to the representation ρ_j).

It is easy to see that for each finite subgroup $\Gamma \subset$ SL(2, $\mathbb{C}$), we have $a_{ij} = a_{ji}$ and a_{ij} must be 0 or 1. Therefore, the directed graph $\mathfrak{G}_\Gamma$ constructed by McKay is a graph [59, Chapter 10].

Theorem 5.2.3 ([43, McKay]) *Let Γ be a finite subgroup of* SL(2, $\mathbb{C}$) *and let $X_\Gamma := \mathbb{C}^2/\Gamma$ be the corresponding Kleinian Singularity. Then the dual graph Δ_Γ of the minimal resolution of X_Γ and the graph $\mathfrak{G}_\Gamma$ without the trivial representation coincide.*

Theorem 5.2.3 is called the *McKay correspondence*. Since the graphs Δ_Γ and $\mathfrak{G}_\Gamma$ coincide, this gives a one to one correspondence between:

1. The set of smooth curves $\{E_j\}$, the components of the exceptional set of the minimal resolution of X_Γ.
2. The set of non-trivial irreducible complex representations of Γ.

In Table 5.3 we show the McKay graphs of the finite subgroups of $SL(2, \mathbb{C})$.

Notice that if we remove the vertex that corresponds to the trivial representation ρ_0 we recover the corresponding Dynkin diagram.

For sake of the reader, in an appendix at the end of the chapter, we include all the irreducible representations of the finite subgroups of $SL(2, \mathbb{C})$.

Example 5.2.4 Let us show how to use the representations to construct the McKay graph for the case of the finite cyclic subgroups of SL(2, $\mathbb{C}$). In this case, we have

$$\Gamma = \left\langle \begin{pmatrix} \zeta_n & 0 \\ 0 & \zeta_n^{-1} \end{pmatrix} \right\rangle,$$

Table 5.3 McKay graphs of finite subgroups of SL(2, $\mathbb{C}$)

Γ	C_n	BD_{2n}	BT	BO	BI
$\mathfrak{G}_\Gamma$	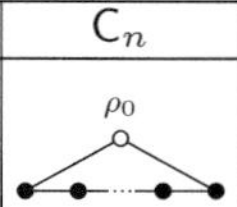	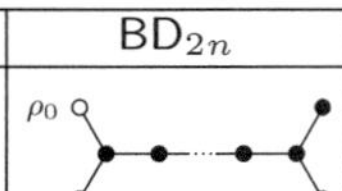	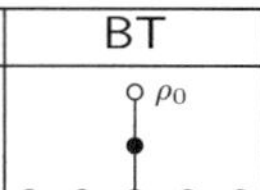	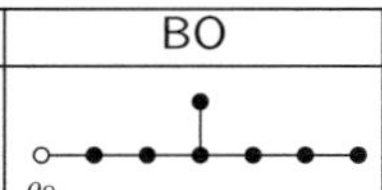	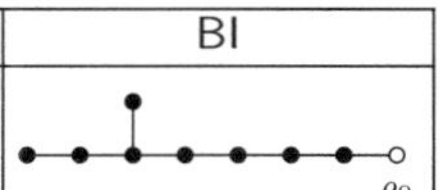

where ζ_n is a primitive root of the unity of order n. The group C_n has n irreducible representations ρ_j, $0 \leq j \leq n-1$ given by $\rho_j(\zeta_l) = \zeta_l^j$. The natural representation Q is given by

$$Q = \rho_1 \oplus \rho_{n-1} \colon \mathsf{C}_n \to \Gamma \subset \mathrm{SL}(2, \mathbb{C})$$
$$\zeta_n \mapsto \begin{pmatrix} \zeta_n & 0 \\ 0 & \zeta_n^{-1} \end{pmatrix}$$

Since,

$$\left(\rho_j \otimes Q\right)(\zeta_n) = \zeta_n^j \begin{pmatrix} \zeta_n & 0 \\ 0 & \zeta_n^{-1} \end{pmatrix} = \begin{pmatrix} \zeta_n^{j+1} & 0 \\ 0 & \zeta_n^{j-1} \end{pmatrix} = \left(\rho_{j+1} \oplus \rho_{j-1}\right)(\zeta_n).$$

Therefore, the McKay graph $\mathfrak{G}_{\mathsf{C}_n}$ is as follows:

1. It has n vertices.
2. The vertex that corresponds to the representation ρ_j is connected to the vertices that correspond to the representations ρ_{j-1} and ρ_{j+1} (everything is modulo n).

So we get the McKay graph given in Table 5.3 corresponding to the group C_n and removing the vertex corresponding to the trivial representation we recover the Dynkin diagram of type A_{n-1} as shown in Table 5.2.

A more geometric approach to the Mckay's correspondence was discovered by Gonzales-Sprinberg and Verdier [18]. As before let Γ be a finite subgroup of $\mathrm{SL}(2, \mathbb{C})$, $X_\Gamma := \mathbb{C}^2/\Gamma$ the corresponding Kleinian singularity and $\tilde{X}_\Gamma$ the minimal resolution of X_Γ. Gonzales-Sprinberg and Verdier constructed for each irreducible, non-trivial representation of Γ a locally free sheaf $\mathcal{M}$ over $\tilde{X}_\Gamma$, such that the first Chern class of $\mathcal{M}$ is the Poincaré dual of a curvette hitting transversely an unique irreducible component E_i of the exceptional set E of $\tilde{X}_\Gamma$. The work of Gonzales-Sprinberg and Verdier was done case by case. Later, Artin and Verdier [2] gave a proof of the McKay correspondence using only properties of Kleinian singularities.

In this survey, we review these constructions and several generalizations. For that, we start reviewing some preliminaries on singularity theory.

5.3 Review on Singularity Theory

In this section, we recall basics on singularity theory and topological invariants of normal surface singularities. The reader may see [7, 25, 45, 48] for more details.

5.3.1 Setting and Notation: Normal Singularities

Throughout this chapter, we denote by (X, x) the germ of a *complex analytic surface singularity*, i.e.,

$$(X, x) = (\{f_1 = f_2 = \ldots = f_m = 0\}, 0) \subset \left(\mathbb{C}^N, 0\right),$$

where $f_l \colon (\mathbb{C}^n, 0) \to (\mathbb{C}, 0)$ are germs of holomorphic functions such that

$$r(p) = \operatorname{rank}\left(\frac{\partial f_l}{\partial z_j}(p)\right) = N - 2,$$

for generic p.

Definition 5.3.1 Let (X, x) be the germ of a complex analytic surface singularity. We say that:

1. the germ (X, x) is *smooth* if $r(x) = N - 2$.
2. the germ (X, x) has an *isolated singularity* if $r(x) < N - 2$ and $r(p) = N - 2$ for $p \neq x$.
3. the germ (X, x) is a *complete intersection* if $m = N - 2$.
4. the germ (X, x) is a *hypersurface* if $m = 1$ and $N = 3$.

For example, any Klenian singularity is an isolated hypersurface singularity. As usual, we denote by $\mathbb{C}\{z_1, \ldots, z_n\}$ the local ring of holomorphic functions on the variables $z_1, \ldots, z_n$.

Definition 5.3.2 Let (X, x) be the germ of a complex analytic complex surface singularity. Its *local ring of functions* denoted by $\mathcal{O}_{X,x}$ is defined as follows:

$$\mathcal{O}_{X,x} = \mathbb{C}\{z_1, \ldots, z_n\}/\langle f_1, \ldots, f_m\rangle.$$

From now on, we will assume that (X, x) is irreducible (equivalently, the ideal $\langle f_1, \ldots, f_m\rangle$ is a prime ideal).

Definition 5.3.3 Let (X, x) be the germ of a complex analytic complex surface singularity. We say that (X, x) is *normal* if and only if $\mathcal{O}_{X,x}$ is integrally closed in its field of fractions (see [25, Definition 4.1.10]). Equivalently, (X, x) is normal if the Riemann extension theorem holds on $X \setminus \{x\}$ (see [48, Section 12.2.6]).

Remark 5.3.4 By Serre's Criterion for normality (see for example [25, Theorem 3.5.16]), any normal surface singularity has at most an isolated singularity. The converse is not true, there are isolated surface singularities that are not normal.

We have the following examples of normal surface singularities:

Quotient Singularities
A *quotient surface singularity* (X, x) is isomorphic to a germ $(\mathbb{C}^2/\Gamma, 0)$, where Γ is a *small* finite subgroup of $\mathrm{GL}(2, \mathbb{C})$, that is, its elements does not have 1 as an eigenvalue, and therefore Γ acts freely on $\mathbb{C}^2 \setminus \{0\}$. If Γ and Γ' are two small finite subgroups of $\mathrm{GL}(2, \mathbb{C})$, then $\mathbb{C}^2/\Gamma$ and $\mathbb{C}^2/\Gamma'$ are analytically isomorphic if Γ and Γ' are conjugate. The classification of these singularities was done by Brieskorn [5] and Riemenschneider [54]. Any quotient surface singularity is a normal surface singularity (for more details see [25, Chapter 7, Section 4]).

Remark 5.3.5 All the finite subgroups of $\mathrm{SL}(2, \mathbb{C})$ are small, thus, Kleinian singularities are a subfamily of quotient singularities.

Hypersurfaces and Complete Intersections
Let $f \colon (\mathbb{C}^3, 0) \to (\mathbb{C}, 0)$ be a holomorphic function. Denote by $(X, x) = (\{f = 0\}, 0)$. If (X, x) has an isolated singularity, then (X, x) is a normal surface singularity. More generally, let (X, x) be a complete intersection. By Serre's Criterion for normality, the germ (X, x) is normal if and only if (X, x) has an isolated singularity.

Remark 5.3.6 Kleinian singularities are the quotient singularities which are hypersurfaces [12, Corollary 5.3].

5.3.2 *The Link*

From now on (X, x) is always a normal surface singularity. Let X be a representative of the germ (X, x), we may assume that $X \setminus \{x\}$ is connected and $(X, x) \subset (\mathbb{C}^n, 0)$. If $\epsilon > 0$ is small enough, the intersection $\Sigma = X \cap \mathbb{S}_\epsilon^{2n-1}$ is called the *link* of (X, x). The link does not depend on ϵ. Moreover, the link Σ is a smooth, compact, connected and oriented 3-manifold. In singularity theory is common to denote the fundamental group of the link by $\pi_1^{\mathrm{loc}}(X, x) = \pi_1(\Sigma)$ and call it the *local fundamental group*. In the case of a normal surface singularity, the local fundamental group satisfies the following characterization.

Proposition 5.3.7 ([35]) *For a normal surface singularity (X, x):*

1. *The group $\pi_1^{\mathrm{loc}}(X, x)$ is finite if and only if (X, x) is a quotient singularity.*
2. *If $\pi_1^{\mathrm{loc}}(X, x)$ is infinite, then the link Σ is an Eilenberg–MacLane space of type $K(\pi_1(\Sigma), 1)$. Furthermore, the local fundamental group is torsion free.*

The first point of Proposition 5.3.7 is Brieskorn's characterization of quotient singularities [5, Satz 2.8].

Example 5.3.8 Some examples of links of normal surfaces are the following:

1. Mumford proved in [44] that the germ (X, x) is smooth if and only if $\pi_1(\Sigma) \cong \{e\}$.
2. Let $(X, x) = (\{xy - z^n = 0\}, 0)$ be the A_{n-1} singularity. Its link is the *lens space* $L(n, -1)$ *that is,* $\Sigma = \mathbb{S}^3/\Gamma$, where $\Gamma = \left\langle \begin{pmatrix} \zeta_n & 0 \\ 0 & \zeta_n^{-1} \end{pmatrix} \right\rangle \leq \mathrm{SL}(2, \mathbb{C})$.
3. Let $(X, x) = (\{x^2 + y^3 + z^5 = 0\}, 0)$ be the E_8 singularity. Its link is the *Poincaré homology* 3*-sphere* $\Sigma = \mathbb{S}^3/\mathsf{BI}$, hence $|\pi_1(\Sigma)| = 120$ and $H_1(\Sigma, \mathbb{Z}) = 0$.

The reader may see [45] for more examples and references.

5.3.3 Resolutions

Let (X, x) be the germ of an analytic complex normal surface singularity. Let

$$\pi : \tilde{X} \to X,$$

be a *resolution of* (X, x), i.e., a proper holomorphic map from a smooth complex surface $\tilde{X}$ to a given representative of (X, x) such that π restricted to the complement of $\pi^{-1}(x)$ is biholomorphic. Sometimes we will require π to be a *good resolution*, which means that the *exceptional divisor* $E := \pi^{-1}(x)$ is a normal crossing divisor and each irreducible component of E is smooth. For any normal surface singularity there is always a good resolution, however it is not unique. The resolution is *minimal* if there is no rational irreducible component of the exceptional divisor E_i with self intersection $E_i^2 = -1$. The minimal resolution always exists and it is unique. Given a good resolution, the *dual graph* is a decorated graph Δ constructed as follows: the set of vertices, say V, is in bijection with the set of irreducible components of E, say $\{E_v\}_{v \in V}$, two vertices u and v are connected by k edges in Δ if and only if the corresponding components E_u and E_v have k intersection points. Each vertex v is decorated with two numbers: the self-intersection E_v^2 in $\tilde{X}$ and the genus $g(E_v)$. We omit the genus when it is equal to zero.

The *intersection matrix* $M = (m_{u,v})_{u,v \in V}$ associated to the dual graph Δ is the intersection matrix of the curves $\{E_v\}_{v \in V}$, i.e., $m_{u,v} = (E_u, E_v)$. It is negative definite. The reader may see [34, 37] for more details about resolutions of singularities.

Example 5.3.9 Here are some examples of singularities and the dual graph of a resolution:

1. The germ (X, x) is smooth if and only if the dual graph of its minimal resolution is empty.

2. For $(X, x) = (\{xy - z^n = 0\}, 0)$ the A_{n-1} singularity, the minimal resolution graph is the following:

3. For $(X, x) = (\{x^2 + y^3 + z^7 = 0\}, 0)$ a good resolution graph is the following:

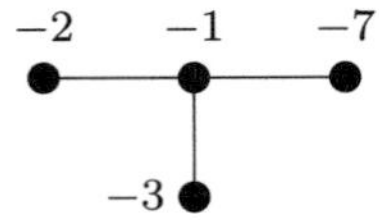

Remark 5.3.10 The resolution graph determines and it is determined by the link of the singularity. In some cases, the topological type of a singularity carries exactly one analytic structure, i.e., the minimal resolution graph determines the analytic structure. If this happens, we say that the singularity is *taut*. For example, the Klenian singularities are taut.

5.3.4 *The Divisor Lattice Structure and the Lipman Cone*

Let (X, x) be the germ of an analytic complex normal surface singularity. Let $\pi : \tilde{X} \to X$ be any resolution. The *divisor class group* is defined as follows:

$$\mathrm{Div}(\tilde{X}) = \left\{ \sum n_i D_i \mid D_i \text{ is a irreducible curve on } \tilde{X} \text{ and } n_i \in \mathbb{Z} \right\}.$$

For a divisor $D = \sum n_i D_i$, we denote its *support* as follows $|D| = \cup_{n_i \neq 0} D_i$.

For $D \in \mathrm{Div}(\tilde{X})$, we define the line bundle $\mathcal{O}_{\tilde{X}}(D)$ (also denoted by $\mathcal{L}(D)$) by its sections: since the resolution is smooth and any divisor in a smooth space is Cartier, therefore locally in an open set U the divisor D is given by the zero locus of some f. Thus, $\mathcal{O}_{\tilde{X}}(D)(U) = f \cdot \mathcal{O}_U$. The reader may see [25, Section 5.2] or [45, Section 2] for more details.

Definition 5.3.11 The *cycles* are divisors supported on the exceptional locus. Thus, any cycle is of the form $D = \sum_{v \in V} n_v E_v$.

Set $L := H_2(\tilde{X}, \mathbb{Z})$, i.e., L is the free abelian group generated by the classes $\{E_v\}_{v \in V}$. Consider L and the intersection form I given by the intersection matrix M. Denote also $L' := H_2(\tilde{X}, \partial \tilde{X}, \mathbb{Z})$. Notice that L' is the dual of L, indeed if for each $v \in V$ we take a transversal disc D_v to E_v (at a generic point of E_v), then their classes form a basis of L'. In these bases, the duality map is $L \otimes L' \to \mathbb{Z}$ is given by

$$(E_v, D_w) = \delta_{vw} \text{ (Kronecker delta).}$$

The map from L to L' in the bases $\{E_v\}$ and $\{D_v\}$ is exactly the map induced by the intersection form. Since the intersection matrix M is non-degenerate, the homological map $L \to L'$ is injective. Thus, by the homological long exact sequence of the pair $(\tilde{X}, \partial\tilde{X})$ (where $\partial\tilde{X} \simeq \Sigma$) we have

$$0 \to L \to L' \to H_1(\partial\tilde{X}, \mathbb{Z}) \to H_1(\tilde{X}, \mathbb{Z}) \to 0. \tag{5.4}$$

Since $H_1(\tilde{X}, \mathbb{Z})$ is free, $H_1(\partial\tilde{X}, \mathbb{Z}) \cong \mathrm{Tors}(H_1(\partial\tilde{X}, \mathbb{Z})) \oplus H_1(\tilde{X}, \mathbb{Z})$, and the quotient L'/L is identified with

$$H := \mathrm{Tors}\big(H_1(\partial\tilde{X}, \mathbb{Z})\big) = \mathrm{Tors}\left(H_1(\Sigma, \mathbb{Z})\right). \tag{5.5}$$

We extend the intersection form $(,)$ from L to $L_{\mathbb{Q}} := L \otimes \mathbb{Q}$, which we denote by $(,)_{\mathbb{Q}}$. Via $(,)_{\mathbb{Q}}$ the lattice $L' \cong \mathrm{Hom}_{\mathbb{Z}}(L, \mathbb{Z})$ can be identified with a lattice of rational cycles

$$\{\ell' \in L_{\mathbb{Q}} : (\ell', \ell)_{\mathbb{Q}} \in \mathbb{Z} \text{ for any } \ell \in L\} \subset L_{\mathbb{Q}}.$$

Hence, we can identify L' with $\bigoplus_{v \in V} \mathbb{Z}\langle E_v^* \rangle$, the lattice generated by the rational cycles $E_v^* \in L_{\mathbb{Q}}$ (with $v \in V$ the set of vertices of the dual graph Δ) defined via

$$(E_u^*, E_v)_{\mathbb{Q}} = -\delta_{u,v} \text{ (Kronecker delta) for any } u, v \in V.$$

Let $\ell_1', \ell_2' \in L_{\mathbb{Q}}$ where $\ell_j' = \sum_v l'_{jv} E_v$ for $j \in \{1, 2\}$. We consider the partial order in $L_{\mathbb{Q}}$ giving by $\ell_1' \geq \ell_2'$ if and only if $l'_{1v} \geq l'_{2v}$ for all $v \in V$. We set $\min\{\ell_1', \ell_2'\} := \sum_v \min\{l'_{1v}, l'_{2v}\} E_v$.

Let $\ell' \in L'$, we denote its class in H (given in (5.5)) by $[\ell']$. The lattice L' has a partition parametrized by H, where for any $h \in H$ we have

$$L'_h := \{\ell' \in L' \mid [\ell'] = h\}.$$

Note that $L'_0 = L$. Given any $h \in H$ we define $r_h := \sum_v l'_v E_v$ as the unique element of L'_h such that $0 \leq l'_v < 1$. We define the *rational Lipman cone* by

$$\mathcal{S}_{\mathbb{Q}} := \{\ell' \in L_{\mathbb{Q}} \mid (\ell', E_v) \leq 0 \text{ for any } v \in V\}.$$

It is a cone generated over $\mathbb{Q}_{\geq 0}$ by E_v^*. We set

$$\mathcal{S}' := \mathcal{S}_{\mathbb{Q}} \cap L' \quad \text{and} \quad \mathcal{S} := \mathcal{S}_{\mathbb{Q}} \cap L.$$

Note that $\mathcal{S}'$ is the monoid of anti-nef rational cycles of L', it is generated over $\mathbb{Z}_{\geq 0}$ by the cycles E_v^*. The Lipman cone $\mathcal{S}'$ also has a natural equivariant partition indexed by H. We denote $\mathcal{S}'_h = \mathcal{S}' \cap L'_h$. The monoid $\mathcal{S} = \mathcal{S}'_0$ has the following properties:

1. if $Z = \sum n_v E_v \in \mathcal{S}$ and $Z \neq 0$, then $n_v > 0$ for all $v \in V$,
2. if $Z_1, Z_2 \in \mathcal{S}$, then $Z_1 + Z_2 \in \mathcal{S}$,
3. if $Z_1, Z_2 \in \mathcal{S}$, then $\min\{Z_1, Z_2\} \in \mathcal{S}$.

Thus, $\mathcal{S} \setminus \{0\}$ has a unique minimal element Z_{min} called the *Artin's fundamental cycle (minimal cycle or numerical cycle)*. Similarly, for any $h \in H$, the set $\mathcal{S}'_h$ has the following properties [45, 46]:

1. If $s_1, s_2 \in \mathcal{S}'_h$, then $s_1 - s_2 \in L$ and $\min\{s_1, s_2\} \in \mathcal{S}'_h$,
2. for any h there exists a unique *minimal cycle* $s_h := \min\{\mathcal{S}'_h\}$,
3. the set $\mathcal{S}'_h$ is a cone with *vertex* s_h in the sense that $\mathcal{S}'_h = s_h + \mathcal{S}'_0$.

5.3.5 *Families of Singularities*

So far, we have introduced some families of normal complex surface singularities, for example the Klenian singularities, the quotient singularities and the complete intersections. One can refine the classification of singularities by means of their resolutions. Let (X, x) be the germ of a normal surface singularity. Let $\pi : \tilde{X} \to X$ be a resolution. We denote by $\mathcal{O}_{\tilde{X}}$ the structure sheaf of $\tilde{X}$, i.e., the sheaf of holomorphic functions on $\tilde{X}$.

Definition 5.3.12 The *geometric genus* p_g of (X, x) is the dimension of $H^1(\tilde{X}, \mathcal{O}_{\tilde{X}})$ as a $\mathbb{C}$-vector space. It is independent of the choice of the resolution. We say that (X, x) has a *rational singularity* if $p_g = 0$.

Remark 5.3.13 By [25, Corollary 7.4.10], every quotient surface singularity is a rational singularity.

Recall that $\tilde{X}$ is a smooth complex surface. We denote by $\omega_{\tilde{X}}$ the sheaf of holomorphic 2-forms. The divisor of any meromorphic section of $\omega_{\tilde{X}}$ is called *the canonical divisor*; it is denoted by $K_{\tilde{X}}$. The rational cycle $Z_K \in L'$ (supported on E) that satisfies $(Z_K, E_v)_{\mathbb{Q}} = -K_{\tilde{X}} \cdot E_v$ for any $v \in V$ is called *the canonical cycle*. (It is independent of the choice of $K_{\tilde{X}}$.) We define the Riemann-Roch expression

$$\chi : L \to \mathbb{Z}, \ \chi(\ell) = -(\ell, \ell - Z_K)_{\mathbb{Q}}/2.$$

By Artin's criterion [1] (X, x) is rational if and only if $\chi(\ell) > 0$ for any $\ell > 0$ with $\ell \in L$.

Definition 5.3.14 We say that (X, x) has a *Gorenstein* singularity if the restriction of $\omega_{\tilde{X}}$ to $\tilde{X} \setminus E$ has a global section that is never vanishing on $\tilde{X} \setminus E$. Equivalently, (X, x) has a Gorenstein singularity if and only if $\mathcal{O}_{X,x}$ is a Gorenstein local ring (see [7] and [15, Section 21.3]).

In the case of Gorenstein singularities, we get $Z_K \in L$, see [45].

Remark 5.3.15 By [45, Section 2], any complete intersection is Gorenstein. In particular, Klenian singularities are Gorenstein.

By Remarks 5.3.13 and 5.3.15, Klenian singularities are rational and Gorenstein. This motivates the following definition.

Definition 5.3.16 We say that (X, x) has a *rational double point* singularity if (X, x) is Gorenstein and rational.

Remark 5.3.17 By [13], the Klenian singularities are all the rational double point singularities. We will use both names as synonymous.

Another important family of singularities is the following:

Definition 5.3.18 The germ (X, x) is called elliptic if (X, x) is not rational but $\chi(\ell) \geq 0$ for any $\ell \in L$. By an equivalent definition, (X, x) is elliptic if $\chi(Z_{min}) = 0$. Both topological criterions are independent of the choice of the resolution.

Definition 5.3.19 Let (X, x) be an elliptic normal surface singularity. Let $\pi : \tilde{X} \to X$ be a resolution. A non-zero effective cycle C supported on E is called *minimally elliptic cycle* if $\chi(C) = 0$ and for any $0 < \ell < C$, one has $\chi(\ell) > 0$.

Once a resolution is fixed, a minimally elliptic cycle always exists and it is unique.

Definition 5.3.20 We say that (X, x) is a *minimally elliptic singularity*, if the geometric genus of (X, x) is one and (X, x) is Gorenstein.

Laufer proved the following topological characterization: an elliptic singularity is minimally elliptic if and only if in the minimal resolution satisfies $C = Z_K = Z_{min}$, see [38] or [48, Theorem 7.2.15].

5.3.6 *The Picard Group and Vanishing Theorems*

Let (X, x) be the germ of a normal surface singularity. Let $\pi : \tilde{X} \to X$ be a resolution. The *Picard group* of $\tilde{X}$, denoted by $\mathrm{Pic}(\tilde{X})$, is the group of isomorphism classes of holomorphic line bundles with operation the tensor product. Denote by $\mathcal{O}^*_{\tilde{X}}$, the subsheaf of $\mathcal{O}_{\tilde{X}}$ consisting of the non-vanishing holomorphic functions. It is well known that $H^1(\tilde{X}, \mathcal{O}^*_{\tilde{X}}) \cong \mathrm{Pic}(\tilde{X})$. Consider the *exponential exact sequence*

$$0 \longrightarrow \mathbb{Z} \longrightarrow \mathcal{O}_{\tilde{X}} \xrightarrow{\exp} \mathcal{O}^*_{\tilde{X}} \longrightarrow 0,$$

The exponential exact sequence induces in cohomology the following long exact sequence

$$H^1(\tilde{X}, \mathbb{Z}) \longrightarrow H^1(\tilde{X}, \mathcal{O}_{\tilde{X}}) \xrightarrow{\exp} H^1(\tilde{X}, \mathcal{O}^*_{\tilde{X}}) \xrightarrow{\delta} H^2(\tilde{X}, \mathbb{Z}) \longrightarrow 0.$$

If (X, x) is a rational singularity, we get $H^1(\tilde{X}, \mathcal{O}_{\tilde{X}}) = 0$. Therefore, $\mathrm{Pic}(\tilde{X}) \cong H^1(\tilde{X}, \mathcal{O}^*_{\tilde{X}}) \cong H^2(\tilde{X}, \mathbb{Z})$. The last isomorphism is given by the first Chern class, i.e., $\delta = c_1$.

Since $H^2(\tilde{X}, \mathbb{Z}) \cong H_2(\tilde{X}, \partial\tilde{X}, \mathbb{Z}) = L'$, we can write $c_1 \colon \mathrm{Pic}(\tilde{X}) \to L'$. We denote by

$$\mathrm{Pic}^{\ell'}(\tilde{X}) := c_1^{-1}(\ell') \qquad \text{for any } \ell' \in L'.$$

Note that $\mathrm{Pic}^{\ell'}(\tilde{X})$ becomes naturally a $\mathrm{Pic}^0(\tilde{X})$-torsor. The following vanishing theorem will be used later.

Theorem 5.3.21 (Lipman's Vanishing [41, Theorem 11.1]) *Let (X, x) be a rational normal surface singularity. Let $\pi : \tilde{X} \to X$ be any resolution. If $\mathcal{L} \in \mathrm{Pic}(\tilde{X})$ with $-c_1(\mathcal{L}) \in \mathcal{S}'$, then $H^1(\tilde{X}, \mathcal{L}) = 0$.*

5.3.7 $R^1\pi_*\mathcal{F}$, and How to Calculate It

Let (X, x) be the germ of a normal complex surface singularity. We denote by V any affine neighborhood of x. Let $\pi : \tilde{X} \to X$ be any resolution. Denote by $\mathcal{F}$ a coherent sheaf on $\tilde{X}$. Under these assumptions, we have the following proposition.

Proposition 5.3.22 ([52, Chapter 4, Section 7]) *For any $n \in \mathbb{N}$ with $n \geq 2$, we get $H^n\left(\pi^{-1}V, \mathcal{F}\right) = 0$. Moreover, $H^1\left(\pi^{-1}V, \mathcal{F}\right)$ is independent of V.*

Definition 5.3.23 We define $R^1\pi_*\mathcal{F}$ as $H^1\left(\pi^{-1}X, \mathcal{F}\right) = H^1\left(\tilde{X}, \mathcal{F}\right)$, viewed as a sheaf supported at the point $x \in X$.

The reader can find more information on [52, Chapter 4, Section 7].

5.4 The Riemann-Hilbert Correspondence

As before, we denote by (X, x) the germ of a normal complex surface singularity or the spectrum of a complete normal local ring over an algebraically closed field of characteristic zero. We denote by $U = \mathrm{Reg}X = X \setminus \{x\}$ the *regular part* of X and by $i \colon U \to X$ the natural inclusion.

5.4.1 Reflexive Modules

We will only work with finite generated modules, and there will be no distinction between $\mathcal{O}_{X,x}$-modules and their Sheafification over (X, x). Let $\mathcal{H}om_{\mathcal{O}_X}(\cdot, \cdot)$ be the sheaf theoretical Hom functor. The dual of an $\mathcal{O}_X$-module M is denoted by $M^{\vee} := \mathcal{H}om_{\mathcal{O}_X}(M, \mathcal{O}_X)$, where $\mathcal{O}_X$ is the structure sheaf of X.

Definition 5.4.1 An $\mathcal{O}_X$-module M is called *reflexive* if the natural homomorphism from M to $M^{\vee\vee}$ is an isomorphism (see [25, Definition 5.1.12]).

We denote by $\mathrm{Cl}(X, x)$ the *local divisor class group* of (X, x) (see [48, Chapter 6]). The group $\mathrm{Cl}(X, x)$ can also be interpreted as the group of isomorphism classes of reflexive sheaves on (X, x) of rank one [56].

Definition 5.4.2 An $\mathcal{O}_X$-module M is called *Cohen-Macaulay* if the depth of each of its stalks M_y is equal to the dimension of the module. If the depth of M_y is equal to the dimension of $\mathcal{O}_{X,y}$, then the module M_y is called *maximal Cohen-Macaulay*; these definitions extend to sheaves if we ask that they hold for every stalk.

A module is *indecomposable* if it can not be written as a direct sum of two non trivial submodules.

By [59, Proposition 1.5] and [24, Section 1] some basic properties of maximal Cohen-Macaulay modules are:

1. If $\mathcal{O}_{X,y}$ is a regular local ring, then any maximal Cohen-Macaulay module over it is free.
2. If $\mathcal{O}_{X,y}$ is a reduced local ring of dimension one, then an $\mathcal{O}_{X,y}$-module M is maximal Cohen-Macaulay if and only if it is torsion free, that is, when the natural homomorphism $M \to M^{\vee\vee}$ is a monomorphism.
3. If $\mathcal{O}_{X,y}$ is normal of dimension two, then an $\mathcal{O}_{X,y}$-module M is maximal Cohen-Macaulay if and only if it is reflexive.
4. A consequence of the previous properties is that reflexive sheaves over regular rings $\mathcal{O}_{X,y}$ of dimension at most 2 are free.
5. Let M be a $\mathcal{O}_X$-module. Then M_x is a reflexive $\mathcal{O}_{X,x}$-module if and only if the adjunction morphism $M \to i_*i^*M$ is an isomorphism.

Recall that (X, x) is a normal surface singularity. Denote by $\mathcal{O}_{X,x}$ the local ring at x. By [25, Corollary 3.5.17], we have that $\mathcal{O}_{X,x}$ is a Cohen-Macaulay ring, i.e., $\dim \mathcal{O}_{X,x} = \operatorname{depth} \mathcal{O}_{X,x} = 2$. Using the Cohen-Macaulay condition one can prove the existence of a dualizing sheaf ω_X, and by abuse of notation we denote also by ω_X its stalk at x, which is the dualizing module of the ring $\mathcal{O}_{X,x}$. The reader may see [25, Definition 5.3.5] or [15, Theorem 21.15] for more details. If (X, x) is Gorenstein then the dualizing sheaf is an invertible trivial sheaf (see for example [15, Section 21.3] or [7, Theorem 3.3.7]). This means that if (X, x) is Gorenstein then the dualizing module coincides with $\mathcal{O}_{X,x}$. The canonical module and Cohen-Macaulay modules have the following properties, which are a special case of [7, Theorem 3.3.10]. The reader may find more details about the dualizing module and the dualizing sheaf on [25, Chapter 5, Section 3] and [7, Chapter 3, Section 3].

Theorem 5.4.3 *Let X be a normal surface. For $t = 0, 1, 2$ and all Cohen-Macaulay $\mathcal{O}_X$-modules M of dimension t one has*

1. $\mathcal{E}xt^{2-t}_{\mathcal{O}_X}(M, \omega_X)$ *is Cohen-Macaulay of dimension t,*
2. $\mathcal{E}xt^{i}_{\mathcal{O}_X}(M, \omega_X) = 0$ *for all $i \neq 2 - t$,*
3. *there exists an isomorphism $M \to \mathcal{E}xt^{2-t}_{\mathcal{O}_X}\left(\mathcal{E}xt^{2-t}_{\mathcal{O}_X}(M, \omega_X), \omega_X\right)$ which in the case $t = 2$ is just the natural homomorphism from M into the ω_X-bidual of M.*

We also have the following properties:

Proposition 5.4.4 ([59, Proposition 1.2]) *Let X be a normal surface. Denote by M be an $\mathcal{O}_X$-module. The following conditions are equivalent:*

1. *The module M is reflexive.*
2. $\mathcal{E}xt^{i}_{\mathcal{O}_X}(\mathbb{C}, M) = 0$ *for $i < 2$.*
3. $H^i_{\mathfrak{m}}(M) = 0$ *for $i \neq 2$.*

Where $H^i_{\mathfrak{m}}$ denotes the i-th local cohomology functor with support on $\mathfrak{m}$.

Theorem 5.4.5 ([7, Theorem 3.5.7]) *Let X be a normal surface. Denote by M an $\mathcal{O}_X$-module of depth t and dimension d. Then,*

1. $H^i_{\mathfrak{m}}(M) = 0$ *for $i < t$ and $i > d$.*
2. $H^t_{\mathfrak{m}}(M) \neq 0$ *and* $H^d_{\mathfrak{m}}(M) \neq 0$.

Corollary 5.4.6 ([7, Corollary 3.5.9]) *Let $(R, \mathfrak{m}, k)$ be a Gorenstein normal local ring of dimension 2. Then, for all finite R-modules M and all integers i there exist natural isomorphism*

$$H^i_{\mathfrak{m}}(M) \cong \mathrm{Hom}_R\left(\mathrm{Ext}^{2-i}_R(M, R), E(k)\right).$$

Where $E(k)$ is the injective hull of k.

Remark 5.4.7 We will use the previous results as follows: Let $(R, \mathfrak{m}, k)$ be a Gorenstein normal local ring of dimension 2. Let M be a finitely generated R-module. Then, by Proposition 5.4.4 the module M is reflexive if and only if $H^i_{\mathfrak{m}}(M) = 0$ for $i \neq 2$. By Theorem 5.4.5 $H^i_{\mathfrak{m}}(M) = 0$ for $i \neq 2$ if and only if $H^0_{\mathfrak{m}}(M) = 0$ and $H^1_{\mathfrak{m}}(M) = 0$. By Corollary 5.4.6 it is enough to prove that $\mathrm{Ext}^1_R(M, R) = 0$ and $\mathrm{Ext}^2_R(M, R) = 0$ to guarantee that M is a reflexive module.

We will also use the Grothendieck duality.

Theorem 5.4.8 ([23, Ch. VII]) *Let (X, x) be the germ of a normal surface singularity. Denote by ω_X its dualizing module for the singularity [7, Theorem 3.3.7]. Let $\pi : \tilde{X} \to X$ be a resolution. Denote by $\omega_{\tilde{X}}$ the canonical divisor. Then, the map π establishes the isomorphism*

$$R\pi_* R\mathcal{H}om_{\mathcal{O}_{\tilde{X}}}\left(-, \omega_{\tilde{X}}\right) \cong R\mathrm{Hom}_{\mathcal{O}_X}(R\pi_*-, \omega_X).$$

5.4.2 Flat Modules and the Riemann-Hilbert Correspondence

Let M be a coherent $\mathcal{O}_X$-module. As usual, denote by $\mathrm{Der}_{\mathbb{C}}(\mathcal{O}_X)$ the set of all derivations over $\mathbb{C}$, the reader may see [42, Chapter 10] for more details. Following [19] a *connection* on M is an $\mathcal{O}_X$-linear map

$$\nabla : \mathrm{Der}_{\mathbb{C}}(\mathcal{O}_X) \to \mathrm{End}_{\mathbb{C}}(M),$$

that is, for any $f \in \mathcal{O}_X$ and any $D \in \mathrm{Der}_{\mathbb{C}}(\mathcal{O}_X)$ we get the following:

$$\nabla_{fD} = f\nabla_D.$$

Moreover, for all $f \in \mathcal{O}_X$, $m \in M$ and $D \in \mathrm{Der}_{\mathbb{C}}(\mathcal{O}_X)$ satisfies the *Leibniz rule*

$$\nabla_D(fm) = D(f)m + f\nabla_D(m).$$

A connection ∇ is called *integrable or flat* if it is a $\mathbb{C}$-Lie-algebra homomorphism. Denote by:

$$\begin{aligned}
\mathrm{Ref}_X &:= \text{the category of reflexive } \mathcal{O}_X\text{-modules,}\\
\mathrm{MCM}_{\mathcal{O}_X} &:= \text{the category of maximal Cohen-Macaulay } \mathcal{O}_X\text{-modules,}\\
\mathrm{Vect}_U &:= \text{the category of holomorphic vector bundles over } U,\\
\mathrm{Ref}_X^{\nabla} &:= \text{the category of pairs } (M, \nabla) \text{ where } M \text{ is a reflexive } \mathcal{O}_X\text{-module}\\
&\quad\ \text{and } \nabla \text{ is an integrable connection,}\\
\mathrm{Vect}_U^{\nabla} &:= \text{the category of pairs } (V, \nabla) \text{ where } V \text{ is a vector bundle over } U\\
&\quad\ \text{and } \nabla \text{ is an integrable connection,}\\
\mathrm{Rep}_{\pi_1(\Sigma)} &:= \text{the category of complex finite dimensional representations of } \pi_1(\Sigma).
\end{aligned}$$

By [19] there is an equivalence

$$\mathrm{Ref}_X^{\nabla} \cong \mathrm{Rep}_{\pi_1(\Sigma)},$$

called *the Riemann-Hilbert correspondence.*

Remark 5.4.9 By [24], we get the following functor

$$\begin{aligned}
i^*\colon \mathrm{Ref}_X \cong \mathrm{MCM}_{\mathcal{O}_X} &\to \mathrm{Vect}_U\\
M \quad &\longmapsto \quad i^*M = M|_U
\end{aligned}$$

with inverse (defined only in the image)

$$\begin{array}{ccc} i^*(\mathrm{Ref}_X) & \rightarrow \mathrm{Ref}_X \cong & \mathrm{MCM}_{\mathcal{O}_X} \\ V & \longmapsto & i_*V \end{array}$$

Analogously, we have the following equivalence of categories

$$\mathrm{Ref}_X^{\nabla} \cong \mathrm{Vect}_U^{\nabla}.$$

Therefore, the categories $\mathrm{Rep}_{\pi_1(\Sigma)}$ and Vect_U^{∇} are equivalent.

In general, the *forgetful functor*

$$\Psi : \mathrm{Ref}_X^{\nabla} \to \mathrm{Ref}_X,$$

is not an equivalence of categories. Moreover, the forgetful functor may not be essentially surjective, i.e., the forgetful functor may not be onto on objects, see [9]. The Riemann-Hilbert (R.H.) correspondence and these forgetful functors fit in the following conmutative diagram:

$$\begin{array}{ccccc} \mathrm{Rep}_{\pi_1(\Sigma)} & \xrightarrow[\cong]{\text{R.H.}} & \mathrm{Ref}_X^{\nabla} & \xrightarrow{\Psi} & \mathrm{Ref}_X \\ & & \cong\downarrow & & \downarrow \\ & & \mathrm{Vect}_U^{\nabla} & \xrightarrow{\Psi} & \mathrm{Vect}_U \end{array} \tag{5.6}$$

From this diagram we get the following remarks:

Remark 5.4.10 Let (X, x) be the germ of a normal surface singularity.

1. If (X, x) is Klenian singularity, by the Riemann-Hilbert correspondence we can interpret the Mckay correspondence as a classification of the objects of the category Ref_X^{∇} in terms of the exceptional divisor of the minimal resolution.
2. By Esnault, the functor Ψ is an equivalence of categories if and only if (X, x) is a quotient surface singularity. Moreover, by Brieskorn characterization, this is the case only when the irreducible elements of $|\mathrm{Obj}(\mathrm{Rep}_{\pi_1(\Sigma)})|$ is finite. If this is the case, we say that (X, x) has finite representation type.
3. In general, the functor Ψ is not onto neither injective on objects, see [9].

For reflexive modules of rank one we can identify the categories as follows: Denote by

Ref_X^1 := the category of reflexive $\mathcal{O}_X$-modules of rank one,

$\mathrm{Ref}_X^{1,\nabla}$:= the category of pairs (M, ∇) where M is a reflexive $\mathcal{O}_X$-module of rank one and ∇ is an integrable connection,

$\mathrm{Rep}^1_{\pi_1(\Sigma)}$:= the category of complex one dimensional representations of $\pi_1(\Sigma)$.

Let ρ be a representation of $\pi_1(\Sigma)$. Thus, we have $\rho\colon \pi_1(\Sigma) \to GL(1,\mathbb{C}) = \mathbb{C}^*$ where $\mathbb{C}^*$ is the group of units of $\mathbb{C}$. Since $\mathbb{C}^*$ is abelian,

$$\mathrm{Hom}(\pi_1(\Sigma), \mathbb{C}^*) = \mathrm{Hom}(\pi_1(\Sigma)_{\mathrm{ab}}, \mathbb{C}^*) = \mathrm{Hom}(H_1(\Sigma,\mathbb{Z}), \mathbb{C}^*). \tag{5.7}$$

By the Universal Coefficient Theorem we get

$$\mathrm{Hom}(H_1(\Sigma,\mathbb{Z}), \mathbb{C}^*) \cong H^1(\Sigma, \mathbb{C}^*). \tag{5.8}$$

Hence, by (5.7) and (5.8) $\rho \in H^1(\Sigma, \mathbb{C}^*)$. Replacing $\mathbb{C}^*$ by $\mathbb{C}/\mathbb{Z}$, the exact sequence

$$0 \to \mathbb{Z} \to \mathbb{C} \to \mathbb{C}/\mathbb{Z} \cong \mathbb{C}^* \to 0,$$

induces the following cohomological long exact sequence

$$\cdots \to H^1(\Sigma,\mathbb{C}) \to H^1(\Sigma,\mathbb{C}/\mathbb{Z}) \xrightarrow{\Phi} H^2(\Sigma,\mathbb{Z}) \to H^2(\Sigma,\mathbb{C}) \to \dots \tag{5.9}$$

By (5.9), the image of Φ is torsion. Hence, it has a factorization through

$$\Phi\colon H^1(\Sigma,\mathbb{C}/\mathbb{Z}) \to \mathrm{Tors}\left(H^2(\Sigma,\mathbb{Z})\right) \cong H, \quad \rho \mapsto \Phi(\rho) \in H.$$

On the other hand, by the equivalence between $\mathrm{Ref}^{1,\nabla}_X$ and $\mathrm{Rep}^1_{\pi_1(\Sigma)}$, let (M_ρ, ∇_ρ) be the pair in $\mathrm{Obj}(\mathrm{Ref}^{1,\nabla}_X)$ associated to ρ. Hence, in this reformulation, the *forgetful morphism* is

$$\Psi\colon H^1(\Sigma,\mathbb{C}/\mathbb{Z}) \to \mathrm{Cl}(X,x), \quad \rho = (M_\rho, \nabla_\rho) \mapsto \Psi(\rho) = M_\rho.$$

Note also that the groups H and $\mathrm{Cl}(X,x)$ are related by the following long exact sequence

$$0 \longrightarrow H^1(\Sigma,\mathbb{Z}) \longrightarrow \mathbb{C}^{p_g} \longrightarrow \mathrm{Cl}(X,x) \xrightarrow{\overline{c}_1} H \longrightarrow 0,$$

where $\overline{c}_1$ assigns to a Weil divisor the homological class of its intersection with the link.

5.4.3 Full Sheaves

Let (X, x) be the germ of a normal surface singularity and $\pi\colon \tilde{X} \to X$ be a resolution. By Remark 5.4.9, there is a one-to-one correspondence between the reflexive $\mathcal{O}_X$-modules and the vector bundles over the regular part. Moreover, there is an isomorphism $\operatorname{Reg} U \cong \tilde{X} \setminus E$. Thus, there should be a natural way to study the reflexive modules over the resolution. This is the notion of full sheaf, introduced by Esnault [16] and generalized by Kahn [28].

Definition 5.4.11 An $\mathcal{O}_{\tilde{X}}$-module $\mathcal{M}$ is called *full* if there is a reflexive $\mathcal{O}_X$-module M such that $\mathcal{M} \cong (\pi^* M)^{\vee\vee}$. We call $\mathcal{M}$ the full sheaf associated to M. We denote by $\mathrm{Full}_{\tilde{X}}$ as the category of full sheaves over $\tilde{X}$.

The following characterization of full sheaves will be very important in the following sections.

Proposition 5.4.12 ([28, Proposition 1.2]) *A locally free sheaf $\mathcal{M}$ on $\tilde{X}$ is full if and only if*

1. *the sheaf $\mathcal{M}$ is generically generated by global sections.*
2. *The natural map $H^1_E(\tilde{X}, \mathcal{M}) \to H^1(\tilde{X}, \mathcal{M})$ is injective, where $H^1_E(\tilde{X}, \mathcal{M})$ denotes the first cohomology group of $\mathcal{M}$ supported on the exceptional divisor E.*

If $\mathcal{M}$ is the full sheaf associated to M, then $\pi_* \mathcal{M} = M$.

In the particular case of a rational singularity the characterization of full sheaves is the following:

Proposition 5.4.13 ([16, Lemma 2.2]) *Suppose that (X, x) is a rational normal surface singularity. A locally free sheaf $\mathcal{M}$ on $\tilde{X}$ is full if and only if*

1. *the sheaf $\mathcal{M}$ is generated by global sections,*
2. *$H^1_E(\tilde{X}, \mathcal{M}) = 0$.*

The following two lemmas will be used later.

Lemma 5.4.14 *If $\mathcal{M}$ is a full $\mathcal{O}_{\tilde{X}}$-module, then $R^1\pi_*(\mathcal{M} \otimes \omega_{\tilde{X}}) = 0$.*

Lemma 5.4.15 *If $\mathcal{M}$ is a full sheaf, then $\pi_* \left(\mathcal{M}^{\vee}\right) = (\pi_* \mathcal{M})^{\vee}$.*

As in the previous section, we denote:

$$\mathrm{Full}_{\tilde{X}} := \text{the category of full shaves over } \tilde{X},$$

$$\mathrm{Full}^1_{\tilde{X}} := \text{the category of full shaves over } \tilde{X} \text{ of rank 1}.$$

Remark 5.4.16 By Proposition 5.4.12 there is a one-to-one correspondence between full sheaves and reflexive $\mathcal{O}_X$-modules. The equivalence of categories is given by

$$\begin{aligned} \mathrm{Ref}_X &\cong \mathrm{Full}_{\tilde{X}} \\ M &\mapsto \left(\pi^* M\right)^{\vee\vee} \end{aligned}$$

with inverse

$$\begin{aligned} \mathrm{Full}_{\tilde{X}} &\cong \mathrm{Ref}_X \\ \mathcal{M} &\mapsto \pi_* \mathcal{M} \end{aligned}$$

Moreover, by Lemma 5.4.15 this isomorphism is compatible with dualization.

Another important notion is the concept of specialty.

Definition 5.4.17 A full $\mathcal{O}_{\tilde{X}}$-module $\mathcal{M}$ on $\tilde{X}$ of rank r is called *special* if $\dim_{\mathbb{C}}(R^1\pi_* \left(\mathcal{M}^{\vee}\right)) = rp_g$. *The defect of specialty of* $\mathcal{M}$ is the number

$$\dim_{\mathbb{C}}(R^1\pi_* \left(\mathcal{M}^{\vee}\right)) - rp_g.$$

This notion was defined by Wunram [58] and Riemenschneider [55] for rational singularities. Later, this notion was generalized for Gorenstein singularities by Fernández de Bobadilla and the second author [4]. Notice that this definition is a generalization of the concept given by Wunram and Riemenschneider and both definitions coincide in the case of a rational singularity.

Remark 5.4.18 By Definition 5.3.23, the speciality condition is equivalent to the following condition $\dim_{\mathbb{C}}(H^1 \left(\tilde{X}, \mathcal{M}^{\vee}\right)) = rp_g$.

Since the definition of being special depends on the resolution, we have a another related notion.

Definition 5.4.19 Let M be a reflexive $\mathcal{O}_X$-module. We say that M is *a special module* if for any resolution the full sheaf associated to M is special.

Remark 5.4.20 Another approach to the study of reflexive modules on normal surface singularities was provided by Iyama and Wemyss [26]. They generalized the notion of speciality as the vanishing of $\mathrm{Ext}^1_{\mathcal{O}_X}(M, \mathcal{O}_X)$. Using this algebraic notion of speciality, they classified all the non-trivial indecomposable special modules on quotient surface singularities in a combinatorial way. For this, they use the vanishing of the first Ext group to reduce the classification problem to a combinatorial problem on the Auslander-Reiten quiver. It is important to remark that their notion of speciality is different to the notion given by Definition 5.4.17.

5.5 The McKay Correspondence for Rational Singularities

In this section we show how to prove the McKay correspondence for rational singularities. This correspondence generalizes the classical McKay correspondence for Klenian singularities. The reader may see [2, 16, 58] for more details.

Let (X, x) be the germ of a rational normal surface singularity. Let $\pi : \tilde{X} \to X$ be its minimal resolution. By Remark 5.4.10, we have to construct a correspondence between:

- the indecomposable reflexive $\mathcal{O}_X$-modules,
- the irreducible components of the exceptional divisor of the minimal resolution.

We will prove that this correspondence is false in general. The correct correspondence is stated using the special modules. First, we have the following lemma by Artin and Verdier. For sake of the reader we write the proof.

Lemma 5.5.1 ([2, Lemma 1.1]) *Let (X, x) be the germ of a rational normal surface singularity. Let $\pi : \tilde{X} \to X$ be its minimal resolution. Let M be a $\mathcal{O}_X$-module. Denote by $\tilde{M} = \pi^* M/\text{tor}$. Then,*

1. $\pi_* \tilde{M} = M$,
2. $R^1\pi_* \tilde{M} = 0$,
3. *the sheaf $\tilde{M}$ is locally free and generated by global sections.*

Proof Consider the following local presentation,

$$\mathcal{O}_X^q \longrightarrow \mathcal{O}_X^p \longrightarrow M \longrightarrow 0. \tag{5.10}$$

Applying the functor π^*- to the previous presentation, we get

$$0 \longrightarrow K_1 \longrightarrow \mathcal{O}_{\tilde{X}}^p \longrightarrow \pi^* M \longrightarrow 0, \tag{5.11}$$

and

$$0 \longrightarrow K_2 \longrightarrow \mathcal{O}_{\tilde{X}}^q \longrightarrow K_1 \longrightarrow 0. \tag{5.12}$$

Applying the functor π_*- to the exact sequences (5.11) and (5.12), we get

$$\begin{array}{l} 0 \longrightarrow \pi_* K_1 \longrightarrow \mathcal{O}_X^p \longrightarrow \pi_*\pi^* M \longrightarrow \\ \longrightarrow R^1\pi_* K_1 \longrightarrow R^1\mathcal{O}_{\tilde{X}}^p \longrightarrow R^1\pi_*\pi^* M \longrightarrow 0 \end{array} \tag{5.13}$$

and

$$
\begin{array}{l}
0 \longrightarrow \pi_*K_2 \longrightarrow \mathcal{O}_X^q \longrightarrow \pi_*K_1 \longrightarrow \\
R^1\pi_*K_2 \longrightarrow R^1\mathcal{O}_{\tilde{X}}^q \longrightarrow R^1\pi_*K_1 \longrightarrow 0
\end{array}
\tag{5.14}
$$

By hypothesis (X, x) is rational, thus $R^1\mathcal{O}_{\tilde{X}}^q = R^1\mathcal{O}_{\tilde{X}}^p = 0$. Therefore, by (5.13) and (5.14) we get $R^1\pi_*\pi^*M = R^1\pi_*K_1 = 0$. Using this vanishes with (5.13) and (5.10) we get

$$
\begin{array}{ccccccc}
0 \longrightarrow \pi_*K_1 \longrightarrow & \mathcal{O}_X^p & \longrightarrow & \pi_*\pi^*M & \longrightarrow 0 \\
& \cong\uparrow & & \uparrow & \\
& \mathcal{O}_X^p & \longrightarrow & M & \longrightarrow 0
\end{array}
$$

Hence, the natural map from M to $\pi_*\pi^*M$ is onto. Moreover, the kernel of this map is all torsion and M is torsion-free. Therefore, this map is injective. Thus,

$$M \cong \pi_*\pi^*M. \tag{5.15}$$

Now, consider the following exact sequence

$$0 \longrightarrow \operatorname{tor}\pi^*M \longrightarrow \pi^*M \longrightarrow \tilde{M} \longrightarrow 0.$$

Applying the functor π_*- to the previous exact sequence, we get

$$0 \longrightarrow \pi_*\operatorname{tor}\pi^*M \longrightarrow \pi_*\pi^*M \xrightarrow{\alpha} \pi_*\tilde{M} \longrightarrow R^1\pi_*\operatorname{tor}\pi^*M \longrightarrow \tag{5.16}$$

Thus, $R^1\pi_*\tilde{M} = 0$. This proves the second assertion of the lemma. Since $\pi_*\operatorname{tor}\pi^*M$ is all torsion and by (5.15) the module $\pi_*\pi^*M$ is torsion free, thus

$$\pi_*\operatorname{tor}\pi^*M = 0.$$

Denote by $s\colon \tilde{M} \to i_*i^*\tilde{M} \cong i_*i^*M \cong M$ the natural map given by adjunction. Hence, $s \circ \alpha\colon M \to M$ is the identity map. Thus, we have constructed a section of the exact sequence (5.16). Therefore, $R^1\pi_*\operatorname{tor}\pi^*M \subset \pi_*\tilde{M}$. Since $\pi_*\tilde{M}$ is torsion free, we get $R^1\pi_*\operatorname{tor}\pi^*M = 0$. This implies, $M \cong \pi_*\tilde{M}$. This proves the first assertion of the lemma.

Now we prove the last assertion of the lemma. Since M is globally generated, we get $\tilde{M}$ is also generated by its global sections. We have to prove that $\tilde{M}$ is locally free. Since $\tilde{X}$ is smooth, we only have to prove that $\tilde{M}$ is reflexive. By Remark 5.4.7 we have to prove that

$$\mathscr{E}xt^t_{\mathcal{O}_{\tilde{X}}}(\tilde{M}, \omega_{\tilde{X}}) = 0 \quad \text{for } t > 0.$$

Since $\tilde{M}$ is locally free in codimension two, $\mathscr{E}xt^t_{\mathcal{O}_{\tilde{X}}}(\tilde{M}, \omega_{\tilde{X}})$ is supported on points of E. By Grothendieck's duality 5.4.8, we have

$$R\pi_* R\mathscr{H}om_{\mathcal{O}_{\tilde{X}}}\left(\tilde{M}, \omega_{\tilde{X}}\right) \cong R\mathrm{Hom}_{\mathcal{O}_X}\left(R\pi_*\tilde{M}, \omega_X\right). \tag{5.17}$$

Since

$$\begin{aligned} \pi_*\tilde{M} &= M, \\ R^j\pi_*\tilde{M} &= 0 \quad \text{for any } j \geq 1, \\ \mathrm{Ext}^j_{\mathcal{O}_X}(M, \omega_X) &= 0 \quad \text{for any } j \geq 1, \end{aligned} \tag{5.18}$$

where the last isomorphism follows by Theorem 5.4.3 because M is reflexive (equivalently, M is maximal Cohen-Macaulay). By (5.17) and (5.18) we get

$$\mathscr{E}xt^t_{\mathcal{O}_{\tilde{X}}}(\tilde{M}, \omega_{\tilde{X}}) = 0 \quad \text{for } t > 0.$$

Thus, $\tilde{M}$ is locally free. This finishes the proof. □

Remark 5.5.2 In general, Lemma 5.5.1 is false if we omit the hypothesis of rationality. We will show an example in the last section of this survey.

Corollary 5.5.3 *Let (X, x) be a rational normal singularity. Let $\pi : \tilde{X} \to X$ be a resolution. Let M be a reflexive $\mathcal{O}_X$-module. Denote by $\mathcal{M}$ the full sheaf associated to M and by $\tilde{M} = \pi^* M/\mathrm{tor}$. Then, $\mathcal{M} \cong \tilde{M}$.*

Proof The natural map from $\tilde{M}$ to $(\pi^* M)^{\vee\vee} = \mathcal{M}$ is an injection. Moreover, by Lemma 5.5.1 the sheaf $\tilde{M}$ is locally free. Thus, the natural map from $\tilde{M}$ to $\mathcal{M}$ is an isomorphism because $\mathcal{M}$ and $\tilde{M}$ are locally free sheaves of the same rank. □

In the particular case of a Klenian singularity the notion of speciality is redundant, any reflexive module is special. This is proved in the following lemma.

Lemma 5.5.4 *Let (X, x) be a Klenian singularity. Let $\pi : \tilde{X} \to X$ be the minimal resolution. Let M be a reflexive $\mathcal{O}_X$-module. Denote by $\mathcal{M}$ the full sheaf associated to M. Then, $\mathcal{M}$ is special.*

Proof By Serre duality, we get $H^1\left(\tilde{X}, \mathcal{M}^{\vee}\right) \cong H^1_E\left(\tilde{X}, \mathcal{M} \otimes \omega_{\tilde{X}}\right)$. Since (X, x) is a rational double point we get that $\omega_{\tilde{X}}$ is a trivial vector bundle (the resolution is *crepant*, see [53, Section 1.3] for more details). Thus,

$$H^1_E\left(\tilde{X}, \mathcal{M} \otimes \omega_{\tilde{X}}\right) \cong H^1_E\left(\tilde{X}, \mathcal{M}\right).$$

By Proposition 5.4.13 we get $H^1_E\left(\tilde{X}, \mathcal{M}\right) = 0$. Therefore, $H^1\left(\tilde{X}, \mathcal{M}^{\vee}\right) = 0$. Hence, $\mathcal{M}$ is special. □

Now we sketch the idea of the proof of the McKay correspondence for rational singularities: Let (X, x) be a rational normal singularity. Let $\pi : \tilde{X} \to X$ be a resolution. Let M be a reflexive $\mathcal{O}_X$-module of rank r. Denote by $\mathcal{M}$ the full sheaf associated to M. Let $\phi_1, \phi_2, \ldots, \phi_r$ be generic sections of $\mathcal{M}$. Consider the following exact sequence given by the sections

$$0 \to \mathcal{O}^r_{\tilde{X}} \overset{(\phi_1,\ldots,\phi_r)}{\longrightarrow} \mathcal{M} \to \mathcal{A}' \to 0. \tag{5.19}$$

One can prove (see [2, Lemma 1.2]) that the sections $\phi_1, \phi_2, \ldots, \phi_r$ generate outside of a closed subscheme D of $\tilde{X}$ of codimension 1. Moreover, $\mathcal{A}' \cong \mathcal{O}_D$. Moreover, the class $[D]$ of the divisor D in $\mathrm{Pic}(\tilde{X})$ is the first Chern class $c_1(\tilde{M})$ of $\tilde{M}$. With this construction of the first Chern class we can prove the following theorem (see [2] for Klenian singularities and [16, 58] for rational singularities).

Theorem 5.5.5 *Let (X, x) be a rational singularity. Let $\pi : \tilde{X} \to X$ be the minimal resolution with exceptional divisor $E = \bigcup_{i=1}^{l} E_i$. Then, there exists a bijection between the following sets:*

1. *The set of non-trivial indecomposable special reflexive $\mathcal{O}_X$-modules up to isomorphism.*
2. *The set of irreducible divisors E_i where E_i is an irreducible component of the exceptional divisor E.*

The correspondence is constructed via the first Chern class: to M_j a reflexive $\mathcal{O}_X$-module, we associate $c_1(\tilde{M}_j) \cdot E_k = \delta_{jk}$. Moreover, the rank of M_j equals the multiplicity r_j of the curve E_j in the fundamental cycle Z_{min}.

Remark 5.5.6 Theorem 5.5.5 is false without the hypothesis of specialty. For example, consider the following groups:

$$\Gamma_1 = \left\langle \begin{pmatrix} \zeta_3 & 0 \\ 0 & \zeta_3^{-1} \end{pmatrix} \right\rangle \quad \text{and} \quad \Gamma_2 = \left\langle \begin{pmatrix} \zeta_3 & 0 \\ 0 & \zeta_3 \end{pmatrix} \right\rangle.$$

Denote by

$$X_1 = \mathbb{C}^2/\Gamma_1 \quad \text{and} \quad X_2 = \mathbb{C}^2/\Gamma_2.$$

Thus, X_1 and X_2 are quotient surface singularities. Moreover, X_1 is a Klenian singularity. In both cases, the links Σ_1 and Σ_2 are lens spaces and $\pi_1(\Sigma_1) \cong \pi_1(\Sigma_2) \cong \mathbb{Z}_3$. By the Riemann-Hilbert correspondence, both singularities have two non-trivial reflexive modules (corresponding to rank one representations of the fundamental group of the links). It is well known that the number of irreducible components of the exceptional divisor of the minimal resolution of X_1, has two components (one for each rank one reflexive module because all the modules are special because X_1 is Klenian). For the singularity X_2, the number of irreducible components of the exceptional divisor of the minimal resolution is one (see [57, pp. 651]) but there are two different reflexive modules. Thus it is impossible to produce a one-to-one correspondence between the components of the exceptional divisor and the non-trivial reflexive modules.

By the previous remark, it is clear that we can not generalize the McKay correspondence in full generality. Related to this problem, Esnault asked the following question (see [16]): Does the rank and the first Chern class determine all the reflexive modules (equivalently, the full shaves)? The answer is no and it was provided by Wunram in [58]. In this paper, Wunram proved that in the singularity $\mathbb{I}_7$ (following the Riemenschneider's notation in [54]) there are two different reflexive modules such that their full shaves have the same rank and the same first Chern class. Hence, it is natural to study not the whole category of reflexive modules but a suitable subcategory of these. This is done in the following section.

5.6 Classification of Reflexive Modules of Rank One over Rational and Minimally Elliptic Singularities

As we have seen the classification of reflexive modules is an open problem and even in quotient surface singularities it is a difficult one. Esnault [16] studied this problem and she classified rank one reflexive modules for quotient singularities (without using the notion of specialty). Later, Nemethi and the second author improved Esnault's result to more classes of singularities. In this section we will see the classification of reflexive modules of rank one over rational and minimally elliptic singularities.

Recall that if (X, x) is rational singularity then the link Σ is a rational homology sphere, thus (see (5.4))

$$H_1(\Sigma, \mathbb{Z}) = \mathrm{Tors}(H_1(\Sigma, \mathbb{Z})) = L'/L = H.$$

Furthermore, we have the isomorphism $c_1 \colon \mathrm{Pic}(\tilde{X}) \to L'$, i.e., all the line bundles are characterized topologically. As in Sect. 5.3.4, for any $\ell' \in L'$ we denote by $\mathcal{O}_{\tilde{X}}(\ell') \in \mathrm{Pic}(\tilde{X})$ the unique line bundle that satisfies $c_1(\mathcal{O}_{\tilde{X}}(\ell')) = \ell'$. For rational surface singularities, the classification of full shaves of rank one is as follows (see [49, Theorem 3.1]):

Theorem 5.6.1 *Let* (X, x) *be a rational normal surface singularity. Let* $\pi : \tilde{X} \to X$ *be any resolution. Then,*

1. *for any* $h \in H$ *the line bundle* $\mathcal{O}_{\tilde{X}}(-s_h)$ *is full.*
2. *If a line bundle* $\mathcal{L} \in \mathrm{Pic}(\tilde{X})$ *is full, then* $c_1(\mathcal{L}) = -s_h$ *for some* $h \in H$. *Hence* $\mathcal{L} = \mathcal{O}_{\tilde{X}}(-s_h)$.

Thus, any full sheaf of rank one is of the form $\mathcal{O}_{\tilde{X}}(-s_h)$ *for some* $h \in H$. *Where* s_h *is the minimal cycle of* $\mathcal{S}'_h$.

This theorem provides a topological classification of reflexive modules of rank one over rational singularities. This can be understood via the following theorem (see [49, Theorem 3.5]):

Theorem 5.6.2 *Let* (X, x) *be a rational normal surface singularity. Let* $\pi : \tilde{X} \to X$ *be any resolution. Then, any full* $\mathcal{O}_{\tilde{X}}$*-sheaf of rank one is flat.*

Remark 5.6.3 In general, the computation of the set $\{s_h\}_{h \in H}$, even for very concrete families of topological types of singularities (families of links) in general is not easy. The reader may see [49, Remark 3.6] for more about this problem and some references.

If the singularity is not rational, then the first Chern class $c_1 : \mathrm{Pic}(\tilde{X}) \to L'$ does not provide an isomorphism anymore, hence the identification of certain line bundles (with special properties) from the class of line bundles with given Chern class is considerably harder. From now on, suppose that (X, x) is the germ of a minimally elliptic singularity (see Definition 5.3.20). In this case, we have the following classification theorem (see [49, Theorem 4.8]):

Theorem 5.6.4 *Let* (X, x) *be a minimally elliptic singularity. Let* $\pi : \tilde{X} \to X$ *be the minimal resolution. Then*

$$\mathrm{Full}^1(\tilde{X}) \subset \left(\bigcup_{h \in H \setminus \{0\}} \mathrm{Pic}^{-s_h}(\tilde{X}) \right) \cup \left(\mathrm{Pic}^{-Z_{min}}(\tilde{X}) \setminus \{\mathcal{O}_{\tilde{X}}(-Z_{min})\} \right) \cup \{\mathcal{O}_{\tilde{X}}\}.$$

Moreover, suppose that $E_v \cong \mathbb{P}^1_{\mathbb{C}}$ *for any* $v \in V$. *Then,*

$$\mathrm{Full}^1(\tilde{X}) = \left(\bigcup_{h \in H \setminus \{0\}} \mathrm{Pic}^{-s_h}(\tilde{X}) \right) \cup \left(\mathrm{Pic}^{-Z_{min}}(\tilde{X}) \setminus \{\mathcal{O}_{\tilde{X}}(-Z_{min})\} \right) \cup \{\mathcal{O}_{\tilde{X}}\}.$$

Remark 5.6.5 By Theorem 5.6.4, the classification of rank one full sheaves cannot be described using only the first Chern classes. For example, consider the minimally elliptic singularity $X = \{x^2 + y^3 + z^7 = 0\} \subset \mathbb{C}^3$. In this case, $H = 0$, hence by Theorem 5.6.4

$$\mathrm{Full}^1(\tilde{X}) = \left(\mathrm{Pic}^{-Z_{min}}(\tilde{X}) \setminus \{\mathcal{O}_{\tilde{X}}(-Z_{min})\} \right) \cup \{\mathcal{O}_{\tilde{X}}\}.$$

Since Σ is an integer homology 3-sphere, we get $H^1(\Sigma, \mathbb{C}/\mathbb{Z}) = 0$. Therefore, the only one dimensional representation of $\pi_1(\Sigma)$ is the trivial one. Thus, $\mathcal{O}_{\tilde{X}}$ is the only flat full sheaf. The non-trivial non-flat full sheaves of rank one constructed in [9] belong to $\mathrm{Pic}^{-Z_{min}}(\tilde{X}) \setminus \{\mathcal{O}_{\tilde{X}}(-Z_{min})\}$.

Now that we already know how to classify all the rank one full shaves, the natural problem to study is to determine which of them are flat bundles, i.e., which of them are coming from the representation theory of the local fundamental group.

First, consider the scenario where the link is a rational homology sphere. Observe that every $\mathrm{Pic}^{\ell'}(\tilde{X})$ constitutes an affine space, specifically a $\mathrm{Pic}^0(\tilde{X}) = \mathbb{C}$ torsor. Consequently, the space of rank one full sheaves is one-dimensional. Nevertheless, $H_1(\Sigma, \mathbb{Z})$ is finite, which implies the set of flat bundles is discrete: for any $h \in H$, there is exactly one flat line bundle in $\mathrm{Pic}^{-s_h}(\tilde{X})$. For $h = 0$ is straightforward, it is just the trivial bundle. For $h \neq 0$, the election is known as *natural line bundles*, see [47, 48].

Definition 5.6.6 Let $\tilde{X}$ be the resolution of a normal surface singularity with a rational homology sphere link. Then, the morphism $c_1 : \mathrm{Pic}(\tilde{X}) \to L'$ has a section (group morphism). For any $\ell' \in L'$, there exists a unique line bundle $\mathcal{L}_{\ell'} \in \mathrm{Pic}^{\ell'}(\tilde{X})$, called a *natural line bundle*, with the following universal property: if $N\ell'$ is an integral cycle for some $N \in \mathbb{Z}_{>0}$, then $(\mathcal{L}_{\ell'})^{\otimes N} = \mathcal{O}_{\tilde{X}}(N\ell')$. This line bundle will be designated as $\mathcal{O}_{\tilde{X}}(\ell')$.

Theorem 5.6.7 ([49, Theorem 5.1]) *Let (X, x) be a minimally elliptic singularity such that its link is a rational homology sphere. Let $\pi : \tilde{X} \to X$ be the minimal resolution.*

Corresponding to $h = [c_1(\mathcal{L})] = 0$, among the rank one full sheaves in

$$\left(\mathrm{Pic}^{-Z_{min}}(\tilde{X}) \setminus \{\mathcal{O}_{\tilde{X}}(-Z_{min})\}\right) \cup \{\mathcal{O}_{\tilde{X}}\}$$

only the trivial sheaf $\mathcal{O}_{\tilde{X}}$ admits a flat connection.

Corresponding to $h = [c_1(\mathcal{L})] \neq 0$, among the rank one full sheaves in $\mathrm{Pic}^{-s_h}(\tilde{X})$ only $\mathcal{O}_{\tilde{X}}(-s_h)$ admits a flat connection.

If the singularity is a cusp singularity, i.e., a minimally elliptic singularity such that that their minimal resolution graph is cyclic (loop) with all $g(E_v) = 0$ [38, 45], we have the following theorem (see [49, Theorem 5.2]):

Theorem 5.6.8 *Let (X, x) be a cusp singularity. Let $\pi : \tilde{X} \to X$ be the minimal resolution. Then, any reflexive module of rank one admits a flat connection.*

Remark 5.6.9 Behnke conjectured (see [3]) that every reflexive module on a cusp singularity admits a flat connection. Later, Gustavsen and Ile [20] extended this conjecture for any log-canonical surface singularity. By Theorem 5.6.8 the conjecture is true for rank one reflexive modules. The conjecture is still an open problem.

The next natural step is to try to study another families of singularities. This will be done in the next section.

5.7 Classification of Special Reflexive Modules on Gorenstein Singularities

As we can see, by Theorem 5.6.4 it is impossible to fully generalize the McKay correspondence for non-rational Gorenstein singularities, i.e., this is because even for fix rank and fix Chern class we have a moduli and hence, the classification problem becomes more difficult. In this section, we will review some results about the classification, deformation theory and moduli space of reflexive modules.

By the results of the previous sections, for rational surface singularities the classification of reflexive modules is well understood (under the hypothesis of specialty on the modules). In the case of a non-rational surface singularity there are some results in concrete families of singularities: a complete classification and description of the Cohen-Macaulay representation type in the case of minimally elliptic singularities by Kahn [28] and Drozd et al. [10], some results on matrix factorizations by Laza et al. in [40] and Laza et al. [39].

Fernández de Bobadilla and the second author in [4] obtained several results for the case of normal Gorenstein singularities. We begin by fixing the notation of this section. By (X, x) we denote a complex analytic normal Gorenstein surface germ. We denote by $\pi : \tilde{X} \to X$ a resolution of singularities. The exceptional set is denoted by E, and its decomposition into irreducible components is $E = \cup_i E_i$.

Since (X, x) is Gorenstein, the canonical cycle is defined by $Z_k := \sum_i -q_i E_i$, where the q_i are the coefficients determined by the canonical divisor $K_{\tilde{X}}$ (see Sect. 5.3.5). We say that $\tilde{X}$ is *small with respect to the Gorenstein form* if Z_k is greater than or equal to 0.

Let (X, x) be a normal surface singularity and $\pi : \tilde{X} \to X$ be a resolution. Let M be a reflexive $\mathcal{O}_X$-module. A crucial theorem in the paper by Fernández de Bobadilla and Romano [4] is the computation of the first cohomology of full sheaves.

Theorem 5.7.1 ([4, Theorem 6.1]) *Let (X, x) be the germ of a normal Gorenstein surface singularity. Let M be a reflexive $\mathcal{O}_X$-module of rank r. Let $\pi : \tilde{X} \to X$ be a small resolution with respect to the Gorenstein form, let Z_k be the canonical cycle at $\tilde{X}$. Let $\mathcal{M}$ be the full $\mathcal{O}_{\tilde{X}}$-module associated to M. Then we have the equality*

$$\dim_{\mathbb{C}}(R^1\pi_*\mathcal{M}) = \dim_{\mathbb{C}}(R^1\pi_*\mathcal{M}^{\vee}) - [c_1(\mathcal{M})] \cdot [Z_k].$$

As a corollary of this theorem, they construct the *minimal adapted resolution* of M. Such resolution is characterized as the minimal resolution $\pi : \tilde{X} \to X$ such that the full $\mathcal{O}_{\tilde{X}}$-module $\mathcal{M}$ associated with M is generated by global sections.

Remark 5.7.2 By Proposition 5.4.13, for rational singularities the minimal resolution is the minimal adapted resolution. If the singularity is not rational, one has to show the existence of such resolution. This was done in [4, Proposition 5.1].

The minimal adapted resolution motives the following definition:

Definition 5.7.3 We say that M is a *special reflexive* module if the associated full sheaf $\mathcal{M}$ at the minimal adapted resolution satisfies $\dim_{\mathbb{C}}(R^1\pi_*\left(\mathcal{M}^{\vee}\right)) = rp_g$, where r is the rank of M.

Remark 5.7.4 If (X, x) is the germ of a Klenian singularity, then any reflexive module is special. Indeed, by Remark 5.7.2 the minimal resolution is the minimal adapted resolution. Since (X, x) is rational, then $p_g = 0$. Now, by Lemma 5.5.4 all full sheaves are special. Thus, any reflexive module is special. Therefore, the notion of special module is a generalization of the notion of special full sheaf.

Remark 5.7.5 Note that not every reflexive $\mathcal{O}_X$-module is special. For example, consider $X = V(x^3 + y^3 + z^3) \subset \mathbb{C}^3$ and $\pi : \tilde{X} \to X$ be the minimal resolution. Let M be the module of Zariski differentials. By [3, p. 216] the sheaf π^*M/tor is locally free, therefore

$$(\pi^*M)^{\vee\vee} \cong \pi^*M/\text{tor}.$$

Since π^*M/tor is generated by global sections, so is $(\pi^*M)^{\vee\vee}$. Therefore, $\pi : \tilde{X} \to X$ is the minimal adapted resolution associated to M. Now, suppose that M is special. By [4, Proposition 4.14] the canonical cycle Z_K should be 0. This is impossible since X is not a rational double point singularity. Thus, M is a non-special reflexive module.

One of the main results of the work of Fernández de Bobadilla and Romano is a determination of special reflexive modules in terms of its first Chern class.

Theorem 5.7.6 ([4, Theorem 7.6]) *Let (X, x) be a normal Gorenstein surface singularity. Let M be a special $\mathcal{O}_X$-module without free factors. Let $\pi : \tilde{X} \to X$ be the minimal resolution adapted to M, and $\mathcal{M}$ the full $\mathcal{O}_{\tilde{X}}$-module associated to M. The module $\mathcal{M}$ (and equivalently M) is determined by its first Chern class in $Pic(\tilde{X})$.*

Using the previous theorem, they classify the special reflexive modules over normal Gorenstein surface singularities as follows:

Theorem 5.7.7 ([4, Theorem 7.10]) *Let (X, x) be a normal Gorenstein surface singularity. Then, there exists a bijection between the following sets:*

1. *The set of special indecomposable non-trivial reflexive $\mathcal{O}_X$-modules up to isomorphism.*
2. *The set of irreducible divisors E over x, such that at any resolution of X where E appears, the Gorenstein form has not either zeros or poles along E.*

5.7.1 Deformation Theory of Full Shaves

In [4] Fernández de Bobadilla and Romano studied the deformation theory of full shaves. We will try to enunciate some of their main results, the reader may check [4, Section 8] for more details.

As before, let (X, x) be the germ of a normal surface singularity and let M be a reflexive $\mathcal{O}_X$-module.

A *deformation* of (X, M) over a base (S, s) consists of a flat deformation $\mathcal{X}$ of X over (S, s) together with an $\mathcal{O}_{\mathcal{X}}$-module which is flat over S and whose fibre over s is isomorphic to M.

Remark 5.7.8 Since reflexivity is an open property in flat families this is a good notion of deformations of reflexive sheaves. The reader may see [4, Lemma 8.3] for more details.

This definition of deformation leads to a deformation functor $\mathbf{Def_{X,M}}$. The functor of deformations fixing the base space X is a sub-functor.

On the other hand, fullness is not an open property in flat families. The correct definition of the deformation functor of full sheaves needs a cohomological condition. Let X and M be as before. Let $\pi : \tilde{X} \to X$ be a resolution and $\mathcal{M}$ be the full $\mathcal{O}_{\tilde{X}}$-module associated with M. A deformation of $(\tilde{X}, X, \mathcal{M})$ is formed by a very weak simultaneous resolution $\Pi : \tilde{X}d \to \mathcal{X}$ of a flat deformation $\mathcal{X}$ of X over S, and an $\mathcal{O}_{\tilde{X}d}$-module $\overline{\mathcal{M}}$ which is flat over S, whose specialization over s is isomorphic to $\mathcal{M}$ and such that $R^1\Pi_*\overline{\mathcal{M}}$ is flat over S. This leads to a deformation functor $\mathbf{FullDef_{\tilde{X},X,\mathcal{M}}}$. One has subfunctors fixing the underlying space and/or the resolution.

Remark 5.7.9 Although there is a bijection between full sheaves and reflexive sheaves (see Remark 5.4.16), one can see that the transformation $\mathbf{FullDef}_{\mathcal{M}} \to \mathbf{Def_M}$ is not an isomorphism of functors. The reader may see [4, Example 10.5]. Nevertheless, the push-forward functor Π_* defines a natural transformation from $\mathbf{FullDef_{\tilde{X},X,\mathcal{M}}}$ to $\mathbf{Def_{X,M}}$ ([4, Proposition 8.10].

Before stating the applications of the deformation theory of reflexive and full sheaves, we need a definition. First, let us recall that a surface singularity (X, x) is of *finite, tame or wild Cohen-Macaulay representation type* if there are at most finite, 1-dimensional or unbounded dimensional families of indecomposable Maximal Cohen-Macaulay $\mathcal{O}_{X,x}$-modules respectively (see [10]). With this definition we have the following theorem:

Theorem 5.7.10 ([4, Theorem 9.3]) *A Gorenstein surface singularity is of finite Cohen-Macaulay representation type if and only if it is a rational double point. Gorenstein surface singularities of tame Cohen-Macaulay representation type are precisely the log-canonical ones. The remaining Gorenstein surface singularities are of wild Cohen-Macaulay representation type.*

The second application is the construction of fine moduli spaces of special modules without free summands of prescribed graph and rank on Gorenstein normal surface singularities. Let $\mathcal{G}$ be the graph of a special reflexive $\mathcal{O}_X$-module on a Gorenstein normal surface singularity X. Ler r be a positive integer. Denote by $\mathbf{Mod}_{\mathcal{G}}^{\mathbf{r}}$ the moduli functor that parametrizes flat families of cohomologically special reflexive modules without free summands of rank r and graph $\mathcal{G}$, over normal base spaces. The main result is:

Theorem 5.7.11 ([4, Theorem 10.8]) *The functor* $\mathbf{Mod}_{\mathcal{G}}^{\mathbf{r}}$ *is represented by an algebraic variety.*

Moreover, in the proof of the above theorem we see that the variety representing the functor has a very nice geometric description: it parametrizes sequences of infinitely near points to x in the singularity X with a given combinatorial type.

Appendix: Irreducible Representations of Finite Subgroups of SL(2, ℂ)

In this appendix we include all the irreducible representations of the finite subgroups of $SL(2, \mathbb{C})$, since they are difficult to find in the literature.

Let $\langle 2, q, r\rangle$ denote the group given by the presentation,

$$\langle 2, q, r\rangle = \langle b, c \mid (bc)^2 = b^q = c^r\rangle$$

and let $\zeta_n = e^{\frac{2\pi i}{n}}$ be a n-th primitive root of unity.

Cyclic Groups: C_n, $n \geq 2$

We identify C_n with the n-th roots of unity. The group C_n has n irreducible representations $\rho_j, 0 \leq j \leq n-1$ given by $\rho_j(\zeta_n) = \zeta_n^j$.

Binary Dihedral Groups: $\mathsf{BD}_{2r} = \langle 2, 2, r\rangle$, $r \geq 2$

It has $r + 3$ irreducible representations. Four one-dimensional representations, denoted by ρ_j, $j = 0, \ldots, 3$ given by:

r even

$$\rho_0(b) = \rho_0(c) = 1, \qquad \rho_1(b) = -1, \rho_1(c) = 1,$$

$$\rho_2(b) = 1, \rho_2(c) = -1, \qquad \rho_3(b) = \rho_3(c) = -1.$$

r odd

$$\rho_0(b) = \rho_0(c) = 1, \qquad \rho_1(b) = -1, \rho_1(c) = 1,$$
$$\rho_2(b) = i, \rho_2(c) = -1, \qquad \rho_3(b) = -i, \rho_3(c) = -1.$$

And $r - 1$ two-dimensional representations, denoted by ρ_{3+t}, $1 \le t \le r - 1$, given for any r by:

$$\rho_{3+t}(b) = \begin{pmatrix} 0 & 1 \\ (-1)^t & 0 \end{pmatrix}, \qquad \rho_{3+t}(c) = \begin{pmatrix} \zeta_{2r}^t & 0 \\ 0 & \zeta_{2r}^{-t} \end{pmatrix},$$

where ρ_4 is the natural representation.

Binary Tetrahedral Group: **BT** = ⟨2, 3, 3⟩

It has 7 irreducible representations, denoted by ρ_j, $j = 0, \dots, 6$. There are three 1-dimensional representations:

$$\rho_0(b) = \rho_0(c) = 1, \quad \rho_1(b) = \zeta_3, \rho_1(c) = \zeta_3^2 \quad \text{and} \quad \rho_2(b) = \zeta_3^2, \rho_2(c) = \zeta_3.$$

There are three 2-dimensional representations:

$$\rho_3(b) = \begin{pmatrix} 0 & 1 \\ -1 & 1 \end{pmatrix}, \qquad \rho_3(c) = \begin{pmatrix} 0 & -\zeta_3 \\ \zeta_3^2 & 1 \end{pmatrix},$$
$$\rho_4(b) = \begin{pmatrix} -\zeta_3^2 & \zeta_3 \\ 0 & -1 \end{pmatrix}, \qquad \rho_4(c) = \begin{pmatrix} -1 & 0 \\ \zeta_3^2 & -\zeta_3 \end{pmatrix},$$
$$\rho_5(b) = \begin{pmatrix} -\zeta_3 & -\zeta_3^2 \\ 0 & -1 \end{pmatrix}, \qquad \rho_5(c) = \begin{pmatrix} 0 & 1 \\ -\zeta_3^2 & \zeta_3 \end{pmatrix}.$$

where ρ_3 is the natural representation. There is one 3-dimensional representation:

$$\rho_6(b) = \begin{pmatrix} -1 & -1 & -1 \\ 1 & 0 & 0 \\ 0 & 0 & 1 \end{pmatrix}, \qquad \rho_6(c) = \begin{pmatrix} -1 & -1 & -1 \\ 0 & 1 & 0 \\ 1 & 0 & 0 \end{pmatrix}.$$

Binary Octahedral Group: **BO** = ⟨2, 3, 4⟩

It has 8 irreducible representations, denoted by ρ_j, $j = 0, \dots, 7$. There are two 1-dimensional representations:

$$\rho_0(b) = \rho_0(c) = 1, \quad \rho_1(b) = 1, \rho_1(c) = -1.$$

There are three 2-dimensional representations:

$$\rho_2(b) = \begin{pmatrix} 0 & 1 \\ -1 & -1 \end{pmatrix}, \qquad \rho_2(c) = \begin{pmatrix} -1 & -1 \\ 0 & 1 \end{pmatrix},$$

$$\rho_3(b) = \frac{1}{2}\begin{pmatrix} 1+\sqrt{2}i & -i \\ -i & 1-\sqrt{2}i \end{pmatrix}, \qquad \rho_3(c) = \frac{1}{2}\begin{pmatrix} -\sqrt{2}-i & -1 \\ 1 & -\sqrt{2}+i \end{pmatrix},$$

$$\rho_4(b) = \begin{pmatrix} 0 & -i \\ -i & 1 \end{pmatrix}, \qquad \rho_4(c) = \begin{pmatrix} i & -\sqrt{2}i \\ -1 & \sqrt{2}-i \end{pmatrix},$$

where ρ_3 is the natural representation. There are two 3-dimensional representations:

$$\rho_5(b) = \begin{pmatrix} 1 & 0 & -1 \\ 0 & 0 & 1 \\ 0 & -1 & -1 \end{pmatrix}, \qquad \rho_5(c) = \begin{pmatrix} 0 & 1 & 1 \\ -1 & -1 & 0 \\ 0 & 1 & 0 \end{pmatrix},$$

$$\rho_6(b) = \begin{pmatrix} -1 & 0 & -1 \\ 1 & 1 & 0 \\ 1 & 0 & 0 \end{pmatrix}, \qquad \rho_6(c) = \begin{pmatrix} 0 & 0 & -1 \\ 1 & 0 & 1 \\ 0 & -1 & 1 \end{pmatrix}.$$

There is one 4-dimensional representation:

$$\rho_7(b) = \begin{pmatrix} -\zeta_3 & 0 & 0 & 0 \\ -1 & -1 & 0 & 0 \\ 0 & 0 & -1 & -1 \\ 0 & 0 & 0 & -\zeta_3^2 \end{pmatrix}, \qquad \rho_7(c) = \begin{pmatrix} 0 & 0 & \zeta_3^2 & \zeta_3^2 \\ 0 & 0 & 0 & \zeta_3 \\ 0 & -\zeta_3 & 0 & 0 \\ \zeta_3^2 & -1 & 0 & 0 \end{pmatrix}.$$

Binary Icosahedral Group: **BI** = ⟨2, 3, 5⟩

It has 9 irreducible representations, denoted by ρ_j, $j = 0, \ldots, 8$. There is one 1-dimensional representation ρ_0:

$$\rho_0(b) = \rho_0(c) = 1.$$

There are two 2-dimensional representations ρ_1, ρ_2, where ρ_1 is the natural representation:

$$\rho_1(b) = \frac{1}{2}\begin{pmatrix} 1+i & 1+i \\ -1+i & 1-i \end{pmatrix}, \qquad \rho_1(c) = \frac{1}{2}\begin{pmatrix} \frac{\sqrt{5}+1}{2}+\frac{\sqrt{5}-1}{2}i & 1 \\ -1 & \frac{\sqrt{5}+1}{2}-\frac{\sqrt{5}-1}{2}i \end{pmatrix},$$

$$\rho_2(b) = \begin{pmatrix} -\zeta_5^2 & -\zeta_5^2-\zeta_5^4 \\ -\zeta_5^4 & -\zeta_5-\zeta_5^3-\zeta_5^4 \end{pmatrix}, \qquad \rho_2(c) = \begin{pmatrix} 1 & -\zeta_5-\zeta_5^3-\zeta_5^4 \\ \zeta_5+\zeta_5^2+\zeta_5^4 & \zeta_5+\zeta_5^4 \end{pmatrix}.$$

There are two 3-dimensional representations ρ_3, ρ_4:

$$\rho_3(b) = \begin{pmatrix} 0 & 0 & 1 \\ 1 & 0 & 0 \\ 0 & 1 & 0 \end{pmatrix} \qquad \rho_3(c) = \begin{pmatrix} -\zeta_5^2-\zeta_5^3 & 1 & -\zeta_5^2-\zeta_5^3 \\ 0 & 0 & -1 \\ -1 & 0 & 0 \end{pmatrix},$$

$$\rho_4(b) = \begin{pmatrix} -\zeta_5^2-\zeta_5^3 & -\zeta_5^2-\zeta_5^3 & -1 \\ 0 & 0 & -1 \\ 1 & -\zeta_5^2-\zeta_5^3 & \zeta_5^2+\zeta_5^3 \end{pmatrix} \qquad \rho_4(c) = \begin{pmatrix} \zeta_5^2+\zeta_5^3 & -1 & -\zeta_5^2-\zeta_5^3 \\ 1 & 0 & 0 \\ 0 & 0 & 1 \end{pmatrix}.$$

There are two 4-dimensional representations ρ_5, ρ_6:

$$\rho_5(b) = \begin{pmatrix} \frac{1}{5}(2\zeta_5-\zeta_5^2+\zeta_5^3+3\zeta_5^4) & \frac{1}{5}(-4\zeta_5-3\zeta_5^2-2\zeta_5^3-\zeta_5^4) & \frac{1}{5}(-2\zeta_5+\zeta_5^2-6\zeta_5^3-3\zeta_5^4) & \frac{1}{5}(-6\zeta_5-7\zeta_5^2-3\zeta_5^3-4\zeta_5^4) \\ 0 & 0 & \zeta_5^3 & 0 \\ \frac{1}{5}(2\zeta_5+4\zeta_5^2+\zeta_5^3+3\zeta_5^4) & \frac{1}{5}(-4\zeta_5+2\zeta_5^2+3\zeta_5^3-\zeta_5^4) & \frac{1}{5}(-7\zeta_5-4\zeta_5^2-6\zeta_5^3-3\zeta_5^4) & \frac{1}{5}(-\zeta_5-2\zeta_5^2+2\zeta_5^3+\zeta_5^4) \\ -\zeta_5^3 & -\zeta_5^4 & -\zeta_5 & -\zeta_5^4 \end{pmatrix},$$

$$\rho_5(c) = \begin{pmatrix} 0 & 0 & \zeta_5^3 & 0 \\ \zeta_5^4 & 0 & 0 & 0 \\ 0 & 0 & 0 & \zeta_5^3 \\ -\zeta_5^4 & -1 & -\zeta_5^2 & -1 \end{pmatrix},$$

$$\rho_6(b) = \begin{pmatrix} -1 & 1 & \frac{1}{5}(-2\zeta_5-4\zeta_5^2-6\zeta_5^3-3\zeta_5^4) & \frac{1}{5}(\zeta_5+2\zeta_5^2-2\zeta_5^3-\zeta_5^4) \\ \zeta_5+\zeta_5^3+\zeta_5^4 & -\zeta_5-\zeta_5^4 & \frac{1}{5}(-3\zeta_5-\zeta_5^2-4\zeta_5^3-2\zeta_5^4) & \frac{1}{5}(4\zeta_5+3\zeta_5^2+2\zeta_5^3+6\zeta_5^4) \\ -\zeta_5-\zeta_5^4 & -\zeta_5^3+\zeta_5^4 & \zeta_5+\zeta_5^4 & -\zeta_5^4 \\ 1 & 0 & 0 & 0 \end{pmatrix},$$

$$\rho_6(c) = \begin{pmatrix} 0 & 0 & -\zeta_5^2 & 0 \\ \zeta_5+\zeta_5^3+\zeta_5^4 & -\zeta_5-\zeta_5^2-\zeta_5^4 & \frac{1}{5}(-4\zeta_5-3\zeta_5^2-2\zeta_5^3-6\zeta_5^4) & \frac{1}{5}(2\zeta_5-\zeta_5^2+\zeta_5^3+3\zeta_5^4) \\ -\zeta_5-\zeta_5^4 & -\zeta_5^3 & \frac{1}{5}(-\zeta_5-2\zeta_5^2-3\zeta_5^3+1\zeta_5^4) & \frac{1}{5}(-2\zeta_5+\zeta_5^2-\zeta_5^3-3\zeta_5^4) \\ \zeta_5^4 & -\zeta_5^2-\zeta_5^4 & \frac{1}{5}(-2\zeta_5+\zeta_5^2-\zeta_5^3-3\zeta_5^4) & \frac{1}{5}(\zeta_5+2\zeta_5^2-2\zeta_5^3-\zeta_5^4) \end{pmatrix}.$$

There is one 5-dimensional representation ρ_7:

$$\rho_7(b) = \begin{pmatrix} 0 & 0 & 0 & 0 & 1 \\ 1 & 0 & 0 & 0 & 0 \\ -1 & -1 & -1 & -1 & -1 \\ 0 & 0 & 1 & 0 & 0 \\ 0 & 1 & 0 & 0 & 0 \end{pmatrix}, \qquad \rho_7(c) = \begin{pmatrix} 0 & 1 & 0 & 0 & 0 \\ 0 & 0 & 0 & 1 & 0 \\ 0 & 0 & 0 & 0 & 1 \\ 0 & 0 & 1 & 0 & 0 \\ 1 & 0 & 0 & 0 & 0 \end{pmatrix}.$$

There is one 6-dimensional representation ρ_8:

$$\rho_8(b) = \begin{pmatrix} \zeta_5+\zeta_5^3+\zeta_5^4 & -\zeta_5^3 & -\zeta_5-\zeta_5^4 & 0 & \zeta_5^3 & \zeta_5+\zeta_5^4 \\ 0 & 0 & -\zeta_5^2 & 0 & 0 & 0 \\ \zeta_5+\zeta_5^3 & \zeta_5+\zeta_5^2+\zeta_5^4 & -\zeta_5-\zeta_5^3 & 0 & \zeta_5^3 & \zeta_5 \\ 0 & 0 & 0 & 0 & 1 & 0 \\ \zeta_5 & \zeta_5^2 & 1 & -\zeta_5^4 & -\zeta_5^2 & \zeta_5+\zeta_5^2+\zeta_5^3 \\ -\zeta_5-\zeta_5^3-\zeta_5^4 & \zeta_5+\zeta_5^2+\zeta_5^3 & \zeta_5+\zeta_5^4 & \zeta_5 & \zeta_5^4 & \zeta_5^2-\zeta_5^4 \end{pmatrix},$$

$$\rho_8(c) = \begin{pmatrix} 0 & 0 & 0 & 0 & 0 & 1 \\ 0 & 0 & \zeta_5^3 & 0 & 0 & 0 \\ -\zeta_5-\zeta_5^3 & -\zeta_5^4 & \zeta_5+\zeta_5^3+\zeta_5^4 & 0 & \zeta_5^4 & -\zeta_5-\zeta_5^3-\zeta_5^4 \\ 0 & 0 & 0 & 0 & -\zeta_5^4 & 0 \\ -2\zeta_5-\zeta_5^2-\zeta_5^3-2\zeta_5^4 & \zeta_5-\zeta_5^2 & \zeta_5+\zeta_5^2+\zeta_5^3+2\zeta_5^4 & \zeta_5^4 & \zeta_5^2 & 1 \\ 0 & 0 & 0 & -\zeta_5 & 0 & 0 \end{pmatrix}.$$

Acknowledgments We thank the referee for his/her comments which greatly improved the clarity of this chapter. The second author is partially supported by UNAM-DGAPA-PAPIIT IN112424 "Complex Kleinian groups".

References

1. Artin, M.: On isolated rational singularities of surfaces. Am. J. Math. **88**, 129–136 (1966). https://doi.org/10.2307/2373050
2. Artin, M., Verdier, J.L.: Reflexive modules over rational double points. Math. Ann. **270**(1), 79–82 (1985). https://doi.org/10.1007/BF01455531
3. Behnke, K.: On Auslander modules of normal surface singularities. Manuscr. Math. **66**(2), 205–223 (1989). https://doi.org/10.1007/BF02568491
4. de Bobadilla, J.F., Velazquez, A.R.: Reflexive modules on normal gorenstein stein surfaces, their deformations and moduli (2019). https://arxiv.org/abs/1812.06543
5. Brieskorn, E.: Rationale Singularitäten komplexer Flächen. Invent. Math. **4**, 336–358 (1968). https://doi.org/10.1007/BF01425318
6. Browder, F.E. (ed.): Mathematical developments arising from Hilbert problems. Proceedings of the symposium in pure mathematics of the American Mathematical Society, held at Northern Illinois University, DeKalb, Illinois, May 1974, *Proc. Symp. Pure Math.*, vol. 28. American Mathematical Society (AMS), Providence, RI (1976)
7. Bruns, W., Herzog, J.: Cohen-Macaulay rings., *Camb. Stud. Adv. Math.*, vol. 39, rev. ed. edn. Cambridge: Cambridge University Press (1998)
8. Cisneros Molina, J.L., Tosun, M.: Finite dimensional Lie algebras in singularities. In: Handbook of geometry and topology of singularities. I, pp. 541–587. Springer, Cham ([2020] ©2020). https://doi.org/10.1007/978-3-030-53061-7_10. https://doi-org.pbidi.unam.mx:2443/10.1007/978-3-030-53061-7_10
9. Dan, A., Romano-Velázquez, A.: Examples of non-flat bundles of rank one. C. R., Math., Acad. Sci. Paris **361**, 965–968 (2023). https://doi.org/10.5802/crmath.459
10. Drozd, Y., Greuel, G.M., Kashuba, I.: On cohen-macaulay modules on surface singularities. Moscow Mathematical Journal, v.3, 397–418, 742 (2003) **3** (2003). https://doi.org/10.17323/1609-4514-2003-3-2-397-418
11. Du Val, P.: On isolated singularities of surfaces which do not affect the conditions of adjunction. I, II, III. Proc. Camb. Philos. Soc. **30**, 453–465, 483–491 (1934)
12. Durfee, A.H.: Fifteen characterizations of rational double points and simple critical points. Enseign. Math. (2) **25**(1–2), 131–163 (1979)
13. Durfee, A.H.: Fifteen characterizations of rational double points and simple critical points. Enseign. Math. (2) **25**, 132–163 (1979)
14. Eisenbud, D.: Homological algebra on a complete intersection, with an application to group representations. Trans. Am. Math. Soc. **260**, 35–64 (1980). https://doi.org/10.2307/1999875
15. Eisenbud, D.: Commutative algebra. With a view toward algebraic geometry, *Grad. Texts Math.*, vol. 150. Berlin: Springer-Verlag (1995)
16. Esnault, H.: Reflexive modules on quotient surface singularities. J. Reine Angew. Math. **362**, 63–71 (1985). https://doi.org/10.1515/crll.1985.362.63
17. Esnault, H., Knörrer, H.: Reflexive modules over rational double points. Math. Ann. **272**, 545–548 (1985). https://doi.org/10.1007/BF01455865. https://eudml.org/doc/164030
18. Gonzalez-Sprinberg, G., Verdier, J.L.: Construction géométrique de la correspondance de McKay. Ann. Sci. École Norm. Sup. (4) **16**(3), 409–449 (1984) (1983)
19. Gustavsen, T.S., Ile, R.: Reflexive modules on normal surface singularities and representations of the local fundamental group. J. Pure Appl. Algebra **212**(4), 851–862 (2008). https://doi.org/10.1016/j.jpaa.2007.07.002
20. Gustavsen, T.S., Ile, R.: Representation theory for log-canonical surface singularities. Ann. Inst. Fourier **60**(2), 389–416 (2010). https://doi.org/10.5802/aif.2526
21. Gustavsen, T.S., Ile, R.: Deformations of rational surface singularities and reflexive modules with an application to flops. Adv. Math. **340**, 1108–1140 (2018). https://doi.org/10.1016/j.aim.2018.10.023
22. Hara, N., Sawada, T., Yasuda, T.: F-blowups of normal surface singularities. Algebra Number Theory **7**(3), 733–763 (2013). https://doi.org/10.2140/ant.2013.7.733

23. Hartshorne, R.: Residues and duality. Lecture notes of a seminar on the work of A. Grothendieck, given at Havard 1963/64. Appendix: Cohomology with supports and the construction of the $f^!$ functor by P. Deligne, *Lect. Notes Math.*, vol. 20. Springer, Cham (1966). https://doi.org/10.1007/BFb0080482
24. Hartshorne, R.: Stable reflexive sheaves. Math. Ann. **254**, 121–176 (1980). https://doi.org/10.1007/BF01467074
25. Ishii, S.: Introduction to singularities, 2nd edition edn. Tokyo: Springer (2018). https://doi.org/10.1007/978-4-431-56837-7
26. Iyama, O., Wemyss, M.: The classification of special Cohen-Macaulay modules. Math. Z. **265**(1), 41–83 (2010). https://doi.org/10.1007/s00209-009-0501-3
27. James, G., Liebeck, M.: Representations and Characters of groups. Cambridge University Press (2001)
28. Kahn, C.P.: Reflexive modules on minimally elliptic singularities. Math. Ann. **285**(1), 141–160 (1989). https://doi.org/10.1007/BF01442678
29. Kapustin, A., Li, Y.: D-branes in landau-ginzburg models and algebraic geometry. Journal of High Energy Physics **2003**(12), 005 (2004). https://doi.org/10.1088/1126-6708/2003/12/005
30. Khovanov, M., Rozansky, L.: Matrix factorizations and link homology. Fundam. Math. **199**(1), 1–91 (2008). https://doi.org/10.4064/fm199-1-1
31. Khovanov, M., Rozansky, L.: Matrix factorizations and link homology. II. Geom. Topol. **12**(3), 1387–1425 (2008). https://doi.org/10.2140/gt.2008.12.1387
32. Klein, F.: Lectures on the icosahedron and the solution of equations of the fifth degree. Dover (1956). English translation from the 1884 original
33. Klein, F.: Lectures on the icosahedron and the solution of equations of the fifth degree. Translated by George Gavin Morrice. With a new introduction and commentaries by Peter Slodowy. Translated by Lei Yang, *CTM, Class. Top. Math.*, vol. 5, reprint of the English translation of the 1884 German original edition edn. Beijing: Higher Education Press (2019)
34. Kollár, J.: Lectures on resolution of singularities, *Ann. Math. Stud.*, vol. 166. Princeton, NJ: Princeton University Press (2007). https://doi.org/10.1515/9781400827800
35. Kollár, J., Némethi, A.: Holomorphic arcs on singularities. Invent. Math. **200**(1), 97–147 (2015). https://doi.org/10.1007/s00222-014-0530-3
36. Lamotke, K.: Regular solids and isolated singularities. Advanced Lectures in Mathematics. Braunschweig - Wiesbaden: Friedr. Vieweg & Sohn. IX, 224 p. (1986)
37. Laufer, H.B.: Normal two-dimensional singularities, *Ann. Math. Stud.*, vol. 71. Princeton University Press, Princeton, NJ (1971). https://doi.org/10.1515/9781400881741
38. Laufer, H.B.: On minimally elliptic singularities. Am. J. Math. **99**, 1257–1295 (1977). https://doi.org/10.2307/2374025
39. Laza, R., O'Carroll, L., Popescu, D.: Maximal Cohen-Macaulay modules over $Y_1^3 + \cdots + Y_n^3$ with few generators. Math. Rep., Buchar. **3**(2), 177–185 (2001)
40. Laza, R., Pfister, G., Popescu, D.: Maximal Cohen-Macaulay modules over the cone of an elliptic curve. J. Algebra **253**(2), 209–236 (2002)
41. Lipman, J.: Rational singularities, with applications to algebraic surfaces and unique factorization. Publ. Math., Inst. Hautes Étud. Sci. **36**, 195–279 (1969). https://doi.org/10.1007/BF02684604
42. Matsumura, H.: Commutative algebra. 2nd ed, *Math. Lect. Note Ser.*, vol. 56. The Benjamin/Cummings Publishing Company, Reading, MA (1980)
43. McKay, J.: Graphs, singularities, and finite groups. In: The Santa Cruz Conference on Finite Groups, *Proc. Sympos. Pure Math.*, vol. 37, pp. 183–186. Amer. Math. Soc., Providence, R.I. (1980)
44. Mumford, D.: The topology of normal singularities of an algebraic surface and a criterion for simplicity. Publ. Math., Inst. Hautes Étud. Sci. **9**, 5–22 (1961). https://doi.org/10.1007/BF02698717. https://eudml.org/doc/103827
45. Némethi, A.: Five lectures on normal surface singularities. In: Low dimensional topology. Proceedings of the summer school, Budapest, Hungary, August 3–14, 1998, pp. 269–351. Budapest: János Bolyai Mathematical Society (1999)

46. Némethi, A.: On the Ozsváth-Szabó invariant of negative definite plumbed 3-manifolds. Geom. Topol. **9**, 991–1042 (2005). https://doi.org/10.2140/gt.2005.9.991
47. Némethi, A.: Graded roots and singularities. In: Singularities in geometry and topology. Proceedings of the Trieste singularity summer school and workshop, ICTP, Trieste, Italy, August 15–September 3, 2005, pp. 394–463. Hackensack, NJ: World Scientific (2007)
48. Némethi, A.: Normal surface singularities, *Ergeb. Math. Grenzgeb., 3. Folge*, vol. 74. Cham: Springer (2022). https://doi.org/10.1007/978-3-031-06753-2
49. Némethi, A., Romano-Velázquez, A.: The classification of reflexive modules of rank one over rational and minimally elliptic singularities (2023). https://arxiv.org/abs/2302.08768
50. Orlov, D.: Landau-Ginzburg models, D-branes and mirror symmetry. Mat. Contemp. **41**, 75–112 (2012)
51. Orlov, D.O.: Triangulated categories of singularities and D-branes in Landau-Ginzburg models. In: Algebraic geometry. Methods, relations, and applications. Collected papers. Dedicated to the memory of Andrei Nikolaevich Tyurin., pp. 227–248. Moscow: Maik Nauka/Interperiodica (2004)
52. Reid, M.: Chapters on algebraic surfaces. In: Complex algebraic geometry. Lectures of a summer program, Park City, UT, 1993, pp. 5–159. Providence, RI: American Mathematical Society (1997)
53. Reid, M.: McKay's correspondence. In: Séminaire Bourbaki. Volume 1999/2000. Exposés 865–879, pp. 53–72, ex. Paris: Société Mathématique de France (2002)
54. Riemenschneider, O.: Die Invarianten der endlichen Untergruppen von GL(2,C). Math. Z. **153**, 37–50 (1977). https://doi.org/10.1007/BF01214732
55. Riemenschneider, O.: Special representations and the two-dimensional McKay correspondence. Hokkaido Math. J. **32**(2), 317–333 (2003). https://doi.org/10.14492/hokmj/1350657526
56. Sakai, F.: Weil divisors on normal surfaces. Duke Math. J. **51**, 877–887 (1984). https://doi.org/10.1215/S0012-7094-84-05138-X
57. Wemyss, M.: The $GL(2, \mathbb{C})$ McKay correspondence. Math. Ann. **350**(3), 631–659 (2011). https://doi.org/10.1007/s00208-010-0572-9
58. Wunram, J.: Reflexive modules on quotient surface singularities. Math. Ann. **279**(4), 583–598 (1988). https://doi.org/10.1007/BF01458530
59. Yoshino, Y.: Cohen-Macaulay modules over Cohen-Macaulay rings, *Lond. Math. Soc. Lect. Note Ser.*, vol. 146. Cambridge (UK): Cambridge University Press (1990)

Chapter 6
Calculations with Characteristic Cycles

David B. Massey

Contents

Abstract We provide many results on how to begin with the characteristic cycle (possibly in a graded, enriched form) of a given complex of sheaves and then calculate the characteristic cycles of related complexes, such as the nearby cycles and vanishing cycles of the original complex.

6.1 Introduction

Throughout this chapter, we must assume that the reader is familiar with basic aspects of the derived category of bounded constructible complexes of sheaves, perverse sheaves, and the nearby and vanishing cycles. Good references for the theory are [8, 17], and [40].

D. B. Massey (✉)
Department of Mathematics, Northeastern University, Boston, MA, USA
e-mail: d.massey@northeastern.edu

© The Author(s), under exclusive license to Springer Nature Switzerland AG 2026

J. L. Cisneros-Molina et al. (eds.), *Handbook of Geometry and Topology of Singularities VIII*, https://doi.org/10.1007/978-3-031-99571-2_6

Suppose that $\mathcal{U}$ is an open subset of $\mathbb{C}^{n+1}$ and, for convenience, assume that $\mathbf{0} \in \mathcal{U}$. Let $f : (\mathcal{U}, \mathbf{0}) \to (\mathbb{C}, 0)$ be a complex analytic function, which is nowhere locally constant, and consider the hypersurface $V(f) = f^{-1}(0)$.

There are several spaces which are typically studied when investigating the local, embedded topology of $V(f)$ at $\mathbf{0}$: the complement $V(f)$ in a small ball around $\mathbf{0}$, the real link of $V(f)$ at $\mathbf{0}$, the complex link of $V(f)$ at $\mathbf{0}$, and the Milnor fiber of f at $\mathbf{0}$.

If $\mathbf{0}$ is a non-isolated critical point of f, then it is important to look at each of these spaces at every critical point, and to know how the topology of $V(f)$ at these nearby critical points is related to the topology at $\mathbf{0}$. This is a question of how local data patches together to give global data. Thus, complexes of sheaves of $\mathbb{Z}$-modules naturally enter the picture. And so, it is beneficial to look at the constant sheaf on the complement of $V(f)$, the Verdier dual of the constant sheaf on $V(f)$, the vanishing cycles of the constant sheaf on $V(f)$ along a generic linear form, and the nearby and vanishing cycles of the constant sheaf on $\mathcal{U}$ along f.

But complexes of sheaves contain so much data that they are not amenable to computation. Hence, researchers consider other objects, which do not contain as much data, but which can be calculated algebraically/geometrically. The characteristic cycle and micro-support of a complex of sheaves are prominent examples of two such objects. We will define these carefully later, but we wish to describe them briefly here.

Let X be a complex analytic space inside $\mathcal{U}$. Let $\mathfrak{S}$ be a complex analytic Whitney stratification of X, with connected strata. We let $\overline{T^*_S\mathcal{U}}$ denote the closure of the conormal variety of S in $\mathcal{U}$, that is, the closure of the set of $(\mathbf{p}, \eta) \in T^*\mathcal{U}$ such that $\mathbf{p} \in S$ and $\eta(T_\mathbf{p}S) \equiv 0$.

Now, let $\mathbf{F}^\bullet$ be a bounded complex of sheaves of $\mathbb{Z}$-modules, which is constructible with respect to $\mathfrak{S}$. Then, as described by Goresky and MacPherson [15], to each stratum S in $\mathfrak{S}$, there are an associated normal slice $\mathbb{N}_S$ and complex link $\mathbb{L}_S$. The isomorphism-types of the hypercohomology modules $\mathbb{H}^*(\mathbb{N}_S, \mathbb{L}_S; \mathbf{F}^\bullet)$ are independent of the choices made in defining the normal slice and complex link; these are the *Morse modules* of S, with respect to $\mathbf{F}^\bullet$. We let $d_S := \dim S$, and $m^k_S(\mathbf{F}^\bullet) := \mathbb{H}^{k-d_S}(\mathbb{N}_S, \mathbb{L}_S; \mathbf{F}^\bullet)$.

The Morse modules $m^k_S(\mathbf{F}^\bullet)$ tell one how the cohomology of $\mathbf{F}^\bullet$ changes as one moves through the stratum S. We say that a stratum S is $\mathbf{F}^\bullet$*-visible* provided that $m^*_S(\mathbf{F}^\bullet) \neq 0$, i.e., provided that there exists k such that $m^k_S(\mathbf{F}^\bullet) \neq 0$. The $\mathbf{F}^\bullet$-invisible strata, the strata which are **not** $\mathbf{F}^\bullet$-visible, are, in a sense, strata that could be omitted from the stratification as far as the cohomology of $\mathbf{F}^\bullet$ is concerned.

The union of the closures of conormal varieties to $\mathbf{F}^\bullet$-visible strata is the *microsupport*, $\mathrm{SS}(\mathbf{F}^\bullet)$, of $\mathbf{F}^\bullet$, as defined by Kashiwara and Schapira in [17], i.e.,

$$\mathrm{SS}(\mathbf{F}^\bullet) := \bigcup_{m^*_S(\mathbf{F}^\bullet)\neq 0} \overline{T^*_S\mathcal{U}}.$$

The microsupport is microlocal data which encodes the directions in which $\mathbf{F}^\bullet$ changes at each point.

Let $c_S(\mathbf{F}^\bullet)$ be the Euler characteristic of the Morse modules, i.e., let $c_S(\mathbf{F}^\bullet) := \sum_{k\in\mathbb{Z}}(-1)^k \operatorname{rank}(m_S^k(\mathbf{F}^\bullet))$. The *characteristic cycle of* $\mathbf{F}^\bullet$ is the cycle

$$\mathrm{CC}(\mathbf{F}^\bullet) = \sum_{S\in\mathfrak{S}} c_S(\mathbf{F}^\bullet)\left[\overline{T_S^*\mathcal{U}}\right],$$

where the square brackets indicate that we are considering the conormal varieties as cycles. Both the microsupport and the characteristic cycle are independent of the stratification $\mathfrak{S}$, and so are intrinsic to the complex $\mathbf{F}^\bullet$ (and the ambient space $\mathcal{U}$).

But the microsupport and the characteristic cycle throw away a large amount of the Morse module data. Why not take the conormal varieties and, instead of weighting them with Euler characteristic data from the Morse modules, consider a formal sum in which the coefficients are the Morse modules themselves?

Thus, we define $\mathrm{gecc}^\bullet(\mathbf{F}^\bullet)$, the *graded, enriched characteristic cycle of* $\mathbf{F}^\bullet$, by defining a "cycle" in degree k as a formal sum of modules times conormal varieties:

$$\mathrm{gecc}^k(\mathbf{F}^\bullet) := \sum_{S\in\mathfrak{S}} m_S^k(\mathbf{F}^\bullet)\left[\overline{T_S^*\mathcal{U}}\right] = \sum_{S\in\mathfrak{S}} H^{k-d_S}(\mathbb{N}_S,\mathbb{L}_S;\mathbf{F}^\bullet)\left[\overline{T_S^*\mathcal{U}}\right],$$

and we use an intersection theory which is a very mild extension of the theory of proper intersection of cycles (**not** cycle classes) inside a complex manifold, as described in [12].

There have been numerous other works on the computations of characteristic cycles: notably, the papers of Ginsburg [13], Briancon et al. [6], and Parusiński and Pragacz [37], plus portions of the books of Kashiwara and Schapira [17] and of Schürmann [40]. However, there are several advantages to the techniques and results presented here.

- The intersection theory that we use is that of properly intersecting cycles inside a complex manifold. For such intersections, there are well-defined intersection cycles, not cycle **classes** (see [12], Chapter 8).
  The fact that we have intersection cycles with fixed underlying analytic sets makes calculations and formulas much easier and algorithmic, and, typically, the amount of genericity that we need in statements is merely that the intersections are proper, which is a relatively simple thing to check.
  It is an interesting aspect of the theory that, using only enough genericity to obtain proper intersections does not yield objects which are as generic as possible, and it is precisely this lack of really generic genericity that makes formulas work so well.
- While we use an easy intersection theory, we use modules in various degrees for the coefficients of our cycles. This graded, enriched intersection theory adds essentially no difficulty to computations, and yet, almost magically, yields results on the levels of modules, instead of merely giving numerical results.

- In addition to the notion of graded, enriched characteristic cycles, our primary new device involved in the calculus of gecc's is the *graded, enriched relative polar curve* (see [35] and Sect. 6.5 of this chapter). This is a substantial generalization the now-classic relative polar curve introduced by Hamm, Lê, and Teissier in 1973 in [16, 19, 42], and [20].
  By giving the "correct" definition of the general polar curve, we are not required to make choices as generically as did Hamm, Lê, and Teissier and, thus, once again, the genericity hypotheses that we need in theorems are simply that certain intersections are proper.
- Our calculation of the gecc of the vanishing cycles uses a generalization of the Lê cycle algorithm that we developed in [28], and so really does allow for explicit calculations in many examples.
- The relative local Euler obstruction, $\mathrm{Eu}_{\mathbf{p}}\, f$, was introduced in [4]. Our contribution to [4] was to give a definition in terms of complexes of sheaves and vanishing cycles—this "definition" works even for non-isolated singularities/critical points. In Sect. 6.3, we recall this formal definition and give a number of important properties. Most of these results appeared in our recent article [36]; nonetheless, we give the proofs here also because they are simple and show the power of using characteristic complexes.

While a large part of the chapter is a unification and summary of some of our earlier work, many of the results here have been sharpened with optimal hypotheses, and many examples are given. Specific new content in this chapter includes:

- In Theorems 6.6.1 and 6.6.2 of Sect. 6.6, we recall our previous formulas involving $\mathrm{gecc}^{\bullet}(\psi_f[-1]\mathbf{F}^{\bullet})$. However, the example that we calculate is new, as is the easy—but interesting—corollary, Corollary 6.6.4, that the shifted nearby cycles of numerical complexes (see Definition 6.2.12) are numerical.
- In Sect. 6.7, Theorem 6.7.1 is an enriched cycle version of one of our earlier results, which relates to the calculation of the graded enriched characteristic cycle of the complement of a hypersurface. However, Corollary 6.7.2, Corollary 6.7.4, Theorem 6.7.6, and Corollary 6.7.8 are new. Furthermore, as we show in Example 6.7.11, Corollary 6.7.10 is a new generalization of the classically-known formula for the number of spheres in the homotopy-type of the complex link of an affine hypersurface.
- In Sect. 6.8, we recall our earlier result, Theorem 6.8.2, and then show, in Corollary 6.8.4, that the shifted vanishing cycles of numerical complexes are numerical. Theorem 6.8.7 is new, but follows quickly from some of our previous results. In Remark 6.8.8, we discuss a general method for calculating the graded enriched characteristic of the vanishing cycles; this uses a cycle approach which is a generalization of our method for calculating Lê cycles (see [28]). In Example 6.8.9, we give an example of how the method of Remark 6.8.8 actually works in practice.

We reiterate that, throughout this chapter, it is important that, when we state that a choice must be made "generically", we actually give effective means of checking that the choice is generic enough. This makes the results much more useful when applying them to specific examples, and we give sample calculations to illustrate this point.

6.2 The Characteristic Cycle and the Graded, Enriched Characteristic Cycle

Throughout this chapter, we fix a base ring R that is a regular, Noetherian ring with finite Krull dimension (e.g., $\mathbb{Z}$, $\mathbb{Q}$, or $\mathbb{C}$). This implies that every finitely-generated R-module has finite projective dimension (in fact, it implies that the projective dimension of the module is at most $\dim R$).

We let $\mathcal{U}$ be an open neighborhood of the origin of $\mathbb{C}^{n+1}$, and let X be a closed, analytic subset of $\mathcal{U}$. We let $\mathbf{z} := (z_0, \dots, z_n)$ be coordinates on $\mathcal{U}$. Having fixed the coordinates, we identify the cotangent space $T^*\mathcal{U}$ with $\mathcal{U} \times \mathbb{C}^{n+1}$ by mapping $(\mathbf{p}, w_0 d_\mathbf{p} z_0 + \cdots + w_n d_\mathbf{p} z_n)$ to $(\mathbf{p}, (w_0, \dots, w_n))$. Let $\pi : T^*\mathcal{U} \to \mathcal{U}$ denote the projection.

Let $\mathfrak{S}$ be a complex analytic Whitney stratification of X, with connected strata. Let $\mathbf{F}^\bullet$ be a bounded complex of sheaves of R-modules on X, which is constructible with respect to $\mathfrak{S}$. For each $S \in \mathfrak{S}$, we let $d_S := \dim S$, and let $(\mathbb{N}_S, \mathbb{L}_S)$ denote *complex Morse data for S in X*, consisting of a normal slice and complex link of S in X; see, for instance, [15] or [33].

A general reference for the remainder of this section is [34].

Definition 6.2.1 For each $S \in \mathfrak{S}$ and each integer k, the isomorphism-type of the module $m^k_S(\mathbf{F}^\bullet) := \mathbb{H}^{k-d_S}(\mathbb{N}_S, \mathbb{L}_S; \mathbf{F}^\bullet)$ is independent of the choice of $(\mathbb{N}_S, \mathbb{L}_S)$; we refer to $m^k_S(\mathbf{F}^\bullet)$ as the *degree k Morse module of S with respect to* $\mathbf{F}^\bullet$.

Remark 6.2.2 The shift by d_S above is present so that perverse sheaves can have non-zero Morse modules in only degree 0.

We also remark that, up to isomorphism, $m^k_S(\mathbf{F}^\bullet)$ can be obtained in terms of vanishing cycles. To accomplish this, select any point $\mathbf{p} \in S$. Consider an analytic function $\tilde{g} : (\mathcal{U}', \mathbf{p}) \to (\mathbb{C}, 0)$ on some open neighborhood of $\mathbf{p}$ in $\mathcal{U}$ such that $d_\mathbf{p}\tilde{g}$ is a nondegenerate covector (in the sense of [15]), and such that $\mathbf{p}$ is a (complex) nondegenerate critical point of $\tilde{g}_{|_{\mathcal{U}' \cap S}}$. Let $g := \tilde{g}_{|_{\mathcal{U}' \cap X}}$. Then, $m^k_S(\mathbf{F}^\bullet)$ is isomorphic to the stalk cohomology $H^k(\phi_g[-1]\mathbf{F}^\bullet)_\mathbf{p}$.

Note that, if $\mathbf{0}$ is a point-stratum, then $m^k_\mathbf{0}(\mathbf{F}^\bullet) \cong H^k(\phi_\mathcal{L}[-1]\mathbf{F}^\bullet)_\mathbf{0}$, where $\mathcal{L}$ is the restriction to X of a generic linear form $\tilde{\mathcal{L}}$.

For any analytic submanifold $M \subseteq \mathcal{U}$, we denote the conormal space

$$\{(\mathbf{p}, \omega) \in T^*\mathcal{U} \mid \omega(T_\mathbf{p}M) \equiv 0\}$$

by $T^*_M\mathcal{U}$, and will typically be interested in its closure $\overline{T^*_M\mathcal{U}}$ in $T^*\mathcal{U}$.

Definition 6.2.3 Suppose that R is an integral domain.

Define $c_S(\mathbf{F}^\bullet) := \sum_{k\in\mathbb{Z}}(-1)^k \operatorname{rank}(m_S^k(\mathbf{F}^\bullet))$, and define the **characteristic cycle of $\mathbf{F}^\bullet$ (in $T^*\mathcal{U}$)** to be the analytic cycle

$$\mathrm{CC}(\mathbf{F}^\bullet) = \sum_{S\in\mathfrak{S}} c_S(\mathbf{F}^\bullet)\left[\overline{T_S^*\mathcal{U}}\right].$$

We write $c_{\mathbf{0}}(\mathbf{F}^\bullet)$ in place of $c_{\{\mathbf{0}\}}(\mathbf{F}^\bullet)$, and let $c_{\mathbf{0}}(\mathbf{F}^\bullet) = 0$ if $\{\mathbf{0}\} \notin \mathfrak{S}$.

The underlying set $|\mathrm{CC}(\mathbf{F}^\bullet)| = \bigcup_{c_S(\mathbf{F}^\bullet)\neq 0} \overline{T_S^*\mathcal{U}}$ is the **characteristic variety of $\mathbf{F}^\bullet$ (in $T^*\mathcal{U}$)**.

Throughout this chapter, whenever we refer to $c_S(\mathbf{F}^\bullet)$ or $\mathrm{CC}(\mathbf{F}^\bullet)$, we assume that the base ring is an integral domain, even if we do not explicitly state this.

Remark 6.2.4 We should remark that there are various conventions for the signs involved in the characteristic cycle. In fact, our definition above uses a different convention than we used in our earlier works. Our definition above is the most desirable considering the graded, enriched characteristic cycle that we will define below. In hopes of avoiding confusion with our earlier work, we have also changed our notation for the characteristic cycle.

Note that, using the above convention, the characteristic cycle is not changed by extending $\mathbf{F}^\bullet$ by zero to all of $\mathcal{U}$.

We give some basic, easy properties of the characteristic cycle concerning how they work with shifting, constant sheaves, distinguished triangles, and the Verdier dual $\mathcal{D}\mathbf{F}^\bullet$. The proofs of the items below are not difficult, we leave them to the reader.

Proposition 6.2.5

1. $\mathrm{CC}(\mathbf{F}^\bullet[j]) = (-1)^j\mathrm{CC}(\mathbf{F}^\bullet)$.
2. *If X is a pure-dimensional (e.g., connected) complex manifold, then*

$$\mathrm{CC}(\mathbf{R}_X^\bullet) = (-1)^{\dim X}[T_X^*\mathcal{U}],$$

   i.e., $\mathrm{CC}(\mathbf{R}_X^\bullet[\dim X]) = [T_X^*\mathcal{U}]$.
3. *If* $\mathbf{A}^\bullet \to \mathbf{B}^\bullet \to \mathbf{C}^\bullet \xrightarrow{[1]} \mathbf{A}^\bullet$ *is a distinguished triangle in* $D_c^b(X)$*, then*

$$\mathrm{CC}(\mathbf{B}^\bullet) = \mathrm{CC}(\mathbf{A}^\bullet) + \mathrm{CC}(\mathbf{C}^\bullet).$$

4. $\mathrm{CC}(\mathbf{F}^\bullet) = \mathrm{CC}(\mathcal{D}\mathbf{F}^\bullet)$.

Exercise 6.2.6 Prove the four items in Proposition 6.2.5.

For calculating the characteristic cycle of the constant sheaf, the following is very useful:

Corollary 6.2.7 *Suppose that Y and Z are closed analytic subsets of X such that $X = Y \cup Z$. Then,*

$$\mathrm{CC}(\mathbf{R}^\bullet_X) = \mathrm{CC}(\mathbf{R}^\bullet_Y) + \mathrm{CC}(\mathbf{R}^\bullet_Z) - \mathrm{CC}(\mathbf{R}^\bullet_{Y\cap Z}).$$

Proof Let $j : Y \hookrightarrow X$, $k : Z \hookrightarrow X$, and $l : Y \cap Z \hookrightarrow X$ denote the respective inclusions. Then, there is a canonical distinguished triangle

$$\mathbf{R}^\bullet_X \to j_*j^*\mathbf{R}^\bullet_X \oplus k_*k^*\mathbf{R}^\bullet_X \to l_*l^*\mathbf{R}^\bullet_X \xrightarrow{[1]} \mathbf{R}^\bullet_X.$$

As the pull-back of the constant sheaf is the constant sheaf, and as the characteristic cycle is unaffected by extensions by zero, the desired conclusion follows immediately from Item 3 of Proposition 6.2.5. □

We also have the following result.

Proposition 6.2.8 *Suppose that $\mathbf{A}^\bullet$ and $\mathbf{B}^\bullet$ are bounded, constructible complexes of sheaves on the d-dimensional analytic space X. Suppose that $\mathfrak{S}$ is a stratification with respect to which both $\mathbf{A}^\bullet$ and $\mathbf{B}^\bullet$ are constructible (which always exists).*

For $0 \le k \le d$, let $\mathrm{CC}_{\ge k}(\mathbf{A}^\bullet)$ and $\mathrm{CC}_{\ge k}(\mathbf{B}^\bullet)$ denote the portions of the characteristic cycles which lie over closures of strata of dimension greater than or equal to k.

Then, $\mathrm{CC}_{\ge k}(\mathbf{A}^\bullet) = \mathrm{CC}_{\ge k}(\mathbf{B}^\bullet)$ if and only if, for all $S \in \mathfrak{S}$ such that $\dim S \ge k$, for all $\mathbf{p} \in S$, there is an equality of Euler characteristics of the stalk cohomology $\chi(\mathbf{A}^\bullet)_\mathbf{p} = \chi(\mathbf{B}^\bullet)_\mathbf{p}$.

In particular, $\mathrm{CC}(\mathbf{A}^\bullet) = \mathrm{CC}(\mathbf{B}^\bullet)$ if and only if, for all $\mathbf{p} \in X$, $\chi(\mathbf{A}^\bullet)_\mathbf{p} = \chi(\mathbf{B}^\bullet)_\mathbf{p}$.

Proof The proof is by downward induction on k. Certainly the result is trivial for $k = d$. Now suppose that $k_0 \ge 0$ and that the statement is true for all k such that $k_0 + 1 \le k \le d$; we wish to show that the statement is true for $k = k_0$.

Let $S_0 \in \mathfrak{S}$ be a stratum of dimension k_0, and let $\mathbf{p}_0 \in S_0$. For each stratum S of dimension greater than or equal to $k_0 + 1$, let $\mathbf{p}_S$ denote a point of S. If we let be $\mathbf{A}^\bullet$ or $\mathbf{B}^\bullet$, then

$$c_{S_0}(\mathbf{F}^\bullet) := \chi(\mathbb{N}_{S_0}, \mathbb{L}_{S_0}; \mathbf{F}^\bullet[-k_0]) = \chi(\mathbb{N}_{S_0}; \mathbf{F}^\bullet[-k_0]) - \chi(\mathbb{L}_{S_0}; \mathbf{F}^\bullet[-k_0]) =$$

$$(-1)^{k_0}\Big\{\chi(\mathbf{F}^\bullet)_{\mathbf{p}_0} - \sum_{S, \dim S \ge k_0+1} \chi\big(\mathbb{L}_{S_0} \cap S\big) \cdot \chi(\mathbf{F}^\bullet)_{\mathbf{p}_S}\Big\}.$$

Note that our inductive hypothesis implies that the summation on the right above is the same whether $\mathbf{F}^\bullet$ equals $\mathbf{A}^\bullet$ or $\mathbf{B}^\bullet$.

Therefore, $c_{S_0}(\mathbf{A}^\bullet) = c_{S_0}(\mathbf{B}^\bullet)$ if and only if $\chi(\mathbf{A}^\bullet)_{\mathbf{p}_0} = \chi(\mathbf{B}^\bullet)_{\mathbf{p}_0}$, and we are finished. □

Corollary 6.2.9 *Suppose that* $\mathrm{CC}(\mathbf{A}^\bullet) = \mathrm{CC}(\mathbf{B}^\bullet)$ *and that we have a complex analytic* $f : X \to \mathbb{C}$. *Then, for all,* $\mathbf{p} \in X$,

$$\chi(\psi_{f-f(\mathbf{p})}\mathbf{A}^\bullet)_\mathbf{p} = \chi(\psi_{f-f(\mathbf{p})}\mathbf{B}^\bullet)_\mathbf{p} \quad \text{and} \quad \chi(\phi_{f-f(\mathbf{p})}\mathbf{A}^\bullet)_\mathbf{p} = \chi(\phi_{f-f(\mathbf{p})}\mathbf{B}^\bullet)_\mathbf{p}.$$

Proof For convenience, we shall assume that $f(\mathbf{p}) = 0$. Let $F_{f,\mathbf{p}}$ denote the Milnor fiber of f at $\mathbf{p}$. Once again, choose a Whitney stratification $\mathfrak{S}$ with respect to which both $\mathbf{A}^\bullet$ and $\mathbf{B}^\bullet$ are constructible and, for each $S \in \mathfrak{S}$, select a $\mathbf{p}_S \in S$.

Then,

$$\chi(\psi_f\mathbf{A}^\bullet)_\mathbf{p} = \chi(F_{f,\mathbf{p}}; \mathbf{A}^\bullet) = \sum_{S\in\mathfrak{S}} \chi(F_{f,\mathbf{p}} \cap S) \cdot \chi(\mathbf{A}^\bullet)_{\mathbf{p}_S}.$$

By the proposition, this also equals $\chi(\psi_f\mathbf{B}^\bullet)_\mathbf{p}$.

The result about the vanishing cycles follows immediately since

$$\chi(\phi_f\mathbf{A}^\bullet)_\mathbf{p} = \chi(\psi_f\mathbf{A}^\bullet)_\mathbf{p} - \chi(\mathbf{A}^\bullet)_\mathbf{p} = \chi(\psi_f\mathbf{B}^\bullet)_\mathbf{p} - \chi(\mathbf{B}^\bullet)_\mathbf{p} = \chi(\phi_f\mathbf{B}^\bullet)_\mathbf{p}.$$

□

The characteristic cycle uses only the Euler characteristic information from the Morse data to strata. While this makes many calculations far easier, it disposes of a great deal of cohomological data. Hence, we define a formal graded "cycle" with module coefficients (actually, isomorphism classes of modules); we shall discuss such "enriched" cycles more generally in Sect. 6.4.

Definition 6.2.10 The **graded, enriched characteristic cycle of $\mathbf{F}^\bullet$ in the cotangent bundle** $T^*\mathcal{U}$ is defined in degree k to be

$$\mathrm{gecc}^k(\mathbf{F}^\bullet) := \sum_{S\in\mathfrak{S}} m_S^k(\mathbf{F}^\bullet)\big[\,\overline{T_S^*\mathcal{U}}\,\big] = \sum_{S\in\mathfrak{S}} H^{k-d_S}(\mathbb{N}_S, \mathbb{L}_S; \mathbf{F}^\bullet)\big[\,\overline{T_S^*\mathcal{U}}\,\big].$$

The underlying set $\left|\mathrm{gecc}^k(\mathbf{F}^\bullet)\right|$ is $\bigcup_{m_S^k(\mathbf{F}^\bullet)\neq 0} \overline{T_S^*\mathcal{U}}$.

The total underlying set $|\mathrm{gecc}^\bullet(\mathbf{F}^\bullet)| := \bigcup_{m_S^*(\mathbf{F}^\bullet)\neq 0} \overline{T_S^*\mathcal{U}}$ is the **microsupport of $\mathbf{F}^\bullet$ (in $T^*\mathcal{U}$)**, and is denoted by $\mathrm{SS}(\mathbf{F}^\bullet)$ (see [17]).

We have the following basic properties.

Proposition 6.2.11

1. $\mathrm{gecc}^k(\mathbf{F}^\bullet[j]) = \mathrm{gecc}^{k+j}(\mathbf{F}^\bullet)$.
2. $\mathrm{supp}\,\mathbf{F}^\bullet = \pi(\mathrm{SS}(\mathbf{F}^\bullet))$.
3. If X is a pure-dimensional (e.g., connected) complex manifold, then

$$\mathrm{gecc}^k(\mathbf{R}_X^\bullet[\dim X]) = \begin{cases} 0, & \text{if } k \neq 0; \\ R[T_X^*\mathcal{U}], & \text{if } k = 0. \end{cases}$$

4. *More generally,* $\mathbf{F}^\bullet$ *is a perverse sheaf if and only if* $\mathrm{gecc}^\bullet(\mathbf{F}^\bullet)$ *is concentrated in degree* 0*, i.e.,* $\mathrm{gecc}^k(\mathbf{F}^\bullet) = 0$ *if* $k \neq 0$*. In particular, the characteristic cycle of a perverse sheaf has only non-negative coefficients.*
5. *Let* ${}^\mu H^k$ *denote the degree k (middle perversity) perverse cohomology (see[17], Section 10.3). Then,* ${}^\mu H^k(\mathbf{F}^\bullet)$ *is a perverse sheaf,* $m_S^k(\mathbf{F}^\bullet) \cong m_S^0({}^\mu H^k(\mathbf{F}^\bullet))$*, and*
$$\mathrm{gecc}^k(\mathbf{F}^\bullet) = \mathrm{gecc}^0\big({}^\mu H^k(\mathbf{F}^\bullet)\big).$$
6. *Suppose that* $\mathbf{A}^\bullet \to \mathbf{B}^\bullet \to \mathbf{C}^\bullet \xrightarrow{[1]} \mathbf{A}^\bullet$ *is a distinguished triangle in* $D_c^b(X)$*. Then, for all k,* $|\mathrm{gecc}^k(\mathbf{B}^\bullet)| \subseteq |\mathrm{gecc}^k(\mathbf{A}^\bullet)| \cup |\mathrm{gecc}^k(\mathbf{C}^\bullet)|$ *and, consequently, the microsupport of each complex is contained in the union of the microsupports of the other two.*
7. *If R is a Dedekind domain, then, for all k and for all* $S \in \mathfrak{S}$*,*
$$m_S^{-k}(\mathcal{D}\mathbf{F}^\bullet) \cong \mathrm{Hom}(m_S^k(\mathbf{F}^\bullet), R) \oplus \mathrm{Ext}(m_S^{k+1}(\mathbf{F}^\bullet), R).$$
In particular, if R is a field, then $\mathrm{gecc}^{-k}(\mathcal{D}\mathbf{F}^\bullet) = \mathrm{gecc}^k(\mathbf{F}^\bullet)$*.*
8. *Suppose that R is a principal ideal domain. Let X and Y be analytic spaces, let* $\pi_1 : X \times Y \to X$ *and* $\pi_2 : X \times Y \to Y$ *denote the projections. Let* $\mathfrak{S}$ *and* $\mathfrak{S}'$ *be Whitney stratifications of X and Y, respectively. Let* $\mathbf{A}^\bullet$ *and* $\mathbf{B}^\bullet$ *be bounded, complexes of sheaves on X and Y, respectively, which are constructible with respect to* $\mathfrak{S}$ *and* $\mathfrak{S}'$*, respectively. Let* $\mathbf{A}^\bullet \overset{L}{\boxtimes} \mathbf{B}^\bullet := \pi_1^*\mathbf{A}^\bullet \overset{L}{\otimes} \pi_2^*\mathbf{B}^\bullet$*.*

   Then, $\mathbf{A}^\bullet \overset{L}{\boxtimes} \mathbf{B}^\bullet$ *is constructible with respect to the product stratification* $\{S \times S' \mid S \in \mathfrak{S},\ S' \in \mathfrak{S}'\}$ *and, for all* $S \in \mathfrak{S}$ *and* $S' \in \mathfrak{S}'$*,*
$$m^k_{S\times S'}\big(\mathbf{A}^\bullet \overset{L}{\boxtimes} \mathbf{B}^\bullet\big) = \bigoplus_{i+j=k} m^i_S(\mathbf{A}^\bullet) \otimes m^j_{S'}(\mathbf{B}^\bullet) \oplus \bigoplus_{i+j=k+1} \mathrm{Tor}\big(m^i_S(\mathbf{A}^\bullet), m^j_{S'}(\mathbf{B}^\bullet)\big).$$
   Consequently,
$$c_{S\times S'}\big(\mathbf{A}^\bullet \overset{L}{\boxtimes} \mathbf{B}^\bullet\big) = c_S(\mathbf{A}^\bullet) \cdot c_{S'}(\mathbf{B}^\bullet).$$

Proof Items 1, 3, and 6 are trivial. Item 2 is the last equality of Proposition 2.5 in [34]. Item 8 follows immediately from formula 5.6 of [40].

To see Item 4, note that the graded, enriched characteristic cycle of a complex being concentrated in degree zero is equivalent to the complex being **pure with shift** 0 (see Definition 7.5.4 of [17]). This is equivalent to the complex being perverse ([17], 9.5.2).

Item 5:

We will use that $\phi_f[-1]$ naturally commutes with ${}^\mu H^k$ ([17], Corollary 10.3.13) and that, if $\mathbf{A}^\bullet$ has $\mathbf{p}$ as an isolated point in its support, then so does the perverse sheaf ${}^\mu H^k(\mathbf{A}^\bullet)$; in this case, the stalk cohomology of ${}^\mu H^k(\mathbf{A}^\bullet)$ at $\mathbf{p}$ is concentrated in degree 0 and

$$H^0\left({}^{\mu}H^k(\mathbf{A}^\bullet)\right)_{\mathbf{p}} \cong H^k(\mathbf{A}^\bullet)_{\mathbf{p}}.$$

Now, let $S \in \mathfrak{S}$. Let $\mathbf{p} \in S$. Let g be as in Remark 6.2.2, so that $\mathbf{p}$ is an isolated point in the support of $\phi_g[-1]\mathbf{F}^\bullet$ and, hence,

$$m_S^k(\mathbf{F}^\bullet) \cong H^k(\phi_g[-1]\mathbf{F}^\bullet)_{\mathbf{p}} \cong H^0\left({}^{\mu}H^k(\phi_g[-1]\mathbf{F}^\bullet)\right)_{\mathbf{p}} \cong$$

$$H^0\left(\phi_g[-1]{}^{\mu}H^k(\mathbf{F}^\bullet)\right)_{\mathbf{p}} \cong m_S^0({}^{\mu}H^k(\mathbf{F}^\bullet)).$$

Item 7:

We begin as in the proof of Item 5. Let $S \in \mathfrak{S}$. Let $\mathbf{p} \in S$. Let g be as in Remark 6.2.2, so that

$$m_S^k(\mathbf{F}^\bullet) \cong H^k(\phi_g[-1]\mathbf{F}^\bullet)_{\mathbf{p}},\ \ m_S^{k+1}(\mathbf{F}^\bullet) \cong H^{k+1}(\phi_g[-1]\mathbf{F}^\bullet)_{\mathbf{p}},$$

and

$$m_S^{-k}(\mathcal{D}\mathbf{F}^\bullet) \cong H^{-k}(\phi_g[-1]\mathcal{D}\mathbf{F}^\bullet)_{\mathbf{p}} \cong H^{-k}(\mathcal{D}\phi_g[-1]\mathbf{F}^\bullet)_{\mathbf{p}}.$$

As the support of $\phi_g[-1]\mathbf{F}^\bullet$ is contained in $\{\mathbf{p}\}$ and as R is a Dedekind domain, there is a natural split exact sequence

$$0 \to \operatorname{Ext}(H^{k+1}(\phi_g[-1]\mathbf{F}^\bullet)_{\mathbf{p}}, R) \to H^{-k}(\mathcal{D}\phi_g[-1]\mathbf{F}^\bullet)_{\mathbf{p}} \to \operatorname{Hom}(H^k(\phi_g[-1]\mathbf{F}^\bullet)_{\mathbf{p}}, R) \to 0.$$

Item 7 follows. □

Note that, for a perverse sheaf $\mathbf{P}^\bullet$ with free Morse modules, $\operatorname{gecc}^\bullet(\mathbf{P}^\bullet)$ is completely determined by $\operatorname{CC}(\mathbf{P}^\bullet)$. This motivates us to define:

Definition 6.2.12 A complex of sheaves $\mathbf{P}^\bullet$ is **numerical** if and only if $\mathbf{P}^\bullet$ is perverse with free Morse modules.

Remark 6.2.13 We cannot use the perhaps more-obvious term "free perverse sheaf" as "free" has a categorical meaning.

Suppose that X is a local complete intersection of dimension d. In the literature on perverse sheaves, one frequently sees that Lê proved that $\mathbb{Z}_X^\bullet[d]$ is a perverse sheaf in [21]. In fact, what he proved in [21] is stronger; he proved that each of the complex links of each of the strata of any Whitney stratification of X have the homotopy-type of a bouquet of spheres in middle dimension. In cohomological terms, this means that Lê proved:

Theorem 6.2.14 ([21]) *If X is a local complete intersection of dimension d, then $\mathbb{Z}^\bullet_X[d]$ is numerical.*

While most of our examples will have to wait until we have developed more machinery, we can calculate "bare-handedly" what happens for curves and some basic complexes of sheaves.

Example 6.2.15 Suppose that X is a curve. The calculations of $\mathrm{CC}(\mathbf{F}^\bullet)$ and $\mathrm{gecc}^\bullet(\mathbf{F}^\bullet)$ reduce to calculating what happens at the discrete set of points where X is singular or where $\mathbf{F}^\bullet$ is not locally constant. Thus, it suffices to analyze the situation where there is a single zero-dimensional stratum.

Hence, we shall assume that $\mathbf{0} \in X$, and that $X - \{\mathbf{0}\}$ is a smooth curve. In a small enough open ball, each irreducible component of X is homeomorphic to an open disk and, hence, corresponds to an irreducible component of the germ of X at $\mathbf{0}$. Let $\{X_\mathfrak{l}\}_{\mathfrak{l}\in\Lambda}$ denote the collection of irreducible components of X and, for each $X_\mathfrak{l}$, let $\mathfrak{m}_\mathfrak{l} := \mathrm{mult}_\mathbf{0} X_\mathfrak{l}$. Let $\mathfrak{m} := \mathrm{mult}_\mathbf{0} X = \sum_\mathfrak{l} \mathfrak{m}_\mathfrak{l}$. Let $e := |\Lambda|$, i.e., let e be the number of irreducible components of X.

Stratify X by using $S_0 := \{\mathbf{0}\}$ and $S_\mathfrak{l} := X_\mathfrak{l} - \{\mathbf{0}\}$ as strata (we are assuming that 0 is not in the indexing set Λ). Let $j : \{\mathbf{0}\} \hookrightarrow X$ and $i : X - \{\mathbf{0}\} \hookrightarrow X$ denote the inclusions. Let $\mathbf{A}^\bullet := \mathbb{Z}^\bullet_X[1]$, $\mathbf{B}^\bullet := i_!i^!\mathbf{A}^\bullet$, $\mathbf{C}^\bullet := i_*i^*\mathbf{A}^\bullet$, and let $\mathbf{I}^\bullet$ be the perverse sheaf given by intersection cohomology with constant $\mathbb{Z}$-coefficients (here, we use the shifts that put all of the possibly non-zero cohomology in non-positive degrees). These are all complexes of sheaves on X, which are the constant sheaf, shifted by 1, on $X - \{\mathbf{0}\}$.

Let $\tilde{f} : (\mathcal{U}, \mathbf{0}) \to (\mathbb{C}, 0)$ be a complex analytic function, where $\tilde{f}$ may vanish identically on some irreducible components of X. Let $f := \tilde{f}_{|X}$. Consider $\mathbf{P}^\bullet := \psi_f[-1]\mathbf{A}^\bullet$ and $\mathbf{Q}^\bullet := \phi_f[-1]\mathbf{A}^\bullet$. These are complexes of sheaves on $V(f) := f^{-1}(0)$.

We wish to calculate the graded, enriched characteristic cycle, and the ordinary characteristic cycle, for each of the six complexes given above.

As $\mathbf{A}^\bullet$, $\mathbf{B}^\bullet$, $\mathbf{C}^\bullet$, and $\mathbf{I}^\bullet$ are the 1-shifted constant sheaf on $X - \{\mathbf{0}\}$, it follows that, if $\mathbf{F}^\bullet$ is any of these four complexes, then, for all $\mathfrak{l} \in \Lambda$, $m^0_{S_\mathfrak{l}}(\mathbf{F}^\bullet) \cong \mathbb{Z}$, and $m^k_{S_\mathfrak{l}}(\mathbf{F}^\bullet) = 0$ for $k \neq 0$. The question is: what is $m^k_{S_0}(\mathbf{F}^\bullet)$?

A normal slice to S_0 is simply $\overset{\circ}{B}_\epsilon \cap X$, where $\overset{\circ}{B}_\epsilon$ is a small open ball around the origin. The complex link to S_0 is $\overset{\circ}{B}_\epsilon \cap X \cap L^{-1}(a)$, where L is a generic linear form and $0 < |a| \ll \epsilon$.

$\mathbf{A}^\bullet$:

We have

$$m^k_{S_0}(\mathbf{A}^\bullet) = \mathbb{H}^k(\overset{\circ}{B}_\epsilon \cap X, \overset{\circ}{B}_\epsilon \cap X \cap L^{-1}(a); \mathbb{Z}^\bullet_X[1]),$$

which is the ordinary degree $k+1$ integral cohomology of the pair consisting of a contractible space modulo $\mathfrak{m}$ points in the space. Hence, $m^0_{S_0}(\mathbf{A}^\bullet) \cong \mathbb{Z}^{(\mathfrak{m}-1)}$, and $m^k_{S_0}(\mathbf{A}^\bullet) = 0$ if $k \neq 0$.

Thus, we find that, if $k \neq 0$, then $\mathrm{gecc}^k(\mathbf{A}^\bullet) = 0$, and

$$\mathrm{gecc}^0(\mathbf{A}^\bullet) = \mathbb{Z}^{(\mathfrak{m}-1)}\left[T^*_{\mathbf{0}}\mathcal{U}\right] + \sum_{\mathfrak{l}} \mathbb{Z}\left[\overline{T^*_{X_{\mathfrak{l}}-\{\mathbf{0}\}}\mathcal{U}}\right].$$

It follows that

$$\mathrm{CC}(\mathbf{A}^\bullet) = (\mathfrak{m}-1)\left[T^*_{\mathbf{0}}\mathcal{U}\right] + \sum_{\mathfrak{l}} \left[\overline{T^*_{X_{\mathfrak{l}}-\{\mathbf{0}\}}\mathcal{U}}\right].$$

$\mathbf{B}^\bullet$:

We have

$$m^k_{S_0}(\mathbf{B}^\bullet) = \mathbb{H}^{k+1}(\mathring{B}_\epsilon \cap X, \mathring{B}_\epsilon \cap X \cap L^{-1}(a); i_!\mathbb{Z}^\bullet_{X-\{\mathbf{0}\}}).$$

Using the long exact sequence for the hypercohomology of a pair, and that

$$\mathbb{H}^*(\mathring{B}_\epsilon \cap X; i_!\mathbb{Z}^\bullet_{X-\{\mathbf{0}\}}) = 0,$$

we find that

$$m^k_{S_0}(\mathbf{B}^\bullet) \cong H^k(\mathring{B}_\epsilon \cap X \cap L^{-1}(a); \mathbb{Z}).$$

Therefore, $m^k_{S_0}(\mathbf{B}^\bullet) = 0$ if $k \neq 0$, and $m^0_{S_0}(\mathbf{B}^\bullet) \cong \mathbb{Z}^{\mathfrak{m}}$.

Thus, we find that, if $k \neq 0$, then $\mathrm{gecc}^k(\mathbf{B}^\bullet) = 0$, and

$$\mathrm{gecc}^0(\mathbf{B}^\bullet) = \mathbb{Z}^{\mathfrak{m}}\left[T^*_{\mathbf{0}}\mathcal{U}\right] + \sum_{\mathfrak{l}} \mathbb{Z}\left[\overline{T^*_{X_{\mathfrak{l}}-\{\mathbf{0}\}}\mathcal{U}}\right].$$

It follows that

$$\mathrm{CC}(\mathbf{B}^\bullet) = \mathfrak{m}\left[T^*_{\mathbf{0}}\mathcal{U}\right] + \sum_{\mathfrak{l}} \left[\overline{T^*_{X_{\mathfrak{l}}-\{\mathbf{0}\}}\mathcal{U}}\right].$$

$\mathbf{C}^\bullet$:

We have

$$m^k_{S_0}(\mathbf{C}^\bullet) = \mathbb{H}^{k+1}(\mathring{B}_\epsilon \cap X, \mathring{B}_\epsilon \cap X \cap L^{-1}(a); i_*\mathbb{Z}^\bullet_{X-\{\mathbf{0}\}}) \cong$$

$$H^{k+1}(\mathring{B}_\epsilon \cap X - \{\mathbf{0}\}, \mathring{B}_\epsilon \cap X \cap L^{-1}(a); \mathbb{Z}).$$

This splits as a direct sum of the degree $k+1$ integral cohomology of pairs consisting of spaces $\overset{\circ}{B}_\epsilon \cap X_\mathfrak{l} - \{\mathbf{0}\}$, which are homotopy-equivalent to circles, modulo $\mathfrak{m}_\mathfrak{l}$ points. As in the $\mathbf{B}^\bullet$ case, one easily calculates that that $m^k_{S_0}(\mathbf{C}^\bullet) = 0$ if $k \neq 0$, and $m^0_{S_0}(\mathbf{C}^\bullet) \cong \mathbb{Z}^{\mathfrak{m}}$.

Thus, we find that $\mathrm{gecc}^\bullet(\mathbf{B}^\bullet) = \mathrm{gecc}^\bullet(\mathbf{C}^\bullet)$ and, of course, that $\mathrm{CC}(\mathbf{B}^\bullet) = \mathrm{CC}(\mathbf{C}^\bullet)$.

$\mathbf{I}^\bullet$:

The axioms of intersection cohomology imply that $\mathbf{I}^\bullet$ is isomorphic to the direct sum of the extensions by zero of the intersection cohomology on each of the $X_\mathfrak{l}$. As each $X_\mathfrak{l}$ is homeomorphic to an open disk, the intersection cohomology complex on $X_\mathfrak{l}$ is isomorphic to $\mathbb{Z}_{X_\mathfrak{l}}[1]$.

Thus, we have

$$m^k_{S_0}(\mathbf{I}^\bullet) = \bigoplus_{\mathfrak{l} \in \Lambda} H^{k+1}(\overset{\circ}{B}_\epsilon \cap X_\mathfrak{l}, \overset{\circ}{B}_\epsilon \cap X_\mathfrak{l} \cap L^{-1}(a); \mathbb{Z}).$$

It follows that $m^k_{S_0}(\mathbf{I}^\bullet) = 0$ if $k \neq 0$, and $m^0_{S_0}(\mathbf{I}^\bullet) \cong \bigoplus_\mathfrak{l} \mathbb{Z}^{(\mathfrak{m}_\mathfrak{l}-1)} \cong \mathbb{Z}^{(\mathfrak{m}-e)}$.

Thus, we find that, if $k \neq 0$, then $\mathrm{gecc}^k(\mathbf{I}^\bullet) = 0$, and

$$\mathrm{gecc}^0(\mathbf{I}^\bullet) = \mathbb{Z}^{(\mathfrak{m}-e)}\left[T^*_{\mathbf{0}}\mathcal{U}\right] + \sum_\mathfrak{l} \mathbb{Z}\left[\overline{T^*_{X_\mathfrak{l}-\{\mathbf{0}\}}\mathcal{U}}\right].$$

It follows that

$$\mathrm{CC}(\mathbf{I}^\bullet) = (\mathfrak{m}-e)\left[T^*_{\mathbf{0}}\mathcal{U}\right] + \sum_\mathfrak{l} \left[\overline{T^*_{X_\mathfrak{l}-\{\mathbf{0}\}}\mathcal{U}}\right].$$

We still wish to look at the complexes $\mathbf{P}^\bullet$ and $\mathbf{Q}^\bullet$.

Let $\Lambda_\subseteq := \{\mathfrak{l} \in \Lambda \mid f(X_\mathfrak{l}) \equiv 0\}$. Hence, $V(f) = \bigcup_{\mathfrak{l} \in \Lambda_\subseteq} X_\mathfrak{l}$, with the convention that, if $\Lambda_\subseteq$ is empty, then this union is taken as meaning the point-set $\{\mathbf{0}\}$. Let $\Lambda_\nsubseteq := \Lambda - \Lambda_\subseteq$. For each $\mathfrak{l} \in \Lambda_\nsubseteq$, let $\eta_\mathfrak{l}$ equal the intersection multiplicity $(X_\mathfrak{l} \cdot V(\tilde{f}))_\mathbf{0}$, and let $\eta := \sum_{\mathfrak{l} \in \Lambda_\nsubseteq} \eta_\mathfrak{l}$. Let $\mathfrak{m}_\subseteq := \sum_{\mathfrak{l} \in \Lambda_\subseteq} \mathfrak{m}_\mathfrak{l}$.

$\mathbf{P}^\bullet$:

By definition, $\mathbf{P}^\bullet = \psi_f[-1]\mathbf{A}^\bullet$ is a complex of sheaves on $V(f)$, but – in our current setting – the support of $\psi_f[-1]\mathbf{A}^\bullet$ will be contained in $\{\mathbf{0}\}$, and the stalk cohomology $H^*(\psi_f[-1]\mathbf{A}^\bullet)_\mathbf{0}$ is isomorphic to $\mathbb{Z}^\eta$ in degree 0 and is zero in other degrees.

Thus, we find that, if $k \neq 0$, then $\mathrm{gecc}^k(\mathbf{P}^\bullet) = 0$, and

$$\mathrm{gecc}^0(\mathbf{P}^\bullet) = \mathbb{Z}^\eta\left[T^*_{\mathbf{0}}\mathcal{U}\right].$$

It follows that

$$\mathrm{CC}(\mathbf{P}^\bullet) = \eta\left[T^*_{\mathbf{0}}\mathcal{U}\right].$$

$\mathbf{Q}^\bullet$:

By definition, $\mathbf{Q}^\bullet = \phi_f[-1]\mathbf{A}^\bullet$ is a complex of sheaves on $V(f)$, and the restriction of $\mathbf{Q}^\bullet$ to $V(f) - \{\mathbf{0}\}$ is isomorphic to the 1-shifted constant sheaf. In addition, the stalk cohomology $H^*(\phi_f[-1]\mathbf{A}^\bullet)_\mathbf{0}$ is isomorphic to $\mathbb{Z}^{(\eta-1)}$ in degree 0 and is zero in other degrees. We have

$$m^k_{S_0}(\mathbf{Q}^\bullet) = \mathbb{H}^k(\overset{\circ}{B}_\epsilon \cap X \cap V(f), \overset{\circ}{B}_\epsilon \cap X \cap V(f) \cap L^{-1}(a); \phi_f[-1]\mathbf{A}^\bullet),$$

for $0 < |a| \ll 1$. This module fits into the hypercohomology long exact sequence of the pair, in which one has the map induced by inclusion

$$\mathbb{H}^k(\overset{\circ}{B}_\epsilon \cap X \cap V(f); \phi_f[-1]\mathbf{A}^\bullet) \to \mathbb{H}^k(\overset{\circ}{B}_\epsilon \cap X \cap V(f) \cap L^{-1}(a); \phi_f[-1]\mathbf{A}^\bullet). \quad (\dagger)$$

The right-hand term above is clearly isomorphic to

$$\bigoplus_{\mathfrak{l} \in \Lambda_\subseteq} \mathbb{H}^k(\overset{\circ}{B}_\epsilon \cap X_\mathfrak{l} \cap L^{-1}(a); \phi_f[-1]\mathbf{A}^\bullet),$$

and $(\dagger)$ can be rewritten as

$$H^{k+1}(\overset{\circ}{B}_\epsilon \cap X, \overset{\circ}{B}_\epsilon \cap X \cap f^{-1}(b); \mathbb{Z}) \to \bigoplus_{\mathfrak{l} \in \Lambda_\subseteq} H^{k+1}(\overset{\circ}{B}_\epsilon \cap X_\mathfrak{l} \cap L^{-1}(a); \mathbb{Z}),$$

where $0 < |b| \ll |a| \ll 1$. Now, one easily finds from the long exact sequence that $m^k_{S_0}(\mathbf{Q}^\bullet) = 0$ for $k \neq 0$, and $m^0_{S_0}(\mathbf{Q}^\bullet) \cong \mathbb{Z}^{(\mathfrak{m}_\subseteq + \eta - 1)}$.

Thus, we find that, if $k \neq 0$, then $\mathrm{gecc}^k(\mathbf{Q}^\bullet) = 0$, and

$$\mathrm{gecc}^0(\mathbf{Q}^\bullet) = \mathbb{Z}^{(\mathfrak{m}_\subseteq + \eta - 1)}\left[T^*_\mathbf{0}\mathcal{U}\right] + \sum_{\mathfrak{l} \in \Lambda_\subseteq} \mathbb{Z}\left[\overline{T^*_{X_\mathfrak{l} - \{\mathbf{0}\}}\mathcal{U}}\right].$$

It follows that

$$\mathrm{CC}(\mathbf{Q}^\bullet) = (\mathfrak{m}_\subseteq + \eta - 1)\left[T^*_\mathbf{0}\mathcal{U}\right] + \sum_{\mathfrak{l} \in \Lambda_\subseteq} \left[\overline{T^*_{X_\mathfrak{l} - \{\mathbf{0}\}}\mathcal{U}}\right].$$

The characteristic cycle calculations for $\mathbf{P}^\bullet$ and $\mathbf{Q}^\bullet$ can be "checked". There is the canonical distinguished triangle

$$(\mathbb{Z}^\bullet_X[1])_{V(f)}[-1] \to \psi_f[-1]\mathbb{Z}^\bullet_X[1] \to \phi_f[-1]\mathbb{Z}^\bullet_X[1] \xrightarrow{[1]} (\mathbb{Z}^\bullet_X[1])_{V(f)}[-1],$$

and so we should find that $\mathrm{CC}(\mathbf{P}^\bullet) = \mathrm{CC}(\mathbf{Q}^\bullet) + \mathrm{CC}(\mathbb{Z}^\bullet_{V(f)})$.

This is easily checked, for $\mathrm{CC}(\mathbb{Z}^\bullet_{V(f)}) = -\mathrm{CC}(\mathbb{Z}^\bullet_{V(f)}[1])$ and, applying our calculation of $\mathrm{CC}(\mathbf{A}^\bullet)$, we find that

$$\mathrm{CC}(\mathbb{Z}^\bullet_{V(f)}[1]) = (\mathrm{m}_{\subseteq} - 1)\left[T^*_{\mathbf{0}}\mathcal{U}\right] + \sum_{\mathfrak{l}\in\Lambda_{\subseteq}} \left[\overline{T^*_{X_{\mathfrak{l}}-\{\mathbf{0}\}}\mathcal{U}}\right].$$

Note that all of the graded, enriched characteristic cycles in this example are concentrated in degree 0. As we stated in Item 4 of Proposition 6.2.11, this is a reflection of the fact that each of the complexes that we considered above are perverse sheaves.

Example 6.2.16 We wish a give an easy example/problem, where the sheaves under consideration are not perverse.

Let $\mathcal{U} := \mathbb{C}^3$, and use x, y, and z as coordinates. Let $X := V(z) \cup V(x, y)$. There are three obvious strata: $S_0 := \{\mathbf{0}\}$, $S_1 := V(x, y) - \{\mathbf{0}\}$, and $S_2 := V(z) - \{\mathbf{0}\}$. Let $j : \{\mathbf{0}\} \hookrightarrow X$ and $i : X - \{\mathbf{0}\} \hookrightarrow X$ denote the inclusions, and consider the complexes of sheaves $\mathbf{A}^\bullet := \mathbb{Z}^\bullet_X[2]$, $\mathbf{B}^\bullet := i_! i^! \mathbf{A}^\bullet$, and $\mathbf{C}^\bullet := i_* i^* \mathbf{A}^\bullet$.

We leave it to the reader to verify that:

$\mathrm{gecc}^k(\mathbf{A}^\bullet) = 0$ if $k \neq -1, 0$,
$\mathrm{gecc}^{-1}(\mathbf{A}^\bullet) = \mathbb{Z}[T^*_{\mathbf{0}}\mathcal{U}] + \mathbb{Z}[T^*_{V(x,y)}\mathcal{U}]$, and $\mathrm{gecc}^0(\mathbf{A}^\bullet) = \mathbb{Z}[T^*_{V(z)}\mathcal{U}]$.

$\mathrm{gecc}^k(\mathbf{B}^\bullet) = 0$ if $k \neq -1, 0$,
$\mathrm{gecc}^{-1}(\mathbf{B}^\bullet) = \mathbb{Z}^2[T^*_{\mathbf{0}}\mathcal{U}] + \mathbb{Z}[T^*_{V(x,y)}\mathcal{U}]$, and $\mathrm{gecc}^0(\mathbf{B}^\bullet) = \mathbb{Z}[T^*_{V(z)}\mathcal{U}]$.

$\mathrm{gecc}^k(\mathbf{C}^\bullet) = 0$ if $k \neq -1, 0, 1$,
$\mathrm{gecc}^{-1}(\mathbf{C}^\bullet) = \mathbb{Z}[T^*_{\mathbf{0}}\mathcal{U}] + \mathbb{Z}[T^*_{V(x,y)}\mathcal{U}]$, $\mathrm{gecc}^0(\mathbf{C}^\bullet) = \mathbb{Z}[T^*_{V(z)}\mathcal{U}]$, and $\mathrm{gecc}^1(\mathbf{C}^\bullet) = \mathbb{Z}[T^*_{\mathbf{0}}\mathcal{U}]$.

Exercise 6.2.17 Let $f : \mathbb{C}^3 \to \mathbb{C}$ be given by $f(x, y, t) = y^2 - x^3 - tx^2$. Thinking of t as a parameter, $V(f)$ is the total space of the classic family of nodes degenerating to a cusp; it is the complex Whitney umbrella after an analytic change of coordinates.

Calculate $\mathrm{gecc}^\bullet(\mathbb{Z}^\bullet_{V(f)}[2])$ and thereby verify that $\mathbb{Z}^\bullet_{V(f)}[2]$ is numerical (recall Definition 6.2.12).

Remark 6.2.18 Before we leave this section, we need to make an important point.

In this chapter, we shall show that many results about characteristic cycles are true for graded, enriched characteristic cycles—with essentially the same proofs; however, because the graded, enriched characteristic cycles contain far more data, it should not be surprising that, for some results, the ordinary characteristic cycle is easier to calculate with. In particular, the additivity of CC for distinguished triangles, as given in Item 3 of Proposition 6.2.5, is extremely useful.

6.3 Characteristic Complexes and the Local Euler Obstruction

The local Euler obstruction, defined by MacPherson in [24] in 1974 has been studied by many researchers (see, for instance, [5, 7, 14, 23], and [11]) and is, at this point, a standard piece of data associated to a singular point of a complex analytic space. The local Euler obstruction is an obstruction to extending a stratified radial vector field to a non-zero lift in the Nash modification.

The relative local Euler obstruction, defined in [4], is an analog of the local Euler obstruction for a complex analytic function $f : X \to \mathbb{C}$ at a point $\mathbf{p}$ which is a stratified isolated critical point of f. This is again defined in an obstruction-theoretic way; it is an obstruction to extending the conjugate of the stratified gradient vector field of f to a non-zero lift in the Nash modification. This relative concept is beginning to be studied by a number of other researchers; see, for instance, [9, 10], [41], and [1].

Our own contribution to [4] appeared in the last section of that paper, where we used derived category techniques to extend the definition of the relative local Euler obstruction to functions with arbitrary critical loci, but—in [4]—we referred to this generalized concept as the *defect* of f. We also gave an algorithm for calculating the defect of f via a process similar to how we defined the Lê cycles, Lê numbers, and their generalizations in [25, 26, 28], and especially in Remark 1.6 of Part IV of [33].

In this section, we recall our earlier characterization of the local Euler obstruction in terms of *characteristic complexes* and recall and re-derive some standard properties of the local Euler obstruction. We then recall our general definition/characterization of the defect of f—which we now take as the general definition of the relative local Euler obstruction—in terms of vanishing cycles and characteristic complexes. Finally, in Theorem 6.3.23, we prove a number of basic properties for the relative local Euler obstruction and give some examples.

Some classical constructions in the study of singular spaces, such as calculating the polar varieties and polar multiplicities of Lê and Teissier and the Nash modification, deal with contributions from only the smooth strata of X. From our point of view, these are results where the underlying complex of sheaves is a *characteristic complex*.

Note that, while our definition of the characteristic cycle in this chapter is not what we used in [4], our definition below of a characteristic complex has been adjusted in such a way that the same complexes here and in [4] are characteristic complexes.

6.3.1 Characteristic Complexes

Definition 6.3.1 Let $X = \bigcup_i X_i$ be the decomposition of X into its irreducible components.

We say that a complex of sheaves $\mathbf{K}^\bullet$ on X is a **characteristic complex for** X provided that

$$\mathrm{CC}(\mathbf{K}^\bullet) = \sum_i (-1)^{\dim X_i} \left[\overline{T^*_{(X_i)_{\mathrm{reg}}} \mathcal{U}}\right].$$

Thus, $\mathbf{K}^\bullet$ is a characteristic complex if and only if there exists a complex analytic Whitney stratification $\mathfrak{S}$ of X, with connected strata, with respect to which $\mathbf{K}^\bullet$ is constructible and such that, for all $S \in \mathfrak{S}$, the Euler characteristic $c_S(\mathbf{K}^\bullet)$ of the Morse modules of S with respect to $\mathbf{K}^\bullet$ is zero **unless** S is an open dense subset of one of the $(X_i)_{\mathrm{reg}}$, in which case

$$c_S(\mathbf{K}^\bullet) = (-1)^{\dim S} = (-1)^{\dim X_i}.$$

Remark 6.3.2 Note that Remark 6.2.2 implies that, if $\mathbf{p}$ is not an isolated point in X and $\mathbf{K}^\bullet$ is a characteristic complex for X, then for a generic choice of (restricted) linear forms $\mathcal{L}$, $\chi(\phi_{\mathcal{L}-\mathcal{L}(\mathbf{p})}\mathbf{K}^\bullet)_\mathbf{p} = 0$, i.e., $\chi(\mathbf{K}^\bullet)_\mathbf{p} = \chi(\psi_{\mathcal{L}-\mathcal{L}(\mathbf{p})}\mathbf{K}^\bullet)_\mathbf{p}$.

Example 6.3.3 Suppose that X is a complex manifold with connected components $\{X_i\}_i$. Then it is immediate that $\mathbb{Z}^\bullet_X$ is characteristic complex for X.

Proposition 6.3.4 *Let $X = \bigcup_i X_i$ be the decomposition of X into its irreducible components. Suppose that, for each i, $\mathbf{K}^\bullet_i$ is a characteristic complex for X_i and let $\widehat{\mathbf{K}}^\bullet_i$ denote the extension by zero of $\mathbf{K}^\bullet_i$ to all of X. Then, $\bigoplus_i \widehat{\mathbf{K}}^\bullet_i$ is a characteristic complex for X.*

Proof This is immediate from Item 3 of Proposition 6.2.5. □

We wish to give a simple, but important, example of characteristic complexes.

Example 6.3.5 Suppose that X pure-dimensional, of dimension > 0, and has a single singular point at the origin in $\mathcal{U}$. Then,

$$\mathrm{CC}\left(\mathbb{Z}^\bullet_X\right) = (-1)^{\dim X} \left[\overline{T^*_{X_{\mathrm{reg}}} \mathcal{U}}\right] + b\left[T^*_\mathbf{0}\mathcal{U}\right],$$

where the final part of Remark 6.2.2 tells us that $b = 1 - \chi(\mathbb{L}_{X,\mathbf{0}})$.

To produce a characteristic complex for X, we must modify $\mathbb{Z}^\bullet_X$ at the origin. This is easy.

If $b = 0$, there is nothing to do; $\mathbb{Z}^\bullet_X$ would be a characteristic complex.

If $b < 0$, let $\mathbf{A}^\bullet$ be the extension by zero to all of X of the constant sheaf $\mathbb{Z}^{-b}$ on the point-set $\{\mathbf{0}\}$. Then, $\mathbb{Z}^\bullet_X \oplus \mathbf{A}^\bullet$ is a characteristic complex for X.

If $b > 0$, let $\mathbf{A}^\bullet$ be the extension by zero to all of X of the shifted constant sheaf $\mathbb{Z}^b[1]$ on the point-set $\{\mathbf{0}\}$. Then, $\mathbb{Z}^\bullet_X \oplus \mathbf{A}^\bullet$ is a characteristic complex for X.

Note that, in each of these cases, the stalk cohomology at the origin of the resulting characteristic complex is $\chi(\mathbb{L}_{X,\mathbf{0}})$.

The following proposition is well-known to experts in the form: CC yields a surjection from the Grothendieck group of constructible complexes to the group of Lagrangian cycles. For completeness, we give the proof, which is basically induction on the construction given in Example 6.3.5.

Proposition 6.3.6 *Let $\mathfrak{S}$ be any Whitney stratification of X, with connected strata. Then, there exists a characteristic complex on X which is constructible with respect to $\mathfrak{S}$. In particular, characteristic complexes exist for all X.*

Proof This proof is contained in Lemma 3.1 of [29]. However, we wish to sketch it here.

Note that Proposition 6.3.4 implies that we need deal only with the case where X is irreducible. Hence, we assume that X is irreducible of dimension d.

Let $\mathfrak{S}$ be a Whitney stratification of X, with connected strata.

For every stratum $S \in \mathfrak{S}$ and every non-negative integer v, let $\mathbf{U}^\bullet_{S,v}$ denote the extension by zero to all of X of $\left(\mathbb{Z}^\bullet_S\right)^v$ so that $c_S(\mathbf{U}^\bullet_{S,v}) = (-1)^{\dim S} v$ (where c_S is the coefficient of $\left[\overline{T^*_S \mathcal{U}}\right]$ in the characteristic cycle). If v is a negative integer, define $\mathbf{U}^\bullet_{S,v} := \mathbf{U}^\bullet_{S,-v}[1]$ so that, again, $c_S(\mathbf{U}^\bullet_{S,v}) = (-1)^{\dim S} v$.

Now we construct a characteristic complex as a direct sum, canceling out conormal cycles over lower-dimensional strata. Let

$$\mathbf{K}^\bullet_d = \mathbf{K}^\bullet_{\geq d} := \mathbf{U}^\bullet_{X_{\mathrm{reg}},1}.$$

Note that $\mathbf{K}^\bullet_d$ is also constructible with respect to $\mathfrak{S}$ and, if $S \in \mathfrak{S}$ has dimension d, then $c_S(\mathbf{K}^\bullet_d) = (-1)^d$.

Now we need cancel out the contributions to the characteristic cycle from lower-dimensional strata.

Let

$$\mathbf{K}^\bullet_{d-1} := \bigoplus_{S \in \mathfrak{S}, \dim S = d-1} \mathbf{U}^\bullet_{S, -c_S(\mathbf{K}^\bullet_d)},$$

so that $\mathbf{K}^\bullet_{\geq d-1} := \mathbf{K}^\bullet_d \oplus \mathbf{K}^\bullet_{d-1}$ has the property that, for $S \in \mathfrak{S}$ of dimension at least $d-1$,

$$c_S\left(\mathbf{K}^\bullet_{\geq d-1}\right) = \begin{cases} (-1)^d, & \text{if } \dim S = d; \\ 0, & \text{if } \dim S = d-1. \end{cases}$$

Continuing in this manner, we produce $\mathbf{K}^\bullet := \mathbf{K}^\bullet_{\geq 0}$ which is a characteristic complex for X. □

Proposition 6.3.7 *Let X and Y be analytic spaces, let $\pi_1 : X \times Y \to X$ and $\pi_2 : X \times Y \to Y$ denote the projections. Let $\mathbf{K}^\bullet_X$ and $\mathbf{K}^\bullet_Y$ be characteristic complexes for X and Y, respectively. Let $\mathbf{K}^\bullet_X \overset{L}{\boxtimes} \mathbf{K}^\bullet_Y := \pi_1^*\mathbf{K}^\bullet_X \overset{L}{\otimes} \pi_2^*\mathbf{K}^\bullet_Y$.*

Then, $\mathbf{K}^\bullet_X \overset{L}{\boxtimes} \mathbf{K}^\bullet_Y$ is a characteristic complex for $X \times Y$.

Proof This is immediate from Item 8 of Proposition 6.2.11. □

6.3.2 The Local Euler Obstruction

Our primary interest in characteristic complexes lies in their relationship with the local Euler obstruction, defined by MacPherson in [24]. We let $\mathrm{Eu}_\mathbf{p} X$ denote the local Euler obstruction of X at $\mathbf{p}$, and first recall some well-known results, which appear either explicitly or implicitly in [24]. Note that we slightly extend the usual definition of the local Euler obstruction to possibly non-pure-dimensional spaces by adding over the irreducible components.

Proposition 6.3.8

1. *The local Euler obstruction is, in fact, local, i.e., if $\mathcal{W}$ is an open neighborhood of $\mathbf{p}$ in X, then $\mathrm{Eu}_\mathbf{p} X = \mathrm{Eu}_\mathbf{p} \mathcal{W}$.*
2. *If $\mathbf{p}$ is a smooth point of X, then $\mathrm{Eu}_\mathbf{p} X = 1$.*
3. *If $(\mathbf{x}, \mathbf{y}) \in X \times Y$, then $\mathrm{Eu}_{(\mathbf{x},\mathbf{y})}(X \times Y) = (\mathrm{Eu}_\mathbf{x} X)\big(\mathrm{Eu}_\mathbf{y} Y\big)$.*
4. *If $\mathbf{p} \in X$ and X_i denotes the local irreducible components of X at $\mathbf{p}$, then $\mathrm{Eu}_\mathbf{p} X = \sum_i \mathrm{Eu}_\mathbf{p} X_i$.*
5. *$\mathrm{Eu}_\mathbf{x} X$ is a constant function of $\mathbf{x}$ along the strata of any Whitney stratification of X (which has connected strata).*

There is also the important result:

Theorem 6.3.9 (Brylinski et al. [7]) *Suppose that $\mathbf{A}^\bullet$ on X is constructible with respect a Whitney stratification $\mathfrak{S}$, and that*

$$\mathrm{CC}(\mathbf{A}^\bullet) = \sum_{S \in \mathfrak{S}} c_S(\mathbf{A}^\bullet) \left[\overline{T^*_S \mathcal{U}}\right].$$

Then, for all $\mathbf{p} \in X$,

$$\chi(\mathbf{A}^\bullet)_\mathbf{p} = \sum_{S \in \mathfrak{S}} (-1)^{\dim S} c_S(\mathbf{A}^\bullet)\, \mathrm{Eu}_\mathbf{p} \overline{S},$$

where we set $\mathrm{Eu}_\mathbf{p}(\overline{S}) = 0$ if $\mathbf{p} \notin \overline{S}$.

From this, we immediately conclude the fundamental relationship between characteristic complexes and the local Euler obstruction:

Corollary 6.3.10 *Let $\mathbf{K}^\bullet$ be a characteristic complex for X. Let $\mathbf{p} \in X$.*
Then,

$$\mathrm{Eu}_{\mathbf{p}} X = \chi\left(\mathbf{K}^\bullet\right)_{\mathbf{p}}.$$

Remark 6.3.11 We note, as in [39], Remark 0.1, that much of what we have written can be described just using the language of constructible functions. Suppose that, for every constructible complex of sheaves $\mathbf{F}^\bullet$, we let

$$\alpha_{\mathbf{F}^\bullet} : X \to \mathbb{Z}$$

denote the constructible function $\alpha_{\mathbf{F}^\bullet}(\mathbf{x}) := \chi(\mathbf{F}^\bullet)_{\mathbf{x}}$.

Then, it is well-known that $\alpha_{()}$ yields a surjection from the Grothendieck group of constructible complexes on X to the group of constructible functions on X.

As $\mathbf{p} \mapsto \mathrm{Eu}_{\mathbf{p}} X$ is a constructible function, our definition of a characteristic complex $\mathbf{K}^\bullet$ was precisely designed so that $\alpha_{\mathbf{K}^\bullet}$ equals the local Euler obstruction function.

As a corollary to Corollary 6.3.10, and using the additivity of the Euler characteristic of hypercohomology over complex stratifications, we recover the formula of [3]:

Corollary 6.3.12 *Let $\mathfrak{S}$ be a complex analytic Whitney stratification, with connected strata, of X. For each $S \in \mathfrak{S}$, let $\mathbf{p}_S$ denote a point in S. Let $\mathbf{p}$ be an arbitrary non-isolated point of X and let $\mathbb{L}_{X,\mathbf{p}}$ denote the complex link of X at $\mathbf{p}$.*
Then,

$$\mathrm{Eu}_{\mathbf{p}} X = \sum_{S \in \mathfrak{S}} \chi(\mathbb{L}_{X,\mathbf{p}} \cap S)\, \mathrm{Eu}_{\mathbf{p}_S} X.$$

Proof For convenience, we assume that $\mathbf{p} = \mathbf{0}$ and that $\mathcal{L}$ is a generic linear form. Let $\mathbf{K}^\bullet$ be a characteristic complex for X. As $\mathbf{p}$ is not an isolated point of X, Remark 6.3.2 implies that $\chi(\mathbf{K}^\bullet)_{\mathbf{0}} = \chi(\psi_{\mathcal{L}}\mathbf{K}^\bullet)_{\mathbf{0}}$.

Thus, using the additivity of the Euler characteristic of hypercohomology over complex stratifications, we find

$$\mathrm{Eu}_{\mathbf{0}} X = \chi\left(\mathbf{K}^\bullet\right)_{\mathbf{0}} = \chi(\psi_{\mathcal{L}}\mathbf{K}^\bullet)_{\mathbf{0}} = \chi\left(\mathbb{H}^*(\mathbb{L}_{X,\mathbf{0}};\, \mathbf{K}^\bullet)\right) =$$

$$\sum_{S} \chi(\mathbb{L}_{X,\mathbf{0}} \cap S)\chi(\mathbf{K}^\bullet)_{\mathbf{p}_S} = \sum_{S \in \mathfrak{S}} \chi(\mathbb{L}_{X,\mathbf{0}} \cap S)\, \mathrm{Eu}_{\mathbf{p}_S} X.$$

□

Before we leave this section, we wish to give a known example of how Corollary 6.3.12 enables one to calculate local Euler obstructions.

Example 6.3.13 Suppose that $\mathbf{p}$ is an isolated singular point of X. Then, every point in the complex link, $\mathbb{L}_{X,\mathbf{p}}$, is a smooth point of X and, hence, the local Euler obstruction of X at each point of $\mathbb{L}_{X,\mathbf{p}}$ is 1. Thus, we conclude from Corollary 6.3.12 that

$$\mathrm{Eu}_{\mathbf{p}}\, X = \chi(\mathbb{L}_{X,\mathbf{p}}).$$

In particular, suppose that X is a curve, and $\mathbf{p} \in X$. Then the complex link $\mathbb{L}_{X,\mathbf{p}}$, is a finite collection of points; the number of points is simply the multiplicity, $\mathrm{mult}_{\mathbf{p}}\, X$, of X at $\mathbf{p}$. Thus, for a curve, we conclude that

$$\mathrm{Eu}_{\mathbf{p}}\, X = \chi(\mathbb{L}_{X,\mathbf{p}}) = \mathrm{mult}_{\mathbf{p}}\, X.$$

Exercise 6.3.14 Suppose that X is 2-dimensional in a neighborhood of $\mathbf{p} \in X$. Let S's denote 1-dimensional Whitney strata which contain $\mathbf{p}$ in their closures. Let $\mathbf{p}_S$ denote an arbitrary point of S near $\mathbf{p}$.

Show that

$$\mathrm{Eu}_{\mathbf{p}}\, X = \chi(\mathbb{L}_{X,\mathbf{p}}) + \sum_S (\mathrm{mult}_{\mathbf{p}}\, \overline{S})(\mathrm{mult}_{\mathbf{p}_S}\, X - 1).$$

6.3.3 *The Relative Local Euler Obstruction*

We now wish to discuss the *relative local Euler obstruction*, as was introduced in [4].

Recall that $\mathcal{U}$ is an open neighborhood of the origin of $\mathbb{C}^{n+1}$, X is a closed, analytic subset of $\mathcal{U}$. We let $\mathbf{z} := (z_0, \ldots, z_n)$ be coordinates on $\mathcal{U}$. We identify the cotangent space $T^*\mathcal{U}$ with $\mathcal{U} \times \mathbb{C}^{n+1}$ by mapping $(\mathbf{p}, w_0 d_{\mathbf{p}} z_0 + \cdots + w_n d_{\mathbf{p}} z_n)$ to $(\mathbf{p}, (w_0, \ldots, w_n))$. Let $\pi : T^*\mathcal{U} \to \mathcal{U}$ denote the projection.

Suppose that we have $\mathbf{p} \in X$ and a complex analytic $f : X \to \mathbb{C}$. We let $\tilde{f}$ be a local extension of f at $\mathbf{p}$ to an open neighborhood of $\mathbf{p}$ in $\mathcal{U}$; we assume now that $\mathcal{U}$ is re-chosen to be this (possibly) smaller open neighborhood of $\mathbf{p}$. We also let $d\tilde{f}$ denote the section of the cotangent bundle to $\mathcal{U}$ given by $d\tilde{f}(\mathbf{x}) = (\mathbf{x}, d_{\mathbf{x}}\tilde{f})$; we let $\mathrm{im}(d\tilde{f})$ denote the image of this section in $T^*\mathcal{U}$.

Thus, in coordinates,

$$\mathrm{im}(d\tilde{f}) = V\left(w_0 - \frac{\partial \tilde{f}}{\partial z_0}, \ldots, w_n - \frac{\partial \tilde{f}}{\partial z_n}\right).$$

Note that π, restricted to $\mathrm{im}(d\tilde{f})$, is an isomorphism onto $\mathcal{U}$, with inverse given by $\mathbf{x} \mapsto (\mathbf{x}, d_{\mathbf{x}}\tilde{f})$. In particular, we have an isomorphism

$$\overline{T^*_{X_{\mathrm{reg}}}\mathcal{U}} \cap \mathrm{im}(d\tilde{f}) \cong \pi\left(\overline{T^*_{X_{\mathrm{reg}}}\mathcal{U}} \cap \mathrm{im}(d\tilde{f})\right).$$

In [30], we gave a name to this last analytic set:

Definition 6.3.15 The **conormal-regular critical locus**, $\Sigma_{\mathrm{cnr}} f$, of f is defined to be

$$\Sigma_{\mathrm{cnr}} f := \pi\left(\overline{T^*_{X_{\mathrm{reg}}}\mathcal{U}} \cap \mathrm{im}(d\tilde{f})\right) = \left\{\mathbf{p} \in \mathcal{U} \mid (\mathbf{p}, d_{\mathbf{p}}\tilde{f}) \in \overline{T^*_{X_{\mathrm{reg}}}\mathcal{U}}\right\}.$$

Below, when we take intersection cycles and numbers, we will always be in the case of proper intersections inside the complex analytic manifold $T^*\mathcal{U}$ or inside $\mathcal{U}$ itself. In this case, there are well-defined intersection cycles (**not** cycles classes modulo rational equivalence); see [12].

Assuming that X is pure-dimensional and that $\mathbf{p}$ is a stratified isolated critical point of f, the relative local Euler obstruction, $\mathrm{Eu}_{\mathbf{p}} f$, is defined in [4] as an obstruction to extending the conjugate of the stratified gradient vector field of f to a non-zero lift in the Nash modification.

We will extend this definition to the case of non-isolated critical points; we are guided by Corollary 5.4 of [4], which states:

Proposition 6.3.16 *Suppose that X is pure-dimensional and that $f : X \to \mathbb{C}$ has a stratified isolated critical point at* $\mathbf{p}$. *Let* $\mathbf{K}^\bullet$ *be a characteristic complex for X. Then,* $(\mathbf{p}, d_{\mathbf{p}}\tilde{f})$ *is an isolated point in the intersection* $\overline{T^*_{X_{\mathrm{reg}}}\mathcal{U}} \cap \mathrm{im}(d\tilde{f})$ *(equivalently,* $\mathbf{p}$ *is an isolated point in* $\Sigma_{\mathrm{cnr}} f$*) and*

$$\mathrm{Eu}_{\mathbf{p}} f = \chi\left(\phi_{f-f(\mathbf{p})}[-1]\mathbf{K}^\bullet\right)_{\mathbf{p}} = (-1)^{\dim X}\left(\overline{T^*_{X_{\mathrm{reg}}}\mathcal{U}} \cdot \mathrm{im}(d\tilde{f})\right)_{(\mathbf{p}, d_{\mathbf{p}}\tilde{f})}.$$

Note that, in the case where X is affine space, this intersection number is the Milnor number of $f - f(\mathbf{p})$ *at* $\mathbf{p}$.

We can use Proposition 6.3.16 as the basis for generalizing the definition of the relative local Euler obstruction to (possibly) non-isolated critical points of functions on spaces which need not be pure-dimensional. In [4], we referred to this as the *defect* $D_{f,X}$.

Definition 6.3.17 Let $\mathbf{K}^\bullet$ be a characteristic complex for X. Then, we define the **relative, local Euler obstruction of** f at $\mathbf{p} \in X$ to be

$$\mathrm{Eu}_{\mathbf{p}} f := \chi\left(\phi_{f-f(\mathbf{p})}[-1]\mathbf{K}^\bullet\right)_{\mathbf{p}}.$$

Note that $\mathrm{Eu}_{\mathbf{p}} f$ is well-defined by Corollary 6.2.9.

Remark 6.3.18 Note that the definition immediately implies that, if c is a constant, then $\mathrm{Eu}_{\mathbf{p}} f = \mathrm{Eu}_{\mathbf{p}}(f + c)$. Also, note that, in [4], we referred to this generalized relative local Euler obstruction as the *defect* $D_{f,X}$.

In terms of constructible functions, what we have done is to define the relative local Euler obstruction to be the constructible shifted vanishing cycle function along f of the local Euler obstruction.

The relative local Euler obstruction is related to the local Euler obstructions of strata and the Euler characteristics of the intersections of the various strata with the Milnor fiber, $F_{f-f(\mathbf{p}),\mathbf{p}}$, of $f - f(\mathbf{p})$ at $\mathbf{p}$ via the following theorem. We proved this theorem, in slightly different terms, in [4], but the proof is very short, and we include it for completeness.

Theorem 6.3.19 *Let $\mathfrak{S}$ be a complex analytic stratification of X such that the local Euler obstruction of X is constant along the strata (e.g., a Whitney stratification with connected strata). For each $S \in \mathfrak{S}$, let $\mathbf{p}_S$ be a point in S. Then, for $\mathbf{p} \in X$,*

$$\mathrm{Eu}_{\mathbf{p}} f = \mathrm{Eu}_{\mathbf{p}} X - \sum_{S \in \mathfrak{S}} \chi(F_{f-f(\mathbf{p}),\mathbf{p}} \cap S)\, \mathrm{Eu}_{\mathbf{p}_S} X.$$

Proof Let $\mathbf{K}^\bullet$ be a characteristic complex for X. Then,

$$\mathrm{Eu}_{\mathbf{p}} f = \chi\big(\phi_{f-f(\mathbf{p})}[-1]\mathbf{K}^\bullet\big)_{\mathbf{p}}$$

which, by the canonical distinguished triangle relating the nearby and vanishing cycles, gives us

$$\mathrm{Eu}_{\mathbf{p}} f = -\chi\big(\mathbf{K}^\bullet[-1]\big)_{\mathbf{p}} + \chi\big(\psi_{f-f(\mathbf{p})}[-1]\mathbf{K}^\bullet\big)_{\mathbf{p}} =$$

$$\mathrm{Eu}_{\mathbf{p}} f - \chi\big(\mathbb{H}^*(F_{f-f(\mathbf{p})};\, \mathbf{K}^\bullet)\big)_{\mathbf{p}} = \mathrm{Eu}_{\mathbf{p}} X - \sum_{S \in \mathfrak{S}} \chi(F_{f-f(\mathbf{p}),\mathbf{p}} \cap S)\, \mathrm{Eu}_{\mathbf{p}_S} X.$$

□

Example 6.3.20 We should note that, given Theorem 6.3.19, our definition of the relative local Euler obstruction in terms of vanishing cycles implicitly appears in 4.6 of [38].

The next two examples contain known results; see, especially, [41].

Example 6.3.21 Consider the case where $f : (\mathcal{U}, \mathbf{0}) \to (\mathbb{C}, 0)$ is a non-locally constant function on an open subset $\mathcal{U} \subseteq \mathbb{C}^{n+1}$. Then, Theorem 6.3.19 tells us that

$$\mathrm{Eu}_{\mathbf{0}} f = 1 - \chi(F_{f,\mathbf{0}}) = -\widetilde{\chi}(F_{f,\mathbf{0}}),$$

where $\widetilde{\chi}$ denotes the Euler characteristic of the reduced cohomology.

In particular, if $\mathbf{0}$ is an **isolated** critical point of $f : (\mathcal{U}, \mathbf{0}) \to (\mathbb{C}, 0)$, then

$$\mathrm{Eu}_{\mathbf{0}} f = (-1)^{n+1} \mu_{\mathbf{0}}(f),$$

where $\mu_{\mathbf{0}}(f)$ is the Milnor number of f at $\mathbf{0}$.

Example 6.3.22 Now suppose that $\mathbf{0}$ is an isolated singular point of X, and we have $f : (X, \mathbf{0}) \to (\mathbb{C}, 0)$. Then, Theorem 6.3.19 tells us that

$$\mathrm{Eu}_{\mathbf{0}} f = \mathrm{Eu}_{\mathbf{0}} X - \chi(F_{f,\mathbf{0}}) = \chi(\mathbb{L}_{X,\mathbf{0}}) - \chi(F_{f,\mathbf{0}}).$$

In particular, if f itself is a generic linear form, then $\mathrm{Eu}_{\mathbf{0}} f = 0$. We shall see in Item 2 below that is true for arbitrarily singular spaces.

As another particular case, suppose that X is a curve and that $\tilde{f}$ is a local extension of f to the ambient affine space such that $\dim_{\mathbf{0}} X \cap V(\tilde{f}) = 0$. Then, we conclude that

$$\mathrm{Eu}_{\mathbf{0}} f = \mathrm{mult}_{\mathbf{0}} X - (X \cdot V(\tilde{f}))_{\mathbf{0}}.$$

We now give a number of basic properties of the relative local Euler obstruction.

Theorem 6.3.23 *Let $X = \bigcup_i X_i$ be the decomposition of X into its irreducible components, and let f_i denote the restriction of f to X_i.*

The relative local Euler obstruction has the following properties:

1. *If f is constant in a neighborhood of $\mathbf{p}$, then $\mathrm{Eu}_{\mathbf{p}} f = \mathrm{Eu}_{\mathbf{p}} X$.*
2. *If $\mathbf{p} \notin \Sigma_{\mathrm{cnr}} f$, then $\mathrm{Eu}_{\mathbf{p}} f = 0$. In particular, if $\mathbf{p}$ is not an isolated point of X, and L is the restriction of a generic linear form to X, then $\mathrm{Eu}_{\mathbf{p}} L = 0$.*
3. *$\mathrm{Eu}_{\mathbf{p}} f = \sum_i \mathrm{Eu}_{\mathbf{p}} f_i$, where we set $\mathrm{Eu}_{\mathbf{p}} f_i = 0$ if $\mathbf{p} \notin X_i$.*
4. *If $\mathbf{p}$ is an isolated point in $\Sigma_{\mathrm{cnr}} f$, then*

$$\mathrm{Eu}_{\mathbf{p}} f = \sum_i (-1)^{\dim X_i} \left(\overline{T^*_{(X_i)_{\mathrm{reg}}} \mathcal{U}} \cdot \mathrm{im}(d\tilde{f})\right)_{(\mathbf{p}, d_{\mathbf{p}}\tilde{f})}.$$

5. *Let $\mathbf{q} \in Y$ and suppose that we have a complex analytic function $g : Y \to \mathbb{C}$. Let $f \boxplus g$ denote the function from $X \times Y$ to $\mathbb{C}$ given by*

$$(f \boxplus g)(x, y) = f(x) + g(y).$$

Then,

$$\mathrm{Eu}_{(\mathbf{p},\mathbf{q})}(f \boxplus g) = \mathrm{Eu}_{\mathbf{p}} f \cdot \mathrm{Eu}_{\mathbf{q}} g.$$

Proof Item 1 follows at once from Theorem 6.3.19 or, alternatively from Corollary 6.3.10 and Definition 6.3.17.

Both Items 2 and 4 essentially follow from the vanishing cycle index theorem of Ginsburg [13], and Lê [22], but those papers require stronger hypotheses. However, the full results appears in 4.5 and 4.6 of Sabbah [38], and also in Corollary 0. 3 of Schürmann [39].

We can also conclude the results from looking ahead to Theorem 6.3.27 in Sect. 6.3.4.

Item 3:

Suppose that, for each i, $\mathbf{K}_i^\bullet$ is a characteristic complex for X_i and let $\widehat{\mathbf{K}}_i^\bullet$ denote the extension by zero of $\mathbf{K}_i^\bullet$ to all of X. Then, by Proposition 6.3.4, $\bigoplus_i \widehat{\mathbf{K}}_i^\bullet$ is a characteristic complex for X. Thus,

$$\mathrm{Eu}_{\mathbf{p}}\, f = \chi\Big(\phi_{f-f(\mathbf{p})}[-1]\bigoplus_i \widehat{\mathbf{K}}_i^\bullet\Big)_{\mathbf{p}} = \sum_i \chi\big(\phi_{f-f(\mathbf{p})}[-1]\widehat{\mathbf{K}}_i^\bullet\big)_{\mathbf{p}} =$$

$$\sum_i \chi\big(\phi_{f_i-f_i(\mathbf{p})}[-1]\mathbf{K}_i^\bullet\big)_{\mathbf{p}} = \sum_i \mathrm{Eu}_{\mathbf{p}}\, f_i.$$

Item 5:

Let us assume, without loss of generality, that $f(\mathbf{p}) = 0$ and $g(\mathbf{q}) = 0$. Let $\mathbf{K}_X^\bullet$ and $\mathbf{K}_Y^\bullet$ be characteristic complexes for X and Y, Then, we know from Proposition 6.3.7 that $\mathbf{K}_X^\bullet \overset{L}{\boxtimes} \mathbf{K}_Y^\bullet$ is a characteristic complex for $X \times Y$.

Now, the derived category version of the Sebastiani-Thom Theorem which we proved in [32] (but, here, using the more-usual definition/shift on the vanishing cycles) tells us that

$$H^k\big(\phi_{f\boxplus g}[-1](\mathbf{K}_X^\bullet \overset{L}{\boxtimes} \mathbf{K}_Y^\bullet)\big)_{(\mathbf{p},\mathbf{q})} \cong H^k\big(\phi_f[-1]\mathbf{K}_X^\bullet \overset{L}{\boxtimes} \phi_g[-1]\mathbf{K}_Y^\bullet\big)_{(\mathbf{p},\mathbf{q})} \cong$$

$$\bigoplus_{i+j=k} H^i(\phi_f[-1]\mathbf{K}_X^\bullet)_{\mathbf{p}} \otimes H^j(\phi_g[-1]\mathbf{K}_Y^\bullet)_{\mathbf{q}} \oplus \bigoplus_{i+j=k+1} \mathrm{Tor}\big(H^i(\phi_f[-1]\mathbf{K}_X^\bullet)_{\mathbf{p}}, H^j(\phi_g[-1]\mathbf{K}_Y^\bullet)_{\mathbf{q}}\big).$$

Item 5 follows. □

Remark 6.3.24 We naturally refer to Item 5 above as the *Sebastiani-Thom property* of the relative local Euler obstruction.

We should also remark that the quantity

$$\Big(\overline{T^*_{(X_i)_{\mathrm{reg}}}\mathcal{U}} \cdot \mathrm{im}(d\tilde{f})\Big)_{(\mathbf{p},d_{\mathbf{p}}\tilde{f})}$$

in Item 5 above can be characterized more geometrically as the number of non-degenerate critical points of a small perturbation of f by a generic linear form L which occur near the origin on $(X_i)_{\mathrm{reg}}$; see Theorem 3.2 of [29] and Theorem 1.1 of [31]. Thus, Item 5 is a significant generalization of Proposition 2.3 of [41], as we do not require that f has a stratified isolated critical point.

Example 6.3.25 Let (x, y, z) be coordinates on $\mathcal{U} := \mathbb{C}^3$, and let (w_0, w_1, w_2) be the corresponding cotangent coordinates. Let

$$X := V(xy) = V(x) \cup V(y) \subseteq \mathbb{C}^3.$$

Let $\tilde{f} : \mathcal{U} \to \mathbb{C}$ be given by $\tilde{f}(x, y, z) := x + y^2 + yz$, and let f be the restriction of $\tilde{f}$ to X.

Note that f does **not** possess a stratified isolated singularity at $\mathbf{0}$, since $\Sigma X = V(x, y) = \{\mathbf{0}\} \times \mathbb{C}$ and $f_{|V(x,y)} \equiv 0$. However, we claim that $(\mathbf{0}, d_\mathbf{0}\tilde{f})$ **is** an isolated point in $\overline{T^*_{X_{\text{reg}}}\mathcal{U}} \cap \operatorname{im} d\tilde{f}$.

We find

$$\overline{T^*_{X_{\text{reg}}}\mathcal{U}} = V(x, w_1, w_2) \cup V(y, w_0, w_2)$$

and

$$\operatorname{im} d\tilde{f} = V(w_0 - 1, w_1 - 2y - z, w_2 - y).$$

Therefore, $\operatorname{im} d\tilde{f}$ does not intersect $V(y, w_0, w_2)$, and intersects $V(x, w_1, w_2)$ in the single point $V(x, y, z, w_0 - 1, w_1, w_2)$.

Hence, Item 4 of Theorem 6.3.23 tells us that

$$\operatorname{Eu}_\mathbf{0} f = (-1)^2 \left(\overline{T^*_{(X)_{\text{reg}}}\mathcal{U}} \cdot \operatorname{im}(d\tilde{f})\right)_{(\mathbf{0}, d_\mathbf{0}\tilde{f})} =$$

$$\Big(\big(V(x, w_1, w_2) + V(y, w_0, w_2)\big) \cdot V(w_0 - 1, w_1 - 2y - z, w_2 - y)\Big)_{(\mathbf{0},1,0,0)} = 1.$$

Example 6.3.26 Suppose that Y is a curve through the origin in $\mathbb{C}^2$. Let $g : \mathbb{C}^2 \to C$ be such that $\dim_\mathbf{0} Y \cap V(g) = 0$. Let $X := Y \times \mathbb{C}$, where we use z as the coordinate on this new copy of $\mathbb{C}$.

Consider the function $f : (X, \mathbf{0}) \to (\mathbb{C}, 0)$ given by $f(x, y, z) = g(x, y) + z^b$, for some positive integer b.

Then, by Item 5 of Theorem 6.3.23—the Sebastiani-Thom property—combined with Example 6.3.21 and the last part of Example 6.3.22—we find that

$$\operatorname{Eu}_\mathbf{0} f = \big(\operatorname{mult}_\mathbf{0} Y - (Y \cdot V(g))_\mathbf{0}\big)(1 - b).$$

6.3.4 Calculating the Relative Local Euler Obstruction

Here, we recall the algorithm and result which is described in Section 6 of [4].

Once again, let $X = \bigcup_i X_i$ be the decomposition of X into its irreducible components, and let f_i denote the restriction of f to X_i. Recall from Item 4 of Theorem 6.3.23 that $\mathrm{Eu}_{\mathbf{p}} f = \sum_i \mathrm{Eu}_{\mathbf{p}} f_i$. Therefore, calculating $\mathrm{Eu}_{\mathbf{p}} f$ boils down to needing to calculate in the case where X is irreducible.

Thus, suppose throughout this section that X is irreducible, embedded in $\mathcal{U}$, an open subset of $\mathbb{C}^{n+1}$.

Recall our previous set-up:

We let $\mathbf{z} := (z_0, \dots, z_n)$ be coordinates on $\mathcal{U}$, we identify the cotangent space $T^*\mathcal{U}$ with $\mathcal{U} \times \mathbb{C}^{n+1}$ by mapping $(\mathbf{p}, w_0 d_{\mathbf{p}} z_0 + \cdots + w_n d_{\mathbf{p}} z_n)$ to $(\mathbf{p}, (w_0, \dots, w_n))$, and let $\pi : T^*\mathcal{U} \to \mathcal{U}$ denote the projection.

Assume that $\mathbf{p} \in X$, that $\tilde{f} : \mathcal{U} \to \mathbb{C}$ is a complex analytic function, and that f is the restriction of $\tilde{f}$ to X. We let $d\tilde{f}$ denote the section of the cotangent bundle to $\mathcal{U}$ given by $d\tilde{f}(\mathbf{x}) = (\mathbf{x}, d_{\mathbf{x}}\tilde{f})$; we let $\mathrm{im}(d\tilde{f})$ denote the image of this section in $T^*\mathcal{U}$.

Thus, in coordinates,

$$\mathrm{im}(d\tilde{f}) = V\left(w_0 - \frac{\partial \tilde{f}}{\partial z_0}, \dots, w_n - \frac{\partial \tilde{f}}{\partial z_n}\right).$$

Our method of calculation requires a "generic" choice of coordinates for the ambient, affine space $\mathcal{U}$. This choice of coordinates is made as follows. Refine a Whitney stratification of X to a stratification $\mathcal{W} := \{W_j\}$ such that $\mathcal{W}$ satisfies Thom's a_f condition and such that $f^{-1}(0)$ is a union of strata of $\mathcal{W}$ (we are not assuming that $\mathcal{W}$ is still a Whitney stratification).

Choose the first coordinate z_0 so that the hyperplane $z_0^{-1}(0)$ transversely intersects, in some neighborhood of the origin, all positive-dimensional strata of $\{W_j\}$. Then, there is an induced stratification (on the germ at the origin) of $X \cap z_0^{-1}(0)$ given by $\{W_j \cap z_0^{-1}(0)\}$. We choose z_1 so that $z_1^{-1}(0)$ transversely intersects, in some neighborhood of the origin, all positive-dimensional strata of $\{W_j \cap z_0^{-1}(0)\}$. We continue in this inductive manner to produce $z_0, \dots, z_{n+1}$. We call such a coordinate choice *prepolar* (at the origin).

Below, we will discuss the Lê-Vogel cycles and numbers with respect to a choice of coordinates; prepolar coordinates are generic enough so that these cycles and numbers are defined and have many desirable properties. Note that, while different sets of prepolar coordinates may yield different Lê-Vogel numbers, the Lê-Vogel numbers do have generic values which are obtained by putting more conditions on the coordinates; however, for the sake of being able to check if our coordinates are "generic enough" for calculations in a given example, "prepolar" is the condition that we want to use.

It is important to notice that $\mathbf{z}$ being prepolar at the origin implies that the coordinates are also prepolar at each point in a neighborhood of the origin, and we assume that we are in such a neighborhood throughout the remainder of this section.

Assuming that the coordinates are prepolar for f at the origin, all of the intersections that we write below are proper in $T^*\mathcal{U}$ (resp., in $\mathcal{U}$) in a neighborhood of $(\mathbf{p}, d_{\mathbf{p}}\tilde{f})$ (resp., $\mathbf{p}$).

The cycle $\left[\overline{T^*_{X_{\mathrm{reg}}}\mathcal{U}}\right]$ can be written as a sum of purely $(n+1)$-dimensional cycles

$$\left[\overline{T^*_{X_{\mathrm{reg}}}\mathcal{U}}\right] = \widehat{\Gamma}^{n+1}_{f,\mathbf{z}} + \widehat{\Lambda}^{n+1}_{f,\mathbf{z}},$$

where no component of $\widehat{\Gamma}^{n+1}_{f,\mathbf{z}}$ is contained in $\mathrm{im}(d\tilde{f})$ (i.e. in $\overline{T^*_{X_{\mathrm{reg}}}\mathcal{U}} \cap \mathrm{im}(d\tilde{f})$) in and every component of $\widehat{\Lambda}^{n+1}_{f,\mathbf{z}}$ is contained in $\mathrm{im}(d\tilde{f})$.

Now, we define $\widehat{\Gamma}^{k}_{f,\mathbf{z}}$ and $\widehat{\Lambda}^{k}_{f,\mathbf{z}}$ by downward induction. If we have defined the purely $(k+1)$-dimensional cycle $\widehat{\Gamma}^{k+1}_{f,\mathbf{z}}$, then the hypersurface

$$\left[V\left(w_{k+1} - \frac{\partial \tilde{f}}{\partial z_{k+1}}\right)\right]$$

properly intersects $\widehat{\Gamma}^{k+1}_{f,\mathbf{z}}$ inside U, and therefore there is a well-defined, purely k-dimensional intersection cycle

$$\widehat{\Gamma}^{k+1}_{f,\mathbf{z}} \cdot \left[V\left(w_{k+1} - \frac{\partial \tilde{f}}{\partial z_{k+1}}\right)\right],$$

which we can decompose as

$$\widehat{\Gamma}^{k+1}_{f,\mathbf{z}} \cdot \left[V\left(w_{k+1} - \frac{\partial \tilde{f}}{\partial z_{k+1}}\right)\right] =: \widehat{\Gamma}^{k}_{f,\mathbf{z}} + \widehat{\Lambda}^{k}_{f,\mathbf{z}},$$

where no component of $\widehat{\Gamma}^{k}_{f,\mathbf{z}}$ is contained in $\mathrm{im}(d\tilde{f})$ and every component of $\widehat{\Lambda}^{k}_{f,\mathbf{z}}$ is contained in $\mathrm{im}(d\tilde{f})$.

As the $\widehat{\Lambda}^{k}_{f,\mathbf{z}}$ are contained in $\mathrm{im}(d\tilde{f})$, the projection, π, maps each $\widehat{\Lambda}^{k}_{f,\mathbf{z}}$ isomorphically onto a cycle in $\mathcal{U}$; we let $\Lambda^{k}_{f,\mathbf{z}} := \pi_*\big(\widehat{\Lambda}^{k}_{f,\mathbf{z}}\big)$ (this is the proper projection of a cycle). We refer to $\Lambda^{k}_{f,\mathbf{z}}$ as the *k-dimensional Lê-Vogel cycle*.

Note that in the case where $X = \mathcal{U}$, $\left[\overline{T^*_{X_{\mathrm{reg}}}\mathcal{U}}\right] = \mathcal{U} \times \{0\}$, and the Lê-Vogel cycles coincide with the Lê cycles of [25, 26], and [28].

Now, $\Lambda^{k}_{f,\mathbf{z}}$ properly intersects the linear subspace $V(z_0 - p_0, \ldots, z_{k-1} - p_{k-1})$ at $\mathbf{p}$, and we define the *k-dimensional Lê-Vogel number of f at* $\mathbf{p}$ to be the intersection number

$$\lambda^{k}_{f,\mathbf{z}}(\mathbf{p}) := \big(\Lambda^{k}_{f,\mathbf{z}} \cdot V(z_0 - p_0, \ldots, z_{k-1} - p_{k-1})\big)_{\mathbf{p}}.$$

Theorem 6.3.27 (Theorem 6.2 of [4]) *Suppose that X is irreducible, and let $\mathbf{K}^\bullet$ be a characteristic complex for X.*

Let $s := \dim_{\mathbf{p}} \Sigma_{\mathrm{cnr}} f$ (where we set $s = -\infty$ if $\mathbf{p} \notin \Sigma_{\mathrm{cnr}} f$), and assume that the coordinates $\mathbf{z}$ are prepolar for f at $\mathbf{0}$.

Then, the Lê-Vogel numbers $\lambda^k_{f,\mathbf{z}}(\mathbf{p})$ are zero if $k > s$, and

$$\mathrm{Eu}_{\mathbf{p}}\, f = \chi(\phi_f[-1]\mathbf{K}^\bullet)_{\mathbf{p}} = (-1)^{\dim X} \sum_{k=0}^{s} (-1)^k \lambda^k_{f,\mathbf{z}}(\mathbf{p}).$$

In particular, when $s = 0$, the only Lê-Vogel number which is possibly non-zero is $\lambda^0_{f,\mathbf{z}}(\mathbf{p})$, and $\lambda^0_{f,\mathbf{z}}(\mathbf{p}) = \left(\overline{T^*_{X_{\mathrm{reg}}}\mathcal{U}} \cdot \mathrm{im}(d\tilde{f})\right)_{(\mathbf{p}, d_{\mathbf{p}}\tilde{f})}$; thus, when $s = 0$,

$$\mathrm{Eu}_{\mathbf{p}}\, f = (-1)^{\dim X} \left(\overline{T^*_{X_{\mathrm{reg}}}\mathcal{U}} \cdot \mathrm{im}(d\tilde{f})\right)_{(\mathbf{p}, d_{\mathbf{p}}\tilde{f})}.$$

The serious weakness of Theorem 6.3.27 is that there is no effective way to obtain the a_f stratification of X which we need in order to know if our coordinates are prepolar or not. The case where $\mathbf{p}$ is an isolated point of $\Sigma_{\mathrm{cnr}} f$ is nice because the resulting formula is independent of the coordinates. However, we can also handle special 1-dimensional cases fairly easily.

Example 6.3.28 Suppose X is irreducible at $\mathbf{p}$, of dimension d, and suppose that $\dim_{\mathbf{p}} \Sigma X \leq 1$. Then, near $\mathbf{p}$, there is a Whitney stratification of X consisting of the connected components of X_{reg}, the connected components of $\Sigma X - \{\mathbf{p}\}$, and $\{\mathbf{p}\}$.

Now, suppose also that $\dim_{\mathbf{p}} \overline{\Sigma(f_{|X_{\mathrm{reg}}})} \leq 1$. Let $\Sigma := \Sigma X \cup \Sigma(f_{|X_{\mathrm{reg}}})$. Then there is an a_f stratification consisting of the connected components of $X_{\mathrm{reg}} - V(f)$, the connected components of $V(f) - \Sigma$, the connected components of $\Sigma - \{\mathbf{p}\}$, and $\{\mathbf{p}\}$.

Then, one easily sees that the requirement that $(z_0, z_1, \dots, z_n)$ be polar is equivalent to requiring that, near $\mathbf{p}$,

- $V(z_0 - p_0)$ contains no irreducible component of Σ;
- for $0 \leq i \leq d-1$, $V(z_0 - p_0, \dots, z_i - p_i)$ transversely intersects X_{reg} near $\mathbf{p}$; and
- for $0 \leq i \leq d-1$, $\Sigma\left(f_{|X_{\mathrm{reg}} \cap V(z_0 - p_0, \dots, z_i - p_i)}\right) = \emptyset$.

Let us consider a specific example.

Let $\mathcal{U} := \mathbb{C}^3$ and let $X := V(y^2 - x^3) \subseteq \mathcal{U}$, where we use coordinates (t, x, y) on $\mathcal{U}$. Thus, X is a cross-product of a cusp and $\mathbb{C}$, and has a Whitney stratification consisting of $\{X - V(x, y), V(x, y)\}$.

We wish to determine $\overline{T^*_{X_{\mathrm{reg}}}\mathcal{U}}$. Using (t, x, y, w_0, w_1, w_2) for coordinates on $T^*\mathcal{U} \cong \mathcal{U} \times \mathbb{C}^3$, one looks at the vanishing of $y^2 - x^3$ and the 2×2 minors of the matrix

$$\begin{pmatrix} 0 & -3x^2 & 2y \\ w_0 & w_1 & w_2 \end{pmatrix},$$

and disposes of those irreducible components which lie over $\Sigma X = V(x, y)$.

We find that, as analytic sets,

$$\overline{T^*_{X_{\mathrm{reg}}}\mathcal{U}} = \overline{V(y^2 - x^3, w_0x^2, w_0y, 2w_1y + 3w_2x^2) - V(x, y)} =$$

$$\overline{V(y^2 - x^3, w_0, 2w_1y + 3w_2x^2) - V(x, y)}.$$

Thus, as cycles,

$$V(y^2 - x^3, w_0, 2w_1y + 3w_2x^2) = \overline{T^*_{X_{\mathrm{reg}}}\mathcal{U}} + mV(x, y, w_0), \qquad (\dagger)$$

for some positive integer m.

In fact, it is easy to show that

$$\overline{T^*_{X_{\mathrm{reg}}}\mathcal{U}} = V(y^2 - x^3, w_0, 2w_1y + 3w_2x^2, 4w_1^2 - 9w_2^2x),$$

but, as this is not defined by a regular sequence, it is somewhat problematic to deal with this in the intersections below, so (†) is more useful.

Let $\tilde{f} : \mathcal{U} \to \mathbb{C}$ be given by $\tilde{f}(t, x, y) = 2y - 3tx + t^3$, and let $f := \tilde{f}_{|X}$. Then one easily checks that

$$\overline{\Sigma(f_{|X_{\mathrm{reg}}})} = V(x - t^2, y - t^3),$$

and (t, x, y) are prepolar coordinates at $\mathbf{0}$. Thus, we may use the method of Theorem 6.3.27 to calculate the relative local Euler obstruction.

We will suppress the reference to the coordinate system (t, x, y) in the subscripts below.

We have

$$\operatorname{im}(d\tilde{f}) = V\Big(w_0 + 3(x - t^2),\ w_1 + 3t,\ w_2 - 2\Big),$$

and one easily finds that

$$\overline{T^*_{X_{\mathrm{reg}}}\mathcal{U}} \cap \operatorname{im}(d\tilde{f}) = V\left(y - t^3, x - t^2, w_0, w_1 + 3t, w_2 - 2\right),$$

which is 1-dimensional. Note for later that this implies that $\widehat{\Lambda}^2_f = 0$.

We wish to proceed with the algorithm described just before Theorem 6.3.27, but we get slightly "tricky" in the first intersection, to avoid the problem that $\overline{T^*_{X_{\text{reg}}}\mathcal{U}}$ is not defined by a regular sequence.

We begin:

$$\widehat{\Gamma}^3_f = \overline{T^*_{X_{\text{reg}}}\mathcal{U}}.$$

Now, we use (†):

$$V(y^2 - x^3, w_0, 2w_1 y + 3w_2 x^2) \cdot V(w_2 - 2) =$$

$$\overline{T^*_{X_{\text{reg}}}\mathcal{U}} \cdot V(w_2 - 2) \ + \ mV(x, y, w_0) \cdot V(w_2 - 2).$$

Thus, we conclude that

$$V(y^2 - x^3, w_0, w_1 y + 3x^2, w_2 - 2) = \widehat{\Gamma}^2_f + mV(x, y, w_0, w_2 - 2), \qquad (\ddagger)$$

where we have used our earlier observation that $\widehat{\Lambda}^2_f$ must equal 0.

We wish to investigate the purely 2-dimensional cycle

$$Y := V(y^2 - x^3, w_0, w_1 y + 3x^2, w_2 - 2).$$

As sets, $V(w_1) \cap Y = V(x, y, w_0, w_1, w_2 - 2)$, which is 1-dimensional. Therefore, we may calculate the cycle structure of Y by looking at the structure where $w_1 \neq 0$.

Via this approach, we find

$$V(y^2 - x^3, w_0, w_1 y + 3x^2, w_2 - 2) =$$

$$V(27y + w_1^3, 9x - w_1^2, w_0, w_2 - 2) + 3V(x, y, w_0, w_2 - 2).$$

Now (‡) implies that

$$\widehat{\Gamma}^2_f = V(27y + w_1^3, 9x - w_1^2, w_0, w_2 - 2).$$

We proceed:

$$\widehat{\Gamma}^2_f \cdot V(w_1 + 3t) = V(27y + w_1^3, 9x - w_1^2, w_0, w_2 - 2) \cdot V(w_1 + 3t) =$$

$$V\left(y - t^3, x - t^2, w_0, w_1 + 3t, w_2 - 2\right),$$

which as a set equals $\overline{T^*_{X_{\text{reg}}}\mathcal{U}} \cap \operatorname{im}(d\tilde{f})$.

Therefore,

$$\widehat{\Gamma}^1_f = 0 \text{ and } \widehat{\Lambda}^1_f = V\left(y - t^3, x - t^2, w_0, w_1 + 3t, w_2 - 2\right).$$

Since $\widehat{\Gamma}^1_f = 0$, we have that $\widehat{\Lambda}^0_f = 0$. Thus there is only one non-zero Lê-Vogel cycle:

$$\Lambda^1_f = \pi_*\big(\widehat{\Lambda}^1_f\big) = \pi_*\big(V(y-t^3, x-t^2, w_0, w_1+3t, w_2-2)\big) = V(y-t^3, x-t^2),$$

with corresponding Lê-Vogel number at the origin:

$$\lambda^1_f(\mathbf{0}) = \big(V(y-t^3, x-t^2)\cdot V(t)\big)_{\mathbf{0}} = 1.$$

Finally, Theorem 6.3.27 tells us that

$$\mathrm{Eu}_{\mathbf{0}}\, f = (-1)^2(-1)\lambda^1_f(\mathbf{0}) = -1.$$

The following exercise is not trivial, but it is also not too difficult.

Exercise 6.3.29 Use Theorem 6.3.19 to verify the calculation above that $\mathrm{Eu}_{\mathbf{0}}\, f = -1$.

6.4 Basics of Enriched Cycles

In Definition 6.2.10, we defined the graded, enriched characteristic cycle. In this section, we wish to describe graded, enriched cycles more generally and carefully. We also describe the associated intersection theory. The intersection theory that we use is a fairly simple extension of the intersection theory of properly intersecting cycles in an analytic manifold, as described in Section 8.2 of [12]. In this case, one obtains intersection cycles (with specific underlying analytic sets), **not rational equivalence classes of cycles**.

Definition 6.4.1 An *enriched cycle*, E, in X is a formal, locally finite sum $\sum_V E_V[V]$, where the V's are irreducible analytic subsets of X and the E_V's are finitely-generated R-modules. We refer to the V's as the *components* of E, and to E_V as the *V-component module of E*. Two enriched cycles are considered the same provided that all of the component modules are isomorphic. The underlying set of E is $|E| := \cup_{E_V \neq 0} V$.

If $C = \sum n_V[V]$ is an ordinary positive cycle in X, i.e., all of the n_v are non-negative integers, then there is a corresponding enriched cycle $[C]^{\mathrm{enr}}$ in which the V-component module is the free R-module of rank n_V. If R is an integral domain, so that rank of an R-module is well-defined, then an enriched cycle E yields an ordinary cycle $[E]^{\mathrm{ord}} := \sum_V(\mathrm{rk}(E_V))[V]$.

If q is a finitely-generated module and E is an enriched cycle, then we let $qE := \sum_V (q \otimes E_V)[V]$; thus, if R is an integral domain and E is an enriched cycle, $[qE]^{\text{ord}} = (\text{rk}(q))[E]^{\text{ord}}$ and if C is an ordinary positive cycle and n is a positive integer, then $[nC]^{\text{enr}} = R^n[C]^{\text{enr}}$.

The (direct) sum of two enriched cycles D and E is given by $(D+E)_V := D_V \oplus E_V$.

There is a partial ordering on isomorphism classes of finitely-generated R-modules given by $M \leq Q$ if and only if there exists a finitely-generated R-module N such that $M \oplus N \cong Q$. This relation is clearly reflexive and transitive; moreover, anti-symmetry follows from the fact that if M and N are Noetherian modules such that $M \oplus N \cong M$, then $N = 0$. This partial ordering extends to a partial ordering on enriched cycles given by: $D \leq E$ if and only if there exists an enriched cycle P such that $D + P = E$. If the base ring R is a PID and $D + P = E$, then D is uniquely determined by P and E, and we write $D = E - P$.

If two irreducible analytic subsets V and W intersect properly in $\mathcal{U}$, then the (ordinary) intersection cycle $[V] \cdot [W]$ is a well-defined positive cycle; we define the enriched intersection product of $[V]^{\text{enr}}$ and $[W]^{\text{enr}}$ by $[V]^{\text{enr}} \odot [W]^{\text{enr}} = ([V]\cdot[W])^{\text{enr}}$. If D and E are enriched cycles, and every component of D properly intersects every component of E in $\mathcal{U}$, then we say that *D and E intersect properly* in $\mathcal{U}$ and we extend the intersection product linearly, i.e., if $D = \sum_V D_V[V]$ and $E = \sum_W E_W[W]$, then

$$D \odot E := \sum_{V,W} (D_V \otimes E_W)([V]\cdot[W])^{\text{enr}}.$$

A *graded, enriched cycle* $E^\bullet$ is simply an enriched cycle E^i for each i in some bounded set of integers. An single enriched cycle is considered as a graded enriched cycle by being placed totally in degree zero. The analytic set V is a *component of* $E^\bullet$ if and only if V is a component of E^i for some i, and the underlying set of $E^\bullet$ is $|E^\bullet| = \cup_i |E^i|$. If R is a domain, then $E^\bullet$ yields an ordinary cycle $[E^\bullet]^{\text{ord}} := \sum_i (-1)^i (\text{rk}(E^i_V))[V]$. If k is an integer, we define the *k-shifted graded, enriched cycle* $E^\bullet[k]$ by $(E^\bullet[k])^i := E^{i+k}$.

If q is a finitely-generated module and $E^\bullet$ is a graded enriched cycle, then we define the graded enriched cycle $qE^\bullet$ by $(qE^\bullet)^i := \sum_V (q \otimes E^i_V)[V]$. The (direct) sum of two graded enriched cycles $D^\bullet$ and $E^\bullet$ is given by $(D^\bullet + E^\bullet)^i_V := D^i_V \oplus E^i_V$. If D^i properly intersects E^j for all i and j, then we say that $D^\bullet$ and $E^\bullet$ *intersect properly* and we define the intersection product by

$$(D^\bullet \odot E^\bullet)^k := \sum_{i+j=k} (D^i \odot E^j).$$

Whenever we use the enriched intersection product symbol, we mean that we are considering the objects on both sides of $\odot$ as graded, enriched cycles, even if we do not superscript by enr or $\bullet$.

Let $\tau : W \to Y$ be a proper morphism between analytic spaces. If $C = \sum n_V[V]$ is an ordinary positive cycle in W, then the proper push-forward $\tau_*(C) = \sum n_V \tau_*([V])$ is a well-defined ordinary cycle.

Definition 6.4.2 If $E^\bullet = \sum_V E^\bullet_V[V]$ is an enriched cycle in W, then we define the **proper push-forward of $E^\bullet$ by τ** to be the graded enriched cycle $\tau^\bullet_*(E^\bullet)$ defined by

$$\tau^j_*(E^\bullet) := \sum_V E^j_V[\tau_*([V])]^{\mathrm{enr}}.$$

The ordinary projection formula for divisors ([**F**], 2.3.c) immediately implies the following enriched version.

Proposition 6.4.3 *Let $E^\bullet$ be a graded enriched cycle in X. Let $W := |E^\bullet|$. Let $\tau : W \to Y$ be a proper morphism, and let $g : Y \to \mathbb{C}$ be an analytic function such that $g \circ \tau$ is not identically zero on any component of $E^\bullet$. Then, g is not identically zero on any component of $\tau^\bullet_*(E^\bullet)$ and*

$$\tau^\bullet_*\big(E^\bullet \odot V(g \circ \tau)\big) = \tau^\bullet_*(E^\bullet) \odot V(g).$$

6.5 The Relative Polar Curve

We will use the notation established in Sect. 6.2: $\mathcal{U}$ is an open neighborhood of the origin in $\mathbb{C}^{n+1}$, X is a closed, analytic subset of $\mathcal{U}$, $\mathfrak{S}$ is a complex analytic Whitney stratification of X, with connected strata, R is a base ring that is a regular, Noetherian ring with finite Krull dimension, $\mathbf{F}^\bullet$ is a bounded complex of sheaves of R-modules on X, which is constructible with respect to $\mathfrak{S}$, $\mathbf{z} = (z_0, \dots, z_n)$ is a set of coordinates on $\mathcal{U}$, we identify the cotangent space $T^*\mathcal{U}$ with $\mathcal{U} \times \mathbb{C}^{n+1}$ by mapping $(\mathbf{p}, w_0 d_\mathbf{p} z_0 + \cdots + w_n d_\mathbf{p} z_n)$ to $(\mathbf{p}, (w_0, \dots, w_n))$, for each $S \in \mathfrak{S}$, $d_S = \dim S$, and $(\mathbb{N}_S, \mathbb{L}_S)$ is the complex Morse data for S in X, consisting of a normal slice and complex link of S in X.

Let $\tilde{f}$ and $\tilde{g}$ be analytic functions from $(\mathcal{U}, \mathbf{0})$ to $(\mathbb{C}, 0)$, and let f and g denote the restrictions of $\tilde{f}$ and $\tilde{g}$, respectively, to X. By refining $\mathfrak{S}$, if necessary, we assume that $V(f)$ is a union of strata.

Definition 6.5.1 Let $\mathfrak{S}(\mathbf{F}^\bullet) := \{S \in \mathfrak{S} \mid \mathbb{H}^*(\mathbb{N}_S, \mathbb{L}_S; \mathbf{F}^\bullet) \neq 0\}$; we refer to the elements of $\mathfrak{S}(\mathbf{F}^\bullet)$ as the $\mathbf{F}^\bullet$*-visible strata of* $\mathfrak{S}$.

Fix a point $\mathbf{p} \in \mathcal{U}$. In [16, 19, 20, 42], Hamm, Teissier, and Lê define and use the relative polar curve (of $\tilde{f}$ with respect to z_0), $\Gamma^1_{\tilde{f}, z_0}$, to prove a number of topological results related to the Milnor fiber of hypersurface singularities. We shall recall some definitions and results here. We should mention that there are a number of different characterizations of the relative polar curve, all of which agree when z_0

is sufficiently generic; below, we have selected what we consider the easiest way of describing the relative polar curve as a set, a scheme, and a cycle.

As a set, $\Gamma^1_{\tilde{f},z_0}$ is the closure of the critical locus of $(\tilde{f}, z_0)$ minus the critical locus of $\tilde{f}$, i.e., $\Gamma^1_{\tilde{f},z_0}$ equals $\overline{\Sigma(\tilde{f}, z_0) - \Sigma\tilde{f}}$, as a set. If z_0 is sufficiently generic for $\tilde{f}$ at p, then, in a neighborhood of p, $\Gamma^1_{\tilde{f},z_0}$ will be purely one-dimensional (which includes the possibility of being empty).

It is not difficult to give $\Gamma^1_{\tilde{f},z_0}$ a scheme structure. If $\Gamma^1_{\tilde{f},z_0}$ is purely one-dimensional at $\mathbf{p}$, then, at points $\mathbf{x}$ near, but unequal to, $\mathbf{p}$, $\Gamma^1_{\tilde{f},z_0}$ is given the structure of the scheme $V\left(\frac{\partial\tilde{f}}{\partial z_1}, \dots, \frac{\partial\tilde{f}}{\partial z_n}\right)$. One can also "algebraically" remove any embedded components of $\Gamma^1_{\tilde{f},z_0}$ at $\mathbf{p}$ by using *gap sheaves*; see Chapter 1 of [28].

In practice, all topological applications of the relative polar curve use only its structure as an analytic cycle (germ), that is, as a locally finite sum of irreducible analytic sets (or germs of sets) counted with integral multiplicities (which will all be non-negative).

If C is a one-dimensional irreducible germ of $\Gamma^1_{\tilde{f},z_0}$ at $\mathbf{p}$, and $\mathbf{x} \in C$ is close to, but unequal to, $\mathbf{p}$, then the component C appears in the cycle $\Gamma^1_{\tilde{f},z_0}$ with multiplicity given by the Milnor number of $\tilde{f}_{|_H}$ at $\mathbf{x}$, where H is a generic affine hyperplane passing through $\mathbf{x}$.

Suppose that M is a complex submanifold of $\mathcal{U}$. Recall:

Definition 6.5.2 The relative conormal space $T^*_{\tilde{f}_{|_M}}\mathcal{U}$ is given by

$$T^*_{\tilde{f}_{|_M}}\mathcal{U} := \{(x, \eta) \in T^*\mathcal{U} \mid \eta(T_xM \cap \ker d_x\tilde{f}) = 0\}.$$

If $M \subseteq X$, then $T^*_{\tilde{f}_{|_M}}\mathcal{U}$ depends on f, but not on the particular extension $\tilde{f}$. In this case, we write $T^*_{f_{|_M}}\mathcal{U}$ in place of $T^*_{\tilde{f}_{|_M}}\mathcal{U}$.

Definition 6.5.3 The **graded, enriched relative conormal cycle**, $\left(T^*_{f,\mathbf{F}^\bullet}\mathcal{U}\right)^\bullet$, **of** f, **with respect to** $\mathbf{F}^\bullet$, is defined by

$$\left(T^*_{f,\mathbf{F}^\bullet}\mathcal{U}\right)^k := \sum_{\substack{S\in\mathfrak{S}(\mathbf{F}^\bullet)\\ f_{|_S}\neq \text{ const.}}} m^k_S(\mathbf{F}^\bullet)\left[\overline{T^*_{f_{|_S}}\mathcal{U}}\right].$$

We now wish to define the graded, enriched relative polar curve. We will consider the image, $\operatorname{im} d\tilde{g}$, of $d\tilde{g}$ in $T^*\mathcal{U}$; this scheme is defined by

$$V\left(w_0 - \frac{\partial\tilde{g}}{\partial z_0}, \dots, w_n - \frac{\partial\tilde{g}}{\partial z_n}\right) \subseteq \mathcal{U} \times \mathbb{C}^{n+1}.$$

We will consider $\operatorname{im} d\tilde{g}$ as a scheme, an analytic set, an ordinary cycle, and as a graded, enriched cycle; we will denote all of these by simply $\operatorname{im} d\tilde{g}$, and explicitly state what structure we are using or let the context make the structure clear.

Note that the projection π induces an isomorphism from the analytic set $\operatorname{im} d\tilde{g}$ to $\mathcal{U}$. We will use the proper push-forward (Definition 6.4.2) of the map π restricted to $\operatorname{im} d\tilde{g}$; we will continue to denote this restriction by simply π.

By our conventions in Sect. 6.4, the graded, enriched cycle $\operatorname{im} d\tilde{g}$ is zero outside of degree 0, and is the enriched cycle $R[\operatorname{im} d\tilde{g}]$ in degree 0.

Definition 6.5.4 If $S \in \mathfrak{S}$ and $f_{|S}$ is not constant, we define the **relative polar set**, $|\Gamma_{f,\tilde{g}}(S)|$, to be $\pi\left(\overline{T^*_{f_{|S}}\mathcal{U}} \cap \operatorname{im} d\tilde{g}\right)$; if this set is purely 1-dimensional, so that $\overline{T^*_{f_{|S}}\mathcal{U}}$ and $\operatorname{im} d\tilde{g}$ intersect properly, we define the (ordinary) **relative polar cycle**, $\Gamma_{f,\tilde{g}}(S)$, to be the cycle $\pi_*\left(\left[\overline{T^*_{f_{|S}}\mathcal{U}}\right] \cdot [\operatorname{im} d\tilde{g}]\right)$.

The **relative polar set**, $|\Gamma_{f,\tilde{g}}(\mathbf{F}^\bullet)|$, is defined by

$$|\Gamma_{f,\tilde{g}}(\mathbf{F}^\bullet)| := \pi\left(\left|(T^*_{f,\mathbf{F}^\bullet}\mathcal{U})^\bullet\right| \cap \operatorname{im} d\tilde{g}\right).$$

Each 1-dimensional component C of $|\Gamma_{f,\tilde{g}}(\mathbf{F}^\bullet)|$ is the image of a component of $\left|(T^*_{f,\mathbf{F}^\bullet}\mathcal{U})^\bullet\right| \cap \operatorname{im} d\tilde{g}$ along which $\left|(T^*_{f,\mathbf{F}^\bullet}\mathcal{U})^\bullet\right|$ and $\operatorname{im} d\tilde{g}$ intersect properly. We give such a component C the structure of the graded, enriched cycle whose underlying set is C and whose graded, enriched cycle structure is given by $\pi^\bullet_*\left((T^*_{f,\mathbf{F}^\bullet}\mathcal{U})^\bullet \odot \operatorname{im} d\tilde{g}\right)$ over generic points in C. We refer to this as the **graded, enriched cycle structure of C in $|\Gamma_{f,\tilde{g}}(\mathbf{F}^\bullet)|$**.

If $|\Gamma_{f,\tilde{g}}(\mathbf{F}^\bullet)|$ is purely 1-dimensional, we say that the **graded, enriched relative polar curve**, $(\Gamma^1_{f,\tilde{g}}(\mathbf{F}^\bullet))^\bullet$, is defined, and is given by

$$(\Gamma^1_{f,\tilde{g}}(\mathbf{F}^\bullet))^\bullet := \pi^\bullet_*\left((T^*_{f,\mathbf{F}^\bullet}\mathcal{U})^\bullet \odot \operatorname{im} d\tilde{g}\right),$$

i.e.,

$$(\Gamma^1_{f,\tilde{g}}(\mathbf{F}^\bullet))^k = \sum_{\substack{S\in\mathfrak{S}(\mathbf{F}^\bullet)\\ f_{|S}\neq \text{ const.}}} m^k_S(\mathbf{F}^\bullet)\,(\Gamma_{f,\tilde{g}}(S))^{\text{enr}}.$$

Remark 6.5.5 If $\tilde{g} = z_0$ is a generic linear form and $S = \mathcal{U}$, then $\Gamma_{f,\tilde{g}}(S)$ is the classical polar curve $\Gamma^1_{\tilde{f},z_0}$ (as a cycle) of Hamm, Lê, and Teissier.

In the notation for the polar curve, we write $\tilde{g}$, not simply g; we do not, in fact, know if $(\Gamma^1_{f,\tilde{g}}(\mathbf{F}^\bullet))^\bullet$ is independent of the extension to $\tilde{g}$. However, when $(\Gamma^1_{f,\tilde{g}}(\mathbf{F}^\bullet))^\bullet$ is defined and has no component on which f is constant, then $(\Gamma^1_{f,\tilde{g}}(\mathbf{F}^\bullet))^\bullet$ is independent of the extension $\tilde{g}$. It is also not difficult to show that the set $\left|\Gamma_{f,\tilde{g}}(\mathbf{F}^\bullet)\right|$ is independent of the extension of g, but we shall not need this result here.

Note that $\overline{T^*_{f_{|S}}\mathcal{U}} \cap \operatorname{im} d\tilde{g}$ is at least 1-dimensional at each point of intersection, and so $\left|\Gamma_{f,\tilde{g}}(\mathbf{F}^\bullet)\right|$ has no isolated points. Also, note that, as $\left|(T^*_{f,\mathbf{F}^\bullet}\mathcal{U})^\bullet\right| \cap \operatorname{im} d\tilde{g}$ is a closed subset of $\operatorname{im} d\tilde{g}$, and π induces an isomorphism from $\operatorname{im} d\tilde{g}$ to $\mathcal{U}$, $\left|\Gamma_{f,\tilde{g}}(\mathbf{F}^\bullet)\right|$ is a closed subset of $\mathcal{U}$.

Finally, if the relative polar set is 1-dimensional, we frequently superscript with a 1 to emphasize that fact.

For the purposes of this chapter, we need to recall the following proposition, which is Proposition 3.13 of [35].

Proposition 6.5.6

1. *There exists a non-zero linear form $\tilde{\mathcal{L}}$ such that $\mathbf{0} \notin \left|\Gamma_{f,\tilde{\mathcal{L}}}(\mathbf{F}^\bullet)\right|$ if and only if for generic linear $\tilde{\mathcal{L}}$, $\mathbf{0} \notin \left|\Gamma_{f,\tilde{\mathcal{L}}}(\mathbf{F}^\bullet)\right|$.*
2. *For generic linear $\tilde{\mathcal{L}}$,*

$$\dim_{\mathbf{0}} V(f) \cap \left|\Gamma_{f,\tilde{\mathcal{L}}}(\mathbf{F}^\bullet)\right| \leq 0 \quad \text{and} \quad \dim_{\mathbf{0}} V(\tilde{\mathcal{L}}) \cap \left|\Gamma_{f,\tilde{\mathcal{L}}}(\mathbf{F}^\bullet)\right| \leq 0.$$

Example 6.5.7 In this example, we will calculate another graded, enriched characteristic cycle. We shall use this as a basis for the next two examples, in which we calculate a graded, enriched relative conormal cycle and a graded, enriched relative polar cycle.

Let $f : \mathbb{C}^3 \to \mathbb{C}$ be given by $f(x, y, t) = y(y^2 - x^3 - t^2x^2)$, and let

$$X := V(f) = V(y) \cup V(y^2 - x^3 - t^2x^2).$$

The singular set of X, ΣX, is the 1-dimensional set $V(x, y) \cup V(x + t^2, y)$. Thus, near the origin (actually, in this specific example, globally),

$$\mathfrak{S} := \{V(y) - V(y^2 - x^3 - t^2x^2), V(y^2 - x^3 - t^2x^2) - V(y), \\ V(x, y) - \{\mathbf{0}\}, V(x + t^2, y) - \{\mathbf{0}\}, \{\mathbf{0}\}\}$$

is a Whitney stratification of X with connected strata. Let $\mathbf{F}^\bullet := \mathbb{Z}^\bullet_X[2]$. We want to calculate the graded, enriched characteristic cycle of $\mathbf{F}^\bullet$.

First, consider the 2-dimensional strata. Let $S_1 := V(y) - V(y^2 - x^3 - t^2x^2)$. Then, $\mathbb{N}_{S_1}$ is simply a point, and $\mathbb{L}_{S_1}$ is empty. Hence,

$$H^{k-2}(\mathbb{N}_{S_1}, \mathbb{L}_{S_1}; \mathbf{F}^\bullet) = H^k(\mathbb{N}_{S_1}, \mathbb{L}_{S_1}; \mathbb{Z})$$

isomorphic to $\mathbb{Z}$ if $k = 0$, and is 0 if $k \neq 0$. The same conclusion holds if S_1 is replaced by $S_2 := V(y^2 - x^3 - t^2x^2) - V(y)$.

Now, consider the 1-dimensional strata. Let $S_3 := V(x, y) - \{\mathbf{0}\}$, and $S_4 := V(x + t^2, y) - \{\mathbf{0}\}$. The normal slice $\mathbb{N}_{S_3}$ is, as a germ, up to analytic isomorphism, three complex lines in $\mathbb{C}^2$, which intersect at a point, and $\mathbb{L}_{S_3}$ is three points.

Similarly, the normal slice $\mathbb{N}_{S_4}$ is, as a germ, up to analytic isomorphism, two complex lines in $\mathbb{C}^2$, which intersect at a point, and $\mathbb{L}_{S_4}$ is two points. Hence, $H^{k-1}(\mathbb{N}_{S_3}, \mathbb{L}_{S_3}; \mathbf{F}^\bullet) = H^{k+1}(\mathbb{N}_{S_3}, \mathbb{L}_{S_3}; \mathbb{Z})$ isomorphic to $\mathbb{Z}^2$ if $k = 0$, and is 0 if $k \neq 0$. Similarly, $H^{k-1}(\mathbb{N}_{S_4}, \mathbb{L}_{S_4}; \mathbf{F}^\bullet) = H^{k+1}(\mathbb{N}_{S_4}, \mathbb{L}_{S_4}; \mathbb{Z})$ isomorphic to $\mathbb{Z}$ if $k = 0$, and is 0 if $k \neq 0$.

Finally, consider the stratum $\{\mathbf{0}\}$. Then, $\mathbb{N}_{\{\mathbf{0}\}}$ is all of X, intersected with a small ball around the origin. The complex link $\mathbb{L}_{\{\mathbf{0}\}}$ is usually referred to as simply the complex link of X at $\mathbf{0}$. Thus, $\mathbb{L}_{\{\mathbf{0}\}}$ has the homotopy-type of a bouquet of 1-spheres (see [21]), and the number of spheres in this bouquet is equal to the intersection number $(\Gamma^1_{f,L} \cdot V(L))_{\mathbf{0}}$, where L is any linear form such that $d_{\mathbf{0}}L$ is not a degenerate covector from strata of X at 0 (see [15]), and the relative polar curve here is the classical one of Lê, Hamm, and Teissier. We claim that we may use $L := t$ for this calculation.

To see this, first note that $V(y^2 - x^3 - t^2x^2)$ is the classic example of a space such that the regular part satisfies Whitney's condition (a) along the t-axis (or, alternatively, this is an easy exercise). Thus, $d_{\mathbf{0}}t$ is not a limit of conormals from S_2. Now, the closures of S_1, S_3, and S_4 are all smooth, and $d_{\mathbf{0}}t$ is not conormal to these closures at the origin.

To find the ordinary cycle $\Gamma^1_{f,t}$, we take the components of the cycle below which are **not** contained in Σf:

$$\begin{aligned} V&\left(\frac{\partial f}{\partial x}, \frac{\partial f}{\partial y}\right) \\ &= V(y(-3x^2 - 2t^2x), 3y^2 - x^3 - t^2x^2) \\ &= V(y, x^2(x + t^2)) + V(x(3x + 2t^2), 3y^2 - x^3 - t^2x^2) \\ &= 2V(x, y) + V(x + t^2, y) + 2V(x, y) + V(3x + 2t^2, 3y^2 - x^3 - t^2x^2). \end{aligned}$$

Thus, $\Gamma^1_{f,t} = V(3x + 2t^2, 3y^2 - x^3 - t^2x^2)$, and

$$(\Gamma^1_{f,t} \cdot V(t))_{\mathbf{0}} = [V(3x + 2t^2, 3y^2 - x^3 - t^2x^2, t)]_{\mathbf{0}} = 2,$$

and $H^{k-0}(\mathbb{N}_{\{\mathbf{0}\}}, \mathbb{L}_{\{\mathbf{0}\}}; \mathbf{F}^\bullet) = H^{k+2}(\mathbb{N}_{\{\mathbf{0}\}}, \mathbb{L}_{\{\mathbf{0}\}}; \mathbb{Z})$ is isomorphic to $\mathbb{Z}^2$ if $k = 0$, and is 0 if $k \neq 0$.

Therefore, we find that $\mathrm{gecc}^k(\mathbf{F}^\bullet) = 0$ if $k \neq 0$, and

$$\mathrm{gecc}^0(\mathbf{F}^\bullet) = \mathbb{Z}\left[\overline{T^*_{S_1}\mathbb{C}^3}\right] + \mathbb{Z}\left[\overline{T^*_{S_2}\mathbb{C}^3}\right] + \mathbb{Z}^2\left[\overline{T^*_{S_3}\mathbb{C}^3}\right] + \mathbb{Z}\left[\overline{T^*_{S_4}\mathbb{C}^3}\right] + \mathbb{Z}^2\left[T^*_{\{\mathbf{0}\}}\mathbb{C}^3\right].$$

Example 6.5.8 We continue with the setting of Example 6.5.7, where

$$X = V(y) \cup V(y^2 - x^3 - t^2x^2)$$

and $\mathbf{F}^\bullet = \mathbb{Z}_X^\bullet[2]$. We had Whitney strata consisting of $\{\mathbf{0}\}$,

$$S_1 = V(y) - V(y^2 - x^3 - t^2x^2), \quad S_2 = V(y^2 - x^3 - t^2x^2) - V(y),$$

$$S_3 = V(x, y) - \{\mathbf{0}\} \text{ and } S_4 = V(x + t^2, y) - \{\mathbf{0}\}.$$

We found that $\mathrm{gecc}^k(\mathbf{F}^\bullet) = 0$ if $k \neq 0$, and

$$\mathrm{gecc}^0(\mathbf{F}^\bullet) = \mathbb{Z}\left[\overline{T^*_{S_1}\mathbb{C}^3}\right] + \mathbb{Z}\left[\overline{T^*_{S_2}\mathbb{C}^3}\right] + \mathbb{Z}^2\left[\overline{T^*_{S_3}\mathbb{C}^3}\right] + \mathbb{Z}\left[\overline{T^*_{S_4}\mathbb{C}^3}\right] + \mathbb{Z}^2\left[T^*_{\{\mathbf{0}\}}\mathbb{C}^3\right].$$

We will calculate $\left(T^*_{x,\mathbf{F}^\bullet}\mathbb{C}^3\right)^\bullet$.

As we said above, we identify $T^*\mathbb{C}^3$ with $\mathbb{C}^3 \times \mathbb{C}^3$, and will use coordinates (w_0, w_1, w_2) for cotangent coordinates, so that (w_0, w_1, w_2) represents $w_0dx + w_1dy + w_2dt$.

Since x is identically zero on $\{\mathbf{0}\}$ and S_3, these two strata are not used in the calculation of $\left(T^*_{x,\mathbf{F}^\bullet}\mathbb{C}^3\right)^\bullet$. For the 1-dimensional stratum S_4, $\left[\overline{T^*_{x_{|S_4}}\mathbb{C}^3}\right]$ is the 4-dimensional cycle $V(x + t^2, y) \subseteq \mathbb{C}^3 \times \mathbb{C}^3$.

The fiber of $T^*_{x_{|S_1}}\mathbb{C}^3$ over any $p \in S_1$ is

$$\begin{aligned}(T^*_{S_1}\mathbb{C}^3)_p + < d_px > &:= \{\omega + ad_px \mid \omega \in (T^*_{S_1}\mathbb{C}^3)_p, a \in \mathbb{C}\} \\ &= \{bd_py + ad_px \mid a, b \in \mathbb{C}\}.\end{aligned}$$

Hence, $\left[\overline{T^*_{x_{|S_1}}\mathbb{C}^3}\right] = V(y, w_2)$.

The fiber of $T^*_{x_{|S_2}}\mathbb{C}^3$ over any $p \in S_2$ which is a regular point of x restricted to S_2 is

$$(T^*_{S_2}\mathbb{C}^3)_p + < d_px >= \{\omega + ad_px \mid \omega \in (T^*_{S_2}\mathbb{C}^3)_p, a \in \mathbb{C}\} =$$

$$\{b\big((-3x^2 - 2t^2x)d_px + 2yd_py - 2tx^2d_pt\big) + ad_px \mid a, b \in \mathbb{C}\}.$$

The form $w_0d_px + w_1d_py + w_2d_pt$ is in this set if and only if the determinant of the following matrix is 0:

$$\begin{bmatrix} w_0 & w_1 & w_2 \\ -3x^2 - 2t^2x & 2y & -2tx^2 \\ 1 & 0 & 0 \end{bmatrix},$$

i.e., if and only if $yw_2 + tx^2w_1 = 0$. It is tempting to conclude that

$$\left[\overline{T^*_{x|_{S_2}}\mathbb{C}^3}\right] = V(y^2 - x^3 - t^2x^2,\ yw_2 + tx^2w_1),$$

but this is not the case; we must eliminate any components of

$$V(y^2 - x^3 - t^2x^2,\ yw_2 + tx^2w_1)$$

which are contained in $V(y)$. Obviously, $V(y^2 - x^3 - t^2x^2, yw_2 + tx^2w_1)$ is purely 4-dimensional, and one easily shows that any component contained in $V(y)$ must, in fact, equal $V(x, y)$ (on the level of sets). Thus, we need to remove any components of $V(y^2 - x^3 - t^2x^2, yw_2 + tx^2w_1)$ which are contained in $V(x, y)$.

Our notation for the resulting scheme (a gap sheaf, see [33], I.1) is

$$V(y^2 - x^3 - t^2x^2,\ yw_2 + tx^2w_1) \neg V(x, y).$$

Note that, as schemes,

$$\begin{aligned} &V(y^2 - x^3 - t^2x^2,\ yw_2 + tx^2w_1)\\ &\quad = V(y^2 - x^3 - t^2x^2,\ yw_2 + tx^2w_1,\ y^2w_2 + ytx^2w_1)\\ &\quad = V(y^2 - x^3 - t^2x^2,\ yw_2 + tx^2w_1,\ (x^3 + t^2x^2)w_2 + ytx^2w_1). \end{aligned}$$

Note the x^2 factor of the last polynomial listed above, and note that, in the analytic set above, if $x = 0$, then y must be 0, i.e., if a point has $x = 0$, the point must be in $V(x, y)$. Hence, using [33], I.1.3.iv, we find that, as cycles,

$$V(y^2 - x^3 - t^2x^2,\ yw_2 + tx^2w_1) \neg V(x, y) =$$

$$V(y^2 - x^3 - t^2x^2,\ yw_2 + tx^2w_1,\ (x + t^2)w_2 + ytw_1).$$

(This last equality need not be true on the level of schemes, since our generators do not form a regular sequence and, hence, there may be embedded subvarieties.)

Therefore, we find that $\left(T^*_{x,\mathbf{F}^\bullet}\mathbb{C}^3\right)^k$ is 0 unless $k = 0$, and

$$\begin{aligned} &\left(T^*_{x,\mathbf{F}^\bullet}\mathbb{C}^3\right)^0\\ &\quad = \mathbb{Z}[V(y, w_2)] + \mathbb{Z}[V(y^2 - x^3 - t^2x^2,\ yw_2 + tx^2w_1,\ (x + t^2)w_2 + ytw_1)]\\ &\qquad + \mathbb{Z}[V(x + t^2, y)]. \end{aligned}$$

Example 6.5.9 We continue with our setting from Examples 6.5.7 and 6.5.8, and consider $X = V(y) \cup V(y^2 - x^3 - t^2x^2)$ and $\mathbf{F}^\bullet = \mathbb{Z}^\bullet_X[2]$. We will calculate $\left(\Gamma^1_{x,t}(\mathbf{F}^\bullet)\right)^\bullet$.

Using the isomorphism $T^*\mathbb{C}^3 \cong \mathbb{C}^3 \times \mathbb{C}^3$ from Example 6.5.8, $\operatorname{im} dt$ is the scheme

$$V\left(w_0 - \frac{\partial t}{\partial x},\ w_1 - \frac{\partial t}{\partial y},\ w_2 - \frac{\partial t}{\partial t}\right) = V(w_0, w_1, w_2 - 1).$$

In Example 6.5.8, we found that $\left(T^*_{x,\mathbf{F}^\bullet}\mathbb{C}^3\right)^k$ is 0 unless $k = 0$, and
$\left(T^*_{x,\mathbf{F}^\bullet}\mathbb{C}^3\right)^0 = \mathbb{Z}[V(y, w_2)]+$
$\mathbb{Z}[V(y^2 - x^3 - t^2x^2, yw_2 + tx^2w_1, (x + t^2)w_2 + ytw_1)] + \mathbb{Z}[V(x + t^2, y)]$.
Let

$$E = V(y^2 - x^3 - t^2x^2, yw_2 + tx^2w_1, (x + t^2)w_2 + ytw_1),$$

on the level of cycles, throughout the remainder of this example.

Thus, $\left(\Gamma^1_{x,t}(\mathbf{F}^\bullet)\right)^k$ is 0 unless $k = 0$ and, to calculate $\left(\Gamma^1_{x,t}(\mathbf{F}^\bullet)\right)^0$, we need first to calculate the three ordinary cycles

$$\pi_*\big(V(y, w_2) \cdot V(w_0, w_1, w_2 - 1)\big),$$

$$\pi_*\big(E \cdot V(w_0, w_1, w_2 - 1)\big),$$

and

$$\pi_*\big(V(x + t^2, y) \cdot V(w_0, w_1, w_2 - 1)\big).$$

Now, $V(y, w_2) \cap V(w_0, w_1, w_2 - 1) = \emptyset$, and so $\pi_*\big(V(y, w_2) \cdot V(w_0, w_1, w_2 - 1)\big) = 0$. In addition, it is trivial that there is an equality of cycles

$$\pi_*\big(V(x + t^2, y) \cdot V(w_0, w_1, w_2 - 1)\big) = V(x + t^2, y).$$

However, the remaining cycle is more difficult to calculate.

It is trivial that, as sets,

$$E \cap V(w_0, w_1, w_2 - 1) = V(x + t^2, y, w_0, w_1, w_2 - 1),$$

but the difficulty in calculating

$$\pi_*\big(E \cdot V(w_0, w_1, w_2 - 1)\big)$$

is due to the fact that $y^2 - x^3 - t^2x^2, yw_2 + tx^2w_1, (x + t^2)w_2 + ytw_1$ is not a regular sequence. To "fix" this, note that, in Example 6.5.8, we saw that, as cycles, there is an equality

$$V(y^2 - x^3 - t^2x^2, yw_2 + tx^2w_1) = C + E,$$

where the underlying set $|C| = V(x, y)$. Therefore,

$$C \cdot V(w_0, w_1, w_2 - 1) + E \cdot V(w_0, w_1, w_2 - 1) =$$

$$V(y^2 - x^3 - t^2x^2, yw_2 + tx^2w_1) \cdot V(w_0, w_1, w_2 - 1) =$$

$$V(y^2-x^3-t^2x^2, yw_2+tx^2w_1, w_0, w_1, w_2-1) = V(x^2(x+t^2), y, w_0, w_1, w_2-1) =$$

$$2V(x, y, w_0, w_1, w_2 - 1) + V(x + t^2, y, w_0, w_1, w_2 - 1).$$

Thus, as cycles,

$$E \cdot V(w_0, w_1, w_2 - 1) = V(x + t^2, y, w_0, w_1, w_2 - 1),$$

and so $\pi_*(E \cdot V(w_0, w_1, w_2 - 1)) = V(x + t^2, y)$.

Finally, we find that

$$\left(\Gamma^1_{x,t}(\mathbf{F}^\bullet)\right)^0 = \pi^0_* \left((T^*_{f,\mathbf{F}^\bullet}\mathcal{U})^\bullet \odot \operatorname{im} dt \right) =$$

$$\mathbb{Z}[V(x + t^2, y)] + \mathbb{Z}[V(x + t^2, y)] = \mathbb{Z}^2[V(x + t^2, y)].$$

We shall discuss the main results on the graded, enriched relative conormal cycle and the graded, enriched relative polar curve in the following sections.

6.6 The Nearby Cycles

The following theorem was our primary motivation for defining the graded, enriched conormal cycle. While we state the theorem in the elegant form given in [34], this theorem, in terms of ordinary cycles, is essentially contained in [6]

Theorem 6.6.1 ([34], Theorem 3.3) *There is an equality of graded enriched cycles given by*

$$\operatorname{gecc}^\bullet(\psi_f[-1]\mathbf{F}^\bullet) = \left(T^*_{f,\mathbf{F}^\bullet}\mathcal{U}\right)^\bullet \odot (V(f) \times \mathbb{C}^{n+1}).$$

If one knows the irreducible components $\{V_j\}_j$ of the underlying set $\mathrm{SS}(\psi_f[-1]\mathbf{F}^\bullet)$, then, by selecting a generic point of each V_j, and taking a normal slice, the calculation of $\mathrm{gecc}^\bullet(\psi_f[-1]\mathbf{F}^\bullet)$ is reduced to calculating the Morse modules of point strata. In other words, if we know $\mathrm{SS}(\psi_f[-1]\mathbf{F}^\bullet)$, then, by taking normal slices, the calculation of $\mathrm{gecc}^\bullet(\psi_f[-1]\mathbf{F}^\bullet)$ reduces to calculating $m^k_{\mathbf{0}}(\psi_f[-1]\mathbf{F}^\bullet)$, i.e., $H^k(\phi_{\mathcal{L}}[-1]\psi_f[-1]\mathbf{F}^\bullet)_{\mathbf{0}}$, where $\mathcal{L}$ is the restriction to $V(f)$ of a generic linear form $\tilde{\mathcal{L}}$.

The next result follows from Theorem 4.2 of [29], but is stated as in Theorem 3.12 of [35].

Theorem 6.6.2

$$m^k_{\mathbf{0}}(\psi_f[-1]\mathbf{F}^\bullet) \cong \left(\left(\Gamma^1_{f,\tilde{\mathcal{L}}}(\mathbf{F}^\bullet)\right)^k \odot V(f)\right)_{\mathbf{0}} = \bigoplus_{\substack{S\in\mathfrak{S}(\mathbf{F}^\bullet)\\ S\not\subseteq V(f)}} m^k_S(\mathbf{F}^\bullet)\otimes R^{\alpha_S},$$

where $\alpha_S := \left(\Gamma^1_{f,\tilde{\mathcal{L}}}(S)\cdot V(\tilde{f})\right)_{\mathbf{0}}$ *and* $\tilde{\mathcal{L}}$ *is a generic linear form. Specifically, the amount of genericity that we need is that* $(\mathbf{0}, d_{\mathbf{0}}\tilde{\mathcal{L}}) \notin \overline{\mathrm{SS}(\psi_f[-1]\mathbf{F}^\bullet) - T^*_{\mathbf{0}}\mathcal{U}}$, *which is equivalent to* $\dim_{\mathbf{0}} |\left(\Gamma^1_{f,\tilde{\mathcal{L}}}(\mathbf{F}^\bullet)\right)^\bullet| \cap V(f) \leq 0$.

Example 6.6.3 We continue with our example from Example 6.5.8:

$$X = V(y)\cup V(y^2-x^3-t^2x^2),$$

$\mathbf{F}^\bullet = \mathbb{Z}^\bullet_X[2]$, and we found that $\left(T^*_{x,\mathbf{F}^\bullet}\mathbb{C}^3\right)^k$ is 0 unless $k=0$, and
$\left(T^*_{x,\mathbf{F}^\bullet}\mathbb{C}^3\right)^0 = \mathbb{Z}[V(y, w_2)]+$
$\mathbb{Z}[V(y^2-x^3-t^2x^2, yw_2+tx^2w_1, (x+t^2)w_2+ytw_1)] + \mathbb{Z}[V(x+t^2, y)]$.
Let $f := x$.

In light of Theorem 6.6.1, we find that $\mathrm{gecc}^\bullet(\psi_f[-1]\mathbf{F}^\bullet)$ is concentrated in degree 0, and that

$$\mathrm{gecc}^0(\psi_f[-1]\mathbf{F}^\bullet) = \mathbb{Z}[V(x, y, w_2)] + \mathbb{Z}^2[V(x,y,t)] + \mathbb{Z}\big[V(y^2-x^3-t^2x^2, yw_2+tx^2w_1, (x+t^2)w_2+ytw_1)\cdot V(x)\big]^{\mathrm{enr}}. \quad (\dagger)$$

The difficulty is in calculating the cycle

$$E := \Big[V(y^2-x^3-t^2x^2, yw_2+tx^2w_1, (x+t^2)w_2+ytw_1)\cdot V(x)\Big].$$

The underlying set $|E|$ is easily found to be $V(x,y,t)\cup V(x,y,w_2)$, and we may find the geometric multiplicity of each component in E by moving to generic points.

At a generic point of $V(x,y,t)$, $w_2\neq 0$ and, at such a point, one easily shows that there is an equality of ideals

$$\langle y^2-x^3-t^2x^2, yw_2+tx^2w_1, (x+t^2)w_2+ytw_1\rangle = \langle yw_2+tx^2w_1, (x+t^2)w_2+ytw_1\rangle$$

and, as $yw_2+tx^2w_1$, $(x+t^2)w_2+ytw_1$ is a regular sequence, one easily calculates that, at a point where $w_2\neq 0$, there are equalities of cycles

$$\Big[V(yw_2+tx^2w_1,\,(x+t^2)w_2+ytw_1)\cdot V(x)\Big]=$$

$$\Big[V(yw_2+tx^2w_1,\,(x+t^2)w_2+ytw_1,\,x)\Big]=[V(y,t^2,x)]=2[V(x,y,t)].$$

This is the component of E with underlying set $V(x,y,t)$.

At a generic point of $V(x,y,w_2)$, neither t nor w_1 is zero, and it follows that $x+t^2$ is not zero. At such a point, one easily shows that there is again an equality of ideals

$$\langle y^2-x^3-t^2x^2,\,yw_2+tx^2w_1,\,(x+t^2)w_2+ytw_1\rangle=\langle yw_2+tx^2w_1,\,(x+t^2)w_2+ytw_1\rangle$$

and one easily calculates that, at a point where neither t nor w_1 is zero, there are equalities of cycles

$$\Big[V(yw_2+tx^2w_1,\,(x+t^2)w_2+ytw_1)\cdot V(x)\Big]=[V(yw_2,\,tw_2+yw_1,\,x)]=$$

$$[V(y,tw_2,x)]+[V(w_2,yw_1,x)]=[V(y,w_2,x)]+[V(w_2,y,x)]=2[V(x,y,w_2)].$$

This is the component of E with underlying set $V(x,y,w_2)$.

Therefore, (†) tells us that

$$\mathrm{gecc}^0(\psi_f[-1]\mathbf{F}^\bullet)=\mathbb{Z}^3[V(x,y,w_2)]\;+\;\mathbb{Z}^4[V(x,y,t)].$$

Note that the component of $\mathrm{gecc}^0(\psi_f[-1]\mathbf{F}^\bullet)$ over the origin agrees with Theorem 6.6.2 and our calculation in Example 6.5.9. For

$$(\mathbf{0},d_{\mathbf{0}}t)\notin\overline{\mathrm{SS}(\psi_f[-1]\mathbf{F}^\bullet)-T^*_{\mathbf{0}}\mathcal{U}}=V(x,y,w_2)$$

and

$$\left(\Gamma^1_{x,t}(\mathbf{F}^\bullet)\right)^0\odot V(t)=\mathbb{Z}^2[V(x+t^2,y)]\odot V(t)=\mathbb{Z}^4[V(x,y,t)].$$

Another example, which we leave as an exercise for the reader, is to recalculate $\mathrm{gecc}^\bullet(\psi_f[-1]\mathbf{F}^\bullet)$ from Example 6.2.15 by using either Theorem 6.6.1 or Theorem 6.6.2.

Recall from Definition 6.2.12 that a complex of sheaves $\mathbf{P}^\bullet$ on X is numerical if and only if $\mathbf{P}^\bullet$ is a perverse sheaf and all of its Morse modules are free, so that the ordinary characteristic cycle of $\mathbf{P}^\bullet$ carries all of the information about $\mathrm{gecc}^\bullet(\mathbf{P}^\bullet)$.

The following corollary is immediate from Theorem 6.6.1.

Corollary 6.6.4 *If $\mathbf{P}^\bullet$ is numerical, then so is $\psi_f[-1]\mathbf{P}^\bullet$.*

6.7 Hypersurface Complements and Restrictions

Let $i : X - V(f) \hookrightarrow X$ and $j : V(f) \hookrightarrow X$ denote the inclusions. Recall that we are assuming that $V(f)$ is a union of strata, and recall the partial ordering on isomorphism classes of finitely-generated R-modules given in Definition 6.4.1.

We would like to give an elegant formula for $\mathrm{gecc}^\bullet(i_!i^!\mathbf{F}^\bullet)$, something along the lines of what we gave for $\mathrm{gecc}^\bullet(\psi_f[-1]\mathbf{F}^\bullet)$ in Theorem 6.6.1. We do not quite do this. However, we do the next best thing; we give a formula for the set $\mathrm{SS}(i_!i^!\mathbf{F}^\bullet)$ and a formula for $m^k_{\mathbf{0}}(i_!i^!\mathbf{F}^\bullet)$.

Once we have these formulas, and so, in principle, know $\mathrm{gecc}^\bullet(i_!i^!\mathbf{F}^\bullet)$, we can use how the graded, enriched characteristic cycle works with Verdier duals to obtain $\mathrm{gecc}^\bullet(i_*i^*\mathbf{F}^\bullet)$. In addition, we can use the additivity of ordinary characteristic cycles over distinguished triangles, and the duality formula, to obtain the characteristic cycles of $j_*j^*[-1]\mathbf{F}^\bullet$ and $j_!j^![1]\mathbf{F}^\bullet$ when the base ring is a domain.

The following result is immediate from Theorem 4.2 B of [29] and, for ordinary cycles, is proved in [6].

Theorem 6.7.1

$$m^k_{\mathbf{0}}(i_!i^!\mathbf{F}^\bullet) \cong \left(\left(\Gamma^1_{f,\tilde{\mathcal{L}}}(\mathbf{F}^\bullet)\right)^k \odot V(\tilde{\mathcal{L}})\right)_{\mathbf{0}} = \bigoplus_{\substack{S\in\mathfrak{S}(\mathbf{F}^\bullet)\\ S\not\subseteq V(f)}} m^k_S(\mathbf{F}^\bullet)\otimes R^{\beta_S},$$

where $\beta_S := \left(\Gamma^1_{f,\tilde{\mathcal{L}}}(S)\cdot V(\tilde{\mathcal{L}})\right)_{\mathbf{0}}$ *and* $\tilde{\mathcal{L}}$ *is a generic linear form. Specifically, the amount of genericity that we need is that*

1. $(\mathbf{0}, d_{\mathbf{0}}\tilde{\mathcal{L}}) \notin \overline{\mathrm{SS}(i_!i^!\mathbf{F}^\bullet) - T^*_{\mathbf{0}}\mathcal{U}}$ *and*
2. $\dim_{\mathbf{0}} |(\Gamma^1_{f,\tilde{\mathcal{L}}}(\mathbf{F}^\bullet))^\bullet| \cap V(\mathcal{L}) \le 0.$

Corollary 6.7.2 *If* $S \in \mathfrak{S}$ *and* $S \not\subseteq V(f)$, *then* $m^k_S(i_!i^!\mathbf{F}^\bullet) \cong m^k_S(\mathbf{F}^\bullet)$.

If $S \in \mathfrak{S}$ *and* $S \subseteq V(f)$, *then* $m^k_S(i_!i^!\mathbf{F}^\bullet) \neq 0$ *if and only if* $m^k_S(\psi_f[-1]\mathbf{F}^\bullet) \neq 0$, *and* $m^k_S(i_!i^!\mathbf{F}^\bullet) \le m^k_S(\psi_f[-1]\mathbf{F}^\bullet)$.

Proof Outside of $V(f)$, the complex $i_!i^!\mathbf{F}^\bullet$ agrees with $\mathbf{F}^\bullet$; this yields the first statement.

A comparison of Theorem 6.7.1 with Theorem 6.6.2 yields the second statement.

□

Definition 6.7.3 Suppose that $E^\bullet$ is a graded, enriched cycle in $T^*\mathcal{U}$ given by $E^k = \sum_{S\in\mathfrak{S}} E^k_S[\overline{T^*_S\mathcal{U}}]$, where E^k_S is a finitely-generated R-module.

Let $(E_{\not\subseteq V(f)})^\bullet$ be the graded, enriched cycle such that $(E_{\not\subseteq V(f)})^k$ is the sum of those $E^k_S[\overline{T^*_S\mathcal{U}}]$ such that $S \not\subseteq V(f)$. Similarly, let $(E_{\subseteq V(f)})^\bullet$ be the graded, enriched cycle such that $(E_{\subseteq V(f)})^k$ is the sum of those $E^k_S[\overline{T^*_S\mathcal{U}}]$ such that $S \subseteq V(f)$.

Let $|E^k|_{\not\subseteq V(f)} := |(E_{\not\subseteq V(f)})^\bullet|$ and $|E^k|_{\subseteq V(f)} := |(E_{\subseteq V(f)})^\bullet|$.

Corollary 6.7.4

$$|\mathrm{gecc}^k(i_!i^!\mathbf{F}^\bullet)| = |\mathrm{gecc}^k(\mathbf{F}^\bullet)|_{\not\subseteq V(f)} \cup |\mathrm{gecc}^k(\psi_f[-1]\mathbf{F}^\bullet)|.$$

Proof This is immediate from Corollary 6.7.2. □

In light of Corollary 6.7.4 and the hypotheses on $\tilde{\mathcal{L}}$ in Theorem 6.7.1, the following proposition is of interest.

Proposition 6.7.5 *Suppose that* $(\mathbf{0}, d_{\mathbf{0}}\tilde{\mathcal{L}}) \notin |\mathrm{SS}(\mathbf{F}^\bullet)|_{\not\subseteq V(f)}$.

Then, $\dim_{\mathbf{0}} |\big(\Gamma^1_{f,\tilde{\mathcal{L}}}(\mathbf{F}^\bullet)\big)^\bullet| \cap V(\mathcal{L}) \leq 0$ *if and only if* $\dim_{\mathbf{0}} |\big(\Gamma^1_{f,\tilde{\mathcal{L}}}(\mathbf{F}^\bullet)\big)^\bullet| \cap V(f) \leq 0$.

Proof Lemma 3.10 of [35] tells us that, if $\dim_{\mathbf{0}} |\big(\Gamma^1_{f,\tilde{\mathcal{L}}}(\mathbf{F}^\bullet)\big)^\bullet| \cap V(\mathcal{L}) \leq 0$, then $\dim_{\mathbf{0}} |\big(\Gamma^1_{f,\tilde{\mathcal{L}}}(\mathbf{F}^\bullet)\big)^\bullet| \cap V(f) \leq 0$.

Suppose then that $\dim_{\mathbf{0}} |\big(\Gamma^1_{f,\tilde{\mathcal{L}}}(\mathbf{F}^\bullet)\big)^\bullet| \cap V(f) \leq 0$. Let $\mathbf{p}(t)$ be an analytic parametrization of an irreducible component C of $|\big(\Gamma^1_{f,\tilde{\mathcal{L}}}(\mathbf{F}^\bullet)\big)^\bullet|$ such that $\mathbf{p}(0) = \mathbf{0}$. Suppose that $C \subseteq V(\tilde{\mathcal{L}})$; we wish to derive a contradiction.

Let $S' \in \mathfrak{S}$ be an $\mathbf{F}^\bullet$-visible stratum such that $C = \pi(\overline{T^*_{f_{|S'}}\mathcal{U}} \cap \operatorname{im} d\tilde{\mathcal{L}})$. Let S denote the stratum of $\mathfrak{S}$ which contains $\mathbf{p}(t)$ for $t \neq 0$. Note that neither S nor S' is contained in $V(f)$, since $\dim_{\mathbf{0}} |\big(\Gamma^1_{f,\tilde{\mathcal{L}}}(\mathbf{F}^\bullet)\big)^\bullet| \cap V(f) \leq 0$. On the other hand, in a neighborhood of the origin, the stratified critical locus of f is contained in $V(f)$.

It follows that, for all $\mathbf{x} \in C - \{\mathbf{0}\}$, the fiber $(\overline{T^*_{f_{|S'}}\mathcal{U}})_{\mathbf{x}}$ is equal to $(\overline{T^*_{S'}\mathcal{U}})_{\mathbf{x}} + \langle d_{\mathbf{x}}\tilde{f}\rangle$. Thus, for $t \neq 0$, there exists a complex number $a(t)$ such that

$$d_{\mathbf{p}(t)}\tilde{\mathcal{L}} + a(t)d_{\mathbf{p}(t)}\tilde{f} \in (\overline{T^*_{S'}\mathcal{U}})_{\mathbf{p}(t)} \subseteq (T^*_S\mathcal{U})_{\mathbf{p}(t)}. \qquad (\dagger)$$

By evaluating at $\mathbf{p}'(t)$, we find that $(\mathcal{L}(\mathbf{p}(t)))' + a(t)(f(\mathbf{p}(t))' \equiv 0$. As $C \subseteq V(\tilde{\mathcal{L}})$, we find that $a(t)(f(\mathbf{p}(t))' \equiv 0$. As $C \not\subseteq V(f)$, we conclude that $a(t) \equiv 0$. From (†), it follows that $d_{\mathbf{p}(t)}\tilde{\mathcal{L}} \in (\overline{T^*_{S'}\mathcal{U}})_{\mathbf{p}(t)}$ and, hence, that $(\mathbf{0}, d_{\mathbf{0}}\tilde{\mathcal{L}}) \in \overline{T^*_{S'}\mathcal{U}}$. This contradicts the fact that $(\mathbf{0}, d_{\mathbf{0}}\tilde{\mathcal{L}}) \notin |\mathrm{SS}(\mathbf{F}^\bullet)|_{\not\subseteq V(f)}$. □

Theorem 6.7.6 *For all k,* $\mathrm{gecc}^k(i_*i^*\mathbf{F}^\bullet) = \mathrm{gecc}^k(i_!i^!\mathbf{F}^\bullet)$.

Proof One uses Theorem 6.7.1 and Corollary 6.7.4, together with the isomorphisms $i_*i^*\mathbf{F}^\bullet \cong \mathcal{D}i_!i^!\mathcal{D}\mathbf{F}^\bullet$ and $\mathcal{D}\psi_f[-1] \cong \psi_f[-1]\mathcal{D}$. We leave the proof as an exercise. □

Remark 6.7.7 Note that $\mathrm{gecc}^k(i_*i^*\mathbf{F}^\bullet)$ and $\mathrm{gecc}^k(i_!i^!\mathbf{F}^\bullet)$ do not depend on any degree of $\mathrm{gecc}^\bullet(\mathbf{F}^\bullet)$, other than the degree k portion. In particular, if $\mathrm{gecc}^\bullet(\mathbf{F}^\bullet)$ is concentrated in degree 0, then so are $\mathrm{gecc}^\bullet(i_*i^*\mathbf{F}^\bullet)$ and $\mathrm{gecc}^\bullet(i_!i^!\mathbf{F}^\bullet)$. Thus, we recover the well-known fact that, if $\mathbf{F}^\bullet$ is a perverse sheaf and i is the inclusion of the complement of the zero locus of a single function, then $i_*i^*\mathbf{F}^\bullet$ and $i_!i^!\mathbf{F}^\bullet$ are also perverse.

The following corollary is immediate from Theorem 6.7.1 and Theorem 6.7.6.

Corollary 6.7.8 *If* $\mathbf{P}^\bullet$ *is numerical, then so are* $i_*i^*\mathbf{P}^\bullet$ *and* $i_!i^!\mathbf{P}^\bullet$.

Recall that we have the closed inclusion $j : V(f) \hookrightarrow X$. Unlike the functors i_*i^* and $i_!i^!$ discussed above, the functors j_*j^* and $j_!j^!$ (with or without shifts) do not take perverse sheaves to perverse sheaves. In other words, $\mathrm{gecc}^k(j_*j^*\mathbf{F}^\bullet)$ and $\mathrm{gecc}^k(j_!j^!\mathbf{F}^\bullet)$ are not determined by a single degree of $\mathrm{gecc}^\bullet(\mathbf{F}^\bullet)$. Of course, given the canonical distinguished triangles relating j_*j^* and $i_!i^!$, and $j_!j^!$ and i_*i^*, we immediately obtain:

Proposition 6.7.9

$$\mathrm{CC}(j_*j^*\mathbf{F}^\bullet) = \mathrm{CC}(\mathbf{F}^\bullet) - \mathrm{CC}(i_!i^!\mathbf{F}^\bullet) = \mathrm{CC}(\mathbf{F}^\bullet) - \mathrm{CC}(i_*i^*\mathbf{F}^\bullet) = \mathrm{CC}(j_!j^!\mathbf{F}^\bullet).$$

Perhaps more interesting is the easy corollary to Theorem 6.7.1:

Corollary 6.7.10 *Suppose that, for all k,* $m^k_{\mathbf{0}}(\mathbf{F}^\bullet) = 0$. *Then,*

$$\begin{aligned} m^k_{\mathbf{0}}(j_*j^*[-1]\mathbf{F}^\bullet) \cong m^k_{\mathbf{0}}(j_!j^![1]) &\cong \left(\left(\Gamma^1_{f,\tilde{\mathcal{L}}}(\mathbf{F}^\bullet)\right)^k \odot V(\tilde{\mathcal{L}})\right)_{\mathbf{0}} \\ &= \bigoplus_{\substack{S\in\mathfrak{S}(\mathbf{F}^\bullet)\\ S\not\subseteq V(f)}} m^k_S(\mathbf{F}^\bullet)\otimes R^{\beta_S}, \end{aligned}$$

where $\beta_S := \left(\Gamma^1_{f,\tilde{\mathcal{L}}}(S)\cdot V(\tilde{\mathcal{L}})\right)_{\mathbf{0}}$ *and* $\tilde{\mathcal{L}}$ *is a generic linear form.*

Proof Let $\mathcal{L} = \tilde{\mathcal{L}}_{|X}$. The corollary follows immediately from Theorems 6.7.1 and 6.7.6, together with applying the functor $\phi_{\mathcal{L}}[-1]$ to the two canonical distinguished triangles

$$\cdots \to j_*j^*[-1]\mathbf{F}^\bullet \to i_!i^!\mathbf{F}^\bullet \to \mathbf{F}^\bullet \xrightarrow{[1]} \cdots$$

and

$$\cdots \to \mathbf{F}^\bullet \to i_*i^*\mathbf{F}^\bullet \to j_!j^![1]\mathbf{F}^\bullet \xrightarrow{[1]} \cdots.$$

□

We wrote that the above corollary is "more interesting" because it lets us conclude the classical result below.

Example 6.7.11 Suppose that we have $f : (\mathcal{U}, \mathbf{0}) \to (\mathbb{C}, 0)$, where $\mathcal{U}$ is connected and f is not identically 0. Consider the constant sheaf $\mathbf{P}^\bullet := \mathbb{Z}^\bullet_{\mathcal{U}}[n+1]$.

Then, for all k, $m^k_{\mathbf{0}}(\mathbf{P}^\bullet) = 0$ and so, by Corollary 6.7.10,

$$m^k_{\mathbf{0}}(j_*j^*[-1]\mathbf{P}^\bullet) \cong \bigoplus_{\substack{S\in\mathfrak{S}(\mathbf{P}^\bullet)\\ S\not\subseteq V(f)}} m^k_S(\mathbf{P}^\bullet) \otimes R^{\beta_S}$$

where $\beta_S := \left(\Gamma^1_{f,\tilde{\mathcal{L}}}(S) \cdot V(\tilde{\mathcal{L}})\right)_{\mathbf{0}}$ and $\tilde{\mathcal{L}}$ is a generic linear form.

The only $\mathbf{P}^\bullet$-visible stratum is $\mathcal{U}$. As $\mathbf{P}^\bullet$ is perverse, the only possibly non-zero $m^k_{\mathcal{U}}(\mathbf{P}^\bullet)$ occurs when $k = 0$, and $m^0_{\mathcal{U}}(\mathbf{P}^\bullet) \cong \mathbb{Z}$.

Therefore, the only possibly non-zero $m^k_{\mathbf{0}}(j_*j^*[-1]\mathbf{P}^\bullet)$ occurs when $k = 0$ and

$$m^0_{\mathbf{0}}(j_*j^*[-1]\mathbf{P}^\bullet) \cong \mathbb{Z}^\beta,$$

where $\beta = \left(\Gamma^1_{f,\tilde{\mathcal{L}}}(\mathcal{U}) \cdot V(\tilde{\mathcal{L}})\right)_{\mathbf{0}}$.

This says that

$$m^0_{\mathbf{0}}(j_*j^*[-1]\mathbf{P}^\bullet) = m^0_{\mathbf{0}}(\mathbb{Z}^\bullet_{V(f)}[n]) = \left(\Gamma^1_{f,L} \cdot V(L)\right)_{\mathbf{0}}$$

for generic linear L, where, since we are in affine space, we have written the more usual L in place of $\tilde{\mathcal{L}}$ and have written $\Gamma^1_{f,L}$ in place of $\Gamma^1_{f,\tilde{\mathcal{L}}}(\mathcal{U})$. Note that $m^0_{\mathbf{0}}(\mathbb{Z}^\bullet_{V(f)}[n])$ is precisely the reduced integral cohomology in degree $n-1$ of the complex link of $V(f)$ at $\mathbf{0}$.

This is the cohomological version of the well-known result that the complex link of an n-dimensional hypersurface has the homotopy-type of a bouquet of $(n-1)$-spheres, where the number of spheres is given by $\left(\Gamma^1_{f,L} \cdot V(L)\right)_{\mathbf{0}}$.

6.8 The Vanishing Cycles

Before we can give a formula for $\mathrm{gecc}^\bullet(\phi_f[-1]\mathbf{F}^\bullet)$, we must first discuss the graded, enriched exceptional divisor in the blow-up of a graded, enriched cycle along an ideal.

Let us recall the notation established thus far. $\mathcal{U}$ is an open neighborhood of the origin of $\mathbb{C}^{n+1}$, X is a closed, analytic subset of $\mathcal{U}$, $\mathbf{z} := (z_0, \dots, z_n)$ are coordinates on $\mathcal{U}$, we identify the cotangent space $T^*\mathcal{U}$ with $\mathcal{U}\times\mathbb{C}^{n+1}$ by mapping $(\mathbf{p}, w_0d_{\mathbf{p}}z_0 + \cdots + w_nd_{\mathbf{p}}z_n)$ to $(\mathbf{p}, (w_0, \dots, w_n))$., and we let $\pi : T^*\mathcal{U} \to \mathcal{U}$ denote the projection.

Consider a graded, enriched cycle $D^\bullet$ in $T^*\mathcal{U}$ given by $D^k := \sum D^k_V[V]$. Let $h_0, \dots, h_m$ be analytic functions on $T^*\mathcal{U}$, and let I be the ideal $\langle h_0, \dots h_m\rangle$. Then, for each V, the blow-up $\mathrm{Bl}_I V$ of V along I is naturally a subspace of $T^*\mathcal{U} \times \mathbb{P}^m \cong \mathcal{U} \times \mathbb{C}^{n+1} \times \mathbb{P}^m$. Let $\mathrm{Ex}_I(V)$ denote the exceptional divisor as a cycle.

Definition 6.8.1 The **graded, enriched blow-up** $\mathrm{Bl}^\bullet_I(D^\bullet)$ **of** $D^\bullet$ **along** I **in** $T^*\mathcal{U}\times\mathbb{P}^m$ is given by $\mathrm{Bl}^k_I(D^\bullet) := \sum_V D^k_V[\mathrm{Bl}_I V]^{\mathrm{enr}}$.

The **graded, enriched exceptional divisor** $\mathrm{Ex}^\bullet_I(D^\bullet)$ **of** $D^\bullet$ **along** I **in** $T^*\mathcal{U}\times\mathbb{P}^m$ is given by $\mathrm{Ex}^k_I(D^\bullet) := \sum_V D^k_V[\mathrm{Ex}_I V]^{\mathrm{enr}}$.

Instead of subscripting the blow-up and exceptional divisor by the ideal I, it is common to subscript by the analytic scheme $V(I)$. In particular, below, we shall blow-up along $\mathrm{im}\, d\tilde{f} \subseteq \mathcal{U}\times\mathbb{C}^{n+1}$; we remind the reader that this is defined by the ideal

$$\left\langle w_0 - \frac{\partial \tilde{f}}{\partial z_0}, \dots, w_n - \frac{\partial \tilde{f}}{\partial z_n}\right\rangle.$$

Let $\tau : \mathcal{U}\times\mathbb{C}^{n+1}\times\mathbb{P}^n \to \mathcal{U}\times\mathbb{P}^n$ denote the projection, and recall that τ_* denotes the proper push-forward. The following is Theorem 3.5 of [34].

Theorem 6.8.2 *There is an equality of closed subsets of X given by*

$$\bigcup_{v\in\mathbb{C}} \mathrm{supp}\,\phi_{f-v}[-1]\mathbf{F}^\bullet \;=\; \pi\big(\mathrm{SS}(\mathbf{F}^\bullet)\cap \mathrm{im}\, d\tilde{f}\big),$$

and, for all k, an equality of graded, enriched cycles given by

$$\sum_{v\in\mathbb{C}} \mathbb{P}\big(\mathrm{gecc}^k(\phi_{f-v}[-1]\mathbf{F}^\bullet)\big) \;=\; \tau_*\big(\mathrm{Ex}_{\mathrm{im}\, d\tilde{f}}(\mathrm{gecc}^k(\mathbf{F}^\bullet))\big).$$

In particular, for all k, there is an equality of sets

$$\bigcup_{v\in\mathbb{C}} \pi\big(|\mathrm{gecc}^k(\phi_{f-v}[-1]\mathbf{F}^\bullet)|\big) \;=\; \pi\big(|\mathrm{gecc}^k(\mathbf{F}^\bullet)|\cap \mathrm{im}\, d\tilde{f}\big).$$

Remark 6.8.3 We should remark that, in the above unions and sum over $v\in\mathbb{C}$, the unions and sum are not merely locally finite, but, in fact, locally over open neighborhoods of points in X, there is only one non-zero (or non-empty) summand (respectively, indexed subset in the union).

The following corollary is immediate from Theorem 6.8.2.

Corollary 6.8.4 *If* $\mathbf{P}^\bullet$ *is numerical, then so is* $\phi_f[-1]\mathbf{P}^\bullet$.

The following result follows at once from Theorem 4.2 of [29], and is used in the proof of Theorem 6.8.2.

Theorem 6.8.5

$$m^k_{\mathbf{0}}(\phi_f[-1]\mathbf{F}^\bullet) \cong m^k_{\mathbf{0}}(\mathbf{F}^\bullet) \oplus \bigoplus_{\substack{S\in\mathfrak{S}(\mathbf{F}^\bullet)\\ S\not\subseteq V(f)}} m^k_S(\mathbf{F}^\bullet)\otimes R^{\delta_S},$$

where $\delta_S := \left(\Gamma^1_{f,\tilde{\mathcal{L}}}(S)\cdot V(\tilde{f})\right)_{\mathbf{0}} - \left(\Gamma^1_{f,\tilde{\mathcal{L}}}(S)\cdot V(\tilde{\mathcal{L}})\right)_{\mathbf{0}}$*, where* $\tilde{\mathcal{L}}$ *is a generic linear form; specifically, we need for the following three conditions to hold:*

1. $(\mathbf{0}, d_{\mathbf{0}}\tilde{\mathcal{L}}) \notin \overline{\mathrm{SS}(\phi_f[-1]\mathbf{F}^\bullet) - T^*_{\mathbf{0}}\mathcal{U}}$;
2. $\dim_{\mathbf{0}} |(\Gamma^1_{f,\tilde{\mathcal{L}}}(\mathbf{F}^\bullet))^\bullet| \cap V(\mathcal{L}) \leq 0$*, and*
3. $\left(\Gamma^1_{f,\tilde{\mathcal{L}}}(S)\cdot V(\tilde{f})\right)_{\mathbf{0}} \geq \left(\Gamma^1_{f,\tilde{\mathcal{L}}}(S)\cdot V(\tilde{\mathcal{L}})\right)_{\mathbf{0}}$.

Before we can state and prove our next result, we need a definition.

Definition 6.8.6 The **algebraic critical locus of** f, $\Sigma_{\mathrm{alg}} f$, is the set of those $x \in X$ such that $f \in \mathfrak{m}^2_{X,x}$, where $\mathfrak{m}_{X,x}$ is the maximal ideal of X at x.

Now we have:

Theorem 6.8.7 *For all k,*

$$\left|\mathrm{gecc}^k(\mathbf{F}^\bullet)\right|_{\subseteq V(f)} \subseteq \left|\mathrm{gecc}^k(\phi_f[-1]\mathbf{F}^\bullet)\right| \subseteq \left|\mathrm{gecc}^k(\mathbf{F}^\bullet)\right|_{\subseteq V(f)} \cup \left|\mathrm{gecc}^k(\psi_f[-1]\mathbf{F}^\bullet)\right|.$$

Moreover, over $\Sigma_{\mathrm{alg}} f$*, this second containment is an equality, i.e.,*

$$\left|\mathrm{gecc}^k(\phi_f[-1]\mathbf{F}^\bullet)\right| \cap \pi^{-1}\left(\Sigma_{\mathrm{alg}} f\right) = \left(\left|\mathrm{gecc}^k(\mathbf{F}^\bullet)\right|_{\subseteq V(f)} \cup \left|\mathrm{gecc}^k(\psi_f[-1]\mathbf{F}^\bullet)\right|\right) \cap \pi^{-1}\left(\Sigma_{\mathrm{alg}} f\right).$$

Proof The first line of containments follows immediately from Theorems 6.8.5 and 6.6.2.

The equality over $\Sigma_{\mathrm{alg}} f$ follows at once from Theorem 6.6.1, Theorem 6.8.2, and Theorem 4.2 of [30]. □

Remark 6.8.8 In this long remark, we wish to address how effectively one can calculate $\mathrm{gecc}^\bullet(\phi_f[-1]\mathbf{F}^\bullet)$, given f and $\mathrm{gecc}^\bullet(\mathbf{F}^\bullet)$. This will require us to discuss much of our work in [27, 33], and [34]. As usual, we will identify $T^*\mathcal{U}$ with $\mathcal{U} \times \mathbb{C}^{n+1}$, and use four different projections: $\tau : \mathcal{U} \times \mathbb{C}^{n+1} \times \mathbb{P}^n \to \mathcal{U} \times \mathbb{P}^n$, $\eta : \mathcal{U} \times \mathbb{P}^n \to \mathcal{U}$, $\nu : \mathcal{U} \times \mathbb{C}^{n+1} \times \mathbb{P}^n \to \mathcal{U} \times \mathbb{C}^{n+1}$, and $\pi : \mathcal{U} \times \mathbb{C}^{n+1} \to \mathcal{U}$.

We shall assume that our base ring R is a PID, and that X has codimension at least 1 in $\mathcal{U}$ (so that our projectivizations below do not totally discard components).

Assume that we have re-chosen $\mathcal{U}$ small enough so that $\mathbb{P}(\mathrm{gecc}^k(\phi_f[-1]\mathbf{F}^\bullet))$ is the only non-zero summand in Theorem 6.8.2. Then, Theorem 6.8.2 gives a nice, elegant algebraic characterization of the projectivized gecc of $\phi_f[-1]\mathbf{F}^\bullet$, in terms of blow-ups and exceptional divisors. The problem is that blow-ups and exceptional divisors are not so easy to calculate.

Suppose that $\mathbf{A}^\bullet$ is a bounded complex of sheaves of modules over R, which is constructible with respect to $\mathfrak{S}$. We shall first describe a general method for "calculating" $\mathrm{gecc}^\bullet(\mathbf{A}^\bullet)$, and then describe in the case of $\mathrm{gecc}^\bullet(\phi_f[-1]\mathbf{F}^\bullet)$ why this really leads to an effective calculation.

First, projectivize and obtain $\mathbb{P}(\mathrm{gecc}^k(\mathbf{A}^\bullet)) = \sum_{S\in\mathfrak{S}} m^k_S(\mathbf{A}^\bullet)[\mathbb{P}(\overline{T^*_S\mathcal{U}})] \subseteq \mathcal{U} \times \mathbb{P}^n$. Recall that our coordinates $\mathbf{z} = (z_0, \ldots, z_n)$ determine our cotangent coordinates $(w_0, \ldots, w_n)$ and, hence, determine projective coordinates $[w_0 : \cdots :$

$w_n]$ on $\mathbb{P}^n$. We assume that $\mathcal{U}$ is small enough and that our coordinates $\mathbf{z}$ are generic enough so that, for all S such that $\overline{T_S^*\mathcal{U}}$ is a component of $\mathrm{SS}(\mathbf{A}^\bullet)$, for all j such that $0 \leq j \leq n$, the intersection of $\mathbb{P}(\overline{T_S^*\mathcal{U}})$ and $\mathcal{U} \times \mathbb{P}^j \times \{\mathbf{0}\}$ in $\mathcal{U} \times \mathbb{P}^n$ is proper, and so is purely j-dimensional. We claim that the proper push-forwards

$$^k\Gamma^j_{\mathbf{A}^\bullet,\mathbf{z}} := \eta_*\big(\mathbb{P}(\mathrm{gecc}^k(\mathbf{A}^\bullet)) \odot \mathcal{U} \times \mathbb{P}^j \times \{\mathbf{0}\}\big)$$

completely determine $\mathbb{P}(\mathrm{gecc}^k(\mathbf{A}^\bullet))$ (and, hence, $\mathrm{gecc}^k(\mathbf{A}^\bullet)$). The $^k\Gamma^j_{\mathbf{A}^\bullet,\mathbf{z}}$ are the **characteristic polar cycles**; we refer the reader to Section 5 of [34].

The characteristic polar cycles determine $\mathbb{P}(\mathrm{gecc}^k(\mathbf{A}^\bullet))$ by downward induction on the dimension of strata of X. Let $d := \dim X$, which we are assuming is at most n.

Consider first a stratum S of dimension d. Then, $\overline{T_S^*\mathcal{U}}$ appears in $\mathrm{gecc}^k(\mathbf{A}^\bullet)$ if and only if $\overline{S}$ is a component of $\left|{}^k\Gamma^d_{\mathbf{A}^\bullet,\mathbf{z}}\right|$. In addition, as

$$\eta_*\Big(m_S^k(\mathbf{A}^\bullet)[\overline{T_S^*\mathcal{U}}] \odot \mathcal{U} \times \mathbb{P}^d \times \{\mathbf{0}\}\Big) = m_S^k(\mathbf{A}^\bullet)\Big[\eta_*\Big([\overline{T_S^*\mathcal{U}}] \odot \mathcal{U} \times \mathbb{P}^d \times \{\mathbf{0}\}\Big)\Big],$$

once we know that $\overline{T_S^*\mathcal{U}}$ appears in $\mathrm{gecc}^k(\mathbf{A}^\bullet)$ and we know $^k\Gamma^d_{\mathbf{A}^\bullet,\mathbf{z}}$, then we can determine $m_S^k(\mathbf{A}^\bullet)$. Note that in this process, we do not actually determine the stratum S, but rather a closed analytic set which agrees with S on an open dense set—but this is enough.

Now, suppose that we know the pieces of the enriched cycle $\mathbb{P}(\mathrm{gecc}^k(\mathbf{A}^\bullet))$ for all of those strata of dimension at least $j+1$. Let us write $D^k_{\geq j+1}$ for the (enriched) sum of these pieces. Then, one can calculate the enriched cycle $\eta_*\big(D^k_{\geq j+1} \odot \mathcal{U} \times \mathbb{P}^j \times \{\mathbf{0}\}\big)$; this cycle is an enriched form of the j-dimensional absolute polar varieties of the strata of dimension at least $j+1$. Now, one can consider the difference (we use that R is a PID here)

$$M^j := {}^k\Gamma^j_{\mathbf{A}^\bullet,\mathbf{z}} - \eta_*\big(D^k_{\geq j+1} \odot \mathcal{U} \times \mathbb{P}^j \times \{\mathbf{0}\}\big).$$

Suppose that S is a stratum of dimension j. Then, one easily sees that $\overline{T_S^*\mathcal{U}}$ appears in $\mathrm{gecc}^k(\mathbf{A}^\bullet)$ if and only if $\overline{S}$ is a component of $\left|M^j\right|$. As above, once we know that $\overline{T_S^*\mathcal{U}}$ appears in $\mathrm{gecc}^k(\mathbf{A}^\bullet)$ and we know M^j, we can determine $m_S^k(\mathbf{A}^\bullet)$ by calculating $\Big[\eta_*\Big([\overline{T_S^*\mathcal{U}}] \odot \mathcal{U} \times \mathbb{P}^j \times \{\mathbf{0}\}\Big)\Big]$.

We have seen above that the characteristic polar cycles determine $\mathrm{gecc}^\bullet(\mathbf{A}^\bullet)$. The question now is: how does one effectively calculate the characteristic polar cycles in the case where $\mathbf{A}^\bullet = \phi_f[-1]\mathbf{F}^\bullet$?

Let us adopt the notation $^k\Lambda^j_{\mathbf{F}^\bullet,\mathbf{z}} := {}^k\Gamma^j_{\phi_f[-1]\mathbf{F}^\bullet,\mathbf{z}}$. From our discussion above, we see that we could reconstruct $\mathrm{gecc}^\bullet(\phi_f[-1]\mathbf{F}^\bullet)$ if we knew the $^k\Lambda^j_{\mathbf{F}^\bullet,\mathbf{z}}$. The result of Corollary 6.8 of [34] gives an algorithm for calculating the $^k\Lambda^j_{\mathbf{F}^\bullet,\mathbf{z}}$, assuming that the coordinates $\mathbf{z}$ are **$\phi_f[-1]\mathbf{F}^\bullet$-isolating**. Let us put off the discussion of what

$\phi_f[-1]\mathbf{F}^\bullet$-isolating means; for now, simply assume that the coordinates are generic enough to make true what we write below.

We work in each degree separately; so, fix k.

Let $\Pi^{n+1} := \mathrm{gecc}^k(\mathbf{F}^\bullet)$. Then, Π^{n+1} properly intersects $V\left(w_n - \dfrac{\partial \tilde{f}}{\partial z_n}\right)$, and we may consider the enriched cycle defined by the intersection

$$\sum_V M_V[V] := \Pi^{n+1} \odot V\left(w_n - \frac{\partial \tilde{f}}{\partial z_n}\right);$$

this enriched cycle may have some components contained in $\mathrm{im} d\tilde{f}$ and some components not contained in $\mathrm{im} d\tilde{f}$. Let $\Pi^n := \sum_{V \not\subseteq \mathrm{im} d\tilde{f}} M_V[V]$ and let $\Delta^n := \sum_{V \subseteq \mathrm{im} d\tilde{f}} M_V[V]$.

Now, proceed inductively: if we have Π^{j+1}, then $V\left(w_j - \dfrac{\partial \tilde{f}}{\partial z_j}\right)$ properly intersects Π^{j+1}, and we define Π^j and Δ^j by the equality

$$\Pi^{j+1} \odot V\left(w_j - \frac{\partial \tilde{f}}{\partial z_j}\right) = \Pi^j + \Delta^j,$$

where no component of Π^j is contained in $\mathrm{im} d\tilde{f}$, and every component of Δ^j is contained in $\mathrm{im} d\tilde{f}$.

Continue with this process until one obtains Π^0 and Δ^0.

Then, for all j, as germs at $\mathbf{p}$, ${}^k\Lambda^j_{f,\mathbf{z}}(\mathbf{F}^\bullet) = \pi_*(\Delta^j)$ and

$$\left[{}^k\Lambda^j_{f,\mathbf{z}}(\mathbf{F}^\bullet) \odot V(z_0 - p_0, \dots, z_{j-1} - p_{j-1})\right]_{\mathbf{p}} \cong$$

$$H^k(\phi_{z_j - p_j}[-1]\psi_{z_{j-1}-p_{j-1}}[-1] \dots \psi_{z_0 - p_0}[-1]\phi_f[-1]\mathbf{F}^\bullet)_{\mathbf{p}},$$

where, when $j = 0$, we mean

$$\left[{}^k\Lambda^0_{f,\mathbf{z}}(\mathbf{F}^\bullet)\right]_{\mathbf{p}} \cong H^k(\phi_{z_0 - p_0}[-1]\phi_f[-1]\mathbf{F}^\bullet)_{\mathbf{p}}.$$

Note that, as we are interested in the end only in the Δ^j, throughout the algorithm above, we may, in each step, discard any components of Π^j which do not intersect $\mathrm{im} d\tilde{f}$.

The above works very well for calculating the germs of ${}^k\Lambda^j_{f,\mathbf{z}}(\mathbf{F}^\bullet)$ at $\mathbf{p}$, and so $\mathrm{gecc}^\bullet(\phi_f[-1]\mathbf{F}^\bullet)$ above a neighborhood of $\mathbf{p}$, as long as the coordinates $\mathbf{z}$ are

$\phi_f[-1]\mathbf{F}^\bullet$*-isolating at* $\mathbf{p}$. In [34], we give two characterizations of $\phi_f[-1]\mathbf{F}^\bullet$-isolating that are relevant here.

Let $s := \dim_{\mathbf{p}} \operatorname{supp} \phi_f[-1]\mathbf{F}^\bullet = \dim_{\mathbf{p}} \pi(SS(\phi_f[-1]\mathbf{F}^\bullet))$. Then, the coordinates $\mathbf{z}$ are $\phi_f[-1]\mathbf{F}^\bullet$-isolating at $\mathbf{p}$ if and only if, for all j such that $0 \leq j \leq s-1$, $\mathbf{p}$ is an isolated point in the support of

$$\phi_{z_j - p_j}[-1]\psi_{z_{j-1}-p_{j-1}}[-1]\ldots\psi_{z_0-p_0}[-1]\phi_f[-1]\mathbf{F}^\bullet.$$

This is equivalent to:
for all j such that $0 \leq j \leq s-1$, there exists an open neighborhood $\mathcal{W}$ of $\mathbf{p}$ in $\mathcal{U}$ such that $\mathbb{P}\big(SS(\phi_f[-1]\mathbf{F}^\bullet)\big)$ properly intersects $\mathcal{W} \times \mathbb{P}^j \times \{\mathbf{0}\}$ inside $\mathcal{W} \times \mathbb{P}^n$ and

$$\dim_{\mathbf{p}} \Big(V(z_0 - p_0, \ldots, z_{j-1} - p_{j-1}) \cap \eta\Big(\mathbb{P}\big(SS(\phi_f[-1]\mathbf{F}^\bullet)\big) \cap \mathcal{W} \times \mathbb{P}^j \times \{\mathbf{0}\}\Big)\Big) \leq 0. \tag{6.1}$$

When $j = 0$, this condition is interpreted as

$$\dim_{\mathbf{p}} \eta\Big(\mathbb{P}\big(SS(\phi_f[-1]\mathbf{F}^\bullet)\big) \cap \mathcal{W} \times \{[1:0:0:\cdots:0]\}\Big) \leq 0.$$

Are either one of these characterizations of $\phi_f[-1]\mathbf{F}^\bullet$-isolating useful? Yes—the latter one is. Theorem 6.8.7 tells us that

$$SS(\phi_f[-1]\mathbf{F}^\bullet) \subseteq \big(SS(\mathbf{F}^\bullet)\big)_{\subseteq V(f)} \cup SS(\psi_f[-1]\mathbf{F}^\bullet).$$

So, if our coordinates are generic enough so that Formula 6.1 above holds with $\mathbb{P}\big(SS(\phi_f[-1]\mathbf{F}^\bullet)\big)$ replaced by $\mathbb{P}\big(SS(\mathbf{F}^\bullet)\big)_{\subseteq V(f)} \cup \mathbb{P}\big(SS(\psi_f[-1]\mathbf{F}^\bullet)\big)$, then the entire process above works.

Now we can give an example.

Example 6.8.9 We continue with our earlier situation:

$$X = V(y) \cup V(y^2 - x^3 - t^2x^2) \subseteq \mathbb{C}^3, \quad \mathbf{F}^\bullet = \mathbb{Z}_X^\bullet[2], \quad \text{and } \tilde{f} := x.$$

This will give us an easy, but nonetheless, illustrative example of the procedure described in Remark 6.8.8.

From Example 6.5.7, we know that $SS(\mathbf{F}^\bullet)_{\subseteq V(f)} = T^*_{V(x,y)}\mathcal{U} \cup T^*_{\mathbf{0}}\mathcal{U}$. From Example 6.6.3, we know that $SS(\psi_f[-1]\mathbf{F}^\bullet)$ is also equal to $T^*_{V(x,y)}\mathcal{U} \cup T^*_{\mathbf{0}}\mathcal{U}$. Therefore, Theorem 6.8.7 tells us that

$$SS(\phi_f[-1]\mathbf{F}^\bullet) \subseteq T^*_{V(x,y)}\mathcal{U} \cup T^*_{\mathbf{0}}\mathcal{U},$$

(in fact, Theorem 6.8.7 tells us that this an equality, though we will not use this stronger fact).

Hence,

$$\operatorname{supp}\phi_f[-1]\mathbf{F}^\bullet = \pi\big(\mathrm{SS}(\phi_f[-1]\mathbf{F}^\bullet)\big) \subseteq \pi\big(T^*_{V(x,y)}\mathcal{U} \cup T^*_{\mathbf{0}}\mathcal{U}\big) = V(x,y).$$

Considering how simple this set is, we could calculate $\mathrm{gecc}^\bullet(\phi_f[-1]\mathbf{F}^\bullet)$ "barehandedly", by applying Theorem 6.8.5 at the origin, and then moving to a generic point on $V(x,y)$, taking a hyperplane slice, and applying Theorem 6.8.5 again.

However, we want to demonstrate the procedure that we described in Example 6.8.8. Hence, we will first determine $\phi_f[-1]\mathbf{F}^\bullet$-isolating coordinates at $\mathbf{0}$, and then go through the graded, enriched cycle calculation from Example 6.8.8.

From the above, we see that $s = \dim_{\mathbf{0}}\operatorname{supp}\phi_f[-1]\mathbf{F}^\bullet \le 1$, and thus our coordinates are $\phi_f[-1]\mathbf{F}^\bullet$-isolating at $\mathbf{0}$ if Formula 6.1 holds for $\mathbf{p} = \mathbf{0}$ and $j = 0$; this is the degenerate case mentioned immediately after Formula 6.1.

It follows that, if we let $(z_0, z_1, z_2) = (t, x, y)$, so that the cotangent coordinates (w_0, w_1, w_2) correspond to $w_0dt + w_1dx + w_2dy$, then $\mathbb{P}^0 = \{[1:0:\cdots:0]\}$ in Formula 6.1 corresponds to the projective class $[dt]$ and
$\mathbb{P}\big(\mathrm{SS}(\phi_f[-1]\mathbf{F}^\bullet)\big) \cap \mathcal{W} \times \{[1:0:0:\cdots:0]\} \subseteq$
$\Big(\mathbb{P}(T^*_{V(x,y)}\mathcal{U}) \cup \mathbb{P}(T^*_{\mathbf{0}}\mathcal{U})\Big) \cap \mathcal{W} \times \{[1:0:0:\cdots:0]\} = \emptyset \cup \{(\mathbf{0}, [1:0:\cdots:0])\}$.

Therefore,

$$\dim_{\mathbf{0}} \eta\Big(\mathbb{P}\big(\mathrm{SS}(\phi_f[-1]\mathbf{F}^\bullet)\big) \cap \mathcal{W} \times \{[1:0:0:\cdots:0]\}\Big) \le 0,$$

and the coordinates (t, x, y) are $\phi_f[-1]\mathbf{F}^\bullet$-isolating at $\mathbf{0}$. Note that this ordering on the coordinates is different from what we used earlier, because we need for t to come first.

We can now proceed with the enriched cycle calculation as described in Example 6.8.8.

As we saw in Example 6.5.7, $\mathrm{gecc}^k(\mathbf{F}^\bullet) = 0$ if $k \neq 0$; thus, we need calculate only in the fixed degree $k = 0$. As we also saw in Example 6.5.7,

$$\mathrm{gecc}^0(\mathbf{F}^\bullet) = \mathbb{Z}\left[\overline{T^*_{S_1}\mathbb{C}^3}\right] + \mathbb{Z}\left[\overline{T^*_{S_2}\mathbb{C}^3}\right] + \mathbb{Z}^2\left[\overline{T^*_{S_3}\mathbb{C}^3}\right] + \mathbb{Z}\left[\overline{T^*_{S_4}\mathbb{C}^3}\right] + \mathbb{Z}^2\left[T^*_{\{\mathbf{0}\}}\mathbb{C}^3\right],$$

where

$$S_1 = V(y) - V(y^2 - x^3 - t^2x^2),\ S_2 = V(y^2 - x^3 - t^2x^2) - V(y),\ S_3 = V(x,y) - \{\mathbf{0}\},$$

and $S_4 = V(x + t^2, y) - \{\mathbf{0}\}$.

Using a computer algebra system to find equations defining $\overline{T^*_{S_2}\mathbb{C}^3}$, we have
$\mathrm{gecc}^0(\mathbf{F}^\bullet) = \mathbb{Z}[V(y, w_0, w_1)] +$
$\mathbb{Z}\Big[V\big(y^2 - x^3 - t^2x^2,\ 2tw_0w_1 - w_0^2 - 2t^2w_2^2x,\ -w_0w_1 + 2tw_1^2 - tw_2^2(2t^2 + 3x),$

$2w_1x+tw_0+3w_2y,\ w_0^2(t^2+x)-t^2w_2^2x^2,\ w_0y+tw_2x^2,\ 2w_1y+w_2(2t^2+3x)x,$
$tw_2y + w_0(t^2+x))\Big]+$
$\mathbb{Z}^2\,[V(x,y,w_0)] + \mathbb{Z}\left[V(y, x+t^2, 2tw_1 - w_0)\right] + \mathbb{Z}^2\,[V(t,x,y)]\,.$
Before we proceed with the algorithm, note that

$$\operatorname{im} d\tilde{f} \;=\; \mathcal{U} \times \{(0,1,0)\} \;=\; V(w_0, w_1 - 1, w_2).$$

We now let $\Pi^3 = \text{gecc}^0(\mathbf{F}^\bullet)$, and calculate

$$\Pi^3 \odot V\left(w_2 - \frac{\partial x}{\partial y}\right) = \Pi^3 \odot V(w_2) =$$

$$\mathbb{Z}\left[V(y, x+t^2, 2tw_1 - w_0, w_2)\right] + \mathbb{Z}^2\,[V(x,y,w_0,w_2)] +$$

$$\mathbb{Z}\left[V(y, x+t^2, 2tw_1 - w_0, w_2)\right] + \mathbb{Z}^2\,[V(t,x,y,w_2)]$$

$$+ \text{ components which do not intersect } \operatorname{im} d\tilde{f}.$$

Therefore, we have

$$\Delta^2 = \mathbb{Z}^2\,[V(x,y,w_0,w_2)] + \mathbb{Z}^2\,[V(t,x,y,w_2)]\,,$$

and

$$\Pi^2 = \mathbb{Z}^2\left[V(y, t^2+x, 2tw_1 - w_0, w_2)\right].$$

6.9 Future Directions

- It is somewhat annoying in Example 6.8.9, and in the general algorithm given in Remark 6.8.8, that, essentially, we first have to know $\mathrm{SS}(\phi_f[-1]\mathbf{F}^\bullet)$ in order to begin the calculation of the cycles ${}^k\Lambda^j_{\mathbf{F}^\bullet,\mathbf{z}}$.

  Why do we have to know $\mathrm{SS}(\phi_f[-1]\mathbf{F}^\bullet)$ first? Solely because we need to know that our coordinates are $\phi_f[-1]\mathbf{F}^\bullet$-isolating. Ideally, we could begin with the calculation of the ${}^k\Lambda^j_{\mathbf{F}^\bullet,\mathbf{z}}$, and check "on-the-fly" that certain intersections are proper, which would then tell us that the coordinate choice is generic enough. This is what happens with the Lê cycles for affine hypersurface singularities; see [28].

  Unfortunately, while we suspect that such a result is true, we have yet to find a proof.

- The characteristic cycle of the intersection cohomology complex is of great importance in representation theory (see [18] and [2]), and yet, aside from the curve case in Example 6.2.15, we have not discussed the calculation of the characteristic cycle (enriched or not) for intersection cohomology complexes (with constant or twisted coefficients). This is because such a calculation is, not surprisingly, hard, and we have no satisfactory results in this area.

What may be surprising is that the calculation of the characteristic cycle of intersection cohomology, with constant coefficients, is closely related to the relative Milnor monodromy of the constant sheaf along a hypersurface containing the singular set. We will describe this relationship briefly.

Suppose that X is an analytic space and, as we are happy to work locally at $\mathbf{0}$, assume that we have an analytic function $f : X \to \mathbb{C}$ such that the singular set of X is contained in $V(f)$, but that f does not vanish on any irreducible component of X.

As before, let $i : X - V(f) \hookrightarrow X$ and $j : V(f) \hookrightarrow X$ denote the inclusions.

Note that, as intersection cohomology $\mathbf{I}^\bullet$, with constant coefficients, on X is a perverse sheaf, the graded, enriched characteristic cycle is concentrated in degree 0. Also, the case that is of concern in representation theory is when the base ring is a field. Consequently, we would be satisfied with calculating $\mathrm{CC}(\mathbf{I}^\bullet)$.

Now, Proposition 6.7.9 tells us that we can calculate $\mathrm{CC}(\mathbf{I}^\bullet)$ if we can calculate $\mathrm{CC}(j_*j^*[-1]\mathbf{I}^\bullet) = -\mathrm{CC}(j_*j^*\mathbf{I}^\bullet)$ and $\mathrm{CC}(i_!i^!\mathbf{I}^\bullet)$. But $i^!\mathbf{I}^\bullet \cong i^*\mathbf{I}^\bullet$ is the restriction of $\mathbf{I}^\bullet$ to a generic subset of the smooth part of X, by our assumptions on f. Thus, $i^!\mathbf{I}^\bullet$ coincides with the restriction of the constant sheaf to a generic smooth subset of X. Hence, $\mathrm{CC}(i_!i^!\mathbf{I}^\bullet)$ can be calculated using Theorem 6.7.1 and its corollary.

We are left with the problem of calculating $\mathrm{CC}(j_*j^*[-1]\mathbf{I}^\bullet)$, which, after slicing, reduces to the problem of calculating the Morse module $m^0_{\mathbf{0}}(j_*j^*[-1]\mathbf{I}^\bullet)$ or, more precisely, reduces to knowing when this Morse module is zero and when it is not.

There is the fundamental distinguished triangle, relating the nearby and vanishing cycles:

$$j_*j^*[-1]\mathbf{I}^\bullet \to \psi_f[-1]\mathbf{I}^\bullet \xrightarrow{\text{can}} \phi_f[-1]\mathbf{I}^\bullet \xrightarrow{[1]},$$

which is actually a short exact sequence in the Abelian category of perverse sheaves, due to the fact that $j_*j^*[-1]\mathbf{I}^\bullet$ is perverse (which uses that $\mathbf{I}^\bullet$ is intersection cohomology).

There is also the dual variation triangle

$$\phi_f[-1]\mathbf{I}^\bullet \xrightarrow{\text{var}} \psi_f[-1]\mathbf{I}^\bullet \to j_!j^![1]\mathbf{I}^\bullet \xrightarrow{[1]},$$

which is also a short exact sequence in the Abelian category of perverse sheaves, due to the fact that $j_!j^![1]\mathbf{I}^\bullet$ is perverse.

This is where the monodromy automorphism $T_f : \psi_f[-1]\mathbf{I}^\bullet \to \psi_f[-1]\mathbf{I}^\bullet$ comes in. It is well-known that $\mathrm{var} \circ \mathrm{can} = \mathrm{id} - T_f$. It follows that, in the Abelian category of perverse sheaves on $V(f)$, $j_* j^*[-1]\mathbf{I}^\bullet \cong \ker\{\mathrm{id} - T_f\}$.

Suppose now that $\tilde{\mathcal{L}}$ is a generic linear form, and that $\mathcal{L}$ is the restriction of $\tilde{\mathcal{L}}$ to $V(f)$. Then, it follows that $\phi_{\mathcal{L}}[-1]j_* j^*[-1]\mathbf{I}^\bullet$, which is a finite-dimensional vector space concentrated in degree 0, is isomorphic to the kernel of the map induced by $\mathrm{id} - T_f$ on $\phi_{\mathcal{L}}[-1]\psi_f[-1]\mathbf{I}^\bullet$, and so is determined by relative Milnor monodromy.

As $\mathbf{I}^\bullet$ agrees with the constant sheaf on the complement of $V(f)$, which is all that $\psi_f[-1]\mathbf{I}^\bullet$ cares about, we can calculate $m^0_{\mathbf{0}}(\psi[-1]\mathbf{I}^\bullet) \cong \phi_{\mathcal{L}}[-1]\psi[-1]\mathbf{I}^\bullet$ via Theorem 6.6.2, in the easy case of the constant sheaf, where the relevant strata are open dense subsets of the smooth parts of the components of X. Moreover, the relative monodromy that we need to analyze is also that from the "easy" constant sheaf case.

It is, of course, our hope to analyze the above relative monodromy, and produce a method for calculating, in principle and in practice, characteristic cycles of intersection cohomology.

References

1. Ament, D. A. H., Nuño-Ballesteros, J. J., Oréfice-Okamoto, B., and Tomazella, J. N. The Euler obstruction of a function on a determinantal variety and on a curve. *Bull. Braz. Math. Soc., new series*, 47 (3):955–970, 2016.
2. Braden, T. On the Reducibility of Characteristic Varieties. *Proc. AMS*, 130:2037–2043, 2002.
3. Brasselet, J.-P., Lê D. T., Seade, J. Euler Obstruction and indices of vector fields. *Topology*, 39:1193–1208, 2000.
4. Brasselet, J.-P., Massey, D., Parameswaram, Seade, J. Euler Obstruction and Functions with Isolated Singularities. *J. London Math. Soc. (2)*, 70:59–76, 2004.
5. Brasselet, J.-P., Seade, J., and Suwa, T. *Vector Fields on Singular Varieties*, volume 1987 of *Lecture Notes in Mathematics*. Springer-Verlag, 2009.
6. Briançon, J., Maisonobe, P., and Merle, M. Localisation de systèmes différentiels, stratifications de Whitney et condition de Thom. *Invent. Math.*, 117:531–550, 1994.
7. Brylinski, J. L., Dubson, A., and Kashiwara, M. Formule de l'indice pour les modules holonomes et obstruction d'Euler locale. *C. R. Acad. Sci., Série A.*, 293:573–576, 1981.
8. Dimca, A. *Sheaves in Topology*. Universitext. Springer-Verlag, 2004.
9. Dutertre, N. and Grulha Jr., N. Some notes on the Euler Obstruction of a Function. In *Journal of Singularities, vol. 10, Proc. 12th Intern. Workshop on Sing., São Carlos, 2012*, pages 82–91, 2014.
10. Ebeling, W. and Gusein-Zade, S. Radial Index and Euler Obstruction of a 1-Form on a Singular Variety. *Geom. Dedicata*, 113:231–241, 2005.
11. Ebeling, W. and Gusein-Zade, S. An Equivariant Version of the Euler Obstruction. *Bull. Braz. Math. Soc.*, pages 1–10, 2016.
12. Fulton, W. *Intersection Theory*, volume 2 of *Ergeb. Math.* Springer-Verlag, 1984.
13. Ginsburg, V. Characteristic Varieties and Vanishing Cycles. *Invent. Math.*, 84:327–403, 1986.
14. Gonzalez-Sprinberg, G. L'obstruction locale d'Euler et le theoreme de MacPherson. *Asterisque*, 82-83:7–32, 1981.
15. Goresky, M. and MacPherson, R. *Stratified Morse Theory*, volume 14 of *Ergeb. der Math.* Springer-Verlag, 1988.

16. Hamm, H. and Lê D. T. Un théorème de Zariski du type de Lefschetz. *Ann. Sci. Éc. Norm. Sup.*, 6 (series 4):317–366, 1973.
17. Kashiwara, M. and Schapira, P. *Sheaves on Manifolds*, volume 292 of *Grund. math. Wissen.* Springer-Verlag, 1990.
18. Kazhdan, D. and Lusztig, G. A topological approach to Springer's representations. *Adv. Math.*, 38:222–228, 1980.
19. Lê, D. T. Calcul du Nombre de Cycles Évanouissants d'une Hypersurface Complexe. *Ann. Inst. Fourier, Grenoble*, 23:261–270, 1973.
20. Lê, D. T. Topological Use of Polar Curves. *Proc. Symp. Pure Math.*, 29:507–512, 1975.
21. Lê, D. T. Sur les cycles évanouissants des espaces analytiques. *C. R. Acad. Sci. Paris, Sér. A-B*, 288:A283–A285, 1979.
22. Lê, D. T. Le concept de singularité isolée de fonction analytique. *Advanced Studies in Pure Math.*, 8:215–227, 1986.
23. Lê, D. T. and Teissier, B. Variétés polaires locales et classes de Chern des variétés singulières. *Annals of Math.*, 114:457–491, 1981.
24. MacPherson, R. Chern classes for singular varieties. *Annals of Math.*, 100:423–432, 1974.
25. Massey, D. The Lê Varieties, I. *Invent. Math.*, 99:357–376, 1990.
26. Massey, D. The Lê Varieties, II. *Invent. Math.*, 104:113–148, 1991.
27. Massey, D. Numerical Invariants of Perverse Sheaves. *Duke Math. J.*, 73(2):307–370, 1994.
28. Massey, D. *Lê Cycles and Hypersurface Singularities*, volume 1615 of *Lecture Notes in Math.* Springer-Verlag, 1995.
29. Massey, D. Hypercohomology of Milnor Fibres. *Topology*, 35:969–1003, 1996.
30. Massey, D. Critical Points of Functions on Singular Spaces. *Top. and Appl.*, 103:55–93, 2000.
31. Massey, D. A Little Microlocal Morse Theory. *Math. Ann.*, 321:275–294, 2001.
32. Massey, D. The Sebastiani-Thom Isomorphism in the Derived Category. *Compos. Math.*, 125:353–362, 2001.
33. Massey, D. *Numerical Control over Complex Analytic Singularities*, volume 778 of *Memoirs of the AMS*. AMS, 2003.
34. Massey, D. Singularities and Enriched Cycles. *Pacific J. Math.*, 215, no. 1:35–84, 2004.
35. Massey, D. Enriched Relative Polar Curves and Discriminants. *Contemp. Math.*, 474:107–144, 2008.
36. Massey, D. Characteristic Cycles and the Relative Local Euler Obstruction. In *A panorama of Singular Varieties: a conference to celebrate the 70th birthday of Lê Dung Tráng in Seville (Spain)*, volume 742 of *Contemp. Math.*, pages 137–156, 2020.
37. Parusiński, A. and Pragacz, P. Characteristic classes of hypersurfaces and characteristic cycles. *J. Alg. Geom.*, 10(1):63–79, 2001.
38. Sabbah, C. Quelques remarques sur la géométrie des espaces conormaux. *Astérisque*, 130:161–192, 1985.
39. Schürmann, J. A general intersection formula for Lagrangian cycles. *Comp. Math.*, 140 (4):1037–1052, 2004.
40. Schürmann, J. *Topology of Singular Spaces and Constructible Sheaves*, volume 63 of *Monografie Matematyczne*. Birkhäuser, 2004.
41. Seade, J., Tibăr, M., and Verjovsky, A. Milnor Numbers and Euler Obstruction. *Bull. Braz. Math. Soc.*, 36, no. 2:275–283, 2005.
42. Teissier, B. Cycles évanescents, sections planes et conditions de Whitney. *Astérisque*, 7-8:285–362, 1973.

Chapter 7
Monodromy Zeta Functions in Singularity Theory

Sabir M. Gusein-Zade

Contents

Abstract We discuss notions of the monodromy zeta functions, their properties, methods of computation and some modifications of them.

S. M. Gusein-Zade (✉)
Faculty of Mechanics and Mathematics, Moscow Center for Fundamental and Applied Mathematics, Moscow State University, Moscow, Russia

National Research University "Higher School of Economics", Moscow, Russia
e-mail: sabirg@list.ru

© The Author(s), under exclusive license to Springer Nature Switzerland AG 2026

J. L. Cisneros-Molina et al. (eds.), *Handbook of Geometry and Topology of Singularities VIII*, https://doi.org/10.1007/978-3-031-99571-2_7

7.1 Introduction

In the simplest context the monodromy zeta function is defined in the following way.

Let $f : (\mathbb{C}^n, 0) \to (\mathbb{C}, 0)$ be a germ of a holomorphic function. The Milnor fibre of f is

$$V_f \cong V_{f,\varepsilon} = f^{-1}(\varepsilon) \cap B_\delta(0)$$

with $0 < |\varepsilon| \ll \delta$, δ is small enough, $B_\delta(0)$ is the (closed) ball of radius δ centred at the origin in $\mathbb{C}^n$. As a differentiable (that is C^∞–) manifold with the boundary

$$\partial V_f \cong \partial V_{f,\varepsilon} = f^{-1}(\varepsilon) \cap S_\delta(0)$$

($S_\delta(0) = \partial B_\delta(0)$ is the sphere of radius δ), the Milnor fibre V_f is well-defined, that is does not depend on the choice of δ and ε. According to the celebrated Milnor Theorem [54], if f has an isolated critical point at the origin, the Milnor fibre V_f is homotopy equivalent to the bouquet of spheres of dimension $n - 1$. In general V_f is homotopy equivalent to a CW-complex of dimension $\leq (n - 1)$.

Let us fix δ small enough. For $\varepsilon_0 > 0$ small enough, the map

$$f : f^{-1}(D_{\varepsilon_0}(0)) \cap B_\delta(0) \to D_{\varepsilon_0}(0),$$

where $D_{\varepsilon_0}(0)$ is the closed disk of radius ε_0 in $\mathbb{C}$ centred at the origin, is a locally trivial fibration outside of $0 \in \mathbb{C}$. Let $h = h_f : V_f \to V_f$ be a monodromy transformation of this fibration, i.e., a monodromy transformation corresponding to the loop $\gamma_0(\tau) = \exp(2\pi i\tau)\varepsilon_0$, $0 \leq \tau \leq 1$. One can assume that the monodromy transformation h is a diffeomorphism of $V_f = f^{-1}(\varepsilon_0) \cap B_\delta(0)$. It is well defined up to isotopy. The monodromy operators are the actions h^* of the monodromy transformation h on the (co)homology groups $H^q(V_f; \mathbb{Z})$.

The monodromy zeta function of the germ f is the rational function in t defined by

$$\zeta_f(t) = \prod_{q \geq 0} \left(\det(\mathrm{id} - t \cdot h^*_{|H^q(V_f;\mathbb{R})})\right)^{(-1)^q}. \tag{7.1}$$

Remark 7.1.1 Sometimes (say, in [1]), in the definition (7.1), one uses the exponent $(-1)^{q+1}$.

In the Singularity Theory, the monodromy zeta function appeared in [54]. However, one can say that it became a standard and widely used notion after the paper [1] by N. A'Campo.

If the germ f has an isolated critical point at the origin, the monodromy zeta function essentially coincides with the characteristic polynomial $\Delta_f(t) = \det\left(\mathrm{id} - t \cdot h^*_{|H^{n-1}(V_f;\mathbb{R})}\right)$:

$$\zeta_f(t) = (1-t) \cdot \left(\Delta_f(t)\right)^{(-1)^{n-1}}.$$

If, moreover, $n = 2$, the classical monodromy zeta function of a function germ $f : (\mathbb{C}^2, 0) \to (\mathbb{C}, 0)$ is equal to $\frac{1-t}{\Delta^C(t)}$ where $\Delta^C(t) = \Delta_f(t)$ is the Alexander polynomial (in one variable) of the algebraic link $C \cap \mathbb{S}^3_\delta(0)$ of the curve germ $(C, 0) = \{f = 0\} \subset (\mathbb{C}^2, 0)$ ($\mathbb{S}^3_\delta(0)$ is the three-dimensional sphere of small radius δ centred at the origin in $\mathbb{C}^2$). This is the main reason why, in the definition (7.1), some people use the exponent $(q-1)$ instead of q. In that case, the monodromy zeta function is the Alexander polynomial divided by $(1-t)$, what looks somewhat better.

Let us briefly outline the content of this article. In Sect. 7.2, we collect the basic notions and classical facts about the monodromy zeta functions for the case of function germs on a complex affine space. In Sect. 7.3, we discuss their appearance in other situations: function germs on singular varieties, polynomial functions, and meromorphic germs. In Sect. 7.4, we discuss relations of monodromy zeta functions with analytic invariants like Poincaré series. Finally, in Sect. 7.5, we discuss an orbifold version of the monodromy zeta function and symmetries for them related with the mirror symmetry.

7.2 Zeta Function of a Transformation and the Classical Monodromy Zeta Function

7.2.1 The Additive Euler Characteristic

Let X be a topological space nice enough, say, homeomorphic to a locally closed union of cells in a finite CW-complex. (Any complex or real semialgebraic variety is of this sort.) The ("usual") Euler characteristic of X is defined as the alternating sum of the dimensions of the cohomology groups:

$$\chi(X) = \sum_{q \geq 0} (-1)^q \dim H^q(X; \mathbb{R}). \tag{7.2}$$

This Euler characteristic is a homotopy invariant of a topological space. However, in a number of computations, it is more convenient to use the so called *additive Euler characteristic* defined as the alternating sum of the dimensions of the cohomology groups with compact support:

$$\chi_c(X) = \sum_{q \geq 0} (-1)^q \dim H_c^q(X; \mathbb{R}) \,. \tag{7.3}$$

It is not a homotopy invariant of a topological space. (It is invariant with respect to homotopy equivalence defined in the category of proper maps.) The main property of this Euler characteristic is its additivity: if Y is a locally closed subset of X (X and Y are nice enough: see above), then

$$\chi_c(X) = \chi_c(Y) + \chi_c(X \setminus Y) \,. \tag{7.4}$$

This permits to use it as a sort of a measure for the notion of the integration with respect to the Euler characteristic: [74] (see also [34]). If the space X is compact, one obviously has $\chi_c(X) = \chi(X)$. If X is a complex quasiprojective variety, that is the difference of two projective varieties, then the additive Euler characteristic $\chi_c(X)$ coincides with the usual one $\chi(X)$: see, e.g., [36, Proposition 3]. In this paper we shall always have in mind the additive Euler characteristic omitting the index c in its notation.

7.2.2 The Zeta Function of a Transformation

For the case when X was the Milnor fibre V_f of an analytic function germ $f : (\mathbb{C}^n, 0) \to (\mathbb{C}, 0)$ and $h = h_f$ was a classical monodromy transformation of the germ f, the notion of the (classical) monodromy zeta function was defined in the Introduction. One can see that the degree of the monodromy zeta function (the degree of the numerator minus the degree of the denominator) is equal to the Euler characteristic of the Milnor fibre V_f. If we want to have a notion of the zeta function of a transformation $h : X \to X$ such that its degree will be equal to the Euler characteristic of X (the additive one!) we should consider the action of h on the cohomology groups of X with compact support. This is impossible, in general: for an arbitrary map h its action on the cohomology groups with compact support is not defined. One can either assume that all the maps under consideration are proper (that is the preimage of a compact set is compact), or (more restrictively) that they are self-homeomorphisms. Since it is always possible to assume that monodromy transformations are self-homeomorphisms, in what follows we shall use the second assumption. As a reason, one can refer to the fact that monodromy transformations of locally trivial fibrations always can be assumed to be self-homeomorphisms of the fibres.

Let h be a self-homeomorphism of a topological space X (nice enough: see above).

Definition 7.2.1 The *zeta function* $\zeta_h(t)$ of the map h is the rational function in t defined by

$$\zeta_h(t) = \prod_{q\geq 0} \left(\det\left(\mathrm{id} - t h^*_{|_{H^q_c(X;\mathbb{R})}}\right)\right)^{(-1)^q}, \tag{7.5}$$

where $h^*_{|\dots}$ is the action of h on the cohomology groups with compact support.

Obviously, one has $\deg \zeta_h(t) = \chi(X)$. If X is compact or if it is a complex quasiprojective variety, one gets the same (rational) function using, in (7.5), the usual cohomology groups instead of those with compact support. One has the following multiplicativity property of the zeta function of a transformation generalizing the additivity property (7.4) of the Euler characteristic.

Proposition 7.2.2 *Let h be a self homeomorphism of a topological space X and let Y be a locally closed subspace of X such that $h(Y) = Y$ (and therefore $h(X \setminus Y) = X \setminus Y$). Then one has*

$$\zeta_h(t) = \zeta_{h_{|Y}}(t) \cdot \zeta_{h_{|X\setminus Y}}(t), \tag{7.6}$$

where $h_{|\dots}$ is the restriction of h to the corresponding subspace.

As we mentioned above, the Euler characteristic and the zeta function of a transformation are defined only for spaces nice enough. For that one may assume that all spaces under consideration are homeomorphic to locally closed unions of cells in CW-complexes. Complex (or real) quasiprojective varieties are of this sort.

7.2.3 Monodromy of a Holomorphic Function Germ and Its Zeta Function

The classical monodromy zeta function of a function germ $f : (\mathbb{C}^n, 0) \to (\mathbb{C}, 0)$ defined in Introduction (Eq. (7.1)) is the zeta function of a monodromy transformation $h = h_f$ of f. Let us describe the monodromy transformation h_f in more details. This turns out to be useful in a number of constructions and computations related with it. As it was written in Introduction, for $0 < |\varepsilon_0| \ll \delta$ and $\delta > 0$ small enough the restriction of f to $f^{-1}(D_{\varepsilon_0}(0)) \cap B_\delta(0)$ ($B_\delta(0)$ is the closed ball of radius δ centred at the origin in $\mathbb{C}^n$, $D_{\varepsilon_0}(0)$ is the closed disk of radius ε_0 centred at the origin in $\mathbb{C}$) is a smooth locally trivial fibration over the punctured disk $D_{\varepsilon_0}(0) \setminus \{0\}$ with the fibres $V_{f,z} = f^{-1}(z) \cap B_\delta(0)$ (diffeomorphic to the Milnor fibre of f). This implies that there exists a family of diffeomorphisms $\varGamma_\tau : V_{f,\varepsilon_0} \to V_{f,\varepsilon_0\cdot\exp(2\pi i\tau)}$, $0 \leq \tau \leq 1$. One defines the classical monodromy transformation h_f as $\varGamma_1 : V_{f,\varepsilon_0} \to V_{f,\varepsilon_0}$. This definition is completely appropriate for the definition of the monodromy zeta function (7.5). However, one can choose the family $\varGamma_t$ satisfying additional properties.

First, one can impose an additional condition on the behaviour of the monodromy transformation on the boundary of the Milnor fibre. For $0 < |\varepsilon_0| \ll \delta$, $\delta > 0$ small enough, the restriction of f to $f^{-1}(D_{\varepsilon_0}(0)) \cap B_\delta(0)$ is a smooth locally trivial fibration over $D_{\varepsilon_0}(0) \setminus \{0\}$. However, its restriction to $f^{-1}(D_{\varepsilon_0}(0)) \cap S_\delta^{2n-1}(0)$ is a locally trivial fibration over the entier disc D_{ε_0} (whose fibre is the boundary of the Milnor fibre of f). Any locally trivial fibration over the disc is trivial. Thus, there is an isomorphism between this fibration and the "direct product fibration" $D_{\varepsilon_0} \times \partial V_f \to D_{\varepsilon_0}$ (a trivialization). (This trivialization is well defined up to a fibre isotopy.) One can demand that the family Γ_τ is constant on the boundary with respect to this trivialization (and therefore $h = \Gamma_1$ is the identity on the boundary of the Milnor fibre). The monodromy transformation obtained this way is used to define some objects related with it and with its zeta function: the variation operator, the Seifert form, etc.

Another particular choice of the monodromy transformation (not compatible with the construction in the previous paragraph) is used for quasihomogeneous functions. Let $f : (\mathbb{C}^n, 0) \to (\mathbb{C}, 0)$ be quasihomogeneous with respect to the (integer, positive) weights $q_1, \ldots, q_n$ and the quasidegree d (i.e., $f(\lambda^{q_1} z_1, \ldots, \lambda^{q_n} z_n) = \lambda^d f(z_1, \ldots, z_n)$). Then the family Γ_τ can be defined by the equation

$$\Gamma_\tau(z_1, \ldots, z_n) = (\exp \frac{2\pi i q_1 \tau}{d} z_1, \ldots, \exp \frac{2\pi i q_n \tau}{d} z_n)$$

and thus the monodromy transformation h_f is given by

$$h_f(z_1, \ldots, z_n) = (\exp \frac{2\pi i q_1}{d} z_1, \ldots, \exp \frac{2\pi i q_n}{d} z_n).$$

This form of the monodromy transformation is used, in particular, in constructions related with Landau-Ginzburg models and mirror symmetry (see Sect. 7.5).

7.2.4 A'Campo Equation for the Classical Monodromy Zeta Function

Let $f : (\mathbb{C}^n, 0) \to (\mathbb{C}, 0)$ be a germ of a holomorphic function. A resolution of the germ f is a holomorphic map $p : (W, D) \to (\mathbb{C}^n, 0)$ ($D = p^{-1}(0)$) from an n-dimensional complex manifold M such that

(1) the map p is proper;
(2) the total transform $p^{-1}(\{f = 0\})$ of the zero level set of f is a normal crossing divisor on M (this implies, in particular, that the exceptional divisor D is a normal crossing divisor as well);
(3) the map p is an isomorphism outside of the zero level set $\{f = 0\}$ of f.

The condition (2) implies (in fact is equivalent to) that, for each point $Q \in M$, on a neighbourhood of Q there exist local coordinates $u_1, \ldots, u_n$ centred at Q such that

$$f \circ p(u_1, \ldots, u_n) = \varphi(u_1, \ldots, u_n) \cdot u_1^{m_1} \cdot \ldots \cdot u_n^{m_n}, \tag{7.7}$$

where m_i are non-negative integers, $\varphi(0, \ldots, 0) \neq 0$.

Remark 7.2.3 The factor $\varphi(u_1, \ldots, u_n)$ is inserted into Eq. (7.7) in order to make it valid for all the point of W including those outside of the total preimage of $\{f = 0\}$. Also factors like that have to be present in a description of a resolution of a collection of germs.

For an integer m, let S_m be the set of points $Q \in D$ such that, in their neighbourhoods, the representation (7.7) contains only one non-zero exponent m_i equal to m, i.e., is of the form $f \circ p(\overline{u}) = \varphi(\overline{u}) \cdot u_1^m$ ($\overline{u} = (u_1, \ldots, u_n)$). This means, in particular, that the points of S_m are smooth points of the normal crossing divisor $p^{-1}(\{f = 0\})$. The following statement describes the monodromy zeta function of the germ f in terms of its resolution.

Theorem 7.2.4 (N.A. Campo: [1]) *One has*

$$\zeta_f(t) = \prod_{m>0} (1 - t^m)^{\chi(S_m)}. \tag{7.8}$$

Remark 7.2.5 In a description of a resolution in [1], the condition (3) (which is essential) was omitted (forgotten). This caused some problems for the authors of [39] in the process of elaboration of the paper.

The right hand side of (7.8) can be rewritten in the following form. For a point $Q \in D$, let $\zeta_Q(t)$ be the classical monodromy zeta function of the germ $f \circ p$ at Q. One can see that $\zeta_Q(t) = 1 - t^m$ for $Q \in S_m$ and $\zeta_Q(t) = 1$ for $Q \notin \bigcup_m S_m$ (that is if, at the point Q, the representation (7.7) contains more than one exponent m_i different from 0). Then one can write (7.8) as an integral with respect to the Euler characteristic [74]

$$\zeta_f(t) = \int_D \zeta_Q(t)^{d\chi}. \tag{7.9}$$

Here the "measure" $d\chi$ is put into the exponent since the operation in the group of rational functions defining the integral is the multiplication.

Equation (7.8) implies that all the zeroes or/and poles of the classical monodromy zeta function are roots of unity. If f has an isolated critical point at the origin, one has

$$\zeta_f(t) = (1 - t) \cdot \left(\det\left(\mathrm{id} - t \cdot h^*_{|_{H^{n-1}(V_f;\mathbb{R})}}\right)\right)^{(-1)^{n-1}}.$$

Therefore all the eigenvalues of $h_f^* = h^*_{|_{H^{n-1}(V_f;\mathbb{R})}}$ are also roots of unity. This is equivalent to the fact that there exist positive integers M and N such that $\left((h_f^*)^M - \mathrm{id}\right)^N = 0$. (One says that the classical monodromy operator h_f^* is *quasi-unipotent*.) If f has a non-isolated critical point at the origin, Eq. (7.8) does not imply that the operator $h^*_{|_{H^{n-1}(V_f;\mathbb{R})}}$ is quasi-unipotent. The Milnor fibre V_f may have non-trivial (co)homology groups not only in the dimensions 0 and $n-1$, but in other dimensions in between these two.

However, the quasi-unipotence does hold in any dimension: for any $q \in \mathbb{Z}_{\geq 0}$ the operator $h^*_{|_{H^q(V_f;\mathbb{R})}}$ is quasi-unipotent. Moreover, all the Jordan blocks of this operator are of size not greater than $q+1$: see, e.g., [16]. Thus, their sizes are always not grater than the number of variables n.

7.2.5 *Zeta Function in Terms of the Newton Diagram*

Let $f : (\mathbb{C}^n, 0) \to (\mathbb{C}, 0)$ be a germ of a holomorphic function and let

$$f(\overline{z}) = \sum_{\overline{k} \in \mathbb{Z}^n_{\geq 0}} a_{\overline{k}} \overline{z}^{\overline{k}} \tag{7.10}$$

be its Taylor decomposition ($\overline{z} = (z_1, \ldots, z_n)$, $\overline{k} = (k_1, \ldots, k_n)$, $\overline{z}^{\overline{k}} = z_1^{k_1} \cdot \ldots \cdot z_n^{k_n}$). The Newton diagram Γ_f of f (with respect to the chosen coordinates in $\mathbb{C}^n$) is the union of the compact faces of the convex hull (in $\mathbb{R}^n \supset \mathbb{Z}^n_{\geq 0}$)) of the set $\{\overline{k} \in \mathbb{Z}^n_{\geq 0} : a_{\overline{k}} \neq 0\} + \mathbb{R}^n_{\geq 0}$ ($\{\overline{k} \in \mathbb{Z}^n_{\geq 0} : a_{\overline{k}} \neq 0\}$ is the set of points corresponding to the monomials in (7.10) with non-zero coefficients): see, e.g., [4, Section 12.7].

An equation for the Milnor number of f in terms of the Newton diagram Γ_f was proved in [50] (under some non-degeneracy conditions: see below). An analogue of this equation for the monodromy zeta function (under the same non-degeneracy conditions) was obtained in [72]. For the germ f with the Taylor decomposition (7.10) and for a face γ of its Newton diagram Γ_f, let f_γ be the sum of monomials in (7.10) with the exponents from γ:

$$f_\gamma(\overline{z}) = \sum_{\overline{k} \in \gamma} a_{\overline{k}} \overline{z}^{\overline{k}}.$$

The germ f is *non-degenerate* (with respect to its Newton diagram Γ_f) if, for any face γ of Γ_f, the system of equations $\partial f_\gamma / \partial z_1 = \ldots = \partial f_\gamma / \partial z_n$ has no solutions in the complex torus $(\mathbb{C}^*)^n = \{\overline{z} \in \mathbb{C}^n : z_1 \cdot \ldots \cdot z_n \neq 0\}$. The Sard Theorem implies that almost all function germs f with the fixed Newton diagram are non-degenerate.

For an s-dimensional polyhedron Δ in $\mathbb{R}^n$ with integer vertices, its integer (s-dimensional) volume is defined by the condition that the volume of the parallelepiped spanned by any basis of the lattice $\mathbb{Z}^n$ intersected with the vector subspace corresponding to the s-dimensional affine subspace containing Δ is equal to 1. (The integer volume of any polyhedron with integer vertices is a multiple of $1/s!$. In what follows, the word "volume" always means "integer volume". Let $I_0 := \{1, 2, \ldots, n\}$. For $I \subset I_0$, let $\mathbb{R}^I := \{\overline{x} = (x_1, \ldots, x_n) \in \mathbb{R}^n : x_i = 0 \text{ for } i \notin I\}$. For a subset $I \subset I_0$ with the number of elements $|I|$ equal to ℓ and for an $(\ell - 1)$-dimensional face γ of $\Gamma_f \cup \mathbb{R}^I$, let $V(\gamma)$ be the volume of γ and let $m(\gamma)$ be the order of the quotient of the lattice $\mathbb{Z}^I$ by the subgroup generated by all the integer elements in the affine space spanned by γ. (The latter is equal to $(\ell - 1)$ times the volume of the cone over γ with the vertex at the origin divided by the volume of γ.) Let $\zeta^{(\ell)}(t)$ be the product over all $I \subset I_0$ with $|I| = \ell$ and over all $(\ell - 1)$-dimensional faces γ of $\Gamma_f \cup \mathbb{R}^I$ of the polynomials $(1 - t^{m(\gamma)})^{(\ell-1)!V(\gamma)}$.

Theorem 7.2.6 ([72]) *If f satisfies the non-degeneracy conditions mentioned above (and thus for almost all f with the fixed Newton diagram Γ_f), one has*

$$\zeta_f(t) = \prod_{\ell=1}^{n} \left(\zeta^{(\ell)}(t)\right)^{(-1)^{\ell-1}}.$$

Remark 7.2.7 In Theorem 7.2.6, it is not assumed that f has an isolated critical point at the origin.

7.2.6 Zeta Function in Terms of a Partial Resolution

Equation (7.9) gives the classical monodromy zeta function of a function germ as the integral with respect to the Euler characteristic over the exceptional divisor D of a resolution of f of the local zeta functions of the lifting of f at the points of D. This statement holds not only for a resolution, but also for other modifications: "partial resolutions".

Let $f : (\mathbb{C}^n, 0) \to (\mathbb{C}, 0)$ be a holomorphic function germ and let $p : (W, D) \to (\mathbb{C}^n, 0)$ $(D = p^{-1}(0))$ be a map from an n-dimensional complex manifold M such that

(1) the map p is proper;
(2) the map p is an isomorphism outside of the zero level set $\{f = 0\}$ of f.

(Thus the condition that the total transform $p^{-1}(\{f = 0\})$ of the zero level set of f is a normal crossing divisor is omitted.) For a point $Q \in D$, let $\zeta_Q(t)$ be the classical monodromy zeta function of the germ $f \circ p$ at Q.

Theorem 7.2.8 ([39]) *One has the equation formally coinciding with (7.9):*

$$\zeta_f(t) = \int_D \zeta_Q(t)^{d\chi}. \tag{7.11}$$

7.2.7 "Finite Set" Interpretation of the Monodromy Zeta Function

The classical monodromy zeta function (as well as any monodromy zeta function from Sect. 7.3) is a finite product

$$\prod_{m\geq 1}(1-t^m)^{s_m} \tag{7.12}$$

with $s_m \in \mathbb{Z}$. There is a way to identity the set of rational functions of the form (7.12) with the *Grothendieck ring of permutations*. This identification is useful, in particular, since this Grothendieck ring has a natural power structure: [44]. In terms of this power structure, one can formulate, in particular, analogues of the Macdonald equation ([51], see also [36, Section 2]) like in [26]. Besides that, similar constructions permit to define versions of the monodromy zeta functions in some situations with presence of additional structures, say, equivariant versions of them: [35, 45].

A *permutation* is a pair (X, h), where X is a finite set and h is a bijection $X \to X$. (One can regard h as an action of the group $\mathbb{Z}$ on X.) Two permutations (X_1, h_1) and (X_2, h_2) are *isomorphic* if there exists a bijection $\varphi : X_1 \to X_2$ such that $h_2(\varphi(x)) = \varphi(h_1(x))$. The set of the isomorphism classes $[(X, h)]$ of permutations is a semiring $\mathbb{S}$ with the addition defined by the disjoint union and the multiplication defined by the Cartesian product (with the diagonal action of $\mathbb{Z}$). (It is a semigroup with the cancelation property: if $a + c = b + c$, then $a = b$.) Let $K_0(\mathbb{S})$ be the corresponding Grothendieck ring. (Elements of $K_0(\mathbb{S})$ are represented by the formal differences $[(X_1, h_1)] - [(X_2, h_2)]$; $[(X_1, h_1)] - [(X_2, h_2)]$ is equal to $[(X_1', h_1')] - [(X_2', h_2')]$ if (and only if) $(X_1, h_1) + (X_2', h_2')$ is isomorphic to $(X_2, h_2) + (X_1', h_1')$.)

A permutation (X, h) can be regarded as a zero-dimensional space (with the discrete topology) endowed with a self-homeomorphism. Thus, the zeta function $\zeta_{X,h}(t)$ is defined. Defining $\zeta_{[(X_1,h_1)]-[(X_2,h_2)]}$ as $\zeta_{X_1,h_1}(t)/\zeta_{X_2,h_2}(t)$, one gets a map from the Grothendieck ring $K_0(\mathbb{S})$ to the set of rational functions of the form (7.12). It is not difficult to understand that this map is a bijection. Thus, monodromy zeta functions can be regarded as elements of the Grothendieck ring $K_0(\mathbb{S})$. (In terms of the rational functions of the form (7.12), the addition in the ring is represented by the multiplication of the functions. A description of the multiplication is somewhat more complicated.)

7.3 Other Monodromy Transformations and Their Zeta Functions

7.3.1 *Zeta Functions for Function Germs on Singular Varieties*

Let $(X, 0) \subset (\mathbb{C}^N, 0)$ be a germ of a complex analytic variety (of dimension n) and let $f : (X, 0) \to (\mathbb{C}, 0)$ be an analytic function germ. One can define the notions of the Milnor fibre, of a monodromy transformation and of the monodromy zeta function of f just in the same way as for a smooth $(X, 0)$, i.e., for $(X, 0) \cong (\mathbb{C}^n, 0)$. In general, the Milnor fibre is not a manifold (if $(X, 0)$ has a non-isolated singularity at the origin) and also is not homotopy equivalent to the bouquet of spheres. In this setting, the monodromy zeta function was studied in a number of papers.

There are some properties of the discussed objects which always hold. First of all, the Milnor fibre is always homotopy equivalent to a CW-complex of dimension $\leq (n-1)$. (This follows from the fact that it is a Stein variety.) Second, the monodromy operators (the actions of a monodromy transformation on the (co)homology groups) are quasi-unipotent (see, e.g., [16]). Finally, the monodromy zeta function is always cyclotomic, i.e., a product/ratio of binomials of the form $(1 - t^m)$.

The situation, most similar to the case $(X, 0) = (\mathbb{C}^n, 0)$, takes place when $(X, 0)$ is an isolated complete intersection singularity (ICIS). In this case the Milnor fibre is homotopy equivalent to a bouquet of spheres of dimension $(n - 1)$ (n is the dimension of the ICIS $(X, 0)$): [46]. A Varchenko type equation for the monodromy zeta function in terms of the Newton diagrams (of f and of equations of $(X, 0)$) was given in [59].

If $(X, 0)$ has an isolated singular point at the origin, one has a direct generalization of the A'Campo equation (7.8). (If the origin is a non isolated singular point of $(X, 0)$, one has to take into account the fact that a resolution of f (which is, at the same time, a resolution of $(X, 0)$) is not, in general, an isomorphism outside of the zero level set of f.)

7.3.2 *Monodromy Zeta Functions of Polynomials*

If $f \in \mathbb{C}[z_1, \ldots, z_n]$ is a polynomial in n variables, the map $f : \mathbb{C}^n \to \mathbb{C}$ is a locally trivial fibration outside of a finite set $\{u_1, \ldots, u_s\} \subset \mathbb{C}$. (This follows, e.g., from [71].) Among the special values $u_1, \ldots, u_s$, there are critical values of the polynomial f (that is the values at critical points). However, there may be other special values. The first example of a polynomial without critical points, but with special values was suggested by S.A. Broughton: [12]. This was the polynomial $f(x, y) = x(xy - 1)$. Its special value is zero. This effect led to the notion of a *critical value at infinity*. For r large enough and for $R >> r$, the map $f : f^{-1}(D_r(0)) \setminus B_R(0) \to D_r(0)$, where $D_r(0)$ is the disc of radius r in $\mathbb{C}$ and $B_R(0)$ is the ball of radius R in $\mathbb{C}^n$, is a smooth locally trivial fibration outside of a finite set. The (differentiable) type of this fibration does not depend on r and R (large enough). A value in $\mathbb{C}$, in whose neighbourhood the described map is not a locally trivial fibration, is called a critical value at infinity of the polynomial f. Some critical values at infinity may coincide with usual critical values.

For a special value u of f, one can consider invariants which reflect non-triviality either of the map $f : \mathbb{C}^n \to \mathbb{C}$ or of the map $f : f^{-1}D_r(0)\setminus B_R(0) \to D_r(0)$. Those which are related with the latter map can be considered as invariants at infinity. The first such invariant (in the general setting) was constructed in [5]. It measures the change of the Euler characteristic and was called the Milnor number at infinity. One can also consider *the Milnor fibre at infinity* of a polynomial and its topology. A detailed description of studies in this direction can be found in [70].

A more fine invariant of a polynomial at infinity than the Milnor number is the monodromy transformation and the corresponding zeta-function. Besides these monodromy transformations corresponding to critical values at infinity, one can consider the monodromy transformation for the infinite value corresponding to the circle $S_r(0) = \partial D_r(0)$ of a large radius r in $\mathbb{C}$. Initially this transformation was defined and studied in [70].

An intention to express invariants like the Milnor fibre at infinity or the monodromy zeta function at infinity in local terms (say, as integrals of local data with respect to the Euler characteristic like in (7.9) or (7.11)) met the problem that, in a neighbourhood of a point at infinity (in the projective space $\mathbb{CP}^n$ or in its modification), a polynomial function is not a holomorphic, but a meromorphic germ (of a special type). This led to the idea to consider topological invariants of meromorphic germs.

7.3.3 Monodromy Zeta Functions of Meromorphic Germs

In Singularity Theory, the first study of meromorphic germs was due to V.Arnold who, in [3], gave classifications of simple meromorphic germs with respect to certain equivalence relations. At the same time, topological study of meromorphic germs (including their monodromy zeta functions) was started in [40]. A summary of studies at the beginning of development of this subject can be found in [43].

A meromorphic germ at the origin in $\mathbb{C}^n$ is a ratio $f = \frac{F}{G}$ of two holomorphic germs F and G on $(\mathbb{C}^n, 0)$. It is useful (and sometimes necessary) to consider the following equivalence relation (first introduced by Arnold in [3]). Two meromorphic germs $f = \frac{F}{G}$ and $f' = \frac{F'}{G'}$ are equal if and only if there exists a germ $U : (\mathbb{C}^n, 0) \to \mathbb{C}$ with $U(0) \neq 0$ such that $F' = UF$, $G' = UG$.

A meromorphic germ $f = \frac{F}{G}$ defines a map from the complement (in a neighbourhood of the origin) of the indeterminacy locus $I_f = \{F = G = O\}$ to the complex projective line $\mathbb{CP}^1$. A natural analogue of the Milnor fibre for a meromorphic germ cannot be defined directly like for a holomorphic one. (The map described above is not, in general, a locally trivial fibration over the complement of a finite set in $\mathbb{CP}^1$. An example can be found in [43, Example 1.1].) However, it is possible to define the Milnor fibre $V_{f,c}$ for a fixed value $c \in \mathbb{CP}^1$. Namely, it was shown in [40] and [41] that, for any value $c \in \mathbb{CP}^1$, there exists $\delta_0 > 0$ (depending on c) such that for any positive $\delta \leq \delta_0$ the sphere $S_\delta(0)$ is transversal to all strata of the level set $f^{-1}(c)$, and the map $f : B_{\delta_0}(0) \setminus I_f \to \mathbb{CP}^1$ is a locally trivial fibration over a punctured neighbourhood of c. The fibre of this fibration was called *the c-Milnor fibre* of the meromorphic germ f. It is a (non-compact in general) $(n - I)$-dimensional complex manifold with boundary. (From formal point of view, in [40], there were defined only the 0-Milnor fibre and the ∞-Milnor fibre, i.e., the Milnor fibres for $c = 0, \infty$. However, there is no difference with any other value c. The reason to consider only these two Milnor fibres there was related with the fact that the main aim of the paper was to express the monodromy zeta functions in terms of Newton diagrams, what was possible only for these two values.)

Since one has a locally trivial fibration over a punctured neighbourhood of a value, one has the monodromy transformation (well defined up to isotopy) and the corresponding monodromy zeta function $\zeta_{f,c}(t)$. Similar constructions and objects for polynomials (i.e., for meromorphic germs of the form $P(z)/z_1^d$) were considered somewhat earlier, in particular, in [60, 65] and [69].

An A'Campo type equation for this zeta function in terms of a resolution of a meromorphic germ was given in [40, Theorem 1]. Therein one also has a Varchenko type equation for the monodromy zeta functions $\zeta_{f,0}(t)$ and $\zeta_{f,\infty}(t)$ in terms of the so-called Newton pair of a meromorphic germ, that is the Newton diagrams of the numerator and of the denominator. (Of course, there is no essential difference between a computation of the c-monodromy zeta function $\zeta_{f,c}(t)$ and of, say, the 0-monodromy zeta function $\zeta_{f,0}(t)$. However, for the A'Campo type equation, one has to consider a resolution of the meromorphic germ $\frac{P-cQ}{Q}$, and for the Varchenko type equation—the Newton pair for $P - cQ$ and Q.) One also has a natural generalization of the A'Campo type equation for an arbitrary modification (a "partial resolution"): [43, Theorem 1.21].

There are natural generalizations of the "Fibration Theorem", of the A'Campo type equation, and of its version for partial resolutions to meromorphic functions on compact complex analytic manifolds: [43, Section 2]. A particular case of this situation is the case of a polynomial function considered as a meromorphic function (of a special form) on the projective space considered in this context in [42]. Let P be a polynomial of degree d in n variables $x_1, \ldots, x_n$. It defines a meromorphic function $f = \frac{\widetilde{P}}{x_0^d}$ on the projective space $\mathbb{CP}^n$ ($\widetilde{P}$ is the homogenization of the polynomial P). For $c \in \mathbb{CP}^1$, the corresponding monodromy zeta function $\zeta_P^c(t)$ can be expressed in the form:

$$\zeta_P^c(t) = \int\limits_{\{\widetilde{P}=0\}\cap\mathbb{CP}^{n-1}_\infty} \zeta_{f,x}^c(t)^{d\chi} \cdot \int\limits_{\{P=c\}} \zeta_{P,x}^c(t)^{d\chi} \text{ for } c \in \mathbb{C}^1 \subset \mathbb{CP}^1;$$

$$\zeta_P^\infty(t) = \int\limits_{\mathbb{CP}^{n-1}_\infty} \zeta_{f,x}^\infty(t)^{d\chi}.$$

Further developments in this direction were carried out, in particular, in [10, 11, 52, 56, 57, 62, 67, 68, 73].

7.4 Zeta Functions and Poincaré Series

7.4.1 *Poincaré Series of a Plane Curve Germ*

Let $(C, 0) \subset (\mathbb{C}^2, 0)$ be an irreducible plane curve germ given by an equation $f = 0$. As it was indicated in Introduction, the classical monodromy zeta function $\zeta_f(t)$ is equal to $(1-t)$ divided by the Alexander polynomial $\Delta^L(t)$ of the algebraic knot $L = C \cap \mathbb{S}^3_\varepsilon(0)$, ε is small enough. This is a topological invariant of the plane curve germ $(C, 0)$. (Moreover, this is a complete topological invariant of it: two (irreducible) plane curve germs $(C_1, 0)$ and $(C_2, 0)$ are topologically equivalent (that is, the triples $(\mathbb{C}^2, C_1, 0)$ and $(\mathbb{C}^2, C_2, 0)$ are homeomorphic) if and only if the Alexander polynomials of the corresponding algebraic knots coincide.)

Let us consider the following (a priori analytic) invariant of the plane curve $(C, 0) \subset (\mathbb{C}^2, 0)$. A (discrete, rank one) valuation on an algebra $\mathcal{A}$ over a field $\mathbb{K}$ is a map $u : \mathcal{A} \to \mathbb{Z}_{\geq 0} \cup \{+\infty\}$ such that

(1) $u(\lambda a) = u(a)$ for $\lambda \in \mathbb{K}^* = \mathbb{K} \setminus \{0\}$;
(2) $u(a + b) \geq \min\{u(a), u(b)\}$;
(3) $u(ab) = u(a) + u(b)$.

(Sometimes, for a valuation, one also demands that $u(a) \neq +\infty$ for $a \neq 0$. In this case one uses the term pre-valuation for a function u satisfying (1)–(3).) The (irreducible) curve germ $(C, 0) \subset (\mathbb{C}^2, 0)$ determines a valuation on the algebra $\mathcal{O}_{\mathbb{C}^2,0}$ of the germs of holomorphic functions in two variables. Let $\varphi : (\mathbb{C}_\tau, 0) \to (\mathbb{C}^2, 0)$ be a parametrization (an uniformization) of the curve $(C, 0)$. (This means that the image of φ is $(C, 0)$ and φ is an isomorphism outside of the origin.) For a function germ $f \in \mathcal{O}_{\mathbb{C}^2,0}$, let

$$f(\varphi(\tau)) = a \cdot \tau^{u(f)} + \text{ terms of higher degree},$$

$a \neq 0$, be the Taylor decomposition of the germ $f \circ \varphi$. If $f(\varphi(\tau)) \equiv 0$, one puts $u(f) = +\infty$. Then $u_C = u(f)$ is a valuation on $\mathcal{O}_{\mathbb{C}^2,0}$ (called *the curve valuation* corresponding to (or defined by) the curve $(C, 0)$).

The valuation u_C defines a filtration on the algebra $\mathcal{O}_{\mathbb{C}^2,0}$. For $k \in \mathbb{Z}_{\geq 0}$, let $J(k) = \{f \in \mathcal{O}_{\mathbb{C}^2,0} : u_C(f) \geq k\}$. We have

$$J(0) = \mathcal{O}_{\mathbb{C}^2,0} \supset J(1) \supset J(2) \supset \dots \tag{7.13}$$

Let $P_C(t)$ be the Poincaré series of the filtration (7.13):

$$P_C(t) = \sum_{k=0}^{\infty} \dim\left(J(k)/J(k+1)\right) \cdot t^k. \tag{7.14}$$

For all k large enough, one has $\dim J(k)/J(k+1) = 1$. Therefore $(1-t) \cdot P_C(t)$ is a polynomial in t.

There exists a generalization of the discussed notion to reducible plane curve germs. Let $(C, 0) \subset (\mathbb{C}^2, 0)$ be a (reduced) plane curve germ and let $(C, 0) = \bigcup_{i=1}^{r}(C_i, 0)$ be its representation as the union of irreducible components. Each component $(C_i, 0)$, $1 \leq i \leq r$, defines a valuation $u_i = u_{C_i}$ on $\mathcal{O}_{\mathbb{C}^2,0}$ and the collection $\{u_i\}$ of valuations defines a multi-index filtration on $\mathcal{O}_{\mathbb{C}^2,0}$: for $\underline{k} \in \mathbb{Z}^r_{\geq 0}$

$$J(\underline{k}) = \{f \in \mathcal{O}_{\mathbb{C}^2,0} : \underline{u}(f) \geq \underline{k}\}, \tag{7.15}$$

where $\underline{u}(f) := (u_1(f), \ldots, u_r(f))$, $\underline{v} = (v_1, \ldots, v_r) \geq \underline{k}$ if and only if $v_i \geq k_i$ for all i. A generalization of the notion of the Poincaré series (7.14) to the case of a multi-index filtration like (7.15) is not obvious. It was given in [15]. Equation (7.15) defines the subspaces (ideals) $J(\underline{k})$ for all $\underline{k} \in \mathbb{Z}^r$ (not only for $\underline{k} \in \mathbb{Z}^r_{\geq 0}$). Let

$$\mathcal{L}(\underline{t}) := \sum_{\underline{k} \in \mathbb{Z}^r} \dim\left(J(\underline{k})/J(\underline{k}+\underline{1})\right) \cdot \underline{t}^{\underline{k}},$$

where $\underline{t} = (t_1, \ldots, t_r)$, $\underline{t}^{\underline{k}} := t_1^{k_1} \cdot \ldots \cdot t_r^{k_r}$, $\underline{1} = (1, \ldots, 1) \in \mathbb{Z}^r$.

Definition 7.4.1 The Poincaré series of the collection $\{u_i\}$ of valuations is

$$P_{\{u_{C_i}\}}(\underline{t}) := \frac{\mathcal{L}(\underline{t}) \cdot \prod_{i=1}^{r}(t_i - 1)}{t_1 \cdot \ldots \cdot t_r - 1}. \tag{7.16}$$

Remark 7.4.2 For a collection of (curve) valuations, $P_{\{u_{C_i}\}}(\underline{t})$ is a power series in r variables $t_1, \ldots, t_r$. In fact, if $r > 1$, $P_{\{u_{C_i}\}}(\underline{t})$ is a polynomial.

7.4.2 Poincaré Series and Topological Invariants of Algebraic Links

The Alexander polynomial in r variables is an invariant of a link with r component which, for the **algebraic** link $L = C \cap \mathbb{S}^3_\varepsilon(0)$ $((C, 0) = \bigcup_{i=1}^{r}(C_i, 0))$ can be defined in terms of monodromy zeta functions in the following way. Let $F_i = 0$ $(F_i : (\mathbb{C}^2, 0) \to (\mathbb{C}, 0))$ be equations of the curve germs $(C_i, 0)$. For $r > 1$, the Alexander polynomial $\Delta^L(t_1, \ldots, t_r)$ is defined by the following property: for $m_i \in \mathbb{Z}_{>0}$, $i = 1, 2, \ldots, r$, the inverse of the zeta function $\zeta_F(t)$ of the function germ $F =$

$\prod_{i=1}^{r} F_i^{m_i}$ is equal to $\Delta^L(t^{m_1}, \ldots, t^{m_r})$. (This property guarantees the uniqueness of the Alexander polynomial $\Delta^L(t_1, \ldots, t_r)$, but does not guarantee its existence. This is a separate statement.) For $r = 1$, the Alexander polynomial is the inverse to the ratio of the corresponding zeta function and $(1 - t)$.

Remark 7.4.3 In the Knot Theory the Alexander polynomial (in one or in several variables) is defined up to multiplication by a monomial. The definition (description) given here fixes a particular representative of it (in fact, defined by the property that its value at the origin is equal to 1).

Theorem 7.4.4 ([13, 37]) *For $r = 1$, the Poincaré series $P_{u_C}(t)$ is equal to the Alexander polynomial $\Delta^L(t)$ divided by $(1-t)$ (i.e., to the inverse of the corresponding monodromy zeta function). For $r > 1$, the Poincaré series $P_{\{u_{C_i}\}}(t_1, \ldots, t_r)$ coincides with the Alexander polynomial $\Delta^L(t_1, \ldots, t_r)$ of the r-component link $L = C \cap \mathbb{S}_\varepsilon^3(0)$.*

This statement, being rather old, up to now has no conceptual proof. All existing proofs, roughly speaking, consist of (independent) computation of the both sides of the equality in some common terms (say, in terms of an embedded resolution of the curve) and comparison of the obtained results.

There were found equations which give the Poincaré series $P_{\{u_{C_i}\}}(\underline{t})$ in terms of some integrals with respect to the Euler characteristic: [13, 38]. They turned out to be rather effective for its computation: [13, 14].

There exists a generalization of Theorem 7.4.4 for the HOMFLY polynomial of an algebraic link suggested by A. Oblomkov and V. Shender in [58] and proved by D. Maulik in [53].

7.4.3 Poincaré Series of Quasihomogeneous Function Germs and Complete Intersections

Another sort of relations between zeta functions and Poincaré series of filtrations (or rather of graduations) was described initially in [19, 20]. This relations include the so-called Saito duality between cyclotomic polynomials (or, more generally, rational functions), that is, functions $\varphi(t)$ represented as finite products of the form

$$\varphi(t) = \prod_{m \geq 1} (1 - t^m)^{s_m}, \tag{7.17}$$

where s_m are integers. Saito duality is defined for functions whose poles and zeroes are roots of unity of a fixed degree d. This means that they are represented in the form

$$\varphi(t) = \prod_{m | d} (1 - t^m)^{s_m}. \tag{7.18}$$

The *Saito dual* to the function $\varphi(t)$ is

$$\varphi^*(t) = \prod_{m|d}(1 - t^m)^{-s_{d/m}}\,. \tag{7.19}$$

This duality was defined by K. Saito to describe the relation between the characteristic polynomials of the classical monodromy operators of quasihomogeneous polynomials dual in the sense of Arnold Strange duality: [63, 64]. (The Arnold Strange duality is a duality on the set of exceptional unimodular singularities: [2, § 4] (there are 14 of them). It is considered to be the first appearance of the mirror symmetry effect in mathematics. After an appropriate change of variables, it becomes a particular case of the BHH-duality described in Sect. 7.5.3.)

Assume that a function f is quasihomogeneous with the quasidegree equal to d. In this case the characteristic polynomial of the classical monodromy operator of f is of the form (7.18). K.Saito noticed that Arnold dual (exceptional unimodular) functions have the same quasidegree and their characteristic polynomials are dual to each other in the sense of (7.19).

Let us describe these relations between the zeta functions and Poincaré series in the forms given in [21] and [22] (in a more general setting than in [19, 20]).

The *reduced monodromy zeta function* of a function germ $f : (\mathbb{C}^n, 0) \to (\mathbb{C}, 0)$ is $\widetilde{\zeta}_f(t) := \zeta_f(t)/(1-t)$. The reduced monodromy zeta function of a function germ can be defined by the equation like (7.1) if, instead of the cohomology groups one uses the cohomology groups reduced modulo a point. If f has an isolated critical point at the origin, the reduced monodromy zeta function essentially coincides with the characteristic polynomial of the monodromy operator.

Let $f : \mathbb{C}^n \to \mathbb{C}$ be a quasihomogeneous polynomial function of quasidegree d with respect to the weights $q_1, \dots, q_n$ (q_i are positive integers, gcd $(q_1, \dots, q_n) = 1$), not necessarily with an isolated critical point at the origin. Let $X = (f^{-1}(0)$ be the zero level set of f. The variety X carries a natural $\mathbb{C}^*$-action:

$$\lambda * (x_1, \dots, x_n) = (\lambda^{q_1}x_1, \dots, \lambda^{q_n}x_n), \quad \lambda \in \mathbb{C}^*\,. \tag{7.20}$$

Let $A = C[x_1, \dots, x_n]/(f)$ be the coordinate ring of the zero level set $(X, 0)$ ((f) is the ideal generated by f). One has a natural grading on the ring A: A_s ($s \in \mathbb{Z}_{\geq 0}$) is the set of polynomials $g \in A$ such that $g(\lambda^{q_1}x_1, \dots, \lambda^{q_n}x_n) = \lambda^s \cdot g(x_1, \dots, x_n)$. Let $P_f(t) = \sum_{s=0}^{\infty} \dim A_s \cdot t^s$ be the Poincaré series of the graded algebra $A = \bigoplus_{s=0}^{\infty} A_s$.

Let $Y := (X \setminus \{0\})/\mathbb{C}^\star$ be the space of orbits of the $\mathbb{C}^\star$-action on $X \setminus \{0\}$. There is a natural stratification of the variety Y defined by the types of the orbits: Y_m is the set of orbits for which the isotropy group is cyclic of order m. (The stratum Y_m can be nonempty only if m is equal to the greatest common divisor of a subset of the weights $q_1, \dots, q_n$.) Let

$$\mathrm{Or}_X(t) := \prod_{m\geq 0}(1 - t^m)^{-\chi(Y_m)}.$$

Theorem 7.4.5 ([21]) *One has*

$$P_f(t) = \widetilde{\zeta}_f^*(t) \cdot \mathrm{Or}_X(t). \tag{7.21}$$

For $n = 2$, the statement reduces to the statement of [37] for irreducible quasihomogeneous plane curve singularities (i.e., plane curve singularities with one Puiseux pair). In this case, the zeta function $\widetilde{\zeta}_f(t)$ is self dual and $Y = Y_1 = \mathrm{pt}$. For isolated hypersurface singularities with $n = 3$, the statement coincides with that of [19]. The number of equal orbit invariants which participate in the equation of [19] is the Euler characteristic of the corresponding 0-dimensional set, the exponent at the term $(1 - t)$ is the Euler characteristic $\chi(Y_1)$ of the open stratum Y_1 of the quotient curve Y where the $\mathbb{C}^\star$-action is free. In [21] Theorem 7.4.5 was formulated (and proved) for f non-degenerate with respect to its Newton diagram in the sense of [72]. The arguments from [22] show that this condition is not necessary.

Now let $f_1, \ldots, f_k$ be quasihomogeneous polynomials in n variables of quasidegrees $d_1, \ldots, d_k$ with respect to the (same) weights $q_1, \ldots, q_n$. Assume that the equations $f_1 = f_2 = \ldots = f_k = 0$ define a complete intersection X in $\mathbb{C}^n$. Again there is a natural $\mathbb{C}^*$-action on $\mathbb{C}^n$ (and on X) defined by (7.20). Let $A = \mathbb{C}[x_1, \ldots, x_n]/(f_1, \ldots, f_k)$ be the coordinate ring of X. One has a natural grading on A described above. Let $P_X(t) := \sum_{s=0}^{\infty} \dim A_s \cdot t^s$ be the Poincaré series of A. One has

$$P_X(t) = \frac{\prod_{j=1}^{k}(1 - t^{d_j})}{\prod_{i=1}^{n}(1 - t^{q_i})}.$$

For $0 \leq j \leq k$, Let $X^{(j)}$ be the complete intersection defined by the equations $f_1 = \ldots = f_j = 0$ ($X^{(0)} = \mathbb{C}^n$, $X^{(k)} = X$). The restriction of the function f_j to the variety $X^{(j-1)}$ ($j = 1, \ldots, k$) defines a locally trivial fibration $X^{(j-1)} \setminus X^{(j)} \to \mathbb{C}^*$. Let $V^{(j)} = f_j^{-1}(1) \cap X^{(j-1)}$ be the (Milnor) fibre of this fibration (the fibre $V^{(j)}$ is not necessarily smooth) and let $h^{(j)} : V^{(j)} \to V^{(j)}$ be the classical monodromy transformation of it. It is easy to see that, for a natural choice of $h^{(j)}$, one has $\left(h^{(j)}\right)^{d_j} = \mathrm{id}$ and therefore the (reduced) monodromy zeta function $\widetilde{\zeta}_j(t)$ of this fibration is of the form

$$\widetilde{\zeta}_j(t) = \prod_{m|d_j} (1 - t^m)^{s_m}, \quad s_m \in \mathbb{Z}_{\geq 0}.$$

Thus its Saito dual with respect to the (quasi)degree d_j is defined:

$$\widetilde{\zeta}_j^*(t) = \prod_{m|d_j} (1 - t^m)^{-s_{d_j/m}}.$$

As above, let $Y := (X \setminus \{0\})/\mathbb{C}^\star$ be the space of orbits of the $\mathbb{C}^\star$-action on $X \setminus \{0\}$ and let $Y_m \subset Y$ be the set of orbits for which the isotropy group is cyclic order m. Let

$$\mathrm{Or}_X(t) := \prod_{m \geq 0} (1 - t^m)^{-\chi(Y_m)}.$$

One has the following generalization of Theorem 7.4.5:

Theorem 7.4.6 ([22]) *One has*

$$P_X(t) = \prod_{j=1}^{k} \widetilde{\zeta}_k^*(t) \cdot \mathrm{Or}_X(t). \tag{7.22}$$

To an irreducible plane curve germ $(C, 0) \subset (\mathbb{C}^2, 0)$ with g Puiseux pairs, one associates a (monomial) curve $\widehat{C}$ in the space $\mathbb{C}^{g+1}$ with the same semigroup of values. Namely, if the semigroup of values of C is generated by $\beta_0, \beta_1, \ldots \beta_g$, then the curve $\widehat{C}$ in the space $\mathbb{C}^{g+1}$ with the coordinates $x_0, x_1, \ldots, x_g$ is defined by the equations $x_0 = t^{\beta_0}, x_1 = t^{\beta_1}, \ldots, x_g = t^{\beta_g}$. The curve $\widehat{C}$ appears to be a complete intersection. In [66]. J. Stevens derived the result of [37] from Theorem 7.4.6.

7.5 Orbifold Monodromy Zeta Function and Mirror Symmetry

7.5.1 Orbifold Euler Characteristic and Orbifold Monodromy Zeta Function

In the framework of the Mirror symmetry, there appears an orbifold version of the monodromy zeta function. This notion was essentially introduced in [32, Definition 5.10]. We shall give its definition in a somewhat more general setting: see [25].

The classical monodromy zeta function is related with the Euler characteristic of the Milnor fibre: the latter one is equal to its degree. The same holds for the orbifold monodromy zeta function. This is a reason to give the definition of the orbifold Euler characteristic first. This notion was introduced in [17] and [18]; see also [6] and [47].

Let X be a topological space (nice enough, say, homeomorphic to a locally closed union of cells in a finite CW-complex) and let G be a finite group acting on X (also with some goodness properties, e.g., with the action of elements of G being cellular maps). The *orbifold Euler characteristic* of the pair (X, G) is defined by the following two equivalent equations:

$$\chi^{\mathrm{orb}}(X, G) = \frac{1}{|G|} \sum_{\substack{(g,h)\in G^2: \\ gh=hg}} \chi(X^{\langle g,h\rangle}) = \sum_{[g]\in \mathrm{Conj}\, G} \chi(X^{\langle g\rangle}/C_G(g)), \qquad (7.23)$$

where $X^{\langle g,\dots\rangle}$ is the fixed point set of the subgroup of G generated by the indicated elements, $\mathrm{Conj}\, G$ is the set of the conjugacy classes of elements of G, g is a representative of the class $[g]$, $C_G(g)$ is the centralizer $\{h \in G : gh = hg\}$ of the element g. (A proof that these two equations are equivalent can be found, e.g., in [47] or [36, Proposition 6].)

Now let G be a finite group acting on the space $\mathbb{C}^n$ (without lost of generality we assume that the action is linear, i.e., induced by a representation of G) and let the function germ $f : (\mathbb{C}^n, 0) \to (\mathbb{C}, 0)$ be G-invariant. One may assume that the monodromy transformation h_f of the germ f is G-invariant. For an element $g \in G$, its *age* [48] (or the *fermion shift number*: [75]) is defined by $\mathrm{age}\,(g) := \sum_{j=1}^n \alpha_j$, where in a certain basis in $\mathbb{C}^n$ one has $g = \mathrm{diag}\,(\exp(2\pi i\, \alpha_1), \dots, \exp(2\pi i\, \alpha_n))$ with $0 \le \alpha_j < 1$.

Remark 7.5.1 The map $\exp(2\pi i\, \mathrm{age}\,(\cdot)) : G \to \mathbb{C}^*$ is a group homomorphism (i.e., a one-dimensional representation). It coincides with the representation of G induced on the nth exterior power of the space $\mathbb{C}^n$. If f is an invertible polynomial (see below) and G is the group G_f of its diagonal symmetries, then it is an element of $G_f^* = \mathrm{Hom}(G, \mathbb{C}^*)$ The definition of the orbifold monodromy zeta function uses only the homomorphism $\exp(2\pi i\, \mathrm{age}\,(\cdot))$. We give the definition of the age itself since it is necessary for definitions of more fine invariants related with the orbifold monodromy zeta function (such as the E-function: see [31, 33]). Moreover, for physically motivated models, one considers the cases when the representation on $\mathbb{C}^n$ takes values in the group $\mathrm{SL}(n, \mathbb{C})$ and therefore the representation $\exp(2\pi i\, \mathrm{age}(\cdot))$ is trivial. However, symmetries of the mirror style discussed below ((7.27) and (7.28)) take place also for non-trivial representations $\exp(2\pi i\, \mathrm{age}(\cdot))$.

For a rational function $\varphi(t)$ of the form $\prod_j (1 - \alpha_j t)^{r_j}$, $r_j \in \mathbb{Z}$, with finitely many factors, its *g-age shift* is defined by

$$(\varphi(t))_g = \varphi(\exp(-2\pi i\, \mathrm{age}\,(g))t) = \prod_j (1 - \alpha_j \exp(-2\pi i\, \mathrm{age}\,(g))t)^{r_j},$$

i.e., all its roots and/or poles are multiplied by $\exp(2\pi i\, \mathrm{age}\,(g)) \in \mathbb{C}^*$.

For a class $[g] \in \mathrm{Conj}\, G$ (the set of conjugacy classes of elements of G), let $g \in G$ be a representative of it. Let $V_f^g = V_f \cap (\mathbb{C}^n)^{\langle g\rangle}$ be the Milnor fibre of the restriction of f to the fixed point set $(\mathbb{C}^n)^{\langle g\rangle}$ and let $\widehat{V}_f^g = V_f^g / C_G(g)$ be the corresponding quotient space (the "twisted sector" in terms, say, of [17]). One may assume that the monodromy transformation is G-equivariant and therefore preserves

V_f^g for each g. Let $\widehat{h}_f^g : \widehat{V}_f^g \to \widehat{V}_f^g$ be the corresponding map (monodromy) on the quotient space. Its zeta function $\zeta_{\widehat{h}_f^g}(t)$ depends only on the conjugacy class of g.

Definition 7.5.2 The *orbifold monodromy zeta function* of the pair (f, G) is defined by

$$\zeta_{f,G}^{\text{orb}}(t) = \prod_{[g]\in \text{Conj}\, G} \left(\zeta_{\widehat{h}_f^g}(t)\right)_g . \tag{7.24}$$

One can see that the degree of $\zeta_{f,G}^{\text{orb}}(t)$ is equal to the orbifold Euler characteristic of (V_f, G). For an abelian G, $\widehat{V}_f^g = V_f^g/G$ and the product in (7.24) runs over all the elements $g \in G$.

Definition 7.5.3 The *reduced orbifold zeta function* $\overline{\zeta}_{f,G}^{\text{orb}}(t)$ is defined by

$$\overline{\zeta}_{f,G}^{\text{orb}}(t) = \zeta_{f,G}^{\text{orb}}(t) \Big/ \prod_{[g]\in \text{Conj}\, G} (1-t)_g . \tag{7.25}$$

The expression in the denominator in the right hand side of (7.25) can be interpreted as the orbifold zeta function of the (unique) transformation on the one-point G-space $\text{pt} = G/G$ (cf. Eq. (7.24)).

7.5.2 *Invertible Polynomials*

One regular way to construct mirror symmetric manifolds uses so-called invertible polynomials.

Definition 7.5.4 A quasihomogeneous polynomial $f(x_1, \ldots, x_n)$ is called *invertible* if

(1) the number of monomials in it (with non-zero coefficients) is equal to the number n of the variables, i.e.,

$$f(x_1, \ldots, x_n) = \sum_{i=1}^{n} c_i \prod_{j=1}^{n} x_j^{E_{ij}} \tag{7.26}$$

with $c_i \neq 0$, $E_{ij} \in \mathbb{Z}_{\geq 0}$;

(2) the matrix $E = (E_{ij})$ is invertible (over $\mathbb{Q}$, i.e., $\det E \neq 0$);

(3) the polynomial f is non-degenerate, i.e., has an isolated critical point at the origin.

Remark 7.5.5 Without loss of generality one may assume that $c_i = 1$ for all i (this can be achieved by rescaling the variables).

Remark 7.5.6 Sometimes, in the definition of an invertible polynomial, the condition (3) is omitted. Some properties of invertible polynomials (e.g., (7.27) and (7.28)) hold without this assumption. However, here we shall consider only non-degenerate polynomials.

Proposition 7.5.7 *A polynomial is invertible if and only if it is the Sebastiani–Thom sum (i.e., sum of polynomials in non-intersecting sets of variables) of polynomials of the form:*

(1) Fermat type: $x_1^{a_1}$, $a_1 \geq 2$;
(2) chain type: $x_1^{a_1}x_2 + x_2^{a_2}x_3 + \cdots + x_m^{a_m}$, $a_m \geq 2$;
(3) loop type: $x_1^{a_1}x_2 + x_2^{a_2}x_3 + \cdots + x_m^{a_m}x_1$ *(if* m *is even, then neither* $a_1 = a_3 = \ldots = a_{m-1} = 1$ *nor* $a_2 = a_4 = \ldots = a_m = 1$*).*

Remark 7.5.8

(a) Usually this statement is attributed to [49]. However, one cannot find there not only a proof of it, but even a formulation. One can consider this statement as a sort of Folklore.
(b) The Fermat type polynomial can be considered as a particular case either of chain type or of loop type.
(c) If $a_1 = 1$ in (1) or $a_m = 1$ in (2), the corresponding polynomial is regular (does not have a critical point) at the origin.

The group of diagonal symmetries of an invertible polynomial $f(x_1, \ldots, x_n)$ is

$$G_f = \{(\lambda_1, \ldots, \lambda_n) \in (\mathbb{C}^*)^n : f(\lambda_1 x_1, \ldots, \lambda_n x_n) = f(x_1, \ldots, x_n)\}.$$

One can show that $|G_f| = |\det E|$. For an "atomic" invertible polynomial, that is for a polynomial of Fermat, chain, or loop type (in m variables), the group G_f is cyclic and has the following description. Let $E^{-1} = (e^{ij})$, e^{ij} are rational numbers (moreover, $e^{ij} \in \frac{1}{\det E}\mathbb{Z}$). Then the group G_f is generated by the element

$$\left(\exp\left(2\pi i e^{1m}\right), \exp\left(2\pi i e^{2m}\right), \ldots, \exp\left(2\pi i e^{mm}\right), \right) \in (\mathbb{C}^*)^m.$$

(For a polynomial of loop type (in m variables), the group G_f is generated by the element

$$\left(\exp\left(2\pi i e^{1j}\right), \exp\left(2\pi i e^{2j}\right), \ldots, \exp\left(2\pi i e^{mj}\right), \right)$$

for any j, $1 \leq j \leq m$.)

The monodromy transformation is a distinguished element of the group G_f (sometimes called *the exponential grading operator*). The representation $\exp(2\pi i \operatorname{age}(\cdot)$ is a character on G, i.e. a homomorphism from G to $\mathbb{C}^*$.

7.5.3 BHH- and BHHT-dualities

The *dual* (or the *transpose*) of the invertible polynomial (7.26) is

$$\widetilde{f}(x_1, \ldots, x_n) = \sum_{i=1}^{n} \prod_{j=1}^{n} x_j^{E_{ji}} .$$

One can show that there is a canonical isomorphism between the group $G_{\widetilde{f}}$ of diagonal symmetries of $\widetilde{f}$ and the group $\mathrm{Hom}(G_f, \mathbb{C}^*)$ of characters of G_f (see details, e.g., in [24]). For a subgroup $G \subset G_f$, its dual $\widetilde{G}$ is the subgroup of $G_{\widetilde{f}}$ consisting of the characters of G_f vanishing (i.e., having the value 1) on G. One can show that, for the subgroup generated by the monodromy transformation $h_f \in G_f$, the dual subgroup of $G_{\widetilde{f}}$ is $\mathrm{Ker}(\exp(2\pi i\, \mathrm{age}(\cdot)) = G_{\widetilde{f}} \cap \mathrm{SL}(n, \mathbb{C})$ (see, e.g., [24]). The pair $(\widetilde{f}, \widetilde{G})$ is called *Berglund–Hübsch–Henningson-* (*BHH-* for short) *dual* to the pair (f, G). In [27, Proposition 5], it was shown that, under the identification of G_f with $G^*_{\widetilde{f}}$, the monodromy operator of f (as an element of G_f) maps to the age representation of $G_{\widetilde{f}}$.

It was conjectured ([8, 9]) that, in the theory of Landau–Ginzburg models, mirror symmetric Calabi–Yau manifolds can be constructed in the following way. Let (f, G) be a pair consisting of an admissible invertible polynomial f and a subgroup G of the group G_f of its diagonal symmetries (with certain "admissibility conditions"). The equation $f = 0$ gives a (singular, in general) hypersurface in the weighted projective space (corresponding to the weights of the quasihomogeneous polynomial f). Let $W_{f,G}$—be a crepant resolution of the quotient $\{f = 0\}/G$, that is, a resolution with the trivial relative canonical class. (Under the admissibility conditions, $W_{f,G}$ is a Calabi-Yau manifold.) The manifold $W_{\widetilde{f},\widetilde{G}}$ is pretended to be mirror symmetric to $W_{f,G}$. (As a support of this statement, in [9], it was shown that BHH-dual pairs have dual (in a certain sense) elliptic genuses.)

In [23] and [25], it was shown that, for BHH-dual pairs (f, G) and $(\widetilde{f}, \widetilde{G})$ (without any assumption of admissibility!), one has

$$\overline{\chi}^{\mathrm{orb}}(V_f, G) = (-1)^n\, \overline{\chi}^{\mathrm{orb}}(V_{\widetilde{f}}, \widetilde{G}) ; \tag{7.27}$$

$$\overline{\zeta}^{\mathrm{orb}}_{f,G}(t) = \left(\overline{\zeta}^{\mathrm{orb}}_{\widetilde{f},\widetilde{G}}(t)\right)^{(-1)^n} . \tag{7.28}$$

In [31], there was proved a more fine (than (7.28)) symmetry, formulated in terms of the mixed Hodge structures on the twisted sector corresponding to elements of G, what goes beyond the aim of the Chapter.

The group S_n of permutations on n elements acts on the space $\mathbb{C}^n$ by permuting the variables. Let f be an invertible polynomial in n variables, let G be a group of diagonal symmetries of f (i.e., $G \subset G_f$), and let S be a subgroup of S_n preserving f and G ($\sigma^{-1} G \sigma = G$ for $\sigma \in S$). The action of S on G (by conjugation) defines the semidirect product $G \rtimes S$ with its action on $\mathbb{C}^n$. (The precise equations for the group structure on $G \rtimes S$ and for its action on $\mathbb{C}^n$ are the following. Elements of $G \rtimes S$ can be represented as pairs $(\underline{\lambda}, \sigma)$ with $\underline{\lambda} = (\lambda_1, \ldots, \lambda_n) \in G$, $\sigma \in S$. The multiplication in $G \rtimes S$ is given by

$$(\underline{\lambda}, \sigma) \cdot (\underline{\mu}, \tau) = (\underline{\lambda}\sigma(\underline{\mu}), \sigma\tau),$$

where, for $\underline{\mu} = (\mu_1, \ldots, \mu_n)$, $\sigma(\underline{\mu}) = (\mu_{\sigma^{-1}(1)}, \ldots, \mu_{\sigma^{-1}(1)})$. The action of the group $G \rtimes S$ on $\mathbb{C}^n$ is defined by

$$(\underline{\lambda}, \sigma)\overline{x} = (\lambda_1 x_{\sigma^{-1}(1)}, \ldots \lambda_n x_{\sigma^{-1}(n)})$$

for $\overline{x} = (x_1, \ldots, x_n) \in \mathbb{C}^n$.)

In this case the dual polynomial $\widetilde{f}$ and the dual subgroup $\widetilde{G} \subset G_{\widetilde{f}}$ are also preserved by S and, for the pair $(f, G \rtimes S)$, one gets the so-called *Berglund–Hübsch–Henningson–Takahashi-* (*BHHT-* for short) *dual* pair $(\widetilde{f}, \widetilde{G} \rtimes S)$: see [30]. In [30], it was shown that, for BHHT-dual pairs, the obvious analogues of the symmetries (7.27) and (7.28) do not hold in general, but may hold only under a special condition on the subgroup $S \subset S_n$ (called "PC": parity condition), namely, for any subgroup $T \subset S$, $\dim(\mathbb{C}^n)^T \equiv n \bmod 2$.

One can conjecture that, under PC, one has symmetries like (7.27) and (7.28), i.e.,

$$\overline{\chi}^{\mathrm{orb}}(V_f, G \rtimes S) = (-1)^n \, \overline{\chi}^{\mathrm{orb}}(V_{\widetilde{f}}, \widetilde{G} \rtimes S)\,; \tag{7.29}$$

$$\overline{\zeta}^{\mathrm{orb}}_{f, G\rtimes S}(t) = \left(\overline{\zeta}^{\mathrm{orb}}_{\widetilde{f}, \widetilde{G}\rtimes S}(t)\right)^{(-1)^n}. \tag{7.30}$$

Equation (7.29) was proved in [28] for polynomials of loop type with S satisfying PC and in [29] for a cyclic subgroup S (also satisfying PC).

From the computations in [61], one can derive the symmetries (7.29) and (7.30) for the following pairs $(f, G \rtimes S)$ (with Fermat type polynomials f):

(1) $f = x_1^4 + \cdots + x_4^4$, $G = \langle J \rangle$, $S = \langle (123) \rangle$, where $J = \exp \frac{\pi i}{2}(1, 1, 1, 1)$ is the exponential grading operator (the monodromy transformation);
(2) $f = x_1^5 + \cdots + x_5^5$, $G = \langle J \rangle$, $S = \langle (12)(34) \rangle$, where $J = \exp \frac{2\pi i}{5}(1, 1, 1, 1, 1)$.

A symmetry, similar to that from [31], was proved in [55] for the pairs $(f, G \rtimes S)$ with $f = x_1^{p-1}x_2 + \cdots + x_p^{p-1}x_1$ (a loop), an arbitrary subgroup G of $G_f \cong \mathbb{Z}_{(p-1)^p+1}$ and $S = \langle (12 \ldots p) \rangle$, where p is an odd prime number. Moreover, from [7] , one can derive a symmetry similar to the symmetry of [55] for pairs $(f, G \rtimes S)$ with $f = x_1^p + \cdots + x_p^p$ (a Fermat polynomial), G is either G_f or $G_f \cap \mathrm{SL}(p, \mathbb{C})$, S satisfies PC, and p is a prime number.

Acknowledgments The work was supported by the grant 24-11-00124 of the Russian Science Foundation.

References

1. A'Campo, N.: La fonction zêta d'une monodromie. Comment. Math. Helv. **50**, 233–248 (1975)
2. Arnold, V.I.: Critical points of smooth functions and their normal forms. (Russian) Usp. Math. Nauk. **30**, no. 5(185), 3–65 (1975); translation in Russ. Math. Surv. **30**. no. 5, 1–75 (1975)
3. Arnold, V.I.: Singularities of fractions and behaviour of polynomials at infinity, (Russian) Tr. Mat. Inst. Steklova **221**, 48–68 (1998); translation in Proc. Steklov Inst. Math. **221**, 40–59 (1998)
4. Arnold, V.I., Guseĭn-Zade, S.M., Varchenko, A.N.: Singularities of Differentiable Maps, Vol. I, The classification of critical points, caustics and wave fronts. Birkhäuser Boston, Boston (1985)
5. Artal Bartolo, E., Luengo, I., Melle-Hernández, A.: Milnor number at infinity, topology and Newton boundary of a polynomial function. Math. Z. **233**, no. 4, 679–696 (2000)
6. Atiyah, M., Segal, G.: On equivariant Euler characteristics. J. Geom. Phys. **6**, no. 4, 671–677 (1989)
7. Basalaev, A.A., Ionov, A.A.: Mirror map for Fermat polynomials with a nonabelian group of symmetries. (Russian) Teoret. Mat. Fiz. **209**, no. 2, 205–223 (2021); translation in Theoret. and Math. Phys. **209**, no. 2, 1491–1506 (2021)
8. Berglund, P., Hübsch, T.: A generalized construction of mirror manifolds. Nucl. Phys. B **393**, 377–391 (1993)
9. Berglund, P., Henningson, M.: Landau–Ginzburg orbifolds, mirror symmetry and the elliptic genus. Nucl. Phys. B **433**, 311–332 (1995)
10. Bodin, A., Pichon, A.: Meromorphic functions, bifurcation sets and fibred links. Math. Res. Lett. **14**, no. 3, 413–422 (2007).
11. Bodin, A., Pichon, A., Seade, J.: Milnor fibrations of meromorphic functions. J. Lond. Math. Soc. (2) **80**, no. 2, 311–325 (2009)
12. Broughton, S.A.: Milnor numbers and the topology of polynomial hyper-surfaces. Invent. Math. **92**, no. 2, 217–241 (1988)
13. Campillo, A., Delgado, F., Gusein-Zade, S.M.: The Alexander polynomial of a plane curve singularity via the ring of functions on it. Duke Math. J. **117**, no. 1, 125–156 (2003)
14. Campillo, A., Delgado, F., Gusein-Zade, S.M.: The Alexander polynomial of a plane curve singularity and integrals with respect to the Euler characteristic. Internat. J. Math. **14**, no. 1, 47–54 (2003)
15. Campillo, A., Delgado, F., Kiyek, K.: Gorenstein property and symmetry for one-dimensional local Cohen-Macaulay rings. Manuscripta Math. **83**, no. 3–4, 405–423 (1994)
16. Clemens, C. H., Jr.: Picard-Lefschetz theorem for families of nonsingular algebraic varieties acquiring ordinary singularities. Trans. Amer. Math. Soc. **136**, 93–108 (1969)
17. Dixon, L., Harvey, J.A., Vafa, C., Witten, E.: Strings on orbifolds. Nuclear Phys. B **261**, no. 4, 678–686 (1985)
18. Dixon, L., Harvey, J., Vafa, C., Witten, E.: Strings on orbifolds. II. Nuclear Phys. B **274**, no. 2, 285–314 (1986)
19. Ebeling, W.: Poincaré series and monodromy of a two-dimensional quasihomogeneous hypersurface singularity. Manuscripta Math. **107**, no. 3, 271–282 (2002)
20. Ebeling, W.: The Poincaré series of some special quasihomogeneous surface singularities. Publ. Res. Inst. Math. Sci. **39**, no. 2, 393–413 (2003)
21. Ebeling, W., Gusein-Zade, S.M.: Poincaré series and zeta function of the monodromy of a quasihomogeneous singularity. Math. Res. Lett. **9**, no. 4, 509–513 (2002)
22. Ebeling, W., Gusein-Zade, S.M.: Monodromies and Poincaré series of quasihomogeneous complete intersections. Abh. Math. Sem. Univ. Hamburg **74**, 175–179 (2004)
23. Ebeling, W., Gusein-Zade, S.M.: Orbifold Euler characteristics for dual invertible polynomials. Mosc. Math. J. **12**, no. 1, 49–54 (2012)
24. Ebeling, W., Gusein-Zade, S.M.: Saito duality between Burnside rings for invertible polynomials. Bull. London Math. Soc. **44**, no. 4, 814–822 (2012)

25. Ebeling, W., Gusein-Zade, S.M.: Orbifold zeta functions for dual invertible polynomials. Proc. Edinb. Math. Soc. (2) **60**, no. 1, 99–106 (2017)
26. Ebeling, W., Gusein-Zade, S.M.: Higher-order spectra, equivariant Hodge–Deligne polynomials, and Macdonald-type equations. Singularities and computer algebra, 97–108, Springer, Cham (2017)
27. Ebeling, W,, Gusein-Zade, S.M.: Enhanced equivariant Saito duality. J. Algebra Appl. **17**, no. 10 (2018).
28. Ebeling, W,, Gusein-Zade, S.M.: On the orbifold Euler characteristics of dual invertible polynomials with non-abelian symmetry groups. Pure Appl. Math. Q. **16**, no. 4, 1099–1113 (2020).
29. Ebeling, W,, Gusein-Zade, S.M.: Dual invertible polynomials with permutation symmetries and the orbifold Euler characteristic. SIGMA Symmetry Integrability Geom. Methods Appl. **16**, Paper No. 051, 15 pp. (2020).
30. Ebeling, W., Gusein-Zade, S.M.: A version of the Berglund-Hübsch-Henningson duality with non-abelian groups. Int. Math. Res. Not. IMRN **2021**, no. 16, 12305–12329 (2021)
31. Ebeling, W., Gusein-Zade, S.M., Takahashi, A.: Orbifold E-functions of dual invertible polynomials. J. Geom. Phys. **106**, 184–191 (2016)
32. Ebeling, W., Takahashi, A.: Mirror symmetry between orbifold curves and cusp singularities with group action. Int. Math. Res. Not. IMRN **2013**, no. 10, 2240–2270 (2013)
33. Ebeling, W., Takahashi, A.: Variance of the exponents of orbifold Landau–Ginzburg models. Math. Res. Lett. **20**, no. 1, 51–65 (2013)
34. Gusein-Zade, S.M.: Integration with respect to the Euler characteristic and its applications. (Russian) Uspekhi Mat. Nauk **65**, no. 3(393), 5–42 (2010); translation in Russian Math. Surveys **65**, no. 3, 399–432 (2010)
35. Gusein-Zade, S.M.: On an equivariant analogue of the monodromy zeta function. (Russian) Funktsional. Anal. i Prilozhen. **47**, no. 1, 17–25 (2013); translation in Funct. Anal. Appl. **47**, no. 1, 14–20 (2013)
36. Gusein-Zade, S.M.: Equivariant analogues of the Euler characteristic and Macdonald type formulas. (Russian) Uspekhi Mat. Nauk **72**, no. 1 (433), 3–36 (2017); translation in Russian Math. Surveys **72**, no. 1, 1–32 (2017)
37. Guseĭn-Zade, S.M., Delgado, F., Campillo, A.: On the monodromy of a plane curve singularity and the Poincaré series of its ring of functions. (Russian) Funktsional. Anal. i Prilozhen. **33**, no. 1, 65–68 (1999); translation in Funct. Anal. Appl. **33**, no. 1, 56–57 (1999)
38. Guseĭn-Zade, S.M., Delgado, F., Campillo, A.: Integration with respect to the Euler characteristic over a function space, and the Alexander polynomial of a plane curve singularity. (Russian) Uspekhi Mat. Nauk **55**, no. 6(336), 127–128 (2000); translation in Russian Math. Surveys **55**, no. 6, 1148–1149 (2000)
39. Gusein-Zade, S.M., Luengo, I., Melle-Hernández, A.: Partial resolutions and the zeta-function of a singularity. Comment. Math. Helv. **72**, no. 2, 244–256 (1997)
40. Gusein-Zade, S.M., Luengo, I., Melle-Hernández, A.: Zeta functions of germs of meromorphic functions, and the Newton diagram. (Russian) Funktsional. Anal. i Prilozhen. **32**, no. 2, 26–35 (1998); translation in Funct. Anal. Appl. **32**, no. 2, 93–99 (1998)
41. Gusein-Zade, S.M., Luengo, I., Melle-Hernández, A.: On atypical values and local monodromies of meromorphic functions. (Russian) Tr. Mat. Inst. Steklova **225**, 168–176 (1999); translation in Proc. Steklov Inst. Math. **225**, no. 2, 156–164 (1999)
42. Gusein-Zade, S.M., Luengo, I., Melle-Hernández, A.: On the zeta-function of a polynomial at infinity. Bull. Sci. Math. **124**, no. 3, 213–224 (2000)
43. Gusein-Zade, S.M., Luengo, I., Melle-Hernández, A.: Bifurcations and topology of meromorphic germs. New developments in singularity theory (Cambridge, 2000), 279–304, NATO Sci. Ser. II Math. Phys. Chem., **21**, Kluwer Acad. Publ., Dordrecht (2001)
44. Gusein-Zade, S.M., Luengo, I., Melle-Hernández, A.: A power structure over the Grothendieck ring of varieties. Math. Res. Lett. **11**, no. 1, 49–57 (2004).
45. Gusein-Zade, S.M., Luengo, I., Melle-Hernández, A.: An equivariant version of the monodromy zeta function. Geometry, topology, and mathematical physics, 139–146, Amer. Math.

Soc. Transl. Ser.2, **224**, Adv. Math. Sci., 61, Amer. Math. Soc., Providence, RI (2008)
46. Hamm, H.A. Exotische Sphären als Umgebungsränder in speziellen komplexen Räumen. (German) Math. Ann. **197**, 44–56 (1972)
47. Hirzebruch, F., Höfer, Th.: On the Euler number of an orbifold. Math. Ann. **286**, no. 1–3, 255–260 (1990)
48. Ito, Y., Reid, M.: The McKay correspondence for finite subgroups of SL(3, $\mathbb{C}$). In: Higher-dimensional complex varieties (Trento, 1994), de Gruyter, Berlin, 221–240, (1996)
49. Kreuzer, M*, Skarke, H*: On the classification of quasihomogeneous functions. Commun. Math. Phys. **150** (1), 137–147 (1992)
50. Kouchnirenko, A.G.: Polyèdres de Newton et nombres de Milnor. (French) Invent. Math. **32**, no. 1, 1–31 (1976)
51. Macdonald, I.G.: The Poincaré polynomial of a symmetric product. Math. Proc. Cambridge Philos. Soc. **58**, no. 4, 563–568 (1962).
52. Matsui, Yu., Takeuchi, K.: Monodromy zeta functions at infinity, Newton polyhedra and constructible sheaves. Math. Z. **268**, no. 1–2, 409–439 (2011)
53. Maulik, D.: Stable pairs and the HOMFLY polynomial. Invent. Math. **204**, no. 3, 787–831 (2016)
54. Milnor, John: Singular points of complex hypersurfaces. Annals of Mathematics Studies, No. 61. Princeton University Press, Princeton, NJ (1968)
55. Mukai, D.: Mirror symmetry of nonabelian Landau-Ginzburg orbifolds with loop type potentials. J. Geom. Phys. **159**, Paper No.103877 (2021)
56. Nguyen, T.Th., Saito, T., Takeuchi, K.: The bifurcation set of a rational function via Newton polytopes. Math. Z. **298**, no. 1–2, 899–916 (2021)
57. Nguyen, T.Th., Takeuchi, K.: Meromorphic nearby cycle functors and monodromies of meromorphic functions (with appendix by T. Saito). Rev. Mat. Complut. **36**, no. 2, 663–705 (2023)
58. Oblomkov, A., Shende, V.: The Hilbert scheme of a plane curve singularity and the HOMFLY polynomial of its link. Duke Math. J. **161**, no. 7, 1277–1303, (2012)
59. Oka, M.: Principal zeta-function of nondegenerate complete intersection singularity. J. Fac. Sci. Univ. Tokyo Sect. IA Math. **37**, no. 1, 11–32 (1990)
60. Parusiński, A.: On the bifurcation set of a complex polynomial with isolated singularities at infinity. Compositio Math. **97**, no. 3, 369–384 (1995).
61. Priddis, N., Ward, J., Williams, M.: Mirror symmetry for nonabelian Landau-Ginzburg models. SIGMA Symmetry Integrability Geom. Methods Appl. **16**, Paper No. 059, 31 pp. (2020).
62. Raibaut, M.: Motivic Milnor fibers of a rational function. Rev. Mat. Complut. **26**, no. 2, 705–734 (2013)
63. Saito, K.: Duality for regular systems of weights: a précis. In: Topological field theory, primitive forms and related topics (Kyoto, 1996), 379–426, Progr. Math., 160, Birkhäuser Boston, Boston, MA (1998)
64. Saito, K.: Duality for regular systems of weights. Asian J. Math. **2**, no. 4, 983–1047 (1998)
65. Siersma, D., Tibăr, M.: Singularities at infinity and their vanishing cycles. Duke Math. J. **80**, no. 3, 771–783 (1995).
66. Stevens, J.: Poincaré series and zeta function for an irreducible plane curve singularity. Bull. London Math. Soc. **37**, no. 3, 399–404 (2005)
67. Takeuchi, K.: On a Bernstein-Sato polynomial of a meromorphic function. Nagoya Math. J. **251**, 715–733 (2023)
68. Takeuchi, K., Matsui, Yu.: Monodromies at infinity of polynomial maps and A-hypergeometric functions. Sugaku Expositions **29**, no. 1, 67–99 (2016)
69. Tibăr, M.: On the monodromy fibration of polynomial functions with singularities at infinity. C. R. Acad. Sci. Paris **324**, 1031–1035 (1997)
70. Tibăr, M.: Polynomials and Vanishing Cycles, Cambridge University Press (2007)
71. Varchenko, A.N.: Theorems on the topological equisingularity of families of algebraic varieties and families of polynomial mappings. (Russian) Izv. Akad. Nauk SSSR Ser. Mat. **36**, no. 5, 957–1019 (1972); translation in Mathematics of the USSR-Izvestiya **6**, no. 5, 949–1008 (1972)

72. Varchenko, A.N.: Zeta-function of monodromy and Newton's diagram. Invent. Math. **37**, 253–262 (1976)
73. Veys, W., Zúñiga-Galindo, W.A.: Zeta functions and oscillatory integrals for meromorphic functions. Adv. Math. **311**, 295–337 (2017)
74. Viro, O.Ya.: Some integral calculus based on Euler characteristic. In: Topology and geometry — Rohlin Seminar, Lecture Notes in Math., **1346**, pp. 127–138, Springer, Berlin (1988)
75. Zaslow, E.: Topological orbifold models and quantum cohomology rings", Comm. Math. Phys. **156**, no 2, 301–331 (1993)

Chapter 8
Singularities of the Minimal Model Program

Christopher Hacon

Contents

C. Hacon (✉)
Department of Mathematics, University of Utah, Salt Lake City, UT, USA
e-mail: hacon@math.utah.edu

© The Author(s), under exclusive license to Springer Nature Switzerland AG 2026

J. L. Cisneros-Molina et al. (eds.), *Handbook of Geometry and Topology of Singularities VIII*, https://doi.org/10.1007/978-3-031-99571-2_8

Abstract In this article we survey results related to the singularities of the minimal model program of complex quasi-projective varieties.

8.1 Introduction

Algebraic geometry is the study of geometric objects defined by polynomial equations. One of the central problems in algebraic geometry is to classify algebraic varieties. Since birational varieties share many properties, it is natural to classify varieties up to birational isomorphism. The main approach to this classification is the minimal model program which has yielded great progress for varieties defined over the complex numbers $\mathbb{C}$ and partial progress for varieties defined over a field of positive characteristic or in mixed characteristics. Moreover, related techniques show promise in a variety of other contexts including for example analytic varieties and foliated varieties.

In this article we will focus on quasi-projective varieties defined over the complex numbers. The geometry of complex surfaces was well understood by the Italian school of algebraic geometry in the early twentieth century. The case of threefolds was then understood in the 1980s by work of Mori, Kawamata, Kollár, Reid, Shokurov and others. The higher dimensional case is now widely understood after work of Birkar, Cascini, Hacon and McKernan [14].

One of the key features of the minimal model program is that even if one is primarily interested in smooth varieties (complex manifolds), certain singularities naturally appear in this context. For example, if X is a smooth projective variety of general type then the canonical model X_{can} of X exists and it has canonical singularities. Minimal models have terminal singularities and in fact a key step in the proof of the minimal model program for threefolds achieved in the 1980s (cf. [95]) was a comprehensive classification of terminal threefold singularities.

It is natural to generalize these results to pairs (X, B) where X is say a smooth variety and B is a simple normal crossings divisor with coefficients in $[0, 1]$ (resp. in $[0, 1)$). The singularities of the corresponding log canonical and log minimal models are log canonical (or lc for short) (resp. Kawamata log terminal (or klt for short)) and divisorially log terminal (dlt) singularities. It is not possible to completely classify these singularities in higher dimensions but many of their features are well understood and play a fundamental role in many related subjects. For technical reasons, several other flavors of singularities naturally appear, such as purely log terminal (plt) singularities. They play a fundamental role in proofs by induction on the dimension. For example, if (X, Δ) is a plt singularity, then we write $\Delta = S + B$ where S is an irreducible divisor and the coefficients of B are < 1. It then follows by adjunction that $(K_X+S+B)|_S = K_S+B_S$ where (S, B_S) is a klt pair. Information on the geometry of (S, B_S) corresponds to information on the geometry of $(X, S + B)$ in a neighborhood of S. This is a key observation, in the proof of many of the crucial results of the minimal model program including the inductive proof of the existence of flips and minimal models for varieties of log general type [14].

The minimal model program and its singularities are well understood in characteristic 0 because of important vanishing theorems of Kodaira, Kawamata-Viehweg, Kollár and others.

The fact that these vanishing theorems fail in positive and mixed characteristics causes many difficulties. On the one hand we are often able to prove positive results especially in low dimension and large characteristics. On the other hand it is clear that new methods are required in general.

The study of singularities of the minimal model program is a vast and rapidly evolving subject and it is impossible to give a concise comprehensive survey (we refer the interested reader to [75] for a recent in depth book on this topic). In this paper we will survey several topics related to singularities of the minimal model program that have either played an important role in the development of the subject or are a particularly active area of current research. Unluckily, we will also omit many significant and very active areas including, for example, results related to foliations, arc spaces, and D-modules to name a few.

8.2 Notations and Conventions

Unless otherwise stated, we will assume that all varieties are quasi-projective, defined over an algebraically closed field k of characteristic 0 (typically $k = \mathbb{C}$). We begin by reviewing some standard notations and conventions that will be regularly used in what follows.

Let X be a normal quasi-projective variety so that X is S_2 and R_1 i.e. smooth in codimension 1. A *prime divisor* is an irreducible reduced subvariety of codimension 1. A divisor D is *principal* if it is given by the zeroes and poles of a rational function, i.e. $D = (f)$ for some $f \in K(X)^\times$. A *Cartier divisor* is a divisor which is locally principal that is locally given by zeroes and poles of a rational function. A $\mathbb{Q}$*-divisor* (resp. $\mathbb{R}$*-divisor*) $D = \sum d_i D_i$ is a formal linear combination of (distinct) prime divisors D_i with coefficients $d_i \in \mathbb{Q}$ (resp. $d_i \in \mathbb{R}$). D is *effective* $D \geq 0$ if all $d_i \geq 0$. An $\mathbb{R}$*-Cartier divisor* D is a linear combination $D = \sum d_i D_i$ where $d_i \in \mathbb{R}$ and D_i are Cartier (similarly for $\mathbb{Q}$-divisors). The *support* of D, denoted by $\mathrm{Supp}(D)$, is the union of the components D_i with non-zero coefficient $d_i \neq 0$. If $r \in \mathbb{R}$, then the *round down* $\lfloor r \rfloor$ is the biggest integer $\leq r$ and the *round up* $\lceil r \rceil$ is the smallest integer $\geq r$. We define the *fractional part* $0 \leq \{r\} := r - \lfloor r \rfloor < 1$. If $D = \sum d_i D_i$ is a formal linear combination of (distinct) prime divisors D_i, then $\lfloor D \rfloor = \sum \lfloor d_i \rfloor D_i$, $\lceil D \rceil = \sum \lceil d_i \rceil D_i$ and $\{D\} = D - \lfloor D \rfloor$.

We will denote by $|D|$ the *complete linear series* corresponding to an $\mathbb{R}$-divisor D. Recall that $G \in |D|$ iff $G \geq 0$ and $G \sim D$ that is $G - D = (f)$ for some rational function $f \in K(X)$. Note that if $G \in |D|$, then $G = \{D\} + G'$ where $G' \in |\lfloor D \rfloor|$. We say that two $\mathbb{R}$-divisors D_1, D_2 are $\mathbb{R}$*-linearly equivalent* $D_1 \sim_{\mathbb{R}} D_2$ (resp. $\mathbb{Q}$*-linearly equivalent* $D_1 \sim_{\mathbb{Q}} D_2$) if $D_1 - D_2 = \sum d_i (f_i)$ where (f_i) are principal divisors for $f_i \in K(X)^\times$ and $d_i \in \mathbb{R}$ (resp. $d_i \in \mathbb{Q}$). We let $|D|_{\mathbb{Q}}$ (resp. $|D|_{\mathbb{R}}$) be the set of effective $\mathbb{R}$-divisors $G \geq 0$ such that $G \sim_{\mathbb{Q}} D$ (resp. $G \sim_{\mathbb{R}} D$). Note that if D is a $\mathbb{Q}$-divisor and $G \in |D|_{\mathbb{Q}}$, then G is a $\mathbb{Q}$-divisor.

We say that an $\mathbb{R}$-Cartier divisor D on a proper normal variety X is *numerically equivalent to* 0 (or *numerically trivial*) $D \equiv 0$ if $D \cdot C = 0$ for any curve $C \subset X$. If $f : X \to Z$ is a proper morphism of quasi-projective varieties, we say that D is *relatively numerically trivial* $D \equiv_f 0$ (or $D \equiv_Z 0$) if $D \cdot C = 0$ for any curve $C \subset X$ contained in a fiber of f. The global notion of numerical triviality $\equiv 0$ agrees with the relative one when $f : X \to \operatorname{Spec}(\mathbb{C})$ is the structure morphism. We say that two $\mathbb{R}$-divisors D_1, D_2 are *f-numerically equivalent* $D_1 \equiv_f D_2$ if their difference is f-numerically trivial $D_1 - D_2 \equiv_f 0$. Clearly if $D_1 \sim_{\mathbb{R}} D_2$, then $D_1 \equiv D_2$.

For any divisor D we consider the corresponding reflexive sheaf of rank 1, $\mathcal{O}_X(D)$ defined by $\mathcal{O}_X(D)(U) = \{f \in K(X) | (D + (f))|_U \geq 0\}$. D is Cartier, iff $\mathcal{O}_X(D)$ is a line bundle (i.e. invertible rank 1 sheaf). The sections $s \in H^0(\mathcal{O}_X(D))$ define effective divisors $(s) \in |D|$ (recall that any section s corresponds locally to a rational function f such that $(s) = (f) + D \geq 0$ locally on X). We have $h^0(D) := \dim H^0(\mathcal{O}_X(D)) = \dim |D| + 1$. The complete linear series $|D|$ defines a morphism on $U = X \setminus \operatorname{Bs}(|D|)$ as follows: Let $s_0, \ldots, s_r$ be a basis of $H^0(\mathcal{O}_X(D))$ then $\phi(x) = [s_0(x) : \ldots : s_r(x)]$ is defined on U. We denote the corresponding rational map $\phi_{|D|} : X \dashrightarrow |D| \cong \mathbb{P}^r$.

A divisor D is *very ample* if $|D|$ defines an embedding $\phi_{|D|} : X \hookrightarrow |D|$. If D is a divisor (resp. $\mathbb{Q}$-divisor) we say that D is an *ample divisor* (resp. an *ample $\mathbb{Q}$-divisor*) if mD is very ample for some $m > 0$ (in particular D must be $\mathbb{Q}$-Cartier). If D is an $\mathbb{R}$-divisor, then we say that D is *ample* if $0 \neq D = \sum r_i D_i$ where each D_i is an ample divisor and each $r_i \geq 0$. A divisor D is *semiample* if $|mD|$ is base point free for some $m > 0$ and hence it defines a morphism $\phi_{|mD|} : X \to |mD|$. Note that if we replace m by a sufficiently big and divisible multiple, then we obtain a surjective morphism of normal varieties $\phi : X \to Z \subset |mD|$ where $\phi_* \mathcal{O}_X = \mathcal{O}_Z$ and $\phi^* \mathcal{O}_Z(1) = \mathcal{O}_X(mD)$. An $\mathbb{R}$-Cartier divisor D on a proper variety X is *nef* if $D \cdot C \geq 0$ for any curve $C \subset X$. It is easy to see that if D is semiample then it is nef. It is well known that if D is nef then $D^{\dim X} \geq 0$ and then D is big (see defn. below) iff $D^{\dim X} > 0$. It is however not true that $D^{\dim X} > 0$ for all big divisors. For example consider S a projective surface and $\nu : S' \to S$ the blow up of a point with exceptional divisor E. If H is ample on S, then $f^*H + nE$ is big for all $n > 0$ but $(f^*H + nE)^2 = H^2 - n^2 < 0$ for $n \gg 0$.

Recall that the *Kodaira dimension* of a $\mathbb{Q}$-divisor D is defined as follows. If $|mD|$ is empty for all $m > 0$ then set $\kappa(D) = -1$ (some authors define it as $\kappa(D) = -\infty$ and some authors just say it is negative). Otherwise we set $\kappa(D) = \max\{\dim \phi_{|mD|}(X) | m \geq 1\}$. Equivalently, if $R(X, D) := \oplus_{m \geq 0} H^0(X, \mathcal{O}_X(mD))$, then $\kappa(D) = \operatorname{tr.deg.}_{\mathbb{C}} R(X, D) - 1 \in \{-1, 1, 0, \ldots, \dim X\}$. We let

$$\operatorname{vol}(D) = \limsup \frac{h^0(X, mD)}{m^n/n!}$$

be the *volume* of D (here $n = \dim X$). It is well known that in fact $\mathrm{vol}(D) = \lim \frac{h^0(X,mD)}{m^n/n!}$, $\mathrm{vol}(D) > 0$ iff $\kappa(D) = \dim X$ i.e. D is *big*. If D is nef, then $\mathrm{vol}(D) = D^{\dim X}$. Fix an ample divisor A, then a $\mathbb{Q}$-divisor D is *pseudo-effective* if for any rational number $\epsilon > 0$, the $\mathbb{Q}$-divisor $D + \epsilon A$ is big.

Lemma 8.2.1 (Negativity Lemma) *Let $\nu : X' \to X$ be a proper birational morphism of normal quasi-projective varieties and $-B$ a ν-nef $\mathbb{Q}$-Cartier divisor on X', then*

1. *B is effective iff so is $\nu_* B$,*
2. *if B is effective then for any point $x \in X$ then either* $\mathrm{Supp}(B) \cap \nu^{-1}(x) = \emptyset$ *or* $\nu^{-1}(x) \subset \mathrm{Supp}(B)$.

Proof This is a well known consequence of the Hodge Index Theorem, see eg. [81, Lemma 3.39]. □

8.3 Singularities of the mmp

There are several flavors of singularities of the minimal model program that are particularly useful for different purposes. Canonical and terminal singularities naturally arise when considering canonical or minimal models of smooth varieties (or of varieties with canonical and terminal singularities). Log canonical and Kawamata log terminal singularities naturally arise when considering log canonical or log minimal models of log canonical or klt pairs. Divisorial log terminal singularities occur on log minimal models of log canonical pairs. Semi log canonical singularities naturally arise in the study of moduli spaces of canonically polarized varieties (or pairs) and purely log terminal singularities are important for technical reasons such as proofs by induction on the dimension. We begin by defining pairs.

Definition 8.3.1 A *sub-pair* (X, B) consists of a normal quasi-projective variety X and an $\mathbb{R}$-divisor $B = \sum b_i B_i$ such that $K_X + B$ is $\mathbb{R}$-Cartier. If $B = \sum b_i B_i$ is effective (i.e. $b_i \geq 0$), then we say that (X, B) is a pair. A sub-pair (X, B) is *log smooth* if X is smooth and the components B_i of $\mathrm{Supp}(B)$ are smooth and intersect transversely.

In order to define the singularities of the minimal model program we use Hironaka's result on the resolution of singularities.

Definition 8.3.2 Let (X, B) be a pair. A *log resolution* of (X, B) is a proper birational morphism $\nu : X' \to X$ such that X' is smooth, the exceptional set $\mathrm{Ex}(\nu)$ is a divisor and $\mathrm{Supp}(B') + \mathrm{Ex}(\nu)$ has simple normal crossings where $K_{X'} + B' = \nu^*(K_X + B)$. We always assume that K_X and $K_{X'}$ are chosen so that $\nu_* K_{X'} = K_X$, and so $\nu_* B' = B$.

The singularities of a pair (X, B) are measured in terms of the coefficients of B'. For technical reasons it is convenient to define the log discrepancies.

Definition 8.3.3 Let $\nu : X' \to X$ be a proper birational morphism and write $K_{X'} + B' = \nu^*(K_X + B)$. For any prime divisor P on X' we define the *log discrepancy* $a(P; X, B) = 1 - \text{mult}_P(B')$. Thus $B' = \sum_{P \subset X'}(1 - a(P; X, B))P$ where the sum runs over all prime divisors P on X'. It is easy to see that for any prime divisor P over X the definition of $a(P; X, B)$ does not depend on the choice of a proper birational morphism $\nu : X' \to X$ such that P is a divisor on X'. The *minimal log discrepancy* (or mld) of a pair (X, B) is given by

$$\text{mld}(X, B) = \inf\{a(P; X, B) | P \text{ is a divisor over } X\}.$$

Lemma 8.3.4 *Let (X, B) be a pair of dimension $\dim X \geq 2$, then either $\text{mld}(X, B) = -\infty$ or $0 \leq \text{mld}(X, B) \leq 1$.*

Proof Let P be any divisor on X not contained in the support of B, then $a(P; X, B) = 1$ and so $\text{mld}(X, B) \leq 1$. It suffices to show that if $\text{mld}(X, B) < 0$, then $\text{mld}(X, B) = -\infty$. Therefore, suppose that $\text{mld}(X, B) < 0$ and let a $\nu : X' \to X$ be a log resolution and P a divisor on X' such that $a := a(P; X, B) < 0$. Consider a sequence of blow ups $X' = X_0 \leftarrow X_1 \leftarrow X_2 \leftarrow \ldots \leftarrow X_n$, where E_i denotes the exceptional divisor of $\nu_i : X_i \to X_{i-1}$, P_i denotes the strict transform of P, ν_1 is given by blowing up a general hypersurface contained in P and ν_i is given by blowing up $P_{i-1} \cap E_{i-1}$. By an easy surface computation, one sees that $a(P_n; X, B) = 1 + ka$ and so $\inf a(P_n; X, B) = -\infty$ and the claim follows. □

Remark 8.3.5 Note that if $\text{mld}(X, B) \geq 0$, then there exists a divisor P over X such that $\text{mld}(X, B) = a(P; X, B)$.

Definition 8.3.6 We say that

1. (X, B) is *Kawamata log terminal* or klt if $\text{mld}(X, B) > 0$.
2. (X, B) is *log canonical* or lc if $\text{mld}(X, B) \geq 0$.
3. (X, B) is *terminal* if $a(P; X, B) > 1$ for every divisor P exceptional over X.
4. (X, B) is *canonical* if $a(P; X, B) \geq 1$ for every divisor P exceptional over X.
5. (X, B) is *purely log terminal (plt)* if $a(P; X, B) > 0$ for every divisor P exceptional over X.

Remark 8.3.7 As we will see later, terminal and canonical singularities are precisely the singularities that appear on terminal and canonical models of smooth varieties and Kawamata log terminal and log canonical singularities are precisely the singularities that appear on log terminal and log canonical models of Kawamata log terminal and log canonical pairs. plt and dlt singularities (Definition 8.3.10) are useful when proving results by induction on the dimension. Semi log canonical singularities (Definition 8.14.1) naturally appear when considering limits of log canonical pairs, for example the stable curves of Deligne and Mumford.

Lemma 8.3.8 *We have the following implications*

1. *terminal singularities are canonical,*
2. *klt singularities are plt,*

3. *plt singularities (in dimension* ≥ 2*) are lc,*
4. *canonical (resp. terminal) singularities are lc (resp. klt),*
5. *if* (X, B) *is canonical and* $\lfloor B \rfloor = 0$*, then* (X, B) *is klt.*

All of the inclusions above are strict.

Proof (1) and (2) are clear from the definitions.

To see (3), first observe that by the proof of Lemma 8.3.4, if (X, B) is plt, then the coefficients of B are ≤ 1. The claim now follows easily from the definitions.

We next check (4). First of all, arguing as in the proof of Lemma 8.3.4, it follows that if (X, B) is a pair and the coefficient of a component B_i of B is $b_i > 1$ (resp. $b_i \geq 1$), then blowing up X along a general codimension 2 point contained in B_i we obtain an exceptional divisor P over X such that $a(P; X, B) \leq 2 - b_i < 1$ (resp. $a(P; X, B) \leq 1$) and so (X, B) is not canonical (resp. is not terminal). Therefore, if (X, B) is canonical (resp. is terminal), we may assume that $a(P; X, B) = b_i \leq 1$ (resp. $a(P; X, B) = b_i < 1$) for all divisors P on X and the same inequality follows for all divisors P over X by definition of canonical (resp. terminal) singularities.

(5) follows similarly.

Example 8.3.13 shows that a cone over an elliptic curve is lc but not plt and the cone over a rational curve is klt but not canonical. The pair $(\mathbb{C}^2, L)$ where L is a line, is plt but not klt. A rational double point is canonical but not terminal see Example 8.3.15. □

Singularities of log smooth pairs are easy to understand by the lemma below.

Lemma 8.3.9 *Let* (X, B) *be a pair. We will denote a log resolution of* (X, B) *by* $\nu : X' \to X$ *and write* $K_{X'} + B' = \nu^*(K_X + B)$ *and* $B' = \sum_{i=1}^k b'_i B'_i$*. We have the following equivalent definitions.*

1. (X, B) *is klt iff there exists a log resolution* ν *such that* $b'_i < 1$ *for* $1 \leq i \leq k$ *iff for any log resolution* ν *we have* $b'_i < 1$ *for* $1 \leq i \leq k$*.*
2. (X, B) *is lc iff there exists a log resolution* ν *such that* $b'_i \leq 1$ *for* $1 \leq i \leq k$ *iff for any log resolution* ν *we have* $b'_i \leq 1$ *for* $1 \leq i \leq k$*.*
3. *A log smooth pair* (X, B) *is terminal (resp. canonical) iff* $b_i < 1$ *(resp.* $b_i \leq 1$*) for* $1 \leq i \leq k$ *and* $b_i + b_j < 1$ *(resp.* $b_i + b_j \leq 1$*) for any* $1 \leq i < j \leq k$ *such that* $B_i \cap B_j \neq \emptyset$*.*
4. *If* (X, B) *is plt then the components of* $\lfloor B \rfloor$ *are disjoint.*

Proof See [81, Corollary 2.31]. □

The following classes of singularities are also very useful.

Definition 8.3.10 Let (X, B) be a pair, then (X, B) is *divisorially log terminal (dlt)* iff $a(P; X, B) \geq 0$ for every divisor P on X, and there exists a log resolution $\nu : X' \to X$ such that $a(P; X, B) > 0$ for every ν-exceptional divisor P on X'.

Remark 8.3.11 By a result of Szabó [81, Theorem 2.44], it is known that (X, B) is dlt iff the coefficients of B are ≤ 1 and there is an open subset $U \subset X$ such that $(U, B|_U)$ is log smooth and $a(P; X, B) > 0$ for every divisor P over X whose center is not contained in U.

Remark 8.3.12 If (X, B) is dlt, then it is a limit of klt pairs. More precisely, fix A ample on X, then for any $\epsilon > 0$ there exists a $\mathbb{R}$-divisor $B_\epsilon \sim_{\mathbb{R}} B + \epsilon A$ such that (X, B_ϵ) is klt (see [81, Proposition 2.43]).

Example 8.3.13 Let X be the cone over a projective curve C of genus g and degree $d > 0$, then X is klt, lc, not-lc iff $g = 0$, $g = 1$ or $g \geq 2$. To see this, observe that blowing up the vertex $P \in X$ we obtain a resolution $\nu : X' \to X$. Let E be the exceptional divisor, then $E \cong C$ and $E^2 = -d$. By adjunction we have $2g - 2 = (K_{X'} + E) \cdot E$. If we let $K_{X'} + bE = \nu^* K_X$, then $(K_{X'} + bE) \cdot E = 0$. Thus $2g - 2 = (1 - b)E^2 = -d(1 - b)$. If $g = 0$, then $b = 1 - \frac{2}{d} < 1$ and hence X is klt. If $g = 1$, then $b = 1$ and so X is lc but not klt. If $g \geq 2$, then $b = 1 + \frac{2g-2}{d} > 0$ and so X is not log canonical.

Remark 8.3.14 Let X be a terminal surface, then X is smooth (see Proposition 8.5.14).

Example 8.3.15 Let $X = \{x^2 + y^2 + z^2 = 0\}$ be the cone over a rational curve of degree 2, then X is canonical but not smooth and hence not terminal (see Remark 8.3.14). Blowing up the vertex we obtain a resolution $\nu : X' \to X$ such that $\nu^* K_X = K_{X'}$.

Theorem 8.3.16 *Let $0 \in X$ be the germ of a normal complex surface singularity. Then*

1. *$0 \in X$ is canonical iff $X \cong \mathbb{C}^2/G$ where G is a finite subgroup of $SL(2, \mathbb{C})$,*
2. *$0 \in X$ is klt iff $X \cong \mathbb{C}^2/G$ where G is a finite subgroup of $GL(2, \mathbb{C})$,*
3. *$0 \in X$ is log canonical iff it is smooth, or a simple elliptic or cusp singularity, or a quotient of these by a finite group.*

Proof See [78, §3]. □

Definition 8.3.17 Let (X, B) be a pair, then a *non-klt place* (resp. a *non-lc place*) of (X, B) is a prime divisor P over X such that $a(P; X, B) \leq 0$ (resp. $a(P; X, B) < 0$). The images of non-klt places (resp. non-lc places) are the *non-klt centers* (resp. the *non-lc centers*). The union of the non-klt centers (resp. the non-lc centers) is the *non-klt locus $nklt(X, B)$* (resp. the *non-lc locus $nlc(X, B)$*).

If (X, B) is log canonical, then the non-klt centers/places are also called *log canonical centers/places*.

Remark 8.3.18 If $\dim X = 2$ then we can drop the assumption that $K_X + B$ is $\mathbb{R}$-Cartier in the definition of lc, klt, terminal and canonical singularities (see Sect. 8.5.4). In [26], the authors study pairs (X, B) without the assumption that $K_X + B$ is $\mathbb{R}$-Cartier.

Remark 8.3.19 There are many interesting examples of klt pairs in the literature. For example it is known that if (A, Θ) is a principally polarized abelian variety and $0 \leq D \sim_{\mathbb{Q}} \Theta$ then (X, D) is lc and in fact klt iff $\lfloor D \rfloor = 0$ (see eg. [42] and references therein). Similarly, by [7], spherical varieties have log terminal singularities and by [84], Richardson varieties are of klt type.

8.4 Conjectures About mld's

One views minimal log discrepancies as a measure of the singularities of a log pair. *The smaller* $\mathrm{mld}(X, B)$ *is, the more singular the pair* (X, B) *is*. One can also define a local version of these measures of singularity.

Definition 8.4.1 (V.V. Shokurov) The *minimal log discrepancy of a log pair* (X, B) *at a proper Grothendieck point* $\eta \in X$ is given by

$$a(\eta; X, B) = \inf\{a(E; X, B) \mid c_X(E) = \eta\},$$

where the infimum is taken over all prime divisors E over X with center on X given by $c_X(E) = \eta$. If η_X is the generic point of X then we define $a(\eta_X; X, B) = 0$.

The following conjecture shows that we expect that big mld's characterize smooth varieties.

Conjecture 8.4.2 ([108]) Let (X, B) be a log pair and let $\eta \in X$ be a Grothendieck point. Then the following inequality holds: $a(\eta; X, B) \leq \mathrm{codim}(\eta)$. Moreover, if $a(\eta; X, B) > \mathrm{codim}(\eta) - 1$, then X is nonsingular at η.

This conjecture follows easily in dimension 2 (as terminal surface singularities are smooth) and in dimension 3 by classification of terminal threefold singularities. The conjecture is open in higher dimensions.

Since smooth closed points not contained in the support of B have $\mathrm{mld}(x; X, B) = \dim X$ and are dense in X, then the first part of the above conjecture for closed points, follows from the conjecture below.

Conjecture 8.4.3 (Semicontinuity for mlds [3, Hypothesis 0.3]) Let (X, B) be a log pair and consider the function $a : X \to \{-\infty\} \cup \mathbb{R}$, defined by $x \to a(x; X, B)$ for all closed points $x \in X$. Then a is lower semicontinuous, that is, for every closed point $x \in X$ there is a neighborhood $x \in U \subset X$ such that $a(x; X, B) = \inf_{x' \in U}\{a(x'; X, B)\}$.

Remark 8.4.4 By [3], it is known that Conjecture 8.4.3 holds if $\mathrm{codim}(\eta) \leq 3$, or if X is a toric variety and B is invariant under the torus action.

Before we state the next conjecture, recall that a subset $\Gamma \subset \mathbb{R}$ satisfies the *ascending chain condition* or ACC, if any non-decreasing sequence $a_1 \leq a_2 \leq \dots$ is eventually constant i.e. $a_i = \sup\{a_i | i \in \mathbb{N}\}$ for $i \gg 0$. A subset $\Gamma \subset \mathbb{R}$ satisfies the *descending chain condition* or DCC if $-\Gamma$ satisfies the ACC. If $B = \sum b_i B_i$, then we say that $B \in \Gamma$ if all $b_i \in \Gamma$.

Conjecture 8.4.5 (ACC for mlds [108]) Let (X, B) be a log pair, $n \in \mathbb{N}$, and $\Gamma \subset [0, 1]$ be a subset satisfying the DCC. Then the set

$$A(\Gamma, n) := \{a(\eta; X, B);\ \dim X = n,\ \ B \in \Gamma\}$$

satisfies the ACC.

Remark 8.4.6 This conjecture is known to hold for exceptional singularities by [45].

One application of these conjectures is that they imply the existence of minimal models; in fact they would imply one of the main conjectures of the minimal model program that is the termination of flips (see Conjecture 8.13.4) and hence the termination of sequences of steps of the minimal model program.

Theorem 8.4.7 (Shokurov [110]) *Assume Conjectures 8.4.5 and 8.4.3 for n-dimensional varieties. Then every sequence of flips in dimension n terminates.*

Remark 8.4.8 Recently, Nakamura and Shibata [100], have shown that some closely related conjectures fail. In particular they show that the precise inversion of adjunction conjecture fails and that the semicontinuity for mlds in families fails [100, 1.3, 6.1, 6.3, 6.5].

8.5 Surface Singularities

For the convenience of the reader we begin by recalling several well known facts about surfaces.

Let $x \in X$ be a smooth point on a smooth surface and $\nu : X' \to X$ be the blow up of $x \in X$, then $E := \nu^{-1}(x)$ is (-1)-curve, i.e. a smooth rational curve such that $E^2 = -1$ (or equivalently $K_X \cdot E = -1$). We have $K_{X'} = \nu^* K_X + E$.

Theorem 8.5.1 (Castelnuovo's Contractibility Criterion) *If X is a smooth surface and $E \subset X$ is a (-1)-curve, then there is a blow up $\nu : X \to \bar{X}$ where ν is the blow up of a closed point on a smooth surface $0 \in \bar{X}$ and $E = \nu^{-1}(0)$.*

Proof See [8, Theorem II.17]. □

Remark 8.5.2 All rational maps of smooth surfaces are given by sequences of blow ups. More precisely we have the following.

If $\nu : X' \to X$ is a proper birational morphism of smooth surfaces, then ν is given by a sequence of blow ups (cf. [8, Theorem II.11]). In particular X' contains a ν-exceptional (-1)-curve.

If $X \dashrightarrow X'$ is a birational morphism of projective smooth surfaces, then there is a common resolution $p : W \to X$ and $q : W \to X'$ (where both p and q are given by sequences of blow ups).

Remark 8.5.3 If C is a curve in a smooth surface X, $\nu : X' \to X$ is the blow up of a point $x \in X$, and $C' = \nu_*^{-1}C$, then $(C')^2 = C^2 - m^2$ where $m = \operatorname{mult}_x(C)$ and hence $(C')^2 \leq C^2$ and equality only holds if ν is an isomorphism near C'. Similarly one sees that $(K_{X'} + C') \cdot C' = (K_X + C) \cdot C - m(m-1)$. It follows that if $N \to C$ is the normalization, then

$$2g(C) - 2 = 2g(N) - 2 + \sum m_i(m_i - 1)$$

where m_i are the multiplicities of points of C (including infinitely close points) [8, Exercise II.20 (2)].

8.5.1 *Minimal Resolutions of Surfaces*

Let X be a normal surface and $\nu : X' \to X$ a resolution with exceptional divisor $E = \sum_{i=1}^k E_i$, then from the Hodge index theorem it follows that the corresponding matrix with entries $E_i \cdot E_j$ is negative definite. In particular, if $C \subset X'$ is a ν-exceptional curve, then $C^2 < 0$.

Definition 8.5.4 Let X be a normal surface and $\nu : X' \to X$ be a resolution with smallest second Betti number $B_2(X')$, then we say that ν is the minimal resolution of X.

Theorem 8.5.5 *Let $\nu : X' \to X$ be a resolution of a normal surface. The following are equivalent.*

1. *ν is the minimal resolution of X,*
2. *X' contains no ν-exceptional (-1)-curves,*
3. *$K_{X'}$ is nef over X,*
4. *if Y is a smooth surface then any morphism $Y \to X$ factors through a morphism $Y \to X'$.*

In particular minimal resolutions are unique.

Proof (1) implies (2). Suppose that X' contains a ν-exceptional (-1)-curve E and let $X' \to X''$ be the contraction of this curve which exists by Castelnuovo's contractibility criterion. Since E is ν-exceptional, there is an induced morphism $X'' \to X$. Since X'' is smooth and $B_2(X'') = B_2(X') - 1$, ν is not the minimal resolution of X.

(2) and (3) are equivalent. Clearly (3) implies (2). Suppose that $K_{X'}$ is not nef over X, then there is a $\nu-$exceptional curve $C \subset X'$ such that $K_{X'} \cdot C < 0$. Since C is exceptional, $C^2 < 0$. Therefore $2g(C) - 2 = (K_{X'} + C) \cdot C \leq -2$. On the other hand, if $N \to C$ is the normalization, then by Remark 8.5.3, we have

$$-2 \geq 2g(C) - 2 = 2g(N) - 2 + \sum m_i(m_i - 1) \geq -2$$

where m_i are the multiplicities of points of C (including infinitely close points). Thus $g(C) = g(N) = 0$ and $C^2 = -1$. It follows that C is ν−exceptional (-1)-curve.

(3) implies (4). Suppose that there is a morphism $\mu : Y \to X$ from a smooth surface that does not factor through X. Let W be a resolution of the graph of $Y \dashrightarrow X$ and $p : W \to Y$, $q : W \to X$ the induced morphisms. By Castelnuovo's contractibility criterion, we may assume that W contains no (-1)-curves that are both p and q exceptional. Since $Y \dashrightarrow X'$ is not a morphism, it follows that there is a (-1)-curve $C \subset W$ that is p-exceptional but not q-exceptional. Let $C' = q_*C$, then $-1 = C^2 \leq (C')^2 < 0$ (since C' is exceptional over X). But then $(C')^2 = -1$ and $C' \cong C$ is a smooth rational curve i.e. C' is a ν-exceptional (-1)-curve which is impossible as $K_{X'}$ is nef over X.

(4) implies (1). Suppose that $\nu : X' \to X$ is not minimal, then there exists a morphism from a smooth surface $\mu : Y \to X$ such that $B_2(Y) < B_2(X)$. By (4), there is a morphism $Y \to X$ and hence $B_2(Y) \geq B_2(X)$ which is impossible.

□

8.5.2 Surface MMP

The minimal model program is particularly simple for log canonical surfaces as there are no flips and abundance holds. We have the following.

Theorem 8.5.6 *Let (X, B) be a $\mathbb{Q}$-factorial log canonical surface and $f : X \to Z$ a projective morphism. Then there exists a sequence of birational maps $f_i : X_i \to X_{i+1}$ for $0 \leq i \leq n-1$ where $X = X_0$, such that if $B_0 = B$ and $B_{i+1} = (f_i)_*B_i$, then*

1. *(X_i, B_i) is $\mathbb{Q}$-factorial and log canonical,*
2. *$\rho(X_{i+1}) = \rho(X_i) - 1$,*
3. *$-(K_{X_i} + B_i)$ is f_i-ample,*
4. *either $K_{X_n} + B_n$ is nef over Z, and then $K_{X_n} + B_n$ is semiample over Z, or there is a Mori fiber space $g : X_n \to W$ over Z so that $\dim W \leq 1$, $-(K_{X_n} + B_n)$ is ample over W and $\rho(X_n/Z) = 1$.*

Proof We refer the reader to [78, Section 11], [79]. □

Corollary 8.5.7 *If (X, B) is a dlt surface, then X is $\mathbb{Q}$-factorial.*

Proof Perturbing B, we may assume that (X, B) is klt (see Remark 8.3.12). Let $\nu : X' \to X$ be a log resolution and $B' = \nu_*^{-1}B + \mathrm{Ex}(\nu)$, then $K_{X'} + B' = \nu^*(K_X + B) + E$ where $E \geq 0$ and $\mathrm{Supp}\,(E) = \mathrm{Ex}(\nu)$. Replacing X' by the output of the $(K_{X'} + B')$-mmp over X, we may assume that E is nef over X. By the Negativity Lemma 8.2.1, $E = 0$ and so $X' \to X$ is an isomorphism. Since the mmp preserves the $\mathbb{Q}$-factorial condition, X is $\mathbb{Q}$-factorial. □

Theorem 8.5.8 *If (X, B) is dlt, then X has rational singularities.*

Recall that a normal variety X has *rational singularities* if given any resolution $\nu : X' \to X$ we have $R\nu_*\mathcal{O}_{X'} = \mathcal{O}_X$, i.e. $R^i\nu_*\mathcal{O}_{X'} = 0$ for $i > 0$ (and of course $\nu_*\mathcal{O}_{X'} = \mathcal{O}_X$).

Proof By Corollary 8.5.7 $(X, 0)$ is klt. Let $\nu : X' \to X$ be the minimal resolution, then $K_{X'} + F = \nu^*K_X$. By the Negativity Lemma 8.2.1, $F \geq 0$. Since $(X, 0)$ is klt, so is (X', F). Since $0 \equiv_X K_{X'} + F - (K_{X'} + F)$ where $-(K_{X'} + F) \equiv_X 0$ is ν-nef and ν-big, then by the Kawamata-Viehweg vanishing theorem 8.9.1, $R^1\nu_*\mathcal{O}_{X'} = 0$. □

8.5.3 Classification of Surface Singularities

Canonical surface singularities are extremely well understood and in fact they can be described by explicit equations [81, Theorem 4.20].

Theorem 8.5.9 *Let* $(0 \in X)$ *be the germ of a canonical singularity (in particular* X *is normal and* $0 \in X$ *is not a smooth point). Then* $(0 \in X)$ *is analytically isomorphic to one of the following singularities*

A_n: $x^2 + y^2 + z^{n+1} = 0$, *for* $n \geq 1$,
D_n: $x^2 + y^2z + z^{n-1} = 0$, *for* $n \geq 1$,
E_6: $x^2 + y^3 + z^4 = 0$,
E_7: $x^2 + y^3 + yz^3 = 0$,
E_8: $x^2 + y^3 + z^5 = 0$.

In order to study surface singularities, it is convenient to consider the dual graphs corresponding to these singularities. Let $(0 \in X, B)$ be the germ of a log canonical surface singularity, $\nu : X' \to X$ the minimal resolution and $E = \sum E_i$ be the set of exceptional curves. We consider the corresponding dual graph Γ whose vertices v_i correspond to the (irreducible) curves C_i and edges between v_i and v_j have multiplicities $C_i \cdot C_j$. Each vertex is labeled by $b_i = -C_i^2$.

Remark 8.5.10 The dual graphs of canonical singularities are ([81, Theorem 4.22]):

1. A_n: a chain of length n,
2. D_n: a tree with 1 fork and 3 branches, of which 2 branches have length 1 and the last branch has length $n - 3$,
3. E_n: a tree with 1 fork and 3 branches, of which one branch has length 1, another one has length 2 and the last branch has length $n - 4$.

Note that each vertex v corresponds to a smooth rational curve C_v. Since $K_{X'} = \nu^*K_X$, then $K_{X'} \cdot C_v = 0$ and so $C_v^2 = -2$.

The dual graphs corresponding to log canonical singularities are also completely classified, see [81, Theorem 4.7], [75, 3.39], and [78, §3]. Note that in this case an irreducible elliptic curve or a nodal rational curve can also appear in the minimal resolution.

8.5.4 Numerically klt and lc Singularities

In the case of surface pairs (X, B), we can relax the assumption that $K_X + B$ be $\mathbb{R}$-Cartier. Let $\nu : X' \to X$ be a log resolution and let $E_1, \ldots, E_k$ be the exceptional curves, then by the Hodge Index Theorem, the $k \times k$ matrix with entries $E_i \cdot E_j$ is negative definite. Therefore, there are uniquely determined real numbers $b_1, \ldots, b_k$ such that

$$\nu_*^{-1} D \equiv_X -\sum b_j E_j \qquad \text{i.e.} \qquad -(\sum b_j E_j)\cdot E_i = \nu_*^{-1} D \cdot E_i, \qquad i = 1, \ldots, k.$$

We can then define the *Mumford pull-back* $\nu^* D := \nu_*^{-1} D + \sum b_j E_j$. It follows easily from the Negativity Lemma 8.2.1, that whenever D is $\mathbb{R}$-Cartier, then the Mumford pull-back agrees with the usual pull-back.

Definition 8.5.11 If X is a normal surface and B is an effective $\mathbb{R}$-divisor and P is a divisor over X, then we pick a resolution $\nu : X' \to X$ such that P is a divisor on X' and we write $K_{X'} + \nu_*^{-1} B + b_i E_i \equiv_X 0$ and we let $\alpha(P; X, B) = 1 - \mathrm{mult}_P(\nu_*^{-1} B + b_i E_i)$ be the numerical log discrepancy of (X, B) along P. Clearly, if $K_X + B$ is $\mathbb{R}$-Cartier, then $\alpha(P; X, B) = a(P; X, B)$. We say that (X, B) is numerically log canonical (resp. numerically klt) if $\alpha(P; X, B) \geq 0$ (resp. $\alpha(P; X, B) > 0$) for all divisors over X. We can similarly define numerically canonical/terminal surface singularities.

Remark 8.5.12 It is easy to see that if $0 \leq B_0 \leq B$, then $\alpha(P; X, B_0) \geq \alpha(P; X, B)$ for any divisor P over X. In particular if (X, B) is numerically lc/klt/canonical/terminal then (X, B_0) is also numerically lc/klt/canonical/terminal. To see this, let $\nu : X' \to X$ be a log resolution of (X, B) (and hence also a log resolution of (X, B_0)) and write $K_{X'} + \nu_*^{-1} B + E \equiv_X 0$ and $K_{X'} + \nu_*^{-1} B_0 + E_0 \equiv_X 0$, then $\nu_*^{-1} B + E - \nu_*^{-1} B_0 - E_0 \equiv_X 0$ and $\nu_*(\nu_*^{-1} B + E - \nu_*^{-1} B_0 - E_0) \geq 0$. By the Negativity Lemma 8.2.1, $\nu_*^{-1} B + E \geq \nu_*^{-1} B_0 + E_0$ and the claim follows.

Definition 8.5.13 If X is a normal surface and $K_{X'} = \nu^* K_X$ where $\nu : X' \to X$ is the minimal resolution, then we say that X has *Du Val singularities* .

Proposition 8.5.14 *If (X, B) is numerically terminal, then X is smooth and $\mathrm{mult}_P(B) < 1$ for any closed point $P \in X$.*

Proof By Remark 8.5.12, $(X, 0)$ is numerically terminal. Consider the minimal resolution $\nu : X' \to X$ and write $K_{X'} = \nu^* K_X + E$, then $E \geq 0$ and its support is the full exceptional locus (by definition of numerically terminal singularities). Since $E \equiv_X K_{X'}$ is exceptional and nef over X, by the Negativity Lemma 8.2.1, $E \leq 0$ and hence $E = 0$ so that $X' \to X$ is an isomorphism and X is smooth. If

$\mathrm{mult}_P(B) \geq 1$, and $\mu : Y \to X$ is the blow up of $P \in X$, then $\alpha(E; X, B) \leq 1$ where E is the exceptional divisor of μ. This is a contradiction so we may assume that $\mathrm{mult}_P(B) < 1$. □

Proposition 8.5.15 *If* $(0 \in X, B)$ *is a numerically canonical germ, then either* $B = 0$*,* X *is Du Val and* K_X *is Cartier, or* X *is smooth and* $\mathrm{mult}_0(B) \leq 1$.

Proof If $B \neq 0$, then $(X, 0)$ is numerically terminal and hence smooth. If $\mathrm{mult}_P(B) > 1$, and $\mu : Y \to X$ is the blow up of $0 \in X$, then $\alpha(E; X, B) < 1$ where E is the exceptional divisor of μ. This is a contradiction so we may assume that $\mathrm{mult}_P(B) \leq 1$.

If $B = 0$ and X is singular, then let $\nu : X' \to X$ be the minimal resolution and write $K_{X'} \equiv_X E$, then $E \geq 0$ (by definition of numerically canonical singularities). Since, as observed above, $K_{X'}$ is nef over X, then by the Negativity Lemma 8.2.1, $E \leq 0$ and hence $E = 0$ so that $(0 \in X)$ is (numerically) a Du Val singularity. Since $K_{X'}$ is nef and big over X, then it is semiample over X by the base point free theorem (see eg. [81, Theorem 3.3] or Theorem 8.5.6). Since $K_{X'} \equiv_X 0$, then $K_{X'} = \nu^* H$ where H is Cartier on X. But then $K_X = \nu_* K_{X'}$ is Cartier. □

Proposition 8.5.16 *If* (X, B) *is a numerically klt germ, then* X *is* $\mathbb{Q}$*-factorial and in particular* (X, B) *is klt.*

Proof See the proof of Corollary 8.5.7. □

Proposition 8.5.17 *If* (X, B) *is a numerically log canonical germ, then* $K_X + B$ *is* $\mathbb{Q}$*-Cartier and in particular* (X, B) *is log canonical.*

Proof We work locally around a point $x \in X$. If $x \in \mathrm{Supp}(B)$, then $(X, 0)$ is numerically klt and hence X is $\mathbb{Q}$-factorial and we are done. Thus, we may assume that $B = 0$ and $(X, 0)$ is not klt. This case is harder, but follows from the classification of these singularities see [78, §3]. Note that by [78, 3.3.4], if X is a rational singularity and N is the least common denominator of all the log discrepancies, then NK_X is Cartier. □

8.6 Terminal Threefold Singularities

Since terminal surfaces are smooth, it follows that terminal varieties are smooth in codimension 2 and in particular terminal 3-folds have only isolated singularities. We also know that terminal singularities are klt and hence rational (see Theorem 8.9.3). Let $(P \in X)$ be the germ of a terminal 3-fold singularity, the index of $(P \in X)$ is the smallest positive integer $r \in \mathbb{N}$ such that rK_X is Cartier. One can construct special examples of terminal 3-fold singularities of index 1 by deforming Du Val surface singularities.

Definition 8.6.1 ([102, §3]) A *cDV singularity* is a 3-fold hypersurface singularity $P \in X = \{F = 0\} \subset \mathbb{A}^4$ defined by an equation of the form

$$F(x, y, z, t) = f(x, y, z) + tg(x, y, z, t)$$

where f is the equation of a Du Val singularity (see Theorem 8.5.9) and g is an arbitrary polynomial.

It turns out that the terminal 3-fold singularities of index 1 are cDV singularities and terminal 3-fold singularities of higher index $r > 1$ are μ_r-quotients of cDV singularities where $\mu_r = \mathbb{Z}/r\mathbb{Z}$.

Lemma 8.6.2 *Let $(P \in X)$ be a terminal 3-fold singularity of index* 1*, then $(P \in X)$ is a cDV singularity.*

Proof See eg. [81, Lemma 5.30]. □

Theorem 8.6.3 ([102, §3]) *Let $(P \in X)$ be a terminal 3-fold singularity of index r, then there exists a cDV singularity $Q \in Y$ and a μ_r-action which is free on the complement of Q such that $Y/\mu_r \cong X$.*

Note that the cover $Y \to X$ is just the canonical cover, i.e. the cyclic cover defined by the isomorphism $(\mathcal{O}_X(K_X))^{[\otimes r]} \cong \mathcal{O}_X$. Here $(\mathcal{O}_X(K_X))^{[\otimes r]} = (((\mathcal{O}_X(K_X))^{\otimes r})^{**}$ is the reflexive hull of $(\mathcal{O}_X(K_X))^{\otimes r}$.

The action of μ_r on the coordinates x, y, z, t can be described explicitely and it follows that every terminal 3-fold singularity $(P \in X)$ has a Q-smoothing, that is there is a flat deformation X_t with $k \geq 1$ singular points such that for $t \neq 0$ all singular points $P_t^i \in X_t$ are terminal cyclic quotient singularities $\mathbb{A}^3/\mu_{r_i}$ where the action is given by

$$\mu_{r_i} \ni \xi : (x, y, z) \to (\xi x, \xi^{-1} y, \xi^{b_i} z)$$

where $0 < b_i \leq r_i/2$ and b_i is coprime with r_i. We say that $\frac{1}{r_i}(1, -1, b_i)$ is the type of the singularity P_t^i and $\{(b_1, r_1), \ldots, (b_k, r_k)\}$ is the basket of singularities. This basket is determined by the type of the singularity $(P \in X)$. For example a singularity of type $cAx/4$ with general elephant of type D_{2k+1} gives rise to the basket $\{(1, 4), (1, 2) \times k\}$. We have the following.

Theorem 8.6.4 (Reid's Riemann Roch [102, §11]) *Let X be a projective 3-fold with terminal singularities, then*

$$\chi(\mathcal{O}_X(mK_X)) = \frac{1}{12} m(m-1)(2m-1)K_X^3 + (2m-1)\chi(\mathcal{O}_X) + l(m)$$

where

$$l(m) = \sum_{i=1}^{k}\sum_{j=1}^{m-1} \frac{\overline{jb_i}(r_i - \overline{jb_i})}{2r_i}$$

and $\overline{jb_i}$ is the residue of jb_i modulo r_i.

The integers b_i, r_i are determined by the singularities of X. Even though this formula appears to be complicated, it has been used very effectively to prove optimal results about the birational geometry of 3-folds. For example J. Chen and M. Chen show the following.

Theorem 8.6.5 ([21]) *Let X be a smooth complex projective 3-fold of general type. Then $h^0(12K_X) > 0$, $|mK_X|$ defines a birational map for $m \geq 126$, and* $\mathrm{vol}(K_X) \geq \frac{1}{63\cdot 126^2}$.

8.7 Birational Models

If $\dim X \geq 3$ it is no longer clear that X has a minimal resolution, however thanks to the minimal model program several natural partial resolutions exist. We have the following (cf. [14, Corollary 1.4.3]).

Theorem 8.7.1 *Let (X, B) be a klt pair and Ω a collection of divisors P over X such that $a(P; X, B) \leq 1$, then there exists a projective birational morphism $\nu : X' \to X$ such that X' is $\mathbb{Q}$-factorial and* $\mathrm{Ex}(\nu) = \cup_{P\in\Omega} P$.

Proof Let $\mu : \tilde{X} \to X$ be a log resolution such that any prime divisor $P \in \Omega$ is a divisor on $\tilde{X}$. Let $K_{\tilde{X}} + \tilde{B} = \mu^*(K_X + B)$ and set $\tilde{\Delta} = \tilde{B}^{\geq 0} + \epsilon \sum_{P\notin\Omega} P$. We run the $(K_{\tilde{X}} + \tilde{\Delta})$-mmp over X. Let $\phi : \tilde{X} \dashrightarrow X'$ be the output of this mmp, in particular X' is $\mathbb{Q}$-factorial. Since $K_{\tilde{X}} + \tilde{\Delta} \equiv_X \tilde{F}$ where $\tilde{F} = -\tilde{B}^{\leq 0} + \epsilon \sum_{P\notin\Omega} P \geq 0$ is supported on the μ-exceptional divisors not contained in Ω, it follows that $K_{X'} + \Delta' \equiv_X F'$ where $\Delta' = \phi_*\tilde{\Delta}$ and $F' = \phi_*\tilde{F} \geq 0$. Since F' is effective and ν-exceptional and $K_{X'} + \Delta'$ is nef over X, it follows from the Negativity Lemma 8.2.1, that $F' = 0$. Thus ϕ contracts all μ-exceptional divisors not in Ω. Since each divisorial contraction is $\tilde{F}$ negative, it follows that ϕ only contracts divisors contained in the support of $\tilde{F}$. Thus $\mathrm{Ex}(\nu) = \cup_{P\in\Omega} P$. □

This leads to the following two useful constructions.

Remark 8.7.2 If $\Omega = \emptyset$, then $\nu : X' \to X$ is a small birational map (i.e. an isomorphism in codimension 1) such that X' is $\mathbb{Q}$-factorial. We say that X' is a *$\mathbb{Q}$-factorialization* of X. Note that this is not uniquely determined and two $\mathbb{Q}$-factorializations are connected by a sequence of flops.

Remark 8.7.3 If $\Omega = \{P|a(P; X, B) \leq 1\}$, then $\nu : X' \to X$ is a birational map such that X' is $\mathbb{Q}$-factorial and $K_{X'} + B' = \nu^*(K_X + B)$ where $B' \geq 0$. Let Q be any divisor exceptional over X', then $a(Q; X', B') = a(Q; X, B) > 1$ and hence (X', B') is terminal. We say that X' is a *terminalization* of (X, B). Note that this is not uniquely determined and two terminalizations are connected by a sequence of flops.

It is much harder to construct interesting partial resolutions of non klt pairs. The following however are some very useful options.

Definition 8.7.4 Let $(X, B = \sum b_i B_i)$ be a pair where $0 \leq b_i \leq 1$ and $\nu : X' \to X$ be a proper, birational morphism and $B' = \nu_*^{-1} B + \mathrm{Ex}(\nu)$, then $\nu : (X', B') \to (X, B)$ is a *log canonical modification* (resp. a *dlt modification*) if (X', B') is log canonical (resp. dlt) and $K_{X'} + B'$ is ν-ample (resp. ν-nef).

Theorem 8.7.5 *Let (X, B) be a pair, then there exist log canonical and dlt modifications of (X, B).*

Proof See [101, Theorem 1.1 and §2]. □

Remark 8.7.6 It is conjectured that we may remove the hypothesis that $K_X + B$ is $\mathbb{R}$-Cartier from Theorem 8.7.5.

Remark 8.7.7 If (X, B) is log canonical and $\nu : X' \to X$ is a dlt modification, then $K_{X'} + B' = \nu^*(K_X + B)$. To see this, write $E = \nu^*(K_X + B) - (K_{X'} + B')$, then E is ν-exceptional and $-E$ is nef over X so that E is effective. If $E \neq 0$, then $(X', B' + E)$ is not log canonical (as $B' = \nu_*^{-1} B + \mathrm{Ex}(\nu)$) and so (X, B) is not log canonical (as $K_{X'} + B' + E = \nu^*(K_X + B)$). This is a contradiction and so $E = 0$.

Remark 8.7.8 Note that the log canonical modification is unique. To see this, let $\mu : \tilde{X} \to X$ be any log resolution and $\tilde{B} = \mu_*^{-1} B + \mathrm{Ex}(\mu)$, then one easily checks that the log canonical modification is defined by $X^{\mathrm{lc}} = \mathrm{Proj}_X(R(K_{\tilde{X}} + \tilde{B}))$ (see [101, Proposition 2.3]).

On the other hand dlt modifications are not unique, however any two dlt modifications are crepant birational, that is, if $\nu : X' \to X$ and $\mu : X'' \to X$ are two dlt modifications, and $p : W \to X', q : W \to X''$ is a common resolution, then $p^*(K_{X'} + B') = q^*(K_{X''} + B'')$. To see this, note that it follows from the proof of [101, Theorem 1.1] that $r : X' \dashrightarrow X^{\mathrm{lc}}$, is the log canonical model of (X', B') over X. Since $K_{X'} + B'$ is nef over X, then r is a morphism and $K_{X'} + B' = r^*(K_{X^{\mathrm{lc}}} + B^{\mathrm{lc}})$. Similarly $s : X'' \to X^{\mathrm{lc}}$ is a morphism and $K_{X''} + B'' = s^*(K_{X^{\mathrm{lc}}} + B^{\mathrm{lc}})$. But then $p^*(K_{X'} + B') = \upsilon^*(K_{X^{\mathrm{lc}}} + B^{\mathrm{lc}}) = q^*(K_{X''} + B'')$ where $\upsilon = r \circ p = s \circ q$.

We remark that by Remark 8.7.2, there is a $\mathbb{Q}$-factorial dlt model.

Finally note that for log canonical models, the exceptional locus has pure codimension one (since $K_{X^{\mathrm{lc}}} + B^{\mathrm{lc}} - \nu^*(K_X + B) = -F$ where F is effective, exceptional and relatively ample). There are however dlt models for which the exceptional locus is not of pure codimension 1. By [28, Lemma 29] a dlt modification such that the exceptional locus has pure codimension one always exists.

Definition 8.7.9 Let $P \in (X, B)$ be a klt singularity and $\nu : X' \to X$ a proper birational morphism. A divisor S on X' is a *Kollár component* if ν is an isomorphism over $X \setminus \{P\}$, and $\nu^{-1}(P)$ is an irreducible divisor S, such that $(X', S + \nu_*^{-1}B)$ is purely log terminal and $-S$ is $\mathbb{Q}$-Cartier and ample over X. Let $K_S + B_S := (K_{X'} + S + \nu_*^{-1}B)|_S$, then (S, B_S) is a klt Fano variety. (The adjunction $K_S + B_S := (K_{X'} + S + \nu_*^{-1}B)|_S$ is discussed in more detail in Sect. 8.8 below.)

Kollár components exist (for any klt singularity $P \in (X, B)$) see [121, Lemma 1].

8.8 Adjunction

Let $(X, S + B)$ be a (sub-)pair where S is a prime divisor not contained in the support of B, then we would like to define and study the "adjunction" $K_S + B_S = (K_X + S + B)|_S$. As a warm up, we consider the case where $(X, S + B)$ is log smooth. In this case it suffices to let $B_S = B|_S$. It is clear that the coefficients of components of B_S are the same as the coefficients of the corresponding components of B and in particular (S, B_S) is klt/lc iff $(X, S + B)$ is plt/lc on a neighborhood of S. In general, it could happen that S is not normal. In this case (S, B_S) can not be a pair (regardless of the choice of B_S) and so we aim to define

$$K_{S^{\mathrm{n}}} + B_{S^{\mathrm{n}}} = (K_X + S + B)|_{S^{\mathrm{n}}}$$

where $n : S^{\mathrm{n}} \to S$ is the normalization. Let $\nu : X' \to X$ be a log resolution, $S' = \nu_*^{-1}S$ and write $K_{X'} + S' + B' = \nu^*(K_X + S + B)$. We let $K_{S'} + B_{S'} = (K_{X'} + S' + B')|_{S'}$. The morphism $S' \to S$ factors through the birational morphism of normal varieties $g : S' \to S^{\mathrm{n}}$ and we let $K_{S^{\mathrm{n}}} + B_{S^{\mathrm{n}}} := g_*(K_{S'} + B_{S'})$. It follows easily from the projection formula that $K_{S^{\mathrm{n}}} + B_{S^{\mathrm{n}}} \sim (K_X + S + B)|_{S^{\mathrm{n}}}$. We have the following important facts.

Theorem 8.8.1 *Let $(X, S+B)$ be a pair where S is a prime divisor not contained in the support of B and $n : S^{\mathrm{n}} \to S$ is the normalization. Then there exists an effective $\mathbb{R}$-divisor $B_{S^{\mathrm{n}}}$ defined as above such that*

1. *$B_{S^{\mathrm{n}}}$ is well defined,*
2. *$K_{S^{\mathrm{n}}} + B_{S^{\mathrm{n}}} \sim (K_X + S + B)|_{S^{\mathrm{n}}}$,*
3. *if (X, S) is log canonical of index m at a closed point $P \in S$, then* $\mathrm{mult}_P(B_{S^{\mathrm{n}}}) = 1 - \frac{1}{m} + \frac{\sum b_i m_i}{m}$ *where $B = \sum b_i B_i$ and $m_i \in \mathbb{N}$.*

Proof With the above notation, let P be any codimension 1 point on S^{n} and P' be the corresponding point on S'. We may assume that that $P' = E|_{S'}$ where E is a ν-exceptional prime divisor on X' intersecting S' transversely. We then have

$$\mathrm{mult}_P(B^{\mathrm{n}}) = \mathrm{mult}_{P'}(B_{S'}) = \mathrm{mult}_E(B') = 1 - a(E; X, B).$$

Note that if we blow up $Q = S' \cap E$, we obtain a new exceptional divisor F such that $a(F; X, B) = a(E; X, B)$. It follows easily that B_{S^n} is well defined. By an easy surface computation it also follows that $a(E; X, B) \leq 1$ and hence $\mathrm{mult}_{P'}(B_{S'}) = \mathrm{mult}_E(B') \geq 0$. It is also easy to see by a surface computation that B_{S^n} is independent of the chosen log resolution.

(3) follows from the classification of lc surface singularities, see [78, §16]. □

We illustrate the above with the following example.

Example 8.8.2 Let X be the cone over a rational curve of degree d, $\nu : X' \to X$ the blow up of the vertex $P \in X$ and $E \subset X'$ the exceptional rational curve with $E^2 = -d$. We have $K_{X'} + (1 - \frac{2}{d})E = \nu^* K_X$ and if L is a line through P, then $\nu^* L = L' + \frac{1}{d}E$ where $L' = \nu_*^{-1} L$. Note in fact that

$$(K_{X'} + (1 - \frac{2}{d})E) \cdot E = (K_{X'} + E) \cdot E - \frac{2}{d} E \cdot E = -2 - \frac{2}{d} \cdot (-d) = 0$$

and similarly $(L' + \frac{1}{d}E) \cdot E = 1 + \frac{1}{d} \cdot (-d) = 0$. It follows that

$$\nu^*(K_X + L)|_{L'} \cong (K_{X'} + L' + (1 - \frac{1}{d})E)|_{L'} \cong K_L + (1 - \frac{1}{d})P.$$

Remark 8.8.3 It is not surprising that given information about the structure of the pair $(X, S + B)$ one can deduce information about the structure of (S^n, B_{S^n}). For example it is easy to see that if $(X, S + B)$ is plt (or lc), then (S^n, B_{S^n}) is klt (or lc). This follows simply by observing that the coefficients of B' are < 1 (or ≤ 1) and hence the coefficients of $B_{S'}$ are also < 1 (or ≤ 1).

Another pleasant feature is that the coefficients of B_{S^n} are controlled in terms of the coefficients of B. This is very useful in proofs by induction on the dimension.

It is more surprising that in fact information on (S^n, B_{S^n}) also translates in to information on $(X, S + B)$ in a neighborhood of S (see eg. Theorems 8.9.4 and 8.9.5). This is extremely important in proofs by induction on the dimension.

Remark 8.8.4 (Dlt Adjunction) Let (X, B) be a dlt pair, then the log canonical centers of B are irreducible components of the strata of $\lfloor B \rfloor$, i.e. if $\lfloor B \rfloor = S_1 + \ldots + S_n$, then the log canonical centers of (X, B) are the irreducible components V of $S_{i_1} \cap \ldots \cap S_{i_k}$ where $1 \leq i_1 < \ldots < i_k \leq n$. It is known that each such component V is normal. By repeatedly applying Theorem 8.8.1, we obtain an adjunction formula $(K_X + B)|_V = K_V + B_V$ where (V, B_V) is a dlt pair and the coefficients of B_V are determined by those of B similarly to (3) of Theorem 8.8.1.

8.9 Vanishing Theorems

One of the reasons why the minimal model program for complex projective varieties is so effective and why the singularities of the minimal model program are so well behaved is because of the Kawamata-Viehweg vanishing theorem and its generalizations.

Theorem 8.9.1 (Relative Kawamata-Viehweg Vanishing) *Let (X, B) be a klt pair, D a $\mathbb{Q}$-Cartier Weil divisor, and $f : X \to Z$ a proper morphism such that $D - (K_X + B)$ is f-nef and f-big, then $R^i f_*\mathcal{O}_X(D) = 0$ for all $i > 0$.*

Proof This follows from [81, Theorem 2.70]. □

To illustrate the power of this vanishing theorem we recall the following easy but important consequence.

Theorem 8.9.2 (Kollár-Shokurov Connectedness Theorem) *Let (X, B) be a sub-klt pair and $f : X \to Z$ a proper morphism such that $f_*\mathcal{O}_X = \mathcal{O}_Z$ and $-(K_X + B)$ is f-nef and big. Suppose that $B = B^+ - B^-$ where $B^+, B^- \geq 0$ are effective without common components and $f_*\mathcal{O}_X(\lceil B^-\rceil) = \mathcal{O}_Z$, then the fibers of the non-klt locus of (X, B) over Z are connected.*

We remark that if $\operatorname{codim}(f(B^-)) \geq 2$, then $f_*\mathcal{O}_X(\lceil B^-\rceil)$ is a torsion free sheaf containing $f_*\mathcal{O}_X = \mathcal{O}_Z$. Since these two sheaves agree in codimension 1 and Z is normal, then they are equal i.e. $f_*\mathcal{O}_X(\lceil B^-\rceil) = \mathcal{O}_Z$.

Proof Let $\nu : X' \to X$ be a log resolution, $K_{X'} + B' = \nu^*(K_X + B)$. Note that

$$\mathcal{O}_Z \subset f_*\nu_*\mathcal{O}_{X'}(\lceil B'^-\rceil) \subset f_*\mathcal{O}_X(\lceil B^-\rceil) = \mathcal{O}_Z,$$

and so $f_*\nu_*\mathcal{O}_{X'}(\lceil B'^-\rceil) = \mathcal{O}_Z$. Since $-(K_{X'} + B')$ is nef and big over Z and the non-klt locus of (X, B) is the image of the non-klt locus of (X', B'), we may replace (X, B) by (X', B') and thus we may assume that (X, B) is log smooth.

We let $A = -\lfloor B\rfloor$, then $A = K_X + \{B\} - (K_X + B)$. We will write $A = E - F$ where $E = \lceil B^-\rceil$ and $F = \lfloor B^+\rfloor$ are effective with no common components. Note that the support of F is the non-klt locus of (X, B). Consider the short exact sequence

$$0 \to \mathcal{O}_X(A) \to \mathcal{O}_X(A + F) \to \mathcal{O}_F(A + F) \to 0.$$

By Theorem 8.9.1, $R^1 f_*\mathcal{O}_X(A) = 0$ and so, since $A + F = E$, $f_*\mathcal{O}_X(E) \to f_*\mathcal{O}_F(E)$ is surjective. Since $E = \lceil B^-\rceil$ is effective and $f_*\mathcal{O}_X(E) \cong \mathcal{O}_Z$, the inclusion $\mathcal{O}_Z = f_*\mathcal{O}_X \to f_*\mathcal{O}_F(E)$ is also surjective. Suppose that a fiber of F over a neighborhood of a point $z \in f(F)$ is disconnected then we may write $F = F_1 \cup F_2$ and it follows that $f_*\mathcal{O}_F(E) = f_*\mathcal{O}_{F_1}(E) \oplus f_*\mathcal{O}_{F_2}(E)$ where neither summand vanishes. This leads to an immediate contradiction. □

The following result shows that dlt singularities behave similarly to smooth varieties at least from the cohomological point of view.

Theorem 8.9.3 *Let (X, B) be a dlt pair, then X has rational singularities.*

Proof See [81, Theorem 5.22] and references therein. □

Theorem 8.9.4 *Let $(X, S + B)$ be a pair such that $\lfloor B \rfloor = 0$ and S is irreducible. Then $(X, S + B)$ is plt on a neighborhood of S iff (S, B_S) is klt where $K_S + B_S = (K_X + S + B)|_S$ is defined by adjunction. In particular S is normal.*

Proof Suppose that $\nu : X' \to X$ is a log resolution, $S' = \nu_*^{-1} S$ and write $K_{X'} + S' + B' = \nu^*(K_X + S + B)$. If $(X, S + B)$ is plt on a neighborhood of S, then $\lfloor B' \rfloor \leq 0$ over a neighborhood of S and so $\lfloor B'|_{S'} \rfloor \leq 0$. Let $\mu = \nu|_{S'}$ and $F = -\lfloor B' \rfloor$. Consider the short exact sequence

$$0 \to \mathcal{O}_{X'}(F - S') \to \mathcal{O}_{X'}(F) \to \mathcal{O}_{S'}(F|_{S'}) \to 0.$$

Since $F - S' = -\lfloor S' + B' \rfloor = K_{X'} + \{B'\} + \nu^*(K_X + B)$ where $(X', \{B'\})$ is klt, then by Theorem 8.9.1, $R^1 \nu_* \mathcal{O}_{X'}(F - S') = 0$ and so $\phi : \nu_* \mathcal{O}_{X'}(F) \to \nu_* \mathcal{O}_{S'}(F|_{S'})$ is surjective. Since F is ν-exceptional, $\nu_* \mathcal{O}_{X'}(F) = \mathcal{O}_X$ and so ϕ factors through $\mathcal{O}_S$. But then $\mathcal{O}_S \cong \nu_* \mathcal{O}_{S'}(F|_{S'})$ which implies that $\nu_* \mathcal{O}_{S'} = \mathcal{O}_S$ and hence S is normal. Since $\mu^*(K_S + B_S) = K_{S'} + B'|_{S'}$, then (S, B_S) is klt.

Suppose now that (S, B_S) is klt, then we know that $\lfloor B' \rfloor|_{S'} = \lfloor B'|_{S'} \rfloor \leq 0$. By Theorem 8.9.2, it follows that no component of $\lfloor B' \rfloor$ intersects fibers $\nu^{-1}(s)$ for $s \in S$. Thus $\nu(\lfloor B' \rfloor) \cap S = \emptyset$ and hence $(X, S + B)$ is plt on a neighborhood of S. □

Theorem 8.9.5 *Let $(X, S + B)$ be a pair such that S is irreducible, $B \in [0, 1]$ and the support of B does not contain S. Then $(X, S + B)$ is lc on a neighborhood of S iff (S^ν, B_{S^ν}) is lc where $S^\nu \to S$ is the normalization and $K_{S^\nu} + B_{S^\nu} = (K_X + S + B)|_{S^\nu}$ is defined by adjunction.*

Proof See [67] (and [43] for a strengthening of this result). □

Remark 8.9.6 It is not the case that the non-klt locus W of a log canonical pair (X, B) is normal, however, it follows from an argument similar to Theorem 8.9.4, that W is seminormal (see [74, Theorem 8.9.1] and [2]).

Theorem 8.9.7 *Let (X, B) be a log canonical pair then (X, B) has finitely many log canonical centers and*

1. *every irreducible component of the intersection of two log canonical centers is a log canonical center, and*
2. *any minimal log canonical center (with respect to inclusion) is normal.*

Proof See [70], [4, Propositions 4.7 and 4.8], [35, Theorem 9.1], [38]. We will assume for convenience that (X, B_0) is klt for some B_0. Let W_1, W_2 be two log canonical centers of (X, B) and W an irreducible component of $W_1 \cap W_2$. We may assume that W_i are minimal centers containing W. By an easy perturbation argument we may in fact assume that W_1, W_2 are the only log canonical centers containing W. Let $\nu : X' \to X$ be a $\mathbb{Q}$-factorial dlt modification so that $(X', B' = \nu_*^{-1}B + \operatorname{Ex}(\nu))$ is dlt and $K_{X'} + B' = \nu^*(K_X + B)$ (see Theorem 8.7.5 and Remark 8.7.7). We may write $\operatorname{Ex}(\nu) = E_1 + E_2$ where the components of E_i map on to W_i. By the Kollár-Shokurov Connectedness Theorem 8.9.2, $E_1 \cap E_2 \neq \emptyset$. Let V be a component of $E_1 \cap E_2$, then V is a log canonical center of (X', B') and $\nu(V) \subset W_1 \cap W_2$. Thus $\nu(V)$ is a log canonical center of (X, B) which is strictly contained in $W_1 \cap W_2$. This is a contradiction and (1) is proved.

For (2), we argue as follows. By an easy perturbation argument, we may assume that W is the unique log canonical center and if $\nu : X' \to X$ is a dlt modification, then $\operatorname{Ex}(\nu) = S$ is an irreducible divisor. In particular $(X', S + \nu_*^{-1}B)$ is plt, S is normal and $(X', \nu_*^{-1}B)$ is klt. Consider the short exact sequence

$$0 \to \mathcal{O}_{X'}(-S) \to \mathcal{O}_{X'} \to \mathcal{O}_S \to 0.$$

Since $-S \equiv_X K_{X'} + \nu_*^{-1}B$, by Theorem 8.9.1, $R^1\nu_*\mathcal{O}_{X'}(-S) = 0$. Therefore, $\nu_*\mathcal{O}_{X'} \to \nu_*\mathcal{O}_S$ is surjective. But then $\nu_*\mathcal{O}_S = \mathcal{O}_W$. Since S is normal, the morphism $S \to W$ factors through the normalization $h : W^{\mathrm{n}} \to W$ and hence $\nu_*\mathcal{O}_S = h_*\mathcal{O}_{W^{\mathrm{n}}} = \mathcal{O}_W$ and so W is normal. □

8.10 Du Bois Singularities

In contrast to klt singularities, log canonical singularities appear to be much harder to study. For instance, log canonical singularities are often not rational or even Cohen Macaulay, however they are Du Bois singularities. The Du Bois condition is a weak version of rationality that was introduced in [114]. The definition of Du Bois singularities is technical and we do not reproduce it here, however we recall the following useful and well known criterion for Du Bois singularities, see [115], and [104].

Theorem 8.10.1 *Let X be a reduced separated scheme of finite type over a field of characteristic zero. Embed X in a smooth scheme Y and let $\pi : \tilde{Y} \to Y$ be a log resolution of X in Y that is an isomorphism outside of X. If E is the reduced pre-image of X in $\tilde{Y}$, then X has Du Bois singularities if and only if the natural map $\mathcal{O}_X \to R\pi_*\mathcal{O}_E$ is a quasi-isomorphism.*

When X is normal and Cohen-Macaulay, there is an even simpler description.

Theorem 8.10.2 ([83]) *Suppose that X is a normal Cohen-Macaulay variety. Let $\nu : X' \to X$ be any log resolution and denote the reduced exceptional divisor of by E. Then X has Du Bois singularities if and only if $\nu_*\omega_{X'}(E) = \omega_X$.*

It turns out that any log canonical pair (X, B) is Du Bois and so is any union of its log canonical centers.

Theorem 8.10.3 ([80, Theorem 1.4]) *Let (X, B) be a log canonical pair. Then X is Du Bois, and if $W \subset X$ is a reduced, closed subscheme that is a union of log canonical centers of (X, B), then W is Du Bois.*

This has many important consequences.

Corollary 8.10.4 ([80, Corollary 1.1]) *If (X, B) is log canonical, then the natural map*

$$H^i(X^{an}, \mathbb{C}) \to H^i(X^{an}, \mathcal{O}_{X^{an}}) \cong H^i(X, \mathcal{O}_X)$$

is surjective for all i.

Corollary 8.10.5 ([80, Corollary 1.1]) *Let $f : X \to T$ be a proper, flat morphism of complex varieties with T connected. Assume that for all $t \in T$ there exists a $\mathbb{Q}$-divisor D_t on X_t such that (X_t, D_t) is log canonical. Then*

1. *$h^i(X_t, \mathcal{O}_{X_t})$ is independent of $t \in T$,*
2. *if X_t is Cohen-Macaulay (resp S_k) for some $t \in T$, then X_t is Cohen-Macaulay (resp S_k) for every $t \in T$.*

An important consequence of Theorem 8.10.1 is that Kodaira vanishing holds for projective varieties with Du Bois singularities (we thank K. Schwede for sharing the arguments below).

Corollary 8.10.6 *Let X be a projective equidimensional scheme of finite type over a field of characteristic zero. If X has Du Bois singularities and L is an ample line bundle, then $H^i(\omega_X \otimes L) = 0$ for $i > 0$.*

Proof By Theorem 8.10.1, if we embed X in a smooth scheme Y and consider a log resolution $\pi : \tilde{Y} \to Y$ where E is the reduced preimage of X, then $\mathcal{O}_X \to R\pi_*\mathcal{O}_E$ is a quasi-isomorphism. By duality

$$R\pi_*\omega_E = R\pi_* R\mathcal{H}om(\mathcal{O}_E, \omega_E) = R\mathcal{H}om(R\pi_*\mathcal{O}_E, \omega^{\cdot}_X) = \omega^{\cdot}_X$$

and so $\omega_X = R^f\pi_*\omega_E$ where $f = \dim E - \dim X$. By [4, Theorem 3.2] or [36], we have $H^i(\omega_X \otimes L) = H^i(R^f\pi_*\omega_E \otimes L) = 0$ for $i > 0$. □

8.11 Bertini Theorems

Let X be a projective variety and $|L|$ a very ample or base point free linear series, then one expects that the singularities of X and a general divisor in $|L|$ behave similarly. For example recall the following Bertini Theorem.

Theorem 8.11.1 *Suppose that X is a smooth projective variety such and $|L|$ is a base point free linear series, then any general divisor $H \in |L|$ is a smooth subvariety of X.*

More generally, we expect that singularities of a divisor $B \in L$ (or of the pair (X, B)) behave in a semi-continuous way. The next result shows that this is indeed the case .

Theorem 8.11.2 ([73, Theorem 4.5.1]) *Let X be a normal $\mathbb{Q}$-Gorenstein projective variety, L a $\mathbb{Q}$-Cartier Weil divisor such that and $V \subset |L|$ a linear series. Assume that for every $x \in X$ there is a $B_x \in V$ such that (X, B_x) is canonical at x, then (X, B) is canonical for any general member $B \in V$.*

Proof Assume that $\dim X \geq 2$. Suppose that $V = F + V^*$ where F is a divisor and $\mathrm{Bs}(V^*)$ has codimension ≥ 2. Let $x \in \mathrm{Bs}(V^*)$, then we claim that $x \notin F$. Suppose that this is not the case, then for any $B^* \in V^*$, $x \in B^* \cap F \neq \emptyset$. Let $B_x = F + B^*_x$, then $(X, F + B^*_x)$ is canonical and in particular log canonical (see Lemma 8.3.8) and so F and B^*_x have no common components. By Theorem 8.9.7 there is a log canonical center $x \in W \subset F \cap B^*_x$ for $(X, B_x = F + B^*_x)$. This is impossible since (X, B_x) is canonical and so we conclude that $\mathrm{Bs}(V^*) \cap F = \emptyset$.

Let $\nu : X' \to X$ be a log resolution of $(X, |V|)$ so that $\mathrm{Ex}(\nu)$ is a divisor, $\nu^*|V| = |M| + \nu^* F + G$ where $|M|$ is base point free and $\nu^* F + G + \mathrm{Ex}(\nu)$ has simple normal crossings support. By what we have seen above $\nu^* F \cap G = \emptyset$. We let $\nu^* B_x = \nu^{-1}_* B_x + \nu^* F + G + N_x$ and $K_{X'} = \nu^* K_X + E$, then

$$K_{X'} + \nu^{-1}_* B_x = \nu^*(K_X + B_x) + E - (\nu^* F + G + N_x)$$

where $E - (\nu^* F + G + N_x) \geq 0$ over a neighborhood of $x \in X$. In particular $E - (\nu^* F + G) \geq 0$. Let $B \in V$ be a general element, then $\nu^* B = \nu^* F + G + \tilde{B}$ where $\tilde{B} = \nu^{-1}_* B$ is smooth and intersects $E - G$ transversely (and is disjoint from $\nu^* F$). Since

$$K_{X'} + \nu^{-1}_* B = \nu^*(K_X + B) + E - (\nu^* F + G)$$

it follows that (X, B) is canonical. □

Remark 8.11.3 We expect that instead of assuming that X is $\mathbb{Q}$-Gorenstein and L is $\mathbb{Q}$-Cartier, it suffices to assume that $K_X + L$ is $\mathbb{Q}$-Cartier, but we do not pursue it here.

A similar result holds for klt and lc singularities (see [73, Theorem 4.8] for a more extensive treatment). Recall the following.

Definition 8.11.4 Let (X, B) be a pair, $V_i \subset |D_i|$ linear series where D_i are Cartier divisors and $v_i \geq 0$ real numbers. Let $\nu : X' \to X$ be a log resolution $\nu : X' \to X$ such that $\nu^* V_i = M_i + F_i$ where M_i is base point free and $F_i = \mathrm{Fix}(\nu^* V_i)$, $\nu^* B + \mathrm{Ex}(\nu) + \sum F_i$ has simple normal crossings. Write $K_{X'} + B' = \nu^*(K_X + B)$. We say that $(X, B + \sum v_i V_i)$ is klt (resp. lc) iff the coefficients of $B' + \sum v_i F_i$ are < 1 (resp. ≤ 1).

Theorem 8.11.5 *Let $(X, B = \sum b_i B_i)$ be a pair, D_i be $\mathbb{Q}$-Cartier Weil divisors, and $0 \le d_i \le 1$ real numbers. If $D_i^g \in |D_i|$ are general, then*

1. *$(X, B + \sum d_i|D_i|)$ is lc iff $(X, B + \sum d_i D_i^g)$ is lc,*
2. *if $d_i < 1$, then $(X, B + \sum d_i|D_i|)$ is klt iff $(X, B + \sum d_i D_i^g)$ is klt,*
3. *if $B = 0$, then $(X, \sum d_i|D_i|)$ is canonical iff $(X, \sum d_i D_i^g)$ is canonical.*

Proof We will focus on the first two statements (see [73, Theorem 4.8] for further details). Let $\nu : X' \to X$ be a log resolution such that $\nu^*|D_i| = M_i + F_i$ where M_i is free and F_i are divisors such that $\nu_*^{-1}B + \mathrm{Ex}(\nu) + \sum F_i$ has simple normal crossings support. If we write $\nu^* D_i^g = \tilde{D}_i^g + F_i$, then $\nu_*^{-1} D_i^g = \tilde{D}_i^g \in M_i$ is a general element and $\tilde{D}_i^g$ is smooth and intersects the support of $\nu_*^{-1}B + \mathrm{Ex}(\nu) + \sum F_i$ transversely. For ease on notation, we also denote the components of $\mathrm{Ex}(\nu)$ by F_j. We have

$$K_{X'} + \nu_*^{-1}B + \sum d_i \nu_*^{-1} D_i^g + \sum e_i F_i = \nu^*(K_X + B + \sum d_i D_i^g)$$

which is klt (resp. lc) iff $b_i, d_i, e_i < 1$ (resp. $b_i, d_i, e_i \le 1$). By our assumptions, this is equivalent to $b_i, e_i < 1$ (resp. $b_i, e_i \le 1$) and hence the claim follows. □

8.12 Deformation Properties of Singularities

Let $f : X \to T$ be a flat projective morphism where $0 \in T$ is a pointed smooth curve, then it is natural to ask how singularities of the fiber X_0 relate to singularities of X (over a neighborhood of $0 \in T$). It is well known that if X_0 is smooth then X is smooth in a neighborhood of X_0, X_0 is S_2 iff X is S_3 in a neighborhood of X_0, and if X_0 is normal then so is X in a neighborhood of X_0.

Remark 8.12.1 It is however not the case that klt singularities are always well behaved from this point of view. Let $f : X \to T$ be a projective morphism from a threefold to a curve and $g : X \to Z$ be a flipping contraction over T (see Definition 8.13.3). Suppose that C is the flipping curve, $f(C) = 0 \in T$, and (X, X_0) is plt, then $X_0 \to Z_0$ is a divisorial contraction and in particular K_{Z_0} is $\mathbb{Q}$-Cartier and Z_0 is klt (see the proof of Proposition 8.13.2(2)), however K_Z can not be $\mathbb{Q}$-Cartier as $X \to Z$ is small (see the proof of Proposition 8.13.2(3)) and hence Z is not klt. The following results show that this problem does not happen if we assume $\mathbb{Q}$-factoriality or canonical singularities.

Proposition 8.12.2 *Let $f : X \to T$ be a flat projective morphism, and (X, B) a pair such that $K_{X_0} + B_0 := (K_X + B)|_{X_0}$ is klt (resp. lc), then (after shrinking T in a neighborhood of 0), we may assume that (X, B) is klt (resp. lc) and so are the fibers (X_t, B_t) for $t \in T$.*

Proof By Theorem 8.9.4, $(X, X_0 + B)$ is plt (near X_0) and hence (X, B) is klt (near X_0). By Theorem 8.11.5 (X_t, B_t) is klt for all but finitely many $t \in T$ and hence for $t \in T$ near 0. □

The following result due to Nakayama [99] and Kawamata [71] is also known as deformation invariance of terminal singularities.

Theorem 8.12.3 *Let $f : X \to T$ be a flat morphism from the germ of an algebraic variety to a germ of a smooth curve. Assume that the central fiber $X_0 = f^{-1}(0)$ has only terminal/canonical singularities. Then so has the total space X as well as any fiber X_t of f.*

Proof A proof of this result can be found in [71]. Here we give a proof for the canonical case that relies on the results of the minimal model program [14] and [50]. Note however that the proof of the existence of flips in [50] relies on the techniques introduced in Siu [116] which in turn inspired the results of [99], [71] and ultimately [50].

Assume that X_0 is canonical. Let $\nu : \tilde{X} \to X$ be a log resolution of (X, X_0) and $\tilde{X}_0 = \nu_*^{-1} X_0$. After running a relative $(K_{\tilde{X}} + \tilde{X}_0)$-minimal model program say $\phi : \tilde{X} \dashrightarrow \bar{X}$, we may assume that $(\bar{X}, \bar{X}_0)$ is plt and $K_{\bar{X}} + \bar{X}_0 = \phi_*(K_{\tilde{X}} + \tilde{X}_0)$ is nef and big over X_0. Let $K_{\bar{X}_0} + \Delta_0 = (K_{\bar{X}} + \bar{X}_0)|_{\bar{X}_0}$ be defined by adjunction (Theorem 8.8.1), then $\Delta_0 \geq 0$. Pick $m > 0$ sufficiently divisible and consider the short exact sequence

$$0 \to \mathcal{O}_{\bar{X}}(K_{\bar{X}} + (m-1)(K_{\bar{X}} + \bar{X}_0)) \to \mathcal{O}_{\bar{X}}(m(K_{\bar{X}} + \bar{X}_0)) \to \mathcal{O}_{\bar{X}_0}(m(K_{\bar{X}_0} + \Delta_0)) \to 0.$$

By relative Kawamata-Viehweg vanishing (cf. Theorem 8.9.1) it follows that $R^1\mu_*\mathcal{O}_{\bar{X}}(K_{\bar{X}} + (m-1)(K_{\bar{X}} + \bar{X}_0)) = 0$ and hence

$$\mu_*\mathcal{O}_{\bar{X}}(m(K_{\bar{X}} + \bar{X}_0)) \to \mu_*\mathcal{O}_{\bar{X}_0}(m(K_{\bar{X}_0} + \Delta_0))$$

is surjective where $\mu : \bar{X} \to X$. Let s_0 be a nowhere vanishing section of $\mathcal{O}_{X_0}(mK_{X_0})$. Since X_0 is canonical and Δ_0 is an effective $\mathbb{Q}$-divisor, then there is an inclusion $\mathcal{O}_{X_0}(mK_{X_0}) \hookrightarrow \mu_*\mathcal{O}_{\bar{X}_0}(m(K_{\bar{X}_0} + \Delta_0))$. Thus s_0 extends to a s section of $\mu_*\mathcal{O}_{\bar{X}}(m(K_{\bar{X}} + \bar{X}_0)) \subset \mathcal{O}_X(m(K_X + X_0))$. This in turn implies that $\mathcal{O}_X(m(K_X + X_0))|_{X_0} = \mathcal{O}_{X_0}(mK_{X_0})$ and hence $K_X + X_0$ is $\mathbb{Q}$-Cartier. It also implies that the inclusion $\mu_*\mathcal{O}_{\bar{X}}(m(K_{\bar{X}} + \bar{X}_0)) \subset \mathcal{O}_X(m(K_X + X_0))$ is an isomorphism on a neighborhood of X_0 and hence that $K_{\bar{X}} + \bar{X}_0 \geq \mu^*(K_X + X_0)$ over a neighborhood of X_0 so that (X, X_0) is canonical. The claim now follows.

□

Remark 8.12.4 We refer the reader to [27] for several related results. For example [27, Theorem 3.2] shows that if the central fiber X_0 is terminal and $\mathbb{Q}$-factorial, then X is terminal and $\mathbb{Q}$-factorial on a neighborhood of X_0.

8.13 Singularities and the Minimal Model Program

The following theorem, known as the cone theorem for log canonical pairs, describes the closure of the convex cone

$$NE(X) = \{\sum c_i C_i \mid C_i \text{ a curve}, \ c_i \geq 0\}/\equiv$$

generated by effective curves on a normal projective variety X up to numerical equivalence. If D is an $\mathbb{R}$-Cartier divisor, then the symbol $\overline{NE}(X)_{D\geq 0}$ denotes all curve classes $[C] \in \overline{NE}(X)$ such that $C \cdot D \geq 0$. When (X, B) is log canonical, the following cone theorem gives us no information about the $(K_X + B)$-positive part of the cone $\overline{NE}(X)$, but it tells us (roughly speaking) that its $(K_X + B)$-negative part is generated by countably many rational curves of bounded degree with respect to $K_X + B$, that can only accumulate around the hyperplane $(K_X + B)^{\perp}$.

Theorem 8.13.1 (Cone Theorem [34, Theorem 1.4]) *Let (X, B) be a projective log canonical pair. Then there are (countably many) rational curves $C_j \subset X$ such that*

1. $0 < -(K_X + B) \cdot C_j \leq 2\dim X$, *and*
$$\overline{NE}(X) = \overline{NE}(X)_{(K_X+B)\geq 0} + \sum \mathbb{R}_{\geq 0}[C_j].$$
2. *For any $\epsilon > 0$ and ample $\mathbb{Q}$-divisor H,*
$$\overline{NE}(X) = \overline{NE}(X)_{(K_X+B+\epsilon H)\geq 0} + \sum_{\text{finite}} \mathbb{R}_{\geq 0}[C_j].$$
3. *Let $F \subset \overline{NE}(X)$ be a $(K_X + B)$-negative extremal face. Then there is a unique morphism $\varphi_F : X \to Z$ such that $(\varphi_F)_*\mathcal{O}_X \cong \mathcal{O}_Z$, Z is projective, and an irreducible curve $C \subset X$ is mapped to a point by φ_F if and only if $[C] \in F$. The map φ_F is called the contraction of F.*
4. *Let F and φ_F be as in (iii), and L a line bundle on X such that $L \cdot C = 0$ for every curve C with $[C] \in F$. Then there is a line bundle L_Z on Z such that $L \cong \varphi_F^* L_Z$.*

If the dimension of a face F of $\overline{NE}(X)$ is one, then we say that F is a ray. In what follows we assume that X is $\mathbb{Q}$-factorial. One could also run the minimal model program for non-$\mathbb{Q}$-factorial varieties, but this leads to several technical difficulties.

Proposition 8.13.2 *Let (X, B) be a projective $\mathbb{Q}$-factorial log canonical pair, R a $(K_X + B)$-negative extremal ray and $\varphi_R : X \to Z$ the associated contraction morphism, then $-(K_X + B)$ is φ_R-ample, $\rho(X/Z) = 1$, the fibers of φ_R are covered by rational curves, and one of the following holds.*

1. $\dim Z < \dim X$. *In this case we say that* φ_R *is a Mori fiber space and* $K_X + B$ *is not pseudo-effective.*
2. $\dim Z = \dim X$ *and* $\dim \mathrm{Ex}(\varphi_R) = \dim X - 1$. *In this case we say that* φ_R *is a divisorial contraction. We have that* $\mathrm{Ex}(\varphi_R)$ *is a prime divisor,* $(Z, (\varphi_R)_*B)$ *is also log canonical,* Z *is* $\mathbb{Q}$*-factorial and* $\mathrm{Ex}(\varphi_R)$ *is an irreducible divisor.*
3. $\dim Z = \dim X$ *and* $\dim \mathrm{Ex}(\varphi_R) < \dim X - 1$. *In this case we say that* φ_R *is a flipping contraction. The* $\mathbb{R}$*-divisor* $K_Z + (\varphi_R)_*B$ *is not* $\mathbb{R}$*-Cartier and hence* $(Z, (\varphi_R)_*B)$ *is not a (log canonical) pair.*

Proof

(1) Since $-(K_X + B)$ is relatively ample, X is dominated by a family of curves C_t such that $(K_X + B) \cdot C_t < 0$. If $K_X + B$ were pseudo-effective, then $[K_X + B] = \lim[E_i]$ where $E_i \geq 0$ are effective $\mathbb{R}$-divisors and $[E_i] \in N^1(X)$ denotes the numerical equivalence class. Since $E_i \geq 0$, then $E_i \cdot C_t \geq 0$ and so $(K_X + B) \cdot C_t = \lim E_i \cdot C_t \geq 0$ which is impossible.

(2) Let E be a prime φ_R-exceptional divisor. By the Negativity Lemma 8.2.1, E is not relatively nef, i.e. $E \cdot R < 0$. Thus $-E$ is φ_R-ample, and one sees that $\mathrm{Supp}(E) = \mathrm{Ex}(\varphi_R)$. Let C be a φ_R-exceptional curve, then $R = \mathbb{R}^{\geq 0}[C]$ and $K_X + B + \lambda E \equiv_Z 0$ where $\lambda = \frac{-(K_X+B)\cdot C}{E \cdot C} \geq 0$. Assume for simplicity that B is a $\mathbb{Q}$-divisor and hence $\lambda \in \mathbb{Q}$. By (4) of Theorem 8.13.1, $K_X + B + \lambda E \sim_{\mathbb{Q},Z} 0$ and $K_Z + B_Z = (\varphi_R)_*(K_X + B)$ is $\mathbb{Q}$-Cartier. Since $\lambda E \geq 0$, we have $a(P; Z, B_Z) \geq a(P; X, B)$ for every divisor P over Z and hence (Z, B_Z) is log canonical.

(3) Suppose that $K_Z + B_Z = (\varphi_R)_*(K_X + B)$ is $\mathbb{R}$-Cartier, then $K_X + B = (\varphi_R)^*(K_Z + B_Z)$ since φ_R is a small birational map. This contradicts the fact that $-(K_X + B)$ is φ_R-ample.

□

Definition 8.13.3 If (X, B) is log canonical $\mathbb{Q}$-factorial and $\varphi : X \to Z$ is a flipping contraction then we define the *flip* $\varphi^+ : X^+ \to Z$ by letting $X^+ = \mathrm{Proj}\, R(K_X + B/Z)$. The existence of X^+ follows from the finite generation of $R(K_X + B/Z)$ which is proved in [11, 14, 57]. Let $\psi : X \dashrightarrow X^+$, then ψ is a small birational map, X^+ is $\mathbb{Q}$-factorial, $\rho(X^+/Z) = 1$, $(X^+, B^+ := \psi_*B)$ is log canonical and $K_{X^+} + B^+$ is φ^+-ample.

Running the Minimal Model Program We start with a projective log canonical pair (X, B) and we aim to produce a sequence of flips and divisorial contractions

$$X = X_0 \overset{\psi_0}{\dashrightarrow} X_1 \overset{\psi_1}{\dashrightarrow} \dots \overset{\psi_{n-1}}{\dashrightarrow} X_n$$

that ends either with a minimal model or a Mori fiber space.

1. If $K_X + B$ is pseudo-effective, then $K_{X_n} + B_n$ is nef. This is a minimal model of (X, B).
2. If $K_X + B$ is not pseudo-effective, then there is a Mori fiber space $(X_n, B_n) \to Z$.

To achieve this, we proceed as follows. If $\overline{NE}(X)_{(K_X+B)<0} = \emptyset$, then $K_X + B$ is nef and we have a minimal model. Otherwise we pick R a $(K_X + B)$-negative extremal ray in $\overline{NE}(X)$. By the Cone Theorem 8.13.1 there is a contraction morphism $\varphi : X \to Z$.

1. If φ is a Mori fiber space, the algorithm terminates.
2. If φ is a divisorial contraction, then we may replace (X, B) by $(Z, \varphi_* B)$ and repeat the procedure.
3. If φ is a flipping contraction, then we replace (X, B) by the corresponding flip (X^+, B^+).

It is not clear that this process terminates. Since the Picard number is a positive integer which is preserved by flips $\rho(X_{i+1}) = \rho(X_i)$ and decreases by one after a divisorial contraction $\rho(X_{i+1}) = \rho(X_i) - 1$, it follows that any sequence of the (X, B) minimal model program contains at most finitely many divisorial contractions. Therefore the termination of the minimal model program follows from the following.

Conjecture 8.13.4 (Termination of Flips) Let (X, B) be a log canonical $\mathbb{Q}$-factorial projective pair. Then there is no infinite sequence of (X, B) flips.

Remark 8.13.5 The above conjecture is very difficult and it is known to hold in full generality only in some particular cases such as $\dim X \leq 3$ [69], for toric varieties and in a few other cases, e.g. [1, 10, 33, 51, 111]. A different approach is to choose our sequence of flips carefully. One such approach is called the minimal model program with scaling. In [14] it is shown that the minimal model program with scaling terminates for klt pairs (X, B) such that either B is big or $K_X + B$ is big or $K_X + B$ is not pseudo-effective.

It is easy to see that steps of the minimal model program always improve the singularities in the sense that the log discrepancies do not decrease and some log discrepancies in fact increase after each flip (or divisorial contraction).

Lemma 8.13.6 *Let (X, B) be a log canonical pair and $\psi : X \dashrightarrow X^+$ be a flip, then $a(E; X, B) \leq a(E; X^+, B^+ = \psi_* B)$ for all divisors E over X and the inequality is strict if the center of E is contained in the flipping locus (or equivalently contained in the flipped locus). Similarly, if $\psi : X \dashrightarrow Z$ is a divisorial contraction, then $a(E; X, B) \leq a(E; Z, B_Z = \psi_* B)$ and the inequality is strict if the center of E is contained in the contracted locus.*

Proof Let $X \to Z$ and $X^+ \to Z$ be the flipping and flipped contractions and $p : W \to X$ and $q : W \to X^+$ be any common resolution. We write $q^*(K_{X^+} + B^+) - p^*(K_X + B) = -F$, then $-F$ is an exceptional divisor which is nef over Z and so $F \geq 0$ by the Negativity Lemma 8.2.1. If E is a divisor on W, then $a(E; X^+, B^+) = a(E; X, B) + \mathrm{mult}_E(F)$ and so $a(E; X, B) \leq a(E; X^+, B^+)$. If the center of E is contained in the flipping (or flipped locus) then $\mathrm{mult}_E(F) > 0$ and the strict inequality holds.

The divisorial contraction case is similar (see also the proof of (3) of Proposition 8.13.2). □

Lemma 8.13.7 *Let (X, B) be a log canonical pair and $\psi : X \dashrightarrow X'$ a step of the minimal model program, then (X', B') is also log canonical. If moreover (X, B) is klt, dlt, plt, or canonical/terminal and $B \wedge \mathrm{Ex}(\psi) = 0$, then (X^+, B^+) is also klt, dlt, plt, or canonical/terminal.*

It is however not always true that canonical/terminal singularities are preserved when $B \wedge \mathrm{Ex}(\psi) \neq 0$.

Proof The statements about lc, klt, dlt, plt, or canonical/terminal singularities such that $B \wedge \mathrm{Ex}(\psi) = 0$ follow immediately from Lemma 8.13.6.

Note that if (X, p) is the cone over a rational curve of degree d and $\nu : X' \to X$ is the blow up of the vertex p and E is the exceptional divisor, then $(X', (1 - \frac{1}{d})E)$ is terminal by Lemma 8.3.9. Since $K_{X'} + (1 - \frac{1}{d})E \equiv_X \frac{1}{d}E$, it is easy to see that ν is a divisorial contraction for $(X', (1 - \frac{1}{d})E)$ and $(X, 0)$ is not canonical. □

8.14 Properness of Moduli Spaces and Semi-log-canonical Singularities

Let $\mathcal{X}^0 \to T^0$ be a family of projective varieties over a curve $T^0 = T \setminus \{0\}$, we would like to compactify this family in a natural way. There are several reasons why this is difficult to achieve. Notice that since $\mathcal{X}^0 \to T^0$ is a projective family, then we can always compactify it $f : \mathcal{X} \to T$ but the central fiber $\mathcal{X}_0$ can have bad singularities even if the fibers $\mathcal{X}_t$ are smooth for all $t \in T$. By Hironaka's resolution of singularities, after blowing $\mathcal{X}$ up, we can assume that $\mathcal{X}$ is smooth and $\mathcal{X}_0$ is a divisor with simple normal crossings support, however different components may appear with different multiplicities $\mathcal{X}_0 = f^{-1}(0) = \sum m_i \mathcal{X}_0^i$. Even if $f : \mathcal{X} \to T$ is a smooth family of curves, this can happen by blowing up points on the central fiber. For example, blow up $p \in \mathcal{X}_0$ to get $\mathcal{X}' \to \mathcal{X}$ and then we blow up q the intersection of the exceptional divisor E' with the strict transform of $\mathcal{X}_0$ to get $f' : \mathcal{X}'' \to \mathcal{X}$. The central fiber is not reduced and in fact it is given by $\mathcal{X}''_0 = (f'')^{-1}_* \mathcal{X}_0 + E'' + 2F''$ where E'' is the strict transform of E' and F'' is the new exceptional divisor. One may argue that we have taken unnecessary blow ups so this could be avoided, however it is easy to come up with other more natural examples where the central fiber is not reduced.

Let $C \to \bar{C}$ and $\tilde{T} \to T$ be degree 2 maps of curves with corresponding involutions σ on C and τ on $\tilde{T}$. Suppose that τ has a fixed point lying over 0 i.e. $\tilde{T} \to T$ is ramified over 0. Let $\mathcal{X} = (C \times \tilde{T})/ < (\sigma, \tau) >$, then general fibers of the induced map $f : \mathcal{X} \to T$ are isomorphic to C whilst $\mathcal{X}_0 \cong 2\bar{C}$, i.e. the central fiber is isomorphic to $\bar{C}$ with multiplicity 2. Of course, if we let $\tilde{\mathcal{X}}$ be the normalization of $\mathcal{X} \times_T \tilde{T}$, then $\tilde{\mathcal{X}} \cong C \times \tilde{T}$ and hence all fibers of $\tilde{\mathcal{X}} \to \tilde{T}$ are reduced (and isomorphic to C).

By semistable reduction, we can always achieve that the special fiber is a reduced divisor with simple normal crossings and general fibers are smooth as follows. Start with any projective family (flat projective morphism of normal varieties) $\mathcal{X}^0 \to T^0$ where $\dim T^0 = 1$. Let $T^0 \subset T$ be a compactification and $f : \mathcal{X} \to T$ a compactification. After a finite sequence of blow ups along smooth centers $\mathcal{X}' \to \mathcal{X}$, we may assume that the general fibers of $f' : \mathcal{X}' \to T$ are smooth and special fibers have simple normal crossings support (but may be non-reduced). By [72], there exists a finite (ramified) base change such that if $\tilde{\mathcal{X}}'$ is the normalization of $\mathcal{X}' \times_T \tilde{T}$ and $\tilde{f}' : \tilde{\mathcal{X}}' \to \tilde{T}$ is the induced morphism, then the fibers of $\tilde{f}'$ are all reduced with simple normal crossings (general fibers are smooth).

There are several problems with the above construction. First of all, even if the fibers of $\mathcal{X}^0 \to T^0$ are smooth, they may not correspond to fibers of $\tilde{f}'$. Even if we avoid extra blow ups so that general fibers of f and $\tilde{f}'$ match, there may be many different choices for the special fibers of $\tilde{f}'$. One example that illustrates the problem is the following. Let $\mathcal{X} = \mathbb{P}^2 \times \mathbb{A}^1$ and consider 3 sections $s_i : \mathbb{A}^1 \to \mathcal{X}$ given by $s_0(t) = [t, 0, 0]$, $s_1(t) = [0 : t : 0]$ and $s_2(t) = [0 : 0 : t]$. We let $\mathcal{X}_1$ be given by blowing up s_1, s_0 and s_2 (in that order) and $\mathcal{X}_2$ be given by blowing up s_2, s_0 and s_1 (in that order). Clearly $\mathcal{X}_1 \times_T T^0 \cong \mathcal{X}_2 \times_T T^0$ where $T = \mathbb{A}^1$ and $T^0 = \mathbb{A}^1 \setminus \{0\}$, however it is easy to see that $\mathcal{X}_1 \ncong \mathcal{X}_2$.

The natural way to resolve these issues is to consider log canonical models and more generally stable models. One should think of stable models as being natural limits of log canonical models. These are the higher dimensional generalizations of Deligne-Mumford stable curves. They are semi-log canonical pairs with ample log canonical class.

Definition 8.14.1 A pair (X, B) is *semi-log canonical* (slc) if

1. $\mathcal{O}_X$ is S_2,
2. X is simple normal crossings in codimension 1,
3. the singular locus of X does not contain components of the support of B,
4. $K_X + B$ is $\mathbb{R}$-Cartier, and
5. if $\nu : X^\nu \to X$ is the normalization and $K_{X^\nu} + B^\nu = \nu^*(K_X + B)$, then each irreducible component of (X^ν, B^ν) is a log canonical pair.

Definition 8.14.2 A projective slc pair (X, B) is a *stable model* if $K_X + B$ is ample.

The properness of the moduli functor of stable models is discussed in detail in [57, §7] and [77]. The properness follows from the following result.

Theorem 8.14.3 ([57, Corollary 1.5]) *Let $f^0 : \mathcal{X}^0 \to T^0$ be a projective morphism, $(\mathcal{X}^0, B^0)$ a log canonical pair, T the germ of a smooth curve, $0 \in T$ a closed point and $T^0 = T \setminus \{0\}$. If $K_{\mathcal{X}^0} + B^0$ is f^0-ample, then there is a finite dominant base change $\theta : \tilde{T} \to T$, a log canonical pair $(\mathcal{X}, B)$ and a projective morphism $(\mathcal{X}, B) \to \tilde{T}$ such that $K_{\mathcal{X}} + B$ is ample over $\tilde{T}$, the restriction of $(\mathcal{X}, B)$ to the pre-image $\theta^{-1}(T^0)$ is isomorphic to $(\mathcal{X}^0, B^0) \times_{T^0} \tilde{T}$.*

Remark 8.14.4 The limits of stable models are also unique. More precisely, let $(X, B) \to T$ and $(X', B') \to T$ be two families of stable models where T the germ of a smooth curve, $0 \in T$ a closed point and $T_0 = T \setminus \{0\}$. If $(X, B) \times_T T_0 \cong (X', B') \times_T T_0$, then $(X, B) \cong (X', B')$ [57, Lemma 7.2].

8.15 Topology of Singularities

The singularities of the minimal model program are surprisingly well behaved from the topological point of view. Typically divisorially log terminal singularities are particularly well understood, log canonical singularities are more subtle and non log canonical singularities exhibit pathological behaviour. In this section we illustrate this from several points of view.

8.15.1 Resolutions of dlt Singularities are Rationally Chain Connected

If (X, B) is a dlt surface singularity and $\nu : X' \to X$ is the minimal resolution, then the positive dimensional fibers of $X' \to X$ are trees of rational curves. The result below shows that a similar pattern holds in all dimensions [49, Theorem 1.2].

Recall that a variety X is *rationally connected* if for very general points $x_1, x_2 \in X$, there exists an irreducible rational curve $C \subset X$ which contains both points x_1, x_2. A variety X is *rationally chain connected* if for very general points $x_1, x_2 \in X$, there exists a connected curve $C \subset X$ which contains both points x_1, x_2 such that every irreducible component of C is a rational curve. Clearly, if X is rationally connected then it is also rationally chain connected, however the cone over an elliptic curve is rationally chain connected but not rationally connected. It is however known that if X is smooth then rational connectedness and rational chain connectedness are equivalent [82, 2.1].

Theorem 8.15.1 *Let (X, B) be a divisorially log terminal pair and $\nu : X' \to X$ any proper birational morphism of normal varieties such that $\nu_*\mathcal{O}_{X'} = \mathcal{O}_X$, then the fibers of ν are rationally chain connected.*

Proof The original proof of the above result is quite technical and involves several ideas that were important in establishing the minimal model program. Assuming the minimal model program [14], we can give a fairly straightforward argument for the $\mathbb{Q}$-factorial case which we sketch below. For a complete proof, we refer the reader to [49, Theorem 1.2].

We may assume that (X, B) is klt (see Remark 8.3.12). Replacing X' by a higher model, we may assume that ν is a log resolution and we write $B' = \nu_*^{-1}B + (1 - \epsilon)\mathrm{Ex}(\nu)$ for some $0 < \epsilon \ll 1$, then $K_{X'} + B' = \nu^*(K_X + B) + F$ where $F \geq 0$ and the support of F equals the exceptional set. We now run the $(K_{X'} + B')$-minimal model program over X [14]. Let $X' = X_0 \dashrightarrow X_1 \dashrightarrow \ldots \dashrightarrow X_n$ be the steps of this mmp, then $K_{X_n} + B_n \equiv_X F_n$ is nef over X. By the Negativity Lemma 8.2.1, $-F_n \geq 0$ and hence $F_n = 0$. Therefore, $X_n \to X$ is small and since we assume that X is $\mathbb{Q}$-factorial, then $X_n = X$. Therefore, it suffices to show by induction that if the fibers of $f_i : X_i \to X$ are rationally chain connected, then the fibers of $f_{i-1} : X_{i-1} \to X$ are also rationally chain connected.

If $\phi_{i-1} : X_{i-1} \to X_i$ is a divisorial contraction, then by [68, Theorem 1], the fibers of ϕ_{i-1} are rationally chain connected and hence so are the fibers of f_{i-1}.

The case of a flip $\phi_{i-1} : X_{i-1} \dashrightarrow X_i$ is similar. Let $g : X_{i-1} \to Z$ be the flipping contraction and $g^+ : X_i \to Z$ the flipped contraction. If the fibers of f_i are rationally chain connected then so are the fibers of $Z \to X$. By [68, Theorem 1], the fibers of $g : X_{i-1} \to Z$ are rationally chain connected and hence so are the fibers of f_{i-1}.

□

Corollary 8.15.2 *Let (X, B) be a divisorially log terminal pair. Then X is rationally chain connected iff it is rationally connected.*

Proof Let $\nu : X' \to X$ be a resolution. By Theorem 8.15.1, since X is rationally chain connected, then so is X'. Since X' is smooth and rationally chain connected, then it is rationally connected see [82, 2.1]. Thus X is also rationally connected. □

These results already fail for log canonical surface singularities as shown, for example, by the cone over an elliptic curve. The following result shows that if X has klt singularities then holomorphic p-forms defined on the non-singular locus of X can be extended to holomorphic p-forms defined on any resolution of singularities of X. Intuitively, one expects that holomorphic p-forms vanish along any rational curve and hence vanish along the fibers of a log resolution $\nu : X' \to X$.

Theorem 8.15.3 (Extension Theorem for Differential Forms on klt Varieties, [41, Theorem 1.4]) *Let X be a complex quasi-projective variety with at most Kawamata log terminal singularities and $\nu : X' \to X$ a log resolution. Then $\nu_* \Omega^p_{X'}$ is a reflexive sheaf for all $p \leq \dim X$.*

We also quote the following interesting closely related result.

Theorem 8.15.4 (Lipman-Zariski Conjecture for klt Spaces, [41, Theorem 6.1]) *Let X be a klt variety such that the tangent sheaf T_X is locally free. Then X is smooth.*

8.15.2 Dual Complexes

We now turn our attention to a generalization of the dual graph of a minimal resolution of a surface. In higher dimensions there are several challenges. First of all, for dimensional reasons, instead of a dual graph, we expect to construct a dual complex. Secondly, and more importantly, there is no natural minimal resolution, however we do have some possible partial resolutions that may be appropriate substitutes, see for example the lc and dlt modifications constructed in Theorem 8.7.5. We would then hope that the dual complexes constructed from different partial resolutions are equivalent.

Definition 8.15.5 Let (X, B) be a dlt pair and $S = \lfloor B \rfloor$, then the corresponding *dual complex* $\mathcal{D}(X, B) = \mathcal{D}(S)$ is defined as follows. If $S = S_0 + \ldots + S_r$ is a decomposition in to irreducible components, then each component S_i corresponds to a vertex v_i. We will construct the d dimensional cell complex $\mathcal{D}(S)$ inductively via a sequence of k-dimensional complexes C_k such that the k-dimensional simplexes W_v corresponding to irreducible components v of the intersection of collections of $k+1$ components of S say $S_I = S_{i_0} \cap S_{i_1} \cap \ldots \cap S_{i_k}$ where $I = \{i_0, \ldots, i_k\}$ (note that S_I is either empty or of pure codimension $k+1$). In particular $C_0 = \cup_{i=0}^r W_i$ (where $W_i = W_{S_i}$). Note that if v is an irreducible component of S_I and $I' \subset I$ where $|I'| = |I| - 1$, then v is contained in a unique irreducible component of $S_{I'}$. Suppose that C_{k-1} has already been constructed together with its attaching maps. We then glue the k-dimensional simplex W_v corresponding to a component $v \subset S_I$ where $I = \{i_0, \ldots, i_k\}$ on to C_k by gluing its facets to the $(k-1)$-dimensional simplexes corresponding to the unique irreducible components w of $S_{I'}$ containing v where $I' \subset I$ and $|I'| = k$. Note that if E is a divisor with simple normal crossings on a smooth variety X, then (X, E) is dlt and hence $\mathcal{D}(E)$ is defined.

Remark 8.15.6 The attaching maps are embeddings, and so the cell complexes $\mathcal{D}(X, B)$ defined above are regular cell complexes. They may however not be simplicial (i.e. the intersection of 2 simplexes is not a single simplex). For example if S is a divisor on a surface X which has 2 components S_1, S_2 meeting transversely at 2 points p, q, then the corresponding dual complex is not simplicial since $W_p \cap W_q = W_{S_1} \cup W_{S_2}$ is not a single simplex. If we blow up say along p, then we obtain 3 divisors E, S_1', S_2' where E is the exceptional curve and S_i' is the strict transform of S_i, and the corresponding dual complex is simplicial.

Definition 8.15.7 Let $0 \in X$ be a point on a normal variety and $\nu : X' \to X$ a resolution such that $\nu^{-1}(0) = E$ is a divisor with simple normal crossings. The simple-homotopy class of the dual complex $\mathcal{D}(\mathrm{Supp}(E))$ is independent of the choice of the resolution ν. We denote it by $\mathcal{DR}(0 \in X)$.

Topological properties of the dual complex often correspond to algebraic properties of the singularity. For example, it follows from [114] that if $(0 \in X)$ is a rational singularity then $H^i(\mathcal{DR}(0 \in X), \mathbb{Q}) = 0$ for $i > 0$.

Kollár has shown that every finite simplicial complex occurs as the dual complex of some isolated singularity [76, Theorem 2]. However, we expect that if the singularity $(0 \in X)$ is mild, then there are restrictions on the possible simplicial complexes that can occur. We have the following [28, Theorems 1 and 2].

Theorem 8.15.8 *Let* $(0 \in X)$ *be an isolated singularity such that* K_X *is* $\mathbb{Q}$*-Cartier,* $\nu : X' \to X$ *be a proper, birational morphism and* $E \subset X'$ *the divisorial part of* $\mathrm{Supp}(\nu^{-1}(0))$. *Assume that* (X', E) *is dlt and* $E \neq \emptyset$.

1. *If ν is a dlt modification, then the dual complex $\mathcal{D}(E)$ is independent of ν, up-to PL homeomorphism. This defines a PL homeomorphism class associated to $(0 \in X)$; denoted by $\mathcal{DMR}(0 \in X)$.*
2. *If $E = \mathrm{Supp}(\nu^{-1}(0))$ then the dual complex $\mathcal{D}(E)$ is simple-homotopy equivalent to $\mathcal{DMR}(0 \in X)$.*
3. *If (X, B) is dlt then $\mathcal{DR}(0 \in X)$ is contractible (and in fact collapsible, see [28, Definition 18]).*

8.15.3 Local Fundamental Groups

Definition 8.15.9 Let $x \in X$ be a point on a normal variety. The link of $L(x, X)$ is the topological space defined by intersecting X with a sphere of radius $0 < \epsilon \ll 1$ centered at x. It is independent of the choice of ϵ and the intersection of (x, X) with a small closed ball centered at x is homeomorphic to the cone over $L(x, X)$.

The *local fundamental group* of $(x \in X)$ is $\pi_1^{\mathrm{loc}}(x, X) = \pi_1(L(x, X))$. Alternatively $\pi_1^{\mathrm{loc}}(x, X)$ is the inverse limit of the fundamental groups of punctured analytic neighborhoods of $x \in X$. The regional fundamental group $\pi_1^{\mathrm{reg}}(x, X)$ is the inverse limit of the fundamental groups of the intersection of analytic neighborhoods of $x \in X$ with the smooth locus of X. There is a natural surjection $\pi_1^{\mathrm{reg}}(x, X) \to \pi_1^{\mathrm{loc}}(x, X)$ (which is an isomorphism if $x \in X$ is an isolated singularity).

By a result of Kapovich and Kollár [66], it is known that any finitely presented group G appears as the fundamental group of the link of some isolated complex singularity $(0 \in X_G)$ i.e. $\pi_1(L(0, X_G)) \cong G$. One expects that mild singularities will impose restrictions on the local fundamental groups. In fact we have the following.

Theorem 8.15.10 ([18]) *Let (X, B) be a klt pair, then for any $x \in X$, the local fundamental group $\pi_1^{\mathrm{loc}}(x, X)$ is finite.*

If G is a finite group, then there exists a quotient singularity $(x \in X) = \mathbb{C}^n/G$, such that $\pi_1^{\mathrm{reg}}(x, X) = G$.

It would then seem that (apart from finiteness) there is no restriction on the possible local fundamental groups of klt singularities. However, the next result shows that if we fix the dimension, then we can bound the "non-abelian" part of $\pi_1^{\mathrm{reg}}(x, X)$ and we can also bound the rank of the abelian part. The bound on the abelian part can actually be made more precise in terms of the regularity of the singularity. Therefore, before stating the next result we make the following.

Definition 8.15.11 Let (X, B) be a log canonical pair such that $x \in X$ is an isolated log canonical center and $\nu : (X', B') \to (X, B)$ a dlt model. We let $\mathcal{D}(X, B; x) = \mathcal{D}(\lfloor B' \rfloor)$. This complex is defined up to simple homotopy equivalence. The regularity $\mathrm{reg}(X, B)$ is the dimension of $\mathcal{D}(X, B; x)$ or equivalently $\mathrm{reg}(X, B) + 1$ is the maximum of the codimension of any strata of $\lfloor B' \rfloor$ (if $\lfloor B' \rfloor = \emptyset$, i.e. (X, B) is klt, then we set $\mathrm{reg}(X, B) = -\infty$). Thus $\mathrm{reg}(X, B) \in \{-\infty, 0, 1, \ldots, \dim X - 1\}$.

If (X, B) is a klt singularity, then we let $\text{reg}(X, B) = \max\{\text{reg}(X, B+\Gamma)\}$ where $(X, B+\Gamma)$ is log canonical and x is an isolated center of log canonical singularities. Since such a divisor always exists, in this case $\text{reg}(X, B) \geq 0$.

Theorem 8.15.12 ([20] and [94]) *Let $n \geq 2$ and r be two positive integers. There exists a positive integer $c(n)$ only depending on n satisfying the following. Let $(x \in X)$ be an n-dimensional klt singularity of regularity r, then there exists a short exact sequence*

$$1 \to A \to \pi_1^{\text{loc}}(x, X) \to N \to 1$$

where A is a finite abelian group of rank at most $r + 1$ and N is a finite group of order at most $c(n)$.

One may also hope that the fundamental groups of log canonical singularities are well behaved, however easy examples show that the situation is quite different from the klt case. For example the cone over an elliptic curve has infinite local fundamental group $\pi_1^{\text{loc}}(x, X) \cong \mathbb{Z} \oplus \mathbb{Z}$. Never-the-less, since log canonical surface singularities are classified, it is not too surprising that the local fundamental groups can be understood in detail. Figueroa and Moraga prove the following.

Theorem 8.15.13 ([30, Theorem 3]) *Let $(x \in X)$ be a log canonical surface singularity. Then, we have a short exact sequence*

$$1 \to N \to \pi_1^{\text{reg}}(x \in X) \to G \to 1$$

where N is a solvable group of length at most 2 and G is a finite group of order at most 6.

Not surprisingly the situation is already more complicated for 3 dimensional log canonical singularities. In this case Figueroa and Moraga show that every surface group appears as a local fundamental group, however free groups with ≥ 3 generators do not occur. In dimension 4, there are even less restrictions since many 3-manifold groups appear and these include all free groups. More precisely, we have the following.

Theorem 8.15.14 ([30, Theorems 6, 7, and 8]) *Let S be a connected 2-dimensional manifold without boundary. Then, there exists an isolated 3-fold log canonical singularity $(x \in X)$ for which $\pi_1^{\text{loc}}(x, X) \cong \pi_1(S)$.*

Let $(x \in X)$ be a log canonical 3-fold singularity. Then, $\pi_1^{\text{reg}}(x, X)$ is not isomorphic to a free group with ≥ 3 generators.

Let S be a connected 3-dimensional manifold without boundary, smoothly embedded in $\mathbb{R}^4$. Then, there exists an isolated 4-fold log canonical singularity $(x \in X)$ for which $\pi_1^{\text{loc}}(x \in X) \cong \pi_1(S)$. In particular every free group appears as the fundamental group of an isolated 4-dimensional lc singularity.

8.15.4 Reductive Quotients

By Nagata's famous answer to Hilbert's 14th problem [98], it is known that there are finitely generated rings whose rings of invariants under the action of an algebraic group G are not finitely generated. It is however known that in characteristic 0, if the group G is reductive then the ring of invariants is indeed finitely generated and hence we can consider quotients of schemes by reductive groups. More precisely, recall that a *reductive group* is an algebraic group G whose connected component at the identity G^0, has no non-trivial connected unipotent normal subgroups.

Definition 8.15.15 Let G be a reductive group acting on a separated scheme X and $\pi : X \to Y$ a surjective G equivariant map to a separated scheme Y. Then Y is a *good quotient* of X, denoted by $Y = X//G$ if

1. π is an affine morphism, and
2. $\pi^* : \mathcal{O}_Y \to (\pi_* \mathcal{O}_X)^G$ is an isomorphism.

Thus, if X is an affine variety (over an algebraically closed field $\mathbb{K}$ of characteristic 0) and G is a reductive group acting on X, then the good quotient $Y = X//G$ is defined by $Y = \mathrm{Spec}(\mathbb{K}[X]^G)$.

One naturally expects that even though Y may be more singular than X, its singularities will not be arbitrary and in fact will be controlled to some extent by the singularities of X. By work of Hochster and Roberts [64] it is known that that reductive quotients of smooth points are Cohen-Macaulay, and by work of Boutot [17] it is known that reductive quotients of rational singularities are also rational. Since klt singularities are rational, one may hope that klt singularities are preserved by reductive quotients. This is well known for finite groups and Schoutens proved this in [103] assuming that the quotient $X//G$ is $\mathbb{Q}$-Gorenstein.

Theorem 8.15.16 ([19, Theorem 1]) *Let X be an affine variety of klt type and G a reductive group acting on X. Then, the (affine invariant-theoretic) quotient $X//G := \mathrm{Spec}(K[X]^G)$ is of klt type.*

Remark 8.15.17 The above theorem has many important applications. For example to the moduli of K-stable Fano varieties. Let $X_{n,v}$ be the good moduli space of K-polystable Fano manifolds of dimension n and volume v. Then $X_{n,v}$ is potentially klt, i.e. there exists an effective divisor $B_{n,v}$ on $X_{n,v}$ so that the pair $(X_{n,v}, B_{n,v})$ has kawamata log terminal singularities [19, Theorem 4].

Remark 8.15.18 Reductive quotients are examples of pure morphisms. In [126], Zhuang proved that if $X \to Y$ is a pure morphism and X is potentially klt, then Y is potentially klt, and in [118], Takagi and Yamaguchi generalized Zhuang's result to the plt setting.

It is still an open question whether potentially log canonical singularities are preserved by reductive quotients.

8.16 Generalized Pairs

Recall the following definition [12, §4.4]

Definition 8.16.1 A projective *generalized pair* consists of

1. a normal projective variety X and a $\mathbb{Q}$-divisor $B \geq 0$ on X,
2. a birational contraction $\nu : X' \to X$ and a nef $\mathbb{Q}$-Cartier divisor M' on X' such that $K_X + B + M$ is $\mathbb{Q}$-Cartier where $M := \nu_* M'$.

Remark 8.16.2 A frequently used alternate definition is a triple $(X, B + \mathbf{M})$ where $B \geq 0$, $\mathbf{M}$ is a nef b-$\mathbb{Q}$-divisor, $K_X + B + \mathbf{M}_X$ is $\mathbb{Q}$-Cartier. Recall that if $\mathbf{M}$ is a nef b-$\mathbb{Q}$-divisor then there exists a birational model $\nu : X' \to X$ and a nef $\mathbb{Q}$-divisor M' on X' such that $\mathbf{M} = \bar{M}'$ and in particular for any birational morphism $\mu : X'' \dashrightarrow X'$, let $p : W \to X'$ and $q : W \to X''$ be a common resolution, then we have $\mathbf{M}_{X''} = q_*(p^* M')$.

It is easy to see that the two definitions of generalized pairs are equivalent, and in what follows we freely switch between the two. Note that if M' is a nef $\mathbb{Q}$-Cartier divisor on X', then $\bar{M}'$ is the nef b-$\mathbb{Q}$-divisor defined as follows: for any birational morphism $\mu : X'' \to X'$, we let $\bar{M}' = \mu^* M'$ and for any morphism $\eta : X'' \to X'''$, we let $\bar{M}'_{X'''} = \eta_* \bar{M}'_{X''}$.

Definition 8.16.3 Let $(X, B + \mathbf{M})$ be a generalized pair and E be any divisor over X, then we let $a(E; X, B+\mathbf{M})$ be the log discrepancy of $(X, B+\mathbf{M})$ which is defined as follows. Let $\nu : X' \to X$ be a resolution such that E is a divisor on X'. Write $B' := \nu^*(K_X + B + \mathbf{M}_X) - (K_{X'} + \mathbf{M}_{X'})$ and let $a(E; X, B+\mathbf{M}) = 1 - \operatorname{mult}_E(B')$. We say that a generalized pair is generalized Kawamata log terminal or gklt (resp. generalized log canonical or glc) if $a(E; X, B+\mathbf{M}) > 0$ (resp. $a(E; X, B+\mathbf{M}) \geq 0$) for all E over X.

Remark 8.16.4 Note that if $(X, B + \mathbf{M})$ is generalized klt and A is ample, then for any $\epsilon > 0$, there exists a $\mathbb{R}$-divisor $B_\epsilon \sim_{\mathbb{R}} B + \mathbf{M}_X + \epsilon A$ such that (X, B_ϵ) is klt.

To see this, note that if $\nu : X' \to X$ is a log resolution such that $\mathbf{M}$ descends to X', then $\mathbf{M}_{X'}$ is nef. Let F be an effective exceptional $\mathbb{Q}$-divisor such that $\nu^* A - F$ is ample. For any $\epsilon \gg 0$, $(X', B' + \epsilon F)$ is sub-klt where $K_{X'} + B' + \mathbf{M}_{X'} = \nu^*(K_X + B + \mathbf{M}_X)$. Since $\mathbf{M}_{X'} + \epsilon(\nu^* A - F)$ is ample we may pick an effective $\mathbb{R}$-divisor $M_\epsilon \sim_{\mathbb{R}} \mathbf{M}_{X'} + \epsilon(\nu^* A - F)$ such that $(X', B' + \epsilon F + M_\epsilon)$ is sub-klt. But then $(X, B_\epsilon := \nu_*(B' + \epsilon F + M_\epsilon))$ is klt.

It is easy to see that many of the results that hold for the usual klt pairs also hold for generalized klt pairs. For example, if $B' + \mathbf{M}_{X'}$ is big, then the generalized pluricanonical ring $R(K_X + B + \mathbf{M}_X)$ is finitely generated. The existence of flips for generalized lc pairs, and the cone and contraction theorems for generalized lc pairs are proved in [44] and [120].

This might look like a technical generalization of the more traditional singularities of the minimal model program, however these singularities arise naturally in a variety of contexts. The first and most natural example of this is the canonical bundle formula.

8.16.1 The Canonical Bundle Formula

Suppose that $f : X \to Z$ is a contraction morphism i.e. a proper morphism of normal varieties such that $f_*\mathcal{O}_X = \mathcal{O}_Z$. If (X, B) is a klt pair (resp. a lc pair), then we can define the *boundary divisor* B_Z for $(X/Z, B)$ as follows. For any codimension 1 point $P \in Z$ we let $a_P = \mathrm{lct}_{\eta_P}(X, B; f^*P)$ be the log canonical threshold of (X, B) with respect to f^*P over the generic point η_P of P. This means that $(X, B + a_P f^*P)$ is lc but not klt over a neighborhood of η_P. Note that Z is normal and hence smooth in codimension 1, thus we may assume that P is Cartier in a neighborhood of η_P so that f^*P makes sense. We then define the discriminant or boundary part

$$B_Z(X, B) := \sum b_Q Q, \qquad \text{where} \qquad b_Q = 1 - a_Q.$$

It is easy to see that $b_Q = 0$ for all but finitely many $Q \in Z$ and hence $\sum b_Q Q$ is a $\mathbb{Q}$-divisor on Z. To see this, we observe that if $\nu : X' \to X$ is a resolution and $K_{X'} + B' = \nu^*(K_X + B)$ is the log pull-back, then $B_Z(X, B) = B_Z(X', B')$, therefore, we may assume that (X', B') is log smooth and the claim follows easily.

Suppose now that $\mu : Z' \to Z$ is a birational morphism and $\nu : X' \to X$ is a resolution such that $f' : X' \to Z'$ is a morphism, and $K_{X'} + B' = \nu^*(K_X + B)$ is the log pull-back, then $\mu_* B_{Z'}(X', B') = B_Z(X, B)$, therefore we have a b-divisor $\mathbf{B}^Z$ such that $\mathbf{B}^Z_{Z'} = B_{Z'}(X', B')$ for any birational model $\mu : Z' \to Z$.

Lemma 8.16.5 *The b-divisor $\mathbf{B}^Z$ is b-$\mathbb{Q}$-Cartier.*

Proof See eg. [6, Theorem 1.1]. □

Definition 8.16.6 If $f : X \to Z$ is a contraction morphism, (X, B) is a pair such that

1. (X, B) is log canonical over the generic point of Z, and
2. $K_X + B \sim_{\mathbb{Q}} f^*L$ for some $\mathbb{Q}$-Cartier divisor L on Z,

then we define the *moduli part* $M_Z(X, B) := L - (K_Z + B_Z(X, B))$. Note that $M_Z(X, B)$ is only defined up to $\mathbb{Q}$-linear equivalence and it defines a b-divisor $\mathbf{M}^Z$ in the obvious way.

Remark 8.16.7 The divisor $M_Z(X, B/Z)$ is determined by the fibration over the generic point $\eta \in Z$ i.e. $(X_\eta, B_\eta/\eta)$ (see eg. [74, Theorem 8.3.7]).

Theorem 8.16.8 *If $f : X \to Z$ is a contraction morphism as in Definition 8.16.6, then $(Z, B_Z + \mathbf{M}^Z)$ is a generalized pair (equivalently $\mathbf{M}^Z$ is a nef $\mathbb{Q}$-Cartier b-divisor).*

If (X, B) is klt (resp. lc), then $(Z, B_Z + \mathbf{M}^Z)$ is a generalized klt (resp. generalized lc).

Proof See [39] and [70]. □

Remark 8.16.9 One can define the boundary and moduli parts for fiber spaces $f : X \to Z$ and generalized pairs $(X, B + \mathbf{M})$ such that $K_X + B + \mathbf{M}_X \sim_{\mathbb{Q}} f^*L$. The theory is similar and the analog of Theorem 8.16.8 holds in this context, see [31]. This is very useful when proving results by induction on the dimension (see eg. [32]).

8.16.2 Kawamata Subadjunction

Theorem 8.16.10 *Let (X, B) be a projective lc pair and W a minimal log canonical center, then W is normal and we have*

$$(K_X + B)|_W \sim_{\mathbb{Q}} K_W + B_W + \mathbf{M}_W, \qquad \text{where} \qquad (W, B_W + \mathbf{M}) \textit{ is a gklt pair.}$$

Proof This follows from [5, 38, 70, 74]. We have already seen in Theorem 8.9.7 that W is normal. The idea is to consider a dlt model $f : X' \to X$ where $K_{X'} + B' = f^*(K_X + B)$ and a minimal strata S of $\lfloor B' \rfloor$ that dominates W. By dlt adjunction, $(K_{X'} + B')|_S = K_S + B_S$ is klt. Assume for simplicity that $S \to W$ is a fiber space (this can be achieved by passing to the Stein factorization, see [38]). Then $K_S + B_S \sim_{\mathbb{Q}} g^*(K_X + B)|_W$ where $g = f|_S$. The claim now follows from Theorem 8.16.8. □

Remark 8.16.11 The analog result for generalized log pairs is proven in [31].

8.16.3 Applications of Generalized Pairs

One natural situation in which generalized pairs arise is in the study of the Iitaka fibration which is defined as follows. Let (X, B) be a klt pair and

$$R(K_X + B) := \oplus_{m \geq 0} H^0(m(K_X + B))$$

the *canonical ring*. Recall that by [14], $R(K_X + B)$ is finitely generated. We let $P_n = \dim H^0(mK_X)$ be the m-th plurigenus. If $|mK_X| \neq \emptyset$, then we let $\phi_m : X \dashrightarrow \mathbb{P}^{P_n - 1} = |mK_X|$ be the m-th *pluricanonical map*. More generally, we let $\phi_{|m(K_X+B)|} : X \dashrightarrow \mathbb{P}^N = |m(K_X + B)|$ be the corresponding rational maps. The *Kodaira dimension* is

$$\kappa(K_X + B) := \text{tr.deg.}_{\mathbb{C}} R(K_X + B) - 1.$$

It has several alternate definitions. We have that if $\kappa(K_X + B) \geq 0$, then $\kappa(K_X + B) = \max\{\dim \phi_{|m(K_X+B)|}(X)\}$ and

$$\phi : X \dashrightarrow Z := \text{Proj}(R(K_X + B))$$

is the Iitaka fibration where $\phi = \phi_{|m(K_X+B)|}$ for any $m > 0$ sufficiently divisible and $\kappa(X) = \dim Z$. Fujino and Mori show that there is a generalized pair $(Z, B_Z + \mathbf{M})$ such that $H^0(m(K_X + B)) \cong H^0(m(K_Z + B_Z + \mathbf{M}_Z))$ for all $m > 0$ sufficiently divisible. Suppose in fact that $K_X + B$ is semiample (we can reduce to this case if (X, B) has a good minimal model), then $|m(K_X + B)|$ is base point free for $m > 0$ sufficiently divisible and defines $\phi : X \to Z$. We then have $K_X + B \sim_{\mathbb{Q},Z} 0$ and so by Theorem 8.16.8, $K_X + B \sim_{\mathbb{Q}} \phi^*(m(K_Z + B_Z + \mathbf{M}_Z))$ where $(Z, B_Z + \mathbf{M})$ is a generalized pair. Therefore, questions about the geometry of (X, B) correspond to questions about the generalized pair $(Z, B_Z + \mathbf{M})$. If the coefficients of B belong to a DCC set I, then it follows from Theorem 8.17.8 that the coefficients of B_Z also belong to a DCC set (that depends on I and $\dim X$). It is much harder to control the properties of $\mathbf{M}$. By work of Fujino and Mori [39], it is known that the Cartier index of $\mathbf{M}_X$ can be bounded in terms of the middle Betti numbers of the canonical cover of the general fiber of $(X, B) \to Z$ (see [119, §3]).

Another situation where generalized pairs arise is for log Fano pairs (X, B) such that (X, B) is klt and $-(K_X + B)$ is nef and big. In this case we let $\mathbf{M} := \overline{(-(K_X + B))}$ so that $(X, B + \mathbf{M})$ is a generalized log Calabi-Yau pair as $K_X + B + \mathbf{M}_X \equiv 0$. This point of view plays an important role in Birkar's work on the boundedness of Fano and log Calabi-Yau varieties and in particular in his resolution of the BAB conjecture [13].

Theorem 8.16.12 ([13]) *Fix $d \in \mathbb{N}$ and $\epsilon > 0$. Then the set of projective varieties X such that $\dim X = d$, (X, B) is ϵ-log canonical for some boundary B, and $-(K_X + B)$ is nef and big, form a bounded family.*

We refer the reader to [12] for further details and applications.

8.17 ACC for LCTs

Definition 8.17.1 Let (X, B) be a lc pair and $D \geq 0$ an $\mathbb{R}$-Cartier divisor, then the *log canonical threshold* of D with respect to (X, B) is

$$\mathrm{lct}(X, B; D) := \sup\{t \geq 0 | (X, B + tD) \text{ is log canonical}\}.$$

For any $x \in X$ we let $\mathrm{lct}_x(X, B; D)$ be the supremum of non negative real numbers t such that $(X, B + tD)$ is log canonical on a neighborhood of $x \in X$.

Fix $n \in \mathbb{N}$ and $I \subset [0, 1]$, then we let $\mathcal{LC}_n(I)$ be the set of all n-dimensional log canonical pairs (X, B) such that $B \in I$. Fix $J \subset \mathbb{R}_{\geq 0}$, then we let

$$\mathrm{LCT}(I, J) = \{\mathrm{lct}(X, B; D) \mid (X, B) \in \mathcal{LC}_n(I),\ D \in J\}.$$

It is well known [85, §9.3] that if X is smooth, $B = 0$, and $\mathrm{mult}_x(D) = m$, then

$$\frac{1}{m} \le \mathrm{lct}_x(X, 0; D) \le \frac{\dim X}{m}.$$

We establish this in the following lemmas.

Lemma 8.17.2 *If X is smooth and B is a $\mathbb{R}$-divisor such that $\mathrm{mult}_x(B) \ge \dim X$, then (X, B) is not klt.*

Proof If $\nu : X' \to X$ is the blow up of $x \in X$, then $\nu^*(B) = mE + \nu_*^{-1}D$ where $m = \mathrm{mult}_x(B) \ge \dim X$ and $K_{X'/X} = -(\dim X - 1)E$. Thus $a(E; X, B) \le 0$ and hence (X, B) is not klt. □

Lemma 8.17.3 *If X is smooth and B is a $\mathbb{R}$-divisor such that $\mathrm{mult}_x(B) \le 1$, then (X, B) is lc.*

Proof We proceed by induction on the dimension. Let H be a general hyperplane passing through $x \in X$, then H is smooth and $\mathrm{mult}_x(B|_H) \le 1$ and so by induction on the dimension $(H, B|_H)$ is log canonical (note that the case $\dim X = 1$ is clear). By Theorem 8.9.5, $(X, B + H)$ is log canonical on a neighborhood of H and the claim follows. □

Remark 8.17.4 If $X = \mathbb{A}^n$ and B is the zero set of a polynomial $f \in \mathbb{C}[z_1, \ldots, z_n]$, then

$$\mathrm{lct}_0(X, 0; B) = \sup\{\lambda > 0 | \frac{1}{|f|^\lambda} \in L^2_{\mathrm{loc}}\}.$$

If B has simple normal crossing support then this can easily be checked. In general we use a log resolution to reduce to the normal crossing support case bearing in mind that $K_{X'/X}$ determines the Jacobian in the change of variables formula (see also Remark 8.20.2).

Next we compute several examples of log canonical thresholds.

Example 8.17.5 If $X = \mathbb{A}^n$, $H_1, \ldots, H_r$ are general hyperplanes through $0 \in X$ and $h_i \in [0, 1)$, then we let $H = \sum h_i H_i$ and $h = \sum h_i = \mathrm{mult}_0(H)$. We claim that $\mathrm{lct}_0(X, 0; H) = n/h$. To see this, note that if $\nu : X' \to X$ is the blow up of $0 \in X$ and E is the exceptional divisor, then ν is a log resolution and $a(E; X, \lambda B) = n - \lambda h$.

Example 8.17.6 Consider the cusp $B = \{x^2 = y^3\} \subset \mathbb{A}^2 = X$. It is easy to see that if $\nu_1 : X_1 \to X$ is the blow up $0 \in X$, then the strict transform $B_1 = (\nu_1^{-1})_* B$ is tangent to the exceptional divisor E_1 at a point p. If we blow up p we obtain $\nu_2 : X_2 \to X_1$ with an exceptional divisor F which intersects $E_2 = (\nu_2^{-1})_* E_1$ and $B_2 = (\nu_2^{-1})_* B_1$ at a point q. If we further blow up q, then we obtain $\nu_3 : X_3 \to X_2$ with an exceptional divisor G which intersects $E_3 = (\nu_3^{-1})_* E_2$, $B_3 = (\nu_3^{-1})_* B_2$ and $F_3 = (\nu_3^{-1})_* F_2$ transversely at 3 distinct points. Thus $\nu : X_3 \to X$ is a log resolution of (X, B). One sees that $\nu^*(B) = B_3 + 2E_3 + 3F_3 + 6G$ and $K_{X_3/X} = E_3 + 2F_3 + 4G$. Therefore $\mathrm{lct}_0(X, 0; B) = 5/6$ (in fact $\mathrm{mult}_G(\nu^*(\lambda B) - K_{X'/X}) = 6\lambda - 4$).

More generally, the following lemma implies that if $B = \{z_1^{a_1} + \ldots + z_n^{a_n} = 0\} \subset \mathbb{A}^n = X$, then

$$\mathrm{lct}_0(X, B) = \min\{1, \frac{1}{a_1} + \ldots + \frac{1}{a_n}\}.$$

Lemma 8.17.7 *Suppose that $f(x_1, \ldots, x_n)$ and $g(y_1, \ldots, y_m)$ are polynomials and $B = \{f(x_1, \ldots, x_n) + g(y_1, \ldots, y_m) = 0\} \subset \mathbb{A}^{n+m} = X$. Then*

$$\mathrm{lct}_0(X, B) = \min\{1, \mathrm{lct}_0(\{f = 0\}, \mathbb{A}^n) + \mathrm{lct}_0(\{g = 0\}, \mathbb{A}^m)\}.$$

Proof See [73, Proposition 8.21]. □

Shokurov realized that the set of log canonical thresholds has interesting subtle structures. In particular he conjectured that log canonical thresholds are well ordered [109]. This was eventually proved by Hacon-M^c^Kernan-Xu.

Theorem 8.17.8 (ACC for Log Canonical Thresholds [46, Theorem 1.1]) *Fix $n \in \mathbb{N}$, $I \subset [0, 1]$ and $J \subset \mathbb{R}_{\geq 0}$ such that I and J satisfy the DCC, then $\mathrm{LCT}_n(I, J)$ satisfies the ACC.*

Theorem 8.17.9 ([46, Theorem 1.11]) *If $I \subset [0, 1]$ is closed under addition and 1 is the only accumulation point of I, then the accumulation points of $LCT_n(I)$ are $LCT_{n-1}(I) - \{1\}$. In particular, if $I \subset \mathbb{Q}$ then the accumulation points of $LCT_n(I)$ are rational numbers.*

We refer the reader to [87] (and references therein) for further results on the structure of accumulation points of log canonical thresholds.

As usual, we expect that local properties of log canonical singularities correspond to global properties of Calabi-Yau pairs. We have the following.

Theorem 8.17.10 ([46, Theorem 1.4]) *Fix $n \in \mathbb{N}$ and a DCC set $I \subset [0, 1]$, then there is a finite subset $I_0 \subset I$ such that if (X, B) is an n-dimensional log pair such that $K_X + B \equiv 0$ and $B \in I$, then $B \in I_0$.*

Remark 8.17.11 One should view Theorems 8.17.8 and 8.17.10 as being local and global versions of the same phenomena. To gain some intuition of why Theorem 8.17.10 implies Theorem 8.17.8, we argue as follows. Suppose that $\mathrm{LCT}_n(I, J)$ is not ACC, then we may pick a strictly increasing sequence $\lambda_i \in \mathrm{LCT}_n(I, J)$. For each i there is a log canonical pair $(X_i, B_i + \lambda_i D_i)$ such that $B_i \in I$ and $D_i \in J$ which is not klt at some point $P_i \in X_i$. Cutting by hyperplanes, we may assume that P_i is a closed point. Let $\nu_i : X_i' \to X_i$ be a dlt model (Theorem 8.7.5), and E_i an exceptional divisor, then we have $K_{X_i'} + \Delta_i = \nu_i^*(K_{X_i} + B_i + \lambda_i D_i)$ where (X_i', Δ_i) is dlt and $\mathrm{Ex}(\nu_i) \subset \lfloor \Delta_i \rfloor$. Pick E_i an irreducible exceptional divisor mapping to P_i, then by adjunction (Theorem 8.8.1) we may write

$$K_{E_i} + \Delta_{E_i} = (K_{X'_i} + \Delta_i)|_{E_i} \equiv 0.$$

It then follows that there exists a DCC set K (determined by I, J, $\{\lambda_i\}$) such that $\Delta_{E_i} \in \Lambda$. By Theorem 8.17.10, there exists a finite subset $\Lambda_0 \subset \Lambda$ such that $\Delta_{E_i} \in \Lambda_0$. This is easily seen to contradict the fact that λ_i is a strictly increasing sequence. We refer the reader to [46, §5].

The converse is more involved. Suppose however that (X, B) is a Calabi-Yau pair where $B = bA$ for some ample Cartier divisor A. The cone Y of X determined by the ample divisor A has a log canonical singularity at the vertex and $b = \mathrm{lct}(X, 0; A)$. By Theorem 8.17.8, b belongs to an ACC set, but since we are assuming that $b \in I$ where I is a DCC set, then b belongs to a finite subset of I.

There are many reasons why Theorem 8.17.8 is an important result, but one of the most compelling ones comes from the following result of Birkar.

Theorem 8.17.12 ([10]) *Assume termination of flips for $\mathbb{Q}$-factorial kawamata log terminal pairs in dimension $n-1$. Let (X, B) be a n-dimensional projective $\mathbb{Q}$-factorial kawamata log terminal pair such that $K_X + B$ is numerically equivalent to a divisor $D \geq 0$ then any sequence of $(K_X + B)$-flips terminates.*

Theorem 8.17.8 also has many important applications to the birational geometry of log pairs. We have already seen an application to Calabi-Yau's, but more importantly there is a strong connection to the boundedness of log canonical models and of Fano varieties.

Theorem 8.17.13 ([46, Theorem 1.3]) *Let $n \in \mathbb{N}$ and $I \subset [0, 1]$ be a DCC set, then there exists a positive constant $\delta > 0$ and an integer $m > 0$ such that if $\mathcal{LC}_n(I)$ is the set of all n-dimensional projective log canonical pairs (X, B) with $B \in I$ and $\kappa(K_X + B) = \dim X$, then*

1. *if $(X, B) \in \mathcal{LC}_n(I)$, then $|m(K_X + B)|$ defines a birational map,*
2. *if $(X, B) \in \mathcal{LC}_n(I)$, then $\mathrm{vol}(K_X + B) \geq \delta$,*
3. *$\{\mathrm{vol}(K_X + B) | (X, B) \in \mathcal{LC}_n(I)\}$ is a DCC set.*

Remark 8.17.14 Theorem 8.17.13 is crucial to proving the boundedness of the KSBA moduli functor of stable models of varieties of general type, see [47, 48, 77]. Similar ideas then also play an important role in the study of Calabi-Yau and Fano varieties (see eg. [13]).

Remark 8.17.15 There is also a close connection between minimal log discrepancies and a variant of log canonical thresholds known as a-log canonical thresholds. If (X, B) is a-lc for $0 \leq a \leq 1$ (i.e. $\mathrm{mld}(X, B) \geq a$), then we let

$$a\text{-lct}(X, B; D) := \sup\{t \geq 0 \mid (X, B + tD) \text{ is } a\text{-lc}\}.$$

It is conjectured that a-lc thresholds also satisfy the ACC property similarly to Theorem 8.17.8. By [15] and [92], it is known that the ACC for MLDs Conjecture 8.4.5 is equivalent to the ACC for a-lcts Conjecture.

8.18 Normalized Volume

Normalized and local volumes are sophisticated invariants of klt singularities. These invariants were introduced quite recently [90] and have already proven to have interesting applications. For example, any klt singularity $x \in (X, B)$ has a canonically determined degeneration to a stable log Fano cone (see Theorem 8.18.8). We refer the reader to [91] and [127] for in depth surveys on these topics.

Definition 8.18.1 Let X be a reduced, irreducible, separated variety over a field k, with function field $K(X)$. A *valuation* $v : K(X)^\times \to \mathbb{R}$ is a non-constant map such that $v(f \cdot g) = v(f) + v(g)$, $v(f+g) \geq \min\{v(f), v(g)\}$, $v(k^\times) = 0$, and $v(0) = +\infty$.

The corresponding *valuation ring* is

$$\mathcal{O}_v = \{f \in K(X)^\times | v(f) \geq 0\} \cup \{0\}.$$

If we have an inclusion of local rings $\mathcal{O}_{X,x} \subset \mathcal{O}_v$, then we say that v is centered at $x \in X$. The center of a valuation, if it exists, is unique. The set of all valuations on $K(X)^\times$ is denoted by Val_X and the set of all valuations centered at $x \in X$ is denoted by $\mathrm{Val}_{X,x}$. We have that $v \in \mathrm{Val}_{X,x}$ iff $v(f) > 0$ for any $f \in \mathfrak{m}_x$.

For any valuation v and $\lambda \in \mathbb{R}^{\geq 0}$ the corresponding valuation ideals are defined by

$$\mathfrak{a}_\lambda(v) := \{f \in \mathcal{O}_{X,x} | v(f) \geq \lambda\} \qquad \text{and} \qquad \mathfrak{a}_{>\lambda}(v) := \{f \in \mathcal{O}_{X,x} | v(f) > \lambda\}.$$

Definition 8.18.2 ([29]) Let $x \in X$ be a closed point on an n-dimensional normal variety and $v \in \mathrm{Val}_{X,x}$, then the *volume* of v is

$$\mathrm{vol}_{X,x}(v) = \limsup_{k \to \infty} \frac{\ell(\mathcal{O}_{X,x}/\mathfrak{a}_k(v))}{k^n/n!}$$

where $\ell(\ldots)$ denotes the length of an Artinian module.

It is known that the above $\limsup$ is an actual limit (see [22, 29, 86]). Next we will define the discrepancy of valuations over X. Before we do this we need to recall several facts about valuations.

Definition 8.18.3 (Quasi Monomial Valuations) Let (X, B) be a pair, $\nu : X' \to X$ a log resolution of (X, B), $E = E_1 + \ldots + E_r$ a divisor such that $E + \nu^{-1}(B) + \mathrm{Ex}(\nu)$ has simple normal crossings, and $\alpha = (\alpha_1, \ldots, \alpha_r) \in (\mathbb{R}_{\geq 0})^r$. Suppose that

$E_1 \cap \ldots \cap E_r \neq \emptyset$ and C is an irreducible component of this intersection. Near the generic point η of C, we may pick local coordinates $z_1, \ldots, z_n$ such that $E_i = \{z_i = 0\}$. For any $f \in \mathcal{O}_{X,\eta}$, we write $f = \sum c_\beta z^\beta \in \hat{\mathcal{O}}_{X,\eta}$ where each c_β is either 0 or a unit. Then the valuation

$$v_\alpha(f) = \min\{\sum_{i=1}^{r} \alpha_i \beta_i \mid c_\beta \neq 0\}$$

is a *quasi-monomial valuation*. If the α_i are rational, this valuation corresponds to a divisorial valuation given by the exceptional divisor for the corresponding weighted blow up. This definition works more generally when $(\eta \in X, E_1 + \ldots + E_r)$ is toroidal (cf. [124, Definition 2.8]). The set of all quasi-monomial valuations where $\eta \in X'$ and $F \subset E$ is a reduced divisor such that (X', F) is toroidal at η is denoted by $\mathrm{QM}_\eta(X', E) \subset \mathrm{Val}_X$ and $\mathrm{QM}(X', E) \subset \mathrm{Val}_X$ is the union of the $\mathrm{QM}_\eta(X', E)$.

By [65], for any log smooth model (X', E) of (X, B) there exists a retraction map $\rho_{X',E} : \mathrm{Val}_X \to \mathrm{QM}(X', E)$ inducing a homeomorphism

$$\mathrm{Val}_X \to \varprojlim_{(X',E)} \mathrm{QM}(X', E).$$

Definition 8.18.4 (Log Discrepancy of a Valuation) If (X, B) is a log canonical pair and v is a valuation, then the log discrepancy of v is defined as follows.

1. If P is a divisor over X and $v = \mathrm{ord}_P$ is the divisorial valuation associated to P, then we let $a(v; X, B) = a(P; X, B)$ be the usual log discrepancy.
2. Let $\nu : X' \to X$ be a log resolution of (X, B) and $E = E_1 + \ldots + E_r$ a divisor such that $E + \nu^{-1}(B) + \mathrm{Ex}(\nu)$ has simple normal crossings, η the generic point of a component of $E_1 \cap \ldots \cap E_r$, and $\alpha = (\alpha_1, \ldots, \alpha_r) \in (\mathbb{R}_{\geq 0})^r$. If $\alpha = (\alpha_, \ldots, \alpha_r) \in (\mathbb{R}_{\geq 0})^r \setminus \{0\}$ and $v_\alpha \in \mathrm{QM}_\eta(X', E)$, then $a(v_\alpha; X, B) = \sum \alpha_i a(E_i; X, B)$.
3. For any real valuation $v \in \mathrm{Val}_X$, the *log discrepancy* is defined by

$$a(v; X, B) = \sup\{a(\rho_{X',E}(v); X, B)\}$$

where (X', E) ranges over all log smooth models over (X, B) and $\rho_{X',E}$ is defined in Definition 8.18.3.

Definition 8.18.5 (Normalized Volume [90]) Let $x \in (X, B)$ be an n-dimensional klt singularity. The normalized volume function $\widehat{\mathrm{vol}}_{(X,B),x} : \mathrm{Val}_{X,x} \to \mathbb{R} \cup \{+\infty\}$ is defined by $\widehat{\mathrm{vol}}_{(X,B),x}(v) = +\infty$ if $a(v; X, B) = +\infty$ and

$$\widehat{\mathrm{vol}}_{(X,B),x}(v) := (a(v; X, B))^n \cdot \mathrm{vol}_{X,x}(v)$$

if $a(v; X, B) < +\infty$. The local volume of $x \in (X, B)$ is then defined as

$$\widehat{\mathrm{vol}}_{(x,X,B)} := \inf\{\widehat{\mathrm{vol}}_{(X,B),x}(v) \mid v \in \mathrm{Val}_{X,x}\}.$$

Note that as $x \in (X, B)$ is klt, then $a(v; X, B) > 0$. By [90, Theorem 1.2], it follows that $\widehat{\mathrm{vol}}_{(x,X,B)} > 0$ and in particular $\widehat{\mathrm{vol}}_{(X,B),x}(v) > 0$ for any $v \in \mathrm{Val}_{X,x}$.

Remark 8.18.6 If $x \in X$ is a smooth point, $B = 0$ and ord_x is the valuation given by the order of vanishing at $x \in X$, then $\widehat{\mathrm{vol}}_{(X,B),x}(\mathrm{ord}_x) = n^n$ (where $n = \dim X$) and $\widehat{\mathrm{vol}}_{(X,B),x}(v) > n^n$ for any valuation $v \in \mathrm{Val}_{X,x}$ that is not a scalar multiple of ord_x ([88, Theorem A.4]).

If $f : (x, X, B) \to (\bar{x}, \bar{X}, \bar{B})$ is a finite Galois morphism between klt singularities such that $f^*(K_{\bar{X}} + \bar{B}) = K_X + B$, then by [88, Theorem 2.7] and [123, Theorem 1.3]

$$\widehat{\mathrm{vol}}_{(x,X,B)} = \deg(f) \cdot \widehat{\mathrm{vol}}_{(\bar{x},\bar{X},\bar{B})}.$$

We say that a valuation v^{m} is a *minimizer* of the normalized volume function $\widehat{\mathrm{vol}}_{(X,B),x}$ if

$$\widehat{\mathrm{vol}}_{(X,B),x}(v^{\mathrm{m}}) = \inf\{\widehat{\mathrm{vol}}_{(X,B),x}(v) \mid v \in \mathrm{Val}_{X,x}\}.$$

Theorem 8.18.7 (Existence of Minimizers [16, 122, 124]) *Let $x \in (X, B)$ be a klt singularity. Then there exists a (unique up to rescaling) minimizer v^{m} of the normalized volume function, and any minimizer is quasi-monomial.*

Theorem 8.18.8 (Local Higher Rank Finite Generation and Stable Degeneration [124]) *Let $x \in (X := \mathrm{Spec}(R), B)$ be a klt singularity. Let $v = v^{\mathrm{m}}$ be a minimizer of the normalized volume function $\widehat{\mathrm{vol}}_{(X,B),x}$. Then the associated graded algebra $\mathrm{gr}_v := \bigoplus_{\lambda \geq 0} \mathfrak{a}_\lambda / \mathfrak{a}_{>\lambda}$ is finitely generated and it induces a degeneration of (X, B) to a K-semistable log Fano cone singularity $(X_0 = \mathrm{Spec}(\mathrm{gr}_v(R)), B_0; \xi_v)$, where ξ_v is the Reeb vector field induced by the grading $\Phi := v(R \setminus \{0\}) \to \mathbb{R}_{\geq 0}$.*

We refer the reader to [124] and references therein for further details. Finally we mention the following recent result of Xu and Zhuang [125, Theorem 1.2].

Theorem 8.18.9 *Fix $n \in \mathbb{N}$ and a finite set $I \subset [0, 1]$. Then 0 is the only accumulation point of the set*

$$\mathrm{Vol}_{n,I} := \{\widehat{\mathrm{vol}}_{(x,X,B)} \mid x \in (X, B) \text{ is klt}, \dim X = n,\ B \in I\}.$$

8.19 F-singularities

Varieties defined over an algebraically closed field of characteristic $p > 0$ behave very differently from varieties defined over a field of characteristic 0. One important issue is that Kodaira vanishing and its generalizations are known to fail already for surfaces, and since this is one of the main tools of the minimal model program, we expect many new features to emerge in positive characteristics. A key observation however is that if $F : X \to X$ is the Frobenius morphism and $\mathcal{A}$ is an ample line bundle on X then $(F^e)^*\mathcal{A} \cong \mathcal{A}^{\otimes p^e}$ and so by Serre vanishing and the projection formula, we have

$$H^i((F^e_*\omega_X) \otimes \mathcal{A}) \cong H^i(F^e_*(\omega_X \otimes \mathcal{A}^{\otimes p^e})) = 0 \qquad \forall i > 0,\ e \gg 0.$$

Note that as additive groups, $H^i(F^e_*(\omega_X \otimes \mathcal{A}^{\otimes p^e})) \cong H^i(\omega_X \otimes \mathcal{A}^{\otimes p^e})$. In order to make effective use of the above observation, it is important to be able to compare $H^0((F^e_*\omega_X) \otimes \mathcal{A})$ and $H^0(\omega_X \otimes \mathcal{A})$. There is a natural inclusion map $\mathcal{O}_X \to F^e_*\mathcal{O}_X$ whose dual is the *Grothendieck trace map* $F^e_*\omega^{\cdot}_X \to \omega^{\cdot}_X$. Taking cohomology we get a homomorphism $\Phi^e : F^e_*\omega_X \to \omega_X$. It is known that the corresponding inverse system $\ldots F^2_*\omega_X \to F_*\omega_X \to \omega_X$ satisfies the Mittag-Leffler condition (see [63, §1] and [89]) and hence if $\Omega = \varprojlim F^e_*\omega_X$, then $H^i(\Omega \otimes \mathcal{A}) = 0$ for all $i > 0$, and $H^0(\Omega \otimes \mathcal{A}) \subset H^0(\omega_X \otimes \mathcal{A})$ is the stable image of $H^0(\Phi^e)$ which we denote by

$$S^0(\omega_X \otimes \mathcal{A}) = \cap_{e>0}\mathrm{Im}\left(H^0(\omega_X \otimes \mathcal{A}^{\otimes p^e}) \to H^0(\omega_X \otimes \mathcal{A})\right)$$

(cf. [105]).

Definition 8.19.1 Let X be a normal variety over an algebraically closed field of characteristic $p > 0$, and (X, B) a pair whose index is not divisible by p. For any integer $e > 0$ such that $(p^e - 1)B$ is integral, we let $\mathcal{L}_{e,B} = \mathcal{O}_X((1-p^e)(K_X+B))$ and $\phi_{e,B} : \mathcal{L}_{e,B} \to \mathcal{O}_X$ the induced map. Note that twisting the above map Φ^e by $\omega_X^\vee$ (and taking reflexive hulls), by the projection formula we obtain a homomorphism $F^e_*\mathcal{O}_X((1-p^e)K_X) \to \mathcal{O}_X$. Precomposing this map with the inclusion $\mathcal{O}_X((1-p^e)(K_X+B)) \hookrightarrow \mathcal{O}_X((1-p^e)K_X)$ we recover $\phi_{e,B}$.

The *non-F-pure ideal* $\sigma(X, B)$ is the unique biggest ideal (respectively the *test ideal* $\tau(X, B)$ is the unique smallest nonzero ideal) $\mathcal{J} \subset \mathcal{O}_X$ such that $\phi_{e,B}(\mathcal{J} \cdot \mathcal{L}_{e,B}) = \mathcal{J}$ for any $e > 0$. If $\sigma(X, B) = \mathcal{O}_X$, we say that (X, B) is *sharply F-pure*. If $\tau(X, B) = \mathcal{O}_X$, we say that (X, B) is *strongly F-regular*.

Remark 8.19.2 It is easy to see that $\sigma(X, B) = \phi_{e,B}(\mathcal{L}_{e,B})$ for any $e \gg 0$.

It is also well known that (X, B) is strongly F-regular if for any effective divisor D there exists an $e > 0$ such that $\phi_{e,B}(\mathcal{L}_{e,B}(-D)) = \mathcal{O}_X$.

Remark 8.19.3 We can drop the condition that $(p^e - 1)B$ is integral and say that (X, B) is sharply F-pure if $\phi_{e,B}(\mathcal{L}_{e,B}) = \mathcal{O}_X$ where $\mathcal{L}_{e,B} = \mathcal{O}_X(\lfloor(1-p^e)(K_X + B)\rfloor)$. Note that then the divisor $B' =: \frac{1}{p^e-1}\lceil(p^e-1)B\rceil$ is a $\mathbb{Q}$-divisor with index not divisible by p such that $B' \geq B$ and (X, B') is sharply F-pure.

Similarly, let (X, B) be a pair such that for any divisor $D \geq 0$ there exists an $e > 0$ such that the map $\mathcal{O}_X(\lfloor(1-p^e)(K_X+B)\rfloor - D) \to \mathcal{O}_X$ is surjective, then there is a pair (X, B') where $B' \geq B$ and the index of B' is not divisible by p such that (X, B') is strongly F-regular. See [106, 4.3].

Lemma 8.19.4 *If X is affine and sharply F-pure, then there exists $e > 0$ and $D \in (1-p^e)K_X$ such that $(X, B := \frac{D}{p^e-1})$ is a sharply F-pure pair.*

Proof Consider the surjection $\phi_e : F^e_*\mathcal{O}_X((1-p^e)K_X) \to \mathcal{O}_X$ and a divisor $D \in |(1-p^e)K_X|$ corresponding to a preimage of $1 \in \Gamma(\mathcal{O}_X)$. Then we have an induced inclusion $\mathcal{O}_X((1-p^e)(K_X+B)) \to \mathcal{O}_X((1-p^e)K_X)$ such that the induced map

$$F^e_*\mathcal{O}_X((1-p^e)(K_X+B)) \to F^e_*\mathcal{O}_X((1-p^e)K_X) \to \mathcal{O}_X$$

is surjective. □

Sharply F-pure and strongly F-regular singularities are positive characteristic analogs of log canonical and klt singularities. Since resolution of singularities in positive characteristic is only known in dimension ≤ 3, in this case log canonical singularities (resp. klt singularities) are singularities such that $a(P; X, B) \geq 0$ (resp. $a(P; X, B) > 0$) for any divisor P over X. Equivalently, we require that for any proper birational morphism $\nu : X' \to X$ with reduced exceptional divisor E, we have that $K_{X'} + \nu^{-1}_* B - \nu^*(K_X + B) \geq -E$.

Lemma 8.19.5 *If (X, B) is log smooth and $B = \sum b_i B_i$ where $0 \leq b_i \leq 1$ (resp. $0 \leq b_i < 1$), then (X, B) is sharply F-pure (resp. strongly F-regular).*

Proof This follows from [62, Corollary 2.5], see also [40, 15.1]. □

Lemma 8.19.6 *If the pair (X, B) is sharply F-pure (resp. strongly F-regular), then (X, B) is log canonical (resp klt).*

Proof *(Cf. [62])* Assume that (X, B) is sharply F-pure and p does not divide the index of B. For any $e \gg 0$, $(p^e-1)(K_X+B)$ is a Weil divisor. Let $\nu : X' \to X$ be a proper birational morphism, E the reduced exceptional divisor, and write

$$\mathcal{F}_e := F^e_*\mathcal{O}_{X'}(\lceil K_{X'} - p^e\nu^*(K_X+B) + p^e\epsilon E\rceil)$$

where $0 < \epsilon \ll 1$. Since $\lceil p^e G\rceil \leq p^e\lceil G\rceil$ for any $\mathbb{R}$-divisor G, by the projection formula applied to the homomorphism $F^e_*\mathcal{O}_{X'}(K_{X'}) \to \mathcal{O}_{X'}(K_{X'})$ we have induced homomorphisms

$$\mathcal{F}_e \to F^e_*\mathcal{O}_{X'}(K_{X'} + p^e\lceil -\nu^*(K_X+B) + \epsilon E\rceil) \to \mathcal{F}_0.$$

By the projection formula we also have

$$\nu_*\mathcal{F}_e = F^e_*\left(\mathcal{O}_X((1-p^e)(K_X+B)) \otimes \nu_*\mathcal{O}_{X'}(\lceil K_{X'} - \nu^*(K_X+B) + p^e\epsilon E\rceil)\right).$$

Note that for any fixed $0 < \epsilon \ll 1$, if $e \gg 0$, then $\nu_*\mathcal{O}_{X'}(\lceil K_{X'} - \nu^*(K_X + B) + p^e\epsilon E\rceil) = \mathcal{O}_X$ and so $\nu_*\mathcal{F}_e = F^e_*\mathcal{O}_X((1 - p^e)(K_X + B))$. We claim that the following diagram commutes

$$\begin{array}{ccc} \nu_*\mathcal{F}_e & \longrightarrow & \nu_*\mathcal{F}_0 \\ {\scriptstyle =}\big\downarrow & & \big\downarrow \\ F^e_*\mathcal{O}_X((1-p^e)(K_X+B)) & \longrightarrow & \mathcal{O}_X \end{array} \tag{8.1}$$

To see this, note that the diagram is clearly defined and commutes on the complement of a closed subset of codimension ≥ 2 and $\mathcal{O}_X$ is reflexive. We have that $\nu_*\mathcal{F}_0 = \nu_*\mathcal{O}_{X'}(\lceil K_{X'} - \nu^*(K_X + B) + \epsilon E\rceil) \subset \mathcal{O}_X$ and equality holds (for $0 < \epsilon \ll 1$) iff $a(P; X, B) \geq 0$ for all ν-exceptional divisors $P \subset X'$. Since (X, B) is sharply F-pure, the bottom row is surjective. By commutativity of the diagram, $\nu_*\mathcal{F}_0 \hookrightarrow \mathcal{O}_X$ is an isomorphism. This concludes the proof. □

Remark 8.19.7 In [60], Hara classifies strongly F-regular singularities (as well as F-pure rational singularities) in dimension 2. It turns out that strongly F-regular surface singularities coincide with klt surface singularities if the characteristic is $p > 5$. These results are strengthened in [58] and used to establish the existence of flips for 3-folds over an algebraically closed field of characteristic $p > 5$ (see also [9, 53–55]).

Theorem 8.19.8 *Let (X, B) be a pair defined over an algebraically closed field k of characteristic zero. Given a model of X over a sufficiently large $\mathbb{Z}$-algebra of finite type A, contained in k, then the generic fiber (X_k, B_k) is klt iff for all but finitely many closed points $s \subset \mathrm{Spec}(A)$, the pair (X_s, B_s) is strongly F-regular.*

Proof See [59, 61, 93, 112, 113], [117, Corollary 3.4] and [62].

Let $\nu : X' \to X$ be a resolution of singularities. We may assume that ν is defined over A. Suppose that for all but finitely many closed points $s \subset \mathrm{Spec}(A)$, (X_s, B_s) is strongly F-regular, then by Lemma 8.19.6, we know that these (X_s, B_s) are in fact klt and hence $a(P_s; X_s, B_s) > 0$ for all divisors P_s on X'_s. By generic smoothness, it follows easily that for any fixed divisor P on X', we have $a(P; X, B) = a(P_s; X_s, B_s)$ for all but finitely many closed points $s \in \mathrm{Spec}(A)$. Thus $a(P; X, B) > 0$ for all $P \in \mathrm{Ex}(\nu) \cup \mathrm{Supp}(B)$. Thus (X, B) is klt.

Conversely, suppose that (X, B) is klt. Let D be a Cartier divisor whose support contains the non log smooth locus of (X, B). Pick $\epsilon > 0$ such that $(X, B + \epsilon D)$ is also klt. It suffices to show that for all but finitely many closed points $s \in \mathrm{Spec}(A)$, there exists an $e > 0$ such that the homomorphism

$$F^e_*\mathcal{O}_{X_s}(K_{X_s} + \lceil -p^e(K_{X_s} + B_s + \epsilon D_s)\rceil) \to \mathcal{O}_{X_s}$$

is surjective. We will replace $B + \epsilon D$ by B to simplify the notation.

Let $\nu : X' \to X$ be a resolution of singularities and write $\nu^*(K_X + B) = K_{X'} + B'$. Let $C \geq 0$ be an effective exceptional $\mathbb{Q}$-divisor such that $-C$ is ample over X and $(X', B' + C)$ is sub-klt. By Lemma 8.19.9, the homomorphism

$$\nu_{s,*}F^e_*\mathcal{O}_{X'_s}(K_{X'_s} + \lceil -p^e(\nu^*(K_{X_s} + B_s) + C_s)\rceil) \\ \to \nu_{s,*}\mathcal{O}_{X'_s}(K_{X'_s} + \lceil -(\nu^*(K_{X_s} + B_s) + C_s)\rceil)$$

is surjective for all but finitely many closed points $s \in \mathrm{Spec}(A)$. Since

$$\lceil K_{X'_s} - \nu^*(K_{X_s} + B_s) - C_s\rceil = \lceil -B'_s - C_s\rceil$$

is effective and ν_s-exceptional, then $\nu_{s,*}\mathcal{O}_{X'_s}(K_{X'_s} + \lceil -(\nu^*(K_{X_s} + B_s) + C_s)\rceil) = \mathcal{O}_{X_s}$ and so the composition

$$\nu_{s,*}F^e_*\mathcal{O}_{X'_s}(K_{X'_s} + \lceil -p^e(\nu^*(K_{X_s} + B_s) + C_s)\rceil) \\ \to F^e_*\mathcal{O}_{X_s}(K_{X_s} + \lceil -p^e(K_{X_s} + B_s)\rceil) \to \mathcal{O}_{X_s}$$

is surjective and the claim follows. □

Lemma 8.19.9 (Hara Surjectivity Theorem [59, 93, 107]) *With the notation of Theorem 8.19.8 Suppose that $\nu : X' \to X$ is a log resolution of singularities, D is a ν-ample $\mathbb{Q}$-divisor with simple normal crossings support. Then the homomorphism*

$$\nu_{s,*}F^e_*\mathcal{O}_{X'_s}(K_{X'_s} + \lceil p^e D_s\rceil) \to \nu_{s,*}\mathcal{O}_{X'_s}(K_{X'_s} + \lceil D_s\rceil)$$

is surjective for all but finitely many closed points $s \in$ Spec(A).

A similar picture is conjectured to hold for log canonical and F-split singularities by work of Mustaţă and Srinivas. We refer the reader to [96] and [97] for further details and the definitions of $\tau(Y_s, a_s^\lambda)$ and $\mathcal{J}(Y, a^\lambda)$.

Conjecture 8.19.10 ([97, Conjecture 1.1]) Let X be a smooth, n-dimensional projective variety over an algebraically closed field k of characteristic zero. Given a model of X over a $\mathbb{Z}$-algebra of finite type A, contained in k, there is a dense set of closed points $S \subset \mathrm{Spec}(A)$ such that the action induced by Frobenius on $H^n(X_s, \mathcal{O}_{X_s})$ is bijective for every $s \in S$.

Conjecture 8.19.11 ([97, Conjecture 1.2]) Let Y be a nonsingular, variety over an algebraically closed field k of characteristic zero, and let $a \subset \mathcal{O}_Y$ be a nonzero ideal on Y. Given a model for Y and the ideal a defined over a $\mathbb{Z}$-algebra of finite type A, contained in k, there is a dense set of closed points $S \subset \mathrm{Spec}(A)$ such that

$$\tau(Y_s, a_s^\lambda) = \mathcal{J}(Y, a^\lambda)_s$$

for all $s \in S$ and all $\lambda \geq 0$. Furthermore, if we have finitely many pairs as above $(Y^{(i)}, a^{(i)})$, and corresponding models over Spec(A), then there is a dense open subset of closed points in Spec(A) such that the equation above holds for each of these pairs.

Theorem 8.19.12 *Conjectures 8.19.10 and 8.19.11 are equivalent.*

Proof See [97, Theorem 1.3] and [96, Theorem 1.3]. □

8.20 Analytic Singularities

The singularities of the minimal model program can also be studied in the analytic context. In fact almost everything we have discussed above translates to analytic varieties. The main reason is that the minimal model program holds (locally over the base) for projective morphisms of analytic varieties, see [23, 24, 37] for precise statements. In particular the following results hold.

1. Inversion of plt and lc adjunction (cf. Theorems 8.9.4 and 8.9.5) hold by [52, Theorems 2.6 and 2.7].
2. The ACC for log canonical thresholds (cf. Theorem 8.17.8) holds by [37] and [56].
3. Kawamata subadjunction (cf. Theorem 8.16.10) holds by [52, Theorem 0.2].

Remark 8.20.1 We note that when studying Kähler varieties, there are often very few interesting divisors or line bundles. To allow for more flexibility, one often works with locally exact (1,1) forms $\phi \in H^{1,1}_{\mathrm{BC}}(X)$. One can define several notions of positivity for these forms including nefness and bigness that agree with the usual notions when ϕ is the class of a divisor on a projective variety. It is then also possible to define generalized pairs $(X, B + \boldsymbol{\beta})$ where $\boldsymbol{\beta}$ is a b-nef (1,1) form (instead of a b-nef $\mathbb{R}$-divisor). Many results that hold for the usual generalized pairs $(X, B + \mathbf{M})$ (cf. Sect. 8.16) also hold for generalized pairs $(X, B + \boldsymbol{\beta})$. We refer the reader to [52] or [25] for precise definitions and statements.

Remark 8.20.2 Another interesting feature in analytic geometry, is that one can consider the singularities of a metric associated to a pluri-subharmonic function. Suppose for example that X is a smooth complex manifold and $\phi \in PSH(X)$ is a pluri-subharmonic function on X. The corresponding multiplier ideal sheaf is

$$\mathcal{J}(\phi) := \{f \in \mathcal{O}_X \mid |f|^2 e^{-2\phi} \in L^1_{\mathrm{loc}}\}.$$

For example, if $D = \sum_{i=1}^r a_i D_i$ is an effective $\mathbb{Q}$-divisor, and $U \subset X$ is an open subset such that each D_i is given by the vanishing of $g_i \in \mathcal{O}_U$, then we let $\phi_D|_U := \sum_{i=1}^r a_i\, log|g_i|$ and it follows that $\mathcal{J}(\phi_D) = \mathcal{J}(D)$ is the usual algebraic multiplier ideal. Therefore, (X, ϕ) is klt if $\mathcal{J}(\phi) = \mathcal{O}_X$ and (X, ϕ) is lc if $\mathcal{J}(\phi^{1-\epsilon}) = \mathcal{O}_X$ for any $0 < \epsilon \ll 1$. This allows for a great deal of flexibility in the analytic context which is key to proving many interesting results.

Acknowledgments The author was partially supported by the NSF research grants no: DMS-1952522 and DMS-2301374 and a grant from the Simons Foundation SFI-MPS-MOV-00006719-07. He would like to thank H. Blum, F. Figueroa, J. Moraga, and K. Schwede for useful comments.

References

1. V. Alexeev, C. Hacon, Y. Kawamata, *Termination of (many) 4-dimensional log flips.* Invent. Math. 168 (2007), no.2, 433–448.
2. F. Ambro, *The locus of log canonical singularities.* 1998. arXiv:9806067.
3. F. Ambro, *On minimal log discrepancies.* 1999. arXiv:9906089.
4. F. Ambro, *Quasi-log varieties*, Tr. Mat. Inst. Steklova 240 (2003), Biratsion. Geom. Linein. Sist. Konechno Porozhdennye Algebry, 220–239; translation in Proc. Steklov Inst. Math. 2003, no. 1 (240), 214–233.
5. F. Ambro, *The moduli b-divisor of an lc-trivial fibration*, Compos. Math. 141 (2005), no. 2, 385–403.
6. F. Ambro, P. Cascini, V. Shokurov, C. Spicer, *Positivity of the Moduli Part*, preprint, arXiv:2111.00423, 2011.
7. V. Alexeev and M. Brion, *Stable reductive varieties. II. Projective case*, Adv. Math. 184 (2004), no. 2, 380–408
8. A. Beauville, *Complex algebraic surfaces*, London Math. Soc. Stud. Texts, 34 Cambridge University Press, Cambridge, 1996, x+132 pp.
9. B. Bhatt, L. Ma, Z. Patakfalvi, K. Schwede, K. Tucker, J. Waldron and J. Witaszek, *Globally +-regular varieties and the minimal model program for threefolds in mixed characteristic*, Publs Math. IHES no. 1 138 (2023) 69–227
10. C. Birkar, *Ascending chain condition for log canonical thresholds and termination of log flips,* Duke Math. J. 136 (2007), no. 1, 173–180.
11. C. Birkar, *Existence of log canonical flips and a special LMMP.* Publ. Math. Inst. Hautes Études Sci. 115 (2012), 325–368.
12. C. Birkar, *Generalised pairs in birational geometry.* EMS Surv. Math. Sci.8(2021), no. 1–2, 5–24.
13. *Singularities of linear systems and boundedness of Fano varieties*, Ann. of Math. (2) 193 (2021), no. 2, 347–405.
14. C. Birkar, P. Cascini, C. D. Hacon and J. McKernan, *Existence of minimal models for varieties of log general type,* J. Amer. Math. Soc. 23(2), 405–468(2010).
15. C. Birkar and V. V. Shokurov, *Mld's vs thresholds and flips.* J. Reine Angew. Math. 638 (2010), 209–234.
16. H. Blum, *Existence of valuations with smallest normalized volume*, Compos. Math. 154 (2018), no. 4, 820–849.
17. J.-F. Boutot. *Singularités rationnelles et quotients par les groupes réductifs.* Invent. Math., 88(1):65–68, 1987.
18. L. Braun, *The local fundamental group of a Kawamata log terminal singularity is finite*, Invent. Math. 226 (2021), no. 3, 845–896.
19. L. Braun, D. Greb, K. Langlois, and J. Moraga, *Reductive quotients of klt singularities*, arXiv:2111.02812, preprint (2021).
20. L. Braun, F. Filipazzi, J. Moraga and R. Svaldi, *The Jordan property for local fundamental groups,* Geom. Topol. 26 (2022), no. 1, 283–319.
21. J. Chen, M. Chen, *Explicit birational geometry of threefolds of general type, I,* Annales scientifiques de l'École Normale Supérieure, Serie 4, Volume 43 (2010) no. 3, pp. 365–394.
22. S. D. Cutkosky, *Multiplicities associated to graded families of ideals,* Algebra Number Theory 7 (2013), no. 9, 2059–2083.

23. O. Das and C. Hacon, *On the minimal model program for Kähler 3-folds*, arXiv:2306.11708, preprint, (2023).
24. O. Das, C. Hacon and M. Paun, *On the 4-dimensional minimal model program for Kähler varieties,* Adv. Math. 443, Paper No. 109615 (2024).
25. O. Das, C. Hacon and J. I. Yanez, *MMP for Generalized Pairs on Kähler 3-folds*, arXiv:2305.00524 (April 2023).
26. T. de Fernex, C. D. Hacon. *Singularities on normal varieties*, Compos. Math. 145 (2009), no. 2, 393–414.
27. T. de Fernex, C. D. Hacon. *Deformations of canonical pairs and Fano varieties.* J. Reine Angew. Math. 651 (2011), 97–126.
28. T. de Fernex, J. Kollár, C. Xu, *The dual complex of singularities.* Adv. Stud. Pure Math., 74 Mathematical Society of Japan, Tokyo, 2017, 103–129.
29. L. Ein, R. Lazarsfeld, K. E. Smith, *Uniform approximation of Abhyankar valuation ideals in smooth function fields*, Amer. J. Math. 125 (2003), no. 2, 409–440.
30. F. Figueroa, J. Moraga, *Fundamental groups of low-dimensional lc singularities.* ArXiv:2302.11790.
31. S. Filipazzi, *On a generalized canonical bundle formula and generalized adjunction.* Ann. Sc. Norm. Super. Pisa Cl. Sci. (5) 21 (2020), 1187–1221.
32. S. Filipazzi, R. Svaldi, *On the connectedness principle and dual complexes for generalized pairs,* Forum Math. Sigma, 11 (2023), e33.
33. O. Fujino, *Termination of 4-fold canonical flips.* Publ. Res. Inst. Math. Sci. 40 (1), 231–237 (2004); Addendum. 41 (1), 252–257 (2005)
34. O. Fujino, *Non-vanishing theorem for log canonical pairs*, Journal of Algebraic Geometry 20 (4), 771–783, 2011-01-03
35. O. Fujino, *Fundamental Theorems for the Log Minimal Model Program*, Journals PRIMS Vol. 47, (2011) No. 3, pp. 727–789
36. O. Fujino, *Vanishing theorems.* arXiv:1202.4200.
37. O. Fujino, *Minimal model program for projective morphisms between complex analytic spaces*, arXiv e-prints , arXiv:2201.11315 (January 2022)
38. O. Fujino, Y. Gongyo, *On canonical bundle formulas and subadjunctions.* Michigan Math. J. 61 (2012), no. 2, 255–264.
39. O. Fujino and S. Mori, *A canonical bundle formula*, J. Differential Geom. 56 (2000), no. 1, 167–188.
40. O. Fujino, K. Schwede, S. Takagi, *Supplements to non-lc ideal sheaves.* RIMS Kokyuroku Bessatsu, B24, 2011, 1–47.
41. D. Greb, S. Kebekus, S. Kovács, T. Peternell, *Differential forms on log canonical spaces*, Publications Mathématiques de l'IHÉS, Volume 114 (2011), pp. 87–169.
42. C. D. Hacon. *Divisors on principally polarized abelian varieties.* Compositio Math. 119 (1999), no.3, 321–329.
43. C. D. Hacon. *On the Log Canonical Inversion of Adjunction.* In: Proceedings of the Edinburgh Mathematical Society 57.1 (2014), pp. 139–143
44. C. D. Hacon, J. Liu, *Existence of flips for generalized lc pairs.* Camb. J. Math. 11 (2023), no.4, 795–828.
45. J. Han, J. Liu, and V. V. Shokurov, *ACC for minimal log discrepancies of exceptional singularities,* arXiv:1903.04338.
46. C. D. Hacon, J. McKernan, C. Xu, *ACC for log canonical thresholds*, Ann. of Math. (2) 180 (2014), no. 2, 523–571.
47. C. D. Hacon, J. McKernan, C. Xu, *Boundedness of varieties of log general type.* Proceedings of Symposia in Pure Mathematics Volume 97.1, 2018
48. C. D. Hacon, J. McKernan, C. Xu, *Boundedness of moduli of varieties of general type.* JEMS Volume 20, Issue 4, 2018, pp. 865–901
49. C. D. Hacon, J. McKernan, *On Shokurov's rational connectedness conjecture*, Duke Math. J. 138 (2007), no. 1, 119–136.

50. C. D. Hacon, J. McKernan, *Flips and flops.* Proceedings of the International Congress of Mathematicians. Volume II, 513–539. Hindustan Book Agency, New Delhi; 2010
51. C. D. Hacon, J. Moraga, *On weak Zariski decompositions and termination of flips,* Math. Res. Lett. 27 (2020), no. 5, 1393–1421.
52. C. Hacon and M. Paun, *On the Canonical Bundle Formula and Adjunction for Generalized Kähler Pairs*, arXiv:2404.12007 (April 2024).
53. C. D. Hacon, J. Witaszek, *On the rationality of Kawamata log terminal singularities in positive characteristic*, Algebr. Geom. 6(5) (2019), 516–529.
54. C. D. Hacon, J. Witaszek, *The minimal model program for threefolds in characteristic 5*, Duke Mathematical Journal (2022), 1–39.
55. C. D. Hacon, J. Witaszek, *On the relative minimal model program for fourfolds in positive and mixed characteristic.* Forum Math. Pi 11 (2023), Paper No. e10, 35 pp.
56. C. Hacon, L. Xie, *ACC for generalized log canonical thresholds for complex analytic spaces*, arXiv:2303.00993, preprint (2023).
57. C. D. Hacon, C. Xu, *Existence of log canonical closures*, Invent. Math. 192 (2013), no. 1, 161–195.
58. C. D. Hacon, C. Xu, *On the three dimensional minimal model program in positive characteristic*, J. Amer. Math. Soc. 28(3) (2015), 711–744.
59. N. Hara, *A characterization of rational singularities in terms of injectivity of Frobenius maps* Sürikaisekikenkyüsho Kökyüroku (1996), no. 964, 138–144.
60. N. Hara, *Classification of two-dimensional F -regular and F -pure singularities.* Adv. Math. 133 (1998), no. 1, 33–53.
61. N. Hara, *Geometric interpretation of tight closure and test ideals*, Trans. Amer. Math. Soc. 353 (2001), 1885–1906.
62. N. Hara and K. Watanabe, *F-regular and F-pure rings vs. log terminal and log canonical singularities*, J. Algebraic Geom. 11 (2002), 363–392.
63. R. Hartshorne, R. Speiser *Local cohomological dimension in characteristic p.* Ann. of Math. (2) 105 (1977), no.1, 45–79.
64. M. Hochster and J. L. Roberts, *Rings of invariants of reductive groups acting on regular rings are Cohen-Macaulay*, Advances in Math. 13 (1974), 115–175
65. M. Jonsson and M. Mustaţă, *Valuations and asymptotic invariants for sequences of ideals*, Ann. Inst. Fourier (Grenoble) 62 (2012), no. 6, 2145–2209 (2013).
66. M. Kapovich and J. Kollár, *Fundamental groups of links of isolated singularities*, J. Amer. Math. Soc., 27(4): 929–952, 2014.
67. M. Kawakita, *Inversion of adjunction on log canonicity.* Invent. Math. 167 (2007), no. 1, 129–133.
68. Y. Kawamata, *On the length of an extremal rational curve*, Invent. Math. 105 (1991), 609–611.
69. Y. Kawamata, *Termination of log flips for algebraic 3-folds.* Int. J. Math. 3 (5), 653–659 (1992)
70. Y. Kawamata, *Subadjunction of log canonical divisors. II.* Amer. J. Math. 120 (1998), no. 5, 893–899.
71. Y. Kawamata, *On the extension problem of pluricanonical forms.* Pragacz, Piotr (ed.) et al., Algebraic geometry: Hirzebruch 70. Proceedings of the algebraic geometry conference in honor of F. Hirzebruch's 70th birthday, Stefan Banach International Mathematical Center, Warszawa, Poland, May 11–16, 1998. Providence, RI: American Mathematical Society. Contemp. Math. 241 (1999), 193–207
72. G. Kempf, F. Knudsen, D. Mumford, and B. Saint-Donat, *Toroidal embeddings.* I, Springer-Verlag, Berlin, 1973, Lecture Notes in Mathematics, Vol. 339
73. *Singularities of pairs*, Algebraic geometry-Santa Cruz 1995, Amer. Math. Soc., Providence, RI, 1997, pp. 221–287.
74. J. Kollár, *Kodaira's canonical bundle formula and adjunction.* Flips for 3-folds and 4-folds, 134–162, Oxford Lecture Ser. Math. Appl., 35, Oxford Univ. Press, Oxford, 2007

75. J. Kollár, *Singularities of the minimal model program*, Cambridge Tracts in Mathematics, vol. 200, Cambridge University Press, Cambridge, 2013. With a collaboration of Sándor Kovács.
76. J. Kollár, *Simple normal crossing varieties with prescribed dual complex,* Algebraic Geometry 1 (2014), 57–68.
77. J. Kollár, *Families of Varieties of General Type* Cambridge Tracts in Mathematics (231), Cambridge University Press, April 2023
78. J. Kollár et al., *Flips and abundance for algebraic threefolds, Société Mathématique de France*, Paris, 1992, Papers from the Second Summer Seminar on Algebraic Geometry held at the University of Utah, Salt Lake City, Utah, August 1991, Astérisque No. 211 (1992).
79. J. Kollár, S. Kovács, *Birational geometry of log surfaces*, Preprint, https://sites.math.washington.edu/~kovacs/pdf/BiratLogSurf.pdf
80. J. Kollár and S. Kovács, *Log canonical singularities are Du Bois*, J. Amer. Math. Soc. 23 (2010), no. 3, 791–813.
81. J. Kollár and S. Mori, *Birational geometry of algebraic varieties*, Cambridge tracts in mathematics, vol. 134, Cambridge University Press, 1998
82. J. Kollár, Y. Miyaoka, S. Mori, *Rationally connected varieties.* J. Alg. Geom. 1 (1992) 429–448.
83. S. J. Kovács, K. E. Schwede, K. E. Smith, *The canonical sheaf of Du Bois singularities*, Advances in Math., 2008.
84. S. Kumar, K. Schwede, *Richardson varieties have Kawamata log terminal singularities,* Int. Math. Res. Not. IMRN (2014), no. 3, 842–864.
85. R. Lazarsfeld, *Positivity in algebraic geometry. II,* Ergebnisse der Mathematik und ihrer Grenzgebiete. 3. Folge. A Series of Modern Surveys in Mathematics [Results in Mathematics and Related Areas. 3rd Series. A Series of Modern Surveys in Mathematics], vol. 49, Springer-Verlag, Berlin, 2004, Positivity for vector bundles, and multiplier ideals.
86. R. Lazarsfeld and M. Mustaţă, *Convex bodies associated to linear series,* Ann. Sci. Éc. Norm. Supér. (4) 42 (2009), no. 5, 783–835
87. J. Liu, F. Meng, L. Xie, *Infinitesimal structure of log canonical thresholds*, arXiv:2209.11369
88. Y. Liu and C. Xu, *K-stability of cubic threefolds*, Duke Math. J. 168 (2019), no. 11, 2029–2073.
89. G. Lyubeznik, *On the vanishing of local cohomology in characteristic* $p > 0$. Compos. Math. 142 (2006), no. 1, 207–221.
90. C. Li, *Minimizing normalized volumes of valuations*, Math. Z. 289 (2018), no. 1–2, 491–513.
91. C. Li, Y. Liu, C. Xu, *A guided tour to normalized volumes*, Progress in Mathematics 333, 167–219, Birkhuser/Springer, Cham, 2020.
92. J. Liu, *Toward the equivalence of the ACC for a-log canonical thresholds and the ACC for minimal log discrepancies,* preprint, https://arxiv.org/pdf/1809.04839 (2018).
93. V. B. Mehta and V. Srinivas, *A characterization of rational singularities,* Asian J. Math. 1 (1997), no. 2, 249–271.
94. J. Moraga, *On a toroidalization for klt singularities.* https://arxiv.org/abs/2106.15019
95. S. Mori, *Flip theorem and the existence of minimal models for 3-folds.* J. Amer. Math. Soc. 1 (1988), no.1, 117–253.
96. M. Mustaţă, *Ordinary varieties and the comparison between multiplier ideals and test ideals II*, Proc. Amer. Math. Soc. Volume 140, Number 3, 2012, 805–810.
97. M. Mustaţă, V. Srinivas, *Ordinary varieties and the comparison between multiplier ideals and test ideals*, Nagoya Math. J. 204 (2011), 125–157
98. M. Nagata, *On the 14th problem of Hilbert.* Sugaku, 12:203–209, 1960/61.
99. N. Nakayama. *Invariance of the plurigenera of algebraic varieties.* RIMS preprint, March 1998.
100. Y. Nakamura, K. Shibata, *A counterexample to the PIA conjecture for minimal log discrepancies*, https://arxiv.org/abs/2404.06164
101. Y. Odaka and C. Xu, *Log-canonical models of singular pairs and its applications*, Math. Res. Lett. 19 (2012), no. 2, 325–334.
102. M. Reid, *Minimal models of canonical 3-folds*, Adv. Stud. Pure Math. 1(1983), 131–180.

103. H. Schoutens, *Log-terminal singularities and vanishing theorems via non-standard tight closure*. J. Algebraic Geom., 14(2):357–390, 2005.
104. K. Schwede, *A simple characterization of Du Bois singularities*, Compos. Math. 143 (2007), no. 4,813–828.
105. K. Schwede, *A canonical linear system associated to adjoint divisors in characteristic* $p >> 0$, J. Reine Angew. Math. 696 (2014), 69–87.
106. K. Schwede and K. Smith, *Globally F -regular and log Fano varieties*. Advances in Mathematics, Vol 224 Issue 3, 863–894, 2010.
107. K. Schwede and K. Smith, *Singularities defined by the Frobenius map*, January 30, 2024, https://github.com/kschwede/FrobeniusSingularitiesBook
108. V.V Shokurov, *Problems about Fano varieties*, Birational Geometry of Algebraic Varieties, Open Problems – Katata 1988, 30–32.
109. V.V Shokurov, *Three-dimensional log perestroikas*, Izv. Ross. Akad. Nauk Ser. Mat. 56 (1992), no. 1, 105–203.
110. V.V Shokurov, *Letters of a bi-rationalist. V. Minimal log discrepancies and termination of log flips*. Tr. Mat. Inst. Steklova 246, 328–351 (2004) (Algebr. Geom. Metody, Svyazi i Prilozh.)
111. V.V Shokurov, *Letters of a bi-rationalist. VII. Ordered termination*. Tr. Mat. Inst. Steklova264, Mnogomernaya Algebraicheskaya Geometriya(2009), 184–208; translation in Proc. Steklov Inst. Math.264(2009), no.1, 178–200.
112. K. Smith, *F-rational rings have rational singularities*. Amer. J. Math. 119 (1997), no. 1, 159–180.
113. K. Smith, *The multiplier ideal is a universal test ideal*, Comm. Algebra 28 (2000), 5915–5929.
114. J. H. M. Steenbrink, *Mixed Hodge structures associated with isolated singularities*, Singularities, Part 2 (Arcata, Calif., 1981), 1983, pp. 513–536.
115. J. H. M. Steenbrink, *Vanishing theorems on singular spaces*, Astérisque 130 (1985), 330–341.
116. Y.-T. Siu, *Invariance of plurigenera*. Invent. Math. 134 (1998), no. 3, 661–673.
117. S. Takagi, *An interpretation of multiplier ideals via tight closure*, J. Algebraic Geom. 13 (2) (2004) 393–415
118. S. Takagi, T. Yamaguchi, *On the behavior of adjoint ideals under pure morphisms*, https://arxiv.org/abs/2312.17537 v2, preprint, (2023).
119. G. T. Todorov, *Effective log Iitaka fibrations for surfaces and threefolds*, Manuscripta Math. 133 (2010), no. 1–2, 183–195.
120. L. Xie, *Contraction theorem for generalized pairs*, https://arxiv.org/abs/2211.10800
121. C. Xu, *Finiteness of algebraic fundamental groups*, Compos. Math. 150 (2014), no. 3, 409–414.
122. C. Xu, *A minimizing valuation is quasi-monomial*, Ann. of Math. (2) 191 (2020), no. 3, 1003–1030.
123. C Xu, Z Zhuang, *Uniqueness of the minimizer of the normalized volume function*, Camb. J. Math. 9 (2021), no. 1, 149–176.
124. C Xu, Z Zhuang, *Stable degenerations of singularities*, preprint arXiv:2205.10915, 2022
125. C Xu, Z Zhuang, *Boundedness of log Fano cone singularities and discreteness of local volumes*, preprint arXiv:2404.17134, 2024
126. Z. Zhuang, *Direct summands of klt singularities*, arXiv:2208.12418, preprint (2022)
127. Z. Zhuang, *Stability of klt singularities*, https://arxiv.org/pdf/2307.10525, preprint (2023), to appear in Handbook of Geometry and Topology of Singularities, volume VII.

Chapter 9
Thom Polynomials for Singularities of Maps

Toru Ohmoto

On the occasion of the 101st anniversary of René Thom's birth

Contents

T. Ohmoto (✉)
Department of Pure and Applied Mathematics, Waseda University, Tokyo, Japan
e-mail: toruohmoto@waseda.jp

© The Author(s), under exclusive license to Springer Nature Switzerland AG 2026

J. L. Cisneros-Molina et al. (eds.), *Handbook of Geometry and Topology of Singularities VIII*, https://doi.org/10.1007/978-3-031-99571-2_9

Abstract This is a gentle introduction to a general theory of universal polynomials associated to classification of map-germs, called *Thom polynomials*. The theory was originated by René Thom in the 1950s and has since been evolved in various aspects by many authors. In a nutshell, this is about intersection theory on certain moduli spaces, say 'classifying spaces of mono/multi-singularities of maps', which provides consistent and deep insights into both classical and modern enumerative geometry with many potential applications.

9.1 Introduction

Thom polynomials are a key component of a general enumerative theory for singularities of real and complex mappings—these are universal cohomological obstructions to the appearance of singular points of prescribed types in given mappings. Roughly speaking, a *singularity* of a differentiable map f means a point where the differential df is degenerate (and the germ of f at such a point) or also a multiple image point of the map. There are a lot of different local types of singularities of maps, and the locus of those singular points carries important information about global topological nature of the source and the target spaces. Historically, the origins lie in classical enumerative geometry founded by pioneers of the nineteenth century, and based on it, the theory was born as a branch of singularity theory of differentiable maps in the mid-twentieth century. In fact, René Thom simultaneously developed singularity theory, cobordism theory, characteristic classes and cohomology operations, armed with his transversality theorem, since his doctor thesis [187]—especially, in 1957, the year before he received the Fields Medal, Thom gave the inaugural lecture for professorship at the University of Strasbourg, in which he manifested there is a deep correspondence from local classification theory of map-germs to global theory on topology of manifolds and maps [189]. This is what people later called Thom polynomial theory.[1] Nowadays, the theory itself has been expanding to much broader fields dealing with various enumeration problems in algebraic geometry, differential topology, complex analytic geometry, representation theory, and their interactions.

Excellent surveys of the theory up to the early 1990s may be found in Kleiman [90, 91, 93] from algebraic geometry side, and in the fourth chapter of [10] written by Vassiliev (Vasil'ev) from topology side. The present article aims to briefly outline subsequent developments with emphasizing new ideas—*equivariant cohomology*, *localization* and *Hilbert schemes of points*—and how those are merged with old traditional methods. As the theory still now continues to develop, this survey may include conjectural (but mostly reliable) statements and should therefore be considered an interim report at this stage. Additionally, content selection is heavily biased by the author's interests—in particular, due to space constraints, we restrict

[1] The author heard this story from B. Teissier at the workshop "Hommage à René Thom, IRMA, Strasbourg, 2016.

our attention only to some basic topics in singularity theory of maps. Consequently, there are several important topics which are not picked up or less mentioned in this survey. For instance, we will not enter advanced topics in contemporary Schubert calculus and enumerative geometry on various moduli stacks, which are deeply connected to geometric representation theory, quantum integrable systems, mirror symmetry, Gromov-Witten and Donaldson-Thomas theories and so on. Also about real Thom polynomials for C^∞-mappings in differential topology, the range of related topics are quite large, e.g., from Gromov's h-principle to low dimensional topology including smooth Poincaré conjecture, so we will only tough on a few basic matters. For these topics not covered in this survey, readers are referred to appropriate references.

A new excellent survey paper on Thom polynomial theory written by Rimányi [160] recently appeared. It differs in style, focus and perspective from this survey and is highly recommended reading. Readers are also referred to Rimányi's website [161], which contains many latest computational results of Thom poylnomials.

We work mainly with the complex analytic category (and real C^∞ category in the last section). It is also possible to take algebraic geometry context with some appropriate modifications. We assume that readers are familiar with basics of differential topology or algebraic geometry, and especially, preliminary elements for the following subjects (e.g., see the very beginning chapters in cited survey papers included in Handbook of Geometry and Topology on Singularities III):

- Thom-Mather theory of singularities of maps (cf. [120]);
- characteristic classes of vector bundles (cf. [24]);
- intersection theory in algebraic geometry (cf. [2]).

Throughout this paper, H_* stands for the Borel-Moore homology and H^* the singular cohomology, where the coefficients are integers unless specifically noted. For complex manifolds, H_* and H^* are identified via the Poincaré duality as usual. In algebraic geometry context, we use Chow groups CH_* (Chow rings CH^* for non-singular varieties) as in [58].

The *codimension* κ *of a map* $f : M \to N$ between manifolds is defined as $\kappa := \dim N - \dim M$ and also the relative dimension to be $d := -\kappa$.

For a complex vector bundle $E \to M$, the Chern class $c_i(E) \in H^{2i}(M)$ is assigned ($c_0(E) = 1$, $c_i(E) = 0$ for $i > \operatorname{rank} E$). We often use the *quotient Chern class* $c_i = c_i(f)$ associated to a map $f : M \to N$, which is defined as the i-th Chern class of the difference bundle $f^*TN - TM$, i.e.,

$$c(f) := 1+c_1(f)+c_2(f)+\cdots = \frac{1 + f^*c_1(TN) + f^*c_2(TN) + \cdots}{1 + c_1(TM) + c_2(TM) + \cdots} \in H^*(M).$$

We also use $\bar{c}_i(f) := c_i(TM - f^*TN)$, i.e.,

$$\bar{c}(f) := 1 + \bar{c}_1(f) + \bar{c}_2(f) + \cdots \quad \text{with} \quad c(f)\bar{c}(f) = 1.$$

Note that every $\bar{c}_i(f)$ is a polynomial in $c_1(f), \cdots, c_i(f)$, vice versa, and the dual version is $c_i(T^*M - f^*T^*N) = (-1)^i\bar{c}_i(f)$.

If f is proper, the pushforward $f_* : H^*(M) \to H^{*+\kappa}(N)$ is defined (precisely, it is the Gysin homomorphism through the Poincaré duality; we simply denote it by f_* unless any confusion occurs; the superscript should be multiplied by 2 in complex case). It satisfies the projection formula

$$f_*(\alpha \cdot f^*\beta) = f_*(\alpha) \cdot \beta.$$

Since f_* is just an additive homomorphism, the *Landweber-Novikov class* with index $I = (i_1, i_2, \cdots, i_k)$

$$s_I(f) := f_*(c^I(f)) = f_*(c_1(f)^{i_1}c_2(f)^{i_2}\cdots c_k(f)^{i_k})$$

takes important role, especially in complex cobordism theory. We often use the same notation s_I to mean its pullback f^*s_I (i.e., drop the letter f^*), for it is easily understood from the context. Also the readers should not confuse the notation with Segre classes of (virtual) vector bundles.

9.1.1 *Two Guiding Examples*

As an initial guide to readers, we first present typical examples of Thom polynomials for classifications of linear maps and for non-linear map-germs. Throughout this survey, these examples will be referred to repeatedly.

9.1.1.1 Thom-Porteous Formula

Consider the classification of linear maps $h : \mathbb{C}^m \to \mathbb{C}^n$ under linear coordinate changes of the source and the target spaces. The orbits are determined by the rank, or equivalently, the dimension of kernel. Note that using $\dim\ker h$ is better than $\operatorname{rank} h$ in the sense that $\dim\ker h$ does not change under the trivial suspension $h : \mathbb{C}^m \to \mathbb{C}^n \rightsquigarrow h \times id_k : \mathbb{C}^{m+k} \to \mathbb{C}^{n+k}$. The corresponding Thom polynomials are described as follows.

Let $E \to M$ and $F \to M$ be complex vector bundles of rank m and n, respectively, over the same base manifold M. Let $\varphi : E \to F$ be a vector bundle morphism, that is a section of the vector bundle $\mathrm{Hom}(E, F) = E^* \otimes F \to M$. We are herewith interested in the loci corresponding to singular linear maps; put

$$\Sigma^k(\varphi) = \{\, x \in M \mid \dim\ker\varphi_x = k \,\}$$

and define the *degeneracy locus* of φ to be the closure

$$\overline{\Sigma^k(\varphi)} = \bigcup_{s \geq k} \Sigma^s(\varphi).$$

The expected codimension of the locus is given by kl where $k = \dim\ker$ and $l := \dim\mathrm{coker} = n - m + k \,(= \kappa + k)$. In particular, for a map $f : M^m \to N^n$, we denote by $\Sigma^k(f)$ the locus for the differential $df : TM \to f^*TN$, called the *first order Thom-Boardman singularity stratum*.

The *Giambelli-Thom-Porteous formula* [58, 151] states that for suitably generic φ, the dual to the degeneracy locus is expressed by

$$\mathrm{Dual}\,[\overline{\Sigma^k(\varphi)}] \;=\; \Delta_{l^k}(c(F-E)) \;=\; \overbrace{\begin{vmatrix} c_l & c_{l+1} & \cdots \\ c_{l-1} & c_l & \cdots \\ \vdots & \vdots & \ddots \end{vmatrix}}^{k} \tag{9.1}$$

where c_i is the i-th component of the quotient Chern class $c(F - E)$. The Schur function of type $l^k = (l, \cdots, l)$ in the right hand side of (9.1) is the *Thom polynomial* $tp(\Sigma^k)$ *for the type* Σ^k. Even for non-generic case, i.e., for an arbitrary bundle map φ, the formula is valid after replacing the left hand side of (9.1) by a certain intersection class, called the *degeneracy loci class*, see Fulton [58]. We remark that there is a well-known identity via conjugacy:

$$\Delta_\lambda(c(F-E)) = \Delta_{\lambda^\vee}(c(E^* - F^*))$$

(cf. [58, Lem.14.5.1]; note that the notation is inconsistent across different references).

In case that $\kappa \geq 0$ and E is trivial, then the largest degeneracy locus represents $c_{\kappa+1}(F)$, i.e., the Chern class of a vector bundle itself is a sort of Thom polynomial.

There are several directions for generalization of the Thom-Porteous formula. When replacing the group by Borel subgroups, we meet the (matrix) Schubert calculus, so this passage goes to geometric representation theory and algebraic combinatorics (Sects. 9.3.1.1 and 9.4.2.1). When one considers non-linear mappings, it goes to singularity theory of maps, see the next example.

9.1.1.2 Fold and Cusp

Recall the *classical Riemann-Hurwitz formula*. Let $f : M \to N$ be a surjective holomorphic map between compact nonsingular complex curves: to each point $x \in M$, we assign $\mu = \mu(f, x)$ so that f is written by $x \mapsto x^{\mu+1}$ in local coordinates centered at x and $f(x)$. Then it holds that

$$\begin{aligned}\int_M \mu(f) &= \deg f \cdot \chi(N) - \chi(M) \\ &= c_1(TN) \frown f_*[M] - c_1(TM) \frown [M] \\ &= c_1(f^*TN - TM) \frown [M] \\ &= c_1(f) \frown [M].\end{aligned}$$

Here we use the fact that the Euler characteristic $\chi(M)$ is the degree of the top Chern class $c_1(TM)$ of the tangent bundle. If f is generic, i.e., it has only simple branches ($\mu = 1$), called the A_1-*singularity*, then the formula counts the number of the singular points. For a holomorphic map $f : M \to N$ between complex surfaces, if it is appropriately generic, local singularities of f are classified into two types, called *fold* and *cusp*, up to local coordinate changes of M and N at the point, whose normal forms are respectively given by

$$A_1 : (x, y) \mapsto (x^2, y), \quad A_2 : (x, y) \mapsto (x^3 + yx, y)$$

(see Fig. 9.1). In general, for map-germs $(\mathbb{C}^m, 0) \to (\mathbb{C}^{m+\kappa}, 0)$ with $\kappa \geq 0$, the symbol A_μ denotes the singularity type whose local algebra is isomorphic to $\mathbb{C}[x]/\langle x^{\mu+1}\rangle$, see Sect. 9.2.1.

The critical locus of f is the closure $\overline{A_1(f)}$, which is formed by fold points and isolated cusp points. In his paper [188], R. Thom showed (precisely saying, in real case, using the Stiefel-Whitey classes) that the cohomology class represented by the critical locus and the number of cusps are expressed by

$$\text{Dual}\,[\overline{A_1(f)}] = c_1(f), \quad \text{Dual}\,[\overline{A_2(f)}] = c_1(f)^2 + c_2(f) \tag{9.2}$$

in $H^*(M)$. The former one is essentially the same as the above classical Riemann-Hurwitz formula. In fact, the formulas (9.2) hold for any generic maps between manifolds of the same dimension ≥ 2. The right hand side, $tp(A_1) := c_1$

Fig. 9.1 Singularities of plane-to-plane maps: fold, cusp and double-folds

and $tp(A_2) := c_1^2 + c_2$, are respectively called *Thom polynomials for A_1 and A_2-singularity type with $\kappa = 0$*. The shape of the polynomial depends only on the singularity type, and the information of the map f itself is only encoded in substituting $c_i(f)$ to variables c_i of the polynomial. In case of $\kappa = 0$, more samples are:

$$\begin{aligned} tp(A_3) &= c_1^3 + 3c_1c_2 + c_3, \\ tp(A_4) &= c_1^4 + 6c_1^2c_2 + 2c_2^2 + 9c_1c_3 + 6c_4, \\ tp(A_5) &= c_1^5 + 10c_1^3c_2 + 25c_1^2c_3 + 10c_1c_2^2 + 38c_1c_4 + 12c_2c_3 + 24c_5 \\ tp(A_6) &= c_1^6 + 15c_1^4c_2 + 55c_1^3c_3 + 30c_1^2c_2^2 + 141c_1^2c_4 + 79c_1c_2c_3 \\ &\qquad + 5c_2^3 + 202c_1c_2 + 55c_2c_4 + 17c_3^2 + 120c_6 \\ &\cdots\ \cdots, \end{aligned} \tag{9.3}$$

and for singularity types of corank greater than one, e.g., $I_{a,b} : (x^a + y^b, xy)$,

$$tp(I_{2,2}) = c_2^2 - c_1c_3, \quad tp(I_{2,3}) = 2c_1c_2^2 - 2c_1^2c_3 + 2c_2c_3 - 2c_1c_4,$$

and so on—a full list up to codimension ≤ 8 with $\kappa = 0$ can be seen in Rimányi [156]. As a catalog of various kinds of Thom polynomials, many computational results have been collected in Rimányi's website 'Thom polynomial Portal' [161].

We may find general formulas which contains $\kappa \geq 0$ as a parameter. The simplest Thom-Porteous formula says that the Thom polynomial for A_1-singularity (i.e., dual to the critical locus) is given by

$$tp(A_1) = \Delta_{(\kappa+1)^1} = c_{\kappa+1} \tag{9.4}$$

and by a result due to Ronga [171] we know

$$tp(A_2) = c_{\kappa+1}^2 + \sum_{i\geq 1} 2^{i-1} c_{\kappa+1-i}c_{\kappa+1+i} \tag{9.5}$$

where we read $c_j = 0$ for $j < 0$. Those lead to a general theory of *Thom series*, see Sect. 9.3.2. In particular, from (9.2), (9.3), (9.4), (9.5) and further computations, Rimányi has conjectured the *Chern basis positivity for A_μ-singularity types* ($\kappa \geq 0$), that means that all coefficients of Chern monomials appearing in $tp(A_\mu)$ are non-negative [156]. That is still open, and sounds mysterious, as the conjecture is unexpectedly related to complex hyperbolic geometry, see Sect. 9.7.3.

Let us look back again at a generic map $f : M \to N$ between surfaces. There is one more stable singularity type, which is a bi-singularity of fold types, denoted by $A_1^2 = A_1A_1$ and called *transverse double-folds*, which corresponds to a *node* of the discriminant curve $D(f) := f(\overline{A_1(f)})$ in the target space N. Assume that f is proper, then the number of nodes of $D(f)$ is half the number counted by

$$\mathrm{Dual}\,[\overline{A_1^2(f)}] = -4c_1(f)^2 - 2c_2(f) + c_1(f)f^*s_1(f), \tag{9.6}$$

see Sect. 9.6.1. Furthermore, the Euler characteristic of $D(f)$ is also universally expressed by

$$\chi(D(f)) = \int_N (c_1(TN)s_1(f) + s_{01}(f) - \frac{1}{2}s_1(f)^2), \tag{9.7}$$

see Sect. 9.6.2. Namely, the way of the appearance of singularities of prescribed type of maps has its own universal rule, that is dominant by local symmetry of the singularity type. Those kinds of enumerative problems are treated in the Thom polynomial theory, and indeed appear in various fields of applications.

Finally we mention old examples about fold and cusps of maps between surfaces (cf. [141, 177, 180, 181])—perhaps, R. Thom must have been familiar with these treasures of French classical geometry since Monge and Poncelet. First, consider a smooth plane curve C of degree $d \geq 2$. Let M be the point-line incidence surface (i.e., it consists of pairs $(p, l) \in C \times \mathbb{P}^{2*}$ with $p \in l$) and $f : M \to N := \mathbb{P}^{2*}$ the projection, then the discriminant $D = D(f)$ is just the projective dual to C. A flex (inflection point) of C corresponds to a cusp of D and a double tangent line to C corresponds to a node of D. Applying (9.2) and (9.6), we recover the Plücker formula:

$$\deg D = d(d-1), \quad \#\mathrm{Cusps} = 3d(d-2), \quad \#\mathrm{Nodes} = d^4 - 2d^3 - 9d^2 + 18d.$$

Second, suppose that we are viewing a surface $M \subset \mathbb{P}^3$ of degree d from a general viewpoint. It is a map projecting M to a screen $N := \mathbb{P}^2$ and $D(f)$ is the apparent contour, and then a handy calculation of (9.2) and (9.6) immediately recovers old formulas:

$$\deg D = d(d-1), \ \#\mathrm{Cusps} = d(d-1)(d-2), \ \#\mathrm{Nodes} = \frac{1}{2}d(d-1)(d-2)(d-3).$$

Like as those, there are many different geometric situations where fold, cusps, double-folds appear naturally, while our formulas enumerate them in exactly the same way—namely, the Thom polynomial unites many such counting problems into a form that is relevant only for the type of local singularity. Furthermore, this theory reveals the intrinsic meaning of classical numerical characters of projective varieties. Although it is little-known, Semple and Roth [181] wrote that for a projective surface there are exactly four basic characters which generate all other characters of Salmon-Cayley-Zeuthen, and afterwards, Roth claimed an analogous statement for a projective three-fold that there are seven basic characters which recover dozens of others. Through computing Thom polynomials [178], it turns out that those basic characters correspond to Landweber-Novikov classes in modern language; there are four s_0, s_1, s_2, s_{01} for a surface, and seven $s_0, s_1, s_2, s_3, s_{01}, s_{11}, s_{001}$ for a three-fold, indeed. This suggests cobordism theory behind (see Sect. 9.6). A more advanced application to enumerative geometry, as well as unexpected applications to other fields, will be reviewed in Sect. 9.7.

9.2 Thom Polynomials for Singularities of Maps

9.2.1 *Classification of Map-Germs*

To begin with, we describe a few basic notions in the Thom-Mather theory, see e.g., [120], [119, Part I], [9, Part I] for details. See also Mather's [112] for algebraic geometry setting.

Let $\mathcal{O}_m$ be the local ring of holomorphic function germs $(\mathbb{C}^m, 0) \to \mathbb{C}$ with the maximal ideal $\mathfrak{m}_m = \{h \in \mathcal{O}_m, f(0) = 0\}$. Put $\mathcal{E}(m, n)$ to be the $\mathcal{O}_m$-module of all homolorphic map-germs $f : (\mathbb{C}^m, 0) \to \mathbb{C}^n$, and also put $\mathcal{E}_0(m, n) := \mathfrak{m}_m \mathcal{E}(m, n)$ which consists of all germs $f : (\mathbb{C}^m, 0) \to (\mathbb{C}^n, 0)$.

Equivalence The group of biholomorphic germs $(\mathbb{C}^m, 0) \to (\mathbb{C}^m, 0)$ is denoted by $\mathrm{Aut}(\mathbb{C}^m, 0)$. There are two different kinds of natural equivalence relations on map-germs: for $\mathcal{G} = \mathcal{A}, \mathcal{K}$, a $\mathcal{G}$-equivalence class of map-germs is called a *$\mathcal{G}$-singularity type* ($\mathcal{G}$-type, for short).

- $\mathcal{A}$-**classification** (right-left equivalence) classifies map-germs up to isomorphisms of the source and the target spaces. The *right-left group* $\mathcal{A}$ $(= \mathcal{A}_{m,n})$ is the direct product $\mathrm{Aut}(\mathbb{C}^m, 0) \times \mathrm{Aut}(\mathbb{C}^n, 0)$, which acts on $\mathcal{E}_0(m, n)$ by

$$(\sigma, \tau).f := \tau \circ f \circ \sigma^{-1}.$$

  Two map-germs $f, g : (\mathbb{C}^m, 0) \to (\mathbb{C}^n, 0)$ are *$\mathcal{A}$-equivalent* if f and g belong to the same $\mathcal{A}$-orbit.
- $\mathcal{K}$-**classification** (contact equivalence) classifies zero loci $f^{-1}(0)$ as schemes up to isomorphisms, i.e., it classifies the local algebra

$$Q(f) := \mathcal{O}_m / \langle f_1, \cdots, f_n \rangle_{\mathcal{O}_m}$$

  where f_i are component functions of f. In other words, the $\mathcal{K}$-equivalence measures the *tangency of the graphs* $y = f(x)$ *and* $y = 0$ *in* $\mathbb{C}^m \times \mathbb{C}^n$; the *contact group* $\mathcal{K}$ $(= \mathcal{K}_{m,n})$ consists of pairs (σ, Φ) of $\sigma \in \mathrm{Aut}(\mathbb{C}^m, 0)$ and map-germs $\Phi : (\mathbb{C}^m, 0) \to GL(n, \mathbb{C})$, and it acts on $\mathcal{E}_0(m, n)$ by

$$((\sigma, \Phi).f)(x) = \Phi(x) f(\sigma(x)).$$

  Two map-germs $f, g : (\mathbb{C}^m, 0) \to (\mathbb{C}^n, 0)$ are *$\mathcal{K}$-equivalent* if $Q(f) = \sigma^* Q(g)$ for some $\sigma \in \mathrm{Aut}(\mathbb{C}^m, 0)$, also equivalently, f and g belong to the same $\mathcal{K}$-orbit. In particular, if $f \sim_{\mathcal{A}} g$, then $f \sim_{\mathcal{K}} g$, i.e., $\mathcal{A}.f \subset \mathcal{K}.f$.

Deformations and Stability Let $f \in \mathcal{E}_0(m,n)$. An *infinitesimal deformation* of f is a vector field-germ along f. The space of infinitesimal deformations is regarded as the 'tangent space' of $\mathcal{E}(m,n)$ at f, and is denoted by

$$\theta(f) = \mathcal{O}_m \otimes \mathbb{C}^n.$$

For the identity map id_m of $\mathbb{C}^m$, $\theta_m := \theta(id_m)$ is the space of germs of vector fields on $\mathbb{C}^m$ at the origin. We set $tf : \theta_m \to \theta(f)$ by $tf(v) := df(v)$ and $\omega f : \theta_n \to \theta(f)$ by $\omega(w) := w \circ f$. For our convenience, we put

$$T\mathcal{A}_e.f = tf(\theta(id_m)) + \omega f(\theta(id_n)), \quad T\mathcal{K}_e.f = tf(\theta(id_m)) + f^*\mathfrak{m}_m\theta(f),$$

which are subspaces of $\theta(f)$ consisting of deformations caused by the actions of source and target isomorphisms and contact group action (not necessarily preserving the origin). Indeed, the subscript e indicates the extended action of reparametrization groups possibly moving the origin, and the above subspaces are the tangent spaces of the orbits.

We say that $f : (\mathbb{C}^m, 0) \to (\mathbb{C}^n, 0)$ is an *(infinitesimally) stable germ* if it holds that

$$\theta(f) = T\mathcal{A}_e.f,$$

namely, any deformation of f is re-produced by deformations of coordinate changes of the source and the target spaces. It is known, as one of major theorems of J. Mather, that for stable germs f and g, $f \sim_{\mathcal{K}} g$ if and only if $f \sim_{\mathcal{A}} g$. Namely, for a stable germ, $\mathcal{A}.f = \{\text{Stable germs}\} \cap \mathcal{K}.f$.

Determinacy We denote the k-jet of $f : (\mathbb{C}^m, 0) \to (\mathbb{C}^n, 0)$ by $j^k f(0)$ and the k-jet space by $J^k(m,n)$ (the space of Taylor coefficients of order $\leq k$ off the constant terms). Let $\mathcal{G} = \mathcal{A}$ or $\mathcal{K}$. A map-germ f is *k-$\mathcal{G}$-determined* if every map-germ g with $j^k g(0) = j^k f(0)$ is $\mathcal{G}$-equivalent to f. For instance, a stable germ $f : (\mathbb{C}^m, 0) \to (\mathbb{C}^n, 0)$ is $(n+1)$-$\mathcal{A}$-determined. Also f is *$\mathcal{G}$-finitely determined* (or *$\mathcal{G}$-finite* for short) if it is k-$\mathcal{G}$-determined for some k. Finite determinacy is equivalent to that the orbit $\mathcal{G}.f$ has finite codimension, that is also equivalent to that *$\mathcal{G}_e$-codimension* $\dim_{\mathbb{C}} \theta(f)/T\mathcal{G}_e.f$ is finite. In particular, every stable germ has $\mathcal{A}_e$-codimension zero.

Jet Space and Jet Extension The *k-jet bundle* $J^k(M,N) \to M \times N$ is the fiber bundle with fiber $J^k(m,n)$ and structure group $\mathcal{A}_{m,n}$. We define the *k-jet extension map*

$$j^k f : M \to J^k(M,N)$$

by assigning to every point $x \in M$ the pair of $(x, f(x))$ with the jet of the germ $f : (M, x) \to (N, f(x))$. Let k be high enough ($k \geq n + 1$). The following properties are equivalent:

1. the germ of $f : M \to N$ at x is stable;
2. $j^k f$ is transverse to the $\mathcal{A}$-orbit at x;
3. $j^k f$ is transverse to the $\mathcal{K}$-orbit at x.

Singularity Locus of Prescribed Type Throughout the present paper, we use the notations $\eta, \xi, \tau, \cdots$ to mean $\mathcal{K}$-invariant subsets of $J^\infty(m, n)$. A particular one is a $\mathcal{K}$-singularity type of mono-germs (also a moduli stratum of $\mathcal{K}$-types). We say that a mono-germ f is of type η if its jet belongs to η. For a holomorphic map $f : M \to N$ ($\dim M = m,\ \dim N = n$), we set

$$\eta(f) := \{\ x \in M \mid \text{the germ } f \text{ at } x \text{ is of type } \eta\ \}$$

and call the closure $\overline{\eta(f)} \subset M$ the *η-type singularity locus* of f. By finite determinacy, we may denote by $\eta \subset J^k(m, n)$ the corresponding $\mathcal{K}^k$-orbit in a jet space of sufficiently high order k. Since η is also $\mathcal{A}$-invariant, there is the sub-bundle of the fiber bundle $J^k(M, N) \to M \times N$ with fiber η, denoted by $\eta(M, N)$:

$$\begin{array}{ccccc} & & J^k(M,N) & \hookleftarrow & \overline{\eta(M,N)} \\ & \overset{j^k f}{\nearrow} & \downarrow & & \\ \overline{\eta}(f) \hookrightarrow M & \xrightarrow[(id,f)]{} & M \times N & & \end{array}$$

By the definition, $\eta(f) = jf^{-1}(\eta(M, N))$ and its closure is

$$\overline{\eta(f)} = jf^{-1}(\overline{\eta(M, N)}) = \bigcup_{\xi \subset \overline{\eta}} \xi(f)$$

where ξ runs over strata of a $\mathcal{K}$-invariant stratification of $\overline{\eta}$, i.e., $\xi = \eta$ or boundary strata in $\overline{\eta} - \eta$. This is a higher jet version of the degeneracy locus $\overline{\Sigma^k(f)}$ in Sect. 9.1.1.1.

We say that f is *generic with respect to* η if the jet extension $j^k f$ is transverse to $\eta(M, N)$ and its intersection with the boundary strata is of lower dimension. In this case, especially, the Zariski closure $\overline{\eta(f)}$ forms a cycle in the source manifold M, and we are interested in the cohomology classes represented by the locus.

9.2.2 Equivariant Cohomology and Poincaré Dual

The $\mathcal{G}$-classification of finitely determined map-germs is essentially reduced to the problem on finite dimensional spaces, i.e., the action of $\mathcal{G}$-equivalence at the level of jets of finite order. So we first review a general framework of *Thom polynomials for Lie group action* [50, 77, 80, 129].

Let G be a complex algebraic Lie group and X a complex algebraic variety on which G acts algebraically. In topology, the Borel construction tells that any family φ of G-orbits in X parametrized by some variety M (precisely, the pair of a fiber bundle $E \to M$ with fiber X and group G and a section of the bundle) is regarded as a certain morphism, called a *classifying map*

$$\bar{\varphi} : M \longrightarrow EG \times_G X$$

where $EG \to BG$ is the universal G-principal bundle and $EG \times_G X = (EG \times X)/G$ is the total space of the associated bundle with fiber X and group G over BG. Any two classifying maps corresponding to isomorphic families are homotopic. The *G-equivariant cohomology of X* (in the sense of Borel) is defined by

$$H_G^*(X) := H^*(EG \times_G X).$$

For a G-morphism $f : X \to Y$, the *pullback* $f_G^* : H_G^*(Y) \to H_G^*(X)$ is defined.

In algebraic geometry context (for reductive G), the universal G-principal bundle is constructed as ind-varieties [34, 186]. This mimics the construction in topology. Roughly saying, take a Zariski open subset U in al linear representation so that G acts on U freely and the quotient variety $U_G := U/G$ exists, then the inductive limit of the quotient map $U \to U_G$ taken over all representations of G (with respect to inclusions) gives an algebraic counterpart to $EG \to BG$. For instance, for $T = \mathbb{C}^* = \mathbb{C} - \{0\}$, we have $U = \mathbb{C}^N - \{0\} \to \mathbb{P}^N = U_T$, and for $G = GL_n$, let U be an open set in $\mathrm{Hom}(\mathbb{C}^n, \mathbb{C}^N)$ consisting of injective linear maps, then U_G is the Grassmaniann of n-planes in $\mathbb{C}^N$; those approximates $EG \to BG$. For quasi-projective X, the diagonal action of G on $X \times U$ is also free and the mixed quotient $X_U := (X \times U)/G$ exists so that the projection $p_U : X_U \to U_G$ is a fiber bundle with fiber X and group G. Then $H_G^*(X) = \varprojlim H^*(X_U)$. On the other hand, there is the dual counterpart, the *i-th equivariant homology group* $H_i^G(X)$, which is constructed in several ways, e.g., a suitable limit of $H_*(X_U)$ through the above Borel construction. It satisfies the following properties:

- it is a covariant functor to $H_G^*(pt)$-modules: for a proper G-morphism $f : X \to Y$, the pushforwad $f_*^G : H_*^G(X) \to H_*^G(Y)$ is defined;
- it has a specialization to the non-equivariant homology $H_*(X)$;
- there exists the *equivariant fundamental class* $[X]_G \in H_{2n}^G(X)$ ($\dim X = n$) which defines the homomorphism $\frown [X]_G : H_G^i(X) \to H_{2n-i}^G(X)$ so that it is isomorphic if X is non-singular (called the *G-equivariant Poincaré dual*).

We remark again that in case that X is non-compact, we work with (equivariant) Borel-Moore homology. For a closed G-subvariety W of non-singular X with the inclusion $\iota : W \hookrightarrow X$, we write the dual to the image $\iota_*^G[W]_G \in H_*^G(X)$ by $\mathrm{Dual}_G[W]_G$ (or simply $[W]_G$ for short). For a G-equivariant proper map $f : X \to Y$ between non-singular X, Y, we denote the pushforwad though Dual_G (the Gysin homomorphism) by the same notation $f_*^G : H_G^*(X) \to H_G^{*+2\kappa}(Y)$.

In particular, we are interested in the case that X is a vector space. Then $EG \times_G X$ is a vector bundle over BG, so its G-equivariant cohomology is isomorphic to $H_G^*(pt) = H^*(BG)$.

Definition 9.2.1 For a G-invariant closed subvariety W in a vector space V with G-action, we define the *Thom polynomial* of W to be the equivariant Poincaré dual

$$tp(W) := \mathrm{Dual}^G[W]_G \in H_G^*(V) = H^*(BG).$$

This notion has implicitly appeared everywhere with different names, e.g., in the setting of torus action, this is known as Joseh polynomial or multidegree etc.

Let V and W as in Definition 9.2.1. Given a vector bundle $\pi : E \to M$ over a complex manifold M with fiber V and structure group G, let $W(E) \to M$ be the fiber bundle with the fiber W and group G. For any section $s : M \to E$, the *degeneracy loci class of type W for s* is defined by

$$m_W(s) := s^*[W(E)] \in H^*(M).$$

We say that a section $s : M \to E$ is *generic with respect to W* if the map s is transverse to every stratum of (a Whitney stratification of) $W(E)$.

Proposition 9.2.2 ([50, 77, 80, 129]) *For a generic section $s : M \to E$ with respect to η, it holds that*

$$\mathrm{Dual}\ [W(s)] = m_W(s) = \rho^* tp(W) \in H^*(M)$$

where $W(s) := s^{-1}(W(E))$ and $\rho : M \to BG$ is the classifying map[2] *for $E \to M$.*

9.2.3 Thom Polynomials for $\mathcal{K}$-Classification

$\mathcal{K}$-equivalence of map-germs has a certain stabilization—essential is the codimension $\kappa = n - m$, rather than m and n. Two germs $f : (\mathbb{C}^{m+s}, 0) \to (\mathbb{C}^{n+s}, 0)$ and $g : (\mathbb{C}^m, 0) \to (\mathbb{C}^n, 0)$ are *stably $\mathcal{K}$-equivalent* if f is $\mathcal{K}$-equivalent to the trivial

---

[2] Precisely saying, the algebraic classifying map is defined over the total space of an affine bundle over M, while this issue does not effect formal computation in cohomology [34, 186].

unfolding $g \times id_s$ with s parameters. The following key lemma is easily checked [29, 126, 171]:

Lemma 9.2.3 *The natural embedding of jet spaces*

$$\Psi : J^k(m,n) \to J^k(m+s, n+s), \quad \Psi(jg(0)) := j(g \times id_s)(0)$$

is transverse to any $\mathcal{K}$-orbits in $J(m+s, n+s)$.

Apply Definition 9.2.1 and Proposition 9.2.2 to the setting

$$G := J^k\mathcal{K}_{m,n}, \quad V := J^k(m,n), \quad W := \overline{\eta}.$$

Note that $J^k(m,n)$ is contractible and $G = J^k\mathcal{K}_{m,n}$ is homotopic to the 1-jets $J^1\mathcal{K}_{m,n} = GL_m \times GL_n$, and thus $H^*_G(V)$ which is identifed with

$$H^*_G(pt) = H^*(BGL_m \times BGL_n) = \mathbb{Z}[a_1, \cdots, a_m, b_1, \cdots, b_n]$$

where a_i and b_j are Chern classes for the source bundle and the target bundle, respectively. Hence, $tp_G(W)$ is a polynomial in a_i's and b_j's. The point is that the stabilization property of $\mathcal{K}$-orbits (Lemma 9.2.3) implies that $tp_G(W)$ is *supersymmetric*, and hence it is written in terms of quotient variables c_i's where $c = 1 + c_1 + c_2 + \cdots = (1 + b_1 + \cdots)/(1 + a_1 + \cdots)$, see e.g., [50, 82, 83] for the detail.

Theorem 9.2.4 ([29, 67, 151, 171, 188]) *For a $\mathcal{K}$-singularity type η of codimension κ, there exists a unique polynomial $tp(\eta) \in \mathbb{Z}[c_1, c_2, \cdots]$ so that for any holomorphic map $f : M \to N$ of relative codimension κ which is generic with respect to η, the singularty locus of type η is expressed by the polynomial evaluated by the quotient Chern class $c_i = c_i(f)$:*

$$\text{Dual}\,[\overline{\eta(f)}] = tp(\eta)(c(f)) \quad \in H^{2\,\text{codim}\,\eta}(M).$$

Definition 9.2.5 We call $tp(\eta)$ *the Thom polynomial of $\mathcal{K}$-type η of mono-germs with codimension κ.*

There is a counterpart in real C^∞ category using the Stiefel-Whitney class and the Pontrjagin class, which we will pick up in Sect. 9.8 later (Theorem 9.8.1).

Remark 9.2.6 (Schur Positivity) By the definition, the universal polynomial $tp(\eta)$ is a linear combination of Chern monomials of degree equal to the codimension of η, while it also has a different expansion. Then a certain *positivity property* is shown by Pragacz and Weber [148, 149], see also a survey [147]. Given a partition $\lambda = (\lambda_1, \cdots, \lambda_l)$, i.e., $\lambda_1 \geq \lambda_2 \geq \cdots \geq \lambda_l \geq 0$, the Schur function with index λ is

$$S_\lambda(E - F) := [s_{\lambda_i + j - i}(E - F)]_{1 \leq i,j \leq l}$$

where $1 + s_1(E - F) + \cdots = \prod(1-\beta)/\prod(1-\alpha)$ and α's and β's are the Chern roots of E and F, respectively. Then we can expand the Thom polynomial with respect to the Schur function basis

$$tp(\eta)(c(f)) = \sum a_\lambda S_\lambda(T^*M - f^*T^*N) \qquad (a_\lambda \in \mathbb{Z}).$$

Since $\overline{\eta(M, N)}$ is a cone bundle, we directly apply the Fulton-Lazarsfeld theory of numerical positivity of polynomials in the Chern classes of ample vector bundles [58, §12] to prove the desired positivity $a_\lambda \geq 0$.

Remark 9.2.7 (Thom Polynomials in Other Cohomology Theory) Instead of using singular cohomology H^* or Chow theory CH^* to define Thom polynomials for singularities of maps, we may consider to use other cohomology theories. For general oriented cohomology theories, note that there is no a priori defined fundamental class (orientation class) for a singular variety, thus there is no obvious unique way to define an *η-singularity loci class*—we need some additional data to find a distinguished class supported on the η-singularity locus. For instance, let us consider complex cobordism MU^*, or more suitably, algebraic cobordism Ω^* in the sense of Levine-Morel [106], and their equivariant version via the Borel construction [69, 70]. Pick up a proper $\mathcal{K}$-equivariant birational map onto an orbit-closure, $\pi_\eta : W_\eta \to \overline{\eta} \subset J^k(m, n)$ with W_η smooth. Then, for an appropriately generic map $f : M \to N$, we may define a *cobordism η-singularity loci class* in $\Omega^*(M)$ represented by a proper map $W_\eta(f) \to \overline{\eta}(f) \subset M$, where $W_\eta(f)$ is the fiber product of π_η and the jet extension map $j^k f$ through the above universal construction. The class depends on the choice of π_η, not only the singularity type η. Then, it is automatic to see that this class is universally expressed in terms of cobordism Chern classes [69, 106], i.e., it lives in

$$\Omega^*_G(pt) = \Omega^*(BGL_m \times BGL_n) = \Omega^*(pt)[[a_1, \cdots, a_m, b_1, \cdots, b_n]].$$

That should be the *cobordism Thom polynomial for a desingularization π_η of η-singularity type*, and it may be supersymmetric provided π_η admits some stabiization property with respect to trivial suspension (see also for multi-singularity case Sect. 9.6.1). In fact, a typical example is the Thom-Porteous formula in algebraic cobordism Ω^*, which has been studied in [71, 72] by using Bott-Samelson resolutions of orbit-closures $\overline{\Sigma^k}$ (historically, in real singularity case (cf. Sect. 9.8), already in the 1970s, McCrory [115] considered $\mathbb{Z}_2$-cobordism Thom polynomials for $\overline{\Sigma^k}$ in connection with cohomology operations).

By the universality of MU^* or Ω^*, a version of Thom polynomial in other oriented cohomology theory (e.g., K-theory) can be deduced from this cobordism Thom polynomial. Namely, there are some choices for 'equivariant fundamental class' in case other than ordinary cohomology theory H^*, and of interest is to find a good one among those choices which behaves nicely. A representation theoretic approach to K-theoretic Thom polynomials in expansion with respect to

Grothendieck polynomials has been studied, especially for A_2-singularity type in detail, by Rimányi and Szenes [166] through iterated residue integrals technique (cf. Sect. 9.3.2 later). They point out a subtle relation of their K-theoretic fundamental class of the orbit-closure (defined using certain characters) with its equivariant resolution, see also Kolokolnikova [97].

9.2.4 Desingularization Method

We make a short remark on a traditional approach, called the *desingularization method*. At the early stages of the theory, the crux of the problem had been recognized as explicitly constructing 'resolution of singularities' for the degeneracy loci of prescribed type of maps (indeed, it is natural from the viewpoint of Remark 9.2.7). This approach directly computes Thom polynomials and essentially does not rely on the existence theorem (Theorem 9.2.4) mentioned above. For instance, many cancellations occur in computation, and consequently we see that the Thom polynomial is written in terms of quotient Chern classes $c_i = c_i(f)$.

Let $f : M \to N$ be a holomorphic map. Before Mather's classification theory of map-germs appeared, Thom considered the following filtration of (locally closed) submanifolds of M. First, set

$$\Sigma^i(f) := \{\, x \in M \mid \dim\ker df(x) = i \,\}$$

and suppose that it is non-singular (i.e., the transversality condition on df is assumed). Then take the restriction map of f to $\Sigma^i(f)$, and set

$$\Sigma^{i,j}(f) := \{\, x \in \Sigma^i(f) \mid \dim\ker d(f|_{\Sigma^i(f)})(x) = j \,\}.$$

Assume again it is non-singular, then take the restriction of f to $\Sigma^{i,j}(f)$ and define $\Sigma^{i,j,k}(f)$ and so on. This is the so-called *Thom-Boardman singularity type* Σ^I (TB singularity type) with symbol $I = (i \ge j \ge k \ge \cdots)$. That is justified by the fact that there are $\mathcal{K}$-invariant algebraic subsets Σ^I in $J^r(m,n)$, defined by Jacobian extension (see e.g., [9]), so that if the jet-extension map $j^r f$ is transverse to every $\Sigma^I(M,N)$, then it holds that $\Sigma^I(f) = (j^r f)^{-1}(\Sigma^I(M,N))$. For instance, the stratum $\Sigma^{1,1,\cdots,1,0}$ (1 repeats k times) in $J^r(m,n)$ coincides with the $\mathcal{K}$-orbit of A_k-singularity type, i.e., the local algebra is isomorphic to $\mathbb{C}[x]/\langle x^{k+1}\rangle$. However, in general, Σ^I is not a single $\mathcal{K}$-orbit, and even they do not form a Whitney stratification.

On the other hand, there is another geometric way to construct some TB singularity loci, by means of *intrinsic derivatives* due to Porteous and Ronga. Let us look at a few examples. Note that

$$J^k(M,N) \simeq \bigoplus_{i=1}^{k} \mathrm{Hom}(\mathrm{Sym}^i TM, TN)$$

where Sym^i means the i-th symmetric power and the structure group is reduced to $GL_m \times GL_n$ (thus $J^k(M, N)$ is a vector bundle). Let p_1 and p_2 denote canonical projection to M and N, respectively. The first intrinsic derivative is a tautological linear map

$$\alpha : T_x M \to T_y N \quad \text{at} \quad \alpha \in J^1(M, N)$$

with $x = p_1(\alpha)$ and $y = p_2(\alpha)$, and the second intrinsic derivative is

$$\beta^* : \ker \alpha \to \mathrm{Hom}(\ker \alpha, \mathrm{coker}\, \alpha) \quad \text{at} \quad (\alpha, \beta) \in J^2(M, N)$$

given by $\mathrm{Sym}^2 \ker \alpha \subset \mathrm{Sym}^2 T_x M \overset{\beta}{\to} T_y N \to \mathrm{coker}\, \alpha$. Then $\Sigma^i(M, N)$ is defined as the set of $\alpha \in J^1(M, N)$ with $\dim \ker \alpha = i$, and $\Sigma^{i,j}(M, N)$ is the set of $(\alpha, \beta) \in J^2(M, N)$ with $\dim \ker \alpha = i$ and $\dim \ker \beta^* = j$.

Let $\pi : G_i \to J^1(M, N)$ be the Grassmann bundle of i-dimensional subspaces $\lambda \subset T_x M$, and ξ_i the tautological bundle over G_i. The vector bundle $\mathrm{Hom}(\xi_i, \pi^* p_2^* TN) \to G_i$ naturally admits a section σ which assigns to $(\alpha, \lambda) \in G_i$ the composed linear map $\lambda \hookrightarrow T_x M \overset{\alpha}{\to} T_y N$, and σ is indeed transverse to the zero section. Hence, the zero locus $Z(\sigma)$ of this section is smooth, and it is birationally mapped to the closure $\overline{\Sigma^i(M, N)}$ via π. Applying the Gysin map of π, we have

$$\pi_*[Z(\sigma)] = \pi_* c_{top}(\mathrm{Hom}(\xi_i, \pi^* p_2^* TN)) = [\overline{\Sigma^i(M, N)}],$$

and a computation yields the Thom-Porteous formula (9.1) mentioned in Sect. 9.1.1.1 [58, 151, 171].

Let $\pi' : G_{i,j} \to J^2(M, N)$ be the flag bundle of $\tau \subset \lambda \subset T_x M$ with $\dim \lambda = i$ and $\dim \tau = j$ and G_i be defined over $J^2(M, N)$ (via pullback). Over the pullback of $Z(\sigma)$ via the natural projection $G_{i,j} \to G_i$, the desingularization of $\overline{\Sigma^{i,j}(M, N)}$ is constructed through a bit technical process using β^* [171]. In particular, a computation of the Gysin image for $\Sigma^{1,1}$ leads to (9.5) mentioned in Sect. 9.1.1.2.

In summary, this traditional approach consists of two steps. The first one is to explicitly construct a tower of some flag bundles

$$\pi : G \to \cdots \to G_{i,j} \to G_i \to J^k(M, N)$$

and a vector bundle $E \to G$ with a section σ defined by intrinsic derivatives such that the zero locus $Z(\sigma)$ is smooth and gives a desingularization (birational map) $\pi : Z(\sigma) \to \overline{\Sigma^I(M, N)}$. The second is to directly compute the Gysin image $\pi_*(c_{top}(E))$ to get the Thom polynomial. Here, one may use a general formula of the Gysin map for the projection of (partial) flag bundles given by Damon [30] with residue expression.

However, this strategy fails quickly. Even for order two, the computation in the second step is not easy at all. For order more than two, it is not known in general whether such a nice desingularization in the first step exists or not, and only a few concrete examples are known in equidimensional case ($\kappa = 0$); e.g., by finding desingularizations, Thom polynomials are computed for the closures of $\Sigma^{1,1,1}$ [151], $\Sigma^{1,1,1,1}$ [62] and $\Sigma^{1,1,1,1,1}$ [191] (those are exactly $tp(A_\mu)$ with $\mu = 3, 4, 5$ in Sect. 9.1.1.2, respectively) and also see [29]. To go further, we need a different perspective, which we will describe in Sect. 9.3.

9.3 Equivariant Localization

9.3.1 *Symmetry of Singularities and Interpolation*

Around the end of the last century, a novel efficient approach to computing Thom polynomials without using any desingularization was proposed by Rimányi [156], that brought a breakthrough in this field.

Let V be a complex affine space with action of a complex algebraic Lie group G. Assume that V consists of finitely many G-orbits ξ, and also that for every orbit ξ, the Euler class of the normal bundle ν_ξ of ξ in V is non-zero, $c_{top}(\nu_\xi) \neq 0$. Let G_ξ be the stabilizer subgroup of ξ, and $\iota_\xi : G_\xi \subset G$ the inclusion. Then it turns out that

$$\bigoplus \iota_\xi^* : H_G^*(V) = H^*(BG) \longrightarrow \bigoplus_\xi H^*(BG_\xi)$$

is an isomorphism. In particular it holds that

$$\iota_\xi^*\, tp(\eta) = \begin{cases} 0 & (\xi \not\subset \overline{\eta}) \\ c_{top}(\nu_\eta) & (\xi = \eta) \end{cases} \tag{9.8}$$

The system of these equations (9.8) has a unique solution modulo torsion [50]. The first equations in (9.8) means that $tp(\eta)$ lives in the kernel of the ring homomorphism $H_G^*(V) \to H_G^*(V - \overline{\eta})$, which is isomorphic to the natural image of

$$H_G^*(V, V - \overline{\eta}) \simeq H_*^G(\overline{\eta}) \quad \text{(equivariant Alexander duality).}$$

The kernel is called the *avoiding ideal associated to* η in Fehér and Rimányi [50, 51] or the *ideal of cohomology classes supported on* $\overline{\eta}$ in Fulton and Pragacz [60]. In this ideal, the above second equation in (9.8) distinguishes the G-fundamental class of $\overline{\eta}$, that is nothing but $tp(\eta)$.

Also note that it suffices to take the maximal torus of G_ξ for each ξ by splitting principle. Once we write $tp(\eta)$ as a homogeneous polynomial in generators of

$H^*(BG)$ with unknown coefficients, (9.8) gives the system of linear equations in the unknowns, and solving it determines the explicit form of $tp(\eta)$. Indeed, usually it is overdetermined.

Today, this is known as the *restriction* or *interpolation method* due to R. Rimányi. Originally it comes from a slightly different and deeper background, *singular Thom-Pontrjagin construction* [155, 167, 190]—namely, it works not only for Lie group action but also for *Lie groupoid action* or *Artin stack with global stratification*, see Sect. 9.6 later.

Below, we demonstrate some typical computations using this method, according to [50, 82, 156], for objects treated in Guiding Examples Sects. 9.1.1.1 and 9.1.1.2.

9.3.1.1 Thom-Porteous Formula, Revisited

We consider the classification of linear maps by linear coordinate changes. Let

$$G = GL_m \times GL_n, \quad V = \mathrm{Hom}\,(\mathbb{C}^m, \mathbb{C}^n)$$

and $\eta = \Sigma^k$, the G-orbit of linear maps with kernel dimension k. To determine the universal form of $tp(\Sigma^k)$, we see that it is enough to use only the last equation in (9.8). Put $l = n - m + k$.

Take a representative h in Σ^k in a standard way, then the tangent space of the orbit at the point h consists of all matrices in the following form:

$$h = \left[\begin{array}{c|c} I_{m-k} & O \\ \hline O & O \end{array}\right] \in \Sigma^k, \quad T_h\Sigma^k = \left\{ \left[\begin{array}{c|c} * & * \\ \hline * & O \end{array}\right] \right\} \subset T_hV = V.$$

Note that the normal space is isomorphic to $\mathrm{Hom}(\ker h, \mathrm{coker}\, h)$. The stabilizer group of $h \in \Sigma^k$, denoted by G_k, consists of pairs of square matrices

$$\left(\left[\begin{array}{c|c} P & O \\ \hline * & A \end{array}\right], \left[\begin{array}{c|c} P & * \\ \hline O & B \end{array}\right] \right), \quad (A, B, P) \in GL_k \times GL_l \times GL_{m-k}.$$

Let ρ_1, ρ_2, ρ_3 be the representations of G_k on $\ker h = \mathbb{C}^k$, $\mathrm{coker}\, h = \mathbb{C}^l$, $\mathrm{Im}\, h = \mathbb{C}^{m-k}$, respectively, and then the normal bundle ν_k of the orbit Σ^k in V is obtained from the universal bundle $EG_k \to BG_k$ with ρ_1 and ρ_2 acting on $\mathrm{Hom}(\mathbb{C}^k, \mathbb{C}^l)$. Put $c(\rho_1) = \prod_{i=1}^k (1 + \alpha_i)$ and $c(\rho_2) = \prod_{j=1}^l (1 + \beta_j)$. Then using (9.8) and the classical resultant, we have

$$\begin{aligned} \iota_k^*\, tp(\Sigma^k) &= c_{top}(\nu_k) = c_{kl}(\rho_1^* \otimes \rho_2) \\ &= \prod (\beta_j - \alpha_i) = \Delta_{l^k}(c(\rho_2 - \rho_1)) \in H^*(BG_k). \end{aligned}$$

Denote by a_i and b_j universal Chern classes for GL_m and GL_n, respectively, and then

$$H^*(BG) = \mathbb{Z}[a_1, \cdots, a_m, b_1, \cdots, b_n].$$

The representation of G_k on $\mathbb{C}^m$ and $\mathbb{C}^n$ are $\lambda_1 = \rho_1 \oplus \rho_3$ and $\lambda_2 = \rho_2 \oplus \rho_3$. Thus

$$\iota_k^* : H^*(BG) \to H^*(BG_k)$$

is determined by $c' \mapsto c(\lambda_1)$ and $c'' \mapsto c(\lambda_2)$, and especially, it sends the quotient variables

$$1 + c_1 + c_2 + \cdots := \frac{1 + b_1 + \cdots + b_n}{1 + a_1 + \cdots + a_m} \longmapsto \frac{c(\lambda_2)}{c(\lambda_1)} = 1 + c_1(\rho_2 - \rho_1) + \cdots.$$

Hence $\iota_k^* \Delta_{l^k}(c) = \Delta_{l^k}(c(\rho_2 - \rho_1))$. Finally it is checked that ι_k^* is *injective* for degree $\leq kl$, thus the consequence is

$$tp(\Sigma^k) = \Delta_{l^k}(c).$$

Moreover, the avoiding ideal associated to Σ^k is explicitly determined, see [50, 60].

This method works for actions of Borel subgroups as well, that leads to application to (matrix) Schubert calculus and quiver representations [26, 27, 48–50] and also other topics such as algebraic combinatorics in configuration problems, e.g., [33, 38, 42–46] and so on. Afterwards, this research direction has been widely expanding, that will be touched in Sect. 9.4.2.1 later.

For the second order Thom-Boardman singularity type $\Sigma^{i,j}$, the interpolation has also been discussed by Fehér and Kümöves [37] based on Ronga's disingularization. It will be treated in a more general framework (Sect. 9.3.2).

9.3.1.2 Fold and Cusp, Revisited

We demonstrate how to compute the Thom polynomial for A_2-singularity type (cusp) for maps between the same dimensional manifolds, i.e., in case of $\kappa = 0$ (see Sect. 9.1.1.2). By the existence theorem of Thom polynomials, it has the form

$$tp(A_2) = Ac_1^2 + Bc_2$$

in quotient Chern classes $c_i = c_i(\text{target} - \text{source})$. Our task is to determine the unknown integer coefficients A, B.

The normal form of stable map-germ of type A_2 is given by a quasi-homogeneous polynomial map

$$A_2 : \mathbb{C}^2 \to \mathbb{C}^2, \quad (x, y) \to (x^3 + yx, y).$$

In jet space of order ≥ 3, the jet-extension of this map is transverse to the $\mathcal{K}$-orbit at the origin, so this map itself is regarded as a normal slice to the $\mathcal{K}$-orbit of A_2-singularity type. The maximal torus T of the stabilizer group gives the torus actions on the source and the target

$$\rho_0 = \alpha \oplus \alpha^{\otimes 2}, \quad \rho_1 = \alpha^{\otimes 3} \oplus \alpha^{\otimes 2}$$

satisfying $A_2 \circ \rho_0 = \rho_1 \circ A_2$, where α is a character $T \to \mathbb{C}^*$. Take the dual tautological line bundle $\ell = \mathcal{O}_{\mathbb{P}^N}(1)$ over a projective space $\mathbb{P}^N$ of large dimension $N \gg 0$ (or the classifying space $BT = \mathbb{P}^\infty$ of the torus T). Define two vector bundles of rank 2

$$E_0\ (= E_0(A_2)) := \ell \oplus \ell^{\otimes 2}, \qquad E_1\ (= E_1(A_2)) := \ell^{\otimes 3} \oplus \ell^{\otimes 2}.$$

Since the normal form of A_2 is invariant under the torus action, we can glue together local patches $id_{U_i} \times A_2 : U_i \times \mathbb{C}^2 \to U_i \times \mathbb{C}^2$. The resulting map $f_{A_2} : E_0 \to E_1$ is a stable map between the total spaces E_0 and E_1 so that the following diagram commutes and the restriction to each fiber is $\mathcal{A}$-equivalent to the normal form of A_2. We call f_{A_2} the *universal map for* A_2.

$$\begin{array}{ccc} E_0 & \xrightarrow{\ f_{A_2}\ } & E_1 \\ & {\scriptstyle p_0}\searrow \quad \swarrow{\scriptstyle p_1} & \\ & \mathbb{P}^N & \end{array}$$

The loci $A_2(f_{A_2})$ and $f(A_2(f_{A_2})))$ are just the zero sections of E_0 and of E_1, respectively.

Put $a = c_1(\ell)$ and then $c(E_0) = (1+a)(1+2a)$ and $c(E_1) = (1+3a)(1+2a)$ in $H^*(\mathbb{P}^N) = \mathbb{Z}[a]/(a^{N+1})$. Note that $H^*(E_0) \simeq H^*(\mathbb{P}^N) \simeq H^*(E_1)$ via p_0^* and p_1^*. Since the A_2-locus in the total space E_0 is the zero section, we see

$$\mathrm{Dual}\,[\overline{A_2(f_{A_2})}] = c_2(p_0^* E_0) = c_2(E_0) = 2a^2.$$

On the other hand, the tangent bundles TE_0 and TE_1 of the total spaces canonically split as $TE_i = p_i^*(E_i \oplus T\mathbb{P}^N)$, thus $f_{A_2}^* TE_1 - TE_0 = p_0^*(E_1 - E_0)$. Therefore, the quotient Chern class for f_{A_2} is written as follows:

$$c(f_{A_2}) := c(f_{A_2}^* TE_1 - TE_0) = c(E_1 - E_0) = \frac{1+3a}{1+a} = 1 + 2a - 2a^2 + \cdots .$$

Hence, $c_1(f_{A_2}) = 2a$, $c_2(f_{A_2}) = -2a^2$, so we have

$$tp(A_2)(f_{A_2}) = Ac_1^2 + Bc_2 = (4A - 2B)a^2.$$

By Theorem 9.2.4, $tp(A_2)(f_{A_2}) = \mathrm{Dual}\,[\overline{A_2}(f_{A_2})]$, thus $(4A-2B)a^2 = 2a^2$, i.e., $2A - B = 1$.

Next we apply tp to the universal map of adjacent singularities. Consider

$$A_1 : \mathbb{C} \to \mathbb{C}, \quad x \mapsto x^2$$

and the associated universal map $f_{A_1} : E_0 \to E_1$, where $E_0 = E_0(A_1) = \ell$ and $E_1 = E_1(A_1) = \ell^{\otimes 2}$ in the same way as above. Obviously, the universal map does not have A_2-singularity: $A_2(f_{A_1}) = \emptyset$, thus by Theorem 9.2.4 again, we have $tp(A_2)(f_{A_1}) = \mathrm{Dual}\,[\emptyset] = 0$. Since $c(f_{A_1}) = \frac{1+2a}{1+a} = 1 + a - a^2 + \cdots$, we have $A - B = 0$.

Finally, solving linear equations in A, B we get $A = B = 1$, thus we conclude that

$$tp(A_2) = c_1^2 + c_2.$$

Note that instead of working over $\mathbb{P}^N$, one can perform the same computation formally as a map from the representation ring $R(T)$ to $H_T^*(pt) = \mathbb{Z}[a]$.

By this method, Rimányi first computed Thom polynomials for $\mathcal{K}$-orbits in equidimensional case $\kappa = 0$ up to codimension 8 in [156], that opened a new stage of this theory. Indeed, it is known that any stable-germs appearing in the so-called Mather's nice dimensions admit quasi-homogeneous normal forms, and also that for such stable-germs, there are only finitely many nearby $\mathcal{K}$-orbits (i.e., no moduli strata nearby), see e.g., [119]. Thus, once fixing κ, there are finitely many $\mathcal{K}$-orbits with non-trivial normal Euler classes *up to certain codimension* (determined by Mather's nice dimension), so the interpolation method works well for computing Thom polynomials for such stable singularity types, at least theoretically.

A more difficult question is to seek for general formulas of Thom polynomials involving κ as a parameter. For A_2-singularity with $\kappa \geq 0$ (i.e., $\Sigma^{1,1}$-type), there was Ronga's result as mentioned in Introduction (Sect. 9.1.1.2), and for A_3-singularity ($\kappa \geq 0$), see Bérczi et al. [18] and Lascoux and Pragacz [102]. Also there are some other partial results due to Pragacz [144–146], Öztürk [137], Öztürk and Pragacz [138] by the interpolation method with combinatorial techniques of Schur functions. A more general and modern treatment for this problem will be discussed in the next section.

9.3.2 Thom Series and Punctual Hilbert Schemes

The next breakthrough came with torus-equivalriant localization technique by Bérczi and Szenes [20]. First, we explain a general framework. Let V be a vector space. Suppose that X is a compact algebraic manifold and $Y \subset X$ a subvariety. Let $E \to Y$ be a subbundle of the trivial bundle $X \times V \to M$ restricted to Y. Let

$\pi_2 : X \times V \to V$ be the projection, $i : E \subset X \times V$ the embedding,and $\phi = \pi_2 \circ i$, as in the diagram

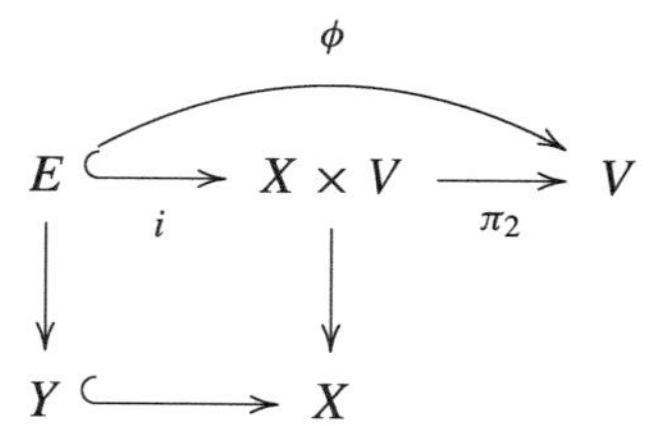

Suppose that the torus T acts on all spaces in the diagram above and that all maps are T-equivariant. Then, as a consequence from the Berline-Vergne-Atiyah-Bott Localization Formula [11] applied to this setup, the following proposition holds. Here $e^T(-)$ is the Euler class, i.e., the top Chern class $c^T_{top}(-)$, and for a proper inclusion $\iota : A \hookrightarrow B$, we denote by $[A \subset B]$ the image of the fundamental class of A, $\iota_*(1) \in H^*_T(B)$.

Proposition 9.3.1 (Bérczi and Szenes [20]) *Assume that the fixed point set $F(X)$ of the T-action on X is finite. Then for the push-forward map $\phi_* : H^*_T(E) \to H^*_T(V) = H^*(BT)$, we have*

$$\phi_*(1) = \sum_{f \in F(Y)} \frac{[Y \subset X]|_f \cdot [E_f \subset V]}{e^T(T_f X)}.$$

Consequently, if ϕ is birational to its image, then the right-hand side of (9) is equal to $[\varphi(E)] \in H^*_T(V) = H^*(BT)$.

If a fixed point f is a smooth point of Y, then $e^T(T_f X) = e^T(T_f Y) \cdot [Y \subset X]|_f$. Hence especially, if Y is non-singular, then

$$\phi_*(1) = \sum_{f \in F(Y)} \frac{[E_f \subset V]}{e^T(T_f Y)}.$$

Proposition 9.3.1 is applied to the following setting associated to a $\mathcal{K}$-singularity type η of finitely determined germs $(\mathbb{C}^m, 0) \to (\mathbb{C}^{m+\kappa}, 0)$, according to Fehér and Rimányi [52, 53]. Namely, we consider

$$V = J^k(m, m+\kappa) = J^k(m, 1) \otimes \mathbb{C}^{m+\kappa}$$

and $X = \mathrm{Gr}^{\mu}$ is a suitable Grassmannian manifold including a certain subvariety Y, whose precise definition will be given later. This may be seen as an equivariant generalization of Damon's partial desingularization [29].

Recall that the $\mathcal{K}$-equivalence classifies ideals I_η, the ideal generated by component functions of map-germs up to isomorphisms of the source space. Namely, a

$\mathcal{K}$-type η is determined by its local algebra $Q(\eta) = \mathcal{O}_{\mathbb{C}^m,0}/I_\eta$, so the dimension of the target space is somewhat auxiliary, i.e., κ is a sort of parameter. Since we are concerned with finitely determined germs η, if $\kappa \geq 0$, the local algebra has finite dimension, and if $\kappa < 0$, it is enough to think of the local algebra modulo sufficiently high degree, i.e., $Q_k(\eta) = \mathcal{O}_{\mathbb{C}^m,0}/I_\eta + \mathfrak{m}_{\mathbb{C}^m,0}^{k+1}$, where k is the determinacy degree of η.

Let $\mathrm{Hilb}^{\mu+1}(\mathcal{O}_{\mathbb{C}^m,0})$ denote the *punctual Hilbert scheme* which parameterizes ideals $I \subset \mathcal{O}_{\mathbb{C}^m,0}$ supported at the origin of colength $\mu + 1$ [73]. It is a projective variety, and especially, it is embedded in a Grassmannian Gr^μ of μ-codimensional subspace of the vector space $J^k(m, 1) = \mathfrak{m}_{\mathbb{C}^m,0}/\mathfrak{m}_{\mathbb{C}^m,0}^{k+1}$. Given a $\mathcal{K}$-type η with $\dim Q(\eta) = \mu + 1$, we assign a stratum of $\mathrm{Hilb}^{\mu+1}(\mathcal{O}_{\mathbb{C}^m,0})$ which consists of all ideals being isomorphic to I_η. We denote the closure of the stratum by

$$Y = Y_\eta \subset \mathrm{Hilb}^{\mu+1}(\mathcal{O}_{\mathbb{C}^m,0}) \subset \mathrm{Gr}^\mu = X.$$

Note that V, X, Y have actions of the maximal torus $T = (\mathbb{C}^*)^m$ of GL_m in a natural way. Apply Proposition 9.3.1 to this setting. Then the information of the fixed points enables us to compute the T-equivariant Poincaré dual to the closure of the $\mathcal{K}$-orbit of type η in $V = J^k(m, n)$, and hence, we get the Thom polynomial $tp(\eta)$ in $c_i = c_i(f)$ by the splitting principle.

At the initial stage of this approach, Bérczi and Szenes [20] considered the case of $\eta = A_\mu$ ($\kappa \geq 0$). In this case, Y_{A_μ} is realized as a particular component of the punctual Hilbert scheme, the *curvilinear Hilbert scheme* $\mathrm{CHilb}^{\mu+1}(\mathcal{O}_{\mathbb{C}^m,0})$, that consists of ideals isomorphic to $\mathbb{C}[x]/\langle x^{\mu+1}\rangle$, and it is a smooth quasi-projective variety. In fact, the authors dealt with the *non-reductive group quotient* $J^k(1, m)/\!\!/\mathrm{Aut}(\mathbb{C}, 0)$, via the action of the group of k-jets of re-parametrization germs of $\mathbb{C}$ at the origin, and compactify it by using the curvilinear Hilbert scheme. Their approach may be seen as a far generalization of Porteous-Gaffney's probe (test curve) method [61, 152] from the equivariant geometry viewpoint. Their main result is the following iterated residue integral expression for the Thom polynomial of type A_μ where $\kappa \geq 0$ is regarded as a parameter.

Theorem 9.3.2 (Bérczi and Szenes [20], Bérczi [13, 16]) *There exists a certain polynomial Q_μ as defined in [20, §6] which depends only on μ and satisfies*

$$tp(A_\mu) = \operatorname*{Res}_{t_1=\infty} \cdots \operatorname*{Res}_{t_\mu=\infty} \frac{\prod_{1\leq i<j\leq\mu}(t_j - t_i) \cdot Q_\mu(t_1, \cdots, t_\mu)}{\prod_{k=1}^{\mu}\prod_{j=1}^{k-1}\prod_{i\leq j}(t_k - t_i - t_j)} \prod_{i=1}^{\mu} \mathrm{C}\left(\frac{1}{t_i}\right) t_i^\kappa \, dt_i$$

where $\mathrm{C}(t^{-1}) = 1 + c_1 t^{-1} + c_2 t^{-2} + \cdots$ *with the quotient Chern classes* c_i.

Here, $\operatorname*{Res}_{t_1=\infty} \cdots \operatorname*{Res}_{t_\mu=\infty} \Psi = (-1)^\mu \int_{|t_1|=R_1} \cdots \int_{|t_\mu|=R_\mu} \Psi \, dt_1 \cdots dt_\mu$ where $0 \ll R_1 \ll R_2 \ll \cdots \ll R_\mu$ for a rational function $\Psi(t_1, \cdots, t_\mu)$.

Example 9.3.3 In Bérczi and Szenes [20], Q_μ for $1 \le \mu \le 6$ are explicitly given: $Q_1 = Q_2 = Q_3 = 1$ and Q_μ ($\mu = 4, 5, 6$) are polynomials of degree $1, 3, 7$, respectively. For example, in case of $\mu = 2$, we have $tp(A_2)$ as the integral of

$$\frac{t_2 - t_1}{t_2 - 2t_1}\left(t_1^\kappa \sum_{i\ge 0} c_i t_1^{-i}\right)\left(t_2^\kappa \sum_{i\ge 0} c_i t_2^{-i}\right) dt_1 dt_2$$

and expanding it, we recover Ronga's formula (9.5) written in Sect. 9.1.1.2. For $\mu = 3$, we have $tp(A_3)$ by integrating

$$\frac{(t_2 - t_1)(t_3 - t_1)(t_3 - t_2)}{(t_2 - 2t_1)(t_3 - 2t_1)(t_3 - t_1 - t_2)} \prod_{i=1}^{3}\left(t_i^\kappa \sum_{i\ge 0} c_i t_i^{-i}\right) dt_1 dt_2 dt_3$$

which gives a simpler expression of a formula in Bérczi et al. [18].

Not only the local algebra of A_n-type but also the local algebra $Q(\eta)$ associated to any $\mathcal{K}$-type η of finitely determined germs $(\mathbb{C}^m, 0) \to (\mathbb{C}^{m+\kappa}, 0)$ with $\kappa \ge 0$, can be dealt with in the same framework. Combining it with the stabilization argument, the following strong consequence is obtained by Fehér and Rimányi [53]:

Theorem 9.3.4 (Fehér and Rimányi [53] Thom series) *Let Q be a local algebra of finite colength $\mu + 1$, then there is a unique formal series*

$$\mathrm{Ts}_Q = \sum_{i_1,\cdots,i_\mu} \psi_{i_1,\cdots,i_\mu} c_{\kappa+1+i_1} \cdots c_{\kappa+1+i_\mu}$$

such that for any $\mathcal{K}$-type η of map-germ with $\kappa \ge 0$ whose local algebra $Q(\eta)$ is isomorphic to Q, the corresponding Thom polynomial $tp(\eta)$ is obtained as Ts_Q *evaluated by the quotient Chern classes $c_{\kappa+1+i} = c_{\kappa+1+i}(f)$ with conventions that $c_0 = 1$ and $c_j = 0$ for $j < 0$. Here the integer coefficients $\psi_{i_1,\cdots,i_\mu}$ are independent from κ.*

In fact, Ts_Q formally consists of infinitely many terms indexed by μ-tuples of integers $(i_1, \cdots, i_\mu)$ whose sum is some constant determined by Q, but fixing κ and applying the conventions on c_j, it is read off as a polynomial ($= tp(\eta)$). This is the most comprehensive existence theorem of universal polynomials for mono-singularity types—Ts_Q is called the *Thom series for the local algebra Q*. For a nice introduction to Thom series with several examples, see Rimányi's survey article [160].

The rule of the appearance of the coefficients $\psi_{i_1,\cdots,i_\mu}$ should be detected. It requires to replace the Grassmannian Gr^n by a finer space. In [85] Kazarian has introduced the concept of *non-associative Hilbert scheme* to be the total space of some (partial) flag bundle which becomes a smooth ambient space into which Y_η is embedded as a closed subvariety. Let $Q = Q(\eta)$ be the local algebra of a $\mathcal{K}$-type η

of map-germs and let $\dim Q = \mu + 1$. In order to construct this ambient space, a key ingredient is a filtration on the nilpotent subalgebra $N := \mathfrak{m}_{\mathbb{C}^m,0}Q$,

$$N = N_1 \supset N_2 \supset \cdots \supset N_{r+1} = 0 \qquad d_i = \dim(N_i/N_{i+1})$$

satisfying $N_i \cdot N_j \subset N_{i+j}$. Of course, $\sum d_i = \mu$. The degree filtration with $N_k := \mathfrak{m}^k_{\mathbb{C}^m,0}Q$ is a typical one, but there may be some other choices. We call the sequence $\underline{d} = (d_1, \cdots, d_r)$ the *dimension vector* associated to the filtration on N. Put $s_k = d_1 + \cdots + d_k$, and for each i with $s_{k-1} \leq i \leq s_k$, put $e(i) := s_k - i$ and $w(i) = k$, and set

$$K_{\underline{d}}(t_1, \cdots, t_\mu) := \frac{\prod_{i=1}^{\mu} t_i^{e(i)} \cdot \prod_{1\leq i<j\leq\mu}(t_j - t_i)}{\prod_{k=1}^{d}\prod_{j=1}^{k-1}\prod_{i\leq j, w(i)+w(j)\leq w(k)}(t_k - t_i - t_j)}.$$

Theorem 9.3.5 (Kazarian [85]) *The coefficients $\psi_{i_1,\cdots,i_\mu}$ arising in* Ts_Q *are determined by the following rational generating function*

$$\sum_{i_1,\cdots,i_\mu} \psi_{i_1,\cdots,i_\mu} t_1^{i_1} \cdots t_\mu^{i_\mu} = K_{\underline{d}}(t_1, \cdots, t_\mu) P_Q(t_1, \cdots, t_\mu)$$

where P_Q is a polynomial defined as in [85, §4]. In other words, the Thom series Ts_Q *has the following residual integral expression:*

$$\mathrm{Ts}_Q = \underset{t_1=\infty}{\mathrm{Res}} \cdots \underset{t_\mu=\infty}{\mathrm{Res}}\ K_{\underline{d}}(t) \cdot P_Q(t) \cdot \prod_{i=1}^{\mu} \mathrm{C}(t_i^{-1})\, t_i^{\kappa}\, dt_i.$$

Obviously, in the case of $\eta = A_\mu$ with the degree filtration, this residual integral is just Theorem 9.3.2. Besides from A_μ-singularity types, for other types, e.g., $\Sigma^{i,j}$, $I_{a,b}$, $III_{a,b}$ etc., the polynomial $P_{Q(\eta)}$, and hence $\mathrm{Ts}_{Q(\eta)}$, are explicitly computed, see [53, 85].

Note that this decomposition (the right hand side) of the generating function of $\psi_{i_1,\cdots,i_\mu}$ is not unique, especially depends on the choice of filtration; K_η is the factor coming from the Gysin map of the projection of the flag bundle (cf. [30]) and P_Q corresponds to the fundamental class of a certain Borel orbit in the ambient space.

Looking back the flow of old and new researches, we observe that the present work of Thom series surely lies on the natural extension of work of Porteous, Ronga, Damon, Gaffney etc. That has not yet been completed, rather say, still only halfway there. In this theory, it is most critical to detect the polynomial P_Q, which comes from the Borel geometry associated to the local algebra, and it essentially relates to tautological integral over punctual Hilbert schemes. Many tasks remain for future work.

9.4 Higher Thom Polyomials

The Thom polynomial $tp(\eta)(c(f))$ is the most basic invariant for the appearance of η-type singular points for a given map $f : M \to N$, while if the locus $\overline{\eta(f)}$ has the expected dimension which is positive, the locus may contain higher codimensional informative cycles other than the fundamental cycle. For instance, if $f : M \to N$ is a Morin map (a stable map which has only local singularities of type A_k), the locus $S_k := \overline{A_k(f)}$ is a submanifold of M, and hence, the pushforward of the Chern class $\iota_* c(TS_k) \in H^*(M)$ is defined (ι is the inclusion of the locus to M). These classes appeared in [7, 105, 123], and especially, Ando called them 'higher Thom polynomials for Morin singularities'. However, the locus for non-Morin maps is singular in general, then this definition does not make sense at all. We need some more tools to build a theory.

9.4.1 *Chern Class for a Singular Variety*

Let X be an irreducible algebraic variety of dimension n. If X is non-singular, it possesses the *total* Chern class

$$c(TX) = 1 + c_1(TX) + \cdots + c_n(TX) \ \in H^*(X).$$

Switching to homology, note that $c_n(TX) \frown [X] = \chi(X)$ and $1 \frown [X] = [X]$. If X is singular, the tangent bundle no longer exists, so $c(TX)$ does not make sense at all. Nevertheless, we may find some alternative to $c(TX)$ in homology—for example, the *Chern-Schwartz-MacPherson class* (CSM class) is a typical one, see Brasselet [24]. It is a total homology class of X such that the degree is topological Euler characteristic of X and the top term is the fundamental class

$$c^{\mathrm{SM}}(X) = \chi(X) \cdot [pt] + \cdots + [X] \ \in H_*(X)$$

together with a nice functorial property (let us note again that if X is non-compact, H_* stands for Borel-Moore homology). Let $\mathcal{F}(X)$ be the abelian group of constructible functions on X, and for a proper morphism $X \to Y$, the *pushforward* $f_* : \mathcal{F}(X) \to \mathcal{F}(Y)$ is defined. It holds that $(f \circ g)_* = f_* \circ g_*$. R. MacPherson introduced the following natural transformation between two covariant functors $\mathcal{F}$ and H_* using the Chern-Mather class and local Euler obstruction, as a solution to the so-called Grothendieck-Deligne conjecture. Also it should be mentioned that the same notion had been found by M. Schwartz in a different context from obstruction theory [24].

Theorem 9.4.1 ([111]) *There is a unique natural transformation*

$$C_* : \mathcal{F}(X) \longrightarrow H_*(X)$$

between these functors so that $C_*(1\!\!1_X) = c(TX) \frown [X]$ *if* X *is non-singular.*

Namely, it holds that

- $C_*(\alpha + \beta) = C_*(\alpha) + C_*(\beta)$ (additive homomorphism)
- $C_* f_*(\alpha) = f_* C_*(\alpha)$ for proper morphisms $f : X \to Y$.

In particular, if $pt : X \to pt$ is proper, we have $pt_* C_*(\alpha) = C_* pt_*(\alpha) = \int_X \alpha$. The *Chern-Schwartz-MacPherson class* of X is defined by $c^{\mathrm{SM}}(X) := C_*(1\!\!1_X)$.

There is also the 'Segre-version' of CSM classes, see Aluffi [1, 2]. Let $\iota : X \hookrightarrow M$ be a closed embedding into non-singular M. Like as the relation between Segre covariance class and Fulton's Chern class, we define the *Segre-Schwartz-MacPherson class* by

$$s^{\mathrm{SM}}(X, M) := c(\iota^* TM)^{-1} \frown c^{\mathrm{SM}}(X) \quad \in H_*(X).$$

We often regard the class $s^{\mathrm{SM}}(X, M)$ as in $H^*(M)$ via the pushforward ι_* and the Poincaré duality, unless specifically mentioned. Also we set for $\alpha \in \mathcal{F}(M)$

$$s^{\mathrm{SM}}(\alpha, M) := c(TM)^{-1} \cdot C_*(\alpha) \in H^*(M).$$

The Segre-SM class has an expected nice property for transverse pullback:

Proposition 9.4.2 *Let* $f : M \to N$ *be a map between complex manifolds, and let* Y *be a closed singular subvariety of* N. *Assume that* f *is transverse to (a Whitney stratification of)* Y. *Then it holds that*

$$f^* s^{\mathrm{SM}}(Y, N) = s^{\mathrm{SM}}(f^{-1}(Y), M) \quad \in \quad H^*(M).$$

Remark 9.4.3 We follow the sign convention of the Segre class as in Fulton's book [58]. Notice that if X is a closed submanifold of M with the normal bundle $\nu = \iota^* TM - TX$, then the Segre-SM class is nothing but the *inverse normal Chern class* for $X \hookrightarrow M$:

$$s^{\mathrm{SM}}(1\!\!1_X, M) = \iota_* c(-\nu) \quad \in H^*(M).$$

Based on equivariant intersection theory [34, 186], an equivariant theory of CSM class is available. Let $\mathcal{F}^G_{inv}(X)$ be the subgroup generated by G-invariant constructible functions in $\mathcal{F}(X)$. Using G-equivariant Chern-Mather class and local Euler obstruction, one gets the following theorem.[3]

Theorem 9.4.4 ([129, 132, 135, 199]) *For G-varieties and proper G-morphisms, there is a unique natural transformation*

$$C_*^G : \mathcal{F}_{inv}^G(X) \to H_*^G(X)$$

so that $C_*^G(1\!\!1_X) = c^G(TX) \frown [X]_G$ *if X is non-singular.*

Each dimensional component of the equivariant CSM class has its support on an invariant algebraic cycle in X. In particular, the lowest and highest terms are as follows: If X is of equidimension n, the top term is the G-equivariant fundamental class: $C_n^G(1\!\!1_X) = [X]_G$. If X is compact, the degree $pt_*^G C_0^G(\alpha)$ is equal to the weighted Euler characteristics $\int_X \alpha$, especially $pt_*^G C_0^G(1\!\!1_X) = \chi(X)$.

Also the Segre-SM classes admit an equivariant verision: for G-embedding $X \hookrightarrow V$ where V is a G-manifold, we define

$$s_G^{\mathrm{SM}}(X, V) := c^G(TV)^{-1} \frown C_*^G(1\!\!1_X) \ \in H_G^*(X)$$

where $c^G(TV)$ is the G-equivariant Chern class of TV. Abusing the notation, we also mean its pushforward into M and take the equivariant Poincaré dual: $s_G^{\mathrm{SM}}(X, V) \in H_G^*(V)$ unless any confusion occurs. It immediately follows from Proposition 9.4.2 that the equvariant SSM class admits the naturality for the transversal pullback via G-morphism.

In particular, if V is a vector space with G-action and $W \subset V$ is a G-subvariety, we define

$$tp^{\mathrm{SM}}(W)\,(= tp^{\mathrm{SM}}(1\!\!1_W)) := s_G^{\mathrm{SM}}(W, V) \quad \in H_G^*(V) \simeq H^*(BG).$$

Its leading term is just the Thom polynomial $tp(W) \in H^*(BG)$. We call $tp^{\mathrm{SM}}(W)$ the *higher Thom polynomial of* $W \subset V$ with respect to the SSM class. This is a universal SSM class in the following sense.

Theorem 9.4.5 ([129, 130, 132, 135]) *Given a vector bundle* $\pi : E \to M$ *over a complex manifold M with fiber V and structure group G, let* $W(E) \to M$ *be the associated cone bundle with the fiber W and group G. Then, for any section*

[3] In [129, p. 122], another construction of C_*^G using inductive limit is also given. Note that if G is not a torus or a finite group, a tiny modification is needed to multiply $T_{U,*}$ by a correction term to guarantee the normalization condition. This does not affect the correctness of Theorem 9.4.4, though.

$s : M \to E$ which is transverse to $W(E)$, it holds that

$$\mathrm{Dual}\; s^{\mathrm{SM}}(W(s), M) = \rho^* tp^{\mathrm{SM}}(W) \in H^*(M)$$

where $W(s) := s^{-1}(W(E))$ and ρ is the classifying map for $E \to M$. The topological Euler characteristic of $W(s)$ is given by

$$\chi(W(s)) = \int_M c(TM) \cdot \rho^* tp^{SM}(W).$$

Remark 9.4.6 (CSM Class of Degeneracy Loci) A prototype of Theorem 9.4.5 is seen in Parusinski and Pragacz [140], see also in Fulton and Pragacz [60]. They actually considered the CSM class $c^{\mathrm{SM}}(\overline{\Sigma^k(\varphi)})$ of degeneracy loci of vector bundle maps $\varphi : E \to F$ over M by using the desingularization method, as the CSM class generalization of the Thom-Porteous formula. More precisely, for all $s \geq k$, they collect Porteous' desingularizations $\pi_s : Z_s \to \overline{\Sigma^s(\varphi)}$ mentioned in Sect. 9.2.4 and compute $\pi_{s*}c^{\mathrm{SM}}(Z_s) \in H^*(M)$ in terms of Chern classes of E and F. It then follows from the additivity (deletion-inclusion principle) of CSM classes to see that

$$c^{\mathrm{SM}}(\overline{\Sigma^k(\varphi)}) = \pi_{k*}c^{\mathrm{SM}}(Z_k) + \sum_{s>k} a_s \pi_{s*}c^{\mathrm{SM}}(Z_s)$$

for some $a_s \in \mathbb{Z}$, and those coefficients a_s are computed using Euler characteristics of several Grassmannians (singular fiber of π_s's). That yields Parusinski-Pragacz's formula.

Remark 9.4.7 (Grothendieck-Riemann-Roch) Note that C_* can be thought of as a *singular* variant of Grothendieck-Riemann-Roch type theorem (GRR) in Grothendieck's original sense. The most general natural transformation in a certain sense is the *Hirzebruch class transformation* given by Brasselet et al. [25], which unites simultaneously

- MacPherson's Chern class transformation,
- Baum-Fulton-MacPherson's Todd class transformation (singular GRR),
- Cappell-Shaneson's L-class transformation

(see surveys in Handbook [24, 204] for more information). G-equivaraint version is also available through the algebraic Borel construction—especially, using localization technique, the torus-equivariant Hirzebruch class transformation has been studied by Weber [200–202], see also Sect. 9.4.2.1 below.

9.4.2 *Universal Segre-Schwartz-MacPherson Classes*

Let us apply Theorem 9.4.5 to the classification of linear maps (Sect. 9.1.1.1) and the $\mathcal{K}$-classification of singularities of maps (Sect. 9.1.1.2).

9.4.2.1 Thom-Porteous Formula, Upgraded

This research direction has been upgraded into modern Schubert calculus and geometric representation theory.

Based on the formulation in [129, 199], Fehér and Rimányi [54] has developed a computational and representation-theoretic aspect on equivariant CSM/SSM classes for orbit-closures of G-action, e.g., generalized Thom-Porteous formula for Σ^k-types (cf. Remark 9.4.6)—in particular, they provide the interpolation characterization. Independently from this work, a different approach to (equivariant) CSM/SSM class of Schubert varieties was initiated by Aluffi and Mihalcea [3, 4], and a more comprehensive study was achieved by Aluffi et al. [5, 6].

In particular, preceding to [54], the torus-equivariant CSM class has been featured by Rimányi and Varchenko [168] in the context of cotangent Schubert calculus, where they use it to identify *cohomological stable envelope* in the sense of A. Okounkov. This is a striking result, indeed. It is soon lifted up to K*-theoretic stable envelope*—the counterpart is the *equivariant motivic Chern classes*, i.e., equivariant version of Brasselet-Schürmann-Yokura's motivic Chern classes [25]. In recent years, this direction has been intensively studied by Rimányi [158, 159], Fehér [36], Fehér et al. [55, 56], and also Aluffi et al. [5, 6], and moreover, for an elliptic cohmology version of stable envelopes, by Rimányi and Weber [169, 170]. For the relation with quantum integrable systems and mirror symmetry, see [163, 163–165] and related references. The theory is still developing rapidly (several new papers have been produced continuously, e.g., [99–101, 150]), and therefore, it is well beyond the scope of this survey paper. For the starting point of this development, the readers should be referred to, e.g., Rimányi [159], Fehér et al. [56], Aluffi et al. [6] and references therein.

9.4.2.2 Fold and Cusp, Upgraded

Let η be a singularity type in $\mathcal{K}$-classification of map-germs with codimension κ. Now we apply Theorem 9.4.5 to the situation with $G = J^k\mathcal{K}_{m,m+\kappa}$ acting on the jet space $V = J^k(m, m+\kappa)$ and the closure $\overline{\eta} \subset V$. Then by entirely the same proof of Theorem 9.2.4, we see that there is a unique universal power series

$$tp^{\mathrm{SM}}(\overline{\eta}) = tp(\eta) + h.o.t \quad \in \mathbb{Z}[[c_1, c_2, \cdots]]$$

so that for any generic map $f : M \to N$ of codimension κ it holds that

$$\mathrm{Dual}\, s^{\mathrm{SM}}(\overline{\eta(f)}, M) = tp^{\mathrm{SM}}(\overline{\eta})(c(f)) \quad \in H^*(M)$$

after substitution by the quotient Chern classes $c_i = c_i(f)$ [129, 135].

Example 9.4.8 The interpolation method of Rimányi is also very effective to compute $tp^{\mathrm{SM}}(\overline{\eta})$ as well as the leading term (Thom polynomial) $tp(\eta)$. For stable singularity type η of maps in case of $\kappa = 0$, we compute the truncated polynomials of $tp^{\mathrm{SM}}(\overline{\eta})$ up to degree 4 as follows:

$$\begin{aligned}
tp^{\mathrm{SM}}(\overline{A_1}) &\equiv c_1 - c_1^2 + c_1^3 - c_1^4 + c_2^2 - c_1c_3 \\
tp^{\mathrm{SM}}(\overline{A_2}) &\equiv c_1^2 + c_2 - (2c_1^3 + 3c_1c_2 + c_3) + 3c_1^4 + 6c_1^2c_2 + 4c_2^2 \\
&\qquad + c_4 - 6c_2^2 + 6c_1c_3 \\
tp^{\mathrm{SM}}(\overline{A_3}) &\equiv c_1^3 + 3c_1c_2 + 2c_3 - (3c_1^4 + 12c_1^2c_2 + 15c_2^2 + 6c_4) + 14c_2^2 - 14c_1c_3 \\
tp^{\mathrm{SM}}(\overline{A_4}) &\equiv c_1^4 + 6c_1^2c_2 + 2c_2^2 + 9c_1c_3 + 6c_4 \\
tp^{\mathrm{SM}}(\overline{I_{2,2}}) &\equiv c_2^2 - c_1c_3
\end{aligned}$$

Furthermore, in cases of $0 \leq \kappa \leq 4$, tp^{SM} for $\mathcal{K}$-types are explicitly computed up to certain codimension, see Rimányi's 'Thom polynomial Portal' [161]. In fact, based on the previous work with Fehér [54] for SSM classes of degeneracy loci, Rimányi [162] has developed the interpolation for computing tp^{SM} in connection with geometric representation theory of Maulik-Okounkov's stable envelopes (cf. Sect. 9.4.2.1). Those higher Thom polynomials are useful, e.g., in classical enumerative geometry: if some η-singular locus is a curve, then we can compute the degree and the genus of its normalization by using the first two terms of $tp^{\mathrm{SM}}(\overline{\eta})$, and those information is needed to recover classical enumerative formulas [124, 178].

Like as this example, we could do compute $tp^{\mathrm{SM}}(\overline{\eta})$ up to certain degree term, but it is not realistic to compute all terms, because, as a theoretical reason, there appear infinitely many strata adjacent to η under the stabilization process ($m \to \infty$), and as a practical reason, if one uses the interpolation method to compute the SSM class, the same difficulty arises as when computing tp, namely moduli of $\mathcal{K}$-orbits appear at certain codimension.

Nevertheless, we may have a chance to find a closed expression of tp^{SM} for some nice invariants. Suppose that we are given a proper stable map $f : M \to N$ with $d = \dim M - \dim N (= -\kappa) \geq 0$. Note that the germ $f : (M, x) \to (N, f(x))$ at every critical point x defines an d-dimensional isolated complete intersection singularity (ICIS). Then we can define a constructible function over M, called the *Milnor number constructible function* associated to f:

$$\mu(f) : M \to \mathbb{Z}, \quad x \mapsto \text{Milnor number of } f \text{ at } x.$$

In fact, $\mu : J^k(m,n) - W_k \to \mathbb{Z}$, where W_k is some $\mathcal{K}$-invariant closed algebraic set, is defined as a $\mathcal{K}$-invariant constructible function such that $\mu(f)$ is obtained by the pulling back μ via the jet extension and the codimension of W_k goes to ∞ ($k \to \infty$) [63, §7]. Thus a universal series $tp^{\mathrm{SM}}(\mu)$ exists. On the other hand, for a proper stable map $f : M \to N$ with connected N and generic fibre F, it holds that

$$f_*(\mathbb{1}_M + (-1)^d \mu(f)) = \chi(F) \cdot \mathbb{1}_N$$

(the Yomdin-Nakai formula [123, 203]). Apply C_* to the both sides of this equality and multiply by $c(TN)^{-1}$, and then, after a small calculation, universal relation among Landweber-Novikov classes s_I suggests the following conjectural formula [123, 133]:

$$s^{\mathrm{SM}}(\mu(f), M) = tp^{\mathrm{SM}}(\mu)(c(f)) = (-1)^d \left(\sum_{i=0}^{\infty} \sum_{j=0}^{d} c_i(f) \bar{c}_j(f) - 1 \right) \tag{9.9}$$

in $H^*(M)$. That is,

$$\begin{aligned} tp^{\mathrm{SM}}(\mu) &= (-1)^d ((1 + c_1 + c_2 + \cdots)(1 + \bar{c}_1 + \cdots + \bar{c}_d) - 1) \\ &= (-1)^{d+1} (1 + c_1 + c_2 + \cdots)(\bar{c}_{d+1} + \cdots). \end{aligned}$$

There is an attempt to give a direct proof using conormal geometry by Liao [110].

Example 9.4.9 In particular, in case of $\dim M = d + 1$ and $\dim N = 1$, the above (9.9) recovers the following classical formula (cf. Fulton [58]): if f has finitely many isolated critical points, then the sum of the Milnor number is expressed by

$$\int_M \mu(f) = (-1)^{d+1} c_{d+1}(TM - f^*TN) \frown [M].$$

As another example, in case of $d\ (= -\kappa) = 0$, we see $\mu = \mathbb{1}_{\overline{A_1}} + \mathbb{1}_{\overline{A_2}} + \mathbb{1}_{\overline{A_3}} + \mathbb{1}_{\overline{A_4}} + \cdots \in \mathcal{F}^{\mathcal{K}}(J(n,n))$ modulo constructible functions with support of codimension greater than 4. Summing up $tp^{\mathrm{SM}}(\overline{\eta})$ in Example 9.4.8, we observe, up to degree four, $tp^{\mathrm{SM}}(\mu) \equiv tp^{\mathrm{SM}}(\overline{A_1}) + \cdots + tp^{\mathrm{SM}}(\overline{A_4}) \equiv c_1 + c_2 + c_3 + c_4$. This is consistent with $tp^{\mathrm{SM}}(\mu) = -(1 + c_1 + \cdots)(\bar{c}_1 + \cdots) = -c(\bar{c} - 1) = c_1 + c_2 + c_3 + c_4 + \cdots$ (cf. [132, p. 223]).

Example 9.4.10 (Milnor Number of Quasi-Homogeneous ICIS) Suppose that we are given a quasi-homogeneous ICIS

$$\eta : (\mathbb{C}^m, 0) \to (\mathbb{C}^n, 0) \qquad (d = m - n \geq 0)$$

with weights $\boldsymbol{w} = (w_1, \cdots, w_m)$ and degrees $\boldsymbol{d} = (d_1, \cdots, d_n)$. Localizing the above conjectural formula (9.9) via the action of $T = \mathbb{C} - \{0\}$, we get a new interpretation or better understanding of a well-known Milnor number formula of Greuel-Hamm [66] and Damon [31]. In the same way as in Sect. 9.3.1.2 we associate the universal map $f_\eta : E_0 \to E_1$, then $c(E_0) = \prod_{i=1}^m (1 + w_i t)$, $c(E_1) = \prod_{j=1}^n (1 + d_j t)$ and $c(E_0 - E_1) = c(E_0)/c(E_1)$. Applying $tp^{\mathrm{SM}}(\mu)$ to f_η, we obtain an expression of the Milnor number of η:

$$\mu_\eta = (-1)^d \left(\frac{c_n(E_1)}{c_m(E_0)} c_d(E_0 - E_1) - 1 \right). \tag{9.10}$$

This is actually the same as the formula of Greuel-Hamm obtained in an entirely different way:

$$\mu_\eta = (-1)^{d+1} + \frac{|\boldsymbol{d}|}{|\boldsymbol{w}|} \sum_{j=0}^{d} (-1)^{d-j} \sigma_{d-j}(\boldsymbol{w}) s_j(\boldsymbol{d})$$

where $|\boldsymbol{w}| = \prod_{i=1}^m w_i$, $|\boldsymbol{d}| = \prod_{i=1}^n d_i$ and $\sigma_k(\boldsymbol{w})$ and $s_k(\boldsymbol{d})$ are respectively the k-th elementary symmetric function in $w_1, \cdots, w_m$ and the k-th complete symetric function in $d_1, \cdots, d_n$, i.e., Chern (resp. Segre) classes.

9.5 Other Classifications of Map-Germs

We give a quick review on Thom polynomial theory for other classifications of map-germs, especially, isolated singularities of functions and hypersurfaces, and $\mathcal{A}$-finitely determined map-germs.

9.5.1 Lagrange and Legendre Thom Polynomials

There is an important class of mappings, called *Lagrange and Legendre maps*, whose singularities correspond to singularities of functions and hypersurfaces, respectively.

Let Q be a complex symplectic manifold and $\pi : Q \to N$ a Lagrange fibration (i.e., every fiber is Lagragian), and $\iota : L \to Q$ a Lagrange immersion. A typical example is $Q = T^*N \to N$, the cotangent bundle equipped with the canonical symplectic structure. The composed map $\pi \circ \iota : L \to N$ is called a Lagrange map; locally such a map is expressed by the gradient maps $(\mathbb{C}^n, 0) \to (\mathbb{C}^n, 0)$, $p \mapsto \frac{\partial g}{\partial p}$, of some generating function $g = g(p)$. There is the counterpart in contact geometry; the composition of a Legendre immersion and the projection of Legendre fibration is a Legendre map. Their singularities are called *Lagrange and Legendre singularities*, respectively, and have thoroughly been explored by Arnold, Zakalyukin and others, see [9]. The corresponding Thom polynomial theory was initiated by Vassiliev in real C^∞-category [10, 197], in which the simplest Thom polynomial for fold singularity corresponds to the so-called the (first) *Maslov class*. The theory is also formulated in complex analytic/algebraic category by Kazarian [78, 79, 81], that we will briefly review below. Also see Mikosz et al. [117, 118]) especially for the positivity question. Some other particular cases had been treated in e.g., [68, 116].

A hypersurface-singularity of dimension d is a $\mathcal{K}$-equivalence class of function-germs $f : (\mathbb{C}^{d+1}, 0) \to (\mathbb{C}, 0)$. It is an isolated singularity if and only if f is $\mathcal{K}$-finitely determined. When ignoring the linear action on the target $\mathbb{C}$, it is the $\mathcal{R}$-classification of function-singularities. The list of typical singularity types starts from the so-called *ADE classification*. Although it might be thought of as a special part (with corank one) of the classification of isolated complete intersection singularities (ICIS), it has own particular nature—there is a different stabilization. Namely, $f : (\mathbb{C}^{d+1}, 0) \to (\mathbb{C}, 0)$ and $g : (\mathbb{C}^{d'+1}, 0) \to (\mathbb{C}, 0)$ ($d \le d'$) are *stablly $\mathcal{G}$-equivalent* if $f(x) + x_{d+2}^2 + \cdots + x_{d'+1}^2$ is $\mathcal{G}$-equivalent to g ($\mathcal{G} = \mathcal{R}$ or $\mathcal{K}$).

Below we focus mainly on the Legendrian case ($\mathcal{G} = \mathcal{K}$). A natural way to study global properties of isolated hypersurface singularities is described as follows. That arises in a family of divisors. Suppose that we are given a locally trivial complex fibration $p : W \to N$ where W and N are complex manifolds of dimension $m + 1$ and n, respectively, and an embedding $\iota : M \hookrightarrow W$ of a non-singular hypersurface. Put

$$\varphi = p \circ \iota : M \to N$$

($\dim M = m$, $\dim N = n$) and L to be the critical locus of φ, i.e., the locus of points $x \in M$ at which $T_x M$ contains the relative tangent space $\ker dp_x$. Note that the map φ has singularities of corank at most one:

$$L = \Sigma^{m-n+1}(\varphi), \quad \Sigma^{m-n+i}(\varphi) = \emptyset \quad (i > 1).$$

In other words, the germ of φ at each critical point is an unfolding of a function-germ. We now assume that M is sufficiently generic so that the 1-jet exptension $d\varphi$ is transverse to the first Thom-Boardman stratum Σ^{m-n+1}, then L becomes a manifold of dimension $n - 1$. In particular, the restricted map

$$\varphi|_L : L \to N$$

is regarded as a Legendre map. Indeed, for $\dim \operatorname{coker} d\varphi_x = 1$ at every $x \in L$, we assign a covector defining the image hyperplane in $T_{\varphi(x)}N$ (up to non-zero constant multiples), then it turns out that the map $\varphi|_L$ is factored via a Legendre immersion: $\varphi|_L : L \to PT^*N \to N$.

Now we put $V = \ker d\varphi$ and $I = \operatorname{coker} d\varphi$; those are vector bundles over L of rank $m - n + 1$ and of rank one, respectively. The map φ yields the germ of a non-linear map $V \to I$ at the zero section over L. Let f denote the map-germ:

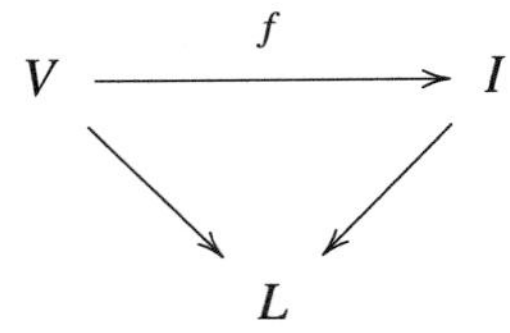

The jet $j^k f(x)$ defines a section of

$$J := \bigoplus_{k \geq 2} \mathrm{Sym}^k V^* \otimes I.$$

This section $L \to J$ is actually determined by the k-th intrinsic derivatives of φ ($k \geq 2$), and is most essential in our argument.

Let η be a stable $\mathcal{K}$-type of isolated hypersurface singularities. Put $\eta(M) \subset L$ to be the locus consisting of the points $x \in L$ at which M is tangent to the fiber with contact type η, in other words, the η-singularity locus of the map $\varphi : M \to N$. Hence, if M is appropriately generic, $[\eta(M)]$ is expressed in $H^*(M)$ by the Thom polynomial $tp(\eta)$ in the quotient Chern classes $\bar{c}_i = c_i(TM - \varphi^* TN)$. However, the geometric setting makes a strong restriction, and especially, Dual $[\eta(M)]$ should be localized into $H^*(L)$. We are interested in representing the class using V, I. More precisely, we start with the above diagram and put

$$u = c_1(I), \quad a_i = c_i(V^* \otimes I - V),$$

where these classes satisfy relations

$$(1 + a_1 + a_2 + \cdots)\left(1 - \frac{a_1}{1+u} + \frac{a_2}{(1+u)^2} - \cdots\right) = 1.$$

Note that when writing $U = V^* \otimes I - V$ formally, this relation follows from the identity

$$U^* \otimes I + U = 0.$$

That allows us to expand the squares of the classes a_i, and hence any polynomial in $u, a_1, a_2, \cdots$ can be expressed as a linear combination of monomials $u^{i_0} a_1^{i_1} a_2^{i_2} \cdots$ with $i_0 \geq 0$ and $i_k = 0$ or 1 for $k > 0$. We define the ring $\mathcal{L}$ of *universal Legendre characteristic classes* to be the quotient ring of the polynomial ring $\mathbb{Z}[u, a_1, a_2, \cdots]$ with $\deg u = 1$ and $\deg a_i = i$ by the ideal obtained by the above relations.

Theorem 9.5.1 ([81] Legendre Thom Polynomials) *Let η be a stable $\mathcal{K}$-type of isolated hypersurface singularities. There is a unique polynomial $P_\eta(u, a_i)$ in $\mathcal{L}$ such that for any generic map $f : V \to I$ over L as above, i.e., any generic section $jf : L \to J$, it holds that*

$$\text{Dual } [\overline{\eta(f)}] = P_\eta(u, a_i) \in H^*(L)$$

evaluated by $u = c_1(I)$ and $a_i = c_i(V^* \otimes I - V)$.

In [81], P_η are explicitly computed for η of codimension up to 6, e.g., a few sample of them are written below (see also Remark 9.5.4).

$$P_{A_2} = \boldsymbol{a_1}$$

$$P_{A_3} = \boldsymbol{3a_2} + ua_1$$

$$P_{A_4} = \boldsymbol{3a_1a_2 + 6a_3} + 4ua_2 + u^2a_1$$

$$P_{D_4} = \boldsymbol{a_1a_2 - 2a_3} - ua_2$$

$$P_{A_5} = \boldsymbol{27a_1a_3 + 6a_4} + u(16a_1a_2 - 12a_3) - 4u^2a_2 + u^3a_1$$

$$P_{D_5} = \boldsymbol{6a_1a_3 - 12a_4} + u(4a_1a_2 - 14a_3) - 4u^2a_2$$

$$\cdots \quad \cdots$$

Return back to the setting with $\varphi = p \circ \iota : M \hookrightarrow W \to N$. Note that V, I are naturally extended to M: we reset V to be the relative tangent bundle $\ker dp$ and I to be the normal bundle of M in W. Then L is the zero locus of a section of $\text{Hom}(V, I)$ on M given by $V \subset TW \to I$, and it is now chosen to be generic. Let $i : L \hookrightarrow M$ be the inclusion, then $[L] = i_*(1) = c_{top}(V^* \otimes I) \in H^*(M)$, and by the projection formula, $[\eta(M)] = [\eta(\varphi)] = i_*([\eta(f)]) = i_*(i^*P_\eta(u, a_i)) = i_*(1)P_\eta(u, a_i)$.

Corollary 9.5.2 ([81] and [118, §10]) *Given a family of divisors, $\varphi = p \circ \iota : M \hookrightarrow W \to N$, as described above, it holds that*

$$\text{Dual } [\overline{\eta(\varphi)}] = (u^r - u^{r-1}c_1 + \cdots + (-1)^r c_r)P_\eta(u, a_i) \in H^*(M)$$

after evaluated by $u = c_1(I)$, $a_i = c_i(V^* \otimes I - V)$ and $c_i = c_i(V)$ with $I = \iota^*TW - TM$ and $V = TW - \varphi^*TN$ and $r := \dim V = m - n + 1$.

For a general Legendre map factored by $L \to PT^*N \to N$, we may take a local chart of PT^*N as $J^1(\mathbb{C}^{n-1}, \mathbb{C}) = T^*\mathbb{C}^{n-1} \oplus \mathbb{C}$ with coordinates (x, p, z) so that the Legendre immersion is locally described using some generating function $z = g(p)$; the z-line $\mathbb{C}$ corresponds to I and the p-space $\mathbb{C}^{n-1}_p$ corresponds to V. We can apply Theorem 9.5.1 to this setting [81, p.731]:

Corollary 9.5.3 ([81]) *Given a generic Legendre map* $\varphi_L : L \to PT^*N \to N$*, it holds that*

$$\text{Dual } [\overline{\eta(\varphi_L)}] = P_\eta(u, a_i) \ \in H^*(L)$$

after evaluated by $u = c_1(I)$, $a_i = c_i(\varphi_L^* TN - TL - I)$ *where* I *is the dual tautological line bundle over* PT^*N*, equivalently, the normal bundle of the contact hyperplane field.*

Remark 9.5.4 (Lagrangian Case) In the above argument, if one takes I to be the trivial line bundle, we cover the case of Lagrange maps, which is tied with the $\mathcal{R}$-classification of function-germs. The universal polynomials of type η for Lagrange singularities, called *Lagrange Thom polynomials*, are obtained by P_η with $u = 0$. For example, in the above sample, the bold font depicts those parts. Lagrange Thom polynomials lie on the polynomial ring generated by $a_1, a_2, \cdots$ with the relation

$$(1 + a_1 + a_2 + \cdots)(1 - a_1 + a_2 - \cdots) = 1$$

which is called the ring of *universal Lagrange characteristic classes*. It is torsion free and has the additive basis of monomials $a_1^{i_1} a_2^{i_2} \cdots a_s^{i_s}$ with $i_k = 0$ or 1. This ring is actually isomorphic to (the limit of) the cohomology ring of the complex Lagrange Grassmannian

$$\Lambda_n^{\mathbb{C}} = Sp_n / U_n$$

i.e., the space of Lagrange subspaces of the symplectic vector space $\mathbb{C}^{2n}$, when one replaces a_i by the ith Chern class of the tautological bundle of rank n over $\Lambda_n^{\mathbb{C}}$. In Legendrian case, the ring $\mathcal{L}$ is isomorphic to the cohomology ring of the twisted complex Lagrange Grassmannian $\tilde{\Lambda}_n^{\mathbb{C}}$, that is the space of Lagrange subspaces in the twisted symplectic fibers of the bundle $\mathbb{C}^n \oplus (\mathbb{C}^n \otimes \xi)$, where ξ is the canonical line bundle over $BU_1 = \mathbb{P}^\infty$.

Positivity of Lagrange and Legendre Thom polynomials with respect to some particular additive basis ('twisted' Schur Q-functions basis and also other basis) has been shown in Mikosz et al. [117, 118]; especially, it contains the list of explicit expansions for ADE-types and P_8, P_9, X_9 up to codimension 7. There is still room for further studies in this direction.

9.5.2 Thom Polynomials for $\mathcal{A}$-Classification

Thom polynomial $tp^{\mathcal{A}}$ for $\mathcal{A}$-classification of map-germs can also be considered. Since for $\mathcal{A}$-equivalence, there is no stabilization of dimension (with respect to suspension), i.e., Lemma 9.2.3 does not hold for $\mathcal{A}$-orbits η, thus in general $tp^{\mathcal{A}}(\eta)$

is just a polynomial in Chern classes of the source bundle and that of the target one, not written in terms of quotient Chern classes.

A relevant geometric setting for $\mathcal{A}$-classification is described as follows. Consider the commutative diagram

$$\begin{array}{ccc} X & \xrightarrow{\quad f \quad} & Y \\ & {\scriptstyle p_0} \searrow \quad \swarrow {\scriptstyle p_1} & \\ & B & \end{array}$$

where X, Y, B are complex manifolds, $p_0 : X \to B$ and $p_1 : Y \to B$ are proper submersions of constant relative dimension m and n, respectively. For each $x \in X$, the germ at x of f restricted to the fiber is defined:

$$f|_{p_0^{-1}(p_0(x))} : (\mathbb{C}^m, 0) \to (\mathbb{C}^n, 0)$$

(in local coordinates of fibers centered at x and $f(x)$, respectively). Given an $\mathcal{A}$-finite singularity type η of map-germs $(\mathbb{C}^m, 0) \to (\mathbb{C}^n, 0)$, the *singularity locus* $\eta(f) \subset X$ and the *bifurcation locus* $B_\eta(f) = p_0(\eta(f)) \subset B$ are defined.

Theorem 9.5.5 ([67, 179]) *Let η be an $\mathcal{A}$-finite singularity type and $f : X \to Y$ as above. If f is appropriately generic,* Dual $[\overline{\eta}(f)] \in H^*(X)$ *is expressed by a universal polynomial $tp^{\mathcal{A}}(\eta)$ in the Chern class $a_i = c_i(T_{X/B})$ and $b_j = c_j(T_{Y/B})$ of relative tangent bundles.* Dual $[\overline{B_\eta(f)}] \in H^*(B)$ *is also expressed by the pushforward $p_{0*}tp^{\mathcal{A}}(\eta)$.*

$$\begin{array}{ccccc} \overline{\eta}(f) & \hookrightarrow & X & \xrightarrow{\quad jf \quad} & J(T_{X/B}, f^*T_{Y/B}) \\ {\scriptstyle p_0} \downarrow & & \downarrow {\scriptstyle p_0} & & \\ \overline{B_\eta}(f) & \hookrightarrow & B & & \end{array}$$

As for Schur basis expansion of $tp^{\mathcal{A}}$ and the positivity, see [149].

Let us look at some computational results. In case of $m = n = 2$, the Thom polynomial of an $\mathcal{A}$-singularity type η is defined to be

$$tp^{\mathcal{A}}(\eta) \in \mathbb{Z}[a_1, a_2, b_1, b_2]$$

where a_i, b_i are Chern classes of relative tangent bundles of source and target, respectively. The $\mathcal{A}$-orbits with $\mathcal{A}$-codimension ≤ 4 are all simple (no nearby moduli). For each of fold, cusp, swallowtail, butterfly and $I_{2,2}^{1,1}$, the $\mathcal{A}$-orbit is an open dense subset of its $\mathcal{K}$-orbit in $J(2, 2)$, that implies that $tp^{\mathcal{A}}$ coincides with

tp for its $\mathcal{K}$-orbit. For other $\mathcal{A}$-type η, the $\mathcal{A}$-orbit has positive codimension in its $\mathcal{K}$-orbit, thus $tp^{\mathcal{A}}(\eta)$ can not be expressed in terms of quotient Chern classes $c_i = c_i(f)$. For instance, $tp^{\mathcal{A}}$ for $\mathcal{A}$-types, lips(beaks) $(x^3 + xy^2, y)$, gulls $(x^4 + xy^2 + x^5, y)$, butterfly $(x^5 + xy + x^7, y)$, are computed by Sasajima and Ohmoto [179] with application to classical projective algebraic geometry:

$$\begin{aligned}
tp^{\mathcal{A}}(\text{lips}) &= -2a_1^3 + 5a_1^2 b_1 - 4a_1 b_1^2 - a_1 a_2 + a_2 b_1 + b_1^3 \\
tp^{\mathcal{A}}(\text{gulls}) &= 6a_1^4 - a_1^2 a_2 - 4a_2^2 - 17a_1^3 b_1 + 4a_1 a_2 b_1 + 17a_1^2 b_1^2 - 3a_2 b_1^2 \\
&\quad -7a_1 b_1^3 + b_1^4 + 2a_1^2 b_2 + 6a_2 b_2 - 4a_1 b_1 b_2 + 2b_1^2 b_2 - 2b_2^2 \\
tp^{\mathcal{A}}(\text{goose}) &= 2a_1^4 + 5a_1^2 a_2 + 4a_2^2 - 7a_1^3 b_1 - 10a_1 a_2 b_1 + 9a_1^2 b_1^2 + 5a_2 b_1^2 \\
&\quad -5a_1 b_1^3 + b_1^4 - 2a_1^2 b_2 - 6a_2 b_2 + 4a_1 b_1 b_2 - 2b_1^2 b_2 + 2b_2^2
\end{aligned}$$

Not only for plane-to-plane germs, the $\mathcal{A}$-Thom polynomials $tp^{\mathcal{A}}$ are also computed in [179] for other $\mathcal{A}$-simple-classifications of $(\mathbb{C}^2, 0) \to (\mathbb{C}^3, 0)$, $(\mathbb{C}^3, 0) \to (\mathbb{C}^3, 0)$, and $(\mathbb{C}^3, 0) \to (\mathbb{C}^4, 0)$. There have been $\mathcal{A}$-simple classifications of parametric curves in plane, 3-space, etc., and those are collectively united by Arnold into a stable form of $\mathcal{A}$-classification of germs $(\mathbb{C}, 0) \to (\mathbb{C}^n, 0)$ with involving n as a parameter, and the corresponding Thom polynomial theory would be a new one, see Pissolato [143].

Remark 9.5.6 (Hurwitz Space) The case of maps between families of curves has intensively been explored by Kazarian and Lando [86, 87] and Kazarian et al. [88, 89] for the study of Hurwitz spaces and genus 0 Gromov-Witten theory (cf. Mumford [122]). Those are actually more advanced ones than ordinary Thom polynomial theory, because the family are allowed to have nodal degeneration and orbifold structures are involved. It also suggests Thom polynomials corresponding to certain classification of function/map-germs over isolated complete intersection singularities.

9.6 Thom Polynomials for Multi-Singularities

At the beginning of this century, another breakthrough occurred. Based on his topological argument using complex cobordism theory MU^*, Kazarian [82, 83] proposed a general framework for universal polynomial expression of the multi-singularity locus of prescribed type for proper holomorphic maps and showed its high potential in application. We briefly describe this conjectural theory and report a progress towards an entire proof.

9.6.1 Multi-Singularity Loci and Hilbert Schemes

Below we use the terminology 'multi-singularity' in the following sense:

Definition 9.6.1 A *multi-singularity type of map-germs with codimension* κ is an ordered set $\underline{\eta} := (\eta_1, \cdots, \eta_r)$ of $\mathcal{K}$-types η_i of mono-germs $(\mathbb{C}^m, 0) \to (\mathbb{C}^n, 0)$ for some m, n with $\kappa = n - m$. In case of $\kappa \leq 0$, we assume that the collection $\underline{\eta}$ contains no submersion-germs.

Note that the entries η_i can be duplicate. We denote by $\mathrm{Aut}(\eta_1, \cdots, \eta_r)$ the subgroup of permutations in $\mathfrak{S}_r$ preserving the types η_i, e.g., $\mathrm{Aut}(A_1, A_1, A_2, A_1, A_2) \simeq \mathfrak{S}_3 \times \mathfrak{S}_2$.

A multi-jet bundle ${}_rJ^k(M, N)$ is defined by the restriction of the cartesian product of r copies of $J^k(M, N) \to M \times N$ over $F(M, r) \times N^r$, where $F(M, r) = M^r \setminus \Delta^r_M$ is the configuration space of ordered distinct r points of M (i.e., the complement of the fat diagonal Δ^r_M) [9, 119]. Put $\pi^r_M : {}_rJ^k(M, N) \to F(M, r)$ and $\pi^r_N : {}_rJ^k(M, N) \to N^r$ to be the canonical projections. For a map $f : M \to N$, the multi-jet extension map ${}_rj^k f$ is defined by

$$ {}_rj^k f : F(M, r) \to {}_rJ^k(M, N), \quad (p_1, \cdots, p_r) \mapsto (j^k f(p_1), \cdots, j^k f(p_r)). $$

Let $\underline{\eta} := (\eta_1, \cdots, \eta_r)$ be a multi-singularity type, and $\dim M = m$ and $\dim N = n\ (= m + \kappa)$. We always assume that k is sufficiently high so that it is greater than the determinacy degree of every entry η_i. Let $W_{\underline{\eta}} \subset {}_rJ^k(M, N)$ be the submanifold consisting of $z = (z_1, \cdots, z_r)$ such that z_i is the k-jet of a mono-germ of type η_i $(1 \leq i \leq r)$ and $\pi^r_N(z)$ lies on the diagonal Δ^r_N of N^r. Let $S = \{p_1, \cdots, p_r\} \subset M$, a set of r distinct points, and suppose that $f(p_1) = \cdots = f(p_r) =: q$ and the germ $f : (M, p_i) \to (N, q)$ is of type η_i for each $1 \leq i \leq r$. The multi-germ $f : (M, S) \to (N, q)$ is *stable* if the multi-jet extension ${}_rj^k f$ (of a representative of the germ f) is transversal to $W_{\underline{\eta}}$ at a point $(p_1, \cdots, p_r) \in F(M, r)$.

Definition 9.6.2 We say that $f : M \to N$ is a *locally stable map* (or simply, a stable map) if for any finite subset S of M such that f maps S to a point $q \in N$, the germ $f : (M, S) \to (N, q)$ is stable.

Let $f : M \to N$ be a proper and locally stable map. Take the preimage of $W_{\underline{\eta}}$ via ${}_rj^k f$ and project it to the first factor M of $F(M, r) \subset M^r$, then the resulting set is written as

$$ \underline{\eta}(f) := \left\{ p_1 \in \eta_1(f) \;\middle|\; \begin{array}{l} \exists\, p_2, \cdots, p_r \in f^{-1}f(p_1) \setminus \{p_1\}\ s.t.\ p_i \neq p_j\ (i \neq j) \\ \text{and the germ of } f \text{ at } p_i \text{ is of type } \eta_i \end{array} \right\}, $$

which becomes a locally closed submanifold of M having the expected dimension. This locus is mapped via f to a locally closed submanifold of N of the same dimension:

$$f(\underline{\eta}(f)) := \left\{ q \in N \;\middle|\; \begin{array}{l} \exists\, p_1, \cdots, p_r \in f^{-1}(q)\ s.t.\ p_i \neq p_j\ (i \neq j) \\ \text{and the germ of } f \text{ at } p_i \text{ is of type } \eta_i \end{array} \right\}.$$

Note that $f : \underline{\eta}(f) \to f(\underline{\eta}(f))$ is a finite-to-one covering whose degree, denoted by $\deg_1 \underline{\eta}$, is equal to the number of η_1 appearing in the tuple $\underline{\eta}$.

For our convenience, we introduce the *source and target multi-singularity loci classes of* f to be

$$m_{\underline{\eta}}(f) := \#\mathrm{Aut}(\eta_2, \cdots, \eta_r)[\overline{\underline{\eta}(f)}], \quad n_{\underline{\eta}}(f) := \#\mathrm{Aut}(\eta_1, \cdots, \eta_r)[\overline{f(\underline{\eta}(f))}] \tag{9.11}$$

(more generally, these classes can be defined for *arbitrary* proper maps [136]). In particular, it holds that

$$f_* m_{\underline{\eta}}(f) = n_{\underline{\eta}}(f).$$

Based on his topological observations supported by many computations, Kazarian claimed in [82, 83] the following statement:

Thom-Kazarian Principle for Multi-Singularities *For any multi-singularity type* $\underline{\eta} = (\eta_1, \cdots, \eta_r)$ *of map-germs of relative codimension* κ*, there is a unique abstract Chern polynomial*

$$R_{\underline{\eta}} = \sum a_I c^I \in \mathbb{Q}[c_1, c_2, \cdots],$$

called the residual polynomial of type $\underline{\eta}$*, such that* $R_{\underline{\eta}}$ *does not depend on the order of entries in* $\underline{\eta}$ *and recursively admits the following property: For a proper morphism* $f : M \to N$ *with* $\dim N - \dim M = \kappa$*, the* $\underline{\eta}$*-singularity loci classes* $m_{\underline{\eta}}(f)$ *and* $n_{\underline{\eta}}(f)$ *are expressed by*

$$m_{\underline{\eta}}(f) = \sum R_{J_1} \cdot f^* f_*(R_{J_2}) \cdots f^* f_*(R_{J_s}) \tag{stp}$$

$$n_{\underline{\eta}}(f) = \sum f_*(R_{J_1}) \cdots f_*(R_{J_s}) \tag{ttp}$$

in rational cohomology rings of M *and* N*, where the summand runs over all possible partitions of the set* $\{1, \cdots, r\}$ *into a disjoint union of non-empty unordered subsets,* $\{1, \cdots, r\} = J_1 \sqcup \cdots \sqcup J_s$ $(s \geq 1)$*, and the subset containing the entry* $1 \in \{1, \cdots, r\}$ *is denoted by* J_1*. Here, for each* $J = \{j_1, \cdots, j_k\}$*,* R_J *stands for the residual polynomial* $R_{(\eta_{j_1}, \cdots, \eta_{j_k})}$ *evaluated by* $c_i = c_i(f)$*, the quotient Chern classes associated to* f*.*

We call the universal polynomials in the right hand side of the formulas stp and ttp the *source and target Thom polynomial of a stable multi-singularity type* $\underline{\eta}$, respectively. Note that the source Thom polynomial $m_{\underline{\eta}}(f)$ is written in quotient Chern classes $c_i(f)$'s and Landweber-Novikov classes $f^*s_I(f)$'s, and the target one is in $s_I(f)$'s, and those polynomials depend only the multi-singularity type $\underline{\eta}$.

For convenience, we introduce a reduced version of multi-singularity Thom polynomials (the last equality is for locally stable maps f):

$$tp(\underline{\eta})(c(f)) := \frac{1}{\#\mathrm{Aut}(\eta_2, \cdots, \eta_r)} m_{\underline{\eta}}(f) = [\overline{\underline{\eta}(f)}],$$

$$tp_{\mathrm{target}}(\underline{\eta})(c(f)) := \frac{1}{\#\mathrm{Aut}(\eta_1, \cdots, \eta_r)} n_{\underline{\eta}}(f) = [\overline{f(\underline{\eta}(f))}].$$

Note that $tp_{\mathrm{target}}(\underline{\eta})(c(f)) := \frac{1}{\deg_1 \underline{\eta}} f_* tp(\underline{\eta})(c(f))$ where $\deg_1 \underline{\eta}$ is the number of entries in $\underline{\eta}$ which is the same type as η_1.

Example 9.6.3 Here are some samples of Thom polynomials for multi-singularities of stable maps in low dimensions [82, 83, 135, 161]. Using the Thom-Kazarian principle, Rimányi's interpolation method works effectively for computing residual polynomials $R_{\underline{\eta}}$. First, in case of $\kappa = 0$, we see

$$\begin{aligned}
tp(A_1) &= c_1 \\
tp(A_2) &= c_1^2 + c_2 \\
tp(A_1^2) &= c_1 s_1 - 4c_1^2 - 2c_2 \\
tp(A_3) &= c_1^3 + 3c_1c_2 + 2c_3 \\
tp(A_1^3) &= \tfrac{1}{2}(c_1 s_1^2 - 4c_2 s_1 - 4c_1 s_2 - 2c_1 s_{01} - 8c_1^2 s_1 + 40c_1^3 + 56c_1c_2 + 24c_3) \\
tp(A_1A_2) &= c_1 s_2 + c_1 s_{01} - 6c_1^3 - 12c_1c_2 - 6c_3 \\
tp(A_2A_1) &= c_1^2 s_1 + c_2 s_1 - 6c_1^3 - 12c_1c_2 - 6c_3.
\end{aligned}$$

Also in case of $\kappa = 1$,

$$\begin{aligned}
tp(A_0^2) &= s_0 - c_1 \\
tp(A_1) &= c_2 \\
tp(A_0^3) &= \frac{1}{2}(s_0^2 - s_1 - 2s_0c_1 + 2c_1^2 + 2c_2) \\
tp(A_0A_1) &= s_{01} - 2c_1c_2 - 2c_3 \\
tp(A_1A_0) &= s_0c_2 - 2c_1c_2 - 2c_3
\end{aligned}$$

$$tp(A_0^4) = \frac{1}{6}(s_0^3 - 3s_0s_1 + 2s_2 + 2s_{01} - 3s_0^2c_1 + 3s_1c_1 + 6s_0c_1^2 + 6s_0c_2 - 6c_1^3 - 18c_1c_2 - 12c_3).$$

Regarding the validity of the above principles, there are some evidences, attempts and solid results, some of which are ongoing projects, as mentioned below. Commonly among those different approaches, it is important to find a nice compactification of the configuration space $F(M, r)$ and to rigorously define multi-singularity loci classes in it—the Hilbert schemes $M^{[l]}$ of 0-dimensional subschemes with colength l of M takes an essential role here.

- *Double and triple-point formulas* (Ronga [172, 173], Fulton and Laksov [59]): For *arbitrary* proper maps $f : M \to N$ with $\kappa > 0$, the *double and triple-point loci classes* $m_{A_0^2}(f)$ *and* $m_{A_0^3}(f)$ are defined, and it is shown that

$$m_{A_0^2}(f) = f^* f_*(1) - c_\kappa(f),$$

$$m_{A_0^3}(f) = f^* f_*(m_{A_0^2}(f)) - 2c_\kappa(f) f^* f_*(1) + 2\left(c_\kappa(f)^2 + \sum_{i=0}^{\kappa-1} 2^i c_{\kappa-i-1}(f) c_{\kappa+i+1}(f)\right).$$

  These are re-written in the form of (stp) using Landweber-Novikov classes. The proofs of the double and triple-point formulas rely on the desingularization method. Namely, desingularizations of the loci are explicitly constructed by blowing-ups along diagonals, and actually the resulting spaces coincide with relative Hilbert schemes associated to f, which are defined in the ordered Hilbert schemes $M^{[[l]]}$ for $l = 2, 3$ (cf. [62, 136]). Following this line, the *quadruple-point formula* was pursued by Dias and Le Barz [32, 104], but it meets a serious difficulty caused from the fact that the Hilbert scheme $M^{[[4]]}$ is singular. On the other hand, once we assume the Thom-Kazarian principle, the quadruple-point formula is explicitly computed by Marangell and Rimányi [114] via the interpolation (also see [157]).
- *Curvilinear multiple-point theory* (Kleiman [92, 94], Colley [28], Ran [153]): The principle is strongly supported by the multiple-point theory for an important particular class of *curvilinear maps*, that is, maps $f : M \to N$ with $\kappa \geq 0$ and corank at most one; its singularity type is denoted by A_μ, i.e., the local ring is isomorphic to $k[x]/\langle x^{\mu+1}\rangle$, and an r-tuple of them is denoted by $A_\bullet = (A_{\mu_1}, \cdots, A_{\mu_r})$ (cf. [84, 121]). That has been established by Kleiman and others around the 80s, where main tools are curvilinear Hilbert schemes and the iteration method [92, 94]. In this case, moduli spaces under consideration become non-singular, and thus concrete computations could be done. For instance, there is a recursive formula for $m_{A_0^r}$ (multiple-point formulas) by Kleiman's iteration method and Colley's program produces $m_{A_\bullet}$ of low codimensions. Note that

those formulas are only applicable to curvilinear maps, i.e., they give wrong answers for non-curvilinear maps.

- *Tautological integrals over curvilinear Hilbert schemes* (Bérczi and Szenes [21]): The target multiple-point formula for generic maps (including non-curvilinear maps) has been established. That corresponds to the formula (ttp) for the type $\underline{\eta} = A_0^r$ where f is a proper locally stable map. Their method relies on torus equivariant localization technique—they use iterated residue integrals with partial desingularization based on the curvilinear Hilbert schemes and non-reductive group quotient $J^k(1, n)/\!\!/\mathrm{Aut}(\mathbb{C}^n, 0)$, that has been developed as a byproduct from their main project for computing Thom polynomials for A_μ-singualrities (Theorem 9.3.2).
- *Functorial approach using algebraic cobordism* (Ohmoto [136]): In contrast to the approaches above, this is a non-constructive approach which proves the existence of universal expression (ttp), and partly (stp), in the Thom-Kazarian principle. In particular, the results ensure the correctness of many formulas counting multi-singular points which are obtained via the interpolation under assuming the principle, including the quadruple-point formula of Marangell and Rimányi [114] for generic maps. Given a multi-singularity type $\underline{\eta} = (\eta_1, \cdots, \eta_r)$, the *source and target multi-singularity loci classes* $m_{\underline{\eta}}(f)$ and $n_{\underline{\eta}}(f)$ are defined for *arbitrary* proper maps $f : M \to N$ by using intersection theory on relative Hilbert schemes so that they coincide with the previously defined ones in (9.11) if f is locally stable. Then, the existence theorem of target multi-singularity Thom polynomials (ttp) for the loci class $n_{\underline{\eta}}(f)$ has been proven, by developing an idea of Kazarian [82] in the context of algebraic geometry—a main feature of the proof is a striking use of algebro-geometric *cohomology operations*, established by Vishik [196] (the universality of Landweber-Novikov operations), on algebraic cobordism theory Ω^* of Levine and Morel [106, 107]. Namely, it turns out that

$$n_{\underline{\eta}} : \Omega^\kappa(N) \to \mathrm{CH}^\ell(N), \quad [f : M \to N] \mapsto n_{\underline{\eta}}(f)$$

is a well-defined cohomology operation, and thus, it is uniquely written as a polynomial of $s_I(f)$'s after tensoring by $\mathbb{Q}$. On the other hand, due to some technical difficulty, the existence of the source multi-singularity Thom polynomials (stp) has been shown, up to now, only in a certainly nice case, i.e., for proper locally stable maps $f : M \to N$ with $\kappa \geq 0$ plus some additional conditions—the proof relies on the existence theorem (ttp) and the interpolation method of Rimányi. For finding an entire proof of the Thom-Kazarian principle for $m_{\underline{\eta}}(f)$, most critical is that the classifying space of multi-singularities has not yet been found properly. That suggests, following Kazarian's view [82], to build a bridge between Thom polynomial theory and motivic homotopy theory.

9.6.2 Higher Thom Polynomials for Multi-Singularity Types

It is natural to expect the Thom-Kazarian principle also holds for the SSM class of the multi-singularity locus of prescribed type, that is, we conjecture that there is a universal polynomial (series) $tp^{\mathrm{SM}}(\overline{\underline{\eta}})$ in variables c_i's and s_I's satisfying

$$s^{\mathrm{SM}}(1\!\!1_{\overline{\underline{\eta}(f)}}, M) = tp^{\mathrm{SM}}(\overline{\underline{\eta}})(c(f))$$

in $H^*(M;\mathbb{Q})$ evaluated by $c_i = c_i(f)$ and $s_I = f^*s_I(f)$ for any proper locally stable maps f. The leading term of the SSM class is the fundamental class $[\overline{\underline{\eta}(f)}]$, thus it is expressed by the reduced multi-singularity Thom polynomial, namely,

$$tp^{\mathrm{SM}}(\overline{\underline{\eta}})(c(f)) = tp(\underline{\eta})(c(f)) + h.o.t.$$

By definition, $c(TM)\cdot tp^{\mathrm{SM}}(\overline{\underline{\eta}})(c(f))$ coincides with the CSM class $C_*(1\!\!1_{\overline{\underline{\eta}(f)}})$. Multiply the both sides of the equality by $\bar{c}(f) = c(f)^{-1} = c(TM - f^*TN)$ and then apply the pushforward f_*, we see

$$s^{\mathrm{SM}}(f_*(1\!\!1_{\overline{\underline{\eta}(f)}}), M) = f_*(\bar{c}(f)\cdot tp^{\mathrm{SM}}(\overline{\underline{\eta}})(c(f)))$$

by using the naturality $f_*C_* = C_*f_*$ of the CSM class transformation, the definition of SSM class and the projection formula. Notice that this total cohomology class in the target space N is written in terms of the Landweber-Novikov classes s_I.

Recall the adjacency relation of multi-singularities both in source and target: the arrow $\underline{\eta} \to \underline{\xi}$ means that the stratum of type $\underline{\xi}$ is contained in the closure of the stratum of $\underline{\eta}$. That yields the set of all multi-singularity types to be a poset (partially ordered set), and we call it the *multi-singularity poset*. Note that this makes sense for each of the source and the target spaces according to the definition of loci $\underline{\eta}(f)$ and $f(\underline{\eta}(f))$ for locally stable maps f; in the source space, the first entry η_1 is distinguished from the others, and in the target space, the order of all entries is ignorable.

For instance, the Hasse diagram of the source multi-singularity poset in case of $\kappa = 0$ starts as follows:

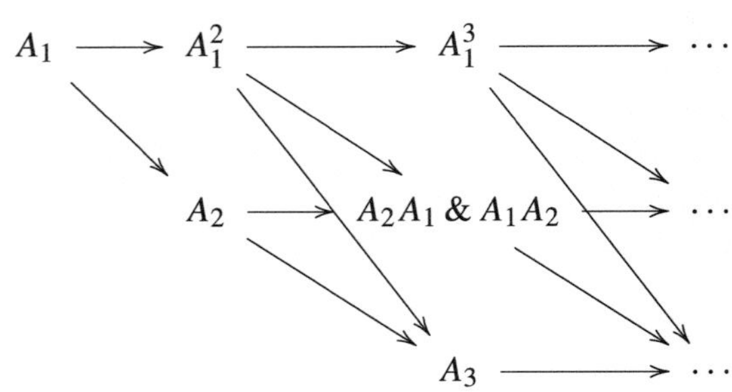

Assuming the Thom-Kazarian principle, the interpolation method of Rimányi yields several computations of the SSM classes for multi-singularity loci provided κ is fixed. Furthermore, in a sophisticated manner inspired by geometric representation theory and the previous work [54, 162], an axiomatic expression of the SSM classes is also studied in a recent paper of Koncki and Rimányi [98]. Generating function formulas may also be considered (cf. [131, 135]).

For instance, in case of $\kappa = 0$, some sample computations are given as follows [124, 135]:

$$\begin{aligned}
tp^{\mathrm{SM}}(\overline{A_1}) &= c_1 - c_1^2 + c_1^3 + \cdots, \\
tp^{\mathrm{SM}}(\overline{A_2}) &= c_1^2 + c_2 - 2c_1^3 - 3c_1c_2 - c_3 + \cdots, \\
tp^{\mathrm{SM}}(\overline{A_1^2}) &= c_1s_1 - 4c_1^2 - 2c_2 - 4c_1^3 - 10c_1c_2 - 4c_3 + c_1s_{01} \\
&\quad + 3c_1^2s_1 + 2c_2s_1 - \tfrac{1}{2}c_1s_1^2 + \cdots, \\
tp^{\mathrm{SM}}(\overline{A_3}) &= c_1^3 + 3c_1c_2 + 2c_3 + \cdots, \\
tp^{\mathrm{SM}}(\overline{A_2A_1}) &= s_1c_1^2 - 6c_1^3 + s_1c_2 - 12c_1c_2 - 6c_3 + \cdots, \\
tp^{\mathrm{SM}}(\overline{A_1A_2}) &= s_2c_1 + s_{01}c_1 - 6c_1^3 - 12c_1c_2 - 6c_3 + \cdots, \\
tp^{\mathrm{SM}}(\overline{A_1^3}) &= \tfrac{1}{2}(s_1^2c_1 - 4s_2c_1 - 2s_{01}c_1 - 8s_1c_1^2 - 4s_1c_2 \\
&\quad + 40c_1^3 + 56c_1c_2 + 24c_3) + \cdots
\end{aligned} \tag{9.12}$$

For instance, assume that $\kappa \le 0$ and let $D(f) := \overline{f(A_1(f))}$ be the discriminant locus. We call $C_*(1\!\!1_{D(f)}) \in H^*(N)$ *the discriminant Chern class of* f. By using the Möbius inversion formula associated to the multi-singularity poset, we find a unique (universal) constructible function $\alpha_{\mathrm{dis}} \in \mathcal{F}(M) \otimes \mathbb{Q}$ so that $f_*\alpha_{\mathrm{dis}} = 1\!\!1_{D(f)}$. For instance, in case of $\kappa = 0$, we compute it as

$$\alpha_{\mathrm{dis}} := 1\!\!1_{\overline{A_1}} - \frac{1}{2}1\!\!1_{\overline{A_1^2}} - \frac{1}{6}1\!\!1_{\overline{A_1^3}} + \frac{1}{2}1\!\!1_{\overline{A_3}} + \cdots \quad \in \mathcal{F}(M) \otimes \mathbb{Q}.$$

The Thom-Kazarian principle indicates that there should be a polynomial $tp^{\mathrm{SM}}(\alpha_{\mathrm{dis}})$ in variables $c_i = c_i(f)$ and $s_I = s_I(f)$ such that

$$C_*(\alpha_{\mathrm{dis}}) = c(TM) \cdot tp^{\mathrm{SM}}(\alpha_{\mathrm{dis}}) \quad \in H^*(M; \mathbb{Q})$$

and also a polynomial $tp^{\mathrm{SM}}(1\!\!1_{D(f)})$ in $s_I = s_I(f)$ such that

$$C_*(1\!\!1_{D(f)}) = c(TN) \cdot tp^{\mathrm{SM}}(1\!\!1_{D(f)}) \quad \in H^*(N; \mathbb{Q}).$$

In particular, the Euler characteristics of the discriminant is universally expressed by

$$\chi(D(f)) = \int_M c(TM)\cdot tp^{\mathrm{SM}}(\alpha_{\mathrm{dis}}) = \int_N c(TN)\cdot tp^{\mathrm{SM}}(1\!\!1_{D(f)}).$$

The naturality of C_* appeals to the inclusion-exclusion principle (indeed, a primitive form of such an argument was often seen in classical computations of elementary characters of projective varieties). For example, in case of $\kappa = 0$, it follows from (9.12) and the above expression of α_{dis} that the low degree terms of the SSM class are given by

$$\begin{aligned} tp^{\mathrm{SM}}(\alpha_{\mathrm{dis}}) &= c_1 + \tfrac{1}{6}(6c_1^2 + 6c_2 - 3c_1s_1) \\ &\quad + \tfrac{1}{6}\begin{pmatrix} c_1^3 + 11c_1c_2 + 6c_3 - 2c_1s_{01} - 5c_1^2s_1 \\ -4c_2s_1 + c_1s_1^2 + 2c_1s_2 \end{pmatrix} + h.o.t. \end{aligned}$$

A reduced divisor D in a complex manifold N is called to be *free* (in the sense of Kyoji Saito) if the sheaf of germs of logarithmic vector fields $\mathrm{Der}_N(-\log D)$ is locally free. As for the CSM class of a free divisor D, the following equality was conjectured by P. Aluffi, and was proved by Liao [109]:

Theorem 9.6.4 (CSM Class of Free Divisors [109]) *If D is locally quasi-homogeneous, it holds that*

$$c^{\mathrm{SM}}(N - D) = c(\mathrm{Der}_N(-\log D)) \quad \in H^*(N).$$

It is well-known that for a stable germ with $\kappa \le 0$, i.e., universal unfolding of ICIS germs, the discriminant is a free divisor (cf. [119]). On the other hand, we have seen that the CSM class of $D = D(f)$ in the ambient space N is expressed using the target universal SSM class for $f_*\alpha_{\mathrm{dis}} = 1\!\!1_{D(f)}$:

$$c^{\mathrm{SM}}(N - D) = c^{\mathrm{SM}}(N) - c^{\mathrm{SM}}(D(f)) = c(TN)(1 - tp^{\mathrm{SM}}(1\!\!1_{D(f)})).$$

Thus we get the following formula:

$$tp^{\mathrm{SM}}(1\!\!1_{D(f)}) = 1 - \frac{c(\mathrm{Der}_N(-\log D(f))}{c(TN)}.$$

This is an unexpected and highly non-trivial identification of the target SSM Thom polynomial $tp^{\mathrm{SM}}(1\!\!1_{D(f)})$ in variables $s_I = s_I(f)$, which measures the difference of Chern classes for locally free sheaves of vector fields and of logarithmic vector fields. That might be comparable to the conjectural formula (9.9) on the Milnor number constructible function.

9.7 Some Topics in Applications

9.7.1 *Application to Enumerative Geometry: Instanton Counting*

The literature on enumerative geometry contain full of counting formulas, including nineteenth century (cf. [91, 141, 142, 177, 180, 181]) as well as modern ones inspired by mathematical physics and applied algebraic geometry. Thom polynomial theory serves as a kind of *taxonomy* or *systematics* to re-organize these counting formulas and discover new species.

Here we present a typical example in this direction—the enumeration of singular curves on surfaces—according to Kazarian [82, §10]. This classical topic has been resurrected since the 1990s due to new interest from string theory, known as *instanton counting*. Consider the linear system $|\mathcal{L}| = \mathbb{P}(H^0(S; \mathcal{L}))$ associated to a line bundle $\mathcal{L}$ on a projective non-singular complex surface S. Let $\underline{\eta} = (\eta_1, \cdots, \eta_r)$ be a combination of types of plane curve singularities $(\mathbb{C}^2, 0) \to (\mathbb{C}, 0)$, e.g., the type of r nodes denoted by $\underline{A_1^r} = (A_1, \cdots, A_1)$, and put n to be the expected codimension of the target multi-singularity locus of type $\underline{\eta}$ (e.g., it is r for $\underline{A_1^r}$ with $\kappa < 0$). Then, a general problem is to determine the number of singular curves in this system which have singular points of the prescribed type $\underline{\eta}$ and pass through $\dim |\mathcal{L}| - n$ common points of S in general position.

The problem is interpreted as counting $\underline{\eta}$-type multi-singular points of a certain map. Let $N = \mathbb{P}^n$ be a generic linear subspace of $|\mathcal{L}|$ which is transverse to the discriminant. Take the trivial fibration $p : N \times S \to N$ and the incident hypersurface $\iota : M \hookrightarrow N \times S$ with the normal bundle I, i.e., M consists of the pairs of a curve in the family and a point lying on this curve. Now $\dim M = n + 1$ and $\dim N = n$ (thus $\kappa = -1$). Let $f = p \circ \iota : M \to N$ be the composed map, that is the family of curves over the sub-linear system N, and put $L = \Sigma^2(f)$, the critical locus of f (recall the setup in Sect. 9.5.1).

Now we can apply target Thom polynomials for $n_{\underline{\eta}}$ to this map $f : M \to N$, since the existence theorem of the universal polynomials has been established in [136]. In this geometric setting, as remarked in Sect. 9.5.1, the quotient Chern class $c(f)$ is rewritten in terms of Legendre characteristic classes in $\mathcal{L}$, so we state the computation in this terminology following [82].

Let $h = c_1(\mathcal{L}) \in H^2(S)$, $1 + c_1 + c_2 = c(T^*S)$, and let t be the divisor class in $N = \mathbb{P}^\ell$. Then

$$u = c_1(I) = [M] = t + h, \quad a = c(T^*S \otimes I - TS) = \frac{(1+u)^2 + (1+u)c_1 + c_2}{1 - c_1 + c_2}.$$

The pushforward $f_* : H^*(L) \to H^*(N)$ is factorized by $f_* = p_* i_*$, where $i : L \to N \times S$ is the embedding and $p : N \times S \to N$ is the projection to the first factor. The class $[L] = i_*(1)$ is easily computed, and the homomorphism p_* commutes with multiplication by t and sends monomials of degree 2 in the classes h, c_1, c_2 to the numbers; $d := p_*(h^2) = \mathcal{L} \cdot \mathcal{L}$, $k := p_*(hc_1) = \mathcal{L} \cdot K_S$, $s := p_*(c_1^2) = K_S \cdot K_S$ and $x := p_*(c_2) = \chi(S)$.

Since the target singularity locus of type $\underline{\eta}$ has codiemsion n, there are finitely many singular curves having isolated singular points of type $\underline{\eta}$ which belong to the family f. Applying the established Thom-Kazarian principle, the degree $\int n_{\underline{\eta}}(f)$ expresses the number of those singular curves; especially, for any $J \subset \underline{\eta}$, the residual class $S_J = f_* R_J \in H^*(\mathbb{P}^n)$ is a linear combination of the numbers d, k, s, x multiplied by t^{n_J} (n_J is the expected codimension of the locus of type J), thus the desired degree is written by a polynomial in d, k, s, x. That is summarized as follows.

Theorem 9.7.1 ([82]) *Let $\underline{\eta} = (\eta_1, \cdots, \eta_r)$ be a multi-singularity type of plane curve-germs with $\mathcal{K}_e$-codimension n. Then it holds that for any projective surface with an ample line bundle $\mathcal{L}$ and any sufficiently generic linear sub-system $\mathbb{P}^n \subset |\mathcal{L}|$, the number $N_{\underline{\eta}}$ of singular curves belonging to the system that possess the combination of singularities of type $\underline{\eta}$ is given by*

$$N_{\underline{\eta}} = \frac{1}{\#\mathrm{Aut}(\underline{\eta})} \sum \int_{\mathbb{P}^n} S_{J_1} \cdots S_{J_s}$$

where the sum runs over all partition of $\{1, \cdots, r\}$ and S_{J_s} denotes the image of the residual polynomial of multi-singularities $\{\eta_j \mid j \in J_s\}$.

A generating function of the numbers $N_{\underline{\eta}}$ is also formulated [82]. Theorem 9.7.1 for the simplest type $\underline{\eta} = A_1^r$ solves the problem on the existence of universal polynomials for counting nodal curves, known as part of the *Göttsche conjecture* [65, 95, 96] (now several different proofs are available, e.g., [13, 17, 108, 154]). Like as this, there are a lot of questions for which Thom polynomial theory may effectively give satisfactory answers. Other examples are:

- higher dimensional analogy to Göttsche conjecture, i.e., counting divisors with a prescribed combination of singularities in a given linear system;
- other type of contact problems, e.g., enumerating multi-tangent hyperplanes to a given hypersurface in projective space etc.;
- computing the degree of bifurcation sets, e.g., event surfaces and its boundary strata when viewing a curve and a surface in 3-space (partly related to Computer Vision);

and so on, see, e.g., Vainsencher [194, 195], Tzeng [192, 193], Rennemo [154], Kazarian [82], Wall [198], Nekarda [124, 125] and references therein.

9.7.2 *Vanishing Topology of Map-Germs*

As a well-known theorem due to J. Milnor, the Milnor fiber of an isolated hypersurface-singularity $f : (\mathbb{C}^{d+1}, 0) \to (\mathbb{C}, 0)$ is homotopic to a bouquet of d-dimensional real spheres; the number $\mu(f)$ of the spheres is called the *Milnor number* of the germ f. The number is interpreted as the number of A_1 (Morse)-singular points appearing in a stable perturbation of the germ f, and also it is computed by the dimension of the Milnor algebra:

$$\mu(f) = \#A_1(f_t) = \dim_{\mathbb{C}} \mathcal{O}_{\mathbb{C}^m,0}/J_f.$$

This picture is generalized to the case of map-germs. For the simplicity, we assume that $m \le n$ throughout this subsection. Let $f : (\mathbb{C}^m, 0) \to (\mathbb{C}^n, 0)$ be a $\mathcal{G}$-finitely determined map-germ, and $\underline{\eta}$ a stable (mono/multi-)singularity type of the expected codimension n in the target. Suppose that there is a *stable perturbation*

$$f_t : U \to \mathbb{C}^n \quad (t \in \Delta \subset \mathbb{C},\ 0 \in U \subset \mathbb{C}^m)$$

so that f_0 is a representative of the germ f and f_t for $t \neq 0$ is a stable map. Note that if η is a mono-singularity type, it is enough to assume that f is $\mathcal{K}$-finitely determined, while for a multi-singularity type, we need the $\mathcal{A}$-finiteness of f. Then $\underline{\eta}(f_t)$ for $t \neq 0$ consists of finitely many isolated points; the number is constant for non-zero t and does not depend on the choice of stable perturbation, and it is usually called a 0-*stable invariant* of the original germ f. We let $\#\underline{\eta}(f_t)$ denote the number of (multi-)singular points in the *target space*, e.g., $\#A_0^3(f_t)$ $(\kappa > 0)$ means the triple-points in the image of a stable perturbation f_t of f_0.

There have been several works for computing 0-stable invariants via colengths of some algebras under certainly nice situations. In general, however, it can be a hard task e.g., in case that the singularity scheme is not Cohen-Macaulay, that often arises.

Now we propose a new topological method localizing Thom polynomials to compute stable invariants for *quasi-homogeneous* map-germs. This new approach provides a number of examples with significantly simpler computation where any corank condition is not assumed.

Let $f : (\mathbb{C}^m, 0) \to (\mathbb{C}^n, 0)$ be a quasi-homogeneous germ with weights $w_1, \cdots, w_m$ and degrees $d_1, \cdots, d_n \in \mathbb{Z}_{>0}$, i.e., there are diagonal representations ρ_0, ρ_1 of $T = \mathbb{C}^*$ in the source and the target space, respectively, such that $f = \rho_1(t) \circ f \circ \rho_0^{-1}(t)$ $(t \in T)$. Suppose that f is $\mathcal{A}$-finitely determined. Then its $\mathcal{A}_e$-versal unfolding $F : (\mathbb{C}^{m+k}, 0) \to (\mathbb{C}^{n+k}, 0)$ is also quasi-homogeneous, so let $r_1, \cdots, r_k$ be the weights of unfolding parameters. We introduce vector bundles over $BT = \mathbb{P}^\infty$

$$E_0 := \oplus_{i=1}^m \mathcal{O}(w_i), \quad E_1 := \oplus_{j=1}^n \mathcal{O}(d_j), \quad E' = \oplus_{i=1}^k \mathcal{O}(r_i),$$

and we have maps between total spaces of these bundles:

$$\begin{array}{ccccccc} & & E_0 & \xrightarrow{\ f\ } & E_1 & & \\ & & \downarrow{\scriptstyle i_0} & & \downarrow{\scriptstyle i_1} & & \\ \underline{\eta}(F) & \hookrightarrow & E_0 \oplus E' & \xrightarrow[F]{} & E_1 \oplus E' & \hookleftarrow & F(\underline{\eta}(F)) \end{array}$$

Given a singularity type $\underline{\eta}$, we apply the Thom polynomial $tp(\underline{\eta})$ for F [82, 136] and compute the intersection of $tp(\underline{\eta})(c(F))$ with the inclusion map i_0 or i_1 in $H^*(BT) = \mathbb{Z}[a]$, that leads to the following theorem.

Theorem 9.7.2 ([135]) *Let $m \le n$ and let $f : (\mathbb{C}^m, 0) \to (\mathbb{C}^n, 0)$ be an $\mathcal{A}$-finitely determined quasi-homogeneous map-germ with weight w_i and degree d_j. Given a stable mono/multi-singularity type $\underline{\eta}$ with the expected codimension n in the target, the 0-stable invariant of f_0 is computed by*

$$\#\underline{\eta}(f_t) = \frac{tp(\underline{\eta})}{\deg_1 \underline{\eta} \cdot w_1 \cdots w_m} = \frac{tp_{\text{target}}(\underline{\eta})}{d_1 \cdots d_n}$$

where numerators stand for the coefficient of a^m and a^n of the source and target Thom polynomials applied to the universal map $f_0 : E_0 \to E_1$, respectively.

In Example 9.4.10 in Sect. 9.4, we have seen that the Greuel-Hamm formula of the Milnor number for quasi-homogeneous $\mathcal{K}$-finite map-germs with $\kappa \le 0$ (i.e., ICIS) can be interpreted as a torus localization of the higher Thom polynomial $tp^{\mathrm{SM}}(\mu)$. As a counterpart in case of $\kappa > 0$, let us consider $\mathcal{A}$-finitely determined map-germs $f : (\mathbb{C}^n, 0) \to (\mathbb{C}^{n+1}, 0)$ where the pair $(n, n+1)$ is in Mather's nice range. The image of f is a hypersurface with non-isolated singularities—it is known that the homotopy type of the image of any stably perturbation f_t is a bouquet of middle dimensional real spheres, whose number is called the *image Milnor number*:

$$\mu_I(f) := (-1)^m(\chi(\mathrm{Im}(f_t)) - 1).$$

The so-called *Mond conjecture* says that the $\mathcal{A}_e$-codimension of $f \;\; \le \mu_I(f)$, with equality if f is quasi-homogeneous [119]. A crucial difficulty lies in how to calculate $\mu_I(f)$. Our method provides a lot of new examples, i.e., we can compute μ_I for $\mathcal{A}$-finite quasi-homogeneous germs in low dimensions without any corank condition, by using the same type theorem as above by replacing tp by higher Thom polynomial tp^{SM} for multi-singularities, see [135] for the detail in case $n = 3$. Furthermore, combining this method with some results on several invariants for $\mathcal{A}$-equivalence, Pallarés Torres and Peñafort Sanchis [139] succeeded to obtain formulas for $\mu_I(f)$ in more cases, e.g., $n = 4$ and 5. Their results have a significant impact on the study of the Mond conjecture [119, §8.8, §8.9].

9.7.3 Positivity and Complex Hyperbolic Geometry

Recently a surprising connection was discovered by Bérczi [12, 14–17] between the positivity question of coefficients of Thom polynomials of A_μ-singularities and higher dimensional versions of the Nevanlinna theory in complex analysis. It was quite unexpected, as there had been no interaction between these two subjects before. The discovery was made possible by a new geometric invariant theory for *non-reductive groups*, established by Bérczi and Kirwan [19].

In complex analytic geometry, one of fascinating themes is the *hyperbolicity* of a complex projective variety X (in the sense of Brody), i.e., any holomorphic curve from $\mathbb{C}$ to X is constant—the famous Kobayashi conjecture says that a generic hypersurface $X \subset \mathbb{P}^{n+1}$ is hyperbolic if $\deg(X)$ is sufficiently large. It is known that a generic hypersurface X is of general type if $\deg(X) \geq n+3$. As a generalization, the Green-Griffiths-Lang (GGL) conjecture claims that for every complex projective algebraic variety X of general type there exists a proper algebraic subvariety of X containing all nonconstant entire holomorphic curves $\gamma : \mathbb{C} \to X$. To attack these conjectures, a key construction is a smooth compactification of the bundle of invariant k-jet differentials, called the *Demailly-Semple tower*, so that global sections of certain line bundles over it give algebraic differential equations which the curves γ should satisfy. Then the problem is reduced to the existence of global sections, thus intersection theory comes up here. By this strategy, there have been several attempts to find effective degree bounds for the conjectures in past decades.

A technical difficulty lies on the fact that the group $J^k\mathrm{Aut}(\mathbb{C}, 0)$ of k-jets of isomoprhism-germs of $(\mathbb{C}, 0)$ is not reductive, thus the geometric invariant theory (GIT) of Mumford can not be applied directly. Recently, Bérczi and Kirwan developed *non-reductive* GIT, and as an application, they obtained a different smooth compactification of the bundle of invariant jet differentials using the algebraic model of Bérczi and Szenes [20] for computing the Thom polynomials of $\mathcal{A}_\mu$-singularity types (Theorem 9.3.2). Namely, the Demailly-Semple tower is replaced by a new fibrewise compactification $\pi : \tilde{X}_k \to X$ of invariant jet differentials, endowed with a tautological line bundle τ on $\tilde{X}_k$, that leads to the discovery/refinement of an iterated residue formula for the tautological integrals.

Theorem 9.7.3 (Bérczi [12, 14–16]) *Let $X \subset \mathbb{P}^{n+1}$ be a smooth projective hypersurface and let $u = c_1(\tau)$ and $h = c_1(\pi^*\mathcal{O}_X(1))$ denote the first Chern classes of the tautological line bundle on $\tilde{X}_\mu$ and the divisor class on X, respectively. For any homogeneous μ polynomial $P = P(u, h)$ of degree $\deg(P) = \dim \tilde{X}_\mu = n + \mu(n-1)$, we have*

$$\int_{\tilde{X}_\mu} P = \int_X \operatorname*{Res}_{t=\infty} \frac{\prod_{m<l}(t_m - t_l) Q_k(t_1, \cdots, t_\mu) P(t_1 + \cdots + t_\mu, h)}{\prod_{m+r \leq l \leq \mu}(t_m + t_r - t_l)(t_1 \cdots t_\mu)^n} \prod_{i=1}^{\mu} \mathrm{S}\left(\frac{1}{t_i}\right) dt_i$$

where $\mathrm{S}(t^{-1}) = 1 + s_1(X)t^{-1} + s_2(X)t^{-2} + \cdots + s_n(X)t^{-n}$ is the total Segre class of X and Q_μ is the polynomial associated to A_μ-singularity type in Theorem 9.3.2.

This theorem is very effective for the so-called *polynomial GGL conjecture*. For instance, in [14, 15] Bérczi proved the polynomial GGL conjecture modulo the *positivity conjecture* of R. Rimányi, posed at the very beginning [156], which claims that there is no negative coefficients in the Chern monomial basis expansion of $tp(A_\mu)$ (this is a different version of positivity from the one in Remark 9.2.6). After improvements by combining it with recent results in complex hyperbolic geometry, finally the following theorems have been obtained:

Theorem 9.7.4 (Bérczi and Kirwan [19]) ***(Polynomial Green-Griffiths-Lang theorem for projective hypersurfaces)*** *Let* $X \subset \mathbb{P}^{n+1}$ *be a generic smooth projective hypersurface of* $\deg(X) \geq 16n^3(5n+4)$. *Then there is a proper algebraic subvariety* $Y \subset X$ *containing all nonconstant entire holomorphic curves in* X.

Theorem 9.7.5 (Bérczi and Kirwan [19]) ***(Polynomial Kobayashi theorem)*** *A generic smooth projective hypersurface* $X \subset \mathbb{P}^{n+1}$ *of* $\deg(X) \geq 16(2n-1)^3(10n-1)$ *is Brody hyperbolic.*

These results have made considerable impact in both (mutually far) communities of singularity theory and complex geometry. The connecting point between two subjects is intersection theory on Hilbert schemes and non-reductive GIT. That also indicates a new future direction for researches in Thom polynomial theory.

9.8 Real Singularities

We have discussed Thom polynomials in complex analytic/algebraic category, but it makes sense for real C^∞-category as well—e.g., the desingularization method in Sect. 9.2.4 commonly works in both categories [10]. There are many related topics in differential topology, but here we focus mainly on a general statement of the existence of real Thom polynomials $tp^{\mathbb{R}}(\eta)$, i.e., the counterpart to Theorem 9.2.4 in the setting of real singularities. Actually, real singularities have their own particular features caused by a finer classification of C^∞ map-germs, real (semi-)algebraic/analytic properties, (co-)orientation problem and complex conjugacy.

9.8.1 *Vassiliev Complex and Real Thom Polynomials*

The classification machinery (with respect to $\mathcal{K}$ and $\mathcal{A}$-equivalence) is the same as the complex case (Sect. 9.2.1), but the real case is more subtle and the classification list becomes larger than the complex case; for instance, there are two different $\mathcal{K}$-orbits with $\kappa = 0$, $I_{2,2} : (x^2 + y^2, xy)$ and $II_{2,2} : (x^2 - y^2, xy)$ as real map-germs, but they are the same in complex case.

Our first task is to define the singularity loci class of prescribed type in (Borel-Moore) homology for real C^∞-maps, that is not straightforward at all. Put $V :=$

$J^k(m,n)$ and $G := J^k\mathcal{G}$, the real algebraic group of k-jets of elements in the equivalence group in real case. Note that G is homotopy equivalent to the product of orthgonal groups $O(m) \times O(n)$. Since the G-action on V is real algebraic, every G-orbit η (or modulus of G-orbits) in V is a locally-closed semi-algebraic subset, and its closure $\overline{\eta}$ in fine topology is also semi-algebraic in general. A semi-algebraic set may not have the $\mathbb{Z}_2$-fundamental cycle in the Borel-Moore homology, while any real algebraic set does [23, Prop. 11.3.1]. Also we may ask for the fundamental cycle with $\mathbb{Z}$-coefficients, that is related to the existence of G-invariant orientation of the normal bundle of η (called the *G-coorientation* of η). In this case, we may replace $\mathcal{G}$ by the one preserving the orientations of the source and the target space, then G is homotopy equivalent to the product of rotation groups $SO(m) \times SO(n)$. For instance, each orbit of $I_{2,2}$ and $II_{2,2}$ ($\kappa = 0$) does not have the fundamental cycle, while the closure of the union of the two orbits forms a real algebraic set, so it has the $\mathbb{Z}_2$-fundamental cycle. Moreover, both orbits are G-coorientable and the union admits a $\mathbb{Z}$-fundamental cycle.

Let us introduce a general framework. Suppose that we are given a G-invariant semi-algebraic Whitney stratification of V. Let Γ_p denote the union of strata of codimension $\geq p$ and put $V_p := V \setminus \Gamma_{p+1}$. Then we have a filtration of invariant open sets

$$V_0 \subset V_1 \subset V_2 \subset \cdots \subset V.$$

It yields the G-equivariant spectral sequence $E_*^{*,*}$, which converges to $H_G^*(V) = H^*(BG)$ with the coefficient ring R (e.g., $\mathbb{Z}_2$ or $\mathbb{Z}$). The E_1-term is

$$E_1^{p,q} := H_G^{p+q}(V_p, V_{p-1}; R) \simeq H_G^q(\Gamma_p \setminus \Gamma_{p+1}; R).$$

In particular, $E_1^{p,0}$ is freely generated by the strata of codimension p (and G-coorientable if $R = \mathbb{Z}$), and $\delta : E_1^{p,0} \to E_1^{p+1,0}$ is the connecting homomorphism of the long exact sequence for the triple (V_{p+1}, V_p, V_{p-1}), which is determined by analyzing the adjacency relation among strata. The cochain complex $(E_1^{*,0}, \delta)$ is called the *Vassiliev complex* (with coefficients in R) associated to the stratification [10, 75, 77, 197]. Intuitively, a p-dimensional cohomology class ω of this cochain complex means a geometric cycle in V defined by a linear combination of strata of codimension p, say $\omega = [\sum n_i \eta_i]$ ($n_i \in R$), and thus it defines an element of $H_G^p(V) = H^p(BG; R)$. This element should be the 'real Thom polynomial' associated to the class ω for C^∞-maps. Originally, such an abstract complex was introduced and computed for the classification of function-germs by Vassiliev in his study on Lagrange and Legendre Thom polynomials [10, 197], see Sect. 9.5.1. For $\mathcal{K}$-classification of C^∞-map-germs with $\kappa = 0$ up to the codimension 8, the $\mathbb{Z}_2$-Vassiliev complex has been determined in Ohmoto [126], and the $\mathbb{Z}$-Vassiliev complex, which is much more difficult, has been computed in Fehér and Rimányi [47] using the interpolation method partly.

For $\mathcal{K}$-classification, the stabilization property (Lemma 9.2.3, as $m \to \infty$ with κ fixed) also holds in real case, that leads to the following theorem. Note that

$$H^*(BO; \mathbb{Z}_2) = \mathbb{Z}_2[w_1, w_2, \cdots], \quad H^*(BSO; \mathbb{Z}[\tfrac{1}{2}]) = \mathbb{Z}[\tfrac{1}{2}][p_1, p_2, \cdots]$$

where w_i and p_i are the Stiefel-Whitney class and the Pontrjagin class, respectively.

Theorem 9.8.1 ([29, 47, 77, 126, 188]) *Fix $\kappa \in \mathbb{Z}$ and consider $\mathcal{K}$-classification of C^∞-map-germs $(\mathbb{R}^*, 0) \to (\mathbb{R}^{*+\kappa}, 0)$. Let $\omega = [\sum \eta_i]$ be a p-dimensional cohomology class for the $\mathbb{Z}_2$-Vassiliev complex. Then there exists a unique polynomial $tp_2^{\mathbb{R}}(\omega) \in \mathbb{Z}_2[w_1, w_2, \cdots]$ such that for any appropriately generic C^∞-map $f : M^m \to N^{m+\kappa}$, the closure of the union of $\eta_i(f)$, denoted by $S_\omega(f)$, forms a $\mathbb{Z}_2$-geometric cycle in M and is expressed by*

$$\mathrm{Dual}_2[S_\omega(f)] = tp_2^{\mathbb{R}}(\eta)(w(f)) \quad \in H^p(M; \mathbb{Z}_2)$$

evaluated by the Stiefel-Whitney class $w_i = w_i(f) = w_i(f^*TN - TM)$. Also for a cohomology class ω for the $\mathbb{Z}$-Vassiliev complex, a similar statement holds: for any generic map $f : M^m \to N^{m+\kappa}$ between orientable manifolds, the locus $S_\omega(f)$ forms a $\mathbb{Z}$-cycle in M and is expressed by a universal polynomial $tp^{\mathbb{R}}(\omega)$ in Pontrjagin classes $p_i = p_i(f) = p_i(f^*TN - TM)$ modulo 2-torsion elements.

We remark that the interpolation method does not work effectively for computing real mod.2 Thom polynomials $tp_2^{\mathbb{R}}$, because the restriction equations mod.2 often become trivial (but see Remark 9.8.2 below). Computing real integer Thom polynomials $tp^{\mathbb{R}}$ in p_i's is a more difficult problem.

Remark 9.8.2 (Borel-Haefliger Theorem) As mentioned above, the union of $\mathcal{K}$-orbits of type $I_{2,2}$ and $II_{2,2}$ with $\kappa = 0$ forms a cycle of the $\mathbb{Z}$-Vassiliev complex; actually its closure in V coincides with the Thom-Boardman stratum $\overline{\Sigma^2}$. Hence, from results by desingularization (which works in both real and complex cases) [151, 171], the integer Thom polynomial in the orientable setting is given as

$$tp^{\mathbb{R}}(I_{2,2} + II_{2,2}) = tp^{\mathbb{R}}(\Sigma^2) = p_1$$

while forgetting the orientation, the mod.2 Thom polynomial is

$$tp_2^{\mathbb{R}}(I_{2,2} + II_{2,2}) = tp_2^{\mathbb{R}}(\Sigma^2) = w_2^2 + w_1 w_3$$

(this is consistent; if M, N are orientable, then $w_1 = w_1(f^*TN - TM) = 0$ and the mod.2 reduction of p_1 gives w_2^2). Also the Thom polynomial for complex singularity type $I_{2,2}$ with $\kappa = 0$ is

$$tp(I_{2,2}) = tp(\Sigma^2) = c_2^2 + c_1 c_3.$$

Note that replacing c_i in $tp(I_{2,2})$ by w_i, we recover $tp_2^{\mathbb{R}}(I_{2,2} + II_{2,2})$. This is generically true. Namely, if a $\mathcal{K}$-orbit $\eta_{\mathbb{C}}$ of complex map-germs admits several real forms η_i such that their complexification is $\eta_{\mathbb{C}}$ (including the case of a single real form η), then it defines a $\mathbb{Z}_2$-Vassiliev cycle $\omega = [\sum \eta_i]$, and the mod.2 Thom polynomial $tp_2^{\mathbb{R}}(\omega)$ is obtained from $tp(\eta_{\mathbb{C}})$ by replacing $c_i \mapsto w_i$ and by taking coefficients mod.2, see Borel-Haefliger [22] and Matszangosz [113] (also [39, 40]).

9.8.2 Global Topology of C^∞ Maps: Miscellaneous

A variant of Vassilev complex for *$\mathcal{A}$-classification of multi-singularities of maps* has been introduced for finding topological invariants for stable maps in low dimensions, called *local first-order Vassilev-type invariants*, e.g., Arnold's $J^\pm$ invariants for plane immersed curves [8], Goryunov's invariants for stable maps from a surface into $\mathbb{R}^3$ [64], and many others (see [128, 134] and references therein). Putting finer semi-global data for mappings under consideration, an *enriched version* of the Vassiliev complex has been considered in Saeki's studies on topology of 4-manifolds via stable maps into low dimensional space [174, 176]. That is also related to Gromov's complexity of C^∞-maps. One of central problems in global topology of C^∞ maps is elimination of singularities of prescribed type—historically it arose in connection with the Poincaré conjecture. In this context, the h-principle and cobordism of singular maps take important roles, where the real Thom polynomial of η is regarded as the primary obstruction for removing η-singular points of generic C^∞ maps $f : M \to N$ via homotopy or surgery, see e.g., [10, 167, 175, 183]. Also 'relative Thom polynomials' for singular Seifert maps have long been considered implicitly (cf. [184]); e.g., Takase [185] has studied oriented bordisms of immersions of 7-manifolds to $\mathbb{R}^8$ in relation to Thom polynomials $tp^{\mathbb{R}}$ with integer coefficients given in [47] (also see [76, 78] for another direction). For an immersion from a C^∞ manifold L into an almost complex manifold M, there are enumerative formulas for complex tangent points or $\mathbb{R}\mathbb{C}$ singularities [103], and elimination problem for complex tangents has recently been discussed [35, 74]. For a singular Lagrange immersion of L into a symplectic manifold M equipped with almost complex structure, there are Thom polynomials for open Whitney umbrellas [127]. Those singularities and complex tangents are characterized by (higher) intrinsic derivatives of the C^∞-bundle map $TL \otimes \mathbb{C} \to TM$, and hence the Thom polynomials may be written in terms of $c_i(f^*TM - TL \otimes \mathbb{C})$. Higher mod.2 Thom polynomials in real case (i.e., real Z_2-counterpart to SSM Thom polynomials) should be based on D. Sullivan's Stiefel-Whitney classes for real algebraic varieties [182]. There had been only a few related works for real Morin maps, such as Fukuda [57] and Nakai [123], but it has potential for further researches, e.g., a recent work of Fehér and Matszangosz [41].

There is still a lot of room for further researches on real Thom polynomials (especially in K-theory, cobordism theory, Chow-Witt groups, and so on), although the story has a long history since 1950s. Unfortunately several interesting topics are dropped and references cited here may be inadequate, thus readers are advised to directly refer to appropriate literature.

Acknowledgments The author is very grateful to the editors of this series, and to the referee for careful reading and comments for improving the draft. On this occasion, the author personally wants to express his deepest gratitude to his teacher Takuo Fukuda, who made a profound study on Isotopy Lemma under René Thom in 1970s, for guiding him to singularity theory in his student time. Also, the author thanks many friends and colleagues, especially Laszlo Fehér, Maxim Kazarian, Richárd Rimányi, Andrzej Weber for helpful discussions on Thom polynomials for a long time over 20 years. This work was partly supported by JSPS KAKENHI Grant Numbers JP17H02838 and JP23H01075.

References

1. P. Aluffi, Characteristic classes of singular varieties, in Topics in cohomological studies of algebraic varieties, Birkhäuser (2005) 1–32. Notes for a cycle of five lectures given at the Banach Center, Warsaw, April 23–27 2002.
2. P. Aluffi, Segre classes and invariants of singular varieties, In ‘Handbook of Geometry and Topology of Singularities, III’, Springer (2022) 419–492.
3. P. Aluffi and L. C. Mihalcea, Chern classes of Schubert cells and varieties, J. Algebraic Geom., **18** (2009), 63–100.
4. P. Aluffi and L. C. Mihalcea, Chern-Schwartz-MacPherson classes for Schubert cells in flag manifolds, Compositio Math. **152** (2016), 2603–2625.
5. P. Aluffi, L. C. Mihalcea, J. Schürmann, C. Su, Positivity of Segre-MacPherson classes, In ‘Facets of Algebraic Geometry: A Collection in Honor of William Fulton’s 80th Birthday. Volume 1’, Cambridge University Press (2022) 1–28.
6. P. Aluffi, L. C. Mihalcea, J. Schürmann, C. Su, Shadows of characteristic cycles, Verma modules and positivity of Chern-Schwartz-MacPherson classes of Schubert cells, Duke Math. J. **172** (2023) 3257–3320.
7. Y. Ando, On the Higher Thom polynomials of Morin Singularities, Publ. RIMS, Kyoto Univ. **23** (1987), 195–207.
8. V. I. Arnol’d, *Topological invariants of plane curves and caustics,* Univ. Lect. Series, vol. 5, AMS (1994).
9. V. I. Arnold, S. Gusein-Zade and A. Varchenko, *Singularities of Differentiable Maps vol. I*, Monographs Math. **82**, Birkhäuser, (1985).
10. V. I. Arnold, V. V. Goryunov, O. V. Lyashko and V. A. Vasil’ev, *Singularity Theory I*, Encyclopaedia of Math. Sci. **6**, Springer-Verlag (1993).
11. M. Atiyah and R. Bott, The moment map and equivariant cohomology, Topology **23** (1984), 1–28.
12. G. Bérczi, Moduli of map germs, Thom polynomials and the Green-Griffiths conjecture, Contributions to Algebraic Geometry (ed. P. Pragacz), EMS (2012), 141–168.
13. G. Bérczi, Tautological integrals on curvilinear Hilbert schemes, Geometry & Topology **21** (2017), 2897–2944.
14. G. Bérczi, Thom polynomials and the Green-Griffiths-Lang conjecture for hypersurfaces with polynomial degree, Internat. Math. Research Notices, (2019), 7037–7092.

15. G. Bérczi, Towards the Green-Griffiths-Lang conjecture via equivariant localisation, Proc. London Math. Soc. **118** (2019), 1057–1083.
16. G. Bérczi, Non-reductive geometric invariant theory and Thom polynomials, (2020), arXiv:2012.06425
17. G. Bérczi, Tautological integrals on Hilbert scheme of points II: geometric subsets, (2023), arXiv: 2303.14812.
18. G. Bérczi, L. Fehér and R. Rimányi, Expressions for resultants coming from the global theory of singularities, Topics in Algebraic and Noncommutative geometry, Contemporary Math. **324**, AMS (2003), 63–69.
19. G. Bérczi and F. Kirwan, Non-reductive geometric invariant theory and hyperbolicity, Invent. Math., **235** (2024) 81–127.
20. G. Bérczi and A. Szenes, Thom polynomials of Morin singularities, Annals of Mathematics **175** (2012), 567–629.
21. G. Bérczi and A. Szenes, Multiple-point residue formulas for holomorphic maps, (2021), arXiv:2112.15502v1.
22. A. Borel and A. Haefliger, La classe d'homologie fondamentale d'un espace analytique, Bull. Soc. Math. France **89** (1961), 461–513.
23. J. Bochnak, M. Coste and J-M. Roy, Real Algebraic Geometry, Ergebnisse der Mathematik und ihrer Grenzgebiete, Folge 3, vol. 36, Springer-Verlag, Berlin (1998).
24. J. P. Brasselet, Characteristic Classes, In 'Handbook of Geometry and Topology of Singularities, III', Springer (2022), 303–417.
25. J. P. Brasselet, J. Schürmann and S. Yokura, Hirzebruch classes and motivic Chern classes for singular spaces, Journal of Topology and Analysis, **2** (1) (2010), 1–55.
26. A.S. Buch and R. Rimányi, A formula for non-equioriented quiver orbits of type A, J. Algebraic Geom. **16** (2007), 531–546.
27. A.S. Buch, L. Fehér and R. Rimányi, Positivity of quiver coefficients through Thom polynomials, Adv. Math. **197** (2005), 306–320.
28. S. J. Colley, Enumerating stationary multiple-points, Advances in Math. **66** (1987), 149–170.
29. J. Damon, *Thom Polynomials for Contact Class Singularities*, Doctoral Dissertation, Harvard University (1972).
30. J. Damon, The Gysin Homomorphism for Flag Bundles, Amer. Jour. Math. **95** (1973), 643–659.
31. J. Damon, Higher multiplicities and almost free divisors and complete intersections, Memoirs, Amer. Math. Soc., **589** (1996).
32. D. Dias and P. Le Barz, *Configuration Spaces over Hilbert Schemes and Applications,* Springer Lecture Notes in Math. **1647** (1996).
33. M. Domokos, L. Fehér and R. Rimányi, Equivariant and invariant theory of nets of conics with an application to Thom polynomials, Jour. Singularities **7** (2013), 1–20.
34. D. Edidin and W. Graham, Equivariant intersection theory, Invent. Math. **131** (1998), 595–634.
35. A. Elgindi, A topological obstruction to the removal of a degenerate complex tangent and some related homotopy and homology groups, Internat. J. Math. **26** (2015), Article No.1550025.
36. L. Fehér, Motivic Chern classes of cones, Singularities and Their Interaction with Geometry and Low Dimensional Topology, Trends in Math., Springer (2021), 181–205.
37. L. Fehér and B. Kömüves, On second order Thom-Boardman singularities, Fund. Math. **191** (2006), 249–264.
38. L. Fehér and A. Juhász, Plücker formulas using equivariant cohomology of coincident root strata, (2023), arXiv:2312.06430
39. L. Fehér and A. Matszangosz, Real solutions of a problem in enumerative geometry, Periodica Mathematica Hungarica (2016), 1–20.
40. L. Fehér and A. Matszangosz, Halving spaces and lower bounds in real enumerative geometry, Algebraic & Geometric Topology **22** (2022), 433–472.

41. L. Fehér and A. Matszangosz, Obstructions for Morin and fold maps: Stiefel-Whitney classes and Euler clarancteristics of singularity loci, (2025), arXiv:2502.07379.
42. L. Fehér, A. Némethi and R. Rimányi, Degeneracy of 2-forms and 3-forms, Canadian Math. Bull. **48** (2005), 547–560.
43. L. Fehér, A. Némethi and R. Rimányi, Coincident root loci of binary forms, Michigan Math. J **54** (2006), 375–392.
44. L. Fehér, A. Némethi and R. Rimányi, The degree of the discriminant of irreducible representations J. Alg. Geom. (2009), 751–780.
45. L. Fehér, A. Némethi and R. Rimányi, Equivariant classes of matrix matroid varieties, Comment. Math. Helvetici **87** (2012), 861–889.
46. L. Fehér and Z. Patakfalvi, The incidence class and the hierarchy of orbits, Central European Jour. Math. **7** (2009), 429–441.
47. L. Fehér and R. Rimányi, Thom polynomials with integer coefficients, Illinois Jour. Math. **46** (2002), 1145–1158.
48. L. Fehér and R. Rimányi, Classes of degeneracy loci for quivers–the Thom polynomial point of view, Duke Math. Jour. **114** (2002), 193–213.
49. L. Fehér and R. Rimányi, Schur and Schubert polynomials as Thom polynomials – cohomology of moduli spaces, Cent. Euro. J. Math. **4** (2003), 418–434.
50. L. Fehér and R. Rimányi, Calculation of Thom polynomials and other cohomological obstructions for group actions, Real and Complex Singularities (Sao Carlos, 2002), Contemp. Math. **354**, A.M.S. (2004), 69–93.
51. L. Fehér and R. Rimányi, Thom polynomial computing strategies. A survey, Adv. Studies in Pure Math. **43** (2007), 45–53.
52. L. Fehér and R. Rimányi, On the structure of Thom polynomials of singularities, Bulletin London Math. Soc. **39** (2007), 541–549.
53. L. Fehér and R. Rimányi, Thom series of contact singularities, Annals of Mathematics **176** (2012), 1381-1426.
54. L. Fehér and R. Rimányi, Chern-Schwartz-MacPherson classes of degeneracy loci, Geom. &Top. **22** (2018), 3575–3622.
55. L. Fehér, R. Rimányi, A. Weber, Characteristic classes of orbit stratifications, the axiomatic approach, International Conference on the Trends in Schubert Calculus (2017), 223–249.
56. L. Fehér, R. Rimányi, A. Weber, Motivic Chern classes and K-theoretic stable envelopes, Proc. London Math. Soc. **122** (2021), 153–189.
57. T. Fukuda, Topology of folds, cusps and Morin singularities, A Fête of Topology, papers dedicated to Itiro Tamura, Academic Press (1988), 331–353 (abstract: RIMS Kokyuroku (1985)).
58. W. Fulton, *Intersection Theory*, 2nd edition, Ergebnisse der Mathematik und ihrer Grenzgebiete, Folge 3, vol. 2, Springer-Verlag, Berlin (1998). [1st edition, (1984)].
59. W. Fulton, D. Laksov, Residual intersections and the double point formula, Real and Complex Singularities (Proc. Conf. Oslo, 1976), P. Holme, ed., Sijthoff & Noordhoff (1977), 171–178.
60. W. Fulton and P. Pragacz, *Schubert varieties and degeneracy loci*, Lecture Notes in Math. **1689** (1998).
61. T. Gaffney, Thom polynomial of $\overline{\Sigma^{1,1,1,1}}$, Proc. Symposia Pure Math. A.M.S. vol **40**-Part 1 (1983), 399–408.
62. T. Gaffney, Punctual Hilbert schemes and resolutions of multiple point singularities, Math. Ann. **295** (1993), 269–289.
63. C. G. Gibson, K. Wirthmüller, A. A. du Plessis and E. J. N. Looijenga, Topological stability of smooth mappings, Springer Lecture Notes in Math. **552** (1976).
64. V. Goryunov, Local invariants of mappings of surfaces into three-space, The Arnold-Gelfand math. seminors, Birkhäuser (1997), 223–255.
65. L. Göttsche, A conjectural generating function for numbers of curves on surfaces, Comm. Math. Phys. **196** (1998), 523–533.

66. G. M. Greuel and H. Hamm, Invarianten quasihomogener vollständiger Durchschnitte, Invent. Math. **49** (1978), 67–86.
67. A. Haefliger and A. Kosinski, Un théorème de Thom sur les singularitiés des applications différentiables, Séminaire H. Cartan, Exp. no. 8 (1956).
68. J. Hayden, Some global properties of singularities, Ph.D. Thesis, Univ. Warwick (1980).
69. J. Heller and J. Malagón-López, Equivariant algebraic cobordism, J. Reine Anguew. Math., **684** (2013), 87–112.
70. J. Hornbostel and V. Kiritchenko, Schubert calculus for algebraic cobordism, J. Reine Anguew. Math., **656** (2011), 59–85.
71. T. Hudson, Thom-Porteous formulas in algebraic cobordism, Dissertation, Univ. Essen (2012).
72. T. Hudson, T. Matsumura and N. Perrin, Stability of Bott–Samelson Classes in Algebraic Cobordism, Schubert Calculus and Its Applications in Combinatorics and Representation Theory, Springer (2020), 281–306.
73. A. Iarrobino, Hilbert Scheme of Points: Overview of Last Ten Years, Proc. Symp. Pure Math. Vol. **46** (1987), 297–320.
74. N. Kasuya and M. Takase, Knots and links of complex tangents, Trans. Amer. Math. Soc. **370** (2018), 2023–2038.
75. M. Kazarian, Hidden singularities and the Vasil'ev homology complex of singularity classes Sbornik Math. **186** (1995), 119–128.
76. M. Kazarian, Relative Morse theory of one-dimensional bundles and cyclic homology, Funct. Anal. Appl. **31** (1997), 16–24.
77. M. Kazarian, Characteristic classes of Singularity theory, Arnold-Gelfand MathematicalSeminars, Birkhäuser (1997), 325–340.
78. M. Kazarian, The Chern-Euler number of circle bundle via singularity theory, Math. Scand., (1998), 207–236.
79. M. Kazarian, On Lagrange and symmetric degeneracy loci Isaac Newton Institute for Mathematical Sciences Preprint Series (2000)
80. M. Kazarian, Classifying spaces of singularities and Thom polynomials. New developments in singularity theory (2001), 117–134.
81. M. Kazarian, Thom polynomials for Lagrange, Legendre, and critical point function singularities, Proc. London Math. Soc. **86** (2003), 707–734.
82. M. Kazarian, Multisingularities, cobordisms and enumerative geometry, Russian Math. Survey **58**:4 (2003), 665–724 (Uspekhi Mat. nauk **58**, 29–88).
83. M. Kazarian, Thom polynomials, Singularity Theory and its application (Sapporo, 2003), Adv. Stud. Pure Math. vol. **43** (2006), 85–136.
84. M. Kazarian, Morin maps and their characteristic classes, preprint (2006).
85. M. Kazarian, Non-associative Hilbert schemes and Thom polynomials, (2017), arXiv:1712.09270.
86. M. Kazarian and S. Lando, Towards an intersection theory on Hurwitz spaces, Izvestiya Math. **68** (2004).
87. M. Kazarian and S. Lando, Thom polynomials for maps of curves with isolated singularities, Proc. Steklov inst. Math. **258** (2007), 87–99.
88. M. Kazarian, S. Lando and D. Zvonkine, Universal cohomological expressions for singularities in families of genus 0 stable maps, Internat. Math. Research Notices (2018), 6817–6843.
89. M. Kazarian, S. Lando and D. Zvonkine, Double Hurwitz numbers and multisingularity loci in genus 0 Internat. Math. Research Notices (2022), 9529–9570.
90. S. L. Kleiman, Problem 15. Rigorous foundation of Schubert's enumerative calculus, Mathematical Developments arising from Hilbert Problems, Proc. Sympos. Pure Math., vol. **28**, Amer. Math. Soc., Providence, R. I. (1976), 445–482.
91. S. L. Kleiman, The enumerative theory of singularities, Real and Complex Singularities (Proc. Conf. Oslo, 1976), P. Holme, ed., Sijthoff & Noordhoff (1977), 297–396.
92. S. L. Kleiman, Multiple point formulas I: Iteration, Acta Math. **147** (1981), 13–49.

93. S. L. Kleiman, Intersection Theory and Enumerative Geometry: A Decade in Review, with the collaboration of A. Thorup, Proc. Symp. Pure Math. Vol. **46** (1987), 321–370.
94. S. L. Kleiman, Multiple point formulas II: The Hilbert scheme, Springer Lecture Notes in Math. **1436** (1990), 101–138.
95. S. L. Kleiman and R. Piene, Enumerating singular curves on surfaces, Algebraic Geometry: Hirzebruch 70, Contemp. Math. A. M. S. **241** (1999), 209–238.
96. S. L. Kleiman and R. Piene, Node polynomials for families: methods and applications, Math. Nachr. **271** (2004), 69–90.
97. N. Kolokolnikova, On the rationality of the singularities of the A_2-loci, (2017), arXiv:1712.03558.
98. J. Koncki and R. Rimányi, The structure of higher Thom polynomials of multisingularities, in preparation (2025).
99. J. Koncki and A. Weber, Twisted motivic Chern class and stable envelopes, Adv. Math. **404** (2022), 108374.
100. J. Koncki and A. Weber, Hecke algebra action on twisted motivic Chern classes and K-theoretic stable envelopes, (2023), arXiv:2301.12746.
101. S. Kumar, R. Rimányi and A. Weber, Elliptic classes of Schubert varieties, Math. Ann. **378** (2020), 703–728.
102. A. Lascoux and P. Pragacz, Thom polynomials and Schur functions: the singularities $A_3(-)$, Publ. RIMS **46** (2010), 183–200.
103. H. F. Lai, Characteristic classes of real manifolds immersed in complex manifolds, Trans. A. M. S., **172** (1972), 1–33.
104. P. Le Barz, Validité de certaines formules de géomeétire énumérative, C. R. Aad. Sci. Paris **289** (1979), 755–758.
105. H. Levine, The Singularities, S_1^q, Illinois J. Math. **8** (1964), 152–168.
106. M. Levine and F. Morel, *Algebraic Cobordism*, Monograph Math., Springer (2007).
107. M. Levine and R. Pandharipande, Algebraic cobordism revisited, Invent. Math. **176** (2009), 63–130.
108. J. Li and Y. Tzeng, Universal polynomials for singular curves on surfaces, Compositio Math. **150** (2014), 1169–1182.
109. X. Liao, Chern classes of logarithmic derivations for free divisors with Jacobian ideal of linear type, J. Math. Soc. Japan **70** (2018) 975–988.
110. X. Liao, Chern classes of vanishing cycles for maps having no blowup in codimension 0 and applications in Thom polynomial theory, preprint (2023)
111. R. MacPherson, Chern classes for singular algebraic varieties, Ann. Math. **100** (1974), 421–432.
112. J. N. Mather, Stable map-germs and algebraic geometry, Manifolds - Amsterdam 1970, Springer Lecture Notes in Math. **197** (1971), 176–193.
113. Á. Matszangosz, Borel-Haefliger type theorems, PhD thesis, Central European Univ., Budapest (2019).
114. R. Marangell and R. Rimányi. The general quadruple point formula, Amer. J. Math. **132** (2010), 867–896. 10.1353/ajm.0.0130.
115. R. McCrory, Cobordism operations and singularities of maps, Bull. A. M. S. **82** (1976), 281–283.
116. R. McCrory, T. Shifrin and R. Varley, The Gauss map of a generic hypersurface in $\mathbb{P}^4$, J. Diff. Geom. **30** (1989), 689–759.
117. M. Mikosz, P. Pragacz and A. Weber, Positivity of Thom polynomials II: the Lagrange singularities, Fundamenta Math. **202** (2009), 65–79.
118. M. Mikosz, P. Pragacz and A. Weber, Positivity of Legendrian Thom polynomials, Jour. Diff. Geom. **89** (2011), 111–132.
119. D. Mond and J. Nuño-Ballesteros, *Singularities of Mappings*, Grundlehren der mathematischen Wissenschaften, vol. 357. Springer (2020).
120. D. Mond and J. Nuño-Ballesteros, Singularities of Mappings, In 'Handbook of Geometry and Topology of Singularities, III', Springer (2022), 81–144.

121. D. Mond and R. Pellikaan, Fitting ideals and multiple points of analytic mappings, Springer Lecture Notes in Math. **1414** (1989), 107–161.
122. D. Mumford, Towards an enumerative geometry of the moduli space of curves, Arithmetic and Geometry, Springer (1983), 271–328.
123. I. Nakai, Elementary topology of stratified mappings, Proc. sympo. "Singularities" (Sapporo 1998), Adv. Stud. Pure Math. 29, Kinokuniya, (2000), 221–243.
124. S. Nekarda, Enumerative Geometry of Bifurcation Surfaces, University of Halle (2025).
125. S. Nekarda and T. Ohmoto, Enumerative geometry of space curve projections using Thom polynomials, preprint (2025).
126. T. Ohmoto, Vassiliev complex for contact classes of real smooth map-germs, Rep. Fac. Sci. Kagoshima Univ. (1994), 1–12.
127. T. Ohmoto, Thom polynomials for open Whitney umbrellas of isotropic mappings, Banach Center Publications **33** Inst. Math. Polish Acd. Sci. (1996) 287–296.
128. T. Ohmoto and F. Aicardi, First order local invariants of apparent contours, Topology, vol. **45** (2006) 27–45.
129. T. Ohmoto, Equivariant Chern classes of singular algebraic varieties with group actions, Math. Proc. Cambridge Phil. Soc. **140** (2006), 115–134.
130. T. Ohmoto, Chern classes and Thom polynomials, Singularities in Geometry and Topology, World Scientific (2007), 464–482.
131. T. Ohmoto, Generating functions of orbifold Chern classes I: Symmetric Products, Math. Proc. Cambridge Phil. Soc. **144** (2008), 423–438.
132. T. Ohmoto, A note on Chern-Schwartz-MacPherson class, IRMA Lectures in Mathematics and Theoretical Physics, European Math. Soc. **20** (2012), 117–132.
133. T. Ohmoto, Mini-Course, 12th International Workshop on Real and Complex Singularities, ICMC-USP, Brazil, July (2012).
134. T. Ohmoto, Vassiliev type invariants for generic mappings, revisited, Contemporary Math., AMS, **569** (2012), 143–160.
135. T. Ohmoto, Singularities of maps and characteristic classes, School on Real and Complex Singularities in São Carlos, 2012, Adv. Studies Pure Math. **68** (2016), 191–265.
136. T. Ohmoto, Universal polynomials for multi-singularity loci of maps, (2024) arXiv: 2406.12166.
137. Ö. Öztürk, On Thom polynomials for $A_4(-)$ via Schur functions, Serdica math. J. **33** (2007), 301–320.
138. Ö. Öztürk and P. Pragacz, On Schur function expansions of Thom polynomials, in "Contributions to Algebraic Geometry", EMS Ser. Congr. Rep. (2012), 443–479.
139. I. Pallarś Torres and G. Peñafort Sanchis, Image Milnor number formulas for weighted-homogeneous map-germs, Results in Math. **76** (2021), no.152.
140. A. Parusiński and P. Pragacz, Chern-Schwartz-MacPherson classes and the Euler characteristic of degeneracy loci and special divisors, Jour. Amer. Math. Soc. **8** (1995), 793–817.
141. R. Piene, Some formulae for a surface in $\mathbb{P}^3$, Springer Lecture Notes in Math. **687** (1978), 196–235.
142. R. Piene, Polar classes of singular varieties, Ann Sci. de l'É.N.S. **11** (1978), 247–276.
143. J. Pissolato, Computation of Thom polynomials for $\mathcal{A}$-singularities of parametric curves germs with degeneracy loci formulas, in preparation (2025).
144. P. Pragacz, Thom polynomials and Schur functions I, (2005), math.AG/0509234
145. P. Pragacz, Thom polynomials and Schur functions: the singularities $I_{2,2}(-)$, Annales de l'Institut Fourier **57** (2007), 1487–1508.
146. P. Pragacz, Thom polynomials and Schur functions: towards the singularities $A_i(-)$, "Real and complex singularities - Sao Carlos 2006", Contemporary Mathematics, AMS **459** (2008) 165–178.
147. P. Pragacz, Positivity of Thom polynomials and Schubert calculus, Adv. Studies Pure Math. Vol. 71 (2016) "Schubert Calculus - Osaka 2012" (2016), 419–451.
148. P. Pragacz and A. Weber, Positivity of Schur function expansions of Thom polynomials, Fundamenta Mathematicae **195** (2007), 85–95.

149. P. Pragacz and A. Weber, Thom polynomials of invariant cones, Schur functions, and positivity, "Algebraic cycles, sheaves, shtukas, and moduli, Trends in Mathematics, (2008), 117–129.
150. S. Promtapan and R. Rimányi Characteristic classes of symmetric and skew-symmetric degeneracy loci, in "Facets of Algebraic Geometry. A Collection in Honor of William Fulton's 80th Birthday", LMS Lecture Note Series **472** (2022), 254–283.
151. I. R. Porteous, Simple singularities of maps, Proceedings of Liverpool Singularities – Symposium I, Lecture Notes in Math. **192** (1971), 286–307.
152. I. R. Porteous, Probing singularities, Proc. Symposia Pure Math. A.M.S. vol **40**-Part 2 (1983), 395–406.
153. Z. Ran, Curvilinear enumerative geometry, Acta Math. **155** (1985), 82–101.
154. J. V. Rennemo, Universal polynomials for tautological integrals on Hilbert schemes, Geometry & Topology **21** (2017), 253–314.
155. R. Rimányi, Generalized Pontrjagin-Thom Construction for Singular Maps, Thesis, Eötvös Loránd University (1997).
156. R. Rimányi, Thom polynomials, symmetries and incidences of singularities, Invent. Math. **143** (2001), 499-521.
157. R. Rimányi, Multiple-point formulas – a new point of view, Pacific Jour. Math. **202** (2002), 475–489.
158. R. Rimányi, Motivic characteristic classes in cohomological Hall algebras, Adv. Math. **360** (2020), 106888.
159. R. Rimányi, $\hbar$-deformed Schubert calculus in equivariant cohomology, K-theory, and elliptic cohomology, in Singularities and Their Interaction with Geometry and Low Dimensional Topology, Birkhauser, (2021)
160. R. Rimányi, Thom polynomials. A primer, survey for lectures in the Newton Institute (2024), arXiv: 2407.13883.
161. R. Rimányi, Thom polynomial Portal, web site, Univ. North Carolina at Chapel Hill, https://rimanyi.web.unc.edu/thom-polynomial-portal/
162. R. Rimányi, Interpolation characterization of higher Thom polynomials, preprint (2025).
163. R. Rimányi and Y. Shou, Bow varieties–geometry, combinatorics, characteristic classes, Comm. in Analysis and Geometry (2022)
164. R. Rimányi, A. Smirnov, A. Varchenko and Z. Zhou, Three dimensional mirror self-symmetry of the cotangent bundle of the full flag variety, SIGMA **15** (2019)
165. R. Rimányi, A. Smirnov, A. Varchenko and Z. Zhou, Three-dimensional mirror symmetry and elliptic stable envelopes, Internat. Math. Research Notices **13** (2022), 10016–10094.
166. R. Rimányi and A. Szenes: Residues, Grothendieck polynomials, and K-theoretic Thom polynomials, Internat. Math. Research Notices **23** (2023), 20039–20075.
167. R. Rimányi and A. Szücs, Pontrjagin-Thom type construction for maps with singularities, Topology **37** (1998), 1177–1191.
168. R. Rimányi and A. Varchenko, Equivariant Chern-Schwartz-MacPherson classes in partial flag varieties: interpolation and formulae, in Schubert Varieties, Equivariant Cohomology and Characteristic Classes, IMPANGA2015 (eds. J. Buczynski, M. Michalek, E. Postingel), EMS (2018), 225–235.
169. R. Rimányi and A. Weber, Elliptic classes of Schubert varieties via Bott-Samelson resolution, J. of Topology, **13** 2020, 1139–1182.
170. R. Rimányi and A. Weber, Elliptic classes on Langlands dual flag varieties, Comm. in Contemp. Math. **24** (2022) 2150014.
171. F. Ronga, Le calcul des classes duales aux singularités de Boardman d'ordre deux, Comment. Math. Helvetici **47** (1972), 15–35.
172. F. Ronga, Le calcul des classes duales aux points doubles d'une application, Compositio. Math. **27** (1973), 223–232.
173. F. Ronga, Desingularisation of the triple points and of the stationary points of a map, Comp. Math. **53** (1984), 197–219.

174. O. Saeki, *Topology of Singular Fibers of Differentiable Maps,* Springer Lecture Notes in Math. **1854** (2004).
175. O. Saeki, R. Sadykov and K. Sakuma, Obstructions to the existence of fold maps, Jour. London Math. Soc. Lect Note Series **81** (2010), 338–354.
176. O. Saeki, O and T. Yamamoto, Singular fibers and characteristic classes, Top. its Appl., vol. **155** (2007), 112–120.
177. G. Salmon, *A treatise on the analytic geometry of three dimensions*, 4th edition, Dublin (1882).
178. T. Sasajima and T. Ohmoto, Classical formulae on projective characters of surfaces and 3-folds, revisited, Proc. JARCS6 (Kagoshima, Nov. 2015), Saitama J. Math. (2017), 141–160.
179. T. Sasajima and T. Ohmoto, Thom polynomials in $\mathcal{A}$-classification I: counting singular projections of a surface, European Math. Soc. Ser. Cong. Rep. (eds. J. Buczynski et al), IMPANGA15 (2018), 203–226.
180. H. Schubert, *Kalkül der abzählenden Geometrie*, Leipzig B.G. Teubner (1879).
181. J. G. Semple and L. Roth, *Introduction to Algebraic Geometry*, Oxford University Press (1949).
182. D. Sullivan, Combinatorial invariants of analytic spaces, Proc. Liverpool Symposium on Singularities, Springer Lecture Notes in Math. **192** (1971), 165–169.
183. A. Szücs, Cobordism of singular maps, Goem. Topol. **12** (2008), 2379–2452.
184. M. Tanabe, Relative Thom polynomials, in preparation, Hokkaido University (2025).
185. M. Takase, Oriented bordism of codimension one immersions of 7-manifolds and relative Thom polynomials, Math. Z **272** (2012), 101–108.
186. B. Totaro, The Chow Ring of a Classifying Space, Proc. Symposia in Pure Math. **67** (1999), 249–281.
187. R. Thom, Espaces fibrés en sphères et carrés de Steenrod, Ann. Sci. I.É.N.S **69** (1951), 109–182.
188. R. Thom, Les singularités des applications différentiables, Ann. Inst. Fourier **6** (1955–56), 43–87.
189. R. Thom, Les ensembles singuliers d'une application différentiable et leurs propriétés homologiques. Coll. Top. Strasbourg (1957).
190. R. Thom, The bifurcation subset of a space of maps, Manifolds - Amsterdam 1970, Springer Lecture Notes in Math. **197** (1971), 202–208.
191. R. Turnbull, The Thom-Boardman singularities Σ^1, $\Sigma^{1,1}$, $\Sigma^{1,1,1}$, $\Sigma^{1,1,1,1}$, $\Sigma^{1,1,1,1,1}$ and their closures, Liverpool University (1989).
192. Y. Tzeng, A proof of the Göttsche-Yau-Zaslow formula, J. Diff. geom. **90** (2012), 439–472.
193. Y. Tzeng, Enumeration of singular varieties with tangency condition, (2017) arXiv:1703.02513.
194. I. Vainsencher, Counting divisors with prescribed singularities, Trans. Amer. Math. Soc. **267** (1981), 399–422.
195. I. Vainsencher, Hypersurfaces with up to six double points, Comm. Algebra **31** (2007), 4107–4129.
196. A. Vishik, Stable and unstable operations in algebraic cobordism, Ann. Sci. l'Ecole Normale Sup. **52** (2019), 561–630. DOI: 10.24033/asens.2393. [arXiv:1209.5793].
197. V. A. Vassiliev, *Lagrange and Legendre characteristic classes*, Adv. Stud. Contemp. Math., **3**, Gordon and Breach, New York (1988).
198. C. T. C. Wall, Geometry of projection-generic space curves, Math. Proc. Camb. Phil.l Soc. **147** (2009), 115–142.
199. A. Weber, Equivariant Chern classes and localization theorem, Jour. Singularities **5** (2012), 153–176.
200. A. Weber, Computing equivariant characteristic classes of singular varieties, RIMS Koyuroku **1868** (2013) 109–129.
201. A. Weber, Equivariant Hirzebruch class for singular varieties, Selecta Math. (N.S.) **22** (2016), 1413–1454.

202. A. Weber, Hirzebruch class and Bialynicki-Birula decomposition, Transform. Groups **22** (2017), 537–557.
203. A. Yomdin, The structure of strata μ = const in a critical set of a complete intersection singularity, Proc. Symposia Pure Math. A.M.S. vol **40**-Part 2 (1983), 663–665.
204. S. Yokura, Motivic Hirzebruch Class and Related Topics, In 'Handbook of Geometry and Topology of Singularities, IV', Springer (2023), 307–410.

Chapter 10
Indices and Residues: From Poincaré-Hopf to Baum-Bott, and Marco Brunella

Maurício Corrêa and José Seade

In memory of Marco Brunella (1964–2012), whose mathematical legacy will endure forever.

Contents

Abstract We study and discuss invariants of vector fields and holomorphic foliations that intertwine the theories of complex analytic singular varieties and singular holomorphic foliations on complex manifolds: two different settings with many points in common.

M. Corrêa (✉)
Università degli Studi di Bari, Bari, Italy,

J. Seade
Instituto de Matemáticas, Unidad Cuernavaca, Universidad Nacional Autónoma de México, Mexico City, Mexico
e-mail: jseade@im.unam.mx

© The Author(s), under exclusive license to Springer Nature Switzerland AG 2026
J. L. Cisneros-Molina et al. (eds.), *Handbook of Geometry and Topology of Singularities VIII*, https://doi.org/10.1007/978-3-031-99571-2_10

Introduction

Holomorphic vector fields on complex manifolds determine holomorphic foliations by complex curves with singularities at the zeros -or singularities- of the vector field. The leaves of this foliation are immersed Riemann surfaces. The most basic invariant of a vector field at an isolated singularity is its Poincaré-Hopf local index, and the Poincaré-Hopf index theorem can be regarded as saying that the Euler characteristic of the manifold, which is a global topological invariant, actually is localized at the singularities of the vector field. A myriad of important lines of research spring from this theorem, permeating all mathematics. Two of them are particularly relevant for us:

- The theory of characteristic classes of vector bundles and singular varieties. This is much related with indices of vector fields and frames; and
- The analogous theory for holomorphic singular foliations on manifolds.

Indeed, the foundational results by R. Bott [15] on indices and residues for holomorphic vector fields with singularities (not necessarily isolated) have profoundly influenced areas ranging from foliation theory to applications in enumerative algebraic geometry [52, 69]. This latter field was notably advanced by Kontsevich [69], who provided a rigorous framework for formulating and verifying predictions in the enumerative geometry of curves, as derived from Mirror Symmetry. In [11] Baum and Bott subsequently generalized Bott's theorem to meromorphic vector fields, and later extended these for higher-dimensional foliations [12]. Whence, we see that the indices and residues associated with foliations are of great significance, and the work of M. Brunella, T. Suwa, M. Soares and others turned this into a rich and fascinating area of mathematics where a lot is happening nowadays. See for instance [87] for an account on this subject.

On the other hand the astounding impact of characteristic classes in manifolds theory of course inspired the search for such invariants in the setting of singular varieties. The work of D. P. Sullivan, M.-H. Schwartz, R. MacPherson, W. Fulton, P. Baum and others laid down the foundations of the theory, and then the work of J.-P. Brasselet, J. Schürmann, L. Maxim, P. Aluffi and others are making this a rich and fascinating theory as well. And as noted for instance in [20, 21, 28, 49], the theory of indices of vector fields and 1-forms comes into the picture playing an important role.

In fact the study of indices and residues is of utmost importance in the local and global study of both, singular holomorphic foliations in complex manifolds and singular complex analytic varieties: two different theories with many points in common. The purpose of this work is to thread a path intertwining these two theories. We use as spearhead the GSV-index.

The first indices of vector fields and 1-forms on singular varieties in the literature appeared in the work of M. H. Schwartz [80] and R. MacPherson [74]. Yet, this was only for specific vector fields or 1-forms, aimed to defining Chern classes for singular varieties. The GSV index was introduced in [60, 81] for vector fields on isolated complex hypersurface singularities, and it began the theory of indices of vector fields and 1-forms on singular varieties.

The definition of the GSV index in 1987 was followed by the radial index (1991, and then again by different authors in 1997, 1998), the homological index (1998), indices for 1-forms (2001), the local Euler obstruction for vector fields (2004) and for 1-forms (2005), the logarithmic index (2005) and several other related invariants. We briefly discuss these invariants in Sect. 10.1; we also comment on their relation with Chern classes for singular varieties, and we refer the reader to [21, 28] for more on this subject.

From the viewpoint of foliations and residues, in his remarkable paper [22], M. Brunella gave in 1997 an interpretation of the GSV index in the setting of holomorphic 1-dimensional foliations in complex surfaces, showing that this index is of great importance in the theory of holomorphic foliations. In fact several of the invariants that we study in the sequel spring from the Baum-Bott residues, so we briefly discuss these in Sect. 10.2. This section also serves as an introduction to the subject of singular holomorphic foliations and Pfaffian systems.

One of the main problems in this field is the famous Poincaré problem, which seeks to bound the degree of invariant varieties of holomorphic foliations in terms of the degree of the foliation. In [22], Brunella showed that the non-negativity of the total sum of the GSV index of the foliation along an invariant curve serves as an obstruction to affirmatively solving the Poincaré problem. He further proved in his book [26] that this invariant plays a significant role for the classification of holomorphic foliations in complex surfaces.

Section 10.3 has Brunella's article [22] as its core; we focus here on 1-dimensional holomorphic foliations on complex surfaces. We briefly discuss the Baum-Bott residues in this setting, and their relations with the Milnor number for foliations, the Camacho-Sad indices, the variations defined by Khanedani and Suwa, and the GSV index as defined by Brunella. We recall Brunella's theorem saying that if we have a holomorphic foliation $\mathcal{F}$ is a complex surface X and S is a nondicritical separatrix at a point p then:

$$\mathrm{GSV}(\mathcal{F}, S, p) \geq 0\,.$$

If $\mathcal{F}$ further is a generalized curve (i.e. there are no saddle-nodes in its resolution) and S is maximal, union of all separatrices at p, then:

$$\mathrm{GSV}(\mathcal{F}, S, p) = 0\,.$$

And the reciprocal from [35]: if $\mathrm{GSV}(\mathcal{F}, S, p) = 0$, then $\mathcal{F}$ is a generalized curve at p and S is the maximal separatrix.

Sections 10.4 and 10.5 look at higher dimensions. These are mostly based on work by Corrêa and Machado [41, 42] and Corrêa et al. [38]. There are two previous articles that somehow paved the way. In [83], Seade and Suwa provide Baum-Bott type formulas for foliations by curves induced by global holomorphic vector fields on complex manifolds with a non-empty boundary. Also, a Bott-type formula for complex orbifolds is studied in [44], and formulas relating residues via good resolutions are obtained. To address the case of a projective variety Y with more complicated singularities, it is natural to consider a log resolution $(X, D) \to Y$ and to relate the residues of holomorphic vector fields on X to the Chern classes of the logarithmic tangent bundle $TX(-\log D)$. This is explained in Sect. 10.5.

In Sect. 10.4, using Aleksandrov's Decomposition Theorem we define, following [42], the GSV index for Pfaff systems ω of dimension k on complex manifolds X, relative to an invariant variety V in X which is a reduced local complete intersection of dimension $n-k$. This is done *la* Brunella and associates an index to each codimension 1 component $S \subset V$ of the singular set of the Pfaff system induced by a twisted form $\omega \in \mathrm{H}^0(X, \Omega_X^{n-k} \otimes \mathcal{L})$. It is proved that these indices localize the Chern class of the line bundle $\mathcal{L} \otimes \det(N_{V/X})^{-1}$ restricted to V. Then a method is given for calculating these GSV-indices, which can be compared with formulas by Suwa [88, Proposition 5.1] and Gómez-Mont [59]. This partially addresses Suwa's question in [88, Remark 5.3.(6)]. As an application, one has that the non-negativity of the GSV-index gives an obstruction to the solution of the Poincaré problem for Pfaff systems, in concordance with Brunella's observation for holomorphic foliations on surfaces. This further yields to a bound for the Poincaré problem similar to those previously established by Esteves and Cruz [46] and by Corrêa and Jardim [37]. Related results were previously obtained in [25, 34, 36, 85]. In particular, this recovers Soares' bound in the case where the invariant hypersurface V is smooth.

In Sect. 10.4.1 we recall Suwa's interpretation in [88] of the GSV and virtual indices for vector fields as residues on complete intersection varieties with isolated singularities, and the fact that if $V = \{f_1 = \cdots = f_k = 0\}$ is a local complete intersection in $\mathbb{C}^n$, with isolated singularity at 0, invariant by a foliation by curves $\mathcal{F}$ singular at most at 0, then the index is a localization of the top Chern class:

$$\mathrm{GSV}(\mathcal{F}, V, 0) = \mathrm{Res}_{c_n}(\mathcal{F}, TU|_V; 0) = \mathrm{Vir}(\mathcal{F}, V, 0).$$

Section 10.5 recalls Aleksandrov's logarithmic index $\mathrm{Ind}_{\log}(\mathcal{F}, D, p)$ defined in [4], inspired by the work of Gómez-Mont, that measures the variation between the Poincaré–Hopf and the homological indices. Then, using this index, we address (following [41, 43]) the problem of establishing a global logarithmic residue theorem for meromorphic vector fields on compact complex manifolds.

One has that if $\mathcal{F}$ is a one-dimensional foliation with isolated singularities and logarithmic along a normal crossings divisor D in a compact complex manifold X. Then

$$\int_X c_n(T_X(-\log D) - T\mathcal{F}) = \sum_{x \in \mathrm{Sing}(\mathcal{F}) \cap (X \setminus D)} \mu_x(\mathcal{F}) + \sum_{x \in \mathrm{Sing}(\mathcal{F}) \cap D} \mathrm{Ind}_{\log}(\mathcal{F}, D, x),$$

where $\mu_x(\mathcal{F})$ is the Milnor number of $\mathcal{F}$ at x. As an application, in [41], the authors prove that if $\mathcal{F}$ is a one-dimensional foliation on $\mathbb{P}^n$, with n odd, and isolated singularities (non-degenerate) along a smooth divisor D, then $\mathrm{Sing}(\mathcal{F}) \subset D$ if and only if Soares' bound [85] for the Poincaré problem is achieved, i.e, if and only if

$$\deg(D) = \deg(\mathcal{F}) + 1.$$

Recently, in [38], a logarithmic version of the Baum–Bott theorem was obtained for more general symmetric polynomials.

In the last section we briefly discuss relations with the adjunction formula and with other invariants, such as the Milnor and Tjurina numbers for foliations.

10.1 Indices of Vector Fields on Isolated Complete Intersection Singularities

Throughout this work, by a complete intersection singularity in $\mathbb{C}^{n+k}$ we mean the germ (say at 0) of a geometric –non necessarily algebraic– complete intersection $(V, 0)$, defined by as many holomorphic equations as its codimension, say k. If V is defined by

$$f = (f_1, \ldots, f_k) : (\mathbb{B}_\varepsilon, 0) \longrightarrow (\mathbb{C}^k, 0),$$

and the critical set of f meets V only at 0, then we say that $(V, 0)$ is an isolated complete intersection singularity, an ICIS for short.

A vector field v on V is a continuous section of the bundle $T\mathbb{C}^{n+1}|_V$ that vanishes at 0 and for each $x \in V^* = V \setminus 0$ the vector $v(x)$ is tangent to V^*. The vector field is holomorphic if the corresponding section is holomorphic. Notice that v must vanish at 0 because this is an isolated singularity of V.

Notice that for each $x \in V^* = V \setminus \{0\}$, the tangent space $T_x V^*$ consists of all vectors in $T_x\mathbb{C}^{n+1}$ which are mapped to 0 by the differential of all the f_i:

$$T_x V^* = \{\zeta \in T_x\mathbb{C}^{n+1} \,|\, df_i(x)(\zeta) = 0, \quad \forall\, i = 1, \ldots, k\}.$$

Example 10.1.1 If V is defined in $\mathbb{C}^{2n}$ by a map $f : (\mathbb{C}^{2n}, 0) \to (\mathbb{C}, 0)$ and we set $V = f^{-1}(0)$, then the Hamiltonian vector field

$$\widetilde{\zeta}(z_1, \ldots, z_{2n}) = \Big(-\frac{\partial f}{\partial z_2}, \frac{\partial f}{\partial z_1}, \ldots, -\frac{\partial f}{\partial z_{2n}}, \frac{\partial f}{\partial z_{2n-1}} \Big)$$

is tangent to V and it is zero only at the origin. Notice that this vector field is actually tangent to all the fibers $f^{-1}(t)$ (Fig. 10.1).

Example 10.1.2 Now let V be defined by the homogeneous polynomial map $f : (\mathbb{C}^n, 0) \to (\mathbb{C}, 0)$ defined by $(z_1, \ldots, z_n) \mapsto z_1^k + \ldots + z_n^k$ for $k \geq 2$. Then the radial vector field $(z_1, \ldots, z_n) = (z_1, \ldots, z_n)$ is tangent to V and we will see later that, unlike the previous example, if $k \geq 3$ this can never be extended to the ambient

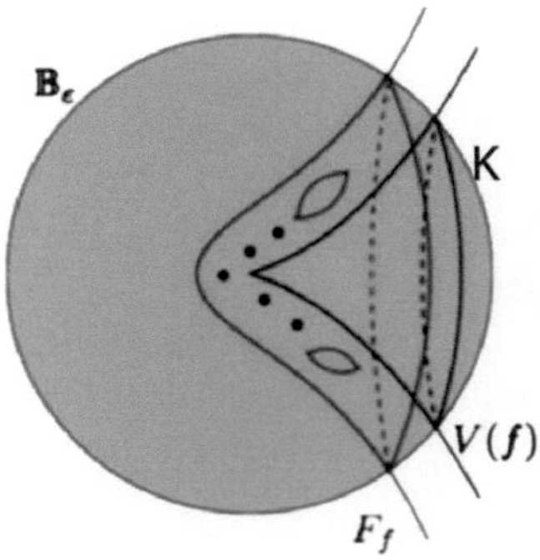

Fig. 10.1 The GSV equals the PH index in a Milnor fiber

space with an isolated singularity at 0 and being tangent to the fibers $f^{-1}(t)$. The case $k = 2$ is special (see Example 10.1.7).

10.1.1 The GSV Index

Given an ICIS $(V, 0)$ as above, in $\mathbb{C}^{n+k}$, defined by $f = (f_1, \ldots, f_k)$ we have that the (complex conjugate) gradient vector fields $\overline{\nabla} f_1, \ldots, \overline{\nabla} f_k$ are linearly independent everywhere and normal to V (for the usual hermitian metric in $\mathbb{C}^{n+k}$). Hence if v is a continuous vector field on V singular only at 0, then the set $\{v(z), \overline{\nabla} f_1, \ldots, \overline{\nabla} f_k)\}$ is a $(k+1)$-frame at each point in $V^* := V \setminus \{0\}$, and up to homotopy it can be assumed to be orthonormal. Hence these vector fields define a continuous map from V^* into the Stiefel manifold of complex orthonormal $(k+1)$-frames in $\mathbb{C}^{n+k}$, denoted $W_k(n+k)$.

Let $\mathbf{K} = V \cap \mathbb{S}_\varepsilon$ be the link of 0 in V. It is an oriented, real manifold of dimension $(2n-1)$ and the above frame defines a continuous map

$$\phi_v = (v, \overline{\nabla} f_1, \ldots, \overline{\nabla} f_k) : \mathbf{K} \longrightarrow W_{k+1}(n+k)\,.$$

The Stiefel manifold $W_k(n+k)$ is diffeomorphic to the homogeneous space $U(n+k)/U(n)$ and therefore the homotopy sequence associated to this fibration implies that $W_k(n+k)$ is $(2n-2)$-connected, while its homology in dimension $(2n-1)$ is isomorphic to $\mathbb{Z}$. Hence, if $\mathbf{K}$ is connected, the map ϕ_v has a well defined degree $\deg(\phi_v) \in \mathbb{Z}$, defined by means of the induced homomorphism $H_{2n-1}(\mathbf{K}) \to H_{2n-1}(W_2(n+1))$ in the usual way.

Definition 10.1.3 Assume $n \geq 2$ or $n = 1$ and V is irreducible. Then **the GSV index** of v at $0 \in V$, $\mathrm{Ind}_{\mathrm{GSV}}(v; V, 0)$, is the degree of the above map ϕ_v.

Remark 10.1.4 Note that the link $\mathbf{K}$ is always connected if $n \geq 2$. For $n = 1$ the link has one connected component for each irreducible component of the plane curve V and in this case one needs to take into account the intersection numbers of the branches. We refer to [21, 3.2] and to Sect. 10.3 for more about this case, which is specially important for the theory of holomorphic foliations.

We recall that one has a Milnor fibration associated to the map f, see [64, 73, 75], and the Milnor fiber F can be regarded as a compact $2n$-manifold with boundary $\partial F = \mathbf{K}$. By the Ehresmann Fibration Lemma there is an ambient isotopy of the sphere $\mathbb{S}_\varepsilon$ taking $\mathbf{K}$ into the boundary ∂F, which can be extended to a collar of $\mathbf{K}$, which goes into a collar of ∂F in F. Hence v can be regarded as a non-singular vector field on ∂F.

We also recall that the Milnor fiber F has the homotopy type of a bouquet of spheres of middle dimension and therefore its Beti numbers are all 0 except $b_0 = 1$ and $b_n = \mu$ for some $\mu \geq 0$

Definition 10.1.5 The number $b_n = \mu$ is **the Milnor number** of V at 0.

Proposition 10.1.6 *The* GSV *index has the following properties:*

(1) *The* GSV *index of v at* 0 *equals the total Poincaré-Hopf index of v in the Milnor fiber:*

$$\mathrm{GSV}(v; V, 0) = PH(v; F)\,.$$

(2) *If v is everywhere transverse to* **K**, *then*

$$\mathrm{GSV}(v; V, 0) = 1 + (-1)^n \mu\,,$$

where n is the complex dimension of V and μ is the Milnor number.

The proof is easy and we refer to [21, 3.2] for details.

We notice that we know from [75] that if V is a hypersurface germ defined by a map f, then its Milnor number μ is the Poincaré-Hopf local index of the gradient vector field ∇f, and therefore it equals the intersection number,

$$\mu = \dim_{\mathbb{C}} \frac{\mathcal{O}_{\mathbb{C}^{n+1},0}}{\left(\frac{\partial f}{\partial z_0}, \ldots, \frac{\partial f}{\partial z_n}\right)}\,. \tag{10.1}$$

For complete intersections the analogous expression is given by the Lê-Greuel formula [61, 70].

Example 10.1.7 Consider the radial vector field v_{rad} in Example 10.1.2 for $k = 2$. The statement (2) in the theorem above implies:

$$\mathrm{GSV}(v_{rad}; V, 0) = 1 + (-1)^n \mu = 1 + (-1)^n\,.$$

Hence if n is odd, then the index is 0 and it is an exercise to show that in this case the vector field can be extended continuously to the Milnor fibers with no singularities.

When the vector field in question is holomorphic, the GSV-index has other interesting properties. We now discuss some of these. To begin, recall that if $\zeta = (\zeta_1, \ldots, \zeta_n)$ is a holomorphic vector field in $\mathbb{C}^n$ with an isolated singularity at 0, then its local Poincaré-Hopf index is the intersection number:

$$\mathrm{PH}(v; 0) = \dim_{\mathbb{C}} \frac{\mathcal{O}_{\mathbb{C}^{n+1},0}}{(\zeta_1, \ldots, \zeta_n)} \tag{10.2}$$

Recall that if ζ is a holomorphic vector field on a complex variety Z (for simplicity, say with isolated singularities), then the solutions of ζ are Riemann surfaces immersed in X, and one has an associated 1-dimensional holomorphic foliation with singularities at the zeroes of ζ.

Definition 10.1.8 Let $(V_i, 0)$, $i = 1, 2$, be germs of isolated hypersurface singularities defined by holomorphic map-germs $(\mathbb{C}^{n+1}, 0) \xrightarrow{f_i} (\mathbb{C}, 0)$, and let v_1, v_2 be germs of holomorphic vector fields on the V_i with an isolated singularity at 0. Then we say that:

1. The germs $(V_1, 0)$ and $(V_2, 0)$ are **topologically equivalent** if there exists an orientation preserving local homeomorphism h of $\mathbb{C}^{n+1}$, such that $f_2 = f_1 \circ h$. In this case we say that h is a topological equivalence between V_1 and V_2.
2. The holomorphic vector fields v_1, v_2 are **topologically equivalent** if there exists a topological equivalence between V_1 and V_2 carrying the leaves of the foliation $\mathcal{F}_1$ of v_1 into the leaves corresponding to v_2.

One has [60, Theorem 4.5]:

Theorem 10.1.9 *Let $(V, 0)$ denote the germ of an isolated hypersurface singularity in $\mathbb{C}^{n+1}$. Then, the* GSV*-index of a holomorphic vector field is a topological invariant. That is, if ζ is a holomorphic vector field on $(V, 0)$ and it is topologically equivalent to another holomorphic vector field ζ' on some other hypersurface germ, then ζ and ζ' have the same* GSV*-index.*

We recall that if $\mathcal{F}$ is a 1-dimensional holomorphic foliation on a complex manifold, then $\mathcal{F}$ is locally free and therefore if p is an isolated singularity of the foliation, then $\mathcal{F}$ is locally generated by a holomorphic vector field. In analogy with Eq. 10.1 one has:

Definition 10.1.10 Let $\zeta = (\zeta_1, \ldots, \zeta_n)$ be a holomorphic vector field tanget to $\mathcal{F}$ in a neighborhood of the isolated singularity p. Then the **Milnor number** of $\mathcal{F}$ at p is the Poincaré-Hopf local index of ζ at p, so it equals:

$$\mu(\mathcal{F}, p) := \dim_{\mathbb{C}} \frac{\mathcal{O}_{\mathbb{C}^{n+1},0}}{(\zeta_1, \ldots, \zeta_n)} \,.$$

It is easy to prove (see [60]) that this invariant equals the PH-local index of any continuous vector field tangent to $\mathcal{F}$ near p. One gets the following corollary to Theorem 10.1.9, which is Theorem A in [29].

Theorem 10.1.11 *The Milnor number of an isolated singularity of a 1-dimensional holomorphic foliation on a smooth complex manifold, is a topological invariant.*

We now remark that, as noticed in [13], given an isolated complex singularity germ $(V, 0)$ in some $\mathbb{C}^m$ and a holomorphic vector field ζ on V with an isolated singularity at 0, there are infinitely many extensions of ζ to a holomorphic vector field in a neighborhood of 0 in $\mathbb{C}^m$. Yet, as noticed in [60, Section 3], if $(V, 0)$ is defined by a single equation f in $\mathbb{C}^m$, and if ζ is a holomorphic vector field on V with an isolated singularity at 0, then an extension $\widetilde{\zeta}$ to the ambient space can have a singular locus of dimension at most 1, because cutting down by a hypersurface can reduce the dimension at most by 1.

One has the following Theorem 3.2 from [60].

Theorem 10.1.12 *Let $(V, 0)$ denote a hypersurface singularity germ in $\mathbb{C}^{n+1}$ defined by a function f with at most an isolated singularity at 0 (which could be a smooth point in V). Let ζ be a holomorphic vector field on V with an isolated singularity at 0. Suppose there exists a holomorphic extension $\widetilde{\zeta}$ of ζ to a neighborhood U of 0 in $\mathbb{C}^{n+1}$ such that $\widetilde{\zeta}$ is tangent to the fibers of the map f. Let Σ be the zero-locus of $\widetilde{\zeta}$, which has dimension at most 1. Then:*

1. *The GSV-index of ζ in V is the intersection number of a Milnor fiber $F_t = f^{-1}(t) \cap U$ and Σ. That is:*

$$\mathrm{GSV}(\zeta; V, 0) = \sum_{p \in U} \dim_{\mathbb{C}} \frac{\mathcal{O}_{\mathbb{C}^{n+1}, p}}{(\widetilde{\zeta}_0, \ldots, \widetilde{\zeta}_n, f - t)},$$

   where the $\widetilde{\zeta}_0$ are the components of $\widetilde{\zeta}$. Hence, in this case the GSV index is non-negative and it is 0 if and only if $\widetilde{\zeta}$ has an isolated singularity at 0.
2. *Localized at 0 the formula becomes:*

$$\mathrm{GSV}(\zeta; V, 0) = \dim_{\mathbb{C}} \frac{\mathcal{O}_{\mathbb{C}^{n+1}, 0}}{(\widetilde{\zeta}_0, \ldots, \widetilde{\zeta}_n, f)} - \dim_{\mathbb{C}} \frac{(f) \cap (\widetilde{\zeta}_0, \ldots, \widetilde{\zeta}_n, f)}{(f\,\widetilde{\zeta}_0, \ldots, f\,\widetilde{\zeta}_n)}.$$

Notice that the number $\dim_{\mathbb{C}} \frac{\mathcal{O}_{\mathbb{C}^{n+1},0}}{(\zeta_0, \ldots, \zeta_n, f)}$ is a reminiscent of the Tjurina number for functions (see Sect. 10.6) and can be called the Tjurina number for vector fields. This appears again in the next section, in Gómez-Mont's formula for the homological index.

For instance, coming back to Example 10.1.2, the Milnor number is $(k-1)^{n+1}$, by Milnor [75]. Hence, if n is odd, then the GSV index of the radial vector field is negative and therefore it can never be holomorphically extended being tangent to the Milnor fibers, not even with singularities.

10.1.2 The Homological Index

Theorem 10.1.12 gives an algebraic formula for the GSV index of holomorphic vector fields under very stringent conditions. Gómez-Mont's search for an algebraic formula in the vein of (10.2) for the GSV index, opened an area of research which is of interest on its own; this all springs from the definition of a homological index in [59]. This has given rise to a series of papers by Gómez-Mont and several co-authors (see [13, 14, 56–58]), including the real analytic setting where it extends to singular hypersurfaces the celebrated index formula of Eisenbud and Levine [51] and Khimshiashvili [67]. There is also a homological index for 1-forms in [50], and Aleksandrov's logarithmic index for vector fields and 1-forms [4, 6] that will be discussed in Sect. 10.5. In [91, 92] M. Zach uses variations of the homological index to study interesting generalizations of the Milnor number to the case of holomorphic maps on singular varieties.

The presentation below is essentially extracted from [21, Chapter 7] and it is a *résumé* of part of Gómez-Mont's work in [59].

We recall that as noted above, if $\zeta := (\zeta_1, \ldots, \zeta_n)$ is a holomorphic vector field in $\mathbb{C}^n$ with an isolated singularity at 0, then its Poincaré-Hopf local index is the intersection number:

$$\mathrm{PH}(\zeta; 0) = \dim_{\mathbb{C}} \frac{\mathcal{O}_{\mathbb{C}^n,0}}{(\zeta_1, \ldots, \zeta_n)} \,.$$

We would like to have a similar expression for the GSV index. Yet we know from Example 10.1.7 that this index can be negative, even for holomorphic vector fields. Therefore an algebraic expression for the GSV index cannot simply be given by the dimension of some algebra. One may notice that for vector fields in $\mathbb{C}^n$ one has the Koszul complex

$$0 \longrightarrow \Omega^n_{\mathbb{C}^n,0} \longrightarrow \Omega^{n-1}_{\mathbb{C}^n,0} \longrightarrow \cdots \longrightarrow \mathcal{O}_{\mathbb{C}^n,0} \longrightarrow 0\,, \tag{10.3}$$

where the arrows are contraction by ζ. The Euler characteristic χ of this complex is $PH(\zeta; 0)$: this is the clue for defining the homological index, though there is an interesting difference. In the smooth case, all the homology of the Koszul complex vanishes except in top dimension, which contributes to χ with the PH index. In the singular case one has a similar complex but the homology groups do not vanish. The miracle is that "almost all of them" have the same dimension, so these cancel down when considering the Euler characteristic (see [59] for the precise statement).

Let $(V, 0) \subset (\mathbb{C}^m, 0)$ be a germ of a complex analytic variety of pure dimension n, which is regular on $V \setminus \{0\}$. So V is either regular at 0 or else it has an isolated singular point at the origin. A vector field ζ on $(V, 0)$ can always be defined as the restriction to V of a vector field $\widehat{v}$ in the ambient space which is tangent to $V \setminus \{0\}$; ζ is holomorphic if $\widehat{v}$ can be chosen to be holomorphic. So we may write $\zeta = (a_1, \cdots, a_m)$ where the a_i are restriction to V of holomorphic functions on a neighborhood of 0 in $(\mathbb{C}^m, 0)$.

It is worth noting that given any space V as above, there are always holomorphic vector fields on V with an isolated singularity at 0. This non-trivial fact is a weak form of a stronger result ([13, p. 19]): in the space $\Theta(V,0)$ of germs of holomorphic vector fields on V at 0, those having an isolated singularity form a connected, dense open subset $\Theta_0(V,0)$. Essentially the same result implies also that every $\zeta \in \Theta_0(V,0)$ can be extended to a germ of a holomorphic vector field in $\mathbb{C}^m$ with an isolated singularity, though it can also be extended with a singular locus of dimension more that 0, a fact that may be useful for explicit computations (as in the sequel).

A (germ of a) holomorphic j-form on V at 0 means the restriction to V of a holomorphic j-form on a neighborhood of 0 in $\mathbb{C}^m$; two such forms in $\mathbb{C}^m$ are equivalent if their restrictions to V coincide on a neighborhood of $0 \in V$. We denote by $\Omega^j_{V,0}$ the space of all such forms (germs); these are the Kähler differential forms on V at 0. So $\Omega^0_{V,0}$ is the local structure ring $\mathcal{O}_{(V,0)}$ of holomorphic functions on V at 0 and each $\Omega^j_{V,0}$ is an $\Omega^0_{V,0}$-module. Notice that if the germ of V at 0 is determined by $(f_1, \cdots, f_k)$ then one has:

$$\Omega^j_{V,0} := \Omega^j_{\mathbb{C}^m,0}/(f_1\Omega^j_{\mathbb{C}^m,0} + df_1 \wedge \Omega^{j-1}_{\mathbb{C}^m,0}, \ldots, f_k\Omega^j_{\mathbb{C}^m,0} + df_k \wedge \Omega^{j-1}_{\mathbb{C}^m,0}), \tag{10.4}$$

where d is the exterior derivative.

Now, given a holomorphic vector field $\widehat{v}$ at $0 \in \mathbb{C}^m$ with an isolated singularity at the origin, and a differential form $\omega \in \Omega^j_{\mathbb{C}^m,0}$, we can always contract ω by ζ in the usual way, thus getting a differential form $i_v(\omega) \in \Omega^{j-1}_{\mathbb{C}^m,0}$. If $\zeta = \widehat{v}|_V$ is tangent to V, then contraction is well defined at the level of differential forms on V at 0 and one gets a complex $(\Omega^\bullet_{V,0}, v)$:

$$0 \longrightarrow \Omega^n_{V,0} \longrightarrow \Omega^{n-1}_{V,0} \longrightarrow \cdots \longrightarrow \mathcal{O}_{V,0} \longrightarrow 0, \tag{10.5}$$

where the arrows are contraction by ζ and n is the dimension of V; of course one also has differential forms of degree $> n$, but those forms do not play a significant role here. We consider the homology groups of this complex:

$$H_j(\Omega^\bullet_{V,0}, v) = \operatorname{Ker}(\Omega^j_{V,0} \to \Omega^{j-1}_{V,0})/\operatorname{Im}(\Omega^{j+1}_{V,0} \to \Omega^j_{V,0}).$$

The first observation in [59] is that if V is regular at 0, so that its germ at 0 is that of $\mathbb{C}^n$ at the origin, and if $\zeta = (a_1, \cdots, a_n)$ has an isolated singularity at 0, then this is the usual Koszul complex. In that case, its homology groups vanish for $j > 0$, while

$$H_0(\Omega^\bullet_{V,0}, v) \cong \mathcal{O}_{\mathbb{C}^n,0}\big/(a_1, \cdots, a_n).$$

In particular the complex is exact if $\zeta(0) \neq 0$. Since the contraction maps are $\mathcal{O}_{V,0}$-module maps, this implies that if V has an isolated singularity at the origin, then the homology groups of this complex are concentrated at 0, and they are finite dimensional because the sheaves of Kähler differentials on V are coherent. Hence it makes sense to define:

Definition 10.1.16 The *homological index* $\mathrm{Ind}_{\mathrm{hom}}(\zeta, 0; V)$ of the holomorphic vector field ζ on $(V, 0)$ is the Euler characteristic of the above complex:

$$\mathrm{Ind}_{\mathrm{hom}}(\zeta, 0; V) = \sum_{i=0}^{n} (-1)^i h_i(\Omega^{\bullet}_{V,0}, v)\,,$$

where $h_i(\Omega^{\bullet}_{V,0}, v)$ is the dimension of the corresponding homology group as a vector space over $\mathbb{C}$.

One has:

Theorem 10.1.17 *If the germ* (V, p) *is a complete intersection, then the* GSV *and the homological indices of every holomorphic local vector field on* V *coincide.*

This theorem was proved by Gómez-Mont in [59] when (V, p) is a hypersurface germ, and the general case is proved in [14]. For simplicity we restrict the discussion to the case where V is a hypersurface. In this case the proof of Theorem 10.1.17 in [59] has three main steps:

Step 1: Finding an algebraic formula for the homological index.
Step 2: Observe that the two indices coincide for one vector field if and only if they coincide for every vector filed.
Step 3: Use the algebraic formula for the homological index and the basic properties of the GSV index, to prove that on every hypersurface germ there exists a tangent holomorphic vector field for which the two indices are 0.

Let us explain first the algebraic formulas for the homological index. Let $\mathbb{B} = \mathbb{B}^{n+1}$ be an open ball around the origin in $\mathbb{C}^{n+1}$. If $f : \mathbb{B}^{n+1} \to \mathbb{C}$ is a holomorphic function with $0 \in \mathbb{C}^{n+1}$ as its only critical point, then the 1-form $\sum_{i=1}^{n+1} \frac{\partial f}{dz_i} dz_i$ vanishes only at $0 \in \mathbb{C}^{n+1}$. Let $\mathcal{I}_f$ be the Jacobian ideal $(\frac{\partial f}{dz_1}, \cdots, \frac{\partial f}{dz_{n+1}}) \subset \mathcal{O}_{\mathbb{C}^{n+1},0}$ of f. Given a vector field ζ tangent to V, with a unique singularity at 0, restriction of a vector field $\widehat{v} = (a_1, \cdots, a_{n+1})$ on $\mathbb{B}$ with a unique singularity at the origin, it is shown in [59, Theorem 2] that the homology groups $H_i(\Omega^{\bullet}_{V,0}, \zeta)$ of the above complex have dimensions:

$$h_0(\Omega^{\bullet}_{V,0}, \zeta) = \dim_{\mathbb{C}} \mathcal{O}_{\mathbb{C}^{n+1},0}/(f, a_1, \cdots, a_{n+1}),$$

$$h_n(\Omega^{\bullet}_{V,0}, \zeta) = \dim_{\mathbb{C}} \mathcal{O}_{\mathbb{C}^{n+1},0}/(f, \mathcal{I}_f),$$

$$h_i(\Omega^{\bullet}_{V,0}, \zeta) = \lambda, \quad \text{for} \quad i = 1, \ldots, n-1,$$

where $0 \leq \lambda \leq \dim_{\mathbb{C}} \mathcal{O}_{\mathbb{C}^{n+1},0}/(f, \mathcal{I}_f)$ is the integer:

$$\lambda = \dim_{\mathbb{C}} \mathcal{O}_{\mathbb{C}^{n+1},0}/(f, a_1, \cdots, a_{n+1}) + \dim_{\mathbb{C}} \mathcal{O}_{\mathbb{C}^{n+1},0}/(\frac{df}{f}(\widehat{v}), a_1, \cdots, a_{n+1})$$
$$- \dim_{\mathbb{C}} \mathcal{O}_{\mathbb{C}^{n+1},0}/(a_1, \cdots, a_{n+1}),$$

noticing that the tangency condition for $\widehat{v}$ means that $\widehat{v}(df)$ is a multiple of f, so that $\frac{df}{f}(\widehat{v})$ is a holomorphic function on $\mathbb{B}$. As a consequence of these computations Gómez-Mont deduced the following expressions for the homological index (Theorem 1 in [59]).

Theorem 10.1.18 *Let $(V, 0)$ be an isolated hypersurface singularity (germ) of dimension n in $\mathbb{C}^{n+1}$, and let ζ be the restriction to V of a holomorphic vector field $\widehat{v}$ on a neighborhood $\mathbb{B}$ of 0 in $\mathbb{C}^{n+1}$, which has an isolated singularity at 0 and is tangent to V.*

(1) *If n is odd, then the homological index of ζ is:*

$$\mathrm{Ind}_{\mathrm{hom}}(\zeta, 0; V) = \dim_{\mathbb{C}} \mathcal{O}_{\mathbb{C}^{n+1},0}/(f, a_1, \cdots, a_{n+1}) - \dim_{\mathbb{C}} \mathcal{O}_{\mathbb{C}^{n+1},0}/(f, \mathcal{I}_f).$$

(2) *If n is even, then :*

$$\mathrm{Ind}_{\mathrm{hom}}(\zeta, 0; V) = \dim_{\mathbb{C}} \mathcal{O}_{\mathbb{C}^{n+1},0}/(a_1, \cdots, a_{n+1})$$
$$- \dim_{\mathbb{C}} \mathcal{O}_{\mathbb{C}^{n+1},0}/(\frac{df}{f}(\widehat{v}), a_1, \cdots, a_{n+1})$$
$$+ \dim_{\mathbb{C}} \mathcal{O}_{\mathbb{C}^{n+1},0}/(f, \mathcal{I}_f).$$

Notice that in the two formulas above, for even and odd dimensions, the following invariant appears:

$$\dim_{\mathbb{C}} \mathcal{O}_{\mathbb{C}^{n+1},0}/(f, a_1, \cdots, a_{n+1}) .$$

This is a reminiscent of the Tjurina number of holomorphic functions. The following name was coined in [32, p. 159]:

Definition 10.1.19 Let $\zeta = (a_1, \cdots, a_{n+1})$ be a holomorphic vector field in $\mathbb{C}^{n+1}$ be a holomorphic vector field in $\mathbb{C}^{n+1}$ with an isolated zero at the origin. Then its **Tjurina (-Gómez-Mont)** number at 0 is:

$$\tau(\zeta) = \dim_{\mathbb{C}} \mathcal{O}_{\mathbb{C}^{n+1},0}/(f, a_1, \cdots, a_{n+1}) .$$

The next step in [59] for identifying the homological and the GSV indices for holomorphic vector fields on hypersurface germs is Step 2 above:

Lemma 10.1.20 *The difference*

$$\mathrm{Ind}_{\mathrm{hom}}(\zeta, 0; V) - \mathrm{Ind}_{\mathrm{GSV}}(\zeta, 0; V) = k\,,$$

is a constant that depends only on V and not on the choice of vector field.

The proof of this lemma is an exercise using that both indices satisfy a certain "Law of conservation of the number". This says roughly that if we perturb slightly a holomorphic vector field ζ at (V, p), the new vector field ξ still has a singularity at p and new singularities arise at smooth points of V, then the index of ζ at p is the index of ξ at p plus the sum of the PH-indices of singularities of ξ at smooth points of V.

The final step for proving Theorem 10.1.17 consists in exhibiting a holomorphic vector on every hypersurface germ (V, p) for which the GSV and homological indices are both 0. When the dimension n of V is even, such an example in [59] is a Hamiltonian vector field as in Example 10.1.1. The case when n is odd is slightly more elaborate. In the two cases the vector field in question extends in an obvious way to the Milnor fibers with no singularities, and therefore the GSV index vanishes. Then Gómez-Mont uses Theorem 10.1.18 to prove that the homological index vanishes as well.

As a consequence we may state:

Corollary 10.1.21 *Let $(V, 0)$ be a hypersurface singularity in $\mathbb{C}^{n+1}$ with an isolated singularity, and let $\zeta = (a_1, \ldots, a_{n+1})$ be a holomorphic vector field tangent to V with an isolated singularity at* 0*. Then the homological index of ζ at* 0*, relative to V, equals the* GSV *index and if n is odd, this is the difference of their Tjurina numbers :*

$$GSV(\zeta, 0, V) = \tau(\zeta) - \tau(V)\,.$$

For n even this is:

$$GSV(\zeta, 0, V) = PH(\zeta, 0) + \tau(V) - \dim_{\mathbb{C}} \frac{\mathcal{O}_{\mathbb{C}^{n+1},0}}{\left(\frac{\zeta(f)}{f}, a_1, \ldots, a_{n+1}\right)}\,.$$

Let us finish this section with two open questions:

Question 1 We know that the GSV index can be negative. We also know from [13] that given any isolated hypersurface germ $(V, 0)$ there are infinitely many holomorphic vector fields on V with an isolated singularity at 0, and those with the smallest homological (or GSV) index are dense. What is this index?

For example, if the singularity is quasi-homogeneous, this bound seems to be given by the index of the radial vector field, which equals the Euler characteristic of the Milnor fiber.

Question 2 Let X be a compact complex analytic variety with isolated singularities which are ICIS. Let ζ be a continuous vector field on X with isolated singularities at the singular points of X and perhaps at some other smooth points of X. Assume further that at each point is X_{sing}, ζ is holomorphic, and define its total homological index in the obvious way. Then it is an exercise to show that this is an invariant of X, independent of the vector field. What is this invariant? For instance, if X is a local complete intersection, this is the 0-degree Fulton class of X, by Seade and Suwa [84].

10.1.3 The Virtual Index

The GSV index inspired also a line of research in the vein of the Baum-Bott's residues, using Chern-Weil theory: this is the virtual index, defined by D. Lehmann, M. Soares and T. Suwa in [71] for vector fields on local complete intersections in complex manifolds. This was originally defined for holomorphic vector fields and it was later extended in [84] for continuous vector fields, which is useful for the study of characteristic classes on singular varieties (see Sect. 10.1.28). The virtual index is defined even if the vector field and the variety have non-isolated singularities; it attaches an index in $\mathbb{Z}$ to each compact connected component of the singular set of the vector field. When the singularity is isolated, this coincides with the GSV index.

The starting point for defining the virtual index is recalling that for a continuous vector field v on a complex m-manifold X^m with an isolated singularity at a point, say 0, the local Poincaré-Hopf index can be regarded as being the localization at 0 of the top Chern class of X. This is explained in detail in [21] using Chern-Weil theory. In the isolated singularity case, which is the most relevant for this article, there is also the article [88] by T. Suwa, where he describes the localization in a particularly nice way, which is relevant for Sect. 10.4, where this is recalled (in 10.4.1).

Here we give an elementary intuitive explanation of the definition of the virtual index from the topological point of view. Let us explain.

The tangent bundle TX has Chern classes $c_1(X), \ldots, c_m(X)$ in the cohomology ring $H^*(X; \mathbb{Z})$. If X is compact, cap product with the fundamental cycle $[X]$ maps these into homology classes in $H_*(X)$; these are called the homology Chern classes of X. The Gauss-Bonnet theorem tell us that the image in $H_0(X) \cong \mathbb{Z}$ of the top Chern class $c_m(X)$ is the Euler characteristic $\chi(X)$, which equals the total index of a vector field on X with isolated singularities, by the theorem of Poincaré-Hopf.

More generally, let v be a continuous vector field on X and assume v vanishes on some analytic subset S of X. We assume further that there is a regular neighborhood $\mathcal{T}$ of S in X such that v is non-singular on $\mathcal{T} \setminus S$. Then V splits the Chern class $c_m(X)$ in two parts: one has support in the interior of $\mathcal{T}$ and another in its exterior (see [21, Chapter 1] for details). By Alexander duality, these can be regarded as homology classes in $H_0(\mathcal{T})$ and $H_0(X \setminus \mathcal{T})$.

Definition 10.1.22 The contribution to $c_m(X)[X] = \chi(X)$ coming from $H_0(\mathcal{T}) \cong H_0(S)$ is *the localization of* $c_m(X)[X]$ at S.

If S is a point p, then the localization of $c_m(X)[X]$ at p is the corresponding local PH-index. We can always perturb v slightly so that it has isolated singularities in $\mathcal{T}$ and the localization of $c_m(X)[X]$ at S is the total PH-index at S, see [21, Chapter 1].

We now want to adapt these ideas to vector fields on singular spaces. Let V be a subvariety of the complex manifold X^m defined by a regular section s of a holomorphic vector bundle E over X of rank k, with $1 \leq k < m$.

If s is transversal everywhere to the zero-section of E, then V is non-singular, the bundle $E|_V$ is isomorphic to the normal bundle of V and the Chern classes of V can be regarded as being those of the virtual bundle $TX|_V - E|_V$. Otherwise V is a singular local complete intersection and one defines its *virtual tangent bundle* by.

$$\tau V := TX|_V - E|_V \,.$$

The total Chern class of τV is defined in the usual way:

$$c_*(\tau V) = c_*(TX|_V) \cdot c_*(E|_V)^{-1} \,,$$

where the term in the extreme right is the inverse of $c_*(E|_V)$ in the cohomology ring.

One gets the *total Fulton class* $c_*^{Fu}(V) = c_0^{Fu}(V) + \ldots. + c_n^{Fu}(V)$ by taking the cap product of $c_*(\tau V)$ with the fundamental cycle $[V]$. Fulton in [54] calls this the canonical class. The components $c_i^{Fu}(V) \in H_{2i}(V)$, for $i = 0, \ldots, n$, are the *Fulton classes* of V.

We remark that in [54] Fulton actually defines these as classes of cycles in the Chow group and calls $c_*^{Fu}(V)$ the canonical class of V. His definition uses the Segre classes and it works for singular varieties in full generality (we refer to [10, 18, 21, 28] for thorough accounts, with different viewpoints, on Chern and their relations with singularity theory).

The idea now is to localize the zero-degree Fulton class (which corresponds to the top dimensional class in cohomology) using a tangent vector field, as we did on manifolds. The problem is that τV is a virtual bundle and the usual methods for localizing Chern classes via topology do not work.

Localization is done in [19, 21, 71] by means of Chern-Weil theory and it works equally well if we replace the isolated singularity p by a compact connected component S of the singular set. We refer to [21] for details.

Let us explain this construction in a special case which is particularly interesting and where it is much simpler. Assume that the regular section s that defines V can be put in a flat family $\{s_t\}_{t \in \mathbb{C}}$ of holomorphic sections of E such that $s_0 = s$ and for $t \neq 0$ the section s_t is transversal to the zero section. Set $V_t := \{x \in X \mid s_t(x) = 0\}$.

We have a flat family of manifolds V_t degenerating to the special fiber V. Over each V_t one has the virtual bundle

$$TX|_{V_t} - E|_{V_t} .$$

For $t \neq 0$ the bundle $E|_{V_t}$ is isomorphic to the normal bundle of V_t, so the Chern classes of $TX|_{V_t} - E|_{V_t}$ are the Chern classes of the tangent bundle TV_t, and these can be regarded in homology. Denote these by $c_{*,t} \in H_*(V_t)$.

Now let $\widetilde{\mathcal{T}}$ be a regular neighborhood of V in X, and let $\varepsilon_o > 0$ be small enough so that whenever $|t| < \varepsilon_o$, the manifold V_t is contained in $\widetilde{\mathcal{T}}$. This exists by the compactness of V. Since $\widetilde{\mathcal{T}}$ has V as a deformation retract, one has well defined homomorphisms:

$$H_*(V_t) \overset{\iota_*}{\to} H_*(\widetilde{\mathcal{T}}) \overset{r_*}{\to} H_*(V) ,$$

induced by the inclusion of V_t in $\widetilde{\mathcal{T}}$ and a retraction to V. Set $\mathcal{S} = r_* \circ \iota_*$. This is a special case of Verdier's specialization [89], and we know that the image $\mathcal{S}(c_{*,t}) \in H_*(V)$ are the Fulton classes of V. Let us summarize this discussion:

Proposition 10.1.23 *The Verdier specialization morphism $\mathcal{S}$ carries the homology Chern classes of the general fiber V_t into the Fulton classes of the special fiber V.*

We now explain how to localize the top Fulton class of V at each connected component of its singular set V_{Sing} using a vector field. Intuitively this is easy: we use the vector field as we did on manifolds, to localize a contribution to the index at a component of the singular set of the vector field, and carry this contribution to the special fiber using the specialization morphism.

For simplicity we do it when the variety V, defined by a regular section of a holomorphic bundle E over X, only has isolated singularities, say $p_1, \ldots, p_r$. The process in general is similar but a little more difficult technically.

For each p_i we may consider a conical neighborhood X_i of p_i in V and "replace it" by a local Milnor fiber $F_{i,t}$. We get a compact manifold $\widehat{V_t}$, and as $|t|$ tends to 0 this degenerates to V. As above, the homology Chern classes of V_t are mapped to the Fulton classes of V. Now, given a vector field on V with isolated singularities, we can take its restriction to the compact manifold with boundary $V \setminus (\bigcup_{i=1}^r X_i \bigcup)$, and extend it to all of $\widehat{V_t}$ with isolated singularities on each local Milnor fiber $F_{i,t}$. This splits $c_n(\widehat{V_t})[V_t]$ into a contribution coming from the smooth part $V \setminus (\bigcup_{i=1}^r X_i \bigcup)$, and a contribution coming from each $F_{i,t}$ (Fig. 10.2).

Definition 10.1.24 The virtual index of v at an isolated singularity p is the localization at p of the Fulton class $c_n(V) \in H_0(V) \cong \mathbb{Z}$ determined by v.

From the above construction we see that this invariant equals the total Poincaré-Hopf index of v at a local Milnor fiber when the singularity (V, p) can be put in a flat family as above. This condition is not necessary and the result id proved in general, either in [19, 21] via Chern-Weil theory, or in [28] topologically. We have:

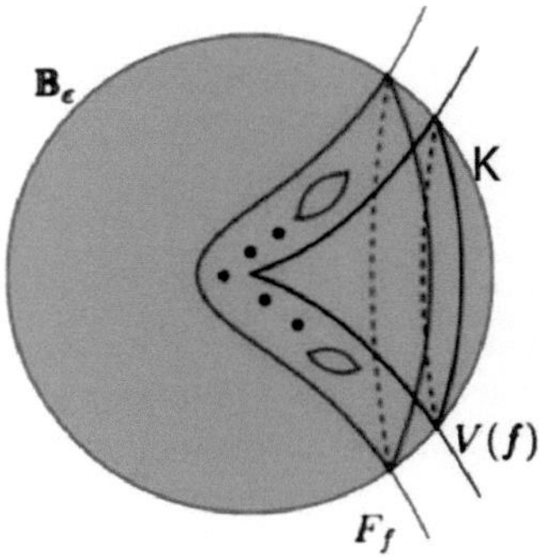

Fig. 10.2 The virtual index is the localization at 0 of the PH index in a Milnor fiber

Theorem 10.1.25 *The GSV and the virtual indices coincide for vector fields on ICIS germs.*

This theorem is discussed again in Sect. 10.4 from the viewpoint of Chern-Weil theory, following Suwa's viewpoint in [88] instead of the original one in [71] (or in [21]).

10.1.4 Radial Index and the Milnor Number

There are other indices of vector fields and 1-forms on singular varieties that play an important role in singularity theory. Two of these are the radial or Schwartz index, and the local Euler obstruction, which plays an important role for MacPherson's construction of Chern classes for singular varieties. This invariant has been extended to an index for vector fields and 1-forms on ICIS singularities. We refer to [21, 28, 49] for thorough accounts on this subject.

The radial (or Schwartz) index springs from the work of M. H. Schwartz in 1965, but she defined it only for a special class of vector fields, those obtained by radial extension. Her goal was defining Chern classes for singular varieties (see [21]). For instance, if p is an isolated singularity in an analytic space X, the only vector fields considered by M. H. Schwartz were the radial ones, i.e., those transversal to each small sphere in the ambient space, centered at the singular point, and in that case the Schwartz index is 1. King and Trotman defined an index in general in [68], a 1991 preprint that unfortunately was not published till 2014, which essentially is the radial index. In the meantime, the same index was discovered simultaneously in [48] and [7]. This is defined in great generality, for real or complex analytic spaces, and we refer to [49] and [21, Ch. 2] for accounts on the subject. In the simplest case, where one considers an isolated singularity germ $(V, 0)$ and a continuous vector field on V with an isolated singularity at 0, this index is defined as:

$$\mathrm{Rad}(v; 0, V) = 1 + d(v, v_{\mathrm{rad}}),$$

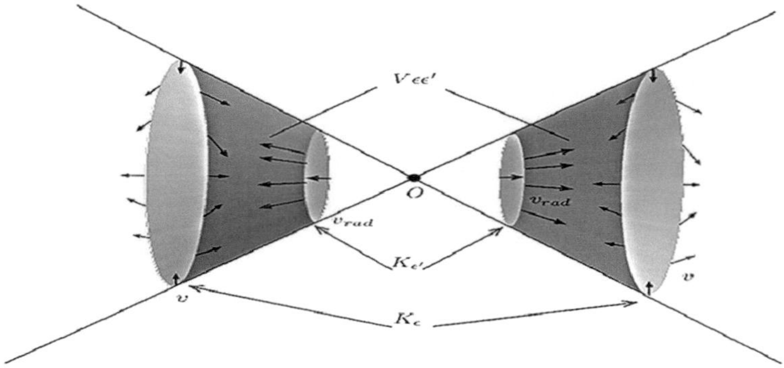

Fig. 10.3 The radial index

where v_{rad} is a radial vector field on V (i.e, it is transversal to the intersection of V with every sufficiently small sphere, pointing outwards) and $d(v, v_{\mathrm{rad}})$ is *the difference between these two vector fields*, as depicted in Fig. 10.3: this is the total index of a vector field ξ on a cylinder bounded by links $\mathbf{K}_{\varepsilon'} = V \cap \mathbb{S}_{\varepsilon'}$ and $\mathbf{K}_{\varepsilon} = V \cap \mathbb{S}_{\varepsilon}$ with $0 < \varepsilon' < \varepsilon$ and ξ restricts to v_{rad} on $\mathbf{K}_{\varepsilon'}$ and to v on $\mathbf{K}_{\varepsilon}$.

When V has non-isolated singularities, the radial index is defined similarly as above but there are two points of view. In [21, Ch. 2] one assigns an index to each connected component Σ of the singular set of a singular variety. For this one takes a regular neighborhood N_Σ of Σ with smooth boundary, and a smaller such neighborhood N'_Σ and use the cylinder bounded by these to define the difference $d(v, v_{\mathrm{rad}})$, where v is the given vector field and v_{rad} is now the unit normal vector field on the boundary $\partial N'_\Sigma$, pointing outwards. In this case we define the radial index of v_{rad} to be the Euler characteristic $\chi(\Sigma)$ and the index of v is $\chi(S) + d(v, v_{\mathrm{rad}})$.

In [48, 49] the definition is more refined as one must consider Whitney stratified vector fields with isolated singularities and one defines a radial index at each singularity of the vector field. The relation between the two definitions is as follows: the above radial index associated to Σ is the sum of the radial indices of the stratified vector field with isolated singularities contained in Σ. In either case the radial index at Σ actually is a class in $H_0(\Sigma; \mathbb{Z}) \cong \mathbb{Z}$, it can be regarded in $H_0(V; \mathbb{Z})$ via the inclusion, and it has the following important properties (see [21, 49]):

Theorem 10.1.26

1. *Let $(V, 0)$ be an n-dimensional ICIS with Milnor number μ, and let v be a continuous vector field on V with an isolated singularity at* 0. *Then their difference is the Milnor number of $(V, 0)$, independently of v:*

$$\mu = (-1)^n \big(\mathrm{GSV}(v; 0, V) - \mathrm{Rad}(v; 0, V)\big) .$$

2. *If V is a compact local complete intersection and v is a continuous vector field on V, singular at the singular set V_{sing} and at finitely many smooth points of V, then its total radial index (defined in the obvious way) is the Euler characteristic $\chi(V)$, which equals the 0-degree Schwartz-MacPherson class of V.*

Remark 10.1.27 (Indices of 1-Forms) The theory of indices of 1-forms on singular varieties is parallel to that for vector fields and it has been developed mostly by W. Ebeling and S. Gusein-Zade in a series of papers. We refer to their survey paper [49] for an account of this, and to [21, Ch. 9] for an introduction to the subject. Given an isolated singularity germ (V, p) in some $\mathbb{C}^m$ and a holomorphic 1-form in a neighborhood of p, its restriction to V is a 1-form on V and its singularities are the points where its kernel is not transversal to V; this includes p. There is a radial index for 1-forms with an isolated singularity at p, and if (V, p) is an ICIS, then there are also GSV and homological indices, and these coincide. The difference between these and the radial index is the local Milnor number. There is a local Euler obstruction as well.

Using 1-forms instead of vector fields has some advantages; perhaps a main one is that instead of imposing the tangency conditions one needs for vector fields, one needs a transversality condition for 1-forms. This allows, for instance, considering indices for collections of 1-forms. The indices we have envisaged so far morally correspond to the top Chern class, while the collections of 1-forms envisaged by Ebeling and Gusein-Zade corespond to different Chern numbers.

Remark 10.1.28 (Milnor Classes) Classical Chern classes of complex manifolds can be defined as being the primary obstructions to constructing frames with appropriate indices at their singularities. These indices are natural extensions of the Poincaré-Hopf index. Similarly, for singular varieties equipped with a Whitney stratification, one may consider appropriate stratified frames and:

- One has a corresponding radial index for frames; the classes one gets are the Schwartz-MacPherson classes (see [21, Ch. 10]).
- If the variety is a local complete intersection in a manifold, one has a corresponding virtual index for frames, and the classes one gets are the Fulton classes (see [21, Ch. 10]).
- In the previous setting, the difference between the Fulton and the Schwartz-MacPherson classes are the Milnor classes of V.
- If V has only isolated singularities, the Milnor class in degree 0 is the sum of the local Milnor numbers (hence the name of Milnor classes). If V is a hypersurface with non-isolated singular set, the 0-degree Milnor class is Parusinski's generalized Milnor number.

Milnor classes arose first in the work of Aluffi [8, 9], though the actual name was coined later (see [28]).

10.2 Singular Holomorphic Foliations, Baum-Bott Residues and Indices

A *regular* (smooth or non-singular) holomorphic foliation $\mathcal{F}$ of dimension k on a complex manifold X is given by an open covering $\{U_\alpha\}$ of X, holomorphic submersions $f_\alpha : U_\alpha \to \mathbb{C}^{n-k}$ and biholomorphisms

$$g_{\alpha\beta} : f_\beta(U_\alpha \cap U_\beta) \subset \mathbb{C}^{n-k} \to f_\alpha(U_\alpha \cap U_\beta) \subset \mathbb{C}^{n-k}$$

such that $f_\alpha = g_{\alpha\beta} \circ f_\beta$, whenever $U_\alpha \cap U_\beta \neq \emptyset$. That is (see Fig. 10.4), the diagramm

$$\begin{array}{ccc} U_\alpha \cap U_\beta & = & U_\alpha \cap U_\beta \\ \downarrow {\scriptstyle f_\beta} & & \downarrow {\scriptstyle f_\alpha} \\ f_\beta(U_\alpha \cap U_\beta) \subset \mathbb{C}^{n-k} & \xrightarrow{g_{\alpha\beta}} & f_\alpha(U_\alpha \cap U_\beta) \subset \mathbb{C}^{n-k} \end{array}$$

commutes.

If we set $f_\alpha = (f_\alpha^1, \dots, f_\alpha^{n-k})$, then the tangent of a fiber passing through $p \in U_\alpha$ is:

$$T_p(f_\alpha^{-1}(f_\alpha(p))) = \ker(df_\alpha)(p) = \{v \in T_pX; i_v(df_\alpha^1 \wedge \cdots \wedge df_\alpha^{n-k})(p) = 0\}.$$

This means that the foliation is locally induced by the decomposable $(n-k)$-form

$$df_\alpha^1 \wedge \cdots \wedge df_\alpha^{n-k} \in H^0(U_\alpha, \Omega_{U_\alpha}^{n-k}),$$

and on $U_\alpha \cap U_\beta \neq \emptyset$ we have

$$df_\alpha^1 \wedge \cdots \wedge df_\alpha^{n-k} = \det(J(g_{\alpha\beta})) \cdot df_\beta^1 \wedge \cdots \wedge df_\beta^{n-k},$$

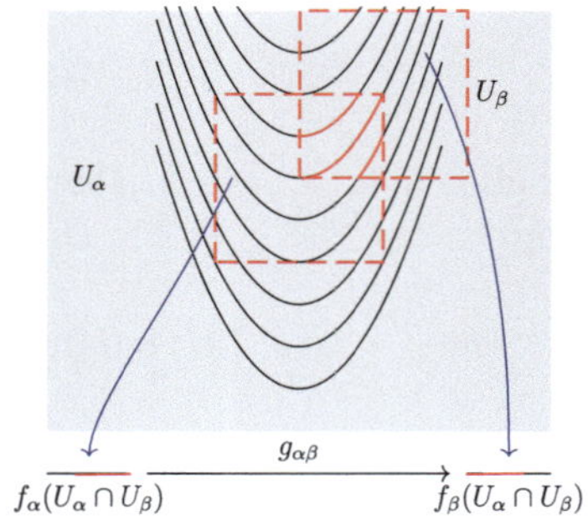

Fig. 10.4 Foliated charts

where $J(g_{\alpha\beta})$ is the jacobian matrix of the biholomorphic map $g_{\alpha\beta}$. Moreover, this gives us a holomorphic vector bundle of rank k defined by

$$T\mathcal{F} := \cup_\alpha \ker(df_\alpha) \subset TX$$

called the tangent bundle of $\mathcal{F}$. On the other hand, $T\mathcal{F}|_{U_\alpha}$ is generated by holomorphic vector fields $v_\alpha^1, \cdots, v_\alpha^k$, such that $v_\alpha^i = \sum_{j=1}^k \Gamma_{ij}^k v_\alpha^j$, where Γ_{ij}^k are holomorphic functions on U_α. That is, the foliation is induced by the decomposable holomorphic k-vector field

$$v_\alpha^1 \wedge \cdots \wedge v_\alpha^k \in H^0(U_\alpha, \wedge^k TU_\alpha).$$

It follows from the Frobenius theorem that a holomorphic subvector bundle $\mathcal{E} \subset TX$ (called a distribution) of rank k is the tangent bundle of a foliation of dimension k if and only if $[\mathcal{E}, \mathcal{E}] \subset \mathcal{E}$. So, a holomorphic regular foliation induces an exact sequence of holomorphic vector bundles,

$$0 \longrightarrow T\mathcal{F} \xrightarrow{\phi} TX \xrightarrow{\pi} N\mathcal{F} \longrightarrow 0,$$

where the quotient bundle $N\mathcal{F}$ is called the normal bundle of $\mathcal{F}$. Notice that $\{\det(J(g_{\alpha\beta}))\}$ is a cocycle of $\wedge^{n-k} N\mathcal{F}$ the determinant of $\mathcal{F}$. Then, the foliation $\mathcal{F}$ induces a nowhere vanishing holomorphic twisted $(n-k)$-form,

$$\omega_{\mathcal{F}} \in H^0(X, \Omega_X^{n-k} \otimes \wedge^{n-k} N\mathcal{F}),$$

with values in the line bundle $\wedge^{n-k} N\mathcal{F}$, given by $\omega_{\mathcal{F}}|_{U_\alpha} = df_\alpha^1 \wedge \cdots \wedge df_\alpha^{n-k}$. Reciprocally, a nowhere vanishing twisted holomorphic $(n-k)$-form $\omega \in H^0(X, \Omega_X^{n-k} \otimes \mathcal{L})$, with values in a line bundle $\mathcal{L}$ on X, induces a regular holomorphic foliation of dimension k, which is locally decomposable and integrable. That is, for all p there is an open neigborhood U and $\omega_1, \ldots, \omega_{n-k} \in H^0(U, \Omega_U^{n-k})$, such that

$$\omega|_U = \omega_1 \wedge \cdots \wedge \omega_{n-k}$$

and

$$d\omega_i \wedge \omega_1 \wedge \cdots \wedge \omega_{n-k} = 0$$

for all $i = 1, \ldots, n-k$. Then, by the Frobenius Theorem, there is a holomorphic map $f = (f^1, \ldots, f^{n-k}) : U \to \mathbb{C}^{n-k}$, such that

$$\ker(\omega|_U) = \{v \in TU; i_v(\omega) = 0\} = \ker(df) = \ker(df^1 \wedge \cdots \wedge df^{n-k}) \subset TU.$$

Therefore, ω induces a holomorphic foliation $\mathcal{F}$ of dimension k, with tangent bundle $T\mathcal{F} = \ker(\omega) \subset TX$. Moreover, for every $g \in H^0(X, \mathcal{O}_X^*)$ the form $\tilde{\omega} = g\omega$, yields the same foliation. In particular, if X is compact, then the foliation $\mathcal{F}$ is represented by a point on the projective space

$$\mathbb{P}H^0(X, \Omega_X^{n-k} \otimes \mathcal{L}).$$

In general, we can define.

Definition 10.2.1 A *regular Pfaff system* of dimension k on a *compact* manifold X is a point $[\omega] \in \mathbb{P}H^0(X, \Omega_X^{n-k} \otimes \mathcal{L})$, for some line bundle $\mathcal{L}$, such that its singular set is $\mathrm{Sing}(\omega) := \{\omega = 0\} = \emptyset$.

So, a regular Pfaff system ω is a regular foliation if and only if ω is *locally decomposable* and *integrable*.

10.2.1 Singular Holomorphic Foliations and Pfaffian Systems

Let X be a complex manifold of dimension n. A singular holomorphic *distribution* $\mathcal{F}$, of dimension k, on X is given by an exact sequence

$$\mathcal{F} : 0 \longrightarrow T\mathcal{F} \xrightarrow{\phi} TX \xrightarrow{\pi} N\mathcal{F} \longrightarrow 0,$$

where $N\mathcal{F}$ is a torsion-free sheaf, called the *normal sheaf* of $\mathcal{F}$, and $T\mathcal{F}$ is a reflexive sheaf, of generic rank k, called the *tangent sheaf* of $\mathcal{F}$.

Definition 10.2.2 A *singular holomorphic foliation* of dimension k is a distribution $\mathcal{F}$, of dimension k, which is involutive, i.e. $[T\mathcal{F}, T\mathcal{F}] \subset T\mathcal{F}$. The *singular set* of the foliation $\mathcal{F}$ is the singular set of its normal sheaf

$$\mathrm{Sing}(\mathcal{F}) := \mathrm{Sing}(N\mathcal{F}) = \{p \in X \mid (N\mathcal{F})_p \text{ is not locally free}\}.$$

Since $N\mathcal{F}$ is torsion-free, the singular set of $\mathcal{F}$ has codimension ≥ 2. The *singular scheme* of $\mathcal{F}$ is defined as follows. Taking the maximal exterior power of the dual morphism $\phi^\vee : \Omega_X^1 \to T\mathcal{F}^*$ we obtain a morphism

$$\wedge^k \phi^\vee : \Omega_X^k \to \det(T\mathcal{F})^*$$

and an induced morphism $\Omega_X^k \otimes \det(T\mathcal{F}) \to \mathcal{O}_X$ whose image is an ideal sheaf $\mathcal{I}_Z$ of a subscheme $Z \subset X$. Moreover, such a morphism induces a global section

$$\nu_{\mathcal{F}} \in H^0(X, \wedge^k TX \otimes \det(T\mathcal{F})^*)$$

which is called the *Pfaff field* induced by $\mathcal{F}$. This means that on a neighborhood U_p of a point p such that $T\mathcal{F}$ is locally free and generated by vector fields $v_1, \ldots, v_k$, then $\nu_{\mathcal{F}}|_{U_p} = v_1 \wedge \cdots \wedge v_k$ and $[v_i, v_j] = \sum_{j=1}^{k} \Gamma_{ij}^k v_k$, for all $i, j, k = 1, \ldots, n$, where $\Gamma_{ij}^k \in \mathcal{O}_{U_p}$ and $v_i, v_j \in H^0(U_p, TU_p)$. Therefore, a singular holomorphic foliation $\mathcal{F}$ of dimension k, induces a regular foliation $\mathcal{F}^0$ of dimension k on $X^0 := X - \mathrm{Sing}(\mathcal{F})$.

The line bundle $K\mathcal{F} := \det(T\mathcal{F})^*$ is called the *canonical bundle* of $\mathcal{F}$. So, By the isomorphism $\wedge^k TX \simeq \Omega_X^{n-k} \otimes \det(TX)$ and the fact that $\det(N\mathcal{F}) \simeq \det(TX) \otimes K\mathcal{F}$ we conclude that the isomorphism

$$\wedge^k TX \otimes K\mathcal{F} \simeq \Omega_X^{n-k} \otimes \det(TX) \otimes K\mathcal{F} \simeq \Omega_X^{n-k} \otimes \det(N\mathcal{F})$$

says that the section $\nu_{\mathcal{F}}$ corresponds to a twisted $(n-k)$-form

$$\omega_{\mathcal{F}} \in H^0(X, \Omega_X^{n-k} \otimes \det(N\mathcal{F}))$$

with values on $\det(N\mathcal{F})$ which yields a *Pfaff system*, of dimension $(n-k)$, associated with the corresponding regular foliation $\mathcal{F}^0$ on the regular part $X^0 = X - \mathrm{Sing}(\mathcal{F})$.

So, in a neighborhood U_p of a point p where $N\mathcal{F}^*$ is locally free, we have

$$\omega_{\mathcal{F}}|_{U_p} = \omega_1 \wedge \cdots \wedge \omega_{n-k},$$

where $\omega_i \in H^0(U_p, \Omega_{U_p}^1)$ satisfies the integrability condition

$$d\omega_j \wedge \omega_1 \wedge \cdots \wedge \omega_{n-k} = 0$$

for all $j = 1, \ldots, n-k$. Therefore,

$$\mathrm{Sing}(\mathcal{F}) = \{\nu_{\mathcal{F}} = 0\} = \{\omega_{\mathcal{F}} = 0\}.$$

Definition 10.2.3 Let X be a compact complex manifold of dimension n. A *Pfaff system* of dimension k on X is an element

$$[\omega] \in \mathbb{P}H^0(X, \Omega_X^{n-k} \otimes \mathcal{L}),$$

where $\mathcal{L}$ is a holomorphic line bundle on X, such that its singular set $\mathrm{Sing}(\omega) := \{\omega = 0\}$ has codimension ≥ 2.

Remark 10.2.4 Note that a holomorphic foliation of dimension k induces a Pfaff system of codimension $n-k$ that is generically decomposable. However, in general, a Pfaff system does not need to be generically decomposable. An example of a non-decomposable Pfaffian system is given by the symplectic 2-form in $\mathbb{C}^{2m}$. In canonical coordinates $(q_1, p_1, \ldots, q_m, p_m) \in \mathbb{C}^{2m}$, this symplectic form is expressed as

$$\omega = \sum_{i=1}^{m} dq_i \wedge dp_i .$$

Example 10.2.5 A holomorphic foliation $\mathcal{F}$ on $\mathbb{P}^n$ of dimension k corresponds to a twisted $(n-k)$-form $\omega \in H^0(\mathbb{P}^n, \Omega^{n-k}_{\mathbb{P}^n}(d+n-k+1)$, where $d \geq 0$ is called the *degree* of $\mathcal{F}$. We can interpret ω as a polynomial differential form

$$\omega = \sum_{1 \leq i_1 < \cdots < i_{n-k} \leq n} A_{i_1 \cdots i_p} dx_{i_1} \wedge \cdots \wedge dx_{i_q}$$

where the $A_{i_1 \cdots i_p}$ are homogeneous of degree $d+1$, and $\iota_R \omega = 0$, where

$$\mathrm{R} = x_0 \frac{\partial}{\partial x_0} + \cdots + x_n \frac{\partial}{\partial x_n}$$

is the radial vector field. The local decomposability translates to the Plücker conditions

$$(i_v \omega) \wedge \omega = 0 \quad \text{for every } v \in \wedge^{n-k-1}(\mathbb{C}^{n+1}),$$

and the integrability condition becomes

$$(i_v \omega) \wedge d\omega = 0 \quad \text{for every } v \in \wedge^{n-k-1}(\mathbb{C}^{n+1}).$$

For more details, we refer the reader to [45, Section 1.3].

Example 10.2.6 Let $\psi : X \to Y$ be a holomorphic fibration, where Y is a complex manifold of dimension $n-k > 0$. Then ψ induces a holomorphic foliation $\mathcal{F}$ on X of dimension k, whose tangent sheaf is the relative tangent sheaf of ψ, and whose singular set corresponds to the singularities of the fibration.

Definition 10.2.7 Let $\omega \in H^0(X, \Omega^{n-k}_X \otimes \mathcal{L})$ be a Pfaff system of dimension k on a complex manifold X. We say that an analytic subvariety $V \subset X$ is *invariant* by ω if $i^*\omega \equiv 0$, where $i : V \hookrightarrow X$ is the inclusion map. If $\mathcal{F}$ is a holomorphic foliation of dimension k on X, then V is invariant by $\mathcal{F}$ if it is invariant by its corresponding Pfaff system

$$\omega_{\mathcal{F}} \in H^0(X, \Omega^{n-k}_X \otimes \det(N\mathcal{F})).$$

Example 10.2.8 Consider the foliation $\mathcal{F}$ in $\mathbb{C}^2$ induced by the holomorphic 1-form $\omega = x(5y^2-9)\,dx - y(5x^2-9)\,dy$. This foliation has 5 singular points given by the singular points of the invariant algebraic curve $C = \{(x^2-y^2)\cdot(x^2+4y^2-9) = 0\}$. See Fig. 10.5.

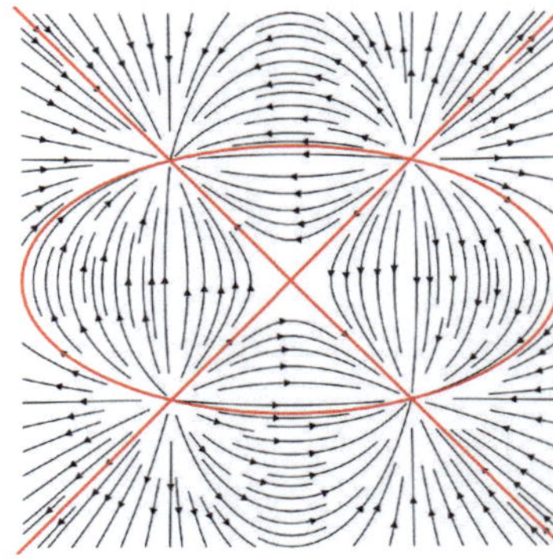

Fig. 10.5 A foliation by lines with 5 singular points

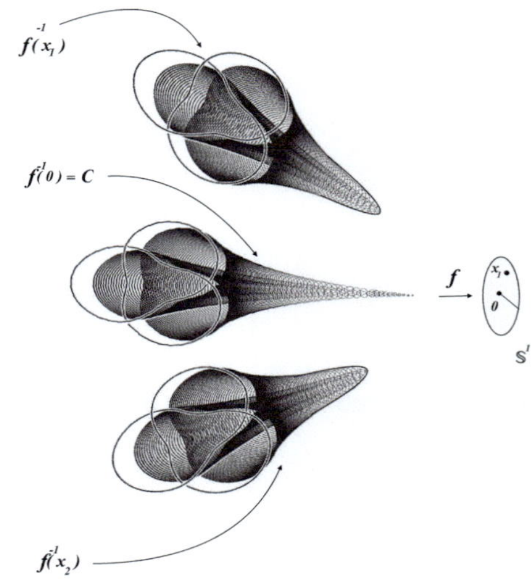

Fig. 10.6 The critical and non-critical levels of the function

Example 10.2.9 In this example, we examine the topology, via the Milnor fibration, of a foliation on $\mathbb{C}^2$ induced by an exact holomorphic 1-form $\omega = df$, i.e., when the foliation admits a holomorphic first integral. Suppose that the foliation is induced by the 1-form $d(x^3 - y^2)$ with an isolated singularity at 0. This foliation is tangent to the fibers of the polynomial function $f : \mathbb{C}^2 \to \mathbb{C}$, given by $f(x, y) = x^3 - y^2$, and it has the invariant (separatrix) curve $f^{-1}(0) := C$ with isolated singularity at 0. Pick the point $0 \in C$ and a small enough ϵ-ball $B_\epsilon(0)$ in $\mathbb{C}^2$ centered at the origin 0, with boundary the 3-sphere $S_\epsilon(x)$. Then by Milnor [75], $B_\epsilon(0) \cap C$ is homeomorphic to the cone on the real link $S_\epsilon(0) \cap C$. In this case, the link $S_\epsilon(0) \cap C \simeq \mathbb{S}^1$ is the trefoil knot. If we take a point $x_1 \in D_\delta(0) - \{0\} \subset \mathbb{C}$, with $|x_1| \ll \delta \ll \epsilon \leq 1$, then on the smooth curve $f^{-1}(x_1)$ we have also that $S_\epsilon(0) \cap f^{-1}(x_1)$ is the trefoil knot, while the leaves of the foliation contained within the ball, which are the Milnor fibers, have the homotopy of a bouquet of 2 circles, $\simeq \mathbb{S}^1 \vee \mathbb{S}^1$. See Fig. 10.6.

10.2.2 Baum–Bott Residues

In this section we recall the Baum–Bott residue theorem for singular foliations on complex manifolds. A starting point is Bott's vanishing theorem presented by R. Bott in the 1970 ICM in Nice (see [16]), where he discussed the state of the art in the problem of "foliating a manifold". As he explains, this can be viewed in two steps:

Step 1: Given a (smooth or complex) n-manifold, construct a k-dimensional subbundle E of the tangent bundle TM; and then,

Step 2: Given a subbundle E of TM as above, can it be deformed to an integrable one?
In the complex case we want E to be holomorphic.
Bott then gives a beautiful answer, which in the complex case can be stated as follows:

Theorem 10.2.10 (Bott's Vanishing Theorem) *Let M be a compact complex analytic n-manifold which admits an integrable holomorphic subbundle E. Let $Q = TM/E$ be the quotient bundle, that one calls the normal bundle. Then the Chern ring TM/E, generated by the real Chern classes of TM/E, vanishes in dimensions greater that* $2 \times \dim_{\mathbb{C}} (TM/E)$. *That is,*

$$Chern^r(TM/E) = 0 \quad \forall r > \dim_{\mathbb{R}} (TM/E) .$$

This provides topological obstructions for the integrability of a distribution on a manifold. It is worth saying that there is also previous work by Bott (e.g. [15]) where he relates the behavior of a holomorphic vector field at its zeroes to the characteristic numbers of complex manifolds and holomorphic bundles.

As an example (from [16]), consider a compact complex n-manifold M with vanishing Euler characteristic. This means that its top Chern class $c_n(M)$ vanishes as well, and also that by the theorem of Poincaré-Hopf, we can construct on M a nowhere vanishing C^∞ vector field v. This spans a smooth trivial 1-dimensional complex sub-bundle E of TM and one has a C^∞ ssplitting $TM \cong E \oplus Q$ where $Q = TM/E$ is the normal bundle. Hence, since E is a trivial bundle, all the Chern numbers of M are degree n products of the Chern classes of Q, evaluated on the fundamental cycle $[M]$.

Now suppose that the vector field V actually is holomorphic, so it is integrable. Then one has a 1-dimensional holomorphic foliation $\mathcal{E}$ with tangent bundle E. Then, by Bott's theorem, all degree n products of Chern classes of Q vanish. Hence, by the above observations, all the Chern numbers of M vanish, so we have:

Corollary 10.2.11 *If a compact complex manifold admits a nowhere vanishing holomorphic vector field, then all its Chern numbers are* 0.

Bott's theorem 10.2.10 was later extended by Baum-Bott to higher dimensional foliations, allowing also singularities. If there are no singularities, then the theorem states that the ring $Chern^r(Q)$, generated by the real Chern classes of the normal bundle, vanishes in dimensions greater that the real codimension of the foliation. In the presence of singularities, the theorem says that each such class, regarded in homology via Alexander duality, localizes as a sum of residues, which are homology classes of the singular set. More precisely:

Theorem 10.2.12 (Baum and Bott [12]) *Let $\mathcal{F}$ be a holomorphic foliation of dimension k on a complex manifold X, and let φ be a homogeneous symmetric polynomial of degree d satisfying $n - k < d \leq n$. Let Z be a compact connected component of the singular set* $\mathrm{Sing}(\mathcal{F})$*. Then, there exists a homology class* $\mathrm{Res}_\varphi(\mathcal{F}, Z) \in \mathrm{H}_{2(n-d)}(Z; \mathbb{C})$ *such that:*

(i) $\mathrm{Res}_\varphi(\mathcal{F}, Z)$ *depends only on φ and on the local behavior of the leaves of $\mathcal{F}$ near Z.*

(ii) Suppose that X is compact, and define $\mathrm{Res}(\varphi, \mathcal{F}, Z) := \alpha_* \mathrm{Res}_\varphi(\mathcal{F}, Z)$*, where α_* is the composition of the maps*

$$\mathrm{H}_{2(n-d)}(Z; \mathbb{C}) \xrightarrow{i_*} \mathrm{H}_{2(n-d)}(X; \mathbb{C}) \quad \text{and} \quad \mathrm{H}_{2(n-d)}(X; \mathbb{C}) \xrightarrow{P} \mathrm{H}^{2d}(X; \mathbb{C}),$$

where i_* is the induced map from the inclusion $i : Z \longrightarrow X$, and P is the Poincaré duality map. Then

$$\varphi(\mathcal{N}_{\mathcal{F}}) = \sum_Z \mathrm{Res}(\varphi, \mathcal{F}, Z).$$

If the foliation has dimension one and isolated singularities, the residues can be expressed in terms of Grothendieck residues (see [11]). More precisely, in this case, for each $p \in \mathrm{Sing}(\mathcal{F})$, an isolated singularity of $\mathcal{F}$, we have

$$\mathrm{Res}(\varphi, \mathcal{F}, p) = \mathrm{Res}_p \begin{bmatrix} \varphi(Jv) \\ v_1, \dots, v_n \end{bmatrix}, \tag{10.6}$$

where $v = (v_1, \dots, v_n)$ is a germ of a holomorphic vector field on (U, p), serving as a local representative of $\mathcal{F}$, Jv is its Jacobian matrix, and

$$\mathrm{Res}_p \begin{bmatrix} \varphi(Jv) \\ v_1, \dots, v_n \end{bmatrix} = \left(\frac{1}{2\pi\sqrt{-1}} \right)^n \int_\Gamma \frac{\varphi(Jv)}{v_1, \dots, v_n},$$

denotes the Grothendieck residue of $\varphi(Jv)$ with respect to $v_1, \dots, v_n$ at p, and Γ is the cycle $\{x \in U; |v_i(x)| = \epsilon_i;\ i = 1, \dots, n\}$.

In general, it is a difficult task to compute the Baum–Bott residues for foliations of higher dimension. In [90], Vishik proved Baum–Bott's Theorem and provided a method to compute the residues, assuming that the foliation has a locally free tangent sheaf. In [17], F. Bracci and T. Suwa studied the behavior of Baum–Bott residues under smooth deformations.

Let us recall how to compute such residues in a particular case. Let $\mathrm{Sing}_{n-k+1}(\mathcal{F}) \subset \mathrm{Sing}(\mathcal{F})$ be the subset of components of pure dimension $n-k+1$. Consider a symmetric homogeneous polynomial φ of degree $n-k+1$. Let $Z \subset \mathrm{Sing}_{n-k+1}(\mathcal{F})$ be an irreducible component. Take a generic point $p \in Z$ such that p is a point where Z is smooth and disjoint from the other irreducible components. Now, consider B_p a ball centered at p, of dimension $n-k+1$ sufficiently small and transversal to Z in p. In [12], under a generic condition (Kupka condition [78]), Baum and Bott proved that

$$\mathrm{Res}(\mathcal{F}, \varphi; Z) = \mathrm{Res}_{\varphi}(\mathcal{F}|_{B_p}; p)[Z],$$

where $\mathrm{Res}_{\varphi}(\mathcal{F}|_{B_p}; p)$ represents the Grothendieck residue at p of the one dimensional foliation $\mathcal{F}|_{B_p}$ on B_p and $[Z]$ denotes the fundamental cycle associated to Z.

In [40], the authors show that Baum–Bott's generic hypothesis is not necessary, and we demonstrate that the above formula always holds as long as the singular set of the foliation has dimension less than or equal to $n-k+1$. More recently, in [65] Kaufmann, Lärkäng, and Wulcan showed that the Baum–Bott residues can be expressed in terms of residue currents.

10.3 Indices for 1-Dimensional Holomorphic Foliations on Surfaces

In his remarkable paper [22], M. Brunella studied various invariants of holomorphic 1-dimensional foliations on complex surfaces that somehow spring from the Baum-Bott theory of residues. He then related all these invariants to establish interesting geometric properties of foliations. These results have been extended to higher dimensions and to other settings by various authors. This is the subject that we study in this section.

10.3.1 Baum-Bott-Brunella Indices

In this subsection we look at the Milnor number of a foliation, already studied earlier (see Definition 10.1.10), as well as at the Camacho-Sad index [31], the variation [66], defined by Khanedani-Suwa, and an invariant that Brunella called Baum-Bott invariant.

Let $\mathcal{F}$ be a holomorphic foliation with isolated singularities on a complex surface X and let p be singular point of $\mathcal{F}$. From the previous section we know that there are two Baum-Bott residues associated to $\mathcal{F}$ at p; these correspond to the Chern numbers c_2 and c_1^2. The one corresponding to c_2, or to the symmetric function given by the determinant, has as residue at p the local Poincaré-Hopf index of a vector field generating $\mathcal{F}$ locally. The local invariant we get is the local multiplicity, or Milnor number $\mu(\mathcal{F}, p)$ of the foliation defined in Definition 10.1.10, which is a topological invariant. Following [22] we denote this invariant by $PH(\mathcal{F}, p) = \mu(\mathcal{F}, p)$.

The second Baum-Bott residue corresponds to the symmetric function given by the trace, or actually to the Chern number $c_1^2(N_{\mathcal{F}})$, where $N_{\mathcal{F}} \in H^1(X, \mathcal{O}^*)$ is the normal bundle of $\mathcal{F}$; this bundle is defined even in the presence of singularities since $\mathcal{F}$ is determined by a morphism of a line bundle into TM.

Following [22] we denote this by $BB(\mathcal{F}, p)$ and call it **the Baum-Bott invariant of $\mathcal{F}$ at p**. This can be studied *la* Godbillon-Vey. Let (z, w) be local coordinates centered at p and let $\omega = F(z, w)dw - G(z, w)dz$ be a holomorphic 1-form that generates $\mathcal{F}$ locally. Let β be a complex valued smooth 1-form of type $(1, 0)$ on a punctured neighborhood $U^* = U \setminus \{p\}$ such that

$$d\omega = \beta \wedge \omega\,.$$

For instance one may take:

$$\beta = \frac{\frac{\partial F}{\partial z} + \frac{\partial G}{\partial w}}{|F|^2 + |G|^2} \left(\overline{F}\, dz + \overline{G}\, dw\right).$$

Then one has (see [22]):

$$\mathrm{BB}(\mathcal{F}, p) \;=\; \frac{1}{(2\pi\, i)^2} \int_{\mathbb{S}^3_\varepsilon} \beta \wedge d\beta\,,$$

where $\mathbb{S}^3_\varepsilon$ is the boundary sphere of a small ball around p with its natural orientation.

If the surface X is compact then one has a global result in the vein of the Poincaré-Hopf index theorem:

$$c_1^2(N_{\mathcal{F}}) = \sum_{p \in \mathrm{sing}\mathcal{F}} \mathrm{BB}(\mathcal{F}, p)\,. \tag{10.7}$$

Now let S be a separatrix of $\mathcal{F}$ at p. That is, a local holomorphic curve containing p which is invariant by $\mathcal{F}$. It is not assumed that this curve is neither smooth nor irreducible. The following invariant is defined in [66]:

Definition 10.3.1 **The variation** of $\mathcal{F}$ at p relative to S is:

$$\mathrm{Var}(\mathcal{F}, S, p) \;=\; \frac{1}{2\pi i} \int_{S \cap \mathbb{S}_\varepsilon} \beta\,,$$

with β as above.

This invariant has the following important global property.

Proposition 10.3.2 (Khanedani-Suwa) *Suppose S is now a compact holomorphic curve in X, invariant by the foliation, then:*

$$\sum_{p \in \mathrm{Sing}(\mathcal{F}) \cap S} \mathrm{Var}(\mathcal{F}, S, p) \;=\; c_1(N\mathcal{F}) \cdot S \,.$$

Now we come to the Camacho-Sad index. This plays a key role for proving in [31] the separatrix theorem: every germ of a holomorphic vector field in $\mathbb{C}^2$ with an isolated singularity at the origen has at least one separatrix S. Yet, this index is not actually defined in $\mathbb{C}^2$ but up in a "desingularization" of the vector field, where one assumes that there is a smooth divisor C, invariant under the induced foliation, with isolated singularities. One attaches a CS-index to each such singularity, which actually is a localization of the Chern class of the normal bundle of C.

Let S be again a separatrix in X for the foliation $\mathcal{F}$ and $p \in S$ a singularity of the foliation. Let ω be a local 1-form on X generating $\mathcal{F}$ locally and f a local function that defines S near p, and we assume it is reduced. Then we know from [86] that there are functions g, ξ and a 1-form η on a neighborhood of p such that

$$g\omega = \xi\, df + f\eta$$

with $k \neq 0$ on $S^* = S \setminus \{p\}$. Notice that on S we have $g\omega = \xi\, df$ and $g \neq 0$ on S^*.

Definition 10.3.3 The Camacho-Sad index of $\mathcal{F}$ at p relative to S is:

$$\mathrm{CS}(\mathcal{F}, S, p) = -\frac{1}{2\pi i} \int_{S \cap \mathbb{S}^3_\varepsilon} \frac{1}{\xi}\eta \,.$$

One has:

Proposition 10.3.4

$$\sum_{p \in \mathrm{Sing}(\mathcal{F}) \cap S} \mathrm{CS}(\mathcal{F}, S, p) \;=\; S \cdot S \;=\; c_1(N_S)[S] \,.$$

Notice that all these formulas have a similar flavor, involving the geometry of the foliation to localize the Chern class of some normal bundle. This contrasts with the GSV index, which (by Proposition 10.1.25) is the localization of a Chern class of the virtual tangent bundle.

10.3.2 *The GSV Index for 1-Dimensional Holomorphic Foliations*

Given a foliation $\mathcal{F}$ on a complex surface X as above, with a singularity at p and a separatrix S through p, S is locally generated by a holomorphic vector field v. Hence S can be regarded itself as being a hypersurface in X with an isolated singularity at p (in fact p can be a regular point of S), equipped with a vector field v. Thus one has the local GSV index of v in S at p, which we take as being its Poincaré-Hopf index in a Milnor fiber of a local defining function f for S (we take this as its definition even if S has several branches).

Brunella in [22] used the foliation to introduce an invariant $Z(\mathcal{F}, S, p)$ in the setting we now envisage, and proved that it coincides with the GSV index. Brunella then established surprising deep relationships between the Baum-Bott, Camacho–Sad, the variation and the GSV indices.

Let X be a compact complex surface, and let $\mathcal{F}$ denote a one-dimensional holomorphic foliation on X. Let C be a reduced curve on X and x a point in C. When C is invariant under $\mathcal{F}$, we say that $\mathcal{F}$ is *logarithmic along* C. Let $f = 0$ be a local equation for C in a neighborhood U_α of x, and let ω_α be a holomorphic 1-form that induces the foliation $\mathcal{F}$ on U_α. Since $\mathcal{F}$ is logarithmic along C, it follows from [79, 86] that there exist holomorphic functions g and ξ defined in a neighborhood of x that do not both vanish identically on C, such that

$$g\,\frac{\omega_\alpha}{f} = \xi\,\frac{df}{f} + \eta, \tag{10.8}$$

where η is an appropriate holomorphic 1-form.

Definition 10.3.5 (Brunella [22]) Let $\mathcal{F}$ be a one-dimensional holomorphic foliation on a compact complex surface X that is logarithmic along a reduced curve $C \subset X$. Given $x \in C$, we define

$$\mathrm{Z}(\mathcal{F}, C, x) = \sum_i \mathrm{ord}_x \left(\frac{\xi}{g}|_{C_i}\right),$$

where $C_i \subset C$ are the irreducible components of C at x, and $\mathrm{ord}_x\left(\frac{\xi}{g}|_{C_i}\right)$ denotes the order of vanishing of $\frac{\xi}{g}|_{C_i}$ at x.

First, we will see that Brunella's Z index coincides with the GSV index.

Proposition 10.3.6 *[22, Page 532] Let* $p \in \mathrm{Sing}(\mathcal{F}) \cap C$. *Then*

$$\mathrm{Z}(\mathcal{F}, C, p) = \mathrm{GSV}(\mathcal{F}, C, p).$$

Proof Let f be a holomorphic function on a neighborhood U of p and defining $C|_U = \{f = 0\}$. So, if ω induces $\mathcal{F}$ on U we have

$$\omega = \frac{\xi}{g} df + \frac{f}{g}\eta.$$

Let $J : \Omega^1_U \to TU$ denote the isomorphism from 1-forms to vector fields in a neighborhood U of p, defined by

$$\omega = P\,dx + Q\,dy \mapsto J(\omega) = Q\frac{\partial}{\partial x} - P\frac{\partial}{\partial y}.$$

The holomorphic vector field $J(\omega)$ defines the foliation $\mathcal{F}$ and can be expressed as the sum of two meromorphic vector fields:

$$J(\omega) = \frac{\xi}{g} J(df) + \frac{f}{g} J(\eta) = v_1 + v_2,$$

with $v_1 = \frac{\xi}{g} J(df)$ and $v_2 = \frac{f}{g} J(\eta)$. Now, observe that

$$J(\omega)|_C = \left(\frac{\xi}{g} J(df) + \frac{f}{g} J(\eta)\right)|_C = \frac{\xi}{g} J(df)|_C = v_1|_C$$

and v_1 is tangent to $C_\epsilon = \{f = \epsilon\}$, for a small ϵ. The restriction $v_1|_{C_\epsilon}$ has poles at the points of $C_\epsilon \cap \{g = 0\}$ and zeros at the points of $C_\epsilon \cap \{\xi = 0\}$. Since the GSV index in C_ϵ is the Poincaré–Hopf index of $v_1|_{C_\epsilon}$, it is the difference between the number of zeroes and the number of poles of $v_1|_{C_\epsilon}$. Therefore,

$$\mathrm{GSV}(\mathcal{F}, C, p) = \frac{1}{2\pi i}\int_{\partial C_\epsilon} \frac{d\xi}{\xi} - \int_{\partial C_\epsilon} \frac{dg}{g} = \frac{1}{2\pi i}\int_{\partial C} \frac{g}{\xi} d\left(\frac{\xi}{g}\right) = \mathrm{Z}(\mathcal{F}, C, p).$$

This proves the proposition.

One has:

Proposition 10.3.7 *[22, Proposition 5] If C is any separatrix at p, then*

$$\mathrm{Var}(\mathcal{F}, C, p) = \mathrm{GSV}(\mathcal{F}, C, p) + \mathrm{CS}(\mathcal{F}, C, p),$$

where $\mathrm{Var}(\mathcal{F}, V, p)$ *denotes the variational Khanedani–Suwa index of $\mathcal{F}$ along V at p [66] and* $\mathrm{CS}(\mathcal{F}, C, p)$ *is the Camacho–Sad index [31].*

Remark 10.3.8 Recently, in [39], the authors generalized the variational index for currents invariant under foliations, with applications to the study of transcendental leaves of foliations and their accumulation on the singular set.

Theorem 10.3.9 (Brunella [22]) *Let $\mathcal{F}$ be a one-dimensional holomorphic foliation on a compact complex surface X, logarithmic along a reduced curve $C \subset X$. Then*

$$\sum_{x \in \mathrm{Sing}(\mathcal{F}) \cap C} \mathrm{GSV}(\mathcal{F}, C, x) = N\mathcal{F} \cdot C - C \cdot C\,,$$

The GSV index is an integer and we already know that it can be negative. In fact, consider Example 10.1.2 and the foliation in $\mathbb{C}^2$ given by the homogeneous vector field $v = w^k \frac{\partial}{\partial z} - z^k \frac{\partial}{\partial w}$, whose leaves are the fibers of the map $f = z^{k+1} + w^{k+1}$. Then $S = f^{-1}(0)$ is a separatrix (with $k+1$ branches) and its GSV index is $1 - k^2$, which is negative whenever $k > 1$. Notice that this foliation has infinitely many separatrices.

Marco Brunella remarked that the non-negativity of the GSV index is the obstruction to the positive solution to "Poincaré problem", as we explain below in Sect. 10.4. The following results from [22] point in that direction and they are interesting on their own.

Recall that a separatrix S of $\mathcal{F}$ at p is nondicritical if there is a sequence of blow-ups based at p, $\pi : \widetilde{X} \to X$, such that π is a resolution of S, $\widehat{S} = \pi^{-1}(S)$ has only normal crossing singularities, and the divisor $\pi^{-1}(p)$ is $\widehat{F}$-invariant, where $\widehat{F}$ is the induced foliation on $\widetilde{X}$ (see [29]). Then (see [22, Propositions 6 and 7]):

Theorem 10.3.10 *If S is a nondicritical separatrix at p then:*

$$\mathrm{GSV}(\mathcal{F}, S, p) \geq 0\,.$$

If S further is a generalized curve (i.e. there are no saddle-nodes in its resolution) and S is maximal, union of all separatrices at p, then:

$$\mathrm{GSV}(\mathcal{F}, S, p) = 0\,.$$

In [35], Cavalier and D. Lehmann proved that the converse is also true, i.e., if

$$\mathrm{GSV}(\mathcal{F}, S, p) = 0,$$

then the foliation $\mathcal{F}$ is a generalized curve at p, and S is the maximal separatrix.

10.4 Indices in Higher Dimensions and the Poincaré Problem

As noted before, indices and residues play a significant role for the study of the famous Poincaré problem in dimensions 2 and higher. In this section we review some of these. The literature is vast and we will focus only on some of the works most related with this article. Let us start by saying a few words about the problem itself.

In [77] H. Poincaré asked whether it is possible to decide if an algebraic differential equation in two variables is algebraically integrable, and he observed that in order to answer this question it is sufficient to know whether one can bound the degree of the generic leaves. He posed a question that in modern language can be stated as follows. Let $\mathcal{F}$ be a holomorphic foliation in $\mathbb{P}^2_{\mathbb{C}}$ and C an irreducible curve which is invariant by $\mathcal{F}$: Is it possible to bound the degree of C in terms of the degree of $\mathcal{F}$?

This problem has been studied by many authors in more generality. One may ask, for instance, if given a foliation $\mathcal{F}$ on $\mathbb{P}^n_{\mathbb{C}}$ of degree d and a hypersurface V in $\mathbb{P}^n_{\mathbb{C}}$ invariant by $\mathcal{F}$, is it possible to bound the degree of V by a number which depends only on d?

The answer in general is obviously negative, even for $n = 2$. For instance the foliation defined by the form $pwdz - qzdw$ has degree 1 and the invariant curve $z^p - w^q = 0$ has degree equal to the maximum of p, q. There are many more easy examples. Yet, there are many positive examples, and as explained in [25] the philosophy is that the failure of having a uniform bound is due to the existence of "bad singularities" of V or of $\mathcal{F}$.

As mentioned before, for $n = 2$, Brunella in [22] established a connection between the GSV-index and both the Khanedani-Suwa variational index [66] and the Camacho-Sad index [31], and he demonstrated that the non-negativity of the GSV-index acts as an obstruction to providing an affirmative answer to the *Poincaré problem* on compact complex surfaces.

In fact, more generally, the non-negativity of the GSV-index provides an obstruction to solving the generalized Poincaré problem for Pfaff systems on projective complex spaces as explained below.

In [42], a GSV-type index for varieties was introduced, in the spirit of Brunella, for varieties invariant under holomorphic Pfaff systems (which may not be locally decomposable) on projective manifolds.

Let U be a germ of an n-dimensional complex manifold. Let D be an analytic reduced hypersurface on U and consider its decomposition into irreducible components

$$D = D_1 \cup \cdots \cup D_{n-k},$$

and suppose that the analytic subvariety $V = D_1 \cap \cdots \cap D_{n-k}$ has pure codimension k. We assume that

$$V = \{z \in U : f_1(z) = \cdots = f_{n-k}(z) = 0\},$$

with $f_1, \ldots, f_{n-k} \in \mathcal{O}_U$ and for each $i \in \{1, \ldots, k\}$,

$$D_i = \{z \in U : f_i(z) = 0\}.$$

Since V is a reduced variety, then the k-form $df_1 \wedge \ldots \wedge df_{n-k}$ is not identically zero on each irreducible component of V. Following Aleksandrov [5], we denote by $\Omega^q_U(\hat{D}_i)$, for $q \geq 1$, the $\mathcal{O}_U$-module of meromorphic differential q-forms with simple poles on the

$$\hat{D}_i = D_1 \cup \cdots \cup D_{i-1} \cup D_{i+1} \cup \cdots \cup D_{n-k},$$

for each $i = 1, \ldots, n-k$. We have the following Aleksandrov's Decomposition Theorem, which is a generalization of Saito's residues [79].

Theorem 10.4.1 (Aleksandrov [5]) *Let $\omega \in \Omega^q_U(D)$ be a meromorphic q-form with simple poles on D. If for each $j = 1, \ldots, k$,*

$$df_j \wedge \omega \in \sum_{i=1}^{k} \Omega^{q+1}_U(\hat{D}_i)$$

then, there exist a holomorphic function g, which is not identically zero on every irreducible component of V, a holomorphic $(q-k)$-form $\xi \in \Omega^{q-k}_U$ and a meromorphic q-form $\eta \in \sum_{i=1}^{k} \Omega^q_U(\hat{D}_i)$ such that the following decomposition holds

$$g\omega = \frac{df_1}{f_1} \wedge \cdots \wedge \frac{df_k}{f_k} \wedge \xi + \eta. \tag{10.9}$$

We can apply this decomposition in our context.

Proposition 10.4.2 *Let $\omega \in \mathrm{H}^0(X, \Omega^{n-k}_X \otimes \mathcal{L})$ be a Pfaff system of dimension k on a complex manifold X, and $V \subset X$ a reduced local complete intersection subvariety of dimension $n-k$ which is invariant by ω. Then for all local representations $\omega_\alpha = \omega|_{U_\alpha}$ of ω, and all local expressions of V in U_α*

$$V \cap U_\alpha = \{z \in U_\alpha : f_{\alpha,1}(z) = \cdots = f_{\alpha,n-k}(z) = 0\},$$

there are holomorphic functions $g_\alpha, \xi_\alpha \in \mathcal{O}(U_\alpha)$ such that

$$g_\alpha \, \omega_\alpha = \xi_\alpha df_{\alpha,1} \wedge \cdots \wedge df_{\alpha,n-k} + \eta_\alpha. \tag{10.10}$$

Moreover, g_α is not identically zero on every irreducible component of V and η_α is given by

$$\eta_\alpha = f_{\alpha,1}\, \eta_{\alpha,1} + \cdots + f_{\alpha,1}\, \eta_{\alpha,n-k},$$

where each $\eta_{\alpha,i} \in \Omega^{n-k}_{U_\alpha}$ *is a holomorphic* $(n-k)$*-form. In particular, the restriction*

$$\frac{\xi_\alpha}{g_\alpha}|_V$$

is well defined.

In this context, we define the GSV index in Brunella's spirit, which coincides with the GSV index when V is a curve invariant under a foliation on a surface.

Definition 10.4.3 Under the assumptions of Proposition 10.4.2, suppose that $S \subset \mathrm{Sing}(\omega) \cap V$ is a codimension one subvariety of V. The GSV-index of ω relative to V in S is defined by

$$\mathrm{GSV}(\omega, V, S) := \sum_j \mathrm{ord}_S \left(\frac{\xi_\alpha}{g_\alpha}|_{V_j} \right),$$

where the sum is taken over all irreducible components V_j of V and $\mathrm{ord}_S \left(\frac{\xi_\alpha}{g_\alpha}|_{V_j} \right)$ denotes the order of vanishing of $\frac{\xi_\alpha}{g_\alpha}|_{V_j}$ along S.

The following result was proved in [42].

Theorem 10.4.4 *Let* X *be a projective manifold, and let* $V \subset X$ *be a reduced, locally complete intersection subvariety of dimension* k*, invariant under a Pfaff system of same dimension induced by a twisted form* $\omega \in \mathrm{H}^0(X, \Omega^{n-k}_X \otimes \mathcal{L})$*. Then the following statements hold:*

(a) The integer number $\mathrm{GSV}(\omega, V, S_i)$ *depends only on the local representatives of* ω*,* V*, and* S_i*;*
(b) If $\mathrm{Sing}(\omega, V) := \mathrm{Sing}(\omega) \cap V$ *has codimension one in* V*, then the following formula is satisfied:*

$$\sum_i \mathrm{GSV}(\omega, V, S_i)[S_i] = c_1([\mathcal{L} \otimes \det(N_{V/X})^{-1}])|_V \frown [V],$$

where S_i *denotes an irreducible component of* $\mathrm{Sing}(\omega, V)$*, and* $N_{V/X}$ *is the normal sheaf of the subvariety* V*.*

The following result offers a method for calculating the GSV-index for Pfaff systems, which can be compared with formulas in [88, Proposition 5.1] and [59]. Moreover, it partially addresses Suwa's question in [88, Remark 5.3.(6)].

Theorem 10.4.5 *With the setup as in Theorem 10.4.4, let* $x \in V$*, and let* $\{f_1 = \cdots = f_{n-k} = 0\}$ *be a local equation for* V *in a neighborhood* U *of* x*. Consider*

$$\omega_{|U} = \sum_{|I|=n-k} a_I \, dZ_I, \quad \textit{where } a_I \in \mathcal{O}(U),$$

as the holomorphic $(n-k)$-form that induces the Pfaff system

$$\omega \in \mathrm{H}^0(U, \Omega_U^{n-k} \otimes \mathcal{L})$$

on U. Then the following formula holds:

$$\mathrm{GSV}(\omega, V, S_i) = \mathrm{ord}_{S_i}(a_I|_V) - \mathrm{ord}_{S_i}(\Delta_I|_V), \tag{10.11}$$

where Δ_I is the $(n-k) \times (n-k)$ minor of the Jacobian matrix $\mathrm{Jac}(f_1, \ldots, f_{n-k})$, *corresponding to the multi-index I.*

More precisely, let $\omega \in \mathrm{H}^0(\mathbb{P}^n, \Omega_{\mathbb{P}^n}^k(d+n-k+1))$ be a holomorphic Pfaff system of dimension k and degree d. Let $V \subset \mathbb{P}^n$ be a reduced complete intersection variety of dimension k and multidegree $(d_1, \ldots, d_{n-k})$, invariant under ω. Suppose that $\mathrm{Sing}(\omega, V)$ has codimension one in V. Then,

$$\sum_i \mathrm{GSV}(\omega, V, S_i) \deg(S_i) = [d+n-k+1-(d_1+\cdots+d_{n-k})] \cdot (d_1 \cdots d_{n-k}),$$

where S_i denotes an irreducible component of $\mathrm{Sing}(\omega, V)$. Therefore, if $\mathrm{GSV}(\omega, V, S_i) \geq 0$ for all i, we have

$$d_1 + \cdots + d_{n-k} \leq d + n - k + 1.$$

This bound is similar to those established by Esteves and Cruz [46] and by Corrêa and Jardim [37]. Related results were previously obtained by Cerveau and Lins Neto [36], Soares [85], Cavalier and Lehmann [34], Brunella and Esteves in [25]. In particular, Soares in [85] gave this bound for a foliation by curves:

$$d_1 + \cdots + d_{n-1} \leq d + n,$$

whenever V is a smooth complete intersection invariant curve.

Example 10.4.6 Consider the Pfaff system $\omega \in \mathrm{H}^0(\mathbb{P}^n, \Omega_{\mathbb{P}^n}^{n-k}(d+n-k+1))$ given by

$$\omega = \sum_{0 \leq j \leq +n-k} (-1)^j d_j f_j \, df_0 \wedge \cdots \wedge \widehat{df_j} \wedge \cdots \wedge df_{n-k},$$

where f_j is a homogeneous polynomial of degree d_j. We can see that

$$d_0 + d_1 + \cdots + d_{n-k} = d + n - k + 1.$$

Suppose that $\deg(f_0) = d_0 = 1$ and that $V = \{f_1 = \cdots = f_{n-k} = 0\}$ is smooth. Then V is invariant by ω, and we have

$$d_1 + \cdots + d_{n-k} = d + n - k.$$

10.4.1 GSV-Indices for Vector Fields as Residues

Suwa in [88] provided an alternative interpretation of the GSV and virtual indices for vector fields as residues on complete intersection varieties with isolated singularities. He considered the residue arising from the localization of the relevant Chern class of the ambient tangent bundle using a frame consisting of v and additional vector fields. Suwa's definition is recalled as follows and does not depend on the choices made.

Let U be a neighborhood of the origin 0 in $\mathbb{C}^m$, and let V be a subvariety (reduced, but not necessarily irreducible) of pure dimension n in U. Assume that V contains 0 and that $V \setminus \{0\}$ is non-singular. We consider a closed ball B around 0 that is sufficiently small so that $R = B \cap V$ has a cone structure over $\partial R = L$, the link of V at 0.

Let E be a C^∞ complex vector bundle of rank l on U, and let $\mathbf{s} = (s_1, \ldots, s_r)$ be a C^∞ r-frame of E on a neighborhood V_0 of L in V, where $r = l - n + 1$. Let ∇_0 be an $\mathbf{s}$-trivial connection for $E|_{V_0}$, and let ∇_1 be a connection for E.

Definition 10.4.7 The residue of $\mathbf{s}$ at 0 with respect to c_n is defined by

$$\mathrm{Res}_{c_n}(\mathbf{s}, E|_V; 0) = \int_R c_n(\nabla_1) - \int_{\partial R} c_n(\nabla_0, \nabla_1).$$

We have an exact sequence

$$0 \longrightarrow TV_0 \longrightarrow TU|_{V_0} \xrightarrow{\pi} NV_0 \longrightarrow 0, \tag{10.12}$$

where TV_0 and TU denote the holomorphic tangent bundles of V_0 and U, respectively, and NV_0 is the normal bundle of V_0 in U. Suppose that V is a complete intersection defined by $f = (f_1, \ldots, f_k)$ in U, where $k = m - n$. In a neighborhood of a regular point of f, we may choose $(f_1, \ldots, f_k)$ as part of the local coordinates on U, so that we have holomorphic vector fields $\frac{\partial}{\partial f_1}, \ldots, \frac{\partial}{\partial f_k}$ away from the critical set of f. These vector fields are linearly independent and normal to the non-singular part of V, so that

$$\partial_f := \left(\pi\left(\frac{\partial}{\partial f_1}\Big|_{V_0}\right), \ldots, \pi\left(\frac{\partial}{\partial f_k}\Big|_{V_0}\right)\right)$$

forms a frame for NV_0. Suppose that we have a C^∞ nonsingular vector field v on V_0.

Theorem 10.4.8 *[88, Theorems 3.4 and 4.4] We have*

$$\mathrm{GSV}(v,0)=\mathrm{Res}_{c_n}((v,\partial_f),TU|_V;0)=\mathrm{Vir}(v,0),$$

where Vir$(v,0)$ *is the virtual index of* v *at* 0.

Since the indices do not depend on the choices made, we have the following consequence.

Corollary 10.4.9 *Let* $V=\{f_1=\cdots=f_k=0\}$ *be a complete intersection, with isolated singularity at* 0*, invariant by a foliation by curves* $\mathcal{F}$ *singular at most at* 0*. Then*

$$\mathrm{GSV}(\mathcal{F},0)=\mathrm{Res}_{c_n}(\mathcal{F},TU|_V;0)=\mathrm{Vir}(\mathcal{F},0).$$

Now, if $n=2$, the formula 10.11 generalizes Suwa's formula for the GSV index. In fact, if $n=2$, then $k=1$, and V is a curve, and ω induces a foliation by curves. Let $x\in V$, and $\{f=0\}$ be a local equation for V in a neighborhood U of x. Consider

$$\omega=a_1dx-a_2dy$$

and its dual vector field

$$v=a_2\frac{\partial}{\partial x}+a_1\frac{\partial}{\partial y}$$

Proposition 10.4.10 *[88, Corollary 5.2] Let* $V=\{f=0\}$ *be a curve, with isolated singularity at* 0*, invariant by a foliation by curves* $\mathcal{F}$ *singular at most at* 0*. Then*

$$\mathrm{Z}(\omega,V,x)=\mathrm{ord}_x(a_1|_V)-\mathrm{ord}_x(\partial_x f|_V)=\mathrm{ord}_x(a_2|_V)-\mathrm{ord}_x(\partial_y f|_V).$$

An interesting application is that the GSV index provides an obstruction to the continuous extension of the normal bundle of the foliation from the regular part of the invariant variety. Indeed, let $(V,0)\subset\mathbb{C}^m$ be an ICIS of dimension $n\geq 2$, and let v be a holomorphic vector field on V, singular only at 0. This vector field defines a one-dimensional holomorphic foliation $\mathcal{F}$ on V singular at 0. On $V\setminus\{0\}$, we have the tangent bundle $T\mathcal{F}$ to the foliation, and we consider the smooth normal bundle $N\mathcal{F}=T(V\setminus\{0\})/T\mathcal{F}$.

Corollary 10.4.11 *[82, Theorem 2] Let* $(V,0)$ *and* $\mathcal{F}$ *be as above. Then the normal bundle to* $\mathcal{F}$ *in* $V\setminus\{0\}$ *extends to* 0 *as a vector bundle (continuous or smooth) if and only if* GSV$(\mathcal{F},0)$ *is a multiple of* $(n-1)!$

10.5 Log Baum–Bott Residues and Aleksandrov's Logarithmic Index

Using techniques from homological algebra, Aleksandrov [4] introduced an algebraic index, inspired by the work of Gómez-Mont, to measure the variation between the Poincaré–Hopf and homological indices. This is defined as follows: Let v be a holomorphic local vector field that induces foliation $\mathcal{F}$. The interior multiplication i_v then generates a complex of logarithmic differential forms:

$$0 \longrightarrow \Omega^n_{X,p}(\log \mathcal{D}) \xrightarrow{i_v} \Omega^{n-1}_{X,p}(\log \mathcal{D}) \xrightarrow{i_v} \cdots \xrightarrow{i_v} \Omega^1_{X,p}(\log \mathcal{D}) \xrightarrow{i_v}$$
$$\mathcal{O}_{X,p} \longrightarrow 0.$$

Let p be an isolated singularity of the foliation $\mathcal{F}$. The i_v-homology groups of the complex $\Omega^\bullet_{X,p}(\log \mathcal{D})$ are finite-dimensional vector spaces. Accordingly, the Euler characteristic

$$\chi(\Omega^\bullet_{X,p}(\log \mathcal{D}), i_v) = \sum_{i=0}^{n}(-1)^i \dim H_i(\Omega^\bullet_{X,p}(\log \mathcal{D}), i_v)$$

of this complex of logarithmic differential forms is well-defined. Since this invariant does not depend on the choice of the local representative v of the foliation $\mathcal{F}$ at p, the *logarithmic index* of $\mathcal{F}$ at the point p is given by

$$\mathrm{Ind}_{\log}(\mathcal{F}, \mathcal{D}, p) := \chi(\Omega^\bullet_{X,p}(\log \mathcal{D}), i_v).$$

In [41, 43], the authors addressed the problem of establishing a global logarithmic residue theorem for meromorphic vector fields on compact complex manifolds. They proved the following result under the assumption that the invariant divisors are of normal crossings type.

Theorem 10.5.1 *Let $\mathcal{F}$ be a one-dimensional foliation with isolated singularities and logarithmic along a normal crossing divisor D in a compact complex manifold X. Then*

$$\int_X c_n(T_X(-\log D) - T\mathcal{F}) = \sum_{x \in \mathrm{Sing}(\mathcal{F}) \cap (X \setminus D)} \mu_x(\mathcal{F}) + \sum_{x \in \mathrm{Sing}(\mathcal{F}) \cap D} \mathrm{Ind}_{\log}(\mathcal{F}, D, x),$$

where $\mu_x(\mathcal{F})$ is the Milnor number of $\mathcal{F}$ at x.

In [41], the authors prove that if $\mathcal{F}$ is a one-dimensional foliation on $\mathbb{P}^n$, with n odd, and isolated singularities (non-degenerate) along a smooth divisor D, then $\mathrm{Sing}(\mathcal{F}) \subset D$ if and only if Soares' bound [85] for the Poincaré problem is achieved, i.e, if and only if

$$\deg(D) = \deg(\mathcal{F}) + 1.$$

More recently, in [38], the authors developed a Baum–Bott residue theory for foliations by logarithmic curves along free divisors:

Theorem 10.5.2 *Let $\mathcal{F}$ be a one-dimensional holomorphic foliation on a complex manifold X, logarithmic along a free divisor $D \subset X$, and let φ be a homogeneous symmetric polynomial of degree n. If $\mathcal{F}$ has only isolated singularities, then*

$$\int_X \varphi(TX(-\log D) - T\mathcal{F}) = \sum_{x \in \mathrm{Sing}(\mathcal{F}) \cap (X \setminus D)} \mathrm{BB}_\varphi(v, x) + \sum_{x \in \mathrm{Sing}(\mathcal{F}) \cap D} \mathrm{Res}_\varphi^{\log}(\mathcal{F}, p),$$

where $\mathrm{BB}_\varphi(v, x)$ is the Baum–Bott residue of $\mathcal{F}$ at x with respect to φ. If, in addition, D has normal crossings at p, then

$$\mathrm{Res}_\varphi^{\log}(\mathcal{F}, D, p) = (2\pi\sqrt{-1})^n \, \mathrm{Res}_p \begin{bmatrix} \varphi(J_{\log}\vartheta) \\ v_1, \ldots, v_n \end{bmatrix}.$$

Also, when $\varphi = \det$, *we have*

$$\mathrm{Res}_{\det}^{\log}(\mathcal{F}, D, p) = (-1)^n (2\pi\sqrt{-1})^n \, \mathrm{Ind}_{\log}(\mathcal{F}, D, p).$$

As an application of this result, we can derive a Baum–Bott-type formula for singular varieties. Consider $T_Y := \mathrm{Hom}(\Omega_Y^1, \mathcal{O}_Y)$ the tangent sheaf of Y, where Ω_Y^1 represents the sheaf of Kähler differentials on Y. A holomorphic foliation on Y is a reflexive subsheaf $T\mathcal{F}$ of T_Y such that $[T\mathcal{F}, T\mathcal{F}] \subset T\mathcal{F}$. Then the following result holds.

Corollary 10.5.3 *Let Y be a compact normal variety, and let $T\mathcal{F} \subset T_Y$ be a one-dimensional holomorphic foliation with isolated singularities. Let φ be a homogeneous symmetric polynomial of degree $n = \dim(Y)$. If $\pi : (X, D) \to (Y, \emptyset)$ is a functorial log resolution of Y, then*

$$\int_X \varphi(T_X(-\log D) - (T\pi^{-1}\mathcal{F})^*)$$

$$= \sum_{x \in \mathrm{Sing}(\mathcal{F}) \cap Y_{\mathrm{reg}}} \mathrm{BB}_\varphi(\mathcal{F}, x) + \sum_{S_\lambda \subset D \cap \mathrm{Sing}(\pi^*\mathcal{F})} \mathrm{Res}_\varphi^{\log}(\pi^*\mathcal{F}, S_\lambda),$$

where $\mathrm{BB}_\varphi(\mathcal{F}, x)$ is the Baum–Bott index of $\mathcal{F}$ at x with respect to φ. In particular, if $D \cap \mathrm{Sing}(\pi^{-1}\mathcal{F})$ is isolated, then

$$\int_X \varphi(T_X(-\log D) - (T\pi^{-1}\mathcal{F})^*) =$$

$$= \sum_{x \in \mathrm{Sing}(\mathcal{F}) \cap Y_{\mathrm{reg}}} \mathrm{BB}_\varphi(\mathcal{F}, x) + \sum_{x \in D \cap \mathrm{Sing}(\pi^{-1}v)} \mathrm{Res}_p \begin{bmatrix} \varphi(J_{\log}(\pi^{-1}v)) \\ v_1, \ldots, v_n \end{bmatrix}.$$

If, in addition, $\pi^{-1}\mathcal{F}$ is induced by a global vector field v and $\mathrm{Sing}(\mathcal{F}) \subset \mathrm{Sing}(Y)$, then the following Poincaré–Hopf type formula holds:

$$\chi(Y) = \sum_{x \in D \cap \mathrm{Sing}(\pi^{-1}v)} \mathrm{Res}_p \begin{bmatrix} \varphi(J_{\log}(\pi^{-1}v)) \\ v_1, \ldots, v_n \end{bmatrix} + \chi(\mathrm{Sing}(Y)).$$

As a consequence of Theorem 10.5.2, we can show that the difference between the Baum-Bott and logarithmic indices can be expressed in terms of the GSV and Camacho-Sad indices. Before stating this result, let us first establish some notation. If $S \subset X$ is an analytic subset and $\mathcal{F}$ is a foliation by curves in X, with isolated singularities, we denote by

$$\mathrm{R}(\mathcal{F}, S) = \sum_{x \in Sing(\mathcal{F}) \cap S} \mathrm{R}(\mathcal{F}, x),$$

where R denotes some residue associated with $\mathcal{F}$ along S. One gets (see [38, Corollary 1.5]:

Theorem 10.5.4 *Let $\mathcal{F}$ be a one-dimensional holomorphic foliation on a compact complex surface X with a reduced invariant curve $S \subset X$. Then*

$$\mathrm{BB}_{c_1^2}(\mathcal{F}, S) - \mathrm{Res}^{\log}_{c_1^2}(\mathcal{F}, S) = 2\mathrm{GSV}(\mathcal{F}, S) - \mathrm{CS}(\mathcal{F}, S).$$

Hence, in particular, the non-negativity of

$$\mathrm{BB}_{c_1^2}(\mathcal{F}, S) - \mathrm{Res}^{\log}_{c_1^2}(\mathcal{F}, S) + \mathrm{CS}(\mathcal{F}, S)$$

is an obstruction to an affirmative answer to the Poincaré problem. Moreover, if $\mathcal{F}$ is a generalized curve along S, then

$$\mathrm{Res}^{\log}_{c_1^2}(\mathcal{F}, S) = 2\mathrm{Res}_{c_1^2}(\mathcal{F}, S).$$

Finally, a new obstruction to giving an affirmative answer to the Poincaré problem for foliations by curves on $\mathbb{P}^n$, n odd, is introduced as a log index as follows(see [38, Corollary 1.4]): let $\mathcal{F}$ be a one-dimensional foliation, with isolated singularities, on $\mathbb{P}^n$ with an invariant *free divisor* D. If n is odd and

then

$$\deg(D) \leq \deg(\mathcal{F}) + n.$$

10.6 Final Remarks

We have explored the intertwining amongst invariants for singular varieties and for singular foliations. There is a lot more in the literature, and in this section we briefly comment on some recent developments related to this work.

The Milnor number permeates this work and it is a topological invariant, fundamental in singularity theory. There are plenty of extensions and applications of this concept in the literature, see for instance [28] for an account on this.

The Tjurina number was introduced by Greuel in [62]; this is the dimension of the base space of a semi-universal deformation of an isolated hypersurface singularity. In fact one has the Milnor and the Tjurina algebras. For a map germ $(\mathbb{C}^n, o) \to (\mathbb{C}, 0)$ these are, respectively:

$$M_f = \frac{\mathcal{O}_{\mathbb{C}^n,0}}{(\frac{\partial f}{\partial z_1}, \cdots \frac{\partial f}{\partial z_n})} \quad \text{and} \quad \mathcal{T}_f = \frac{\mathcal{O}_{\mathbb{C}^n,0}}{(f, \frac{\partial f}{\partial z_1}, \cdots \frac{\partial f}{\partial z_n})}.$$

Their dimensions over $\mathbb{C}$ are the corresponding Milnor and Tjurina numbers, $\mu(f)$ and $\tau(f)$. There are deep relations between these two invariants (see for instance [47, 63]).

The Tjurina algebra for holomorphic vector fields, to our knowledge, appears first in Gómez-Mont's formula in [59] for the homological index (see Definition 10.1.19), though he does not use this name. The actual name of Tjurina number for foliations was coined in [32, p. 159], where F. Cano, N. Corral and R. Mol use the intersection properties of polar curves for 1-dimensional foliations in complex surfaces, to study properties about non-dicritical singularities. In particular, they use the formula in [59] to give a polar interpretation of the GSV-index and use this to study the Poincaré problem for foliations in $\mathbb{P}^2_{\mathbb{C}}$.

In [53] the authors study relations between the Milnor and Tjurina numbers of singular foliations in the complex plane with respect to a balanced divisor of separatrices, an interesting concept introduced in [55]. In the non-dicritical case, where there are only finitely many separatrices of the foliation, a balanced divisor just means the union of all separatrices. For dicritical singularities this is a finite set of separatrices which includes all the isolated separatrices and some separatrices from the ones associated to dicritical components. A balanced divisor provides a certain control of the algebraic multiplicity of the foliation and of its separatrices. The authors find in [53] important relations between various indices of a singular foliation, such as the Baum-Bott, Camacho-Sad, GSV variational indices and the Milnor and Tjurina numbers.

In [30], Camacho, Lins Neto and Sad posed the problem of determining whether non-trivial minimal sets exist for foliations on $\mathbb{P}^2$. This question remains an open problem to this day. In [72], Lins Neto proved the non-existence of minimal sets for codimension one foliations on $\mathbb{P}^n$, with $n \geq 3$. Motivated by Lins Neto's work, Marco Brunella in [24] initiated the investigation of this problem in higher dimensions for codimension one holomorphic foliations with ample normal bundles. More precisely, Brunella proposed the following conjecture: *let $\mathcal{F}$ be a codimension-one foliation on a projective manifold of dimension $n \geq 3$. If $\mathcal{F}$ possesses an ample normal bundle, then every leaf of $\mathcal{F}$ necessarily accumulates on its singular set.* In the remarkable work [27] by Brunella and Perrone, the authors, utilizing Baum-Bott residues, proved this conjecture for projective manifolds with Picard rank one. In [2], Adachi and Brinkschulte completely solved Brunella's conjecture by employing the localization of the first Atiyah class, folowing [1], of the determinant of the normal bundle of the foliation. A second proof of Brunella's conjecture was provided by Adachi, Biard, and Brinkschulte in [3], where the authors employed a residue formula for meromorphic connections. This approach was inspired by a residue formula proved by Pereira in [76]. Recently, in [39], the authors generalized the variational index for currents invariant under foliations, with applications to the study of transcendental leaves of foliations and their accumulation on the singular set. This result represents a higher-dimensional generalization of the result previously established by Brunella for foliations by curves, as proven in [23].

The adjunction formulas for foliations have played an important role in the classification of regular foliations on projective surfaces, as given by Brunella, and also in foliated birational geometry, see [26]. When the curve is invariant under the foliation, the difference $K_{\mathcal{F}} - K_C$ is the total sum of the GSV indices along C. More recently, Cascini and Spicer [33] obtained an adjunction formula for foliations on algebraic varieties, where a correction term serves as a generalization of Brunella's GSV index whenever the divisor is invariant under the foliation. We observe that Theorem 10.4.4 also provides an adjunction formula

$$K\mathcal{F}|_V - K_V = \det(N\mathcal{F})|_V + \det(N_{V/X})^{-1} = \sum_i \mathrm{GSV}(\omega, V, S_i)[S_i],$$

since $\det(N\mathcal{F})|_V = -K_X|_V + K\mathcal{F}|_V$ and $\det(N_{V/X}) = -K_X|_V + K_V$.

In fact the GSV index was originally defined for vector fields on hypersurface singularities. The extension to complete intersections was given in [84] where this is used to prove an adjunction formula for local complete intersections.

References

1. M. Abate, F. Bracci, T. Suwa, and F. Tovena, *Localization of Atiyah classes*, Rev. Mat. Iberoam. **29** (2013), no. 2, 547–578.
2. M. Adachi and J. Brinkschulte, *Dynamical aspects of foliations with ample normal bundle*, Indiana Univ. Math. J. **72** (2023), no. 5, 1931–1947.
3. M. Adachi, S. Biard, and J. Brinkschulte, *A residue formula for meromorphic connections and applications to stable sets of foliations*, J. Geom. Anal., 33 (2023), 338.
4. A. G. Aleksandrov. *The Index of Vector Fields and Logarithmic Differential Forms*. Funct. Anal. Appl. 39 (2005), 245–255.

5. A.G. Aleksandrov, *Multidimensional residue theory and the logarithmic De Rham Complex*, Journal of Singularities, **5** (2012), 1–18
6. A. G. Aleksandrov. *The index of differential forms on complete intersections*. Funct. Anal. Appl. 49 (2015), 1–14.
7. M. Aguilar, J. Seade and A. Verjovsky, *Indices of vector fields and topological invariants of real analytic singularities*, Crelle's, J. Reine u. Ange. Math. 504 (1998), 159–176.
8. P. Aluffi. *MacPherson's and Fulton's Chern Classes of Hypersurfaces*. I.M.R.N. (1994), 455–465.
9. P. Aluffi, *Chern classes for singular hypersurfaces*, Trans. Amer. Math. Soc. 351 (1999), 3989–4026.
10. P. Aluffi. *Segre classes and invariants of singular varieties*. In "Handbook of Geometry and Topology of Singularities III", ed. J. L. Cisneros Molina *et al*, Springer Verlag 2022, p. 419–491.
11. P. Baum, R. Bott, *On the zeros of meromorphic vector fields*, Essay on Topology and Related Topics, Springer-Verlag, New York, 1970, 29–47.
12. P. Baum, R. Bott, *Singularities of holomorphic foliations*, J. Differential Geom. 7 (1972), 279–342.
13. Ch. Bonatti and X. Gómez-Mont, *The index of a holomorphic vector field on a singular variety*, Astérisque 222 (1994), 9–35.
14. H.-C. G. von Bothmer, W. Ebeling and X. Gómez-Mont, *An algebraic formula for the index of a vector field on an isolated complete intersection singularity*, Ann. Inst. Fourier (Grenoble) 58 (2008), no. 5, 1761–1783.
15. R. Bott, *A residue formula for holomorphic vector fields*, J. Differential Geom. 4 (1967), 311–332.
16. R. Bott, *On topological obstructions to integrability.* Actes Congr. Int. Math., Nice 1970, Tome 1, 27–35 (1971).
17. F. Bracci, T. Suwa, *Perturbation of Baum-Bott residues*, Asian J. Math. 19, no. 5 (2015), 871–886.
18. J.-P. Brasselet, *Characteristic Classes*. In "Handbook of Geometry and Topology of Singularities III", ed. J. L. Cisneros Molina *et al*, Springer Verlag 2022, p. 303–417.
19. J.-P. Brasselet, D. Lehmann, J. Seade and T. Suwa, *Milnor classes of local complete intersections*, Transactions A. M. S. 354 (2001), 1351–1371.
20. J.-P. Brasselet, D. T. Lê, J. Seade. *Euler obstruction and indices of vector fields*. Topology 39 (2000), 1193–1208.
21. J.-P. Brasselet, J. Seade, T. Suwa. *Vector fields on singular varieties*. Lecture Notes in Mathematics 1987 (2009). Springer-Verlag, Berlin.
22. M. Brunella, *Some remarks on indices of holomorphic fields*, Publicacions Matemátiques, **41** (1997), 527–544.
23. M. Brunella, *Inexistence of invariant measures for generic rational differential equations in the complex domain*, Bol. Soc. Mat. Mexicana (3) **12** (1) (2006), 43–49.
24. M. Brunella, *On the dynamics of codimension one holomorphic foliations with ample normal bundle*, Indiana Univ. Math. J., **57**(7) (2008), 3101–3113.
25. M. Brunella, L. G. Mendes, *Bounding the degree of Solutions to Pfaff Equations*, Publicacions Matemátiques, **44**, (2000), 593–604.
26. M. Brunella, *Birational Geometry of Foliations*, IMPA Mathematical Publications, Instituto de Matemática Pura e Aplicada (IMPA), Rio de Janeiro, 2004.
27. M. Brunella, C. Perrone, *Exceptional singularities of codimension one holomorphic foliations*, Publ. Mat. 55 (2011), 295–312.
28. Callejas-Bedregal, R., Morgado, M. F. Z. and Seade, J., *Milnor number and Chern classes for singular varieties: an introduction*. In "Handbook of Geometry and Topology of Singularities III", ed. J. L. Cisneros Molina *et al*, Springer Verlag 2022, p. 494–563.
29. C. Camacho, A. Lins, P. Sad, *Topological invariants and equidesingularization for holomorphic vector fields*. J. Differ. Geom. 20, 143–174 (1984).

30. C. Camacho, A. Lins Neto, P. Sad, *Minimal sets of foliations on complex projective spaces*, Publications Mathématiques de l'Institut des Hautes Études Scientifiques, No. 68 (1988), 187–203.
31. C. Camacho, P. Sad, *Invariant varieties through singularities of holomorphic vector fields*, Ann. of Math. **115** (1982) , 579–595. nd topology of singularities III. Ed. J. L. Cisneros-Molina *et al.* Springer. 493–564 (2022).
32. F. Cano, N. Corral and R. Mol, *Local polar invariants for plane singular foliations*, Expo. Math. 37 (2019), 145–164
33. P. Cascini, C. Spicer, *Foliation adjunction*, arXiv:2309.10697, 2024.
34. V. Cavalier, D. Lehmann, *On the Poincaré inequality for one-dimensional foliations*, Compositio Mathematica, 142(2) (2006), 529–540.
35. V. Cavalier and D. Lehmann, *Localisation des résidus de Baum–Bott, courbes généralisées et K-théorie (I: feuilletages dans $\mathbb{C}^2$)*, Commentarii Mathematici Helvetici, **76** (2001), 665–683.
36. D. Cerveau and A. Lins Neto, *Holomorphic foliations in $\mathbb{P}^2_{\mathbb{C}}$ having an invariant algebraic curve,* Ann. Inst. Fourier 41 (1991), 883–903.
37. M. Corrêa Jr, M. Jardim, *Bounds for sectional genera of varieties invariant under Pfaff fields*, Illinois Journal of Mathematics, vol. **56** (2) (2012), 343–352.
38. M. Corrêa, F. Lourenço, and D. Machado, *Log Baum-Bott Residues for Foliations by Curves*, 2024. arXiv:2411.04015v1.
39. M. Corrêa, A. Fernández-Pérez, and M. G. Soares, *Brunella–Khanedani–Suwa Variational Residues for Invariant Currents*, Journal of Singularities **23** (2021), 107–115.
40. M. Corrêa and F. Lourenço, *Determination of Baum-Bott residues of higher codimensional foliations*, Asian J. Math., 23 (3) (2019), 527–538.
41. M. Corrêa, D. Machado, *Residue formulas for logarithmic foliations and applications*, Trans. Amer. Math. Soc., 371 (2019), 6403-6420.
42. M. Corrêa and D. Machado, *GSV–index for holomorphic Pfaff systems*, Documenta Mathematica, vol. 25, pp. 1011–1027, 2020.
43. M. Corrêa, D. Machado, *Global residue formula for logarithmic indices of one-dimensional foliations*, Ann. Math., 32 (2024), no. 4, 1095–1118.
44. M. Corrêa, A. M. Rodríguez, M. G. Soares, *A Bott type residues formula for complex orbifolds*, International Mathematics Research Notices, (10), 2889–2911, 2016.
45. F. Cukierman and J. V. Pereira, *Stability of holomorphic foliations with split tangent sheaf*, Amer. J. Math., **130** (2008), no. 2, 413–439.
46. J. D. A. Cruz, E. Esteves, *Regularity of subschemes invariant under Pfaff fields on projective spaces*. Commentarii Mathematici Helvetici **86** (2011), 947–965.
47. A. Dimca, G.-M. Greuel, *On 1-forms on isolated complete intersection on curve singularities,* J. of Singularities 18 (2018), 114–118.
48. W. Ebeling and S. Gusein-Zade, *On the index of a vector field at an isolated singularity.* The Arnoldfest (Toronto, ON, 1997), 141–152, Fields Inst. Commun., 24, Amer. Math. Soc., Providence, RI, 1999.
49. W. Ebeling and S. Gusein-Zade, *Indices of vector fields and 1-forms*. In Handbook of geometry and topology of singularities III. Ed. J. L. Cisneros-Molina *et al.* Springer. 493–564 (2022).
50. W. Ebeling, G. Gusein-Zade, J. Seade. *Homological index for 1-forms and a Milnor number for isolated singularities*. International Journal of Maths. 15 (2004), no. 9, 895–905.
51. D. Eisenbud and H. Levine, *An algebraic formula for the degree of a C^∞ map germ*, Ann. Math. (2) 106 (1977), no. 1, 19–44.
52. G. Ellingsrud and S. Stromme, *Bott's formula and enumerative geometry*, J. Amer. Math. Soc. 9 (1996), no. 1, 175–193.
53. A. Fernández-Pérez, E. R. García Barroso, N. Saravia-Molina. *On Milnor and Tjurina numbers of foliations*; arXiv:2112.14519.
54. W. Fulton, *Intersection Theory*, Springer-Verlag (1984).
55. Y. Genzmer, *Rigidity for dicritical germ of foliation in $\mathbb{C}^2$*. Int. Math. Res. Notices 19 (2007), 14 p.

56. L. Giraldo, X. Gómez-Mont and P. Mardesic, *On the complex formed by contracting differential forms with a vector field on a hypersurface singularity*, Bol. Soc. Mat. Mexicana (3) 7 (2001), no. 2, 211–221.
57. L. Giraldo, X. Gómez-Mont and P. Mardesic, *A law of conservation of number for local Euler characteristics*, Complex manifolds and hyperbolic geometry (Guanajuato, 2001), 251–259, Contemp. Math., 311, Amer. Math. Soc., Providence, RI, 2002.
58. L. Giraldo, X. Gómez-Mont and P. Mardesić, *On the index of vector fields tangent to hypersurfaces with non-isolated singularities*, J. London Math. Soc. (2) 65 (2002).
59. X. Gómez-Mont. *An algebraic formula for the index of a vector field on a hypersurface with an isolated singularity.* J. Algebraic Geom. 7 (1998), 731–752.
60. X. Gómez-Mont, J. Seade, A. Verjovsky. *The index of a holomorphic flow with an isolated singularity*. Math. Ann. 291 (1991), 737–751.
61. G.-M. Greuel, *Der Gauß-Manin-Zusammenhang isolierter Singularitäten von vollständigen Durchschnitten*, Math. Ann. **214** (1975), 235–266.
62. G.M. Greuel. *Dualität in der lokalen Kohomologie isolierter Singularitäten*. Math. Ann. 250 (2) (1980), 157–173.
63. G.M. Greuel, Ch. Lossen, E. Shustin, *Introduction to Singularities and Deformations*. Springer Monographs in Mathematics, 2007.
64. H. Hamm. *Lokale topologische Eigenschaften komplexer Räume. Math. Ann.*, 191 (1971), 235–252.
65. L. Kaufmann, R. Lärkäng, and E. Wulcan, *Baum–Bott residue currents*, arXiv preprint, arXiv:2302.08887, 2023.
66. B. Khanedani, T. Suwa, *First variation of holomorphic forms and some applications*, Hokkaido Math. J. **26** (1997) , 323–335.
67. G.N. Khimshiashvili, *On the local degree of a smooth mapping*, Comm. Acad. Sci. Georgian SSR 85, No. 2 (1977), 309–312 (in Russian).
68. H. King and D. Trotman, *Poincaré-Hopf theorems on stratified sets*, preprint 1996. (Published as "Poincaré-Hopf theorems on singular spaces", Proc. Lond. Math. Soc. (3) 108, No. 3, 682–703 (2014).)
69. M. Kontsevich, *Enumeration of rational curves via torus actions*, The Moduli Space of Curves (Progress in Mathematics, Vol. 129), Birkhäuser, 1995, 335–368.
70. D.T. Lê, *Computation of the Milnor number of an isolated singularity of a complete intersection*, Funct. Anal. Appl. **8** (1974), 45–49.
71. D. Lehmann, M. Soares and T. Suwa, *On the index of a holomorphic vector field tangent to a singular variety*, Bol. Soc. Bras. Mat. 26 (1995), pp 183–199.
72. A. Lins Neto, *A note on projective Levi flats and minimal sets of algebraic foliations*, Annales de l'Institut Fourier, Vol. 49, No. 4 (1999).
73. E. Looijenga. *Isolated Singular Points on Complete Intersections*. Cambridge Univ. Press, Cambridge, London, New York, New Rochelle, Melbourne, Sydney, 1984.
74. R. MacPherson, *Chern classes for singular varieties*, Ann. of Math **100** (1974), 423–432.
75. J. W. Milnor. *Singular points of complex hypersurfaces*. Annals of Mathematics Studies, 61, Princeton University Press, Princeton, N.J., 1968.
76. J. V. Pereira, *Closed meromorphic 1-forms*, Handbook of Geometry and Topology of Singularities V: Foliations, 447–499. Springer, Cham (2024).
77. H. Poincaré, *Sur l'integration algebrique des equations différentielles du premier ordre et du premier degré*, Rendiconti del Circolo Matematico di Palermo **5** (1891), 161–191; 11 (1897), 193–239.
78. R. Rosas, *Transversely product singularities of foliations in projective spaces*, Comptes Rendus Mathématique, **361** (2023), no. G11, 1785–1787.
79. K. Saito, *Theory of logarithmic differential forms and logarithmic vector fields*, J. Fac. Sci. Univ. Tokyo **27**(2) (1980), 265–291.
80. M.-H. Schwartz, *Classes caractéristiques définies par une stratification d'une variété analytique complexe*, C.R. Acad. Sci. Paris 260 (1965), 3262-3264, 3535–3537.

81. J. Seade, *The index of a vector field on a complex surface with singularities*, in "The Lefschetz Centennial Conf.", ed. A. Verjovsky, Contemp. Math. **58**, Part III, Amer. Math. Soc. (1987), 225–232.
82. J. Seade, *Remarks on contact structures and vector fields on isolated complete intersection singularities*, The Proceedings of the IX Workshop on Real and Complex Singularities, edit. M. Saia *et al*, Contemporary Mathematics, AMS. 459, 2008, 195–205.
83. J. Seade, T. Suwa. *A residue formula for the index of a holomorphic flow*. Mathematische Annalen 304 (1996), 621–634.
84. J. Seade, T. Suwa. *An adjunction formula for local complete intersections*. Int. J. Maths. 9, 1998, 759–768.
85. M. Soares, *The Poincaré problem for hypersurfaces invariant by one–dimensional foliations,* Invent. Math. 128 (1997), 495–500.
86. T. Suwa, *Indices of holomorphic vector fields relative to invariant curves on surfaces*, Proc. Amer. Math. Soc. **123** (1995) , 2989–2997.
87. T. Suwa, *Indices of Vector Fields and Residues of Singular Holomorphic Foliations*, Actualités Mathématiques, Hermann Paris, 1998.
88. T. Suwa, GSV-*Indices as Residues*, Journal of Singularities **9** (2014), 206–218.
89. J.-L. Verdier. *Spécialisation des classes de Chern*. Astérisque 82-83 (1981), 149–159.
90. M. S. Vishik, *Singularities of analytic foliations and characteristic classes*, Funct. Anal. Appl. 7 (1973), 1–15.
91. M. Zach. *A generalization of Milnor's formula*. Math. Annalen 382, 901–942.
92. M. Zach. *Some Lê-Greuel type formulae on stratified spaces*. Preprint 2024, arXiv:2411.02682.

Chapter 11
Monodromy in Integral Geometry and PDE

V. A. Vassiliev

Contents

Abstract Many important functions have integral representations, and their analytic properties (first of all the ramification of analytic continuations) are determined by the monodromy of integration cycles. We demonstrate this approach on two classical problems: the Archimedes'–Newton problem on volumes of plane sections, and the sharpness problem of hyperbolic PDE's.

V. A. Vassiliev (✉)
Weizmann Institute of Science, Rehovot, Israel

© The Author(s), under exclusive license to Springer Nature Switzerland AG 2026

J. L. Cisneros-Molina et al. (eds.), *Handbook of Geometry and Topology of Singularities VIII*, https://doi.org/10.1007/978-3-031-99571-2_11

11.1 Introduction

Many important special functions have integral representations, i.e., they are defined by the values of integrals, whose integration contours and/or forms depend on parameters that serve as arguments of the function. In formal terms, we usually have an analytic map $\pi : X \to Y$ of real analytic manifolds and a differential form or measure on X, which can be integrated along the fibers of this map. The integral function is defined on the manifold Y: its value at a point $y \in Y$ is equal to the integral of this form along the fiber $\pi^{-1}(y)$.

An integral function of this kind is regular analytic at a majority of values of the argument. It has singularities at the critical values of the map π: they are caused by the degeneration of integration contours. The study of these functions implies the questions about their qualitative properties: regularity, algebraicity, asymptotic behavior near these singular points, etc. These properties of integral functions are closely related to the ramification of their analytic continuation into the complex domain, which is conveniently studied by methods of monodromy theory of analytic maps. In particular, the local features of these functions are effectively described in the terms of the local monodromy groups of the corresponding singularities acting on the integration cycles.

We will discuss two illustrative sample problems of this kind and show how the monodromy considerations help in their solution. These are the Archimedes–Newton–Arnold problem on volumes cut off by planes from spatial bodies, and the problem of sharp fronts in the theory of hyperbolic partial differential equations.

In the simplest cases, the monodromy action is described by the usual Picard–Lefschetz theory, therefore our main reference within this series of handbooks is [12]. In addition, this study requires some extensions of these methods, first of all the generalized Pham formulas (see [27, 28, 33]) and the monodromy theory of hyperplane sections of singular varieties, see Sects. 11.3.2 and 11.4 below.

Finally, let me mention the important area of integral geometry and theory of special functions studying integrals of multivalued forms: the Gelfand–Aomoto generalized hypergeometric functions [2, 3, 18, 19] as a perfect representative. Its problems are also effectively studied by a special extension of Picard–Lefschetz theory, see in particular [33, 37]; but the exposition of this theory is beyond the scope of the present article.

11.2 Algebraicity of Volume Functions

11.2.1 Problems

Definition 11.2.1 Let K be a compact subset in $\mathbb{R}^n$ bounded by a smooth hypersurface. The corresponding *volume function* V_K is a two-valued function on the space of all affine hyperplanes in $\mathbb{R}^n$: its two values on such a hyperplane X are equal to the volumes of the intersections of K with two half-spaces of $\mathbb{R}^n$ separated by that hyperplane. The body K is called *algebraically integrable* if this function is algebraic.

The Archimedes' calculation of spherical sections implies that the balls in $\mathbb{R}^3$ are algebraically integrable. It is easy to check that all ellipsoids in odd-dimensional spaces also satisfy this condition.

On contrary, Newton ([25], Lemma XXVIII) proved that there are no algebraically integrable domains with convex infinitely smooth boundaries in $\mathbb{R}^2$:

> "Nulla extat figura ovalis cujus area, rectis pro lubitu abscissam possit per aequationes numero terminorum ac dimensionum finitas generaliter inveniri".

Based on these results, V.I. Arnold conjectured (see [7], Problems 1987-14, 1988-13, and 1990-27) that

(1) there are no algebraically integrable bodies (convex or not) with smooth boundaries in even-dimensional spaces,
(2) all irreducible algebraically integrable bodies with smooth boundaries in odd-dimensional spaces are exhausted by ellipsoids.

11.2.2 First Remarks

The condition of the smoothness of boundaries is essential for these problems: for example, the triangle in $\mathbb{R}^2$ is obviously integrable. (This is an argument of Leibnitz, who attacked the Newton's theorem because Newton did not specified exact conditions for the "figura ovalis" mentioned in his statement. However, analyzing Newton's proof, we can find that he actually used infinite smoothness, convexity and boundedness, see e.g. [4]). Moreover, for any finite r there are examples of convex bounded algebraically integrable domains with boundaries of class C^r in even-dimensional spaces.

Further, the condition of algebraicity of the boundary can be added without loss of generality to its infinite smoothness in the statements of Arnold's conjectures and Newton's theorem. Indeed, it follows from the projective duality and the Tarski-Seidenberg lemma, that if a body with smooth boundary is algebraically integrable then its boundary is also semialgebraic. Thus, we can and will assume that the boundary of K is a smooth connected bounded component of a hypersurface distinguished by a polynomial equation $F(x) = 0$ in $\mathbb{R}^n$.

11.2.3 What Is Known Towards Arnold's Conjectures

Theorem 11.2.2 (See [34]) *The first conjecture is true: a compact body (convex or not) with infinitely smooth boundary in an even-dimensional space cannot be algebraically integrable. Moreover, the analytic continuation of the volume function defined by any such body is infinitely-valued.*

Concerning the second conjecture, there are many obstructions to the integrability of odd-dimensional bodies, which are formulated in terms of the local geometry of the algebraic closure $\{x : F(x) = 0\}$ of the boundary set ∂K in $\mathbb{C}^n$ and can be effectively investigated by methods of the local monodromy theory, see Sects. 11.3, 11.4 below. In particular, if this algebraic closure hypersurface has a non-singular parabolic point, the singularity of whose *generating function* (see Definition 11.3.1 on page 502) is isolated, then the analytic continuation of the volume function is also infinitely valued, and hence this function is not algebraic.

Another class of local obstructions to algebraic integrability follows from some singular strata of these hypersurfaces. In particular, if the complex hypersurface $\{F = 0\}$ has a *standard cuspidal edge* (see Definition 11.3.8 on page 506) and n is odd then the body bounded by any smooth component of this hypersurface in $\mathbb{R}^n$ cannot be algebraically integrable; see Sect. 11.3.2 below.

The obstructions mentioned in the previous two paragraphs appear at particular points of an overwhelming majority of algebraic hypersurfaces of any fixed degree $d \geq 3$, and hence prohibit their algebraic integrability. Nevertheless, there is the following example.

Theorem 11.2.3 (See [35]) *Let n be odd, m be even, and $\varepsilon \in (0, 1)$; let $\mathbb{R}^{n+m} \equiv \mathbb{R}^n \oplus \mathbb{R}^m$ be the Euclidean space with orthogonal coordinates $x_1, \dots, x_n$ in $\mathbb{R}^n$ and $y_1, \dots, y_m$ in $\mathbb{R}^m$. The volume function defined by the body bounded by the hypersurface with the equation*

$$\left(\sqrt{x_1^2 + \cdots + x_n^2} - 1\right)^2 + y_1^2 + \cdots + y_m^2 = \varepsilon^2 . \tag{11.1}$$

(i.e. by the ε-neighborhood of the unit $(n-1)$-dimensional sphere $S^{n-1} \subset \mathbb{R}^n \subset \mathbb{R}^{n+m}$) is then a finite-valued analytic function on the space of affine hyperplanes in $\mathbb{R}^{n+m}$.

See Proposition 11.3.9 on page 506 on the monodromy theoretical explanation of this theorem.

Remark 11.2.4 Conjecturally, this volume function is even algebraic: to prove this, it remains to check that this function has only finite growth at its singular points.

11.2.4 Outline of the Proof of Theorem 11.2.2

11.2.4.1 The Convex Case

First, consider the case where the compact body K is strictly convex. Any linear function $L : \mathbb{R}^n \to \mathbb{R}$ then has only two critical points on ∂K. By Sard's lemma we can choose this function so that both these points are Morse. Let $m < M$ be their critical values. Define the function $v : [m, M] \to \mathbb{R}$, whose value $v(t)$ equals the volume of the set $K \cap L^{-1}((-\infty, t])$. In particular, $v(m) = 0$, and $v(M)$ is equal to the volume of the entire body K. Consider the analytic continuation of this function into the complex domain, and especially the action on it of two *simple loops* in $\mathbb{C}^1$ (see [12], p. 432) associated with the segments connecting a chosen value $t_0 \in (m, M)$ with points m and M; see Fig. 11.1

Lemma 11.2.5 *Let the number n be even. For any* $t_0 \in (m, M)$*, the analytic continuation of the volume function over the simple loop associated with the oriented segment* $[t_0, M]$ *(respectively,* $[t_0, m]$*) moves the value* $v(t_0)$ *to* $2v(M) - v(t_0)$ *(respectively, to* $-v(t_0)$*). In particular, the composition of these two loops adds to the function v the constant function equal identically to* $2v(M)$*.*

Iterating this composition (maybe a negative number of times), we can increase the value $v(t_0)$ by any integer multiple of the number $2v(M)$. Therefore, this analytic continuation takes infinitely many values at the same point t_0 and cannot be an algebraic function.

Proof of Lemma 11.2.5 is a reflection of the similar fact about the monodromy of integration cycles. Indeed, denote by A the hypersurface in $\mathbb{C}^n$ distinguished by our equation $F(x) = 0$, so that $\partial K \subset A \cap \mathbb{R}^n$. For any complex affine hyperplane $X \subset \mathbb{C}^n$ consider the group

$$H_n(\mathbb{C}^n, A \cup X). \tag{11.2}$$

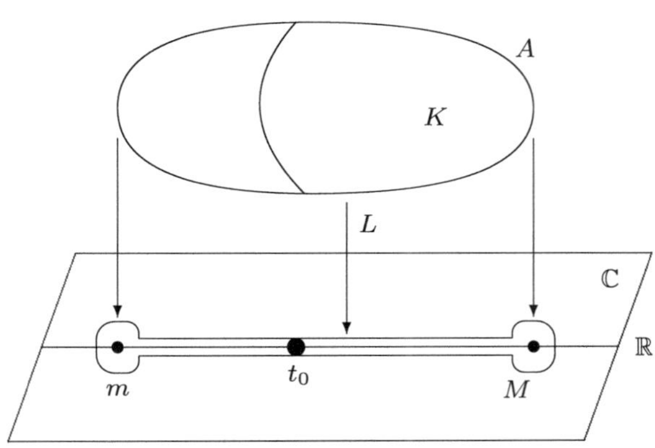

Fig. 11.1 Ramification of the volume function

It easily follows from the Stokes' formula that the integrals of the holomorphic volume form

$$dx_1 \wedge \cdots \wedge dx_n \tag{11.3}$$

along the elements of this group are well-defined, in particular they define a linear function

$$H_n(\mathbb{C}^n, A \cup X) \to \mathbb{C}. \tag{11.4}$$

Let $P^{\mathbb{C}}$ be the space of all affine hyperplanes in $\mathbb{C}^n$, and $P^{\mathbb{C}}_{reg} \subset P^{\mathbb{C}}$ be the subset in it consisting of planes "in general position" with respect to the affine variety A. To formally define this set, let us fix an arbitrary regular complex semialgebraic stratification of the subvariety in $\mathbb{CP}^n$, consisting of A and the "infinitely distant" hyperplane $\mathbb{CP}^{n-1}_\infty \equiv \mathbb{CP}^n \setminus \mathbb{C}^n$. $P^{\mathbb{C}}_{reg}$ is the space of hyperplanes, whose projective closures are transversal (see e.g. [20]) to this stratification. (More canonically, one must consider all such regular stratifications of $A \cup \mathbb{CP}^{n-1}_\infty$ and take the union of corresponding spaces of transversal hyperplanes. For practical computations, however, it is usually sufficient to use only one reasonable stratification.) The set $P^{\mathbb{C}} \setminus P^{\mathbb{C}}_{reg}$ of hyperplanes not in general position with $A \cup \mathbb{CP}^{n-1}_\infty$ will be denoted by $\Sigma(A)$

By *first Thom-Mather isotopy theorem* (see [23] or [29]), we have a locally trivial fiber bundle over $P^{\mathbb{C}}_{reg}$ whose fiber over the point X is the pair $(\mathbb{C}^n, A \cup X)$. In particular, for any choice of the distinguished hyperplane $\tilde{X} \in P^{\mathbb{C}}_{reg}$, the monodromy representation of the group $\pi_1(P^{\mathbb{C}}_{reg}, \tilde{X})$ in the group $H_n(\mathbb{C}^n, A \cup \tilde{X})$ is well defined.

The volume function V_K considered in the Archimedes-Newton problem is the restriction of a function defined in the complex domain in the terms of this fiber bundle. Namely, let $X \in P^{\mathbb{C}}_{reg}$ be an arbitrary generic affine hyperplane, and $\varkappa$ be an arbitrary element of the homology group (11.2). Let U be a simply-connected neighborhood of the point X in $P^{\mathbb{C}}_{reg}$. The function

$$v_\varkappa : U \to \mathbb{C} \tag{11.5}$$

is then defined as follows: its value at any point $X' \in U$ is equal to the integral of the differential form (11.3) along the element of the group $H_n(\mathbb{C}^n, A \cap X')$ obtained from $\varkappa$ by the covering homotopy over an arbitrary path in U connecting X and X'. This function $v_\varkappa$ is analytic in U and can be analytically continued to entire $P^{\mathbb{C}}_{reg}$. Its ramification is defined by the monodromy action of the group $\pi_1(P^{\mathbb{C}}_{reg}, X)$ on the element $\varkappa$. If X is the complexification of a real affine hyperplane in $\mathbb{R}^n$ and $\varkappa$ is the homology class of one of two parts of $K \subset \mathbb{R}^n$ lying on one side of this hyperplane, then in all the *real* points of a neighborhood of the point X in $P^{\mathbb{C}}_{reg}$ (i.e., in its points corresponding to real hyperplanes), the function $v_\varkappa$ coincides with one of two branches of the real volume function.

Now let us consider our linear function L as the restriction to $\mathbb{R}^n$ of a linear function $\mathbb{C}^n \to \mathbb{C}$, which will be denoted by the same letter L. Consider the one-dimensional complex subset in $P^{\mathbb{C}}$ consisting of hyperplanes $X(t) \equiv L^{-1}(t), t \in \mathbb{C}$. In a neighborhood of the real line $\mathbb{R}^1 \subset \mathbb{C}^1$, the set of points t such that $X(t) \notin P^{\mathbb{C}}_{reg}$ consists of finitely many real points, including m and M. The simple loops considered in Lemma 11.2.5[1] define elements of the group $\pi_1(P^{\mathbb{C}}_{reg}, X(t_0))$.

Lemma 11.2.6 *Let n be even, t_0 be a point of the interval (m, M) such that $X(t_0) \in P^{\mathbb{C}}_{reg}$, and let $\varkappa_-$ and $\varkappa_+$ be the elements of the group $H_n(\mathbb{C}^n, A \cup X(t_0))$ defined by the sets $K \cap L^{-1}((-\infty, t_0])$ and $K \cap L^{-1}([t_0, +\infty))$ with the standard orientation of $\mathbb{R}^n$. Then*

(1) the monodromy over the simple loop associated with the oriented segment $[t_0, m]$ moves the cycles $\varkappa_-$ and $\varkappa_+$ respectively to $-\varkappa_-$ and $\varkappa_+ + 2\varkappa_-$;
(2) the monodromy over the simple loop associated with the segment $[t_0, M]$ moves the same cycles to $\varkappa_- + 2\varkappa_+$ and $-\varkappa_+$, respectively.

Proof of Lemma 11.2.6 First of all, we will not care about the points t of the real interval (m, M) such that $t \notin P^{\mathbb{C}}_{reg}$: the "non-genericity" of the corresponding planes $X(t)$ occurs far away from our ramified cycles.

Let B be a small ball in $\mathbb{C}^n$ centered at the minimum point of $L|_{\partial K}$, and δ be a very small positive number. The relative homology group $H_n(B, A \cup L^{-1}(m + \delta))$ is then isomorphic to $\mathbb{Z}$ and is realized by the piece

$$K \cap L^{-1}([m, m + \delta]) \tag{11.6}$$

of the body K. Indeed, by the exact sequence of the pair $(B, (A \cup L^{-1}(m + \delta)) \cap B)$ this group is isomorphic to $H_{n-1}((A \cup L^{-1}(m + \delta)) \cap B)$. The latter group is by Mayer-Vietoris exact sequence isomorphic to the homology group $H_{n-2}(B \cap A \cap L^{-1}(m + \delta))$, whose space $B \cap A \cap L^{-1}(m + \delta)$ is the *Milnor fiber* (see e.g. [12]) of a Morse critical point of the restriction of the function L to the $(n-1)$-dimensional manifold A. This group is equal to $\mathbb{Z}$ and is generated by the vanishing sphere $\partial K \cap L^{-1}(m+\delta)$. It is easy to see that the relative cycle (11.6) is the preimage of this vanishing cycle under the boundary operations of the exact sequence of the pair and Mayer-Vietoris as above.

Statement (1) of Lemma 11.2.6 concerning the case $t_0 = m + \delta$ and the action on the cycle $\varkappa_-$ now follows immediately from the Picard–Lefschetz formula. Indeed, the rotation around the critical value of a Morse critical point moves the even-dimensional cycle vanishing at this point to minus itself; on the other hand the monodromy operators commute with both the boundary and the Mayer-Vietoris maps producing the $(n-2)$-dimensional vanishing cycle from the n-dimensional cycle (11.6). For all other choices of $t_0 \in (m, M)$ the same statement of our lemma follows by continuity over the segment $[m + \delta, t_0]$.

---

[1] And, strictly speaking, avoiding all points $t \in (m, M)$ such that $X(t) \notin P^{C}_{reg}$.

The action of the same simple loop on the cycle $\varkappa_+$ follows from the fact that the cycle $\varkappa_- + \varkappa_+$ does not depend on t_0 and hence is preserved by our monodromy operators.

The proof of statement (2) of Lemma 11.2.6 on the action of the simple loop associated with the segment $[t_0, M]$ is symmetric to the proof of statement (1).

Lemma 11.2.5 follows immediately from Lemma 11.2.6. In particular, we obtain a proof of Theorem 11.2.2 in the case of convex bodies.

11.2.4.2 General Case

Now let K be an arbitrary bounded domain in $\mathbb{R}^n$ with infinitely smooth semialgebraic boundary. If the body K is algebraically integrable, then we can associate a finite-dimensional Weyl group with it. The strong properties of Weyl groups will then lead us to a contradiction.

Indeed, let us again choose a linear function $L : (\mathbb{C}^n, \mathbb{R}^n) \to (\mathbb{C}, \mathbb{R})$, whose restriction to ∂K is *strictly* Morse. Let $m = m_1 < m_2 < \cdots < m_N = M$ be all critical values of this restriction. Let also $u_0, \ldots, u_N$ be all corresponding critical points, i.e. the points of tangency of ∂K with hyperplanes $L^{-1}(m_j)$.

Then, for each $j = 1, \ldots, N$ and a sufficiently small $\delta > 0$, an analytic complex-valued function φ_j is defined on the interval $(m_j, m_l + \delta)$. Namely, consider a small ball $B_j \subset \mathbb{C}^n$ centered at the point u_j for any j. If δ is small enough with respect to the size of this ball, then for any $t \in (m_j, m_j + \delta)$ the homology group

$$H_n(B_j, A \cup X(t)) \tag{11.7}$$

is again isomorphic to $\mathbb{Z}$ by the Morse lemma (although the relative cycle $\Gamma_j(t)$ that generates it cannot be realized inside $\mathbb{R}^n$ if the Morse index of the function $L|_{\partial K}$ is not equal to 0). Let us somehow fix the orientations of all these relative n-cycles $\Gamma_j(t)$ depending continuously on t and define the function $\varphi_j : (m_j, m_j + \delta) \to \mathbb{C}$ by the integrals of the volume form (11.3) along the corresponding cycles $\Gamma_j(t)$.

Lemma 11.2.7 *If the body K is algebraically integrable then analytic continuations of all these functions to the entire line $\mathbb{C}^1$ are algebraic (and hence finite-valued analytic) functions.*

Proof For any point $t \in (m_j, m_j + \delta)$ we have three important elements of the group $H_n(\mathbb{C}^n, A \cup X(t))$: the first is defined by a generator of the group (11.7), the second is the class of the domain $K \cap L^{-1}((-\infty, t])$ with the orientation induced from a fixed orientation of $\mathbb{R}^n$, and the third is obtained from a similar element of the group $K \cap L^{-1}((-\infty, t'])$, where $t' \in (m_j - \delta, m_j)$, by the monodromy over the small arc connecting the points t' and t in the upper half of $\mathbb{C}^1$.

Proposition 11.2.8 (See Lemma 3.3 on p. 125 of [33]) *These three homology classes are dependent: the first of them is equal (up to a sign defined by the orientation) to the difference of the second and the third ones.* □

Therefore, the function defined by the integrals of the form (11.3) along the cycles of the first type is equal to the difference of two functions, defined by integrals over the other two. If the body K is algebraically integrable, then the last two functions are algebraic, and so their difference is also algebraic. □

We can assume that $L^{-1}(0) \in P^{\mathbb{C}}_{reg}$. Let $\mathfrak{O}$ denote the space of germs of holomorphic functions at $0 \in \mathbb{C}^1$. For any $j = 1, \dots, N$ consider a path $\frown_j \subset \mathbb{C}^1$ connecting 0 with a point t_j of the interval $(m_j, m_j + \delta)$ and passing through the upper half-plane of $\mathbb{C}^1$ along the real axis. Let γ_j be the germ of the holomorphic function at 0 obtained from φ_j by the analytic continuation along the corresponding path $\frown_j$. Obviously, it does not depend on the choice of the point t_j. Consider the subgroup in $\mathfrak{O}$ consisting of integer linear combinations of these germs γ_j. It is the image of a homomorphism of the integer lattice $\mathbb{Z}^N$ of all formal linear combinations of these germs to a $\mathbb{C}$-linear space, so it is also an integer lattice $\mathbb{Z}^T$, $T \leq N$.

Now let us construct a reflection group acting on this lattice $\mathbb{Z}^T$. Consider the continuations of the functions γ_l along N simple loops in $\mathbb{C}^1$ associated with our N paths $\frown_j$. It follows from the Picard–Lefschetz formula that these continuations map the group $\mathbb{Z}^T$ to itself; this action can be expressed in the terms of a bilinear form $\langle \cdot, \cdot \rangle$ on it, which is defined as follows. For any j, take the basic element in the group $H_{n-2}(A \cap X(m_j + \delta) \cap B_j)$ obtained from the cycle Γ_j as in Sect. 11.2.4.1 (i.e. by the composition of the boundary and Mayer-Vietoris maps). Consider the image of this element in the group $H_{n-2}(A \cap X(m_j + \delta))$ under the identical embedding of cycles. Transporting it over the corresponding path $\frown_j$ in $\mathbb{C}^1$ makes it an element of $H_{n-2}(A \cap X(0))$. The pairing $\langle \gamma_l, \gamma_k \rangle$ is then defined as the intersection index of two cycles in the $(n-2)$-dimensional complex manifold $A \cap X(0)$, that are obtained in this way from the relative cycles Γ_l and Γ_k.

Proposition 11.2.9 *The analytic continuation of the function germ γ_l along the simple loop in $\mathbb{C}^1$ associated with the path $\frown_k$ is equal to*

$$\gamma_l + (-1)^{n(n-1)/2} \langle \gamma_l, \gamma_k \rangle \gamma_k. \tag{11.8}$$

In particular, the analytic continuations along all N simple loops associated with paths $\frown_j$ preserve the lattice $\mathbb{Z}^T$.

Proof This statement follows from the Picard-Lefschetz formula concerning the action of our loops on the integration cycles.

Remark 11.2.10 This action on the lattice $\mathbb{Z}^T$ does not depend on the representation of the elements of this lattice by formal linear combinations of germs γ_j. This implies that the kernel of the map $\mathbb{Z}^N \to \mathbb{Z}^T$ belongs to the kernel of the bilinear form $\langle \cdot, \cdot \rangle$.

If n is even, then the self-intersection index of any vanishing cycle (and hence also any number $\langle \gamma_l, \gamma_l \rangle$) is equal to $2(-1)^{(n-2)(n-3)/2}$. So by (11.8) the monodromy along the simple loop corresponding to the l-th critical value moves the basic element γ_l to minus itself. Therefore, the action of the simple loop associated with the path $\frown_l$ acts on the lattice $\mathbb{Z}^T$ as a reflection in the hyperplane orthogonal to the element γ_l in the sense of our bilinear form $\langle \cdot, \cdot \rangle$. If K is algebraically integrable then the group generated by all these reflections should be finite, in particular to be a Weyl group.

The well-known classification of Weyl groups (see e.g. [10]) implies that they have no non-trivial elements which are invariant under the action of all reflections. But in our case we have such an element: it is the function identically equal to the volume of the body K. Indeed, this function belongs to the lattice $\mathbb{Z}^T$, since the homology class of the body K in any group $H_n(\mathbb{C}^n, A \cup X)$, $X \in P^{\mathbb{C}}_{reg}$, is equal to the sum of all (appropriately oriented) cycles Γ_j: this equality follows by induction from Proposition 11.2.8. Also, this constant function is obviously non-trivial and invariant under analytic continuations. This contradiction proves Theorem 11.2.2.

11.3 Local Monodromy and Local Obstructions to the Algebraicity of the Volume Function

Parabolic points of the complex hypersurface $A \supset \partial K$ (i.e. its non-singular points at which the second fundamental form is degenerate) are a generous source of monodromy operators acting on the group (11.2), and thus of the ramification of the analytic continuation of the volume function.

This ramification is described by the classical local monodromy theory of function singularities. Indeed, the hypersurface A is locally defined by a smooth function with a critical point (*generating function*, see Definition 11.3.1 on page 502), the space of cutting hyperplanes X can be identified with a deformation of this singularity, the hyperplanes "in general position" with respect to A are exactly the non-discriminant values of this deformation, the integration cycles of the volume function correspond to the homology classes of Milnor fibers (generally speaking, modulo their boundaries), and the local monodromy is the action of arbitrary loops in the complement of this discriminant on these homology classes.

For example, consider the case $n = 2$, where the parabolic points are called *inflection points* of the curve A. At the simplest inflection points (of type A_2) this curve can be defined locally in affine coordinates x, y by the equation $y = f(x)$, where $f(x) = x^3 + O(x^4)$; see Fig. 11.2 for the corresponding real picture.

Consider the family of affine hyperplanes (lines) $X(t)$ defined in these coordinates by the equation $y = \varepsilon x + t$, where ε is some fixed small positive number. This family has two exceptional values of the parameter t, corresponding to the tangencies of $X(t)$ with A. In the considered neighborhood, the group $H_2(\mathbb{C}^2, A \cup X(t_0))$ with non-exceptional t_0 is represented by two relative cycles Δ_1 and Δ_2: they are shown by the dashed domains in Fig. 11.2. The local monodromy group is generated by two simple loops in the space $\mathbb{C}^1$ of the parameter t, corresponding to the segments connecting the non-exceptional value t_0 with two exceptional values. If

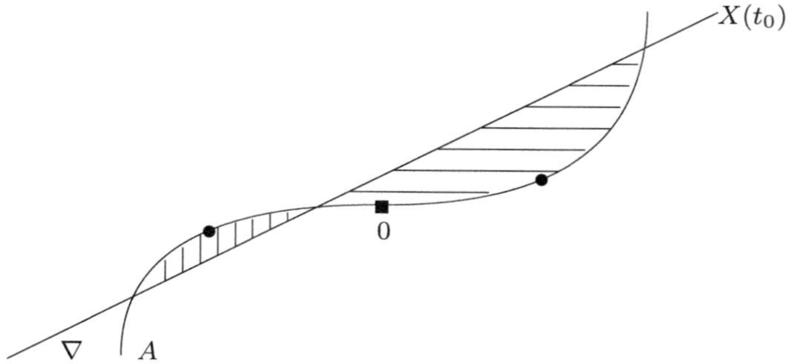

Fig. 11.2 Monodromy of hyperplane sections close to parabolic points

the integration cycle $\varkappa$ of the volume function is represented by one of these cycles, say Δ_1, then its orbit under the local monodromy action consists of six elements $\pm\,\Delta_1,\ \pm\,\Delta_2$ and $\pm\,(\Delta_1+\Delta_2)$. Correspondingly, the ramification of the volume function along the loops in a small neighborhood of the line $\{y=0\}$ is only six-valued. (An important fact is that the volume function defined by each vanishing cycle Δ_i is not equal identically to zero). These six values are not sufficient to prove the non-algebraicity of the volume function (which is proved in the case of even n by methods of global monodromy, see the previous section), but the analog of this consideration in the case of odd $n>2$ gives us the infinite ramification and hence the non-algebraicity: see Theorem 11.3.5 below.

It can also happen that the integration cycle $\varkappa$ is represented in our neighborhood by the domain marked by ∇ in Fig. 11.2. In this case the analytic continuation of the volume function along the local loops is only tri-valued.

Now we move on to higher dimensions and more complicated parabolic points.

11.3.1 Generating Functions, Generating Families, and Their Monodromy Groups

Let a be a non-singular point of the hypersurface $A \in \mathbb{C}^n$. Let us choose local affine coordinates $x_1,\dots,x_n$ at a in such a way that the hyperplane $\{x_n=0\}$ is tangent to A. Then A can be locally defined by an equation of the form

$$x_n = f(x_1,\dots,x_{n-1}), \tag{11.9}$$

where f is a holomorphic function with a singularity at the origin: $f(0)=0$, $df(0)=0$. Furthermore, any hyperplane in $\mathbb{C}^n$ which is close to the tangent plane of A at a can be defined by a linear equation of the form

$$x_n = \lambda_1 x_1 + \cdots + \lambda_{n-1}x_{n-1} + \lambda_n. \tag{11.10}$$

The parameters $\lambda_1, \ldots, \lambda_n$ of these equations form a local affine coordinate system in the space $P^{\mathbb{C}}$ of all affine hyperplanes of $\mathbb{C}^n$.

Definition 11.3.1 The function germ f defining the germ of the hypersurface A by the Eq. (11.9) is called the *generating function* of A at the point a. The point a is called *parabolic* if the singularity of the corresponding generating function is not Morse.

The family of functions in the variables $x_1, \ldots, x_{n-1}$ depending on parameters $\lambda_1, \ldots, \lambda_n$ and defined by the formula

$$F(x_1, \ldots, x_{n-1}; \lambda_1, \ldots, \lambda_n) \equiv f(x_1, \ldots, x_{n-1}) - \lambda_1 x_1 - \cdots - \lambda_{n-1} x_{n-1} - \lambda_n \tag{11.11}$$

is called the *generating family* of A. The functions $F(\cdot, \lambda) : \mathbb{C}^{n-1} \to \mathbb{C}, \lambda \in \mathbb{C}^n$, of this family are also denoted by f_λ (in particular $f_0 \equiv f$). The hyperplane defined by the Eq. (11.10) is denoted by X_λ, where $\lambda = (\lambda_1, \ldots, \lambda_n)$.

So, the generating family (11.11) is a deformation (see [5]) of the generating function f. The *discriminant* $\Sigma(F)$ of this deformation (i.e. the set of parameters $\lambda = (\lambda_1, \ldots, \lambda_n)$ such that the corresponding function $f_\lambda \equiv F(\cdot, \lambda)$ has a critical point with zero critical value) is exactly the set of hyperplanes (11.10) tangent to A close to the point a.

Now let $B \subset \mathbb{C}^n$ be a small ball centered at the point a, and U_δ be a very small neighborhood of the plane $X_0 \equiv \{x_n = 0\}$ in $P^{\mathbb{C}}$. Let $\bar{\lambda}$ be a point of the set $U_\delta \cap P^{\mathbb{C}}_{reg}$, so that the hyperplane $X_{\bar{\lambda}}$ is transversal to A, and $\varkappa$ be an arbitrary element of the group $H_n(\mathbb{C}^n, A \cup X_{\bar{\lambda}})$.

Definition 11.3.2 The *localization* $\varkappa_a$ of the class $\varkappa$ at a is its image under the standard map

$$H_n(\mathbb{C}^n, A \cup X_{\bar{\lambda}}) \to H_n(B, B \cap (A \cup X_{\bar{\lambda}} \cup \partial B)), \tag{11.12}$$

that is, the reduction modulo the complement of the ball B. The *standard reduction* of the latter group is the homomorphism

$$\text{Red} : H_n(B, B \cap (A \cup X_{\bar{\lambda}} \cup \partial B)) \to \tilde{H}_{n-2}(B \cap A \cap X_{\bar{\lambda}}, \partial B) \tag{11.13}$$

defined as the composition of the maps

$$\begin{aligned} & H_n(B, B \cap (A \cup X_{\bar{\lambda}} \cup \partial B)) \to H_{n-1}(B \cap (A \cup X_{\bar{\lambda}}), \partial B) \to \\ & \to H_{n-1}(B \cap (A \cup X_{\bar{\lambda}}, B \cap (X_{\bar{\lambda}} \cup \partial B)) \equiv \\ & \equiv H_{n-1}(B \cap A, B \cap (A \cup X_{\bar{\lambda}} \cup \partial B)) \to \tilde{H}_{n-2}(B \cap A \cap X_{\bar{\lambda}}, \partial B) \end{aligned} \tag{11.14}$$

from appropriate exact sequences of triples.

Remark 11.3.3 We denote by $\tilde{H}_*$ the homology group reduced modulo the fundamental cycle in the case of relative chain complexes (as in the previous formula) and modulo a point in the case of absolute complexes (as in the next formula).

Proposition 11.3.4 *All maps* (11.14) *are isomorphisms, and so their composition* (11.13) *is an isomorphism. Moreover, the similar map of homology groups that are not reduced modulo* ∂B*,*

$$red : H_n(B, B \cap (A \cup X_{\bar{\lambda}})) \to \tilde{H}_{n-2}(B \cap A \cap X_{\bar{\lambda}}) \tag{11.15}$$

is also an isomorphism.

Proof This follows easily from the exact sequences of the corresponding triples and pairs, see e.g. §II.2.5 in [33].

The space $B \cap A \cap X_{\bar{\lambda}}$ involved in the last term of (11.13) is diffeomorphic to the Milnor fiber (see [24]) $f_{\bar{\lambda}}^{-1}(0) \cap B$ of the generating family (11.11): this diffeomorphism is defined by the coordinates $x_1, \dots, x_{n-1}$ on $X_{\bar{\lambda}} \sim \mathbb{C}^{n-1}$.

The group $\pi_1(U_\delta \cap P^{\mathbb{C}}_{reg})$ acts by monodromy operators on the group $H_n(\mathbb{C}^n, A \cup X_{\bar{\lambda}})$ containing the integration cycle $\varkappa$. To prove the non-algebraicity of the function $\varkappa$, it is sufficient to show that the linear function (11.4) takes infinitely many values on the orbit of this action.

Theorem 11.3.5 (See [33], §III.4) *Suppose that n is odd, the singularity of the generating function of the hypersurface A at some its point a is isolated,* $\bar{\lambda} \in U_\delta \cap P^{\mathbb{C}}_{reg}$ *is a non-discriminant parameter value of the corresponding generating family, and* $\varkappa \in H_n(\mathbb{C}^n, A \cup X_{\bar{\lambda}})$ *is a cycle such that the homology class* $Red(\varkappa_a) \in \tilde{H}_{n-2}(B \cap A \cap X_{\bar{\lambda}}, \partial B)$ *is non-trivial. Then the analytic continuation of the integral function* $v_\varkappa$ *(see* (11.5)*) along the paths in* $U_\delta \cap P^{\mathbb{C}}_{reg}$ *is infinitely-valued, in particular it is not algebraic.*

Proof Let μ be the Milnor number of the singularity of the generating function at a. By Sard's lemma, for almost all points $\bar{\lambda} \in U_\delta$ the corresponding functions $f_{\bar{\lambda}}$ are strictly Morse, i.e. they have exactly μ different critical points in B, and the corresponding critical values are all different. We can assume that the point $\bar{\lambda}$ from the statement of theorem satisfies this condition, since the non-triviality condition from this statement is preserved along the paths in $U_\delta \cap P^{\mathbb{C}}_{reg}$. Consider the complex line in $P^{\mathbb{C}}$ consisting of the functions $f_{\bar{\lambda}} - c$, $c \in \mathbb{C}$. It intersects the discriminant variety of the generating family (11.11) at exactly μ points corresponding to critical values of $f_{\bar{\lambda}}$. Connect the point $\bar{\lambda}$ in this line by non-intersecting paths with all these intersection points, and consider μ vanishing cycles $\Delta_i \in \tilde{H}_{n-2}(A \cap X_{\bar{\lambda}})$ defined by these paths. Since the class $\text{Red}(\varkappa_a)$ is non-trivial, the intersection index $\langle \varkappa, \Delta_i \rangle$ is not equal to zero for at least one of these vanishing cycles. By the Picard–Lefschetz formula, the k-fold iteration of the simple loop corresponding to the path, along which this cycle vanishes, increases the initial cycle $\varkappa$ by $\pm k\langle \varkappa, \Delta_i \rangle$ times the cycle equal to the preimage of the cycle Δ_i under the map (11.15).

The integral function on $P^{\mathbb{C}}_{reg}$ defined by integrals of the form (11.3) along the cycles obtained by covering homotopy from this preimage cycle is not identically equal to zero. Indeed, it is surely not zero close to the instance of vanishing, when A and X_t are almost tangent. Therefore the analytic continuation of the integral function $v_{\varkappa}$ is infinite-valued at generic points of $U_\delta \cap P^{\mathbb{C}}_{reg}$. □

Corollary 11.3.6 *Suppose that n is odd, $A \subset \mathbb{C}^n$ is the algebraic closure of the boundary ∂K of the compact semi-algebraic body $K \subset \mathbb{R}^n$ with smooth boundary (that is, the set of common complex zeros of all non-trivial polynomials vanishing on ∂K), and the generating function of the hypersurface A at some its smooth (maybe non-real) point a has a non-Morse but isolated singularity. Then the body K is not algebraically integrable.*

Proof The body K is bounded, so there exists a real hyperplane $Y \subset K$ which is tangent to ∂K at a single point y_0 and has no other intersections with the closure of K. Moreover, we can choose the tangency point so that this tangency is simple (i.e. the second fundamental form of ∂K is non-degenerate. Let $\tilde{Y}$ be a real hyperplane very close and parallel to Y, and intersecting K. Let $\varkappa_0 \in H_n(\mathbb{C}^n, A \cup \tilde{Y}^{\mathbb{C}})$ be the class of the real segment cut by $\tilde{Y}$ from K; here $\tilde{Y}^{\mathbb{C}}$ is the complexification of $\tilde{Y}$.

We can assume that K is connected, and hence the hypersurface A is irreducible. The set of non-parabolic points of A is path-connected, so there exists a path in A that connects the point y_0 with a non-parabolic point in an arbitrarily small neighborhood of the point a. By moving the hyperplane $\tilde{Y}^{\mathbb{C}}$ in $P^{\mathbb{C}}_{reg}$ very close and parallel to the tangent planes of A at this path, we continuously move the cycle $\varkappa_0$ to the preimage $\text{red}^{-1}(\Delta_i)$ of some vanishing cycle Δ_i from the previous proof. The Dynkin diagram of any isolated function singularity is connected (see [15]), and the Milnor number of the generating function at a is greater than 1, so there is another vanishing cycle Δ_j with non-zero intersection index with Δ_i. Now we are in the conditions of the previous theorem. □

11.3.2 Ramification Occurring from Tangencies with Non-maximal Strata

We are interested in the monodromy action of the group $\pi_1(P^{\mathbb{C}}_{reg})$ on the homology groups (11.2). The space $P^{\mathbb{C}}$ is simply-connected, so this fundamental group is generated by simple loops embracing the set of nongeneric hyperplanes at its smooth strata of complex codimension 1. The simplest such strata consist of hyperplanes having a simple tangency with A at its regular non-parabolic points. Rotation around these strata acts on the groups (11.2) according to the classical Picard–Lefschetz formula, as shown in the previous examples.

Another important discriminant subset in $P^{\mathbb{C}}$ consists of *asymptotic hyperplanes* of A, i.e. of hyperplanes whose closures in $\mathbb{CP}^n$ are not in general position with the

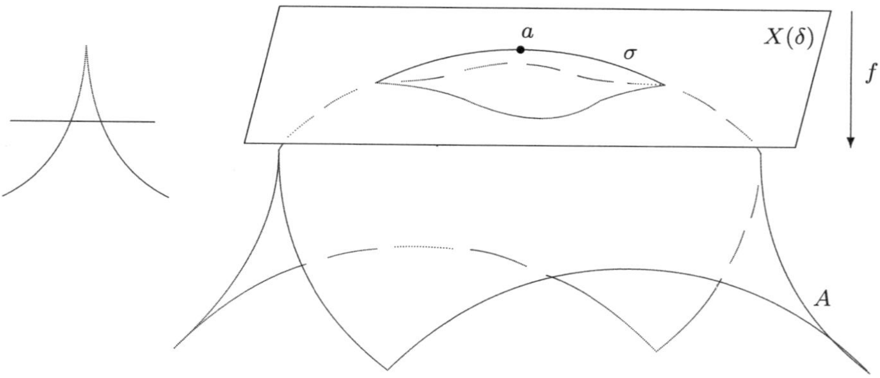

Fig. 11.3 Morse function at a singular stratum

set $\bar{A} \cup (\mathbb{CP}^n \setminus \mathbb{C}^n)$ of "infinitely distant" points of the projective closure of A. They will be discussed shortly in Sect. 11.4.3 below.

The algebraic variety A can also have singular points. In this case, the set of nongeneric hypersurfaces includes those tangent to different strata of regular complex stratifications of A. Say, the real part of the variety $A \in \mathbb{C}^3$ can have a cuspidal edge, i.e., locally look like the lower surface in the right part of Fig. 11.3. In this picture, σ denotes the set of singular points of A; the intersection of A with a plane transversal to σ at an any of its points is a curve with a semicubical singularity shown in the left part of the figure.

The discriminant subset of $P^{\mathbb{C}}$ consisting of points not in general position with a stratum σ of A is a hypersurface in $P^{\mathbb{C}}$ (unless σ is pathologically flat; in the latter case this subset does not contribute to the group $\pi_1(P^{\mathbb{C}}_{reg})$). Generic points of this hypersurface are described in the following definition.

Definition 11.3.7 (Cf. [20]) A linear function $f : \mathbb{C}^n \to \mathbb{C}$ has a *Morse critical point* at the point a of the stratum σ of the algebraic variety A if

(1) its restriction to the manifold σ has a Morse singularity at a, and
(2) for any larger stratum τ approaching σ and any sequence of points $b_i \in \tau$ converging to a and such that the planes tangent to τ at b_i (considered as the points in the associated Grassmann bundle $G^{\dim \tau}(T_*\mathbb{C}^n)$) converge to a plane in $T_a\mathbb{C}^n$, this limit plane is not in the hyperplane $\{df|_a = 0\} \subset T_a\mathbb{C}^n$.

A hyperplane in $\mathbb{C}^n$ tangent to the stratum σ at its point a is said to be a *generic discriminant hyperplane* if it

(a) is the zero set of a linear function with a Morse critical point at a, and
(b) its closure in $\mathbb{CP}^n$ is transversal to the stratified variety $A \cup \mathbb{CP}^{n-1}_\infty$ everywhere except for a.

For example, any line passing through the cusp point of the left curve of Fig. 11.3 is generic, except for the vertical line, whose equation does not satisfy condition (2).

It is easy to see that the hyperplanes tangent to arbitrary strata of A and generic in this sense form a $(n-1)$-dimensional semialgebraic hypersurface in $P^{\mathbb{C}}$, and all hyperplanes tangent to A but not satisfying these conditions fill in a subvariety of dimension $\leq n-2$.

Let $Y \subset \mathbb{C}^n$ be a generic discriminant hyperplane in the sense of Definition 11.3.7, and $f : \mathbb{C}^n \to \mathbb{C}$ be the linear function vanishing on Y. A small circle in $P^{\mathbb{C}}_{reg}$ embracing the discriminant variety $\Sigma(A) \equiv P^{\mathbb{C}} \setminus P^{\mathbb{C}}_{reg}$ at the point $\{Y\}$ can be realized by the family of hyperplanes $X(\delta e^{i\tau})$, $\tau \in [0, 2\pi]$, defined by equations $f(x) = \delta e^{i\tau}$, where δ is a sufficiently small real number.

The study of the monodromy action of such small circles (corresponding to different strata) in the group $H_n(\mathbb{C}^n, A \cup X(\delta))$ is a crucial problem in the computation of the monodromy action of the group $\pi_1(P^{\mathbb{C}}_{reg})$ (which is generated by simple loops consisting of such circles and paths connecting the distinguished point of $P^{\mathbb{C}}_{reg}$ with some points of these circles) in the group (11.2).

The zoo of possible singularities of A is boundless, and the problem of computing all possible monodromy operators of this kind is desperate (but see [27, 28] for many important special cases where exact formulas for these operators have been found). A general simplification that can be done here is to reduce this problem at a point of a positive-dimensional stratum σ to the similar problem in its transversal slice. By the analogy with the main theorem of the Stratified Morse Theory of [20], it is natural to call this reduction the *stratified Picard–Lefschetz theory*.

This reduction will be described in Sect. 11.4; let us formulate here two corollaries of these calculations.

First of all, they give additional obstructions to the algebraic integrability of spatial bodies in odd-dimensional spaces. A complex algebraic hypersurface A of degree ≥ 3 can have no parabolic points that prevent this integrability as in Sect. 11.3. For example, this happens if A is projective dual to a smooth algebraic hypersurface. However, in this case A necessarily has singular strata of codimension 1. The most common of them are as follows.

Definition 11.3.8 The hypersurface $A \subset \mathbb{C}^n$ has a *standard cuspidal edge* at its point a if it is irreducible at a, the germ of the projective dual hypersurface at the corresponding point is smooth, and the generating function of this germ has a critical point of type A_2.

Theorem 11.3.9 (See [33]) *If n is odd and the algebraic closure in $\mathbb{C}^n$ of the boundary of the compact semi-algebraic body $K \subset \mathbb{R}^n$ has standard cuspidal edges, then this body is not algebraically integrable.* □

In the next example, we take for A the hypersurface in $\mathbb{C}^{n+m}$ given by (11.1).

Proposition 11.3.10 (See [35]) *Under the conditions of Theorem 11.2.3, for any complex hyperplane $X \in P^{\mathbb{C}}_{reg}$ in $\mathbb{C}^{n+m}$,*

(a) the group $H_{n+m}(\mathbb{C}^{n+m}, A \cup X)$ is isomorphic to $\mathbb{Z}^6$;
(b) the subgroup of this group generated by relative cycles not approaching the singular set of A is isomorphic to $\mathbb{Z}^4$;

(c) *the latter group* $\mathbb{Z}^4$ *is freely generated by four vanishing cycles associated with four critical points of the restriction of the function* $x_1 + y_1$ *to* A*; and*
(d) *the monodromy representation of the group* $\pi_1(P^{\mathbb{C}}_{reg})$ *in this group* $\mathbb{Z}^4$ *acts by permutations of these basic vanishing cycles. Namely, it realizes only four permutations:* $\binom{1234}{2143}$, $\binom{1234}{4321}$, $\binom{1234}{3412}$ *and the identical one.*

(This subgroup of the permutation group has its own name: the *Klein four-group.*)

All integration cycles occurring in the definition of the volume function related with this variety A belong to this sublattice $\mathbb{Z}^4$, because A has no real singular points. Therefore the volume function defined by this surface is finitely-valued; this proves Theorem 11.2.3.

The discriminant variety of $P^{\mathbb{C}}$, consisting of hyperplanes which are not generic with respect to this variety A, consists of four irreducible components: their points are the hyperplanes which are respectively

(1) tangent to the variety A at its smooth points,
(2) tangent to the unique stratum of singular points of A,
(3) asymptotic to A at the generic points of the variety $\mathbb{CP}^{n-1}_{\infty} \cap \bar{A}$, where $\bar{A}$ is the projective closure of A;
(4) asymptotic to the singular set of A.

The rotations around the first two strata act trivially on the group $\mathbb{Z}^4$ of potential integration cycles, and the rotations around the remaining two strata act as reflections in two mutually orthogonal planes.

11.4 Stratified Picard–Lefschetz Theory: Variation Operator

In this section we show how to reduce the study of local monodromy operators caused by tangencies with positive-dimensional singular strata to the similar problem in the transversal slices of these strata, see the discussion about Fig. 11.3 in the previous section. The main notion of this study is the *variation operator* (see [6, 13, 27]) which acts from a relative localized homology group associated with a singular point to the analogous absolute homology group.

We essentially considered some special case of this operator in the proof of Theorem 11.3.5. Indeed, we calculated the monodromy action on a globally defined cycle $\varkappa$ and saw that this action is in fact local. The difference $M(\varkappa) - \varkappa$ between the result of the monodromy and the initial cycle is realized by a cycle concentrated near the singular point; moreover this difference depends only on the localization $\varkappa_a$ of the cycle $\varkappa$ at this singular point, i.e. on its class in the relative homology group in the right-hand part of (11.12). So, the (global) monodromy operator is the sum of the identity operator and the composition of three operators

$$i \circ \text{Var} \circ j,$$

where j is the localization operator (11.12), i is defined by the inclusion of absolute cycles realized in a neighborhood of the singular point into the globally defined homology group, and Var is some operator from a homology group modulo the complement of the ball B into the homology group of the intersection with that ball. This operator is well defined for arbitrary relative homology classes mod ∂B, not necessarily for the localizations of globally defined ones. Its general construction is as follows.

11.4.1 Variation Operator of Hyperplane Sections

Let A be a complex algebraic subvariety in $\mathbb{C}^n$, $a \in A$, and $L : \mathbb{C}^n \to \mathbb{C}^1$ be a linear function such that $L(a) = 0$. For any $t \in \mathbb{C}^1$, denote by $X(t)$ the hyperplane $L^{-1}(t)$. Let B_ε be a small ball in $\mathbb{C}^n$ centered at a. If $\delta > 0$ is small enough and $D_\delta \subset \mathbb{C}^1$ is a disk of radius δ centered at 0, then the diagrams of spaces

$$\begin{array}{ccc} B_\varepsilon & \supset & A \cap B_\varepsilon \\ \cup & & \cup \\ X(t) \cap B_\varepsilon & \supset & A \cap X(t) \cap B_\varepsilon \end{array} \tag{11.16}$$

form a locally trivial fiber bundle over the punctured disc $D_\delta \setminus 0$.

Suppose that the intersections

$$\begin{array}{ccc} \partial B_\varepsilon & \supset & A \cap \partial B_\varepsilon \\ \cup & & \cup \\ X(t) \cap \partial B_\varepsilon & \supset & A \cap X(t) \cap \partial B_\varepsilon \end{array} \tag{11.17}$$

of these diagrams with the boundary of B_ε form a locally trivial and thus trivializable fiber bundle over the entire disk D_δ.

Then there are a lot of (closely related with each other) *variation operators* from the relative (modulo ∂B_ε) homology groups of some elements of the first diagram to the absolute homology groups. For example, let us describe the variation operator

$$\mathrm{Var} : \tilde{H}_*(A \cap X(\delta) \cap B_\varepsilon, \partial B_\varepsilon) \to \tilde{H}_*(A \cap X(\delta) \cap B_\varepsilon). \tag{11.18}$$

The spaces $A \cap X(\delta e^{i\tau}) \cap B_\varepsilon$, $\tau \in [0, 2\pi]$, form a locally trivial fiber bundle over the segment $[0, 2\pi]$; let E be the space of this bundle. The fixed trivialization of the bundle of diagrams (11.17) defines a family of homeomorphisms $\omega_\tau : A \cap X(\delta) \cap \partial B_\varepsilon \to A \cap X(\delta e^{i\tau}) \cap \partial B_\varepsilon$, which are identical maps for τ equal to 0 and 2π. By the Borsuk's lemma, this family can be extended to a continuous family of maps

$$\bar{\omega}_\tau : A \cap X(\delta) \cap B_\varepsilon \to A \cap X(\delta e^{i\tau}) \cap B_\varepsilon$$

such that ω_0 is the identity map. For any relative cycle ∇ in $A \cap X(\delta) \cap B_\varepsilon$ modulo $A \cap X(\delta) \cap \partial B_\varepsilon$, the cycle $\bar{\omega}_{2\pi}(\nabla) - \nabla$ is an absolute cycle in $A \cap X(\delta) \cap B_\varepsilon$. Its class in $\tilde{H}_*(A \cap X(\delta) \cap B_\varepsilon)$ depends only on the relative homology class of ∇. By definition, it is the image of this class $\{\nabla\}$ under the variation operator associated with our variety A, point a and linear function L.

Example 11.4.1 Suppose that a is a regular point of a positive-dimensional stratum of a Whitney stratification of the variety A, and the restriction of the function L to this stratum is non-singular at a. Then the variation operator is equal to zero: indeed, the fiber bundle can be trivialized over the entire disk D_δ.

Example 11.4.2 Suppose that $A \subset \mathbb{C}^{m+1}$ is the graph of a smooth function $\varphi : (\mathbb{C}^m, 0) \to (\mathbb{C}, 0)$ with a critical point at 0, i.e. A is defined by the equation $z_{m+1} = \varphi(z_1, \dots z_m)$ and $L \equiv z_{m+1}$. The varieties $A \cap X(\delta e^{i\tau}) \cap B_\varepsilon$ can then be identified by the coordinates $z_1, \dots, z_m$ with the Milnor fibers $\varphi^{-1}(\delta e^{i\tau}) \cap B_\varepsilon$ of the function φ. If the critical point of φ at 0 is isolated, then our assumption about the triviality of the bundle of diagrams (11.17) is satisfied, and the corresponding variation operator is the usual variation operator of a smooth function singularity considered in [6].

In this case the groups (11.18) are nontrivial only in the dimension $m - 1$, both of them are free Abelian of rank equal to the *Milnor number* $\mu(\varphi)$ of the critical point, and the variation operator is an isomorphism, see [6].

In the simplest case, the critical point of the function φ is Morse. Its Milnor number is then equal to 1, so there are isomorphisms $\tilde{H}_{m-1}(A \cap X(\delta) \cap B_\varepsilon) \simeq \mathbb{Z} \simeq \tilde{H}_{m-1}(A \cap X(\delta) \cap B_\varepsilon, \partial B_\varepsilon)$. The group

$$\tilde{H}_{m-1}(A \cap X(\delta) \cap B_\varepsilon)$$

is generated by the class of the vanishing cycle of our singularity and is uniquely defined up to the sign. Its self-intersection index with respect to the complex orientation of the Milnor fiber is 0 if m is even, 2 if $m \equiv 1 (\text{mod } 4)$, and -2 if $m \equiv 3 (\text{mod } 4)$. Let us choose an arbitrary generator Δ of this group $H_{m-1}(A \cap X(\delta) \cap B_\varepsilon)$, and choose a generator ∇ of $\tilde{H}_{m-1}(A \cap X(\delta) \cap B_\varepsilon, \partial B_\varepsilon)$ in such a way that their intersection index $\langle \nabla, \Delta \rangle$ is equal to 1.

One formulation of the Picard–Lefschetz theorem is as follows.

Theorem 11.4.3 *The variation operator of a Morse function singularity $\mathbb{C}^m \to \mathbb{C}$ takes ∇ to $(-1)^{m(m+1)/2}\Delta$.*

The usual Picard–Lefschetz formula follows immediately from this one.

Remark 11.4.4 Besides the operator (11.18), many similar operators consisting of elements of diagrams (11.16), (11.17) occur in essentially the same way, for example the variation operators

$$H_*(B_\varepsilon, B_\varepsilon \cap (A \cup X_0 \cup \partial B_\varepsilon)) \to H_*(B_\varepsilon, B_\varepsilon \cap (A \cup X_0))$$

or

$$H_*(B_\varepsilon \cap (A \cup X_0), \partial B_\varepsilon \cap (A \cup X_0)) \to H_*(B_\varepsilon \cap (A \cup X_0)).$$

All of them are related with each other by appropriate boundary or Mayer-Vietoris operations, commuting with these variation operators.

11.4.2 Stratified Picard-Lefschetz Theory

Consider again the situation described in Sect. 11.3.2. Let the linear function $f : \mathbb{C}^n \to \mathbb{C}$ have a Morse critical point (in the sense of Definition 11.3.7) at a point a of some stratum $\sigma \subset A$ of positive dimension. Again, let B_ε be a small ball around a in $\mathbb{C}^n$, δ be a very small positive number, and $X \equiv X(\delta)$ be the hyperplane $f^{-1}(\delta)$. The main characters of the local monodromy action near the generic discriminant point $f^{-1}(0) \in P^{\mathbb{C}}$ are

- the absolute homology group

$$\tilde{H}_*(A \cap X \cap B_\varepsilon), \tag{11.19}$$

- the relative homology group

$$\tilde{H}_*(A \cap X \cap B_\varepsilon, A \cap X \cap \partial B_\varepsilon), \tag{11.20}$$

- the obvious map

$$J : \tilde{H}_*(A \cap X \cap B_\varepsilon) \to \tilde{H}_*(A \cap X \cap B_\varepsilon, A \cap X \cap \partial B_\varepsilon) \tag{11.21}$$

  (reduction modulo the boundary), and
- the variation homomorphism

$$Var : \tilde{H}_*(A \cap X \cap B_\varepsilon, A \cap X \cap \partial B_\varepsilon) \to \tilde{H}_*(A \cap X \cap B_\varepsilon) \tag{11.22}$$

  defined by the family of varieties $A \cap X(\delta e^{i\tau}) \cap B_\varepsilon$, $\tau \in [0, 2\pi]$.

It turns out that this data set is completely defined by the analogous set arising in the section of our picture by a generic hyperplane through a, and hence (by induction over the dimensions) by a similar picture for a section by a generic plane of dimension $n - \dim \sigma$.

Namely, let $T \subset \mathbb{C}^n$ be a complex affine hyperplane through a transversal to the stratum σ and such that the restriction of the function f to T again has a Morse critical point on the variety $A \cap T$ at the point a of the subvariety $\sigma \cap T$. Denote by J_- and Var_- the analogs of the operators (11.21) and (11.22) acting (in different directions) between the groups

$$\tilde{H}_*(A \cap X \cap T \cap B_\varepsilon) \quad \text{and} \quad \tilde{H}_*(A \cap X \cap T \cap B_\varepsilon, A \cap X \cap T \cap \partial B_\varepsilon). \qquad (11.23)$$

Then we have the following statements.

Theorem 11.4.5 *Suppose that the coefficient group of all homology groups* (11.19), (11.20) *and* (11.23) *allows division by 2. Then for any $i \geq 1$ there are*

(1) *an isomorphism*

$$\Sigma : \tilde{H}_{i-1}(A \cap X \cap T \cap B_\varepsilon) \xrightarrow{\sim} \tilde{H}_i(A \cap X \cap B_\varepsilon) \qquad (11.24)$$

(2) *and an isomorphism*

$$\tilde{H}_i(A \cap X \cap B_\varepsilon, A \cap X \cap \partial B_\varepsilon)$$
$$\simeq \tilde{H}_{i-1}(A \cap X \cap T \cap B_\varepsilon) \oplus \tilde{H}_i(A \cap T \cap B_\varepsilon, A \cap T \cap \partial B_\varepsilon) \qquad (11.25)$$

such that

(3) *the operator* (11.21) *acts into the subgroup of $\tilde{H}_i(A \cap X \cap B_\varepsilon, A \cap X \cap \partial B_\varepsilon)$ corresponding by the isomorphism* (11.25) *to the first summand of its right part; namely for any element $\alpha \in H_{i-1}(A \cap X \cap T \cap B_\varepsilon)$ the element $\Sigma(\alpha) \in \tilde{H}_i(A \cap X \cap B_\varepsilon)$ is mapped by* (11.21) *to*

$$- \rightleftarrows (2\alpha + Var_- \circ J_-(\alpha)), \qquad (11.26)$$

where

$$\rightleftarrows : \tilde{H}_{i-1}(A \cap X \cap T \cap B_\varepsilon) \to \tilde{H}_i(A \cap X \cap B_\varepsilon, A \cap X \cap \partial B_\varepsilon)$$

is the injection defined by the first summand of the right part of (11.25), *and*

(4) *the variation operator* (11.22) *is equal to zero on the second summand of* (11.25) *and maps any element $\rightleftarrows(\alpha)$ of the first summand to $\Sigma(\alpha)$.*

Remark 11.4.6

1. A similar statement holds for an arbitrary coefficient group, but the formula for item 3 is then more complicated. In particular, the usual Picard–Lefschetz formula follows from these formulas by the induction over dimensions. For the purposes of integral geometry, however, the homology with real coefficients is sufficient.
2. As in Remark 11.4.4, all these statements can be reformulated for other groups related to these by different boundary operators.
3. Analogous reduction results are proved in [32] also for (middle perversity) intersection homology groups.

The stabilization map (11.24) is defined by a kind of suspension operation. For example, in the situation shown in Fig. 11.3 it maps the generator of the group $\tilde{H}_0(A \cap X \cap T)$, i.e. the difference of two intersection points of the line and the semicubical curve on the left, into the 1-homology class of the (real part of) the intersection set $A \cup X$ on the right. In general, let $\theta : \mathbb{C}^n \to \mathbb{C}$ be a linear form such that the hyperplane T is its zero set; consider the complex 1-parametric family of hyperplanes $T_w = \theta^{-1}(w)$, $w \in \mathbb{C}^1$. This family has two exceptional points w_- and w_+, for which the set $A \cap X_\delta \cap T_w$ degenerates. (In the case of Fig. 11.3, these exceptional planes are complexifications of real planes passing through two points of the set $X(\delta) \cap \sigma$). For any other value of w close to 0 the intersection of this set with the ball B_ε is topologically the same as for $w = 0$, i.e. for $T_w = T$. In particular, these sets form a locally trivial bundle over the complement of these two points $\{w_-, w_+\}$ of the complex line. We can assume that these values $w_\pm$ are real: $w_- < 0 < w_+$. Then for any cycle $\Delta \in \tilde{H}_{i-1}(A \cap X \cap T \cap B_\varepsilon)$ the cycle $\Sigma(\Delta)$ is swept out by the copies of this cycle Δ transmitted by covering homotopy into all fibers over the segment $[w_-, w_+]$; their boundary positions in the degenerate fibers $T_{w_\pm}$ can be contracted there and we get a cycle. Similarly, the operation $\rightleftarrows$ maps such a cycle Δ into the half-difference of two relative chains swept out by the one-parametric families of copies of the cycle Δ over the intervals $[w_+, +\infty)$ and $(-\infty, w_-]$.

11.4.3 On the Monodromy and Ramification of Volume Functions at the Asymptotic Hyperplanes

Suppose that the closure of a hyperplane $Y \in P^{\mathbb{C}}$ is not in general position with respect to the stratified variety $A \cup \mathbb{C}^{n-1}_\infty$ at some "infinite" point $a \in \bar{A} \cap \mathbb{C}^{n-1}_\infty$, and we want to study the ramification of the volume function at the hyperplanes neighboring to Y. This study requires the considerations in a neighborhood of the manifold $\mathbb{C}^{n-1}_\infty$, where the integration form is singular. Let us denote this manifold by S. The analog of the localization (11.12) in this case reduces the integration cycle to an element of the group

$$H_n(B \setminus S, (A \cup X \cup \partial B) \setminus S),$$

where B is a small ball in $\mathbb{CP}^n$ centered at the point a and $X \in P^{\mathbb{C}}_{reg}$ is very close to Y. These groups and the action of variation operators on them can also be studied by the methods of monodromy theory, see e.g. § III.7 of [33]. The first and most important examples of such groups and operators were considered in [27, 28].

11.5 Real Cycle

There are many other problems with volume functions besides their algebraicity. For example, let a be an arbitrary boundary point of the body $K \subset \mathbb{R}^n$ as in Sect. 11.2, and let Y be the hyperplane tangent to ∂K at this point. Suppose for simplicity that this hyperplane has no other tangency points with the algebraic closure of ∂K. What can be said about the asymptotic properties and the local ramification of the analytic continuation of the volume function V_K in a neighborhood of the point $\{Y\}$ in $P^{\mathbb{C}}$?

This function is regular analytic in the set $P^{\mathbb{C}}_{reg} \cap \mathbb{R}^n$, i.e. in the complement of the discriminant of the generating family of A at the point a. Its properties are very different in different local components of this complement, so the enumeration of these components is one of the first problems of this study. See [36] for the solution of this problem for different types of local tangency of Y and A (or, which is the same, of the types of their generating functions).

For any such component, this behavior of the volume function depends very much on the homology class of the localized integration cycle in the group $H_n(B, A \cup X \cup \partial B)$, where $B \in \mathbb{C}^n$ is a small ball centered at a and X is the complexification of a hyperplane of this component. By Proposition 11.3.4, this class is completely characterized by the corresponding relative homology class $\Pi \in \tilde{H}_{n-2}(A \cap X \cap B, A \cap X \cap \partial B)$ of the Milnor fiber of this generating function.

Namely, it is the homology class of the set of all real points of the manifold $A \cap X \cap B$. It can be defined for any non-discriminant parameter value of a real deformation of an analytic real function singularity, not necessarily of a generating function of a hypersurface. This definition is as follows.

Let $f : (\mathbb{C}^m, \mathbb{R}^m, 0) \to (\mathbb{C}, \mathbb{R}, 0)$ be an isolated real function singularity, B_ε be a small ball centered at the origin in $\mathbb{C}^n$, and $\tilde{f}$ be some very small real non-discriminant perturbation of f. The set of real points of the Milnor fiber $V_0 \equiv \tilde{f}^{-1}(0) \cap B_\varepsilon$ is a $(m-1)$-dimensional smooth submanifold in it. Suppose that an orientation of $\mathbb{R}^m$ is chosen, then V_0 is canonically oriented as the boundary of the set $f^{-1}((-\infty, 0]) \cap \mathbb{R}^m$. In particular it defines an element Π of the group $\tilde{H}_{m-1}(V_0, \partial V_0)$. For some reasons, which will be explained in Sect. 11.6, this class is called the *even Petrovskii class* of the perturbation $\tilde{f}$. By Poincaré duality, this class is determined by its intersection indices with elements of the group $\tilde{H}_{m-1}(V_0)$.

To describe it in these terms, let us fix a basis in the latter group.

We can assume that $\tilde{f}$ is a strictly Morse function, then it has exactly $\mu(f)$ critical points in B_ε, where $\mu(f)$ is the Milnor number of f, and all their critical values are different. The set of these critical values is invariant under the complex conjugation. Let us choose a system of non-intersecting paths in $\mathbb{C}^1$ connecting the non-critical value 0 with all these critical values in such a way that

1. The paths to the real critical values lie in the upper half-plane of $\mathbb{C}^1$ in the domain where the imaginary part is less than the absolute values of all non-real critical values;
2. The paths to any two complex conjugate imaginary critical values are symmetric to each other with respect to the real line.

The orientations of all vanishing cycles appearing from real critical points can be canonically derived from the chosen orientation of $\mathbb{R}^m$. Indeed, let α be a real critical value of $\tilde{f}$. By the Morse lemma, we can choose the local coordinates near the corresponding critical point so that $\tilde{f}$ is locally expressed by the formula

$$\tilde{f} = z_1^2 + \cdots + z_r^2 - z_{r+1}^2 - \cdots - z_m^2 + \alpha. \tag{11.27}$$

For small $\rho > 0$ a cycle in $f^{-1}(\alpha - \rho)$, which vanishes at this point, is realized by the sphere S_α^{m-1} with radius $\sqrt{\rho}$ in the real m-dimensional space, which is defined as the real span of the vectors

$$i\partial/\partial z_1, \ldots, i\partial/\partial z_r, \partial/\partial z_{r+1}, \ldots, \partial/\partial z_m. \tag{11.28}$$

Let us orient this space by the following condition: the sequence of vectors (11.28) defines the positive orientation of this space if and only if the sequence $\partial/\partial z_1, \ldots, \partial/\partial z_m$ defines the positive orientation of $\mathbb{R}^m$. This also orients the sphere S_α^{m-1} by the condition that at each of its points the m-frame (grad Re$(\tilde{f})$, a positive tangent frame of this sphere) defines the positive (just defined) orientation of this plane. It is easy to check that this orientation does not depend on the choice of the local coordinates z_i in which $\tilde{f}$ has the canonical form (11.27). Then we transport this oriented cycle to the Milnor fiber V_0 over the path in $\mathbb{C}^1$ connecting the points $\alpha - \rho$ and 0 as above, and obtain an orientation of the corresponding vanishing cycle.

Further, let us choose the orientations of cycles vanishing at imaginary critical points in such a way that the complex conjugation of the Milnor fiber $V(0)$ sends homology classes of any two cycles corresponding to complex conjugate critical points to each other.

Let us somehow number the critical values $u_1, \ldots, u_{\mu(f)}$ of $\tilde{f}$, and denote by Δ_j the oriented vanishing cycle in V_0 defined by the path chosen above to the critical value u_j for any $j = 1, \ldots, \mu(f)$.

These $\mu(f)$ cycles Δ_j form a basis of the group $\tilde{H}_{m-1}(V_0) \simeq \mathbb{Z}^{\mu(f)}$, see e.g. [12].

Theorem 11.5.1 (See [22, 33]) *If a critical value u_j of $\tilde{f}$ is positive (respectively, negative), then*

$$\langle \Delta_j, \Pi \rangle = (-1)^{m(m-1)/2} \chi(S^{m-r-1}) + \sum_k \langle \Delta_j, \Delta_k \rangle \tag{11.29}$$

(*respectively,*

$$\langle \Delta_j, \Pi \rangle = (-1)^{m(m+1)/2} \chi(S^{r-1}) + \sum_k \langle \Delta_j, \Delta_k \rangle\ ) \tag{11.30}$$

(where $\chi(S^p)$ is equal to 0 if p is odd and to 2 if p is even); in both cases, summing over all indices k such that the critical values u_k of $\tilde{f}$ are real and placed between 0 and u_j.

If u_j and $u_{j'}$ are two complex conjugate critical values, then $\langle \Delta_j, \Pi \rangle = (-1)^{m-1}\langle \Delta_{j'}, \Pi \rangle$, and both these numbers are equal mod 2 to $\langle \Delta_j, \Delta_{j'} \rangle$.

Proposition 11.5.2 (See [31]) *If the regular analytic function germ $f : \mathbb{R}^m \to \mathbb{R}$ has a* simple *(i.e. of type A_k, D_k, E_6, E_7, or E_8) singularity at the origin, then any component of the complement of the discriminant of any of its versal deformations contains a strict Morsification whose $\mu(f)$ critical points are all real.*

Therefore in these cases Theorem 11.5.1 gives us a complete computation of the even Petrovskii cycle in the terms of Morse indices of critical points and intersection indices of vanishing cycles. On the other hand, there are strong methods to compute the intersection indices of all vanishing cycles (especially in the case of real function singularities), see [1, 14, 16, 21].

11.6 Hyperbolic PDE with Constant Coefficients

The most known equation of this class is the wave equation

$$\frac{\partial^2 u}{\partial x_1^2} = c^2 \sum_{j=2}^{n} \frac{\partial^2 u}{\partial x_j^2} \tag{11.31}$$

which participates in the Cauchy problems in the half-space $\{x_1 \geq 0\}$. Its *principal fundamental solution* (which is the main ingredient of the universal solution method of all Cauchy problems for this equation) is a distribution with support in this half-space. This fundamental solution is a regular analytic function outside its *wavefront*, which is the conic hypersurface in $\mathbb{R}^n$ given by the equation

$$c^2 x_1^2 = \sum_{j=2}^{n} x_j^2, \quad x_1 \geq 0. \tag{11.32}$$

If n is even and greater than 2, then the principal fundamental solution is equal to zero everywhere outside the wavefront.

If n is odd then it is equal to zero in the "larger"(=exterior) part of the complement of the wavefront, but is non-trivial in the interior (where the left part of (equation 11.32) is greater than the right). When the argument x approaches the smooth part of the wavefront from this interior domain, the fundamental solution behaves asymptotically as a half-integer power of the distance from the wavefront, in particular cannot be extended to a regular analytic function at its points.

Both these types of asymptotic behavior have analogs for arbitrary hyperbolic equations of any degree, see [8, 22, 26]. All such equations have principal fundamental solutions, whose singular sets (wavefronts) are conic surfaces in $\mathbb{R}^n_+$. In the intersection of a neighborhood of a point of the wavefront with some component of the complement of this front, the fundamental solution is a regular analytic function, which either can be extended to a regular function in the entire neighborhood of this point (then we say that this component is a *local lacuna*, and the front is *sharp* from the side of this component), or there is no such regular extension: the latter situation is called the *diffusion of waves*.

These types of the analytic behavior can be explained in the terms of integral representations: the values of fundamental solutions of hyperbolic equations outside the wavefront are equal to integrals of certain differential forms (Herglotz–Petrovskii–Leray forms) along certain cycles (Petrovskii cycles) in the complexified projectivization of the space of momenta, see Sect. 11.6.5 below. These cycles depend on the point of the complement of the wavefront, and the ramification of the analytic continuation of these integrals is controlled by the monodromy of these cycles in the complement of the complexified wavefront.

In the case of general hyperbolic equations of higher order, the wavefronts can have complicated singularities, in particular they can locally divide the space into more than two local parts; at such points of the front the question of the sharpness from the side of each local component of the complement of the front can be asked (and solved). Namely, the classification of typical wavefront singularities of strictly hyperbolic operators coincides with the classification of critical points of smooth functions, and the local geometry of these wavefronts repeats that of discriminants of versal deformations of these functions; see e.g. [36] and references there. Similar to Sect. 11.3.1, the connection between singularities of wavefronts and functions is provided by generating functions.

The dichotomy diffusion/sharpness is only the simplest characterization of the asymptotic behavior of the wave function: the next question is the character of the diffusion when it appears. This problem can also be studied by the methods of singularity theory, see e.g. [30]. Moreover, most of these results can be extended to hyperbolic equations with non-constant coefficients, see [17] and [30].

11.6.1 Hyperbolic Polynomials

Let $F : \mathbb{C}^n \to \mathbb{C}$ be a homogeneous polynomial of degree d, and ϑ be a nonzero vector in $\mathbb{R}^n$. Denote by $A(F)$ the cone in $\mathbb{C}^n$ defined by the equation $F = 0$, and by Re $A(F)$ the cone $A(F) \cap \mathbb{R}^n$.

Definition 11.6.1 The homogeneous polynomial F of degree d is *hyperbolic* with respect to the vector ϑ if $F(\vartheta) \neq 0$ and any line in $\mathbb{R}^n$ parallel to ϑ intersects the cone Re $A(F)$ at exactly d points (counted with their multiplicities). The

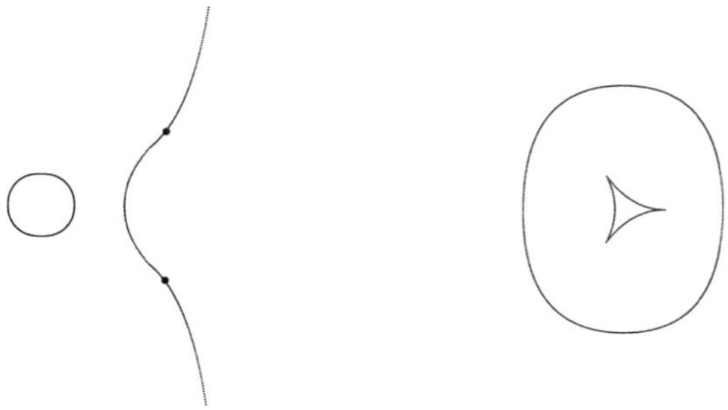

Fig. 11.4 Standard cubic curve and its projective dual

polynomial F is *strictly hyperbolic* with respect to ϑ if it is hyperbolic and has no singular points in $\mathbb{R}^n$ other than the origin.

The set of all vectors ϑ such that F is hyperbolic with respect to them is called the *hyperbolicity domain* of the polynomial F.

Proposition 11.6.2 (See [8]) *Every hyperbolic homogeneous polynomial is equal to a* real *polynomial* $(\mathbb{C}^n, \mathbb{R}^n) \to (\mathbb{C}, \mathbb{R})$ *multiplied by a nonzero constant.*

Example 11.6.3 The quadratic function $x_1^2 - x_2^2 - \cdots - x_n^2$ in $\mathbb{R}^n$ is strictly hyperbolic with respect to any vector directed into the cone where this function is positive.

Example 11.6.4 A non-singular cubic curve in $\mathbb{RP}^2$ can consist of two or one components (and looks like either the union of the two curves in the left part of Fig. 11.4, or just one of them). A degree 3 polynomial in $\mathbb{R}^3$ defining such a curve is not hyperbolic in the case of a one-component curve. In the case of two components it is hyperbolic with respect to all vectors directed inside the "figura ovalis" of the cubic curve.

Proposition 11.6.5 (See [8]) *The set of real zeros of a strictly hyperbolic polynomial of degree d always consists of* $[(d+1)/2]$ *components that meet only at the origin.*

If d is even, then all *these* $d/2$ *components are homeomorphic to the standard cone defined by the equation from Example 11.6.3. If d is odd, then it has* $[d/2]$ *such conic components and one more component that is homeomorphic to a hyperplane.*

If the polynomial F is hyperbolic with respect to the vector ϑ*, then it is also hyperbolic with respect to any other vector from the same component of the set* $\mathbb{R}^n \setminus \mathrm{Re}A(F)$.

For any homogeneous polynomial $F : \mathbb{R}^n \to \mathbb{R}$ *with no singular points outside 0, its hyperbolicity domain is either empty or consists of two opposite convex cones.*

The latter set is surrounded by the "most interior" component of the set $\mathrm{Re}A(F)$.

11.6.2 Hyperbolic Operators

Let $F \equiv F(D)$ be a linear partial differential operator in $\mathbb{R}^n$ of order d with constant coefficients, i.e. a degree d polynomial in variables $D_j \equiv \partial/i\partial x_j,\ j = 1, \dots, n$.

The *Cauchy problem* in the half-space $\mathbb{R}^n_+ = \{x : x_1 \geq 0\}$ for this operator is posed as follows. For any function ϕ on this half-space and any d functions $\chi_0, \dots, \chi_{d-1}$ on the hyperplane $\mathbb{R}^{n-1} = \{x : x_1 = 0\}$, the problem is to find a function u with support in this half-space, such that

$$F(u) = \phi \quad \text{in } \mathbb{R}^n_+,$$

$$\frac{\partial^k u}{\partial x_1^k} = \chi_k \quad \text{in } \mathbb{R}^{n-1}, \qquad k = 0, \dots, d-1.$$

The function ϕ is called the *right-hand side* of the Cauchy problem, and the functions χ_k are its *initial data.*

A *fundamental solution* of the operator F is a distribution Ψ in $\mathbb{R}^n$ that satisfies the condition $F\Psi \equiv \delta_0$, where δ_0 is the Dirac delta function concentrated at the origin.

The operator F is called *hyperbolic* in the half-space $\mathbb{R}^n_+$ if either of the following two conditions holds:

(a) F admits a fundamental solution whose support is a proper cone in the half-space $\mathbb{R}^n_+$. (A closed cone in a half-space is *proper* if its intersection with the boundary of that half-space consists of only one point: its vertex.)
(b) Every Cauchy problem in $\mathbb{R}^n_+$ for this operator has a unique solution, and its value at any point y depends only on the values of the initial data χ_k and the right-hand side ϕ of this problem at the points x, which lie in some proper cone in the half-space $\{x : x_1 \leq x_1(y)\}$ with vertex at the point y.

Proposition 11.6.6 (See e.g. [8]) *The above conditions (a) and (b) are equivalent. Every operator F with constant coefficients can have at most one fundamental solution that satisfies these conditions.* □

This fundamental solution of a hyperbolic operator F is called its *principal fundamental solution* and is denoted by E_F.

The implication (a) $\Rightarrow$ (b) between these conditions is realized by the convolution of the data of the Cauchy problem with the principal fundamental solution.

11.6.3 Hyperbolic Operators and Hyperbolic Polynomials

Let us identify the partial derivatives $\partial / i \partial x_j$ in the "physical space" with the coordinate functions ξ_j of the dual space $\check{\mathbb{R}}^n$ (called the *space of momenta*). Then the differential operator F of order d becomes a polynomial F_ξ of degree d in the variables $\xi_1, \ldots, \xi_n$.

Let us denote by $\bar{F}$ (respectively, $\bar{F_\xi}$) the principal (of highest order of highest degree) homogeneous part of the operator F (respectively, the polynomial F_ξ) obtained from it by forgetting all its lower terms.

Theorem 11.6.7 (See [8])

1. *If the operator F is hyperbolic in the half-space $\mathbb{R}^n_+$, then the principal (of highest degree) homogeneous part $\bar{F_\xi}$ of the corresponding polynomial F_ξ is a hyperbolic polynomial in $\check{\mathbb{R}}^n$ with respect to the vector $\partial / \partial \xi_1$; in particular, it is proportional to a polynomial with real coefficients.*
2. *Conversely, if this principal part $\bar{F_\xi}$ is a* strictly *hyperbolic polynomial with respect to the vector $\partial / \partial \xi_1$, then the operator F is hyperbolic in $\mathbb{R}^n_+$.*
3. *If the operator F is homogeneous, $F = \bar{F}$, then it is hyperbolic if and only if the polynomial F_ξ is hyperbolic.*

□

We will further consider only strictly hyperbolic polynomials and operators.

11.6.4 Wave Front

All singular points of the principal fundamental solution of a hyperbolic operator F lie on some conical semialgebraic surface in $\mathbb{R}^n_+$, called its wavefront and denoted by $W(F)$. In the case of a strictly hyperbolic operator it is the cone projectively dual to the cone of zeros of the principal symbol of the operator.

More precisely, denote by $A(F)$ the conical hypersurface in $\check{\mathbb{C}}^n$ on which the principal part $\bar{F_\xi}$ of the polynomial F_ξ vanishes. Let Re $A(F)$ be the set of its real points, $A(F) \cap \check{\mathbb{R}}^n$.

The spaces $\check{\mathbb{R}}^n$ and $\check{\mathbb{C}}^n$ will be considered as spaces dual to $\mathbb{R}^n$ and $\mathbb{C}^n$ by means of the pairing

$$\langle x, \xi \rangle = \sum x_j \xi_j . \tag{11.33}$$

In particular, each point $x \in \mathbb{C}^n \setminus 0$ defines its orthogonal hyperplane $X(x) \subset \check{\mathbb{C}}^n$. The *wavefront* $W(F)$ of the strictly hyperbolic operator F is defined as the cone consisting of the origin and all points $x \in \mathbb{R}^n_+ \setminus 0$ such that the hyperplane $X(x) \cap \check{\mathbb{R}}^n$ orthogonal to x is tangent to the cone Re $A(F)$ along some line.

Remark 11.6.8 For any (not strictly) hyperbolic operator its wavefront is defined in a more complicated way (see [8]); in any case it lies in the set of points x such that the hyperplane $X(x)$ is not in the general position with respect to the cone Re $A(F)$.

It follows immediately from the definitions and from Proposition 11.6.5 that the wavefront of a strictly hyperbolic operator of degree d consists of $[(d + 1)/2]$ components, each of which is projectively dual to some component of the cone Re $A(F)$. The component dual to the boundary of the hyperbolicity domain is "exterior" one, i.e. it separates all other components from the half-space $\mathbb{R}^n_- = \{x : x_1 \leq 0\}$, and the whole part of $\mathbb{R}^n_+$ separated by this component is convex according to the last statement of Proposition 11.6.5.

For example, the projectivization of the wavefront of the operator from Example 11.6.4 (see Fig. 11.4 left) has the form shown in Fig. 11.4 (right); the singular points of the wave front correspond to the inflation points of the curve $\{F = 0\} \subset \check{\mathbb{RP}}^2$.

For $n = 4$, the projective wavefront is a surface in $\mathbb{RP}^3$; its typical singularities are the standard cuspidal edges (see Definition 11.3.8 and Fig. 11.3) and *swallowtails*, i.e. surfaces like the one shown in Fig. 11.5

Similar to Sect. 11.3.1, these surfaces appear as discriminants of generating families of projectivizations of $A(P) \cup \check{\mathbb{R}}^n$: the cuspidal edges correspond to critical points of type A_2 of generating functions, and the swallowtails to points of type A_3.

Theorem 11.6.9 (See [8]) *The principal fundamental solution of a hyperbolic operator F coincides with a regular analytic function in any connected component of the set* $\mathbb{R}^n \setminus W(F)$*, and is identically equal to zero outside the convex hull of the wavefront* $W(F)$. □

Fig. 11.5 Swallowtail

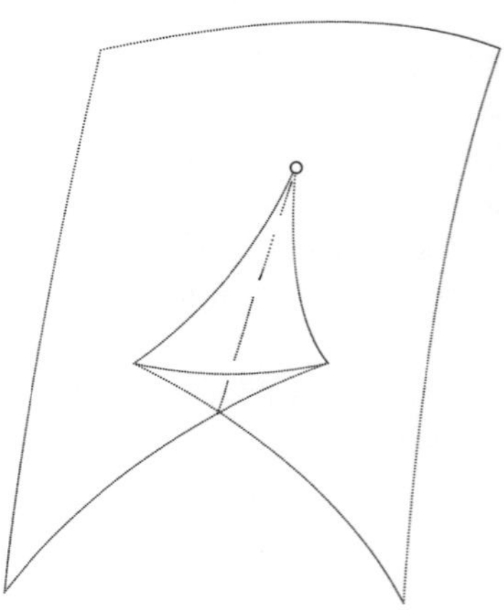

11.6.5 Herglotz–Petrovskii–Leray Integral Representations

Theorems 11.6.7, 11.6.9 and other essential properties of hyperbolic equations follow from the fact that their fundamental solutions have integral representations and thus can be studied by the methods of the monodromy theory.

For any point $x \in \mathbb{R}^n \setminus W(F)$ the value of the fundamental solution (or at least each of its sufficiently high partial derivatives) at this point is given by an explicit integral formula, the *Herglotz–Petrovskii–Leray integral* along some cycle

$$t(\gamma(x)) \in H_{n-1}(\check{\mathbb{CP}}^{n-1} \setminus PX(x) \cup PA(F)) \tag{11.34}$$

where $PX(x)$ and $PA(F)$ are hypersurfaces in $\check{\mathbb{CP}}^{n-1}$ corresponding to the conic subsets $X(x)$ and $A(F)$ in $\check{\mathbb{C}}^n$. More precisely, there is a homology class

$$\gamma(x) \in H_{n-2}(PX(x) \setminus PA(F)),$$

called the *Petrovskii cycle*. The class (11.34) is defined by the *Leray tube* around it, i.e. by the cycle swept out by boundaries of small complex discs in $\check{\mathbb{CP}}^{n-1}$ transversal to the hyperplane $PX(x)$ at the points of a cycle representing the class $\gamma(x)$.

For example, in the case of even n, the Petrovskii cycle is defined by the Leray tube in $X(x) \setminus A(F)$ around the set of real points of the submanifold $A(F) \cap X(x)$; in the case of odd n the definition is a bit more complicated. See e.g. §IV.6 of [33] for the construction of this cycle, and also for the formulas of the Herglotz–Petrovskii–Leray differential forms.

The *Petrovskii condition* for the point $x \in \mathbb{R}^n_+ \setminus W(F)$ (and for the whole component of the complement of the wavefront containing this point) is the condition that the Petrovskii cycle $\gamma(x)$ is homologous to zero. If it is satisfied, then in this component the principal fundamental solution coincides with a holomorphic function defined in the whole of $\mathbb{C}^n$ (and even with a polynomial if the operator F is homogeneous). (In the terminology of this theory, this means that this component is a *lacuna*.)

Atiyah et al. [8] have introduced a local version of the Petrovskii condition, which allows us to explain and predict the *local* regularity of solutions.

Namely, if we approach the point y of the wavefront from the side of a component of its complement along a path $\{x(\tau)\}$, $\tau \in [\varepsilon, 0]$, all whose points corresponding to $\tau > 0$ lie in this component and $x(0) = y$, then the family of sets $PX(x(\tau)) \cup PA(F)$ topologically degenerates at the last instant (corresponding to the point $x(0) = y$).

If nevertheless the Petrovskii cycles corresponding to the points of this path with $\tau > 0$ tend to some cycle in $PX(y) \setminus PA(F)$, then near the point y the restriction of the fundamental solution to this component coincides with an analytic function defined in an entire neighborhood of y. The latter situation is called the *local holomorphic sharpness* at the point y from the side of our component.

The topological condition from the previous paragraph can also be formulated as follows.

Definition 11.6.10 ([8]) The *local Petrovskii condition* is satisfied at the point $y \neq 0$ of the wavefront $W(F)$ from the side of a local (at this point) component L of the set $\mathbb{R}^n \setminus W(F)$, if for some (and then any) point $x \in L$ arbitrarily close to y the homology class of the cycle $\gamma(x)$ in the group $H_{n-2}(PX(x) \setminus PA(F))$ belongs to the image of the group $H_{n-2}(PX(y) \setminus PA(F))$ under the homomorphism defined by the natural projection $PX(y) \to PX(x)$ (which sends any fixed compact subset of $PX(y) \setminus PA(F)$ into the set $PX(x) \setminus PA(F)$ with sufficiently small τ).

Theorem 11.6.11 ([8]) *The local Petrovskii condition implies the sharpness at the point y from the side of the local component L.*

Proof Let $\bar{\gamma}$ be a compact $(n-2)$-dimensional cycle in $PX(y) \setminus PA(F)$ whose homology class in $H_{n-2}(PX(y) \setminus PA(F))$ is mapped to $\gamma(x)$ by this projection. The projections of this cycle $\bar{\gamma}$ to all hyperplanes PX sufficiently close to $PX(y)$ define a family of cycles in all manifolds $PX \setminus PA(F)$ for such X. The Herglotz–Petrovskii–Leray integrals along the Leray tubes around all these cycles are well defined, and their values define a holomorphic function in a neighborhood of the point y in $\mathbb{C}^n$. In the component L this function coincides with the integrals along the cycles $t(\gamma(x))$, which express all sufficiently high partial derivatives of the principal fundamental solution. □

The converse implication is true for all points of the wavefronts of almost all strictly hyperbolic equations, namely for all points corresponding to isolated singularities of generating functions.

Theorem 11.6.12 (See [31] or § IV.8 in [33]) *If F is a strictly hyperbolic operator, $y \in W(F) \setminus 0$, and the projective hyperplane $PX(y) \subset \check{\mathbb{CP}}^{n-1}$ is tangent to the hypersurface $PA(F)$ only at a discrete set of points in a neighborhood of the real space $\check{\mathbb{RP}}^{n-1}$, then the sharpness at y from the side of an arbitrary local component of $\mathbb{R}^n \setminus W(F)$ implies the local Petrovskii condition.*

The assumptions of this theorem are satisfied for all points of the wavefronts of almost all strictly hyperbolic operators.

A proof of this theorem is based on the local monodromy theory and is very similar to the proof of Theorem 11.3.5. Indeed, suppose for simplicity that the plane $PX(y)$ has only one tangency point $a \in \check{RP}^{n-1}$ with the hypersurface $PA(F)$. Then the local Petrovskii condition can be reformulated as the triviality of a certain localization of the Petrovskii cycle, namely of the homology class of the corresponding relative cycle in the group

$$H_{n-2}(B \cap PX(x) \setminus PA(F), \partial B \cap PX(x) \setminus PA(F)),$$

where B is a small ball in $\check{\mathbb{CP}}^{n-1}$ centered at the non-transversality point a, and $x \approx y$ is a point in the examined component. If this relative class is non-trivial then by moving the plane $PX(x)$ along appropriate loops in the set of hyperplanes, that are close to $PX(y)$ but transversal to $PA(F)$, we can increase the Petrovskii cycle by a non-trivial vanishing cycle such that the Herglotz–Petrovskii–Leray integral along its tube is not equal to zero; in particular the ramification of the analytic continuation of our integral function is non-trivial in an arbitrarily small neighborhood of the point y in $\mathbb{C}^n$.

The problem of the local regularity (sharpness) near non-singular points of wavefronts of strictly hyperbolic equations was solved long before [ABG]. (Such points correspond via projective duality to the non-parabolic points of the set of zeros of the principal homogeneous part of the operator.) The answer is called the *Davydova–Borovikov condition* and is formulated in terms of the signature of the second fundamental form of the wavefront. Since the sharpness holds or does not hold simultaneously for all points of the front lying on one and the same ray, we can "projectivize" the problem: we discuss the sharpness at a point y^* of the projectivization $W^*(F) \subset \mathbb{RP}^{n-1}$ of the wavefront from the side of some local component of its complement in $\mathbb{RP}^{n-1}$. Let us fix the local affine coordinates $z_0, z_1, \dots, z_{n-2}$ in $\mathbb{RP}^{n-1}$ with the origin at the point y^* in such a way that the hyperplane $\{z_0 = 0\}$ is tangent to the front at this point, and the vector $\partial/\partial z_0$ is directed into the suspicious component L^*. Near this point, the front is defined by a condition of the form $z_0 = g(z_1, \dots, z_{n-2})$, where g is a smooth function, $g(0) = 0$, $dg(0) = 0$, and the quadratic part of g is non-degenerate.

Definition 11.6.13 A local component L^* of the complement of the wavefront satisfies the *Davydova–Borovikov condition* at the point y^*, if the positive inertia index of this quadratic part $d^2g(0)$ is even.

Theorem 11.6.14 *A local component of the complement of the wavefront of a strictly hyperbolic operator near its non-singular point is a local lacuna* (*i.e. the wavefront is sharp from the side of this component at this point*) *if and only if this component satisfies the Davydova–Borovikov condition.*

The "only if" part was proved in 1945 in the PhD thesis [11] of Petrovskii's student A.M. Davydova. The more difficult part "if" was proved at the end of the 1950s by I.M. Gelfand's student V.A. Borovikov [9] with the help of explicit and complicated analytic calculations. However, using the topological local Petrovskii (–Atiyah–Bott–Gårding) condition and Theorem 11.6.11, it can be proved without any formulas: we only need to check that the Davydova–Borovikov condition is equivalent to the triviality of the localized Petrovskii class.

L. Gårding has also investigated the behavior of the principal fundamental solution close to the cuspidal edges of the wavefront. In particular, he found all the local lacunas near such edges. The answer is as follows.

The "larger" component of the complement of the wavefront near the cuspidal edge is never a local lacuna. If n is even, then we have no sharpness from the side of any component. In the remaining case of odd n and interior component there are two subcases, depending on the parity of the inertia indices of the second fundamental form of the wavefront at the neighboring non-singular points. In one of these cases we obviously cannot have the sharpness, because by the Davydova–Borovikov condition it fails already at the non-singular points of the wavefront adjacent to the cuspidal edge. And only in the remaining case do we have the sharpness.

Gårding has also found the complete list of local lacunas close to the next class of singularities of wavefronts: the swallowtails A_3, see Fig. 11.5.

A complete list of local lacunas close to all simple (i.e. of classes A_k, D_k, E_6, E_7, E_8) and many non-simple singularities of wavefronts can be found in [31] and [33], see § V in [33]. In particular, there is the following geometric criterion of the local sharpness near the simple singularities.

Theorem 11.6.15 (See [33], §V.7) *A local component of the complement of the wavefront of a strictly hyperbolic operator with constant coefficients in $\mathbb{R}^n$ at its* stable *simple singularity is a local lacuna (i.e., the principal fundamental solution is sharp from the side of this component) if and only if*

(a) the Davydova–Borovikov condition is satisfied at all points of all $(n-1)$-dimensional strata of the wavefront bounding this component, and
(b) the boundary of this component contains no cuspidal edges near which this component behaves as the "larger" component of the complement.

The *stability* condition in this theorem means that the generating family of the singularity is a versal deformation of the corresponding generating function; this condition is satisfied for all simple singularities of wavefronts of almost all strictly hyperbolic operators.

The necessity of both conditions (a) and (b) follows from the results of Davydova and Gårding discussed above. Their sufficiency follows from special topological properties of simple singularities, especially from the properness of the Lyashko–Looijenga map, see e.g. [12].

Acknowledgments This work was supported by the Absorption Center in Science of the Ministry of Immigration and Absorption of the State of Israel.

References

1. N. A'Campo, *Le groupe de monodromie du déploiement des singularités isolées de courbes planes. I*, Math. Ann. 213:1 (1975), 1–32.
2. K. Aomoto, *Les équations aux différences linéaires et les intégrales des fonctions multiformes,* J. Fac. Sci. Univ. of Tokyo 22 (1975), 271–297.
3. K. Aomoto, *On the structure of integrals of power products of linear functions,* Sci. papers, Coll. Gen. Ed., Univ. Tokyo., 26 (1977), 49–61.
4. V. I. Arnold, *Kepler's second law and the topology of Abelian integrals*, Kvant, 1987, 12, 17–21. Engl. transl. in Kvant Selecta: Algebra and Analysis II, Math. World, vol. 15, Amer. Math. Soc., Providence, RI, 1999, 131–140.

5. V. I. Arnold, A. N. Varchenko, and S. M. Gusein-Zade, *Singularities of differentiable maps*, vol.1: *Classification of critical points, caustics, and wavefronts*. "Nauka", Moscow, 1982; Engl. transl.: Birkhäuser, Basel, 1985.
6. V. I. Arnold, A. N. Varchenko, and S. M. Gusein-Zade, *Singularities of differentiable maps*, vol.2: *Monodromy and asymptotics of integrals*, "Nauka", Moscow, 1984; Engl. transl.: Birkhäuser, Basel, 1988.
7. V.I. Arnold a.o. *Arnold's problems*. Springer and Phasis, 2004.
8. M. F. Atiyah, R. Bott, and L. Gårding, *Lacunas for hyperbolic differential operators with constant coefficients.*, Acta Math., 124 (1970), 109–189 and 131 (1973), 145–206.
9. V. A. Borovikov, *Fundamental solutions of linear partial equations with constant coefficients*, Trudy Moskov. Mat. Obshch. 8 (1959), 199–257. English transl.: Amer. Math. Soc. Translations, Ser.2, vol. 25, 11–76, 1963.
10. N. Bourbaki, *Groupes et algèbres de Lie, Chapitres 4, 5, 6*, Hermann, Paris, 1968.
11. A. M. Davydova, *A sufficient condition for the absence of a lacuna for a partial differential equation of hyperbolic type*, Ph.D.Thesis, Moscow State Univ., 1945, 43 pp. (In Russian).
12. W. Ebeling, *Distinguished bases and monodromy of complex hypersurface singularities*, Handbook of Singularities, vol. 1, 427–466.
13. D. Fotiadi, M. Froissart, J. Lascoux, F. Pham, *Applications of an isotopy theorem,* Topology, 4 (1965), 159–191.
14. A. M. Gabrielov, *Intersection matrices for some singularities*, Functional Anal. Appl., 7 (1973), 182–193.
15. A. M. Gabrielov, *Bifurcations, Dynkin diagrams and modality of isolated singularities*, Functional Anal. Appl., 8 (1974), 94–98.
16. A. M. Gabrielov, *Polar curves and intersection matrices of singularities*, Invent. Math., 54:1 (1979), 15–22.
17. L. Gårding, *Sharp fronts of paired oscillatory integrals*, Publ. Res. Inst. for Math. Sci. Kyoto Univ., 12 (1977), 53–68.
18. I. M. Gelfand, *A general theory of hypergeometric functions*, Sov. Math. Doklady, 33, 573–577.
19. I. M. Gelfand and G. E. Shilov, *Generalized Functions, vol. 1: Properties and Operations*, Moscow, Fizmatlit, 1959. English translation: Academic Press, New York, 1964.
20. M. Goresky and R. MacPherson, *Stratified Morse theory*, Springer-Verlag, Berlin and New York, 1986.
21. S. M. Gusein-Zade, *Intersection matrices for some singularities of functions of two variables*, Functional Anal. Appl., 8 (1974), 10-13.
22. J. Leray, *Un prolongement de la transformation de Laplace qui transforme la solution unitaire d'un opérateur hyperbolique en sa solution élémentaire (Probleme de Cauchy, IV)*, Bull. Soc. Math. France, 90 (1962), 39–156.
23. Mather, J.: *Notes on topological stability*, Harvard University, 1970 and Bull. Amer. Math. Soc. N.S. 49 (2012), no. 4, 475-506.
24. J. Milnor, *Singular points of complex hypersurfaces*, Princeton Univ. Press and Univ. of Tokyo Press, Princeton, NJ, and Tokyo: 1968. (Ann. of Math. Studies, 61)
25. I. Newton, *Philosophiae Naturalis Principia Mathematica*, London, 1687.
26. I. G. Petrovsky, *On the diffusion of waves and the lacunas for hyperbolic equations*, Matem. Sbornik, 17(59) (1945), 289–370.
27. F. Pham, *Formules de Picard–Lefschetz généralisées et ramification des intégrales*, Bull. Soc. Math. France, 93 (1965), 333–367.
28. F. Pham, *Introduction à l'étude topologique des singularités de Landau*, Gauthier-Villars, Paris, 1967.
29. D. Trotman, *Stratification Theory*, Handbook, Vol. I, 231–260.
30. A. N. Varchenko, *Normal forms of nonsmoothness of solutions of hyperbolic equations*, Mat. USSR Izv. 30:3 (1988), 615–628.
31. V. A. Vassiliev, *Sharpness and the local Petrovskiĭ condition for strictly hyperbolic equations with constant coefficients,* Math. USSR Izv. 28 (1987), 233–273.

32. V. A. Vassiliev, *Stratified Picard–Lefschetz theory*, Selecta Math., New Series. 1:3 (1995), 597–621.
33. V.A. Vassiliev, *Applied Picard–Lefschetz theory*, Amer. Math. Soc., Providence, RI, 2002 , xii+324 pp.
34. V. A. Vassiliev, *Newton's lemma XXVIII on integrable ovals in higher dimensions and reflection groups*, Bull. Lond. Math. Soc., 47:2 (2015), 290–300, arXiv: 1407.7221
35. V. A. Vassiliev, *Algebroidally integrable bodies*, Arnold Math. J., 6:2 (2020), 291–309, arXiv: 2003.04665
36. V.A. Vassiliev, *Real function singularities and their bifurcation sets*, Handbook Vol VII, 2024.
37. V. A. Vassiliev, I. M. Gelfand, and A. V. Zelevinskiĭ, *General hypergeometric functions on complex Grassmannians*, Funct. Anal. and its Appl., **21**:1 (1987), 19–31.

Chapter 12
Topology of Hyperplane Arrangements via Real Structure

Masahiko Yoshinaga

Contents

M. Yoshinaga (✉)
Department of Mathematics, Osaka University, Toyonaka, Osaka, Japan
e-mail: yoshinaga@math.sci.osaka-u.ac.jp

© The Author(s), under exclusive license to Springer Nature Switzerland AG 2026
J. L. Cisneros-Molina et al. (eds.), *Handbook of Geometry and Topology of Singularities VIII*, https://doi.org/10.1007/978-3-031-99571-2_12

Abstract This is a survey on the topology of hyperplane arrangements. We mainly focus on the relationship between topology of complexified real arrangements and the real structure, such as adjacent relations of chambers and stratifications related to real structures.

12.1 Introduction

Let $\mathcal{A} = \{H_1, H_2, \dots, H_n\}$ be a set of hyperplanes in $\mathbb{R}^\ell$. On one hand, $\mathcal{A}$ provides a stratification of the space $\mathbb{R}^\ell$. On the other hand, the complexification gives a complex hyperplane arrangement in $\mathbb{C}^\ell$. The complement $M(\mathcal{A}) = \mathbb{C}^\ell \setminus \bigcup_{H\in\mathcal{A}} H \otimes \mathbb{C}$ is a connected affine variety, which is a real 2ℓ-dimensional manifold. The topology of $M(\mathcal{A})$ is known to be closely related to the structure of $\mathcal{A}$ [38]. In this note, we discuss the relationship between the topology of $M(\mathcal{A})$ and its real structure, focusing on the adjacent relations of chambers.

12.2 Hyperplane Arrangements over General Fields

12.2.1 Braid Arrangements and Graphical Arrangements

As a typical example, we start this note with the *braid arrangements*. Let $V = \mathbb{K}^\ell$. Let $H_{ij} = \{(x_1, \dots, x_\ell) \in \mathbb{K}^\ell \mid x_i - x_j = 0\}$ for $1 \le i < j \le \ell$. The collection of these hyperplanes is called the *braid arrangement* denoted by

$$\mathrm{Br}(\ell) = \{H_{ij} \mid 1 \le i < j \le \ell\}.$$

Example 12.2.1 (Braid Arrangements over $\mathbb{R}$) If $\mathbb{K} = \mathbb{R}$, the complement of the braid arrangement $\mathrm{Br}(\ell)$ consists of finitely many connected components called *regions*, or *chambers*. Each chamber is expressed as

$$\{(x_1, \dots, x_\ell) \mid x_{\sigma(1)} < \dots < x_{\sigma(\ell)}\},$$

where $\sigma \in \mathfrak{S}_\ell$ a permutation. Therefore, there are exactly $\ell!$ chambers.

Example 12.2.2 (Braid Arrangements over $\mathbb{F}_q$) If $\mathbb{K} = \mathbb{F}_q$, the complement of the braid arrangement $\mathrm{Br}(\ell)$

$$\mathbb{F}_q^\ell \setminus \bigcup_{1\le i<j\le\ell} H_{ij}$$

consists of ordered tuples $(x_1, \dots, x_\ell)\ell$ mutually distinct elements in $\mathbb{F}_q$. Therefore, there are exactly $q(q-1)\dots(q-\ell+1)$ elements.

Example 12.2.3 (Braid Arrangements over $\mathbb{C}$. [2, 18, 23]) If $\mathbb{K} = \mathbb{C}$, the complement of the braid arrangement

$$X_\ell = \{(x_1, \ldots, x_\ell) \mid x_i \in \mathbb{C}, x_i \neq x_j (i \neq j)\}$$

has a natural fibration

$$p : X_\ell \longrightarrow X_{\ell-1}, (x_1, \ldots, x_\ell) \longmapsto (x_1, \ldots, x_{\ell-1})$$

with the fiber $\mathbb{C} \setminus \{\ell \text{ points}\}$. Using this fibration, one can prove X_ℓ is a $K(\pi, 1)$-space. Also, the fibration is homologically trivial, hence one can apply Leray-Hirsch theorem to obtain the Poincaré polynomial

$$\mathrm{Poin}(X_\ell, t) = (1+t)(1+2t)\ldots(1+(\ell-1)t).$$

Even though the nature of the problems in Examples 12.2.1, 12.2.2, and 12.2.3 is different, the results (the numbers of chambers, points, and the Poincaré polynomial) look similar. This phenomenon is explained (in Sect. 12.2.2) by the fact that these invariants are determined solely by the intersection poset and the characteristic polynomial which are the same for any field $\mathbb{K}$.

12.2.2 Definitions and Classical Results

Let $V = \mathbb{K}^\ell$ be a linear space. A subset H of V of the form

$$H = \{x \in V \mid \alpha(x) = b\},$$

where $\alpha : V \to \mathbb{K}$ is a non-zero linear form and $b \in \mathbb{K}$, is called an affine hyperplane. A set $\mathcal{A} = \{H_\lambda\}_{\lambda \in \Lambda}$ of hyperplanes is called an *(affine) hyperplane arrangement*, or simply an *arrangement*. An arrangement $\mathcal{A}$ is called *central* if $\bigcap_{H \in \mathcal{A}} H \neq \emptyset$ and contains the origin $\mathbf{0}$. An arrangement $\mathcal{A}$ is called *essential* if there exist $H_1, \ldots, H_\ell \in \mathcal{A}$ such that $\dim(H_1 \cap \cdots \cap H_\ell) = 0$.

12.2.3 Intersection Posets, the Möbius Function, and Chambers

Definition 12.2.4 Let $\mathcal{A}$ be an arrangement. Denote the set of non-empty intersections of hyperplanes by

$$L(\mathcal{A}) = \left\{ \bigcap_{H \in \mathcal{B}} H \neq \emptyset \,\middle|\, \mathcal{B} \subset \mathcal{A} \right\}.$$

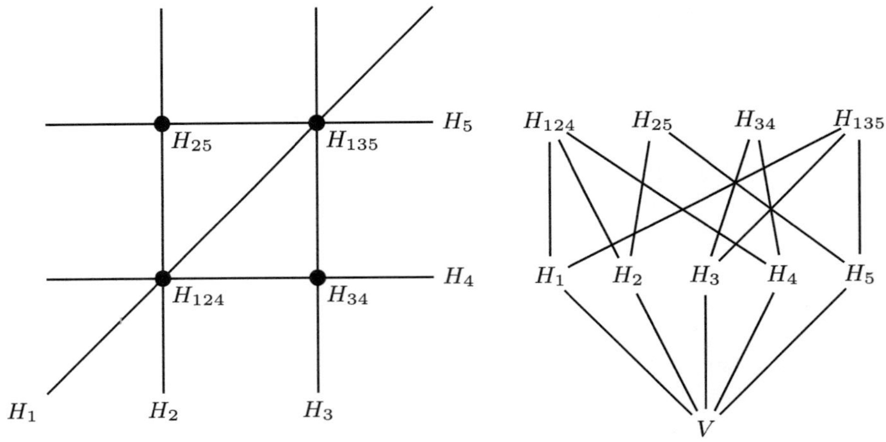

Fig. 12.1 A line arrangement and the intersection poset

Note that if $\mathcal{B} = \emptyset$, $\bigcap_{H \in \mathcal{B}} H = V$. We consider $L(\mathcal{A})$ as a poset equipped with the reverse inclusion order (Fig. 12.1), which is called the intersection poset. We also consider the set of all k-dimensional intersections,

$$L_k(\mathcal{A}) = \{X \in L(\mathcal{A}) \mid \dim X = k\}.$$

Remark 12.2.5 The intersection poset $L(\mathcal{A})$ is graded in the sense that lengths of maximal chains are equal.

We also note that $\mathcal{A}$ is essential if and only if $L_0(\mathcal{A}) \neq \emptyset$.

Let $X \in L(\mathcal{A})$. The arrangement $\mathcal{A}$ naturally defines an arrangement on X

$$\{X \cap H \mid H \in \mathcal{A}, H \not\supset X\}$$

which we denote by $\mathcal{A}^X$ and call the restriction. We also define

$$\mathcal{A}_X := \{H \in \mathcal{A} \mid H \supset X\}$$

which is called the localization at X. It is clear that $L(\mathcal{A}^X) \simeq L(\mathcal{A})_{\geq X}$ and $L(\mathcal{A}_X) \simeq L(\mathcal{A})_{\leq X}$.

If the arrangement $\mathcal{A}$ is defined in $\mathbb{R}^\ell$, then the complement $\mathbb{R}^\ell \setminus \bigcup_{i=1}^n H_i$ is a disjoint union of open convex subsets. Each connected component is called a *chamber*. We denote the set of chambers by $\mathrm{ch}(\mathcal{A})$. We also denote the set of *bounded chambers* by $\mathrm{bch}(\mathcal{A})$.

12.2.4 Characteristic Polynomials

Define the Möbius function $\mu : L(\mathcal{A}) \to \mathbb{Z}$ by

$$\mu(X) = \begin{cases} 1 & \text{if } X = V \\ -\sum_{Z<X} \mu(Z) & \text{if } X \neq V. \end{cases} \tag{12.1}$$

Definition 12.2.6 The characteristic polynomial $\chi(\mathcal{A}, t)$ is defined as

$$\chi(\mathcal{A}, t) = \sum_{X \in L(\mathcal{A})} \mu(X) t^{\dim X}.$$

The characteristic polynomial has other expressions.

Proposition 12.2.7 *Let $\mathcal{A} = \{H_1, \ldots, H_n\}$ be an arrangement.*

(1)

$$\chi(\mathcal{A}, t) = \sum_{\substack{I \subset [n] \\ H_I \neq \emptyset}} (-1)^{\# I} t^{\dim H_I},$$

where $H_I = \bigcap_{i \in I} H_i$.

(2)

$$\chi(\mathcal{A}, t) = \begin{cases} t^{\dim V}, & \text{if } \mathcal{A} = \emptyset, \\ \chi(\mathcal{A} \setminus \{H\}, t) - \chi(\mathcal{A}^H, t), & \text{if } H \in \mathcal{A} \neq \emptyset. \end{cases} \tag{12.2}$$

As applications of Proposition 12.2.7, we have the following.

Proposition 12.2.8

(1) (Zaslavsky [71]) Let $\mathcal{A} = \{H_1, \ldots, H_n\}$ be an arrangement in $V = \mathbb{R}^\ell$. Then

$$\# \mathrm{ch}(\mathcal{A}) = (-1)^\ell \chi(\mathcal{A}, -1).$$

Furthermore, if $\mathcal{A}$ is essential, then

$$\# \mathrm{bch}(\mathcal{A}) = (-1)^\ell \chi(\mathcal{A}, 1).$$

(2) (Crapo-Rota) Let $\mathcal{A} = \{H_1, \ldots, H_n\}$ be an arrangement in $V = \mathbb{F}_q^\ell$. Then

$$\# M(\mathcal{A}) = \chi(\mathcal{A}, q).$$

Example 12.2.9 Let $\mathcal{A} = \{H_1, H_2, H_3, H_4, H_5\}$ be 5 lines in $\mathbb{R}^2$ as shown in Fig. 12.1. Then $\mu(H_i) = -1$ for each $i \in [5]$. By the definition of Möbius function, we also have $\mu(H_{25}) = \mu(H_{34}) = 1$ and $\mu(H_{124}) = \mu(H_{135}) = 2$. Hence, we have

$$\chi(\mathcal{A}, t) = t^2 - 5t + 6 = (t-2)(t-3).$$

Also, we have $\chi(\mathcal{A}, -1) = 12 = \#\mathrm{ch}(\mathcal{A})$, and $\chi(\mathcal{A}, 1) = 2 = \#\mathrm{bch}(\mathcal{A})$.

12.2.5 Chromatic Polynomial

An interesting class of arrangements is the so-called *graphical arrangement*. In this class, the characteristic polynomial is directly connected to enumerative problems [54, 55].

Let $([\ell], E)$ be a finite simple graph, where $[\ell] = \{1, \dots, \ell\}$ is the vertex set, and $E \subset \{\{i, j\} \mid i, j \in [\ell], i \neq j\}$ is the set of edges.

Definition 12.2.10 Let $G = ([\ell], e)$ be a graph and n be a positive integer. A map $c : [\ell] \longrightarrow [n]$ is called a *vertex coloring* (with n-colors) of G if $c(i) \neq c(j)$ whenever $\{i, j\} \in E$. The number of vertex colorings with n-colors is denoted by

$$\chi_G(n) := \#\{c : [\ell] \longrightarrow [n] \mid c \text{ is a coloring}\}.$$

As we will see below, the function $\chi_G(n)$ is known to be a polynomial in n, called the *chromatic polynomial*. Furthermore, it is the characteristic polynomial of the following graphical arrangement.

Definition 12.2.11 Let $G = ([\ell], e)$ be a graph. Define the *graphical arrangement* $\mathcal{A}_G$ by

$$\mathcal{A}_G = \{H_{ij} \mid \{i, j\} \in E\}.$$

Note that if G is the complete graph, then $\mathcal{A}_G$ is equal to $\mathrm{Br}(\ell)$. Hence $\mathcal{A}_G$ is a subarrangement of the braid arrangement.

Proposition 12.2.12 *Let $G = ([\ell], e)$ be a graph. Then $\chi_G(t) = \chi(\mathcal{A}_G, t)$.*

Proof The proof is by induction on the number of edges. In the case $E = \emptyset$, $\mathcal{A}_G$ is the empty arrangement. Then it is clear that $\chi_G(t) = \chi(\mathcal{A}_G, t) = t^\ell$. Suppose that $E \neq \emptyset$. Choose an edge $e = \{i, j\} \in E$ and consider the corresponding hyperplane $H_e = H_{ij}$. Compare the set of colorings of graphs G, the deletion $G \setminus e$, and the contraction G/e (Fig. 12.2). Let $c : [\ell] \longrightarrow [n]$ be a coloring of the deletion $G \setminus e$. The set of colorings of $G \setminus e$ is decomposed into two subsets according to $c(i) \neq c(j)$ or $c(i) = c(j)$. If $c(i) \neq c(j)$, it is equivalent to a coloring of G, and if

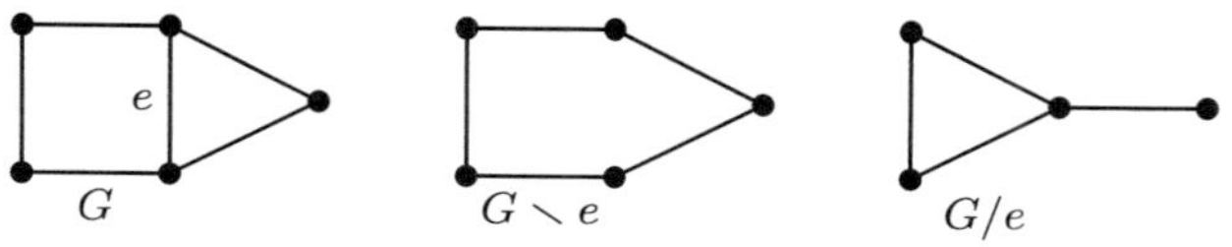

Fig. 12.2 The deletion $G \smallsetminus e$ and the contraction G/e

$c(i) = c(j)$, it is equivalent to a coloring of G/e. Thus we have

$$\{\text{Coloring of } G \smallsetminus e\} = \{\text{Coloring of } G\} \sqcup \{\text{Coloring of } G/e\},$$

This leads to the following recursive formula

$$\chi_G(n) = \chi_{G \smallsetminus e}(n) - \chi_{G/e}(n). \tag{12.3}$$

We also note that the graphical arrangement of the contraction G/e can be identified with the restriction of $\mathcal{A}_G$ on H_e, in other words, $\mathcal{A}_{G/e} = (\mathcal{A}_G)^{H_e}$. Thus we have

$$\begin{aligned}
\chi_G(t) &= \chi_{G \smallsetminus e}(t) - \chi_{G/e}(t) \\
&= \chi(\mathcal{A}_{G \smallsetminus e}, t) - \chi(\mathcal{A}_{G/e}, t) \\
&= \chi(\mathcal{A}_G \setminus \{H_e\}, t) - \chi((\mathcal{A}_G)^{H_e}, t) \\
&= \chi(\mathcal{A}, t).
\end{aligned}$$

□

As we will see, the coefficients of the characteristic polynomial are equal to the Betti numbers of complement of a complex hyperplane arrangement. Recently, a generalization, *characteristic quasi-polynomials* has also been intensively studied in connection with topology of toric arrangements in $(\mathbb{C}^\times)^\ell$ [28, 33, 36].

12.2.6 Orlik-Solomon Algebras

Let $\mathcal{A} = \{H_1, \dots, H_n\}$ be an arrangement. Let $E = \bigcup_{i=1}^n \mathbb{Z}e_i$ be a free $\mathbb{Z}$-module generated by $e_1, \dots, e_n$ and let $\wedge E$ be its exterior algebra. For an (ordered) subset $I = \{i_1, i_2, \dots, i_p\} \subset [n]$, denote $e_I = e_{i_1} \wedge \cdots \wedge e_{i_p}$. We also denote

$$\partial e_I = \sum_{\alpha=1}^{p} (-1)^{\alpha-1} e_{i_1} \wedge \cdots \wedge \widehat{e_{i_\alpha}} \wedge \cdots \wedge e_{i_p}.$$

The Orlik-Solomon algebra is defined as the quotient of $\wedge E$ by the following Orlik-Solomon ideal.

Definition 12.2.13

(1) A subset $I \subset [n]$ is called dependent if $H_I \neq \emptyset$ and $\operatorname{codim} H_I < \#I$, where $H_I = \bigcap_{i \in I} H_i$.
(2) The Orlik-Solomon ideal $\mathcal{I}(\mathcal{A})$ of $\wedge E$ is the ideal generated by the following two types of elements.

$$\mathcal{I}(\mathcal{A}) = \langle e_I \mid I \subset [n], H_I = \emptyset \rangle + \langle \partial e_I \mid I \subset [n] \text{ is dependent} \rangle .$$

(3) The Orlik-Solomon algebra $\mathrm{OS}^\bullet(\mathcal{A})$ is defined as follows.

$$\mathrm{OS}^\bullet(\mathcal{A}) = \wedge E / \mathcal{I}(\mathcal{A}).$$

Since the Orlik-Solomon ideal is generated by homogeneous elements, $\mathrm{OS}^\bullet(\mathcal{A})$ is a graded algebra. The Hilbert series of the Orlik-Solomon algebra $\mathrm{Hilb}(\mathrm{OS}^\bullet(\mathcal{A}), t) = \sum_k \operatorname{rank}_{\mathbb{Z}} \mathrm{OS}^k(\mathcal{A}) \cdot t^k$ is related to the characteristic polynomial by the following formula.

$$\mathrm{Hilb}(\mathrm{OS}^\bullet(\mathcal{A}), t) = (-t)^\ell \chi(\mathcal{A}, -\frac{1}{t}).$$

Example 12.2.14 Let $\mathcal{A}$ be as in Example 12.2.9. Then, the Orlik-Solomon ideal is

$$\mathcal{I}(\mathcal{A}) = \langle e_{23}, e_{45}, e_{24} - e_{14} + e_{12}, e_{35} - e_{15} + e_{13} \rangle.$$

It is easily seen that $1 \in \mathrm{OS}^0(\mathcal{A})$, $e_1, e_2, e_3, e_4, e_5 \in \mathrm{OS}^1(\mathcal{A})$ and

$$e_{25}, e_{34}, e_{12}, e_{24}, e_{13}, e_{35} \in \mathrm{OS}^2(\mathcal{A})$$

form a basis of $\mathrm{OS}^\bullet(\mathcal{A})$.

12.2.7 Cohomology Ring of the Complement of Complex Arrangements

Let $\mathcal{A} = \{H_1, \ldots, H_n\}$ be an arrangement in $V = \mathbb{C}^\ell$. Fix a defining equation α_i for each hyperplane. Then

$$\omega_i = \frac{d \log \alpha_i}{2\pi\sqrt{-1}} = \frac{1}{2\pi\sqrt{-1}} \frac{d\alpha_i}{\alpha_i}$$

is clearly a closed 1-form on $M(\mathcal{A})$ and gives an element in $H^1(M(\mathcal{A}), \mathbb{Z})$.

Proposition 12.2.15 (Orlik-Solomon [37])

(1) *The map* $e_i \longmapsto \omega_i$ *induces an isomorphism* $\mathrm{OS}^\bullet(\mathcal{A}) \xrightarrow{\simeq} H^\bullet(M(\mathcal{A}), \mathbb{Z})$ *of graded* $\mathbb{Z}$*-algebras.*
(2) *The Poincaré polynomial of the complement* $M(\mathcal{A})$ *is equal to* $(-t)^\ell \chi(\mathcal{A}, -1/t)$. *In other words, let* $b_i = b_i(M(\mathcal{A}))$, *then* $\chi(\mathcal{A}, t) = t^\ell - b_1 t^{\ell-1} + \cdots + (-1)^\ell b_\ell$.

Combining Propositions 12.2.15 and 12.2.8, we have the following.

Corollary 12.2.16 *Let* $\mathcal{A} = \{H_1, \ldots, H_n\}$ *be an essential arrangement in* $\mathbb{R}^\ell$. *Then*

$$\#\mathrm{ch}(\mathcal{A}) = \sum_{i=0}^{\ell} b_i(M(\mathcal{A})),$$

$$\#\mathrm{bch}(\mathcal{A}) = \sum_{i=0}^{\ell} (-1)^{\ell-i} b_i(M(\mathcal{A})),$$

The defining equations $\alpha_1, \ldots, \alpha_n$ give an affine linear map

$$\boldsymbol{\alpha} = (\alpha_1, \ldots, \alpha_n) : \mathbb{C}^\ell \longrightarrow \mathbb{C}^n \tag{12.4}$$

which induces the map of complements

$$\boldsymbol{\alpha}|_{M(\mathcal{A})} : M(\mathcal{A}) \longrightarrow (\mathbb{C}^\times)^n. \tag{12.5}$$

Note that $\mathcal{A}$ is essential if and only if $\boldsymbol{\alpha}$ is injective. The description of the cohomology ring in Proposition 12.2.15 implies the surjectivity of the pull-back of this map. More generally, we can prove injectivity (resp. surjectivity) of the induced map on the homology (resp. cohomology) for inclusions of affine spaces as follows.

Proposition 12.2.17 *Let* $\mathcal{A} = \{H_1, \ldots, H_n\}$ *be an arrangement in* $V = \mathbb{C}^\ell$, *and* $M = M(\mathcal{A})$ *be its complement. Let* $F \subset V$ *be a k-dimensional affine subspace which is not contained in* H_i *for any* $1 \leq i \leq n$. *Then the inclusion* $\iota : F \cap M \hookrightarrow M$ *induces the surjection* $\iota^* : H^j(M, \mathbb{Z}) \to H^j(M \cap F, \mathbb{Z})$ *of cohomology groups and the injection* $\iota_* : H_j(F \cap M, \mathbb{Z}) \hookrightarrow H_j(M, \mathbb{Z})$ *of homology groups* ($j \geq 0$). *If F is generic with respect to* $\mathcal{A}$ *(and the hyperplane at infinity), it is isomorphic for* $j \leq k$.

Proof The inclusions $F \cap M \hookrightarrow M \hookrightarrow (\mathbb{C}^\times)^n$ induce ring homomorphisms $\wedge E \to H^*(M, \mathbb{Z}) \to H^*(F \cap M, \mathbb{Z})$. By Proposition 12.2.15, the composition of these homomorphisms is surjective, hence $\iota^* : H^j(M, \mathbb{Z}) \to H^j(M \cap F, \mathbb{Z})$ is surjective. The injectivity of ι_* is its dual assertion. □

12.3 Chambers and Adjacency Relation

Let $\mathcal{A} = \{H_1, \ldots, H_n\}$ be an arrangement of affine hyperplanes in $\mathbb{R}^\ell$. For simplicity, we assume $\mathcal{A}$ is essential. In this section, we summarize the basic notions related to real stratifications.

First let

$$\mathcal{F} : F^0 \subset F^1 \subset \cdots \subset F^{\ell-1} \subset F^\ell = \mathbb{R}^\ell$$

be a generic flag. Associated with the flag, let us define

$$\mathrm{ch}^k_{\mathcal{F}}(\mathcal{A}) = \{C \in \mathrm{ch}(\mathcal{A}) \mid C \cap F^k \neq \emptyset, C \cap F^{k-1} = \emptyset\} \tag{12.6}$$

for $k = 0, \ldots, \ell$, where $F^{-1} = \emptyset$. Then we have the following.

Proposition 12.3.1 ([61, Proposition 2.3.2])

$$\#\,\mathrm{ch}^k_{\mathcal{F}}(\mathcal{A}) = b_k(M(\mathcal{A})). \tag{12.7}$$

(Fig. 12.3).

Proof The proof is by induction on ℓ. If $\ell = 1$, clearly we have $\#\,\mathrm{ch}^0_{\mathcal{F}}(\mathcal{A}) = 1, \#\,\mathrm{ch}^1_{\mathcal{F}}(\mathcal{A}) = n$, which are equal to $b_0(M(\mathcal{A}))$ and $b_1(M(\mathcal{A}))$, respectively. Suppose $\ell > 1$. Since the subspace $F^{\ell-1}$ is generic, $L(\mathcal{A} \cap F^{\ell-1})$ is isomorphic to

$$L(\mathcal{A})_{\leq \ell-1} = \{X \in L(\mathcal{A}) \mid \mathrm{codim}\, X \leq \ell - 1\}.$$

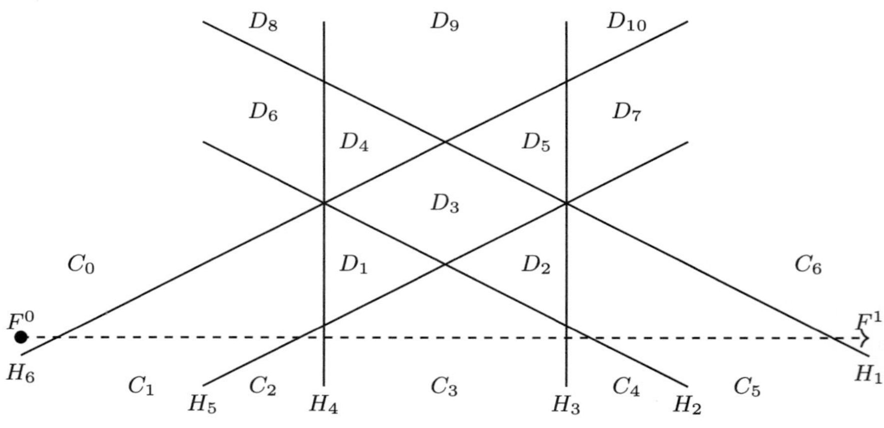

Fig. 12.3 A flag and associate sets of chambers: $\mathrm{ch}^0_{\mathcal{F}}(\mathcal{A}) = \{C_0\}, \mathrm{ch}^1_{\mathcal{F}}(\mathcal{A}) = \{C_1, \ldots, C_6\}$, $\mathrm{ch}^2_{\mathcal{F}}(\mathcal{A}) = \{D_1, \ldots, D_{10}\}$

Thus, by Definition 12.2.6 and Proposition 12.2.15, we have $b_k(M(\mathcal{A})) = b_k(M(\mathcal{A} \cap F^{\ell-1}))$ for $0 \le k \le \ell - 1$. Then by Proposition 12.2.16, we have

$$\#\{C \in \mathrm{ch}(\mathcal{A}) \mid C \cap F^{\ell-1} \neq \emptyset\} = \sum_{k=0}^{\ell-1} b_k(M(\mathcal{A})).$$

Hence $\#\,\mathrm{ch}^{\ell}_{\mathcal{F}}(\mathcal{A}) = b_\ell(M(\mathcal{A}))$. □

Remark 12.3.2 For simplicity, in this article, we pose an extra assumption on the flag. We consider only flags satisfying the following property: F^{k-1} does not separates 0-dimensional intersections of the arrangement $\mathcal{A} \cap F^k$. In other words, $F^{\ell-1}$ does not separates 0-dimensional intersections of $\mathcal{A}$, $F^{\ell-2}$ does not separates 0-dimensional intersections of the induced arrangement $\mathcal{A} \cap F^{\ell-1}$ on $F^{\ell-1}$, and so on. We call such a flag "near to the hyperplane at infinity."

Each hyperplane $H \in \mathcal{A}$ determines two regions. We choose a "positive side" for each $H \in \mathcal{A}$. This leads to a *sign vector* $\sigma(x) \in \{\pm, 0\}^{\mathcal{A}}$ for every point $x \in \mathbb{R}^\ell$. The set of all points $x \in \mathbb{R}^\ell$ having the same sign vector forms a *cell*. Every cell is relatively open and convex. The set $\mathbb{R}^\ell$ is decomposed into a disjoint union of cells. The cells form the *face poset* $\mathcal{F}(\mathcal{A})$, ordered by the inclusion of their topological closures. Note that the maximal cells are the chambers of $\mathcal{A}$ (Fig. 12.4).

Next, we define the adjacency graph of $\mathcal{A}$.

Definition 12.3.3 For two chambers $C_1, C_2 \in \mathrm{ch}(\mathcal{A})$, let $\mathrm{Sep}(C_1, C_2) = \{H \in \mathcal{A} \mid H \text{ separates } C_1 \text{ and } C_2\}$ be the set of separating hyperplanes. C_1 and C_2 are said to be adjacent if $\#\,\mathrm{Sep}(C_1, C_2) = 1$. The *adjacency graph* of $\mathcal{A}$ is the graph $\Gamma(\mathcal{A})$ with the vertex set $\mathrm{ch}(\mathcal{A})$ and the edges corresponding to adjacent pairs of chambers (Fig. 12.5).

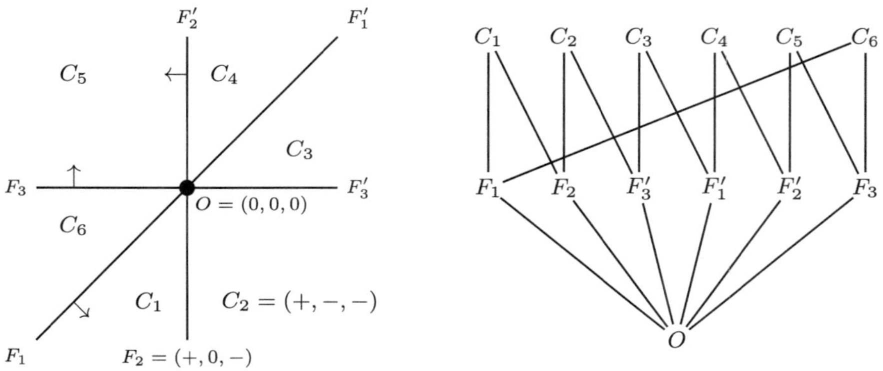

Fig. 12.4 Sign vectors and the face poset

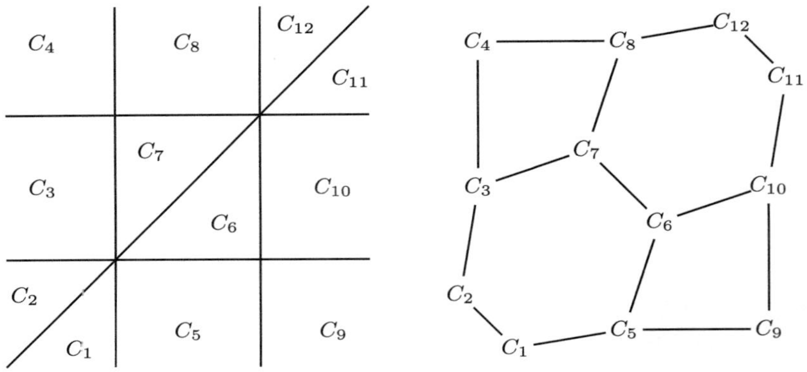

Fig. 12.5 Adjacency graph

12.4 Complexified Complement

Let $x \in \mathbb{R}^\ell$ and $v \in T_x\mathbb{R}^\ell$ be a tangent vector at x. Since we can canonically identify $T_x\mathbb{R}^\ell \simeq \mathbb{R}^\ell$, we can regard v as an element of $\mathbb{R}^\ell$. We identify the total space of the tangent bundle $T\mathbb{R}^\ell$ with $\mathbb{C}^\ell$ by the map: $v \in T_x\mathbb{R}^\ell \longmapsto x + \sqrt{-1} \cdot v \in \mathbb{C}^\ell$ (Fig. 12.6).

Let $H \subset V = \mathbb{R}^\ell$ be a hyperplane defined by a linear form $\alpha \in V^*$. Then the complex point $x + \sqrt{-1} \cdot v$ is contained in the complexified hyperplane $H_\mathbb{C} = H \otimes \mathbb{C}$ if and only if

$$\alpha(x + \sqrt{-1} \cdot v) = \alpha(x) + \sqrt{-1} \cdot \alpha(v) = 0,$$

which is equivalent to both the real part (x) and the imaginary part (v) being contained in H.

Let $\mathcal{A} = \{H_1, \ldots, H_n\}$ be a central arrangement in $V = \mathbb{R}^\ell$. Using the above description, one can describe the complexified complement $M(\mathcal{A}) = \mathbb{C}^\ell \setminus \bigcup_{i=1}^n H_\mathbb{C}$ as follows (see also Fig. 12.6).

Fig. 12.6 Vectors in the complexified complement

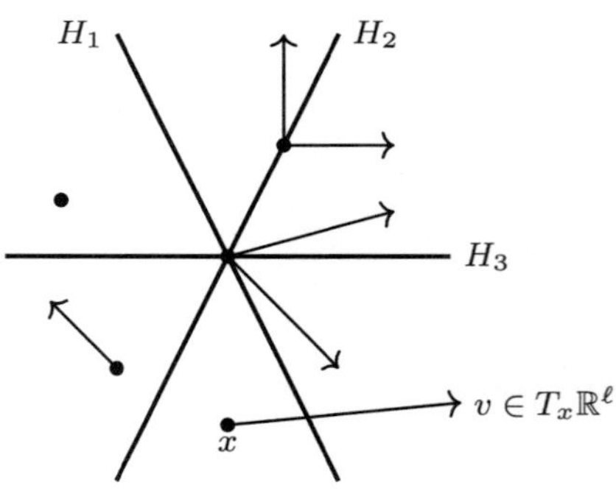

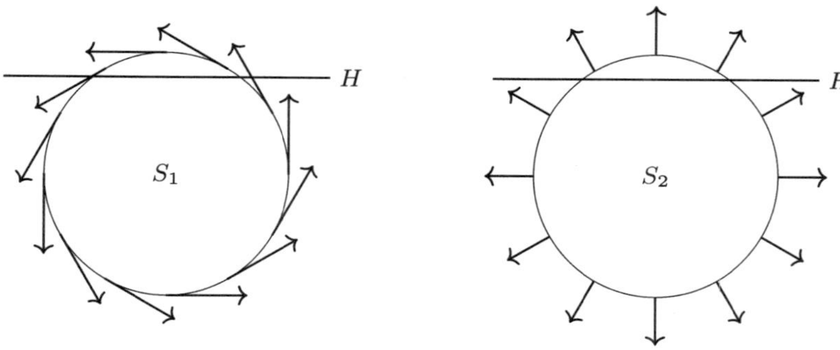

Fig. 12.7 Linked and unlinked embedded circles

$$\{x + \sqrt{-1} \cdot v \mid \text{ if } x \in H_i \text{ for some } 1 \leq i \leq n, \text{ then } v \notin T_x H_i\}.$$

Example 12.4.1 Let S_1 be the set of tangent vectors of a circle in $\mathbb{R}^2$ (left of Fig. 12.7), and S_2 be the set of radial vectors of a circle in $\mathbb{R}^2$ (right of Fig. 12.7). Then S_1 gives a nontrivial cycle in the complement of the complexified line H, while S_2 gives a trivial cycle.

12.5 Fundamental Groups

12.5.1 Commutator Relation

We start with two lines $\mathcal{A} = \{H_1, H_2\}$ in $\mathbb{R}^2$ (left of Fig. 12.8). The complement $M(\mathcal{A})$ is homeomorphic to

$$\{(z_1, z_2) \in \mathbb{C}^2 \mid z_1 \neq 0, z_2 \neq 0\} = \mathbb{C}^\times \times \mathbb{C}^\times.$$

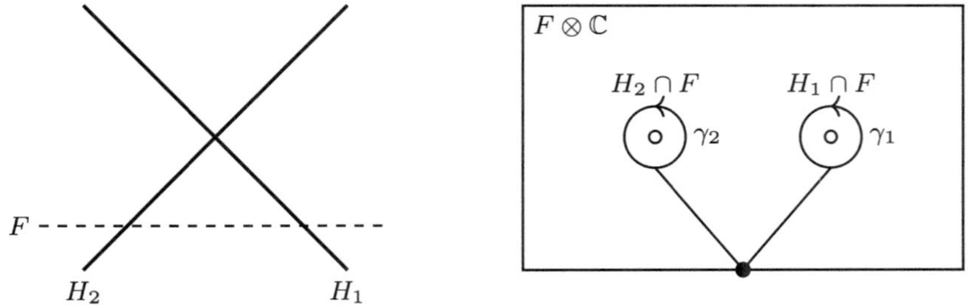

Fig. 12.8 Two lines $\mathcal{A} = \{H_1, H_2\}$

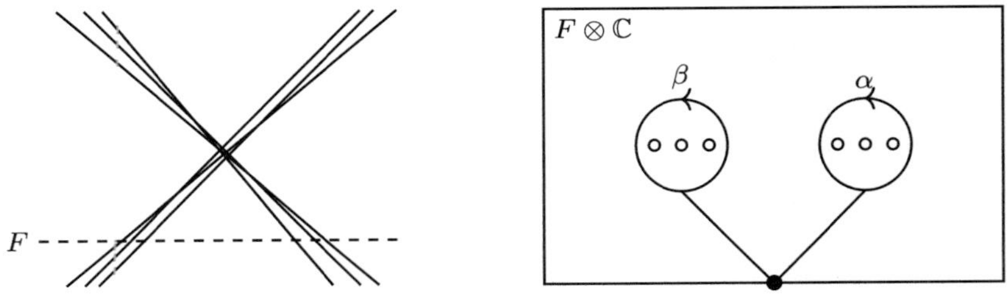

Fig. 12.9 Commutative loops $\alpha\beta = \beta\alpha$

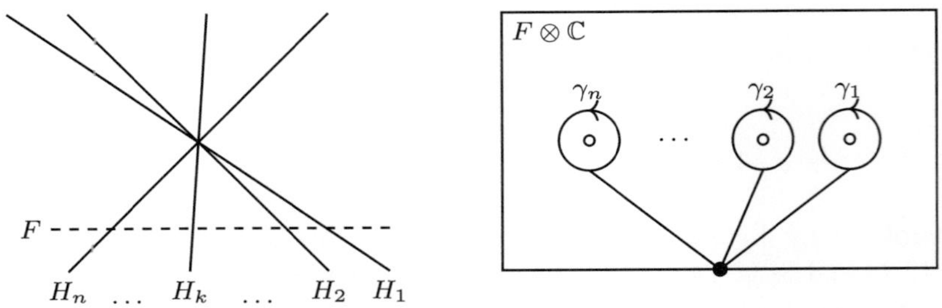

Fig. 12.10 Commutative relation $[\gamma_1 \cdots \gamma_k, \gamma_{k+1} \cdots \gamma_n] = 1$

Hence, $M(\mathcal{A})$ is homotopy equivalent to the 2-torus $(S^1)^2$. The fundamental group is abelian and isomorphic to $\mathbb{Z}^2$. To describe the generators, we first fix a generic line F and take meridians as in the right of Fig. 12.8. In particular, γ_1 and γ_2 are commutative.

Next, we consider the line arrangement as in Fig. 12.9. This arrangement can be considered as a small perturbation of the above $\mathcal{A} = \{H_1, H_2\}$. Hence, the commutativity of γ_1 and γ_2 is preserved. More systematically, we have the following.

Proposition 12.5.1 *Let $\mathcal{A} = \{H_1, \dots, H_n\}$ be an arrangement of n lines intersecting at a point. Let $\gamma_1, \dots, \gamma_n$ be meridians as in Fig. 12.10. Then for each $1 \le k < n$, $\gamma_1\gamma_2 \dots \gamma_k$ and $\gamma_{k+1} \dots \gamma_n$ are commutative. In other words, we have the following cyclic relation.*

$$\gamma_1\gamma_2 \cdots \gamma_n = \gamma_2 \cdots \gamma_n\gamma_1 = \gamma_3 \cdots \gamma_n\gamma_1\gamma_2 = \cdots = \gamma_n\gamma_1 \cdots \gamma_{n-1} \tag{12.8}$$

Roughly speaking, the cyclic relations (12.8) generate all relations in the fundamental group of $M(\mathcal{A})$ for any arrangement.

12.5.2 Presentations

Let $\mathcal{A} = \{H_1, \ldots, H_n\}$ be an affine arrangement in $\mathbb{R}^\ell$. By a version of the Lefschetz hyperplane section theorem [24], if F is a generic affine subspace of complex dimension 2, the inclusion $i : F \cap M(\mathcal{A}) \hookrightarrow M(\mathcal{A})$ induces an isomorphism

$$i_* : \pi_1(F \cap M(\mathcal{A})) \xrightarrow{\simeq} \pi_1(M(\mathcal{A})).$$

Thus, when we consider the fundamental groups, it is enough to consider line arrangements in $\mathbb{R}^2$. There are many ways to give presentations of the fundamental group. Among others (e.g., [9, 27, 49, 50, 56]), we give two of them.

Now, let $\mathcal{A} = \{H_1, \ldots, H_n\}$ be a line arrangement in $V = \mathbb{R}^2$. Let $F^0 \subset F^1$ be a generic flag in V. Let $f(x)$ be a polynomial of degree one such that $F^1 = \{f = 0\}$, and α_i $(i = 1, \ldots, n)$ be the defining equation of H_i. Choose an orientation of F^1. For simplicity, we assume the following (Fig. 12.3).

- All intersections of $\mathcal{A}$ are contained in the half-space $\{f > 0\}$.
- The point F^0 is contained in the half-space $H_i^- := \{\alpha_i < 0\}$ for $i = 1, \ldots, n$.
- On F^1, the order is $F^0 < H_n \cap F^1 < \cdots < H_2 \cap F^1 < H_1 \cap F^1$, with respect to the orientation of F^1.

The set of generators is

$$\{\gamma_X \mid X \text{ is a 1-dimensional face}\}. \tag{12.9}$$

At each point of multiplicity m, we impose the following relations (Fig. 12.11).

$$\gamma_{Y_i} = (\gamma_{X_1} \cdots \gamma_{X_{i-1}})\gamma_{X_i}(\gamma_{X_1} \cdots \gamma_{X_{i-1}})^{-1}, \text{ for } i = 1, \ldots, m, \tag{12.10}$$

$$[\gamma_{X_1} \cdots \gamma_{X_i}, \gamma_{X_{i+1}} \cdots \gamma_{X_m}] = 1, \text{ for } i = 1, \ldots, m-1. \tag{12.11}$$

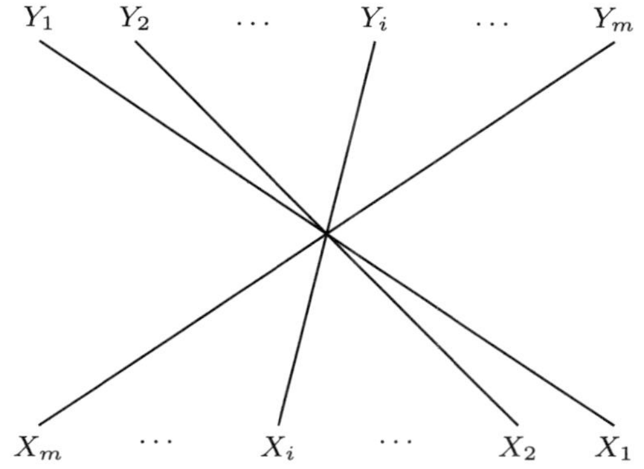

Fig. 12.11 1-Dimensional faces around a point of multiplicity m

Theorem 12.5.2 ([19, 45]) *For a line arrangement $\mathcal{A} = \{H_1, \dots, H_n\}$ in $\mathbb{R}^2$, the fundamental group $\pi_1(M(\mathcal{A}))$ is presented by the generators (12.9) and relations (12.10) and (12.11).*

Note that one can reduce the number of generators to n by using the relations (12.10). This gives a minimal presentation of the fundamental group.

Next, we give a minimal presentation directly. The relations are indexed by

$$\mathrm{ch}^2_{\mathcal{F}}(\mathcal{A}) = \{C \in \mathrm{ch}(\mathcal{A}) \mid C \cap F^1 = \emptyset\}.$$

To each $D \in \mathrm{ch}^2_{\mathcal{F}}(\mathcal{A})$ we will associate a relation.

- Let $\{i \in [n] \mid \alpha_i(D) < 0\} = \{i_1, i_2, \dots, i_p\}$ with $i_1 < \cdots < i_p$.
- Let $\{i \in [n] \mid \alpha_i(D) > 0\} = \{j_1, j_2, \dots, j_q\}$ with $j_1 < \cdots < j_q$.
- Define the relation $R(D)$ as follows.

$$\gamma_1 \cdots \gamma_n = \gamma_{i_1} \cdots \gamma_{i_p} \gamma_{j_1} \cdots \gamma_{j_q}$$

Theorem 12.5.3 ([65]) *The fundamental group $\pi_1(M(\mathcal{A}))$ is presented as*

$$\langle \gamma_1, \dots, \gamma_n \mid R(D), D \in \mathrm{ch}^2_{\mathcal{F}}(\mathcal{A}) \rangle \tag{12.12}$$

This can be proved by using Proposition 12.6.13 in Sect. 12.6.6, see [65] for details.

Example 12.5.4 Let $\mathcal{A} = \{H_1, \dots, H_6\}$ be the line arrangement as in Fig. 12.3 (see also Fig. 12.12). The presentation in Theorem 12.5.2 is reduced to the presentation

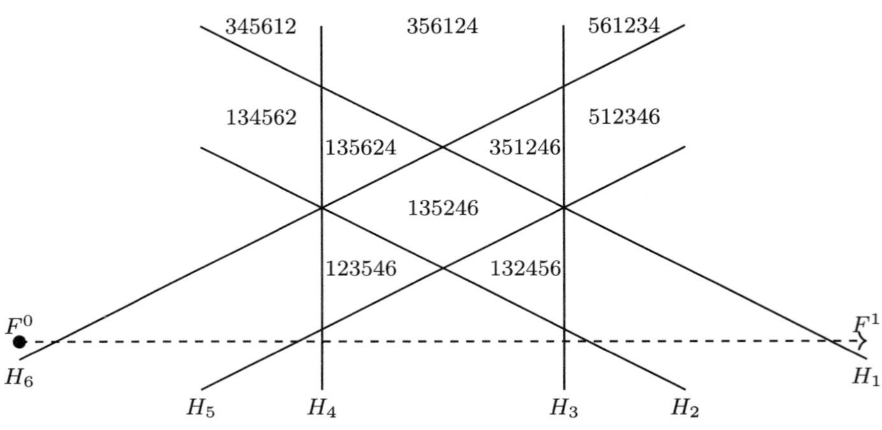

Fig. 12.12 Chamber relations

$$\left\langle \gamma_1, \ldots, \gamma_6 \;\middle|\; \begin{array}{l} [\gamma_2, \gamma_3] = [\gamma_4, \gamma_5] = [\gamma_2, \gamma_5] = \\ [\gamma_1\gamma_3, \gamma_5] = [\gamma_1, \gamma_3\gamma_5] = [\gamma_2\gamma_4, \gamma_6] = [\gamma_2, \gamma_4\gamma_6] = \\ [\gamma_1, \gamma_6] = [\gamma_1, \gamma_2\gamma_4\gamma_2^{-1}] = [\gamma_1\gamma_3\gamma_1^{-1}, \gamma_6] = 1 \end{array} \right\rangle.$$

The presentation in Theorem 12.5.3 is as follows.

$$\left\langle \gamma_1, \ldots, \gamma_6 \;\middle|\; \begin{array}{l} 123456 = \\ 132456 = 123546 = 135246 = 135624 = 351246 = \\ 134562 = 345612 = 356124 = 561234 = 512346 \end{array} \right\rangle$$

Remark 12.5.5 Let us compare the fundamental groups of an arrangement $\mathcal{A}$ and its generic perturbation $\mathcal{A}'$. The minimal homogeneous presentation of $\pi_1(M(\mathcal{A}'))$ is obtained from that of $\pi_1(M(\mathcal{A}))$ by adding several relations determined by the new chambers. Since $\pi_1(M(\mathcal{A}'))$ is isomorphic to the free abelian group $\mathbb{Z}^n$, we have that $\pi_1(M(\mathcal{A}))$ is a group generated by n elements, and adding $\frac{n(n-1)}{2} - b_1$ relations makes it the free abelian group of rank n (Fig. 12.13).

Remark 12.5.6 CW complexes associated with the minimal presentations in Theorems 12.5.2 and 12.5.3 are known to be homotopy equivalent to the complement $M(\mathcal{A})$. A natural question is whether or not the CW complex associated with any minimal presentation of $\pi_1(M(\mathcal{A}))$ is homotopy equivalent to $M(\mathcal{A})$. An affirmative answer to this question yields the following: let $\mathcal{A}_1$ and $\mathcal{A}_2$ be line arrangements in $V = \mathbb{C}^2$. If $\pi_1(M(\mathcal{A}_1)) \simeq \pi_1(M(\mathcal{A}_2))$, then $M(\mathcal{A}_1)$ and $M(\mathcal{A}_2)$ are homotopy equivalent.

12.5.3 Deligne Groupoid

In this section, we construct a category $\mathrm{Gal}^+(\mathcal{A})$ of chambers and paths (galleries), and the Deligne groupoid $\mathrm{Gal}(\mathcal{A})$. These were first introduced by Deligne [11] for simplicial arrangements and later Paris formulated for any real arrangements [43].

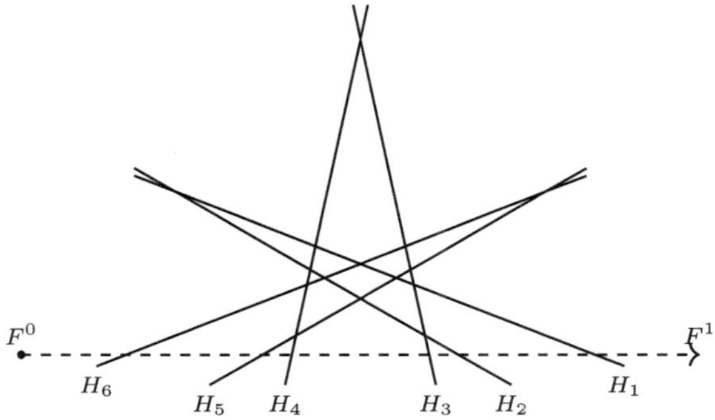

Fig. 12.13 A generic perturbation of Fig. 12.12

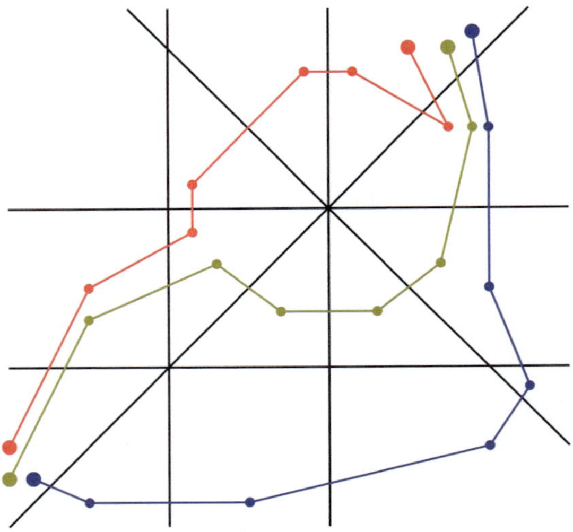

Fig. 12.14 Geodesic flips and equivalent paths

A path of length k is a sequence $(C_0, C_1, \dots, C_k)$ of chambers such that C_{i-1} and C_i are adjacent ($i = 1, \dots, k$). A path $(C_0, \dots, C_k)$ is said to be *geodesic* if $\#\operatorname{Sep}(C_0, C_k) = k$. We construct the category of galleries by identifying geodesics.

Definition 12.5.7 Two paths $(C_0, \dots, C_k)$ and $(D_0, \dots, D_k)$ are related by a *geodesic flip* if there exist indices $0 \leq i < j \leq k$ such that

- $C_0 = D_0, \dots, C_i = D_i$ and $C_j = D_j, \dots, C_k = D_k$.
- both $(C_i, C_{i+1}, \dots, C_j)$ and $(D_i, D_{i+1}, \dots, D_j)$ are geodesic. We denote $\sim$ as the equivalence relation generated by finitely many geodesic flips (Fig. 12.14).

From the adjacency graph, one can define the following categories of galleries.

Definition 12.5.8 $\operatorname{Gal}^+(\mathcal{A})$ is the category such that

- The set of objects is $\operatorname{obj}(\operatorname{Gal}^+(\mathcal{A})) = \operatorname{ch}(\mathcal{A})$.
- The set of morphisms from C to C' is

$$\operatorname{Hom}^+(C, C') = \left\{ (C_0, C_1, \dots, C_n) \,\middle|\, \begin{array}{l} n \geq 0, C_i \in \operatorname{ch}(\mathcal{A}), \\ C_0 = C, C_n = C' \end{array} \right\} \Big/ \sim,$$

where $\sim$ is the equivalence relation as above.

By inverting all morphisms (taking the free groupoid [35]), we obtain the category $\operatorname{Gal}(\mathcal{A})$. To formulate the inverse of the morphisms, we consider signed paths

$$(C_0 *_{\varepsilon_1} C_1 *_{\varepsilon_2} \cdots *_{\varepsilon_k} C_k),$$

where $\varepsilon_i \in \{\pm\}$. If all the signs are $+$, we identify $(C_0 *_+ C_1 *_+ \cdots *_+ C_k)$, with the path $(C_0, C_1, \ldots, C_k)$. If all the signs are $-$, we identify $(C_0 *_- C_1 *_- \cdots *_- C_k)$, with the path $(C_k, C_{k-1}, \ldots, C_1)^{-1}$.

Definition 12.5.9 The Deligne groupoid $\mathrm{Gal}(\mathcal{A})$ is the category such that

- The set of objects is $\mathrm{obj}(\mathrm{Gal}(\mathcal{A})) = \mathrm{ch}(\mathcal{A})$.
- The set of morphisms from C to C' is

$$\mathrm{Hom}(C, C') = \left\{ (C_0 *_{\varepsilon_1} C_1 *_{\varepsilon_2} \cdots *_{\varepsilon_n} C_n) \,\middle|\, \begin{matrix} n \geq 0, C_i \in \mathrm{ch}(\mathcal{A}), \\ C_0 = C, C_n = C' \end{matrix} \right\} \Big/ \sim,$$

  where $\sim$ is the equivalence relation generated by geodesic flips as well as the following:

$$\begin{aligned}(\cdots *_{\varepsilon_k} C_k *_{\varepsilon_{k+1}} \cdots) &\sim (\cdots *_{\varepsilon_k} C_k *_+ D *_- C_k *_{\varepsilon_{k+1}} \cdots) \\ &\sim (\cdots *_{\varepsilon_k} C_k *_- D *_+ C_k *_{\varepsilon_{k+1}} \cdots),\end{aligned}$$

  where D is a chamber adjacent to C_k.

By the construction, all morphisms of $\mathrm{Gal}(\mathcal{A})$ are isomorphisms, hence $\mathrm{Gal}(\mathcal{A})$ is a groupoid. The Deligne groupoid $\mathrm{Gal}(\mathcal{A})$ recovers the fundamental group of $M(\mathcal{A})$.

Theorem 12.5.10 *Let* $C \in \mathrm{ch}(\mathcal{A})$. *Then*

$$\mathrm{Hom}(C, C) \simeq \pi_1(M(\mathcal{A}), p_C), \tag{12.13}$$

where, p_C *is an arbitrary point in* C.

The strategy of the proof of Theorem 12.5.10 is as follows. Choose a point $p_{C'} \in C'$ for each chamber C'. We give a map $\mathrm{Hom}(C, C) \longmapsto \pi_1(M(\mathcal{A}), p_C)$ by associating a curve in $M(\mathcal{A})$ to a signed paths $(C = C_0 *_{\varepsilon_1} C_1 *_{\varepsilon_2} \cdots *_{\varepsilon_k} C_k = C)$ as in Fig. 12.15. This correspondence gives the isomorphism (12.13).

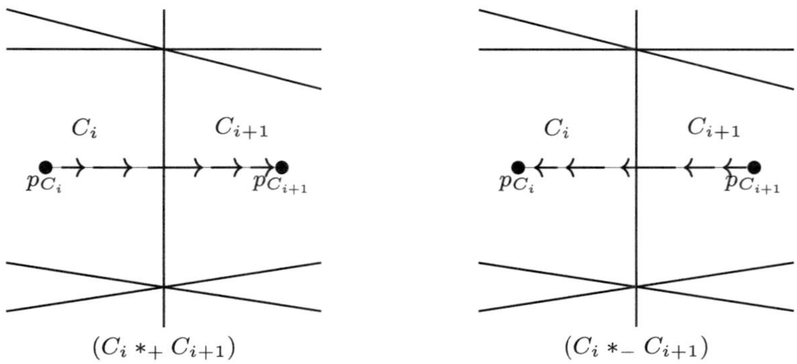

Fig. 12.15 Paths corresponding to $(C_i *_+ C_{i+1})$ and $C_i *_- C_{i+1}$

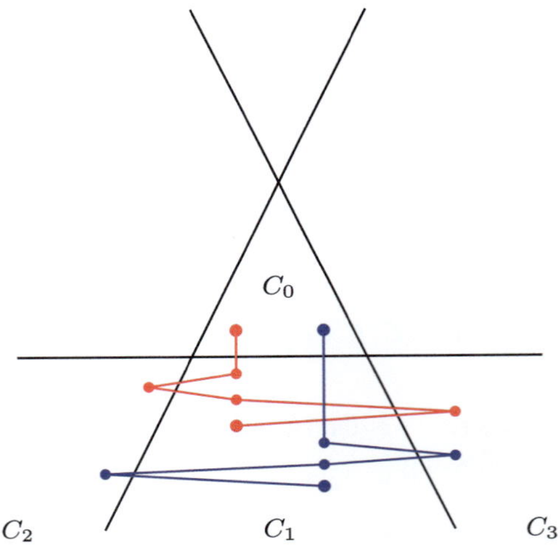

Fig. 12.16 $C_0C_1C_2C_1C_3C_1$ (red) and $C_0C_1C_3C_1C_2C_1$ (blue) are equivalent

Remark 12.5.11 Let $\mathcal{A}$ be a central and simplicial arrangement in $\mathbb{R}^\ell$. Deligne [11] proved that the localization functor $\rho : \mathrm{Gal}^+(\mathcal{A}) \longrightarrow \mathrm{Gal}(\mathcal{A})$ is faithful. This property was one of the crucial steps in his proof of the $K(\pi, 1)$ property.

However, in non simplicial cases, ρ is not faithful [61] (see Fig. 12.16). Non faithfulness seems to have a relationship with non vanishing of the second homotopy group [48].

12.6 Homotopy Type and Minimality

12.6.1 Salvetti Complex

Let $\mathcal{A} = \{H_1, \dots, H_n\}$ be an arrangement in $\mathbb{R}^\ell$. Choose a defining equation α_i for each hyperplane H_i. The hyperplane H_i decomposes $\mathbb{R}^\ell$ into two half-spaces, $H_i^+ = \{x \in \mathbb{R}^\ell \mid \alpha_i(x) > 0\}$ and $H_i^- = \{x \in \mathbb{R}^\ell \mid \alpha_i(x) < 0\}$. Each point $x \in \mathbb{R}^\ell$ determines a sign vector

$$(\mathrm{sign}(\alpha_i(x)) \mid i = 1, \dots, n) \in \{\pm, 0\}^n.$$

Let $X, Y : \mathcal{A} \longrightarrow \{\pm, 0\}$ be sign vectors. The *composition* $X \circ Y : \mathcal{A} \longrightarrow \{\pm, 0\}$ of X and Y is defined by

$$(X \circ Y)(H) = \begin{cases} X(H), & \text{if } X(H) \neq 0 \\ Y(H), & \text{if } X(H) = 0. \end{cases}$$

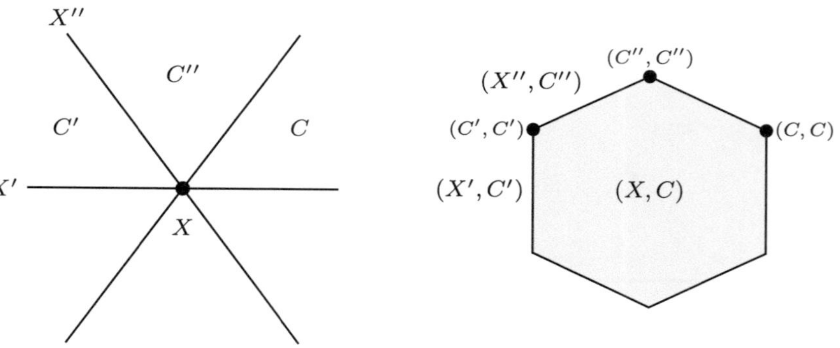

Fig. 12.17 Salvetti complex

Let

$$\mathrm{Sal}(\mathcal{A}) = \{(X, C) \in \mathcal{F}(\mathcal{A}) \times \mathrm{ch}(\mathcal{A}) \mid X \le C\}.$$

Define the partial order on $\mathrm{Sal}(\mathcal{A})$ as follows (Fig. 12.17):

$$(X', C') \le (X, C) \text{ if and only if } X \le X' \text{ and } X' \circ C = C'.$$

The Salvetti complex $\Delta_{\mathrm{Sal}}(\mathcal{A})$ is the regular CW complex that has this $\mathrm{Sal}(\mathcal{A})$ as its face poset.

Theorem 12.6.1 ([12, 43, 50]) *$\Delta_{\mathrm{Sal}}(\mathcal{A})$ is homotopy equivalent to $M(\mathcal{A})$.*

We illustrate the proof of Theorem 12.6.1 via nerve lemma.

Lemma 12.6.2 *Let $H^+ \subset \mathbb{R}^\ell$ be a half-space determined by a hyperplane H. Let $F \subset \mathbb{R}^\ell$ be an affine space. Let $d(x, F)$ be the Euclidean distance from a point x to F. Then*

$$U = \{x \in H^+ \mid d(x, F) < d(x, H)\}$$

is an open convex subset.

Proof Let $x_1, x_2 \in \mathbb{R}^\ell$. From the triangle inequality, we can easily prove

$$d(tx_1 + (1-t)x_2, F) \le t d(x_1, F) + (1-t) d(x_2, F)$$

for $0 \le t \le 1$. On the other hand, if $x_1, x_2 \in H^+$, then

$$d(tx_1 + (1-t)x_2, H) = t d(x_1, H) + (1-t) d(x_2, H).$$

Thus $x_1, x_2 \in U$ implies $tx_1 + (1-t)x_2 \in U$. □

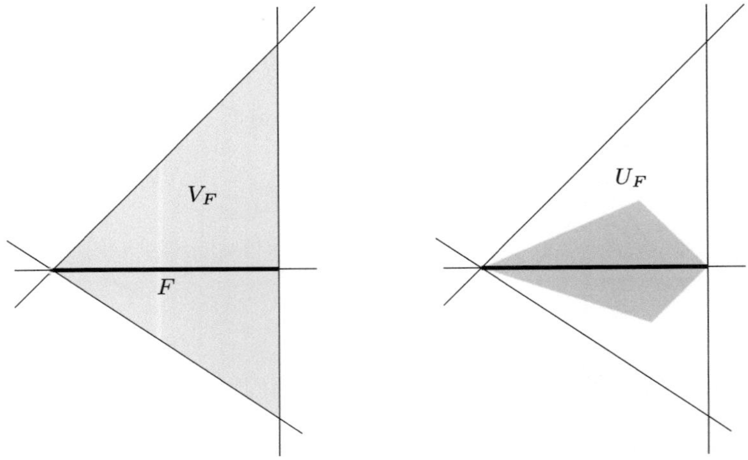

Fig. 12.18 V_F and U_F

Let $F \in \mathcal{F}(\mathcal{A})$ be a face with the sign vector $\sigma(F) = (\varepsilon_1, \ldots, \varepsilon_n) \in \{+, -, 0\}^n$. Define the neighborhood V_F as the intersection of half-spaces containing F,

$$V_F = \bigcap_{\varepsilon_i \neq 0} H^{\varepsilon_i}.$$

Next, define the subset U_F as

$$U_F = \{x \in V_F \mid d(x, F) < d(x, H_i) \text{ for } 1 \leq i \leq n \text{ with } \varepsilon_i \neq 0\}$$

(Fig. 12.18). Note that for a chamber C, $U_C = C$. Since U_F is an intersection of finitely many open convex subsets as in Lemma 12.6.2, U_F is open and convex. The collection $\{U_F \mid F \in \mathcal{F}(\mathcal{A})\}$ provides an open covering of $\mathbb{R}^\ell$. We can also prove the following.

Lemma 12.6.3 *Let $F_1, F_2 \in \mathcal{F}(\mathcal{A})$. Then $U_{F_1} \cap U_{F_2} \neq \emptyset$ if and only if F_1 and F_2 are comparable.*

Now we shall construct an open covering of $M(\mathcal{A})$. Let $(X, C) \in \mathrm{Sal}(\mathcal{A})$. Denote by $\widetilde{C} \in \mathrm{ch}(\mathcal{A}_X)$ the open convex cone corresponding to the unique chamber of $\mathcal{A}_X = \{H \in \mathcal{A} \mid H \supset X\}$ containing C. Note that $\widetilde{C}$ is an open convex cone. Associated with $(X, C) \in \mathrm{Sal}(\mathcal{A})$, let us define

$$Z_{(X,C)} := U_X + \sqrt{-1} \cdot \widetilde{C} = \{x + \sqrt{-1}y \mid x \in U_X, y \in \widetilde{C}\}. \tag{12.14}$$

Then clearly, $Z_{(X,C)} \simeq U_X \times \widetilde{C}$ is an open convex subset in $M(\mathcal{A})$. We can also prove the following.

Lemma 12.6.4

(1) *If* (X_1, C_1) *and* (X_2, C_2) *are not comparable in* $\mathrm{Sal}(\mathcal{A})$, *then* $Z_{(X_1,C_1)} \cap Z_{(X_2,C_2)} = \emptyset$.
(2) *If* $(X_1, C_1) \leq \cdots \leq (X_k, C_k)$ *in* $\mathrm{Sal}(\mathcal{A})$. *Then*

$$Z_{(X_1,C_1)} \cap \cdots \cap Z_{(X_k,C_k)} = \left(\bigcap_{i=1}^{k} U_{X_i}\right) \times \widetilde{C_k}.$$

In particular, $Z_{(X_1,C_1)} \cap \cdots \cap Z_{(X_k,C_k)} \neq \emptyset$ *if and only if* $(X_1, C_1), \ldots, (X_k, C_k)$ *form a chain in* $\mathrm{Sal}(\mathcal{A})$.

By Lemma 12.6.4 and the nerve lemma [25, Cor. 4G.3], [30, Theorem 15.21], the complement $M(\mathcal{A})$ is homotopy equivalent to $\Delta_{\mathrm{Sal}}(\mathcal{A})$. Thus we obtain Theorem 12.6.1.

12.6.2 Delucchi-Falk Complex

Delucchi and Falk [14] introduced a complex which can be described by using only the metric structure of the adjacency graph $\Gamma(\mathcal{A})$. The distance of two chambers $C_1, C_2 \in \mathrm{ch}(\mathcal{A})$ is

$$d(C_1, C_2) = \#\{H \in \mathcal{A} \mid H \text{ separates } C_1 \text{ and } C_2\}.$$

Definition 12.6.5 Define the order on $\mathrm{DF}(\mathcal{A}) := \mathrm{ch}(\mathcal{A}) \times \mathrm{ch}(\mathcal{A})$ by

$$(C, D) \leq (C', D') \Longleftrightarrow d(C', D') = d(C', C) + d(C, D) + d(D, D').$$

Theorem 12.6.6 ([14]) *The order complex (Delucchi-Falk complex) of* $\mathrm{DF}(\mathcal{A})$ *is homotopy equivalent to* $\Delta_{\mathrm{Sal}}(\mathcal{A})$.

Proof The strategy is to construct a poset map $\mathrm{Sal}(\mathcal{A}) \longrightarrow \mathrm{DF}(\mathcal{A})$ that induces the homotopy equivalence of order complexes. Let $(X, C) \in \mathrm{Sal}(\mathcal{A})$. Let $C' \in \mathrm{ch}(\mathcal{A})$ be the opposite chamber of C with respect to X. Then $(X, C) \longmapsto (C', C)$ gives a map $\mathrm{Sal}(\mathcal{A}) \longrightarrow \mathrm{DF}(\mathcal{A})$, which induces a homotopy equivalence between order complexes. (See [14] for details.) □

Remark 12.6.7 Notions in Sects. 12.6.1 and 12.6.2 are generalized to oriented matroids [5].

12.6.3 Minimality

Let X be a finite CW complex. In general, we have the following inequality

$$\#\{k\text{-dimensional cells}\} \geq b_k(X). \tag{12.15}$$

This means that the number of k-dimensional cells of a CW complex is bounded below by the k-th Betti number. A finite CW complex X is called *minimal* if the equality $\#\{k\text{-dimensional cells}\} = b_k(X)$ holds for every $k \geq 0$. This is equivalent to the condition that the boundary maps of the chain complex are vanishing. Hence, the k-dimensional cells of X form a basis of the homology of $H_k(X, \mathbb{Z})$. In general, a manifold does not have the homotopy type of a minimal CW complex. However, for the complement of a complex hyperplane arrangement, it is known to be minimal.

Proposition 12.6.8 (Dimca-Papadima [16], Randell [47]) *Let $\mathcal{A}$ be an essential hyperplane arrangement $\mathcal{A}$ in $\mathbb{C}^\ell$. Then the complement $M = M(\mathcal{A})$ is homotopic to an ℓ-dimensional minimal CW complex.*

Proof *(Proof of Proposition 12.6.8)* The proof is by induction on ℓ. If $\ell = 1$, then $M(\mathcal{A})$ is clearly homotopic to a bouquet of n circles.

Next, let $\ell > 1$. Let $F \subset \mathbb{C}^\ell$ be a generic hyperplane. By the affine Lefschetz hyperplane section theorem [24] (we will see later the idea of the proof of the affine Lefschetz hyperplane section theorem. See Sect. 12.6.4), $M(\mathcal{A})$ is homotopy equivalent to a space obtained by attaching ℓ-dimensional cells to $M(\mathcal{A}) \cap F$. The number of ℓ-cells is clearly $\operatorname{rank} H_\ell(M, M \cap F)$. Consider the homology long exact sequence of the pair of spaces $(M, M \cap F)$. We have

$$H_\ell(M \cap F) \xrightarrow{(\alpha)} H_\ell(M) \xrightarrow{(\beta)} H_\ell(M, M \cap F) \xrightarrow{(\gamma)} H_{\ell-1}(M \cap F) \xrightarrow{(\delta)} H_{\ell-1}(M)$$

Since $M \cap F$ is homotopy equivalent to an $(\ell - 1)$-dimensional CW complex, we have $H_\ell(M \cap F) = 0$. By Proposition 12.2.17, (δ) is isomorphic. Therefore, both (α) and (γ) are zero map, we have (β) is isomorphic, and $b_\ell(M) = \operatorname{rank} H_\ell(M)$ is equal to the number of ℓ-cells. □

12.6.4 Attaching Maps

The homotopy type of the minimal CW complex depends on the attaching maps of cells. The affine Lefschetz hyperplane section theorem [24] tells us that there exist continuous maps

$$\sigma_i : (D^\ell, \partial D^\ell) \longrightarrow (M, M \cap F),$$

$(i = 1, \dots, b_\ell(M))$, such that M is homotopy equivalent to

$$(M \cap F) \cup \bigcup_{i=1}^{b_\ell(M)} D^\ell.$$

This fact can be proved using Morse theory way as follows. Let $Q = \prod_{H \in \mathcal{A}} \alpha_H$ be the defining equation of $\mathcal{A}$. Let $F = \{f = 0\}$, where $f : \mathbb{C}^\ell \longrightarrow \mathbb{C}$ is a defining equation of the generic hyperplane F. Consider

$$\varphi = \left| \frac{f(x)^2}{Q(x)^{1/n}} \right| : M(\mathcal{A}) \longrightarrow \mathbb{R}_{\geq 0} \tag{12.16}$$

By Varchenko conjecture (solved in [39, 53]), φ has non-degenerate critical points with Morse index ℓ, and the number of critical points is equal to

$$(-1)^\ell \chi(M \setminus F) = b_\ell(M(\mathcal{A})).$$

(Note that, to apply solutions to Varchenko conjecture, we might have to perturb the exponents generically.) Recall that the gradient flow is a map $\phi : M \times \mathbb{R} \longrightarrow M$ satisfying

$$\frac{\partial}{\partial t}\phi(x, t) = -\operatorname{grad}\varphi(x). \tag{12.17}$$

For each critical point $p \in \operatorname{Crit}(\varphi)$, define the stable cell $\mathcal{S}_p$ and unstable cell $\mathcal{U}_p$ as

$$\mathcal{S}_p = \{x \in M \mid \lim_{t \to \infty} \phi(x, t) = p\}$$

$$\mathcal{U}_p = \{x \in M \mid \lim_{t \to -\infty} \phi(x, t) = p\}.$$

Then the boundary of $\mathcal{U}_p$ is attached to F. These cells form the minimal CW complex. For the explicit description of the cells, there are two main problems:

(a) Describe the set of critical points.
(b) Describe the unstable cells.

The problem (a) is an algebraic problem. Therefore, it might be possible for individual arrangements. However, (b) is a more transcendental problem. It would be difficult to solve (b) even for a fixed arrangement.

In the real arrangement case, let

$$\operatorname{ch}_F(\mathcal{A}) = \{C \in \operatorname{ch}(\mathcal{A}) \mid C \cap F = \emptyset\}.$$

Using the structure of chambers, one can solve the following weaker versions of (a) and (b) which are enough to describe the homotopy type.

(a') Describe the set of stable cells on which critical points exist.
(b') Describe the homotopy type of unstable cells.

Proposition 12.6.9 ([61, Theorem 4.3.1]) *The Morse function $\varphi : M \to \mathbb{R}_{\geq 0}$ in (12.16) has a critical point p_C in every chamber $C \in \mathrm{ch}^{\ell}_{\mathcal{F}}(\mathcal{A})$. Furthermore, the stable cell $\mathcal{S}_{p_C}$ of p_C is the chamber C itself.*

Proof Let $C \in \mathrm{ch}_F(\mathcal{A})$. By the definition, $|\varphi(x)| \to \infty$ as $x \in C$ approaches the boundary. Hence $\varphi|_C : C \to \mathbb{R}$ has at least one minimal point $p_C \in C$. By the Cauchy-Riemann equation, p_C is a critical point of $\varphi : M \to \mathbb{R}$. Now we have $b_\ell(M)$ critical points. However, by the Varchenko conjecture, there are no other critical points.

The gradient flow of (12.17) preserves real points, hence we have

$$C \subseteq \mathcal{S}_{p_C}.$$

However, both are ℓ-dimensional, thus we have $C = \mathcal{S}_{p_C}$ (Fig. 12.19). □

Proposition 12.6.10 ([61, Corollary 5.1.5]) $X = M(\mathcal{A}) \setminus \bigsqcup_{C \in \mathrm{ch}_F(\mathcal{A})}$ *is diffeomorphic to $(M(\mathcal{A}) \cap F) \times D^2$. In particular, X is homotopy equivalent to $M \cap F$.*

By these results, we conclude as follows.

Proposition 12.6.11 *For each $C \in \mathrm{ch}_F(\mathcal{A})$, the unstable cell $\sigma_C : (D^\ell, \partial D^\ell) \to (M, M \cap F)$ has the following properties.*

(i) σ_C is transversal to C and $\sigma_C(D^\ell) \cap C$ is a point.
(ii) $\sigma_C(D^\ell) \cap C' = \emptyset$ for $C' \in \mathrm{ch}_F(\mathcal{A}) \setminus \{C\}$.

Furthermore, (i) and (ii) characterize the homotopy type of the map $\sigma_C : (D^\ell, \partial D^\ell) \to (M, M \cap F)$.

One can construct an explicit continuous map $\sigma_C : (D^\ell, \partial D^\ell) \to (M, M \cap F)$ satisfying (i) and (ii) in Proposition 12.6.11, see [61] for details. For 2-dimensional

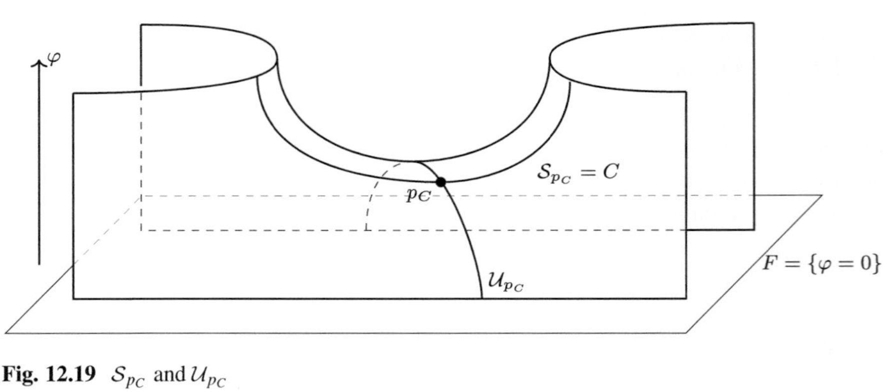

Fig. 12.19 $\mathcal{S}_{p_C}$ and $\mathcal{U}_{p_C}$

case, we will describe the dual stratification in Sect. 12.6.6. We also note that in 2-dimensional case, one can describe the handle decomposition of the 4-manifold $M(\mathcal{A})$ by refining minimal CW decomposition [59].

12.6.5 Minimality via Salvetti Complex

To avoid problems (a) and (b) in the previous section, an alternative way to describe the minimal CW complex is to start from the Salvetti complex. As we saw, the Salvetti complex is a large regular CW complex. For instance, the number of 0-cells is equal to $\#\mathrm{ch}(\mathcal{A})$. However, using discrete Morse theory [30, 31], one can prove that the Salvetti complex $\Delta_{\mathrm{Sal}}(\mathcal{A})$ is homotopy equivalent to a minimal CW complex. The idea is to construct an acyclic matching on the face poset of the complex. The remaining cells are called critical cells. Note that the collapse of cells along an acyclic matching does not change the homotopy type. It is sufficient to construct an acyclic matching such that the number of critical cells is equal to the sum of Betti numbers. This can be done in several ways.

- Salvetti and Settepanella [51] constructed an acyclic matching using a generic flag.
- Delucchi [13], Lofano and Paolini [34] constructed a matching using the Euclidean distance between a generically fixed base point and chambers. This also proves the minimality for locally finite affine arrangements.

These techniques are applicable to many other situations. For example, the minimality was proven for Salvetti complexes for oriented matroids [15], toric arrangements [10], 2-arrangements [1] and so on.

Remark 12.6.12 The minimal CW complex described in [61] and the one based on the Salvetti complex are very different. More precisely, the homotopy types of individual cells are different. It would be interesting to ask whether starting from Delucchi-Falk complex, one can describe both of them or not.

12.6.6 Minimal Stratification

Let $\mathcal{A} = \{H_1, \dots, H_n\}$ be a line arrangement in $V = \mathbb{R}^2$. Fix a flag $F^0 \subset F^1$ and defining equation α_i of H_i as in Sect. 12.5.2. For the sake of simplicity, we also set $\alpha_0 = -1$. Define the subset $S_i \subset M(\mathcal{A})$ $(i = 1, \dots, n)$ as

$$S_i = \left\{ x \in M(\mathcal{A}) \,\middle|\, \frac{\alpha_i(x)}{\alpha_{i-1}(x)} \in \mathbb{R}_{<0} \right\}. \tag{12.18}$$

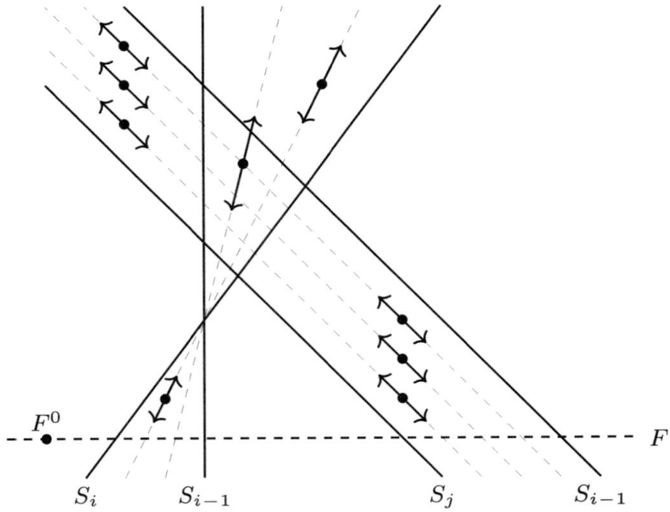

Fig. 12.20 S_i and S_j $(i > j)$

Note that since $\alpha_0 = -1$, $S_1 = \{x \in M(\mathcal{A}) \mid \alpha_1(x) \in \mathbb{R}_{>0}\} = \alpha_1^{-1}(\mathbb{R}_{>0})$. Note also that if $x \in \mathbb{R}^2$, then $x \in S_i$ if and only if $\alpha_i(x)$ and $\alpha_{i-1}(x)$ have different signs, which means, roughly speaking, x is in between H_i and H_{i-1}. More precisely, using the tangent vector description (Sect. 12.4), the real part x of $x + \sqrt{-1} \cdot v \in S_i$ is in between H_i and H_{i-1} and the imaginary part v is tangent to the line passing through $H_i \cap H_{i-1}$ (Fig. 12.20). Each S_i is a 3-dimensional submanifold of $M(\mathcal{A})$. Furthermore, S_i and S_j $(i \neq j)$ intersect transversally, and $S_i \cap S_j$ is a union of chambers (Fig. 12.20). The submanifolds S_i $(i = 1, \dots, n)$ provide a stratification dual to minimal CW complex, in the sense that the number of codimension k strata is equal to $b_k(M(\mathcal{A}))$ for $k = 0, 1, 2$.

Proposition 12.6.13 ([65]) *Let* $U = M(\mathcal{A}) \setminus \bigcup_{i=1}^n S_i$ *and* $S_i^\circ = S_i \setminus \bigcup_{C \in \mathrm{ch}_F(\mathcal{A})} C$. *Then,*

(1) $S_1, \dots, S_n$ *are* 3*-dimensional contractible manifolds.*
(2) U *is a* 4*-dimensional contractible manifold.*
(3) $M(\mathcal{A})$ *is decomposed as follows.*

$$M(\mathcal{A}) = U \sqcup \bigsqcup_{i=1}^n S_i^\circ \sqcup \bigsqcup_{C \in \mathrm{ch}_F(\mathcal{A})} C.$$

From this stratification, one can obtain the presentation of the fundamental group in Proposition 12.5.3.

12.7 Local Systems Cohomology Groups

12.7.1 Rank One Local Systems

Let G be an abelian group. Recall that a locally constant sheaf (or a local system), with local fiber G, on M is a sheaf $\mathcal{L}$ on M such that for any point $p \in M$, there exists an open neighborhood U of p such that the restriction $\mathcal{L}|_U$ is isomorphic to the constant sheaf associated to G. A local system is determined by a group homomorphism $\rho : \pi_1(M, p_0) \longrightarrow \mathrm{Aut}(G)$.

In this section, we mainly consider the rank one local system over a commutative ring R. Namely, $\mathcal{L}$ is a local system with local fiber isomorphic to R as an R-module. Such a local system is determined by a homomorphism

$$\rho : \pi_1(M, p_0) \longrightarrow R^\times,$$

where $R^\times$ is the group of invertible elements. Since $R^\times$ is abelian, the map ρ factors through the homology group, $\pi_1(M) \to H_1(M, \mathcal{A}) \to R^\times$, which is determined by assigning $\rho(\gamma_i) \in R^\times$ for each meridian (that is an oriented small circle turning around H_i) γ_i of the hyperplane H_i.

For our purposes, it is convenient to realize a local system as a representation of the Deligne groupoid $\mathrm{Gal}(\mathcal{A})$. Let C_0 and C_1 be adjacent chambers separated by a hyperplane H_i. Then the sequence $\gamma_i = (C_0C_1C_0) \in \mathrm{Hom}(C_0, C_0)$ is corresponding to the meridian of H_i. We have $\rho(\gamma_i) \in R^\times$. To describe the representation of $\mathrm{Gal}(\mathcal{A})$, we need $\rho((C_0, C_1))$ and $\rho((C_1, C_0))$. Since they can be considered as the square root of the monodromy $\rho(\gamma_i) \in R^\times$, we introduce the formal variable x_i satisfying $x_i^2 = \rho(\gamma_i)$. More formally, define the extended ring $\widetilde{R}$ as follows.

$$\widetilde{R} := R[x_a \mid a \in R^\times]/(x_a^2 - a; a \in R^\times). \tag{12.19}$$

Lemma 12.7.1

(1) The natural map $i : R \longrightarrow \widetilde{R}$ *is injective.*
(2) $x_a \in \widetilde{R}^\times$.

Proof

(1) Let $r \in R$. Suppose $r \in \mathrm{Ker}\, i$. Then there exist $a_1, \dots, a_m \in R^\times$ and $f_1, \dots, f_m \in R[x_1, x_2, \dots, x_m]$, where $x_i = x_{a_i}$, such that

$$r = \sum_{s=1}^{m} f_s(x_1, \dots, x_m) \cdot (x_s^2 - a_s).$$

Let $A_k = R[x_1, \ldots, x_k]/(x_s^2 - a_s; s = 1, \ldots, k)$, for $k = 1, \ldots, m$. Then r is contained in the kernel of $R \longrightarrow A_m$. Note that

$$A_s = A_{s-1}[x_s]/(x_s^2 - a_s).$$

Since the ideal is generated by a monic polynomial, it is clear that the natural map $i_s : A_{s-1} \longrightarrow A_s$ is injective.

(2) This is obvious from $x_a \cdot (x_a a^{-1}) = 1$ in $\widetilde{R}$.

□

From the above lemma, the multiplication map $R \longrightarrow Rx_a, r \longmapsto rx_a$ gives an isomorphism of R-modules.

Let $\mathcal{F} = (F^0 \subset \cdots \subset F^{\ell-1})$ be a generic flag as in Remark 12.3.2. Let us consider $F^0 = p_0$ to be the base point of $M = M(\mathcal{A})$. Let $\rho : \pi_1(M, p) \longrightarrow R^\times$ be a group homomorphism which is determined by $\rho(\gamma_i) = a_i \in R^\times$. Now we construct the representation of $\mathrm{Gal}(\mathcal{A})$. Let $C \in \mathrm{ch}(\mathcal{A})$. Then, define the submodule $\rho(C) \subset \widetilde{R}$ by

$$\rho(C) := R \cdot \prod_{H_i \in \mathrm{Sep}(C_0, C)} x_{a_i},$$

where C_0 is the chamber containing $p_0 = F^0$.

Proposition 12.7.2 *Denote the category of R-modules by* Mod_R. *The correspondence* $\rho : \mathrm{ch}(\mathcal{A}) \longrightarrow \mathrm{Mod}_R : C \longmapsto \rho(C)$ *and* $\rho : \mathrm{Hom}(C, C') \ni (C, C') \longmapsto x_{a_i}$, *where C and C' are adjacent and separated by H_i, gives a representation of the groupoid* $\mathrm{Gal}(\mathcal{A})$.

Proof It is sufficient to show that the image $\rho(\gamma)$ of $\gamma \in \mathrm{Hom}(C, C)$ gives a automorphism of $\rho(C)$. Since $R^\times$ and $\widetilde{R}$ are abelian, each x_{a_i} is multiplied an even number of times. In other words, the action of $\rho(\gamma)$ on $\rho(C)$ is a multiplication by $\prod_i x_{a_i}^{2k_i}$ $(k_i \in \mathbb{Z})$, which is equal to $\prod_i a_i^{k_i} \in R^\times$. □

Definition 12.7.3 Let $C, C \in \mathrm{ch}(\mathcal{A})$. Define $\Delta(C, C') \in \widetilde{R}$ by

$$\Delta(C, C') = \prod_{H_i \in \mathrm{Sep}(C, C')} x_{a_i} - \prod_{H_i \in \mathrm{Sep}(C, C')} x_{a_i}^{-1}.$$

12.7.2 The Degree Map and Twisted Cochain Complex

To describe the twisted cochain complex, we need the notion of the degree map [3, 61, 63]

$$\deg : \mathrm{ch}_{\mathcal{F}}^k(\mathcal{A}) \times \mathrm{ch}_{\mathcal{F}}^{k+1}(\mathcal{A}) \longrightarrow \mathbb{Z}.$$

Let $B = B^k \subset F^k$ be a k-dimensional ball of sufficiently large radius such that every 0-dimensional intersection $X \in L_0(\mathcal{A} \cap F^k)$ is contained in the interior of B. Let $C' \in \mathrm{ch}_{\mathcal{F}}^{k+1}(\mathcal{A})$. Then, consider the vector field $U^{C'}$ on B, that is $U^{C'}(x) \in T_x F^k$ for $x \in B$, satisfying the following conditions.

- $U^{C'}(x) \neq 0$ for $x \in \partial(\overline{C} \cap B)$ with $C \in \mathrm{ch}_{\mathcal{F}}^k(\mathcal{A})$.
- Let $x \in \partial B$, then $U^{C'}(x)$ directs to the inside of B.
- If $x \in H$ for some $H \in \mathcal{A}$, then $U^{C'}(x) \notin T_x H$ and directs the side containing C'.

Roughly speaking $U^{C'}$ is a vector field on B directing towards the chamber C'.

Definition 12.7.4 Define $\deg(C, C') \in \mathbb{Z}$ by the degree of the Gauss map on the boundary $\partial(\overline{C} \cap B)$

$$\deg(C, C') := \deg\left(\left. \frac{U^{C'}}{|U^{C'}|} \right|_{\partial(\overline{C} \cap B)} : \partial(\overline{C} \cap B) \longrightarrow S^{k-1} \right).$$

The next result was first formulated for $R = \mathbb{C}$ in [61]. Recently, the case $R = \mathbb{Z}$ was considered in [58].

Proposition 12.7.5 ([61]) *Let $\rho : \pi_1(M(\mathcal{A})) \longrightarrow R^\times$ be an R-rank one local system. Let*

$$\mathcal{C}^k(\mathcal{A}, \rho) := \bigoplus_{C \in \mathrm{ch}_{\mathcal{F}}^k(\mathcal{A})} \rho(C).$$

Define the map $\nabla_\rho : \mathcal{C}^k(\mathcal{A}, \rho) \longrightarrow \mathcal{C}^{k+1}(\mathcal{A}, \rho)$ by

$$\nabla_\rho(y) = \sum_{C' \in \mathrm{ch}_{\mathcal{F}}^{k+1}} \deg(C, C') \cdot y \cdot \Delta(C, C'),$$

where $y \in \rho(C)$. Then $(\mathcal{C}^\bullet(\mathcal{A}, \rho), \nabla_\rho)$ gives a cochain complex such that

$$H^k(\mathcal{C}^\bullet(\mathcal{A}, \rho), \nabla_\rho) \simeq H^k(M(\mathcal{A}), \mathcal{L}_\rho).$$

Remark 12.7.6 In [34, 51], similar cochain complexes which compute local system cohomology groups are described. However, the relationship with Proposition 12.7.5 is not clear. Proposition 12.7.5 has been applied to the computation of the first Betti number of the Milnor fiber [57, 66, 68], complex rank one local systems [64, 67, 69], and Aomoto complexes [3, 60]. Recently, Sugawara [58] used it to settle an integral version of Cohen-Dimca-Orlik type vanishing theorem [8]. See also for an alternative approach [32].

12.8 Homotopy Groups

12.8.1 $K(\pi, 1)$ Arrangements

Recall that a connected topological space X is called $K(\pi, 1)$ if the higher homotopy groups are vanishing, namely, $\pi_k(X) = 0$ for $k \geq 2$. Naturally, we call a complex arrangement $\mathcal{A}$ a $K(\pi, 1)$ arrangement if the complement $M(\mathcal{A})$ is a $K(\pi, 1)$ space. As we saw in Example 12.2.3, the braid arrangement is $K(\pi, 1)$. Based on this and other examples, Brieskorn [6] conjectured the $K(\pi, 1)$-ness of certain arrangements related to reflection groups. The $K(\pi, 1)$-ness has been settled in many cases, for example,

- Simplicial arrangements [11],
- Many 2-dimensional real arrangements [20],
- Many reflection groups [7],
- Complex reflection groups [4],
- Affine Weyl arrangements [40],

and so on. See [21, 22, 44] for surveys on this topic.

12.8.2 Localization and Hurewicz Map

We recall several basic facts about homotopy groups of $M(\mathcal{A})$. The following is essentially due to M. Oka [42, Lemma 1.1.].

Proposition 12.8.1 *Let A be an arrangement in* $\mathbb{C}^\ell$. *Let* $X \in L(\mathcal{A})$ *and* $i : M(\mathcal{A}) \hookrightarrow M(\mathcal{A}_X)$ *be the natural inclusion. Then* i *induces the split surjective homomorphism*

$$i_* : \pi_k(M(\mathcal{A})) \longrightarrow \pi_k(M(\mathcal{A}_X))$$

of groups for $k \geq 1$. *In other words, there exists a group injection* $f : \pi_k(M(\mathcal{A}_X)) \longrightarrow \pi_k(M(\mathcal{A}))$ *such that* $i_* \circ f : \pi_k(M(\mathcal{A}_X)) \longrightarrow \pi_k(M(\mathcal{A}_X))$ *is the identity.*

Proof Let $p \in X \setminus \bigcup_{H \in \mathcal{A} \smallsetminus \mathcal{A}_X} H$. Take a small convex open neighborhood U of p in such a way that $B \cap H = \emptyset$ for $H \in \mathcal{A} \smallsetminus \mathcal{A}_X$. Then we have inclusions $j : U \cap M(\mathcal{A}) \hookrightarrow M(\mathcal{A})$ and $i : M(\mathcal{A}) \hookrightarrow M(\mathcal{A}_X)$. Then $i \circ j : U \cap M(\mathcal{A}) \longrightarrow M(\mathcal{A})$ is a homotopy equivalence. Hence the composition of induced homomorphism

$$\pi_k(M(\mathcal{A}_X)) \xrightarrow{j_*} \pi_k(M(\mathcal{A})) \xrightarrow{i_*} \pi_k(M(\mathcal{A}_X))$$

is isomorphic. □

Corollary 12.8.2

(1) The fundamental group $\pi_1(M(\mathcal{A}_X))$ embeds into $\pi_1(M(\mathcal{A}))$.
(2) If $\mathcal{A}$ is $K(\pi, 1)$, so is the localization $\mathcal{A}_X$.

One of the difficulties of studying the homotopy groups of $M(\mathcal{A})$ is the following vanishing of Hurewicz maps. One can not detect elements of homotopy groups as elements in the homology groups.

Proposition 12.8.3 (Randell [46]) *Let $\mathcal{A}$ be an arrangement in $\mathbb{C}^\ell$. Then for any $k \geq 2$, the Hurewicz map*

$$h : \pi_k(M) \longrightarrow H_k(M, \mathbb{Z})$$

is vanishing.

Proof Recall (12.5) that $M(\mathcal{A})$ can be embedded into $(\mathbb{C}^\times)^n$. Note that $(\mathbb{C}^\times)^n$ is a $K(\mathbb{Z}^n, 1)$-space, so we have $\pi_k((\mathbb{C}^\times)^n) = 0$, for $k \geq 2$. By the functoriality and naturality of Hurewicz maps, the composition $\boldsymbol{\alpha}_* \circ h : \pi_k(M(\mathcal{A})) \longrightarrow H_k((\mathbb{C}^\times)^n)$ is vanishing. By the injectivity (Proposition 12.2.17) of $\boldsymbol{\alpha}_*$, we conclude that the Hurewicz map $h : \pi_k(M) \longrightarrow H_k(M, \mathbb{Z})$ is vanishing. □

12.8.3 Non-$K(\pi, 1)$ Arrangements

There are many non-$K(\pi, 1)$ arrangements. Hattori [26] proved that if $\mathcal{A} = \{H_1, \dots, H_n\}$ is a generic arrangement in $\mathbb{C}^\ell$ ($n \geq \ell$), then $M(\mathcal{A})$ is homotopy equivalent to the ℓ-skeleton of the standard CW structure of n-torus $T^n = (S^1)^n$, which clearly has $\pi_\ell(M(\mathcal{A})) \neq 0$. Papadima and Suciu [41] studied the structure of higher homotopy groups in a generalized setting. One of the most general results concerning non-vanishing of homotopy groups is the following. The proof in this article was due to M. Falk and S. Papadima.

Proposition 12.8.4 ([62]) *Let $\mathcal{A}$ be an essential arrangement in $V = \mathbb{C}^\ell$ with $\ell \geq 3$. Let $F \subset V$ be a k-dimensional generic subspace with $k \geq 2$. Then the arrangement $\mathcal{A} \cap F$ is not $K(\pi, 1)$. More precisely, $\pi_k(M(\mathcal{A}) \cap F) \neq 0$.*

Proof It is sufficient to prove the case $k = \ell - 1$. Let F be a generic hyperplane. Then we will prove that $\pi_{\ell-1}(M(\mathcal{A}) \cap F) \neq 0$.

Recall (Proposition 12.6.8) that $M(\mathcal{A})$ is homotopy equivalent to a space obtained by attaching ℓ-cells to $M(\mathcal{A}) \cap F$. Suppose that $\pi_{\ell-1}(M(\mathcal{A}) \cap F) = 0$. Then the attaching maps $\partial D^\ell = S^{\ell-1}$ of ℓ-cells must be trivial, and we have

$$M(\mathcal{A}) \simeq (M(\mathcal{A}) \cap F) \vee \bigvee_{i=1}^{b_\ell(M)} S^\ell.$$

Thus we have a map $i : \bigvee_{i=1}^{b} S^\ell \longrightarrow M(\mathcal{A})$ (where $b = b_\ell(M(\mathcal{A}))$) that induces the surjection $i^* : H^\ell(M(\mathcal{A})) \longrightarrow H^\ell(\bigvee_{i=1}^{b} S^\ell) = \mathbb{Z}^b$. Consider the following diagram.

$$\begin{array}{ccc} \wedge^\ell H^1(M(\mathcal{A})) & \xrightarrow{(p)} & \wedge^\ell H^1(\bigvee_{i=1}^{b} S^\ell) = 0 \\ {\scriptstyle (q)}\downarrow & & \downarrow{\scriptstyle (r)} \\ H^\ell(M(\mathcal{A})) & \xrightarrow{(s)} & H^\ell(\bigvee_{i=1}^{b} S^\ell) = \mathbb{Z}^b \end{array}$$

Then, by Orlik-Solomon's presentation of the cohomology ring (Proposition 12.2.15), the map (q) is surjective. Note that $(s) = i^*$ is also surjective. Hence, the composition $(s) \circ (q)$ is also surjective, which contradicts $(r) \circ (p) = 0$. □

Proposition 12.8.5 ([21, 26]) *Let $\mathcal{A}$ be an essential arrangement in $\mathbb{C}^\ell$. If there exists a subgroup $\Gamma \subset \pi_1(M(\mathcal{A}))$ such that $\Gamma \simeq \mathbb{Z}^{\ell+1}$, then $\mathcal{A}$ is not $K(\pi, 1)$.*

Proof Let K be an ℓ-dimensional CW complex which is homotopy equivalent to $M = M(\mathcal{A})$. Suppose K is $K(\pi, 1)$. Then the universal cover $\widetilde{K}$ is contractible space, on which Γ acts freely. Thus $K' := \widetilde{K}/\Gamma$ is a $K(\Gamma, 1)$ space. Since K' is ℓ-dimensional, $H_{\ell+1}(K') = 0$. However, this contradicts the fact that $H_{\ell+1}(\Gamma, \mathbb{Z}) = H_{\ell+1}((S^1)^{\ell+1}, \mathbb{Z}) = \mathbb{Z}$. □

Example 12.8.6 Let $\mathcal{A} = \{H_1, \ldots, H_6\}$ be the arrangement in Fig. 12.3. See Example 12.5.4 for presentations of $\pi_1(M(\mathcal{A}))$. One can directly check that $\gamma_2, \gamma_5, [\gamma_3, \gamma_4]$ are commutative with each other and generate a subgroup isomorphic to $\mathbb{Z}^3$. Hence $M(\mathcal{A})$ is not $K(\pi, 1)$.

12.8.4 Twisted Hurewicz Maps and Homotopy Groups

In this section, we recall the notion of the twisted Hurewicz map. Let X be a CW complex, $\mathcal{L}$ be a local system on X. Let $f : (S^k, *) \longrightarrow (X, x_0)$ be a continuous map from the sphere S^k with $k \geq 2$. Then the pull-back $f^*\mathcal{L}$ is a local system on S^k. Since S^k is simply connected, $f^*\mathcal{L}$ is a trivial local system. Hence the set of global sections $\Gamma(S^k, f^*\mathcal{L})$ is isomorphic to that of the local sections $\mathcal{L}_{x_0}$ at x_0.

Since S^k (equipped with an orientation) is closed, the continuous map $f : (S^k, *) \longrightarrow (X, x_0)$ together with a global section $\sigma \in \Gamma(S^k, f^*\mathcal{L})$ gives a twisted cycle $[f] \otimes \sigma$ on X. Thus we have a map

$$h_{\mathcal{L},x_0} : \pi_k(X, x_0) \otimes \mathcal{L}_{x_0} \longrightarrow H_k(X, \mathcal{L}),$$

which we call the twisted Hurewicz map. Obviously, when $\mathcal{L}$ is the trivial local system $\mathbb{Z}$, it is the classical Hurewicz map.

As we have already seen in Proposition 12.8.3, the classical Hurewicz map does not carry any information on the higher homotopy groups. However, the twisted Hurewicz map can detect some nontrivial elements in the homotopy groups. Actually, in the setting of Proposition 12.8.4, the twisted Hurewicz map become surjective.

Proposition 12.8.7 ([62]) *Let $\mathcal{A}$ be an essential arrangement in $V = \mathbb{C}^\ell$ with $\ell \geq 3$. Let $F \subset V$ be a k-dimensional generic subspace with $k \geq 2$. Let $\mathcal{L}$ be a generic complex rank one local system. Then the twisted Hurewicz map*

$$h_{\mathcal{L},x_0} : \pi_k(M(\mathcal{A}) \cap F, x_0) \otimes \mathcal{L}_{x_0} \longrightarrow H_k(M(\mathcal{A}) \cap F, \mathcal{L})$$

is surjective (and $H_k(M(\mathcal{A}), \mathcal{L})$ is nonzero).

The proof of Proposition 12.8.7 is based on the vanishing theorem of local system homology groups [8, 17, 29, 52]. For certain generic local systems, $H_k(M(\mathcal{A}), \mathcal{L})$ vanishes except for $k = \ell$ and the dimension of the remaining cohomology is equal to the Euler characteristic of $M(\mathcal{A})$. The point of the proof is that the non-vanishing degrees of $H_k(M(\mathcal{A}), \mathcal{L})$ and $H_k(M(\mathcal{A}) \cap F, \mathcal{L}|_{M(\mathcal{A})\cap F})$ differ by one if F is a generic hyperplane. To cancel the twisted cycle of $H_{\ell-1}(M \cap F, \mathcal{L})$, the twisted Hurewicz map needs to be surjective.

We now discuss an explicit construction of a sphere that is non-trivial in the homotopy group. The setting is as follows. Let $\mathcal{A} = \{H_1, \dots, H_n\}$ be a central and essential arrangement in $V = \mathbb{R}^\ell$ with $\ell \geq 3$. Let $S = S^{\ell-1}$ be the unit sphere centered at the origin. Consider the shift of the sphere S into imaginary direction. Let $\theta : S \ni x \longmapsto \theta(x) \in T_x V$ be a vector field. Using the description of $M(\mathcal{A})$ in Sect. 12.4, we consider the sphere

$$S(\theta) = \{x + \sqrt{-1} \cdot \theta(x) \mid x \in S\}. \tag{12.20}$$

Then $S(\theta)$ is embedded into $M(\mathcal{A})$ if and only if

$$\theta(x) \notin T_x H_i \tag{12.21}$$

for any $x \in H_i$, $H_i \in \mathcal{A}$. Since $\ell \geq 3$, $H_i \cap S$ is connected. Therefore, θ determines a half-space $H_i^{\varepsilon_i(\theta)}$, and a sign vector $(\varepsilon_1(\theta), \varepsilon_2(\theta), \dots, \varepsilon_n(\theta)) \in \{\pm\}^n$ (Fig. 12.21).

The (non)triviality of $[S(\theta)] \in \pi_{\ell-1}(M(\mathcal{A}))$ is checked as follows.

Proposition 12.8.8 ([70]) $[S(\theta)] = 0$ *in $\pi_{\ell-1}(M(\mathcal{A}))$ is if and only if the intersection of half-spaces $\bigcap_{i=1}^n H_i^{\varepsilon_i(\theta)}$ is non-empty.*

The "if" part is relatively easy. Indeed, if the intersection of half-spaces $\bigcap_{i=1}^n H_i^{\varepsilon_i(\theta)} = C$ is a chamber, one can extend the vector field θ to that of the ball $\widetilde{\theta} : D^\ell \longrightarrow TV$ in such a way that the condition (12.21) is fulfilled.

The proof of "only if" part needs the twisted Hurewicz map. Suppose $\bigcap_{i=1}^n H_i^{\varepsilon_i(\theta)} = \emptyset$ (see Fig. 12.21 for an example). Let $\mathcal{L}$ be the complex rank

Fig. 12.21 Shifted sphere

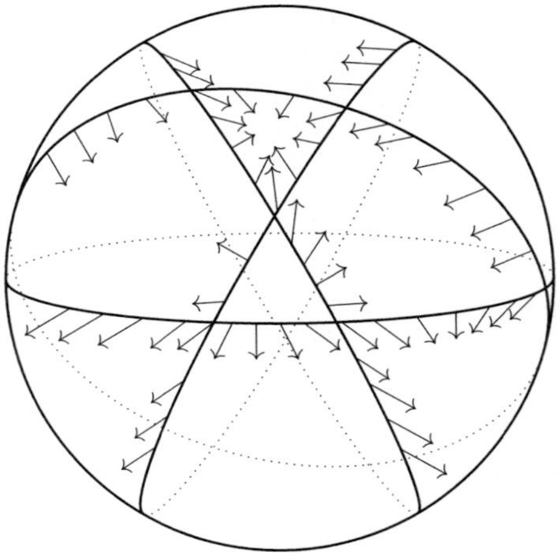

one local system defined by $\rho : \pi_1(M(\mathcal{A})) \longrightarrow \mathbb{C}^\times$, $\gamma_i \longmapsto e^{2\pi\sqrt{-1}/n}$, where γ_i is the meridian of H_i. Then the image of $[S(\theta)]$ by the twisted Hurewicz map

$$h_{\mathcal{L},x_0} : \pi_{\ell-1}(M(\mathcal{A}), x_0) \otimes \mathcal{L}_{x_0} \longrightarrow H_{\ell-1}(M(\mathcal{A}), \mathcal{L})$$

is non-zero. More precisely, one can construct an explicit twisted Borel-Moore cycle $[W] \in H^{BM}_{\ell+1}(M(\mathcal{A}), \mathcal{L}^\vee)$ such that the twisted intersection number between $[W]$ and $[S(\theta)] \in H_{\ell-1}(M(\mathcal{A}), \mathcal{L})$ is nonzero. See [70] for details.

Acknowledgments This work was partially supported by JSPS KAKENHI (Grant No. JP23H00081). The author thanks anonymous referee(s) for useful comments and corrections.

References

1. K. Adiprasito, Combinatorial stratifications and minimality of 2-arrangements. *J. Topol.* **7** (2014), no. 4, 1200–1220.
2. V. I. Arnold, The cohomology ring of the colored braid group, *Mat. Zametki* **5** (1969) 227–231 : *Math. Notes* **5** (1969) 138–140.
3. P. Bailet, M. Yoshinaga, Vanishing results for the Aomoto complex of real hyperplane arrangements via minimality. *Journal of Singularities*, **14** (2016) 74–90.
4. D. Bessis, Finite complex reflection arrangements are $K(\pi, 1)$ *Ann. of Math.* (2) **181** (2015), no.3, 809–904.
5. A. Björner, M. Las Vergnas, B. Sturmfels, N. White, G. Ziegler, Oriented matroids. Second edition Encyclopedia Math. Appl., 46 Cambridge University Press, Cambridge, 1999. xii+548 pp.
6. E. Brieskorn, Sur les groupes de tresses. Séminaire Bourbaki 1971/72. Lecture Notes in Math. **317**, Springer Verlag, 1973, 21–44.

7. R. Charney, M. Davis, The $K(\pi, 1)$-problem for hyperplane complements associated to infinite reflection groups. *J. Am. Math. Soc.* **8** 597–627
8. D. Cohen, A. Dimca, P. Orlik, Nonresonance conditions for arrangements. *Ann. Inst. Fourier* **53** (2003), 1883–1896.
9. D. C. Cohen, A. Suciu, The braid monodromy of plane algebraic curves and hyperplane arrangements. Comment. Math. Helvetici **72** (1997), no. 2, 285–315.
10. G. d'Antonio, E. Delucchi, Minimality of toric arrangements *Journal of the European Mathematical Society* **17** (3), 483–521.
11. P. Deligne, Les immeubles des groupes de tresses généralisés, *Invent. Math.* **17** (1972), 273–302.
12. E. Delucchi, Combinatorics of covers of complexified hyperplane arrangements. Arrangements, Local systems and singularities. Progress in Mathematics Vol. 283, Birkhäuser, 2009, pp. 1–38.
13. E. Delucchi, Shelling-type orderings of regular CW-complexes and acyclic matchings of the Salvetti complex. Int. Math. Res. Not. 2008, no. 6, 39pp.
14. E. Delucchi, M. J. Falk, An equivariant discrete model for complexified arrangement complements. *Proc. Amer. Math. Soc.* **145** (2017), no.3, 955–970.
15. E. Delucchi, S. Settepanella, Combinatorial polar orderings and recursively orderable arrangements. *Adv. Appl. Math.* **44** No. 2 (2010), 124–144.
16. A. Dimca, S. Papadima, Hypersurface complements, Milnor fibers and higher homotopy groups of arrangements. *Ann. of Math. (2)* **158** (2003), no. 2, 473–507.
17. H. Esnault, V. Schechtman, E. Viehweg, Cohomology of local systems on the complement of hyperplanes, *Invent. Math.* **109** (1992), 557–561.
18. E. Fadell, L. Neuwirth, Configuration spaces. *Math. Scand.* **10** (1962) 111–118.
19. M. Falk, Homotopy types of line arrangements. *Invent. Math.* **111** (1993), no. 1, 139–150.
20. M. Falk, $K(\pi, 1)$ arrangements. *Topology* **34** (1995), no. 1, 141–154.
21. M. Falk, R. Randell, On the homotopy theory of arrangements, In: Complex analytic singularities, Adv. Stud. Pure Math., vol. 8, Math. Soc. Japan, Tokyo, 1986, pp. 101–124.
22. M. Falk, R. Randell, On the homotopy theory of arrangements, II, In: Arrangements Tokyo 1998, Adv. Stud. Pure Math., vol. 27, Math. Soc. Japan, Tokyo, 2000, pp. 93–125
23. R. H. Fox, L. Neuwirth, The braid groups. *Math. Scand.* **10** (1962) 119–126.
24. H. A. Hamm, D. T. Lê, Un théorème de Zariski du type de Lefschetz. Ann. Sci. École Norm. Sup. **6** (1973), 317–355.
25. A. Hatcher, Algebraic topology, Cambridge University Press, Cambridge, 2002.
26. A. Hattori, Tolopogy of $\mathbb{C}^n$ minus a finite number of affine hyperplanes in general position. *J. Fac. Sci. Univ. Tokyo* **22** (1975), 205–219.
27. M. Ishikawa, Y. Koda, H. Naoe, Presentation of the fundamental groups of complements of shadows. *Essays in geometry–dedicated to Norbert A'Campo*, 557–588. IRMA Lect. Math. Theor. Phys., 34
28. H. Kamiya, A. Takemura, H. Terao, Periodicity of hyperplane arrangements with integral coefficients modulo positive integers, *J. Algebraic Combin.* **27** (2008), no. 3, 317–330.
29. T. Kohno, Homology of a local system on the complement of hyperplanes. *Proc. Japan Acad. Ser. A Math. Sci.* **62** (1986), 144–147.
30. D. Kozlov, Combinatorial Algebraic Topology, Vol. 31 of Algorithms and Computation in Mathematics, Springer (2008)
31. D. Kozlov, Organized collapse: an introduction to discrete Morse theory, volume 207 of Graduate Studies in Mathematics. American Mathematical Society, Providence, RI, 2020
32. Y. Liu, L. Maxim, B. Wang, Cohomology of $\mathbb{Z}$-local systems on complex hyperplane arrangement complements Preprint arXiv:2209.13193
33. Y. Liu, T. N. Tran, M. Yoshinaga. G-Tutte polynomials and abelian Lie group arrangements. *Int. Math. Res. Notices* **2021**, no. 1 (2021): 150–188.
34. D. Lofano, G. Paolini, Euclidean matchings and minimality of hyperplane arrangements. *Discrete Math. 344* (2021), no.3, Paper No. 112232, 22 pp.

35. S. Mac Lane, Categories for the working mathematician. Second edition. Graduate Texts in Mathematics, 5. Springer-Verlag, New York, 1998.
36. L. Moci, A Tutte polynomial for toric arrangements. *Trans. Amer. Math. Soc.* **364**, no. 2 (2012) 1067–88.
37. P. Orlik, L. Solomon, Combinatorics and topology of complements of hyperplanes. Invent. Math. **56** (1980), no. 2, 167–189.
38. P. Orlik, H. Terao, Arrangements of Hyperplanes. Grundlehren Math. Wiss. **300**, Springer-Verlag, New York, 1992.
39. P. Orlik, H. Terao, The number of critical points of a product of powers of linear functions, *Invent. Math.* **120** (1995), 1–14.
40. G. Paolini, M. Salvetti, Proof of the $K(\pi, 1)$ conjecture for affine Artin groups *Invent. Math. 224* (2021), no. 2, 487–572.
41. S. Papadima, A. Suciu, Higher homotopy groups of complements of complex hyperplane arrangements. *Adv. Math.* **165** (2002), no. 1, 71–100.
42. L. Paris, The Deligne complex of a real arrangement of hyperplanes. *Nagoya Math. J.* **131** (1993) 39–65.
43. L. Paris, On the fundamental group of the complement of a complex hyperplane arrangement, Arrangements—Tokyo 1998, Adv. Stud. Pure Math. 27, Kinokuniya, Tokyo, 2000. 257–272.
44. L. Paris, $K(\pi, 1)$ conjecture for Artin groups *Ann. Fac. Sci. Toulouse Math.* (6) **23** (2014), no. 2, 361–415.
45. R. Randell, The fundamental group of the complement of a union of complex hyperplanes, *Invent. Math.* **69** (1982), 103–108. Correction, *Invent. Math.* **80** (1985), 467–468.
46. R. Randell, Homotopy and group cohomology of arrangements, *Topology Appl.* **78** (1997) 201–213.
47. R. Randell, Morse theory, Milnor fibers and minimality of hyperplane arrangements. *Proc. Amer. Math. Soc.* **130** (2002), no. 9, 2737–2743.
48. K. Saito, Second homotopy classes associated with non-cancellative monoids. Preprint (arXiv:2312.16951)
49. M. Salvetti, Arrangements of lines and monodromy of plane curves. Compositio Math. **68** (1988), 103–122.
50. M. Salvetti, Topology of the complement of real hyperplanes in $\mathbb{C}^N$. *Invent. Math.* **88** (1987), no. 3, 603–618.
51. M. Salvetti, S. Settepanella, Combinatorial Morse theory and minimality of hyperplane arrangements. *Geom. Topol.* **11** (2007), 1733–1766.
52. V. Schechtman, H. Terao, A. Varchenko, Local systems over complements of hyperplanes and the Kac-Kazhdan conditions for singular vectors, *J. Pure Appl. Algebra* **100** (1995), 93–102.
53. R. Silvotti, On a conjecture of Varchenko, *Invent. Math.* **126** (1996), 235–248.
54. R. P. Stanley, Enumerative Combinatorics, vol. 1, 2nd ed., Cambridge Stud. Adv. Math., vol. 49, Cambridge University Press, Cambridge, 2011.
55. R. P. Stanley, An introduction to hyperplane arrangements, (Geometric Combinatorics. Geometric Combinatorics, IAS/Park City Math. Ser., vol. 13, 2007, Amer. Math. Soc.: Amer. Math. Soc. Providence, RI), 389–496
56. A. Suciu, Fundamental groups of line arrangements: enumerative aspects. *Advances in algebraic geometry motivated by physics (Lowell, MA, 2000)*, 43-79, Contemp. Math., 276, Amer. Math. Soc., Providence, RI, 2001.
57. A. Suciu, Hyperplane arrangements and Milnor fibrations. *Ann. Fac. Sci. Toulouse Math. (6)* **23** (2014), no. 2, 417–481
58. S. Sugawara, $\mathbb{Z}$-local system cohomology of hyperplane arrangements and a Cohen-Dimca-Orlik type theorem. *Internat. J. Math.* **34** (2023), no. 8, Paper No. 2350044, 15 pp.
59. S. Sugawara, M. Yoshinaga, Divides with cusps and Kirby diagrams for line arrangements. *Topology and its Applications* **313** (2022), Paper No. 107989.
60. M. Torielli, M. Yoshinaga, Resonant bands, Aomoto complex, and real 4-nets. *J. Singul.* **11** (2015), 33–51.

61. M. Yoshinaga, Hyperplane arrangements and Lefschetz's hyperplane section theorem. *Kodai Math. J.* **30** (2007) no. 2, 157–194.
62. M. Yoshinaga, Generic section of a hyperplane arrangement and twisted Hurewicz maps *Topology Appl.* **155** (2008), no. 9, 1022–1026.
63. M. Yoshinaga, The chamber basis of the Orlik-Solomon algebra and Aomoto complex. *Arkiv för Matematik*, **47** (2009), 393–407.
64. M. Yoshinaga, Minimality of hyperplane arrangements and basis of local system cohomology. *Singularities in geometry and topology*, 345–362, IRMA Lect. Math. Theor. Phys., 20, Eur. Math. Soc., Zürich, 2012.
65. M. Yoshinaga, Minimal stratifications for line arrangements and positive homogeneous presentations for fundamental groups. *Configuration Spaces: Geometry, Combinatorics and Topology*, 503-533, CRM Series, 14, Ed. Norm., Pisa, 2012.
66. M. Yoshinaga, Milnor fibers of real line arrangements. *J. Singul.* **7** (2013), 220–237.
67. M. Yoshinaga, Resonant bands and local system cohomology groups for real line arrangements. *Vietnam J. Math.* **42** (2014), no. 3, 377–392.
68. M. Yoshinaga, Double coverings of arrangement complements and 2-torsion in Milnor fiber homology. *Eur. J. Math.* **6** (2020) 1097–1109.
69. M. Yoshinaga, q-deformation of Aomoto complex. *Rev. Roumaine Math. Pures Appl.* **69** (2024), no. 2, 305–323.
70. M. Yoshinaga, A construction of homotopically non-trivial embedded spheres for hyperplane arrangements. arXiv:2405.20010
71. T. Zaslavsky, Facing up to arrangements: face-count formulas for partitions of space by hyperplanes. *Mem. Amer. Math. Soc.* **1** ,issue 1, no. 154 (1975), vii+102 pp.

Index

© The Author(s), under exclusive license to Springer Nature Switzerland AG 2026
J. L. Cisneros-Molina et al. (eds.), *Handbook of Geometry and Topology of Singularities VIII*, https://doi.org/10.1007/978-3-031-99571-2

V

W

Z

MIX
Papier aus verantwortungsvollen Quellen
Paper from responsible sources
FSC® C105338

If you have any concerns about our products,
you can contact us on
ProductSafety@springernature.com

In case Publisher is established outside the EU,
the EU authorized representative is:
Springer Nature Customer Service Center GmbH
Europaplatz 3, 69115 Heidelberg, Germany

Printed by Libri Plureos GmbH
in Hamburg, Germany